Produced Water

Kenneth Lee · Jerry Neff
Editors

Produced Water

Environmental Risks and Advances in Mitigation Technologies

Editors
Kenneth Lee
Centre for Offshore Oil, Gas and Energy Research (COOGER)
Fisheries and Oceans Canada
Bedford Institute of Oceanography
Dartmouth, NS, Canada

Jerry Neff
Neff & Associates LLC
Duxbury, MA, USA

ISBN 978-1-4614-0045-5 e-ISBN 978-1-4614-0046-2
DOI 10.1007/978-1-4614-0046-2
Springer New York Dordrecht Heidelberg London

Library of Congress Control Number: 2011933551

Printed on acid-free paper

Springer is part of Springer Science+Business Media (www.springer.com)

Preface

The "International Produced Water Conference: Environmental Risks and Advances in Mitigation Technologies" was held in St. John's, Newfoundland, Canada, on October 17–18, 2007, and brought together international stakeholders of the offshore oil and gas industry (scientists, environmental managers, regulators, industry and other non-governmental organizations) to share their concerns, knowledge and expertise on the discharge of produced water at sea. In particular, this conference enabled Canadian scientists to present and discuss the latest findings of environmental effects monitoring and research programs conducted by industry and government (e.g. Environmental Studies Research Funds – ESRF, Program of Energy Research and Development – PERD) for both regulatory compliance and the improvement of policies and regulations. Following the meeting, a decision was made by the conference sponsor, ESRF, to publish a book highlighting the major scientific findings that would assist in the decision-making process related to oil and gas exploration and development, both in Canada's frontier lands as well as globally.

The chapters cover a wide range of topics including:

- chemical composition and characterization,
- fate and transport in the environment,
- biological effects,
- monitoring technologies, and
- predictive modelling, and risk assessment and management.

The chapters cover the major scientific findings reported by the participants of the conference. In addition, a number of additional presentations were added to cover other recent scientific findings of interest identified by the conference session chairs and participants. Controversial papers were not avoided, in order to bring emerging issues of concern forward for discussion and evaluation for additional research. It is our hope that this collection of scientific manuscripts will aid in our evaluation of the potential impacts of produced water and the development of mitigation strategies and regulations to ensure the protection of our marine environment.

We take the opportunity to dedicate this publication to one of our colleagues and the author of a chapter in this volume: the late Maynard G. Brandsma. We hope that the research and information contained herein, on various aspects of produced water, forms a fitting memento to his memory.

Dartmouth, Nova Scotia Kenneth Lee
Duxbury, Massachusetts Jerry Neff

Contents

Contributors

Catherine Andrews Science Branch, Fisheries and Oceans Canada, Northwest Atlantic Fisheries Centre, St. John's, NL, Canada

Antônio S. Araujo Department of Chemistry, UFRN – Federal University of Rio Grande do Norte, Natal, RN, Brazil

Torgeir Bakke Norwegian Institute for Water Research (NIVA), Oslo, Norway

Bent Barman Skaare Department of Environmental Technology, Institute for Energy Technology, Kjeller, Norway

Monique T. Binet Centre for Environmental Contaminants Research, CSIRO Land and Water, Kirrawee, NSW, Australia

Mary C. Boatman U.S. Department of the Interior, Bureau of Ocean Energy Management, Regulation, and Enforcement, Herndon, VA, USA

Neil Bose Australian Maritime Hydrodynamics Research Centre, University of Tasmania, Launceston, TAS, Australia

Monica Boudreau Fisheries and Oceans Canada, Gulf Fisheries Centre, Moncton, NB, Canada

Maynard G. Brandsma Brandsma Engineering, Durango, CO, USA

Les Burridge Fisheries and Oceans Canada, St. Andrews Biological Station, St. Andrews, NB, Canada

Lionel Camus Akvaplan-niva, FRAM Centre, Tromsø, Norway

JoLynn Carroll Akvaplan-niva, FRAM Centre, Tromsø, Norway

Zhi Chen Department of Building Civil and Environmental Engineering, Concordia University, Montreal, PQ, Canada

Susan E. Cobanli Centre for Offshore Oil, Gas and Energy Research (COOGER), Fisheries and Oceans Canada, Bedford Institute of Oceanography, Dartmouth, NS, Canada

Susan Codi King Australian Institute of Marine Science, Townsville, QLD, Australia

Claire Conwell Cawthron Institute, Nelson, New Zealand

Simon Courtenay Fisheries and Oceans Canada, Gulf Fisheries Centre, Moncton, NB, Canada

Salve Dahle Akvaplan-niva, FRAM Centre, Tromsø, Norway

Elisabeth M. DeBlois Elisabeth DeBlois Inc., St. John's, NL, Canada; Jacques Whitford, St. John's, NL, Canada

L.A. dos Santos Petrobras Research and Development Center (CENPES/PDEDS/TTRA), Cidade Universitária, Ilha do Fundão, Rio de Janeiro, Brazil

Linda Fancey Science Branch, Fisheries and Oceans Canada, Northwest Atlantic Fisheries Centre, St. John's, NL, Canada

Boyd French Oceans Ltd, St. John's, NL, Canada

Tone K. Frost Statoil ASA, Trondheim, Norway

Irene T. Gabardo Department of Environmental Monitoring and Assessment, Petrobras Research Center/CENPES, Rio Janeiro, RJ, Brazil

Marthe Monique Gagnon Department of Environment and Agriculture, Curtin University, Perth, WA, Australia

Charles W. Greer National Research Council of Canada, Montreal, QC, Canada

Mary Haasch USEPA Mid-Continent Ecology Division, Molecular and Cellular Mechanisms Research, Duluth, MN, USA

Dounia Hamoutene Science Branch, Northwest Atlantic Fisheries Institute, Fisheries and Oceans Canada, St John's, NL, Canada

Jacqueline Hanlon Oceans Ltd., St. John's, NL, Canada

A.D. Hart CSA International, Inc., Stuart, FL, USA

Kelly Hawboldt Faculty of Engineering and Applied Science, Memorial University of Newfoundland, St. John's, NL, Canada

Libby Howitt Apache Energy Ltd., West Perth, WA, Australia

Tahir Husain Faculty of Engineering and Applied Science, Memorial University of Newfoundland, St. John's, NL, Canada

Per Erik Iversen Climate and Pollution Agency (Klif), Oslo, Norway

A. Jacinto Jr Petrobras Research and Development Center (CENPES/PDEDS/TTRA), Cidade Universitária, Ilha do Fundão, Rio de Janeiro, Brazil

Greg Janes Petro-Canada, East Coast Operations, St. John's, NL, Canada

Ståle Johnsen Statoil Research Centre, Trondheim, Norway

Jan Kihle Department of Environmental Technology, Institute for Energy Technology, Kjeller, Norway

Thomas King Centre for Offshore Oil and Gas and Energy Research (COOGER), Fisheries and Oceans Canada, Bedford Institute of Oceanography, Dartmouth, NS, Canada

Lars-Henrik Larsen Akvaplan-niva, FRAM Centre, Tromsø, Norway

B.A. Law Department of Fisheries and Oceans, Centre for Offshore Oil, Gas and Energy Research (COOGER), Fisheries and Oceans Canada, Bedford Institute of Oceanography, Dartmouth, NS, Canada

Kenneth Lee Centre for Offshore Oil, Gas and Energy Research (COOGER), Fisheries and Oceans Canada, Bedford Institute of Oceanography, Dartmouth, NS, Canada; Department of Building Civil and Environmental Engineering, Concordia University, Montreal, PQ, Canada

Monica Lyons Fisheries and Oceans Canada, St. Andrews Biological Station, St. Andrews, NB, Canada

G. Mabrouk Science Branch, Northwest Atlantic Fisheries Institute, Fisheries and Oceans Canada, St John's, NL, Canada

A. Mansour Science Branch, Northwest Atlantic Fisheries Institute, Fisheries and Oceans Canada, St John's, NL, Canada

Ann Mathieu Oceans Ltd., St John's, NL, Canada

Marcel V. Melo Petrobras Research and Development Center (CENPES/PDEDS/TTRA), Cidade Universitária, Ilha do Fundão, Rio de Janeiro, Brazil

Wynnann Melvin Oceans Ltd., St. John's, NL, Canada

T.G. Milligan Department of Fisheries and Oceans, Bedford Institute of Oceanography, Dartmouth, NS, Canada

Abdullah Mofarrah Faculty of Engineering and Applied Science, Memorial University of Newfoundland, St. John's, NL, Canada

Julie Mondon School of Life & Environmental Sciences, Deakin University, Warrnambool, VIC, Australia

Jochen Müeller National Research Centre for Ecotoxicology, Cooper Plains, QLD, Australia

Mark Myers Northwest Fisheries Science Center, National Oceanic and Atmospheric Administration, Seattle, WA, USA

Jerry Neff Neff & Associates, LLC, Duxbury, MA, USA

Ingunn Nilssen Statoil Research Centre, Rotvoll, Trondheim, Norway

Haibo Niu Centre for Offshore Oil, Gas and Energy Research (COOGER), Fisheries and Oceans Canada, Bedford Institute of Oceanography, Dartmouth, NS, Canada

Gro Harlaug Olsen Akvaplan-niva, FRAM Centre, Tromsø, Norway

C. Parrish Ocean Sciences Centre, Memorial University of Newfoundland, St John's, NL, Canada

Jerry F. Payne Science Branch, Fisheries and Oceans Canada, Northwest Atlantic Fisheries Centre, St. John's, NL, Canada

O.A. Pereira Jr Petrobras Research and Development Center (CENPES/PDEDS/TTRA), Cidade Universitária, Ilha do Fundão, Rio de Janeiro, Brazil

Eduardo B. Platte Department of Environmental Monitoring and Assessment, Petrobras Research Center/CENPES, Rio Janeiro, RJ, Brazil

Fernando H. Pulgatti Department of Statistics, Institute of Mathematics, UFRGS – Federal University of Rio Grande do Sul, Porto Alegre, RS, Brazil

Mark Reed Division for Marine Environmental Technology, SINTEF Materials and Chemistry, Trondheim, Norway

Brian J. Robinson Centre for Offshore Oil, Gas and Energy Research (COOGER), Fisheries and Oceans Canada, Bedford Institute of Oceanography, Dartmouth, NS, Canada

Barry R. Ruddick Dalhousie University, Halifax, NS, Canada

Henrik Rye Division for Environmental Technology, SINTEF Materials and Chemistry, Trondheim, Norway

S. Samuelson Science Branch, Northwest Atlantic Fisheries Institute, Fisheries and Oceans Canada, St John's, NL, Canada

T.C. Sauer ARCADIS-BBL, Cary, NC, USA

Jennifer L. Stauber Centre for Environmental Contaminants Research, CSIRO Land and Water, Kirrawee, NSW, Australia

Christopher T. Taggart Dalhousie University, Halifax, NS, Canada

Terje Torsvik Center for Integrated Petroleum Research, UNIFOB, Bergen, Norway

John H. Trefry Department of Marine & Environmental Systems, Florida Institute of Technology, Melbourne, FL, USA

Robert P. Trocine Department of Marine & Environmental Systems, Florida Institute of Technology, Melbourne, FL, USA

John A. Veil Veil Environmental, LLC, Annapolis, MD, USA

Brian Veitch Faculty of Engineering and Applied Science, Memorial University of Newfoundland, St. John's, NL, Canada

Ann Mari Vik Green Climate and Pollution Agency (Klif), Oslo, Norway

H. Volkoff Departments of Biology and Biochemistry, Memorial University of Newfoundland, St John's, NL, Canada

Francine M. Wight Petro-Canada, East Coast Operations, St. John's, NL, Canada

Urban P. Williams Petro-Canada, East Coast Operations, St. John's, NL, Canada

Trevor Winton Sinclair Knight Merz, Perth, WA, Australia

Gary Wohlgeschaffen Centre for Offshore Oil, Gas and Energy Research (COOGER), Fisheries and Oceans Canada, Bedford Institute of Oceanography, Dartmouth, NS, Canada

Ming Yang TUV NEL Ltd, Scottish Enterprise Technology Park, East Kilbride, Glasgow, UK

Philip A. Yeats Department of Fisheries and Oceans, Bedford Institute of Oceanography, Dartmouth, NS, Canada

C. William Yeung National Research Council of Canada, Montreal, QC, Canada

Lin Zhao Department of Building Civil and Environmental Engineering, Concordia University, Montreal, PQ, Canada

Shiqian Zhu Bioanalytical Systems, Inc., West Lafayette, IN, USA

About the Editors

Kenneth Lee, Ph.D., is the executive director of Fisheries and Oceans Canada's Centre for Offshore Oil, Gas and Energy Research (COOGER). This national Centre of Expertise is responsible for the planning and implementation of national and international research programs with government and academia to provide scientific knowledge and advice pertaining to the potential environmental impacts associated with the development of Canada's offshore oil and gas, and ocean renewable energy sectors.

As the current leader of the Offshore Environmental Impacts Program of the Panel of Energy Research and Development (PERD) under Natural Resources Canada, Dr. Lee is responsible for the coordination of multidisciplinary studies to assess the potential environmental risk associated with operations of the offshore oil and gas industry in Canada's frontier regions. Dr. Lee has also conducted numerous research studies on oil spill countermeasures, including large-scale field trials, to develop and evaluate the efficacy of natural attenuation, oil dispersant use, bioremediation and surf-washing. His research has also provided tools for the monitoring of environmental effects and habitat recovery following oil spill incidents. He served as co-chair of a working group to establish operational guidelines for marine oil spill bioremediation for the International Maritime Organization (IMO). Dr. Lee frequently serves as a science advisor on national and international oil spill response teams. During the recent Deepwater Horizon oil spill in the Gulf of Mexico, Dr. Lee was a member of the Joint Analysis Group (JAG) for surface and subsurface oceanographic, oil and dispersant data collected under the Unified Command and is currently serving on the National Research Council's Committee on the Effects of the Deepwater Horizon Mississippi Canyon-252 Oil Spill on Ecosystem Services.

Dr. Lee has published over 300 articles in scientific journals, technical reports and books. He has served as the chair of the Canadian National Committee for the

Scientific Committee Oceanic Research (CNC/SCOR) and is currently an active committee member on NATO's Science for Peace and Security Program. Dr. Lee is the recipient of the Fisheries and Oceans Canada Prix d'Excellence (Science) for his research contributions to environmental issues associated with offshore oil and gas activities, in addition to the Government of Canada Federal Partners on Technology Transfer (FPTT) Leadership Award for the development of marine oil spill countermeasures.

Jerry M. Neff, Ph.D., was an assistant/associate professor in the Biology Department of Texas A&M University from 1972 to 1980, where he did research on the aquatic toxicity of crude and refined oil, produced water and drilling muds. He continued this research at Battelle Memorial Institute between 1980 and 2005 as senior research leader at the Duxbury, Massachusetts, marine research laboratory. Currently, he is principal of Neff & Associates LLC and consults on environmental challenges related to offshore oil and gas operations and oil spills.

He is an internationally recognized authority on the fate and effects of petroleum hydrocarbons, oil well drilling fluids and produced waters in marine, freshwater and terrestrial environments. During the past 38 years, he has performed more than 150 research and monitoring programs on these and related subjects and assisted in environmental damage assessments following 18 oil spills in the USA, Europe and the Middle East for government and oil industry clients.

Dr. Neff has published more than 250 articles in scientific journals and books, including

- Neff, J.M. 2002. *Bioaccumulation in Marine Organisms. Effects of Contaminants from Oil Well Produced Water*. Elsevier, Amsterdam. 452 pp.
- Neff, J.M. 1979. *Polycyclic Aromatic Hydrocarbons in the Aquatic Environment: Sources, Fates, and Biological Effects*. Applied Science Publishers, London. 266 pp.
- Neff, J.M. 2010. *Fates and Effects of Water Based Drilling Muds and Cuttings in Cold Water Environments*. Report to Shell Exploration & Production, Anchorage AK. 310 pp.

Part I
Overview of Produced Water Fates and Effects

Chapter 1
Produced Water: Overview of Composition, Fates, and Effects

Jerry Neff, Kenneth Lee, and Elisabeth M. DeBlois

Abstract Produced water (formation and injected water containing production chemicals) represents the largest volume waste stream in oil and gas production operations on most offshore platforms. In 2003, an estimated 667 million metric tons (about 800 million m^3) of produced water were discharged to the ocean from offshore facilities throughout the world. There is considerable concern about the ocean disposal of produced water, because of the potential danger of chronic ecological harm. Produced water is a complex mixture of dissolved and particulate organic and inorganic chemicals in water that ranges from essentially freshwater to concentrated saline brine. The most abundant organic chemicals in most produced waters are water-soluble low molecular weight organic acids and monocyclic aromatic hydrocarbons. Concentrations of total PAH and higher molecular weight alkyl phenols, the main toxicants in produced water, typically range from about 0.040 to about 3 mg/L. The metals most frequently present in produced water at elevated concentrations, relative to those in seawater, include barium, iron, manganese, mercury, and zinc. Upon discharge to the ocean, produced water dilutes rapidly, often by 100-fold or more within 100 m of the discharge. The chemicals of greatest environmental concern in produced water, because their concentrations may be high enough to cause bioaccumulation and toxicity, include aromatic hydrocarbons, some alkylphenols, and a few metals. Marine animals near a produced water discharge may bioaccumulate metals, phenols, and hydrocarbons from the ambient water, their food, or bottom sediments. The general consensus of the International Produced Water Conference was that any effects of produced water on individual offshore production sites are likely to be minor. However, unresolved questions regarding aspects of produced water composition and its fate and potential effects on the ecosystem remain. Multidisciplinary scientific studies are needed under an ecosystem-based management (EBM) approach to provide information on the environmental fates (dispersion, precipitation, biological and abiotic transformation) and effects of chronic, low-level exposures to the different chemicals in produced water.

J. Neff (✉)
Neff & Associates, LLC, Duxbury, MA 02332, USA

K. Lee, J. Neff (eds.), *Produced Water*, DOI 10.1007/978-1-4614-0046-2_1,

1 Introduction

Produced water often is generated during the production of oil and gas from onshore and offshore wells. Formation water is seawater or freshwater that has been trapped for millions of years with oil and natural gas in a geologic reservoir consisting of a porous sedimentary rock formation between layers of impermeable rock within the earth's crust (Collins 1975). When a hydrocarbon reservoir is penetrated by a well, the produced fluids may contain this formation water, in addition to the oil, natural gas, and/or gas liquids. Freshwater, brine/seawater, and production chemicals sometimes are injected into a reservoir to enhance both recovery rates and the safety of operations; these surface waters and chemicals sometimes penetrate to the production zone and are recovered with oil and gas during production (Neff 2002; Veil et al. 2004). Produced water (formation and injected water containing production chemicals) represents the largest volume waste stream in oil and gas production operations on most offshore platforms (Stephenson 1991; Krause 1995). Produced water may account for 80% of the wastes and residuals produced from natural gas production operations (McCormack et al. 2001).

The ratio of produced water to oil equivalents (WOR) or the ratio of water to gas (WGR) produced from a well varies widely from essentially zero to more than 50 (98% water and 2% oil). The WGR usually is higher than the WOR. The average worldwide WOR is about 2–3 (Chapter 29, this volume). The volume of produced water generated usually increases as oil and gas production decreases (WOR and WGR increase) with the age of the well (Henderson et al. 1999). In nearly depleted fields, production may be 98% produced water and 2% fossil fuel (Stephenson 1992; Shaw et al. 1999). Mean WOR and WGR for oil and gas production from Federal offshore waters (>4.8 km from shore) of the USA are 1.04 and 86, respectively (Clark and Veil 2009). On the Canadian East Coast, the average WOR was 2 during the life of the Cohasset oil field (1992–1999), the first offshore production in Atlantic Canada (Ayers and Parker 2001). Oil and gas production from the Hibernia field on the Grand Banks is relatively dry, with a WOR of about 1 in September 2007 (Reuters 2007). The gas is reinjected for enhanced oil recovery.

In 2003, an estimated 667 million metric tons (about 800 million m^3) of produced water were discharged offshore throughout the world, including 21.1 million tons to offshore waters of North America, mostly the US Gulf of Mexico, and 358–419 million tons to offshore waters of Europe, mostly the North Sea (OGP 2004; Garland 2005). These are underestimates of actual discharges, because reporting of production to OGP (2004) ranged from 11 to 99% in the seven regions of the world monitored.

For example, the estimated total volume of produced water generated in US Federal offshore waters in 2007 during production of 75.7 million m^3 (476 million barrels) of oil and 2.8 billion ft^3 of natural gas was 93.4 million m^3, or 256,000 m^3/day (Clark and Veil 2009). About 22,000 m^3/day of this produced water was reinjected for enhanced recovery or disposal, and about 234,000 m^3/day was treated and discharged to the ocean.

Produced water production on the Norwegian continental shelf was 135 million m^3 in 2009, a reduction of about 10% from 173 million m^3 produced in 2008 (KLIF 2010). About 85% of the water was discharged to the ocean each year, the remainder was injected.

Off the coast of Atlantic Canada, produced water discharge from the Hibernia field increased from 17,000 m^3/day in July 2007 to 20,300 m^3/day in September 2007 as oil production declined (Reuters 2007). In 2009, the Venture field on the Canadian Scotian Shelf was discharging 100–600 m^3/day of produced water (personal communication, ExxonMobil).

There is considerable concern over the ocean disposal of produced water from production operations, because discharge is continuous during production, discharge volumes are increasing in most mature offshore production areas, and the concentrations of many potentially toxic organic compounds and metals are higher in treated produced water than in the receiving waters, raising concerns about chronic ecological harm.

2 Chemical Composition of Produced Water

Produced water is a complex mixture of dissolved and particulate organic and inorganic chemicals. The physical and chemical properties of produced water vary widely depending on the geologic age, depth, and geochemistry of the hydrocarbon-bearing formation, as well as the chemical composition of the oil and gas phases in the reservoir, and process chemicals added during production. Because no two produced waters are alike, region specific studies are needed to address the environmental risks from its discharge.

Produced water contains a variety of naturally occurring compounds that were dissolved or dispersed from the geologic formations and migration pathways in which the produced water resided for millions of years. These chemicals include inorganic salts, metals, radioisotopes, and a wide variety of organic chemicals, primarily hydrocarbons.

2.1 Salinity and Inorganic Ions

The salt concentration (salinity) of produced water may range from a few parts per thousand (‰) to that of a saturated brine (~300‰; see Chapter 19, this volume), compared to a salinity of 32–36‰ for seawater (Rittenhouse et al. 1969; Large 1990; Table 1.1). Most produced waters have salinities greater than that of seawater and, therefore, are denser than seawater (Collins 1975). Hibernia produced water has a salinity of 46–195‰ (Ayers and Parker 2001).

Produced water contains the same salts as seawater, with sodium and chloride the most abundant ions (Table 1.1). The most abundant inorganic ions in high-salinity produced water are, in order of relative abundance sodium, chloride, calcium,

Table 1.1 Concentrations (mg/kg or parts per million) of several elements and inorganic ions in produced waters of different geologic ages compared with average concentrations in 35‰ seawater (Collins 1975)

		Produced water	
Element/ion	Seawater	Highest concentration (age[a])	Range of mean concentrations
Salinity	35,000	–	<5,000–>300,000,000
Sodium	10,760	120,000 (J)	23,000–57,300
Chloride	19,353	270,000 (P)	46,100–141,000
Calcium	416	205,000 (P)	2,530–25,800
Magnesium	1,294	26,000 (D)	530–4,300
Potassium	387	11,600 (D)	130–3,100
Sulfate	2,712	8,400 (T)	210–1,170
Bromide	87	6,000 (J)	46–1,200
Strontium	0.008	4,500 (P)	7–1,000
Ammonium	–	3,300 (P)	23–300
Bicarbonate	142	3,600 (T)	77–560
Iodide	167	1,410 (P)	3–210
Boron	4.45	450 (T)	8–40
Carbonate	–	450 (M)	30–450
Lithium	0.17	400 (J)	3–50

[a]D, Devonian; J, Jurassic; M, Mississippian; P, Pennsylvanian; T, Tertiary

magnesium, potassium, sulfate, bromide, bicarbonate, and iodide. Concentration ratios of many of these ions are different in seawater and produced water, possibly contributing to the aquatic toxicity of produced water (Pillard et al. 1996).

Sulfate and sulfide concentrations usually are low, allowing barium and other elements that form insoluble sulfates and sulfides to be present in solution at high concentrations. Produced water from sour oil/gas wells may contain high concentrations of sulfide and elemental sulfur. For example, produced water from an offshore California well contained 48–216 mg/L sulfide and 0.6–42 mg/L sulfur (Witter and Jones 1999). If seawater, that naturally contains a high concentration of sulfate (~2,712 mg/L), is injected into the formation to enhance oil and gas recovery and mixes with the formation water, barium and calcium may precipitate as scale in the production pipes and the concentration of dissolved barium in the produced water decreases (Stephenson et al. 1994). Any radium radioisotopes in the produced water co-precipitate with the barium scale. Some Brazilian offshore produced waters contain more than 2,000 mg/L sulfate, a concentration high enough to promote barium and calcium scale formation (Chapter 3, this volume). Hibernia produced water, recovered from a reservoir on the Grand Banks off Newfoundland, Canada, in 2000, contained 248–339 mg/L SO_4, low enough to reduce the likelihood of producing large amounts of barium and calcium scale (Ayers and Parker 2001).

Ammonium ion may be present in some produced waters at elevated concentrations, possibly eliciting inhibitory (toxic) and/or stimulatory (e.g. eutrophication) responses from resident biota (Anderson et al. 2000, Chapter 6, this volume).

Hibernia produced water contains about 11 mg/L NH_3 (Chapter 6, this volume). Brazilian produced water contains 22–800 mg/L NH_3 (Chapter 3, this volume). However, concentrations of nitrate and phosphate often are low in produced waters (Hibernia produced water contains about 0.35 mg/L P and 0.02 mg/L NO_3), decreasing the likelihood of eutrophication in the receiving waters (Johnsen et al. 2004).

A large zone of hypoxic (dissolved oxygen <2.0 mg/L) water develops in nearshore bottom water over an area of more than 17,000 km^2 off Louisiana each summer (Rabalais 2005; Veil et al. 2005; Bierman et al. 2007). The hypoxia is caused primarily by the discharge of large volumes of water containing high concentrations of primary nutrients from the Mississippi River. However, as there are approximately 287 oil and gas production platforms in the hypoxic area (Veil et al. 2005), many of which discharge treated produced water, a comprehensive monitoring program was performed in this area to determine if the discharged production water ($\sim$81,000 m^3/day) was contributing significant amounts of nutrients to the Louisiana nearshore waters (Veil et al. 2005; Bierman et al. 2007). Produced water from 50 platforms, discharging $\sim$280,000 m^3/day to the hypoxic zone, was analyzed for nutrients (Table 1.2). Produced water from most gas platforms contained higher concentrations of BOD and all nutrients except ammonia. The ratio of estimated annual nutrient loading from all the platforms in the hypoxic zone to the annual nutrient loading from the Mississippi River ranged from 0.00003 for nitrate to 0.07 for ammonia. The investigators concluded that produced water discharges contributed very little of the organic loading contributing to oxygen depletion in bottom waters of the hypoxic zone off Louisiana (Rabalais 2005; Bierman et al. 2007).

2.2 Total Organic Carbon

The concentration of total organic carbon (TOC) in produced water ranges from less than 0.1 to more than 11,000 mg/L and is highly variable from one well to another (Tables 1.2 and 1.3). Produced water from Hibernia has a TOC concentration of approximately 300 mg/L (Ayers and Parker 2001). Produced water from wells off

Table 1.2 Mean biological oxygen demand and concentrations (mg/L) of several primary nutrients in produced water from 50 platforms discharging to the hypoxic zone in the Gulf of Mexico off the coast of Louisiana; mass loadings are concentration × discharge volume in kg/day (from Veil et al. 2005; Bierman et al. 2007)

Parameter	Mostly oil	Mostly gas	Oil and gas	Mass loading
No. platforms	6	20	24	50
Biological oxygen demand (BOD)	595	1,444	642	16,330
Total organic carbon (TOC)	551	888	297	6,400
Nitrate (NO_3^-)	1.14	2.71	1.94	31.0
Nitrite (NO_2^-)	0.05	0.05	0.05	1.40
Ammonia (NH_4)	92	57	85	2,160
Orthophospate (PO_4^{3-})	0.34	0.61	0.30	10.2
Total phosphorous	0.62	0.86	0.61	17.0

Table 1.3 Concentration ranges (mg/L or parts per million) of several classes of naturally occurring organic chemicals in produced water worldwide (from Neff 2002)

Chemical class	Concentration range
Total organic carbon	≤0.1–>11,000
Total organic acids	≤0.001–10,000
Total saturated hydrocarbons	17–30
Total benzene, toluene, ethylbenzene, and xylenes (BTEX)	0.068–578
Total polycyclic aromatic hydrocarbons (PAH)	0.04–3.0
Total steranes/triterpanes	0.14–0.175
Ketones	1.0–2.0
Total phenols (primarily C_0-C_5-phenols)	0.4–23

Louisiana contain 67–620 mg/L dissolved TOC (DOC) and 5–127 mg/L particulate TOC (POC) (Veil et al. 2005). A large fraction of the DOC may be in colloidal suspension (Means et al. 1989).

2.3 Organic Acids

The organic acids in produced water are mono- and di-carboxylic acids (—COOH) of saturated (aliphatic) and aromatic hydrocarbons. Much of the TOC in produced water consists of a mixture of low molecular weight carboxylic acids, such as formic, acetic, propanoic, butanoic, pentanoic, and hexanoic acids (Somerville et al. 1987; Means and Hubbard 1987; Barth 1991; Røe Utvik 1999; Table 1.3). The most abundant organic acid usually is formic or acetic acid and abundance typically decreases with increasing molecular weight (Fisher 1987; MacGowan and Surdam 1988; Table 1.4). Strømgren et al. (1995) found 43–817 mg/L total C_1 through C_5 organic acids and 0.04–0.5 mg/L total C_8 through C_{17} organic acids in three samples of North Sea produced water. Several samples of produced water from North Sea, US Gulf of Mexico, and California platforms contained 60–7,100 mg/L total low molecular weight aliphatic organic acids (Table 1.4, MacGowan and Surdam 1988; Jacobs et al. 1992; Flynn et al. 1995; Røe Utvik 1999). Small amounts of aromatic acids also may be present in produced water (Rabalais et al. 1991; Barman Skaare et al. 2007). Produced water from coastal waters of Louisiana contained low concentrations of aliphatic and aromatic acids (Table 1.5). Aliphatic acids were more abundant than benzoic and methylbenzoic acids (Rabalais et al. 1991).

These low molecular weight organic acids are readily biosynthesized and biodegraded by bacteria, fungi, and plants, and so represent nutrients for phyto- and zoo plankton growth. Organic acids are produced by hydrous pyrolysis or microbial degradation of hydrocarbons in the hydrocarbon-bearing formation (Borgund and Barth 1994; Tomczyk et al. 2001; Barman Skaare et al. 2007).

Many crude oils, particularly those that have been biodegraded in the formation, contain high concentrations of naphthenic acids (cycloalkane and/or benzene

Table 1.4 Concentrations (mg/L = ppm) of low molecular weight organic acids in produced water from four production facilities on the Norwegian continental shelf (Røe Utvik 1999), in the Gulf of Mexico off the Texas and Louisiana coast, and in the Santa Maria Basin off the California coast (MacGowan and Surdam 1988)

Organic acid	Formula	Offshore USA	Norwegian North Sea
Formic acid	$CHOOH$	ND–68	26–584
Acetic acid	CH_3COOH	8–5,735	Not determined
Propanoic acid	CH_3CH_2COOH	ND–4,400	36–98
Butanoic acid	$CH_3(CH_2)_2COOH$	ND–44	ND–46
Pentanoic acid	$CH_3(CH_2)_3COOH$	ND–24	ND–33
Hexanoic acid	$CH_3(CH_2)_4COOH$	Not determined	ND
Oxalic acid	$COOHCOOH$	ND–495	Not determined
Malonic acid	$CH_2(COOH)_2$	ND–1,540	Not determined
Total measured organic acids	–	98–7,160	62–761

NA: not analyzed. ND: not detected

Table 1.5 Range of concentrations (mg/L) of aliphatic and aromatic acids in produced water from seven treatment facilities in coastal Louisiana (from Rabalais et al. 1991)

Chemical	Pass Furchon	Bayou Rigoud	5 Other facilities
Aliphatic acids	8.5–120	1.8–78.0	7.9–75.0
Benzoic acid	0.92–15.0	0.13–16.0	1.2–13.0
C_1-Benzoic acid	1.6–11.0	0.089–14.0	1.6–16.0
C_2-Benzoic acid	0.42–2.3	0.043–2.7	0.29–3.8

carboxylic acids with one or more saturated 5- or 6-ring carbon or aromatic structures (Barman Skaare et al. 2007; Grewer et al. 2010). Naphthenic acids are slightly water-soluble and, when abundant in the crude oil, also are present in the associated produced water. Heavy crude oils, bitumens, and process water associated from the oil sands of Alberta, Canada contain high concentrations of hundreds of naphthenic acids with 8–30 carbons. Several process waters from Syncrude and Suncor contain 24–68 mg/L total naphthenic acids (Holowenko et al. 2002).

Produced water from the Troll C platform on the Norwegian continental self contains highly variable concentrations and compositions of naphthenic acids, representing different degrees of anaerobic biodegradation of crude oil in different parts of the reservoir (Barman Skaare et al. 2007). The most abundant napththenic acids in Troll produced water included a series of alkylated benzoic acids, salicylic acid (2-hydroxybenzoic acid), and a variety of naphthoic acids and their ring-reduced analogues. These organic acids were produced by anaerobic biodegradation of aromatic hydrocarbons in the crude oil in the reservoir. Anaerobic bacteria may be abundant in the oil/gas reservoir if formation temperature is below about 100°C (Chapter 18, this volume). Naphthenic acids in crude oil and produced water are of concern because their acidity contributes to corrosion of production pipe and they contribute to the toxicity of produced water (Thomas et al. 2009).

2.4 Petroleum Hydrocarbons

Petroleum hydrocarbons, organic chemicals consisting of just carbon and hydrogen, are the chemicals of greatest environmental concern in produced water. Petroleum hydrocarbons are classified into two groups: saturated hydrocarbons and aromatic hydrocarbons. The solubility of petroleum hydrocarbons in water decreases as their size (molecular weight) increases; aromatic hydrocarbons are more water-soluble than saturated hydrocarbons of the same molecular weight. The hydrocarbons in produced water appear in both dissolved and dispersed (oil droplets) form.

Existing oil/water separators, such as hydrocyclones, are quite efficient in removing oil droplets, but not dissolved hydrocarbons, organic acids, phenols, and metals from produced water. Thus, much of the petroleum hydrocarbons discharged to the ocean in properly treated produced water are dissolved low molecular weight aromatic hydrocarbons and smaller amounts of saturated hydrocarbons. Because there are no treatment procedures that are 100% effective, treated produced water still contains some dispersed oil (droplet size ranging from 1 to 10 μm) (Johnsen et al. 2004). The droplets contain most of the higher molecular weight, less soluble saturated and aromatic hydrocarbons (Faksness et al. 2004).

2.4.1 BTEX and Benzenes

The most abundant hydrocarbons in produced water are the one-ring aromatic hydrocarbons, benzene, toluene, ethylbenzene, and xylenes (BTEX), and low molecular weight saturated hydrocarbons. BTEX may be present in untreated produced water from different sources at concentrations as high as 600 mg/L (Table 1.3). Produced water also contains small amounts of C_3- and C_4-benzenes (Table 1.6). Benzene usually is most abundant and concentration decreases with

Table 1.6 Concentrations (mg/L) of BTEX and selected C_3- and C_4-benzenes in produced water from four platforms in the US Gulf of Mexico and from three offshore production facilities in Indonesia (from Neff 2002)

Compound	7 Gulf of Mexico produced waters	3 Indonesian produced waters
Benzene	0.44–2.80	0.084–2.30
Toluene	0.34–1.70	0.089–0.80
Ethylbenzene	0.026–0.11	0.026–0.056
Xylenes (3 isomers)	0.16–0.72	0.013–0.48
Total BTEX	0.96–5.33	0.33–3.64
Propylbenzenes (2 isomers)	NA	ND–0.01
Methylethylbenzenes (3 isomers)	NA	0.031–0.051
Trimethylbenzenes (3 isomers)	NA	0.056–0.10
Total C_3-benzenes	0.012–0.30	0.066–0.16
Methylpropylbenzenes (5 isomers)	NA	ND–0.006
Diethylbenzenes (3 isomers)	NA	ND
Dimethylethylbenzenes (6 isomers)	NA	ND–0.033
Total C_4-benzenes	ND–0.12	ND–0.068

NA: not analyzed. ND: not detected

increasing alkylation (Dórea et al. 2007; Chapters 3 and 24, this volume). Because BTEX are extremely volatile, they are lost rapidly during produced water treatment by air stripping and during initial mixing of the produced water plume in the ocean (Terrens and Tait 1996).

Saturated hydrocarbons, because of their low solubilities, nearly always are present at low concentrations in produced water (Table 1.3), unless the produced water treatment system is not working properly. Produced water from the US Gulf of Mexico and offshore Thailand contained 0.6–7.8 mg/L total C_{10}- through C_{34}-*n*-alkanes (Neff 2002). The shorter chain-length alkanes, C_{10} through C_{22}, were more abundant than the longer ones. Most of the alkanes probably were associated with droplets.

2.4.2 Polycyclic Aromatic Hydrocarbons

Polycyclic aromatic hydrocarbons (PAH) are defined as hydrocarbons containing two or more fused aromatic rings. These are the petroleum hydrocarbons of greatest environmental concern in produced water because of their toxicity and persistence in the marine environment (Neff 1987, 2002). Concentrations of total PAH in produced water typically range from about 0.040 to 3 mg/L (Tables 1.3 and 1.7) and consist primarily of the most water-soluble congeners, the 2- and 3-ring PAH, such as naphthalene, phenanthrene, and their alkyl homologues (Table 1.7, Fig. 1.1). Higher molecular weight, 4- through 6-ring PAH rarely are detected in properly treated produced water. Because of their low aqueous solubilities, they are associated primarily with dispersed oil droplets (Faksness et al. 2004; Johnsen et al. 2004). Burns and Codi (1999) reported that 5–10% of the total PAH in produced water from the Harriet A platform on the Northwest Shelf of Australia were in the "dissolved" fraction. The dissolved fraction contained mainly alkylnaphthalenes and traces of alkylphenanthrenes. The particulate (droplet) fraction also contained high concentrations of naphthalenes and phenanthrenes, and contained almost all the dibenzothiophenes, fluoranthenes/pyrenes, and chrysenes in the produced water.

2.4.3 Phenols

Concentrations of total phenols in produced water usually are less than 20 mg/L (Table 1.3). Measured concentrations of total phenols in produced waters from the Louisiana Gulf coast and the Norwegian Sector of the North Sea range from 2.1 to 4.5 mg/L and 0.36 to 16.8 mg/L, respectively (Neff 2002; Johnsen et al. 2004). The most abundant phenols in these produced waters are phenol, methylphenols, and dimethylphenols. The abundance of alkyl phenols usually decreases logarithmically with increasing number of alkyl carbons (Boitsov et al. 2007; Fig. 1.2). Long-chain alkylphenols with seven to nine alkyl carbons are the most toxic phenols, exhibiting strong endocrine disruption. They are quite rare in produced water from the Norwegian continental shelf (Fig. 1.2). The concentration of 4-*n*-nonylphenol (the most toxic alkylphenol) in produced waters from six Norwegian platforms

Table 1.7 Concentrations (μg/L = parts per billion: ppb) of individual polycyclic aromatic hydrocarbons (PAH) or alkyl congener groups in produced water from the Scotian Shelf and Grand Banks, Canada, the US Gulf of Mexico, and the North Sea

Compound	Gulf of Mexico[a]	North Sea[a]	Scotian Shelf[b]	Grand Banks[c]
Naphthalene	5.3–90.2	237–394	1,512	131
C_1-Naphthalenes	4.2–73.2	123–354	499	186
C_2-Naphthalenes	4.4–88.2	26.1–260	92	163
C_3-Naphthalenes	2.8–82.6	19.3–81.3	17	97.2
C_4-Naphthalenes	1.0–52.4	1.1–75.7	3.0	54.1
Acenaphthylene	ND–1.1	ND	1.3	2.3
Acenaphthene	ND–0.10	0.37–4.1	ND	ND
Biphenyl	0.36–10.6	12.1–51.7	ND	ND
Fluorene	0.06–2.8	2.6–21.7	13	16.5
C_1-Fluorenes	0.09–8.7	1.1–27.3	3	23.7
C_2-Fluorenes	0.20–15.5	0.54–33.2	0.35	4.8
C_3-Fluorenes	0.27–17.6	0.30–25.5	ND	ND
Anthracene	ND–0.45	ND	0.26	ND
Phenanthrene	0.11–8.8	1.3–32.0	4.0	29.3
C_1-Phenanthrenes	0.24–25.1	0.86–51.9	1.30	45.0
C_2-Phenanthrenes	0.25–31.2	0.41–51.8	0.55	37.1
C_3-Phenanthrenes	ND–22.5	0.20–34.3	0.37	24.4
C_4-Phenanthrenes	ND–11.3	0.50–27.2	ND	13.2
Fluoranthene	ND–0.12	0.01–1.1	0.39	0.51
Pyrene	0.01–0.29	0.03–1.9	0.36	0.94
C_1-Fluoranthenes/ Pyrenes	ND–2.4	0.07–10.3	0.43	5.8
C_2-Fluoranthenes/ Pyrenes	ND–4.4	0.21–11.6	ND	9.1
Benz(a)anthracene	ND–0.20	0.01–0.74	0.32	0.60
Chrysene	ND–0.85	0.02–2.4	ND	3.6
C_1-Chrysenes	ND–2.4	0.06–4.4	ND	6.3
C_2-Chrysenes	ND–3.5	1.3–5.9	ND	18.8
C_3-Chrysenes	ND–3.3	0.68–3.5	ND	6.7
C_4-Chrysenes	ND–2.6	ND	ND	4.2
Benzo(b)fluoranthene	ND–0.03	0.01–0.54	ND	0.61
Benzo(k)fluoranthene	ND–0.07	0.006–0.15	ND	ND
Benzo(e)pyrene	ND–0.10	0.01–0.82	ND	0.83
Benzo(a)pyrene	ND–0.09	0.01–0.41	ND	0.38
Perylene	0.04–2.0	0.005–0.11	ND	ND
Indeno(1,2,3-cd)pyrene	ND–0.01	0.022–0.23	ND	ND
Dibenz(a,h)anthracene	ND–0.02	0.012–0.10	ND	0.21
Benzo(ghi)perylene	ND–0.03	0.01–0.28	ND	0.17
Total PAHs	40–600	419–1,559	2,148	845

[a]Neff (2002)
[b]Thebaud (DFO-COOGER unpublished data)
[c]Hibernia (DFO-COOGER unpublished data)
ND: not detected

ranged from 0.001 to 0.012 mg/L. Five other samples did not contain detectable concentrations of nonylphenol. The concentrations of C_6- through C_9-alkylphenols are highly correlated with the concentration of dispersed oil droplets in produced water (Faksness et al. 2004).

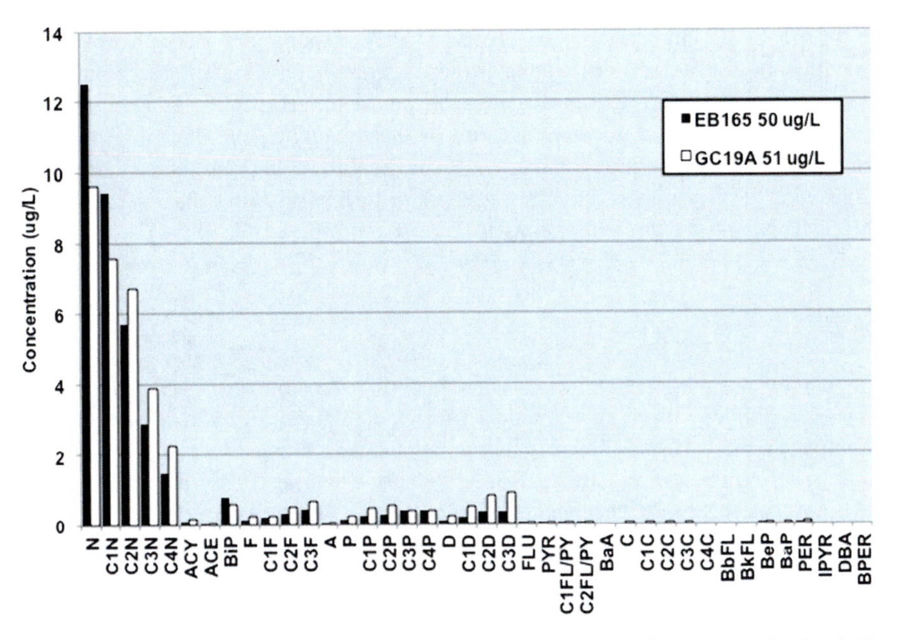

Fig. 1.1 Concentrations (μg/L) of individual PAH in produced water from two production facilities in the US Gulf of Mexico. The mean concentration of total PAH in each produced water is included in the legend. The names of individual PAH (x axis) are in Table 1.7 (based on data in Chapter 24, this volume)

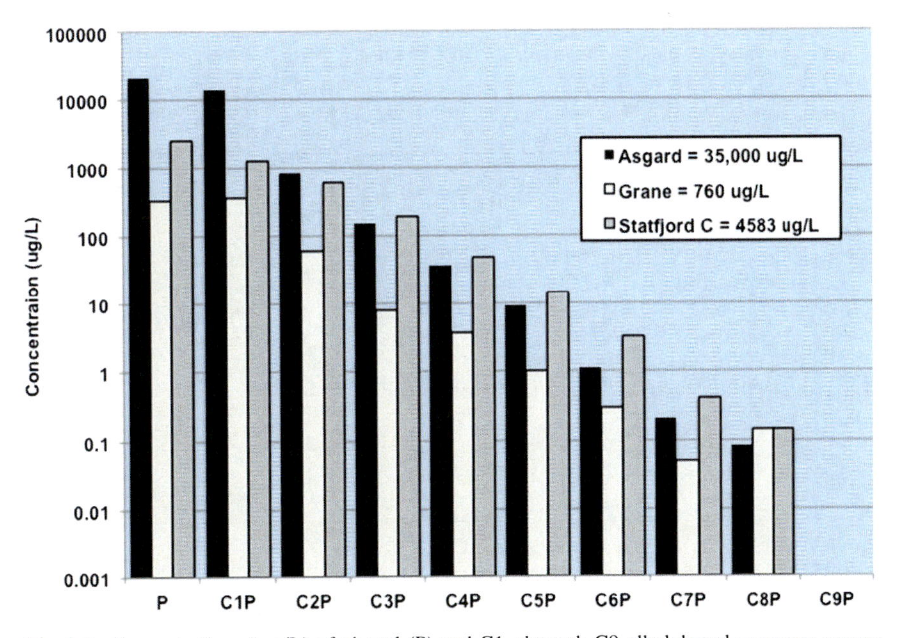

Fig. 1.2 Concentrations (μg/L) of phenol (P) and C1- through C9-alkylphenol congener groups (x axis) in produced water discharged from three production facilities on the Norwegian outer continental shelf. The mean concentration of total phenols in each of the three produced waters is included in the legend (data from Boitsov et al. 2007)

Alkylphenol ethoxylate surfactants (APE), containing octylphenols and nonylphenols, are sometimes used in the production system to facilitate the pumping of viscous or waxy crude oils. If the surfactant degrades, some alkylphenols may dissolve in the produced water. Because of the toxicity of the more highly alkylated phenols as endocrine disruptors, alkylphenol ethoxylate surfactants have been replaced in applications where the surfactant or its degradation products may reach the environment in significant amounts (Getliff and James 1996).

2.5 Metals

Produced water may contain several metals in dissolved or microparticulate forms. The type, concentration, and chemical species of metals in produced waters from different sources is variable, depending on the age and geology of the formations from which the oil and gas are produced (Collins 1975) and the amount and inorganic chemical composition of flood water injected into the hydrocarbon reservoir. A few metals may be present in produced waters from different sources at concentrations substantially higher (1,000-fold or more) than their concentrations in clean natural seawater. The metals most frequently present in produced water at elevated concentrations, relative to those in seawater, include barium, iron, manganese, mercury, and zinc (Neff et al. 1987; Table 1.8). Usually, only

Table 1.8 Concentration ranges (μg/L = ppb) of several metals in seawater and in produced water from the Scotian Shelf and the Grand Banks, Canada, compared to produced water discharged to northwestern Gulf of Mexico and the Norwegian sector of the North Sea

Metal	Seawater	Gulf of Mexico[a]	North Sea[b]	Scotian Shelf[c]	Grand Banks[d]
Arsenic	1–3	0.5–31	0.96–1.0	90	<10
Barium	3–34	81,000–342,000	107,000–228,000	13,500	301–354
Cadmium	0.001–0.1	<0.05–1.0	0.45–1.0	<10	<0.02–0.04
Chromium	0.1–0.55	<0.1–1.4	5–34	<1–10	<1
Copper	0.03–0.35	<0.2	12–60	137	<5
Iron	0.008–2.0	10,000–37,000	4,200–11,300	12,000–28,000	1,910–3,440
Lead	0.001–0.1	<0.1–28	0.4–10.2	<0.1–45	0.09–0.62
Manganese	0.03–1.0	1,000–7,000	NA	1,300–2,300	81–565
Mercury	0.00007–0.006	<0.01–0.2	0.017–2.74	<10	NA
Molybdenum	8–13	0.3–2.2	NA	NA	<1
Nickel	0.1–1.0	<1.0–7.0	22–176	<0.1–420	1.7–18
Vanadium	1.9	<1.2	NA	NA	<0.1–0.6
Zinc	0.006–0.12	10–3,600	10–340	10–26,000	<1–27

[a]Combined results from seven platforms (Neff 2002)
[b]Combined results from 12 platforms (Neff 2002)
[c]SOEP/DFO
[d]Combined results from Hibernia and Terra Nova (DFO-COOGER unpublished data)
NA: not analyzed

a few of these metals are present at elevated concentrations in a particular produced water sample. Concentrations of dissolved and particulate barium, iron, and manganese in produced water from the Hibernia facility off Newfoundland are substantially higher than concentrations in clean seawater (Chapter 6, this volume).

Because formation water is anoxic, iron and manganese may be present in solution at high concentrations. When formation waters containing high concentrations of these metals are brought to the surface and exposed to the atmosphere, the iron and manganese precipitate as iron and manganese oxyhydroxides. Several other metals in produced water may co-precipitate with iron and manganese and be dispersed, adsorbed to, or complexed with very fine, solid hydrous Fe and Mg oxides in the receiving waters (Lee et al. 2005a; Azetsu-Scott et al. 2007). Zinc and possibly lead in produced water could be derived in part from galvanized steel structures in contact with the produced water or with other waste streams that may be treated in the oil/water separator system.

2.6 Radioisotopes

Naturally occurring radioactive material (NORM) is present in produced water in many parts of the world. The most abundant NORM radionuclides in produced water are the natural radioactive elements, radium-226 and radium-228 (^{226}Ra and ^{228}Ra). Radium is derived from the radioactive decay of uranium-238 and thorium-232 associated with certain rocks and clays in the hydrocarbon reservoir (Reid 1983; Kraemer and Reid 1984; Michel 1990). ^{226}Ra (half-life 1,601 years) is an α-emitting daughter of uranium-238 and uranium-234, ^{228}Ra (half-life 5.7 years) is a β-emitting daughter of thorium-232.

The concentration of radionuclides, such as radium, in environmental media is measured as the rate of radioactive decay (number of disintegrations per minute), usually as picocuries/L (pCi/L) or becquerels/L (Bq/L). A pCi is equivalent to 2.22 disintegrations per minute (dpm) or 0.037 becquerels (Bq). One pCi is equivalent to 1 pg of ^{226}Ra or 0.037 pg of ^{228}Ra (Neff 2002).

The surface waters of the ocean have ^{226}Ra and ^{228}Ra activities of 0.027–0.04 pCi/L (0.027–0.04 pg/L: parts/quadrillion) and 0.005–0.012 pCi/L (0.000018–0.00004 pg/L), respectively (Santschi and Honeyman 1989; Nozaki 1991; Table 1.9). Produced water from some onshore and offshore production facilities worldwide contain very high ^{226}Ra and ^{228}Ra activity (Jonkers et al. 1997), relative to activity in nearshore and ocean waters. Concentrations of total ^{226}Ra plus ^{228}Ra in produced water from oil, gas, and geothermal wells along the US Gulf of Mexico coast range from less than 0.2 pCi/L to 13,808 pCi/L (Kraemer and Reid 1984; Fisher 1987; Neff et al. 1989; Hart et al. 1995; Table 1.9). There is no correlation between the concentrations of the two radium isotopes in produced water from the Gulf of Mexico, because of their different origins in the geologic formation. However, there is a good correlation between ^{228}Ra and ^{226}Ra activity in produced

Table 1.9 Mean or range of activities of ^{226}Ra and ^{228}Ra in produced water from different locations, activities are pCi/L (1 pCi = 0.037 Bq)

Location	Radium-226	Radium-228	Reference
Ocean water (background)	0.027−0.04	0.005–0.012	Santschi and Honeyman (1989), Nozaki (1991)
Worldwide	0.05–32,400	8.1–4,860	Jonkers et al. (1997)
Scotian Shelf	1.2	9.2	Nelson (2007)
Grand Banks	33.0	229.7	Nelson (2007)
Texas	0.1−5,150	NA	Fisher (1987)
Louisiana Gulf Coast	ND−1,565	ND−1,509	Kraemer and Reid (1984)
Offshore US Gulf of Mexico	91.2−1,494	162−600	Hart et al. (1995)
Santa Barbara Channel, CA	165	137	Neff (1997)
Cook Inlet, AK	<0.4 − 9.7	NA	Neff (1991)
North Sea	44.8	105	Stephenson et al. (1994)
Dutch North Sea	<54–8,154	<27–540	NRPA (2004)
Norwegian continental shelf	ND–432	ND–567	NRPA (2004)
S. Java Sea, Indonesia	7.6–56.5	0.6−17.7	Neff and Foster (1997)
Offshore Brazil	0.5–294	<2–183	Chapter 3, this volume

NA: not analyzed. ND: not detected

water from the Norwegian continental shelf, probably because most Norwegian produced water contains very low radium activity, with relatively few, mostly from the Sleipner and Njord fields, containing relatively high ^{228}Ra and ^{226}Ra activity (NRPA 2004).

Radium activity in produced water from offshore production areas other than the northern Gulf of Mexico often is low (Table 1.9), with mean ^{226}Ra and ^{228}Ra activity usually less than 200 pCi/L. Produced water from some production facilities in the North Sea, particularly the Dutch sector, contain high ^{226}Ra activity. Preliminary studies of produced water from platforms discharging to Atlantic Canada show low radium isotope activity, still several orders of magnitude higher than activity in natural seawater. However, due to high natural dispersion, only background activity can be detected in seawater samples collected near production platforms located on the Grand Banks and the Scotian Shelf (Nelson 2007). Slightly elevated radium isotope activity has been detected near production facilities on the Norwegian continental shelf (NRPA 2004).

Several other radionuclides may be present at low activity in produced water. The only radioisotope that sometimes is present in produced water at a higher activity than that of ^{226}Ra and ^{228}Ra is ^{210}Pb, a daughter of ^{226}Ra (Table 1.10). Mean ^{210}Pb activity in treated produced water from four platforms off Louisiana ranged from 5.60 ± 5.50 pCi/L to 12.50 ± 2.60 pCi/L (Hart et al. 1995). Another daughter of ^{226}Ra, ^{210}Po, is present at low activity in produced water from the North Sea, probably because of its short half-life, 138.4 days (NRPA 2004). Activity of parents of the radium isotopes, ^{238}U and ^{232}Th, are low, as is the short-lived ^{224}Ra (half-life, 3.66 days).

Table 1.10 Activity (pCi/L except where noted) of radium isotopes and a few of their parents and daughters in produced water and seawater (from Jonkers et al. 1997 and NRPA 2004)

Radionuclide	Seawater	Produced water
^{226}Ra	0.027–0.04	0.054–32,400
^{228}Ra	0.005–0.03	8.1–4,860
^{224}Ra	0.0002–0.008[a]	13.5–1,080
^{238}U	1.1	0.008–2.7
^{232}Th	0.003	0.008–0.027
^{210}Pb	0.026–0.12[b]	1.35–5,130
^{210}Po	0.018–0.068[b]	0.005–0.17

[a]Rasmussen (2003)
[b]Cherry et al. (1987)

2.7 Production Chemicals

Large numbers of specialty additives (treatment chemicals) are available for use in the production system of a well to aid in recovery and pumping of hydrocarbons, to protect the system from corrosion, to facilitate the separation of oil, gas, and water, and prevent methane hydrate (ice) formation in gas production systems. These include biocides, scale inhibitors, emulsion-breakers, and gas-treating chemicals (Table 1.11). Many of these chemicals are more soluble in oil than in produced water and remain in the oil phase. Others are water-soluble, remain in the produced water, and are disposed with it (Tables 1.11 and 1.12). Concentrations of most production chemicals are low in treated produced water (Table 1.12). Corrosion inhibitors, scale inhibitors, and gas treatment chemicals (glycol and methanol) may be high in production systems with these problems.

The point in the production stream where the chemical is added influences the amount that may be discharged or reinjected with the produced water. Treatment chemicals are used to solve specific problems and are not added if there is no demonstrated need. Environmental problems may arise if the more toxic treatment chemicals, such as biocide or corrosion inhibitor, are used at a frequency or concentration greater than needed to solve the problem. Environmental concerns

Table 1.11 Production chemicals used on North Sea oil and gas platforms and the estimated amounts in produced water discharged to the ocean (from Johnsen et al. 2004)

Chemical	Typical use concentration (ppm, v/v)	Phase association of chemical	Amount discharged to North Sea (t/y)
Scale inhibitor	3–10	Water	1,143
Corrosion inhibitor	25–100	Oil	216
H_2O/O_2 scavenger	5–15	Water	22
Biocide	10–200	Water	81
Emulsion breaker	10–200	Oil	9
Coagulants and flocculants	<3	Water	197
Gas treatment chemicals	Variable	Water	2,846

Table 1.12 Concentrations (mg/L) of several production treatment chemicals in produced water discharged in the Gullfaks and Statfjord fields in the North Sea with trade names of commercial formulations included (from Karman et al. 1996)

Chemical	Concentration	Chemical	Concentration
Process biocide (MB554)	1.2	Well treatment scale inhibitor (S432)	6.8
Water injection biocide (Kathon OM)	<0.0001	Antifoam (AF 119)	0.1
Corrosion inhibitor (PK 6050)	1.5	Flocculent (ML 2317 W)	2.4
Corrosion inhibitor (VN 6000 K)	0.3	Glycol	7.7
Process scale inhibitor (SP 250)	2.1	Methanol	0.3
Process scale inhibitor (SP 2945)	0.2		

associated with the use of treatment chemicals are effectively managed through the use of best management practices such as the Offshore Chemical Selection System (OCSS), regulatory compliance effluent toxicity testing protocols, or use of the DREAM model to estimate environmental impact factors (EIF) for individual chemicals (Chapters 9 and 27, this volume). Production chemicals with a high EIF can be replaced with less toxic alternatives or managed in such as way as to reduce amounts discharged to the ocean.

3 Produced Water Treatment

3.1 Regulation of Produced Water Discharge

Most environmental regulatory agencies in countries that have significant offshore oil and gas production place limits on the concentrations of petroleum (usually measured as total oil and grease) that can be present in produced water destined for ocean discharge. Table 1.13 summarizes examples of limits imposed by different countries on total oil and grease or total petroleum hydrocarbon concentration in produced water destined for ocean discharge. Different countries have proposed different standard methods for measuring oil in produced water (Chapter 2, this volume).

The different methods measure different fractions of the total organic chemicals in produced water and, therefore, give different results (Chapters 2 and 29, this volume). For example, in the USA, total oil and grease is defined as those materials that are extracted by *n*-hexane, not evaporated at 70°C, and capable of being

Table 1.13 Monthly average and daily maximum concentrations (mg/L) of total oil and grease permitted by several countries for produced water destined for ocean disposal (from Veil 2006)

Country	Monthly average	Daily maximum
Canada	30	60
USA	29	42
OSPAR (NE Atlantic)	30	–
Mediterranean Sea	40	100
Western Australia	30	50
Nigeria	40	72
Brazil	–	20

weighed (gravimetric) or quantified by infrared analysis (IR). In the OSPAR countries, total (dispersed) oil is defined as the sum of the concentrations of compounds extractable with *n*-pentane, not adsorbed on Florisil, that can be quantified by gas chromatography/flame ionization detection (GC/FID) with retention times between those of *n*-heptane (C_7H_{16}) and *n*-tetracontane ($C_{40}H_{82}$), excluding toluene, ethylbenzene, and xylenes. Gravimetric and IR methods tend to measure many nonpolar organic chemicals in addition to petroleum hydrocarbons, whereas GC/FID, if properly performed, measures mainly semi-volatile aliphatic and aromatic hydrocarbons. None of the methods quantitatively measure the low molecular weight, volatile aromatic hydrocarbons, such as BTEX and naphthalene, that contribute to the aquatic toxicity of produced water.

Current regulatory guidelines for produced water discharge in Canada are based on total petroleum hydrocarbon concentration, measured by IR. Under the 2002 Offshore Waste Treatment Guidelines, the hydrocarbon concentration of produced water must be reduced to acceptable levels prior to discharge into the ocean (NEB 2002). The minimum regulatory standard for the treatment and/or disposal of wastes associated with the routine operations of drilling and production installations offshore of Canada is a 30-day weighted average of oil in discharged produced water of 30 mg/L and a 24-h arithmetic mean of oil in produced water not to exceed 60 mg/L. Similar limits on total petroleum hydrocarbon concentrations in produced water destined for ocean disposal have been promulgated by environmental regulatory agencies in most other countries with offshore oil and gas production (Table 1.13).

3.2 Produced Water Treatment

Produced water intended for ocean disposal usually is treated on the platform or at a shore treatment facility to meet these regulatory limits. The objectives of oil/water/gas treatment on an offshore platform are to produce stabilized crude oil and gas for pipeline or tanker transport to shore facilities, and to generate a produced water that meets discharge requirements (if discharged to the ocean) or is suitable for reinjection into the producing formation or another geologic formation (Bothamley 2004). The discharge requirements for ocean disposal are based on best available technology for treating the produced water. Research is being performed in several countries to determine if the regulatory limits are sufficiently protective of the marine environment.

It is necessary to treat produced water before ocean discharge to avoid the harmful effects that the chemicals in the wastewaters may have on the receiving environment. Treatment removes solids and dispersed non-aqueous liquids from the wastewater, including dispersed oil, suspended solids, scales, and bacterial particles, as well as most volatile hydrocarbons and corrosive gases, such as CO_2 and H_2S. Experience by the offshore oil industry with produced water treatment for ocean disposal has shown that, if dispersed oil is removed, concentrations of volatile and dissolved hydrocarbons are reduced to acceptable levels (Ayers and Parker 2001). If the treated wastewater is intended for disposal to freshwater, recycling for steam generation for the various thermal enhanced oil recovery (EOR) technologies, or for reinjection into the formation, most of the dissolved salts and metals also should be removed. Salt removal is not necessary if the discharge is to the ocean.

The oil/gas/water mixture may be processed through separation devices to separate the three phases from each other. The types of equipment used on many platforms to remove oil and grease from produced water include mechanical and hydraulic gas floatation units, skimmers, coalescers, hydrocyclones, and filters (Otto and Arnold 1996; Chapter 29, this volume). Higher molecular weight, more toxic PAH, alkylphenols, and naphthenic acids are associated almost exclusively with dispersed oil droplets in produced water (Faksness et al. 2004). Efficiency of removal of these toxic chemicals can be improved by removing droplets with high-speed centrifuges and membrane filters, capable of removing particles in the 0.01–2 μm range (Chapter 29, this volume). Chemicals may be added to the process stream to improve the efficiency of oil/gas/water separation. The combination of mechanical and chemical treatment is effective in removing volatile compounds and dispersed oil from the produced water, but is ineffective in removing dissolved organics, ions, and metals. However, even with the most advanced equipment, the oil/water separation is not 100% efficient.

Produced water from offshore oil and gas wells is treated to remove volatile hydrocarbons, dispersed petroleum, and suspended solids to the extent afforded by current wastewater treatment technology. The worldwide average concentration of total petroleum hydrocarbons in produced water discharged offshore in 2003 was 21 mg/ L, with a range for different geographic regions from 14 to 39 mg/L (OGP 2004; Fig. 1.3). The quality of the produced water discharged is a function primarily of the efficiency of the treatment technologies and the strictness and degree of enforcement of environmental discharge regulations.

Within Canada, the formulation of regulatory guidelines for produced water discharge is an adaptive process that promotes the development of improved environmental effects monitoring (EEM) programs and takes into consideration the level of environmental risk, the Best Available Technology (BAT) for mitigative measures, and the socio-economic benefits.

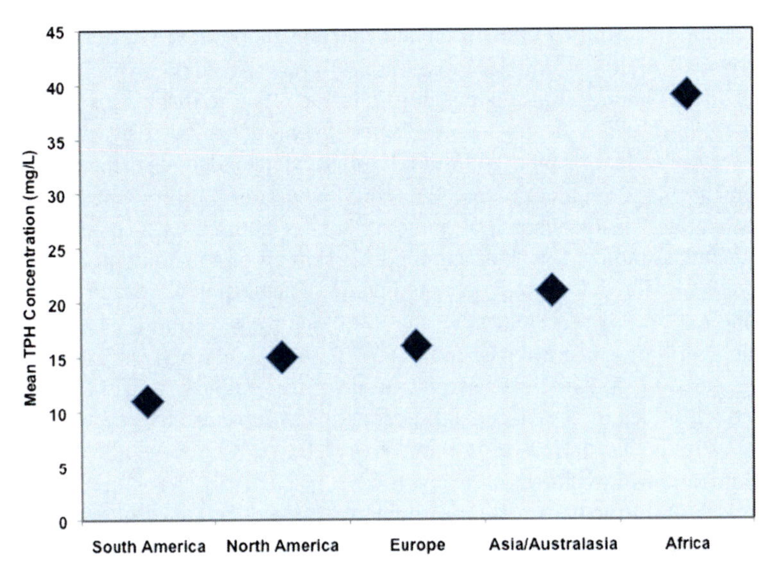

Fig. 1.3 Mean concentrations of total petroleum hydrocarbons (TPH) in produced water discharged offshore in several regions of the world, based on reports to OGP (2004) – no data were available for the former Soviet Union

4 Fate of Produced Water Following Discharge into the Ocean

4.1 Plume Dispersion Models

Treated produced water on offshore platforms may be discharged above or below the sea surface if regulatory compliance concentrations are achieved. The location of subsurface discharge pipes may range in depth from 10 to 100 m. Saline produced waters usually are as dense as or denser than seawater and disperse below the sea surface, diluting rapidly upon discharge into well-mixed marine waters. Low salinity produced water may form a plume on the sea surface and dilute more slowly, particularly if discharged at or above the sea surface (Nedwed et al. 2004). Dispersion modelling studies of the fate of produced water differ in specific details, but all predict a rapid initial dilution of discharges by 30- to 100-fold within the first few tens of meters of the outfall. This is followed by a slower rate of dilution at greater distances (Terrens and Tait 1993; Brandsma and Smith 1996; Strømgren et al. 1995; Smith et al. 2004). When discharge volumes of buoyant or neutral-density produced water are very high, dilution may be slower. Factors that affect the rate of dilution of produced water include discharge rate and height above or below the sea surface, ambient current speed, turbulent mixing regime, water column stratification, water depth, and difference in density (as determined by temperature and total dissolved solids concentration) and chemical composition between the produced water and ambient seawater (Chapters 9 and 11, this volume).

Brandsma and Smith (1996) modeled the fate of produced water discharged under typical Gulf of Mexico conditions. They used two discharge rates: 115.7 m^3/day, which is the median flow rate for offshore discharges to the Gulf of Mexico, and 3,975 m^3/day, which is the maximum allowable discharge rate from a single discharge pipe to the Gulf of Mexico under the general National Pollutant Discharge Elimination System (NPDES) permit. The effluent was a hypersaline brine that was discharged at a temperature of about 29°C. Therefore, it was denser than the ambient seawater and tended to sink. For a median produced water discharge rate of 115.7 m^3/day, the predicted concentration of produced water in the plume 100 m down-current from the discharge ranged from 0.043 to 0.097%, depending on ambient current speed (Table 1.14). At the higher discharge rate, dilutions at 100 m down-current from the discharge ranged from 0.18 to 0.32% produced water, depending on current speed. High-volume discharges into high current speeds in the North Sea (10,000 m^3/day) or Bass Strait off southeast Australia (14,000 m^3/day) were diluted to 1.3 and 0.47%, respectively at 100 m from the outfall (Table 1.14). Brandsma (2001) estimated a rapid decrease of plume centerline concentration to ~2% at 100 m from the high-volume (6,359 m^3/day) discharge from platform Irene off Santa Barbara, California.

High dilution rates for produced water discharges to well-mixed receiving waters appear to be the norm. Terrens and Tait (1994) used the Offshore Operators Committee (OOC) model to estimate dilution of a 14,000 m^3 per day discharge from the Halibut platform in Bass Strait, Australia. The produced water was dynamically indistinguishable from the receiving waters due to the high degree of initial dilution (dilution range of 100:1 to 252:1). At 6 km, the dilution factors were about 13,000:1 for suspended oil droplets and 18,000:1 for dissolved oil. Skåtun (1996) used a BJET model to study the near-field mixing of a warm (32°C), high salinity (84‰) produced water released from a platform in the Gulf of Mexico. The plume had a dilution factor of 400:1 when the current speed was 15 cm/s at 103 m downstream from the release point. The physical dispersion models of projected discharges from the Sable Offshore Energy Project (SOEP) wells (SOEP 1996), located on the Scotian Shelf of Canada also indicated rapid dilution of the

Table 1.14 Predicted dilutions of produced water in the receiving waters at different discharge rates and current speeds (from Brandsma and Smith 1996)

Location	Discharge rate (m^3/day)	Current speed (cm/s)	Concentration at 100 m (% PW)
Gulf of Mexico	115.8	3.3	0.043
Gulf of Mexico	115.8	9.5	0.073
Gulf of Mexico	115.8	25.3	0.097
Gulf of Mexico	3,977.8	3.3	0.18
Gulf of Mexico	3,977.8	9.5	0.32
Gulf of Mexico	3,977.8	25.3	0.32
North Sea	10,000	22	1.3
Bass Strait	14,000	26	0.47

discharge to non-acutely toxic levels within very short distances from discharge points.

The OOC model was designed to estimate only short-term fate and other models described above only considered the near-field mixing. However, it also is important to study the long-term, far-field transport of produced water. Hodgins and Hodgins (1998) studied the dispersion of produced water from the Terra Nova FPSO (floating production, storage, and offloading facility) off the east coast of Canada using the UM3 (Three-dimensional Updated Merge) model coupled with a particle tracking-based far-field model. For the maximum discharge of 18,000 m^3/day, the estimated worst-case initial dilution was 5:1. As the pooled effluent near the hull of the FPSO was carried away by the ambient currents, the far-field model predicted a minimum secondary dilution of 5:1, and this yielded a combined total dilution of 25:1 after the plume had been dispersed a few hundred meters from the FPSO. The same modelling concept was applied to the White Rose development off the east coast of Canada (Hodgins and Hodgins 2000). The near-field UM3 model estimated an initial bulk dilution of 35:1 for a discharge of produced water with a density of 728 kg per m^3 at a maximum rate of 30,000 m^3/day from a 36 cm diameter pipe at 5 m below sea surface (Hodgins and Hodgins 2000). The far-field dispersion simulation showed that the 1% impact line (concentration greater than 0.1 mg L^{-1} or dilution < 400:1 for at least 1% of the time) extended to 1.8–3.2 km. Similar results were described by AMEC (2006) that predicted near, immediate and far-field dilution rates with the US EPA Visual Plumes model (Baumgartner et al. 1994), the USACE CDFate model (Chase 1994), and an advection/diffusion model. The models estimated a dilution of 70:1 at 500 m from the discharge and 400:1 at 2 km from the discharge for a maximum flow rate of 6,400 m^3/day of produced water from the Deep Panuke facility off the coast of Nova Scotia.

The ASA™ MUDMAP model (Spaulding 1994) was run by Burns et al. (1999) to study the dispersion of produced water from the Harriet oil field on the Northwest Shelf of Australia. Based on an averaged daily produced water discharge of 8,000 m^3/day and a total oil concentration ranging from 5.9 to 16 mg/L, the model predicted that the plume had an oil concentration range from 0.006 mg/L near the discharge to 0.00016 mg/L at about 8 km down-current.

Zhao et al. (2008) integrated a random walk-based particle tracking model with the Princeton Ocean Model (POM), which enables the fast prediction of future dispersion and risks of produced water discharges. For a produced water discharge of 21,200 m^3/day from the Hibernia platform off the east coast of Canada, the model predicted a Pb concentration of 0.002 μg/L (about the concentration in clean seawater: Table 1.8) at about 5.3 km south of the platform. An overestimation of dilution and underestimation of chemical concentration may result from the approach of Zhao et al. (2008) due to omission of near-field buoyant jet behaviors. The accuracy of the model can be improved by coupling a near-field model with the particle-tracking model using the method described by Zhang (1995) and Niu et al. (Chapter 11, this volume).

Mukhtasor et al. (2004) estimated a mean concentration of 0.5% of the initial concentration of the produced water plume at about 225 m from the discharge in a hypothetical study of the produced water discharge from the Terra Nova FPSO. The 95 percentile concentration from their analysis at the same location was about 2.25% the initial concentration. Similar approaches were used by Niu et al. (2009a) to study the effects of surface waves on the dispersion of produced water. Because the models of Mukhtasor et al. (2004) and Niu et al. (2009a) can only be used for a limited number of discharge conditions, Niu et al. (2009b; Chapters 11 and 12, this volume) expanded the same approach into a probabilistic-based steady-state model (PROMISE) that was coupled to a MIKE3 model to study dispersion over a wide range of non-steady state discharge and environmental conditions. Niu et al. (2009b; Chapters 11 and 12, this volume) performed a validation study of the PROMISE/MIKE3 model and reported good agreement between model predictions and laboratory measurements.

As is evident in this overview, during the last two decades, a significant amount of effort has been put forward to model the dispersion of produced water plumes in the marine environment. Researchers from different disciplines have approached the problem from different perspectives and developed models with various degrees of sophistication. Produced water plume dispersion models have evolved from the simple steady state near-field, short-term dilution models to comprehensive coupled hydrodynamic dispersion models that predict both the near and far-field dispersion processes in three-dimensional non-steady state conditions. Although the dispersion models are now able to simulate the near-field mixing process very well and predict the far-field mixing process reasonably well, they are still limited in their capacity to predict the fate of the various chemical components of produced water.

4.2 Chemical Fate/Transport Models

In the majority of physical transport models, the chemicals in the produced water stream are treated as passive tracers. Similarly, dye injection tracer studies of produced water plumes (e.g., DeBlois et al. 2007) are based on the same assumptions. The drawback of using dyes is that they may not become fully integrated within the produced water plume, resulting in an inaccurate prediction of the transport of various contaminants that react chemically and separate from the plume. Based on recent studies by Lee et al. (2005a) and Azetsu-Scott et al. (2007), consideration also must be given to the chemical transformations that may influence the subsequent transport, fate, and effects of the contaminants of concern in the produced water following its discharge.

Berry and Wells (2004, 2005) first used the CORMIX model (Doneker and Jirka 2007) to determine the exposure pathways and potential compartment interactions and then employed a Level III fugacity model (Mackay 1991) to study the distribution of benzene and naphthalene among environmental media such as water, suspended particles, fish, and sediments from produced water discharges from the Thebaud platform on Scotian shelf. They predicted that the averaged water column bulk concentrations of benzene and naphthalene over a 1 km by 1 km area are 5.28 $\times 10^{-6}$ μg/g and 8.49 $\times 10^{-7}$ μg/g (~5.28 and 0.85 ng/L (parts per trillion), respectively) for the maximum discharge rate of 211.7 m^3/day. These concentrations are close to the measured background concentrations of benzene and naphthalene in ocean waters (~1 ng/L: Neff 2002). Because the CORMIX model predicted that the produced water plume may not be fully mixed within the selected compartments, the fugacity model may underestimate the concentration within the produced water plume and overestimate the concentration outside of the plume.

Smith et al. (1996), used a coupled model to simulate the transport of produced water from the Pertamina/Maxus operation area in Java Sea, Indonesia. The CORMIX model (Doneker and Jirka 2007) was first used to predict the effluent dispersion and the results were then used in the PISCES model (Turner et al. 1995) to study the partitioning, degradation, and volatilization of chemicals in the produced water plume. The model predicted that the concentration of mercury was about 0.0055–522 μg/L at 500 m from the discharge and the concentration of arsenic was about 5–12 μg/L. At 3000 m downstream, the mean mercury and arsenic concentrations decreased to 65 μg/L and 3 μg/L, respectively. These concentrations are about 10,000 times higher for mercury, and equivalent to the arsenic, concentrations in clean seawater (Table 1.8).

The use of separate dispersion, fate, and chemical models cannot account for the dynamic changes of chemical concentrations following discharge. Murray-Smith et al. (1996) applied the particle-based model, TRK, to the Clyde platform in the UK sector of the North Sea. The model combined physical dispersion with first order degradation to simulate biodegradation and other removal processes. The study found that the initial dilution was rapid, with a minimum dilution factor of between 300 and 3,000 within 100 m from the discharge. Further away, the physical dilution was less rapid and other removal processes, such as precipitation and biodegradation may have become important. The model predicted an overall dilution (including biodegradation) of plume chemicals of 1,000 to 16,000-fold at 1 km from the discharge.

Reed et al. (1996) described a PROVANN model for the simulation of three-dimensional transport, dilution, and degradation of chemicals associated with produced water discharges from one or more simultaneous sources. The model included various transport processes such as adsorption/dissolution kinetics, entrainment and dissolution of oil droplets, volatilization, degradation, and deposition from the water column. For two platforms off Trondheim, Norway, the simulation showed that the naphthalene concentration at the edge of the plume at about 40 km downstream after 50 days was extremely low (~0.00006 μg/L). The simulation showed that inclusion of degradation is clearly an important factor in modelling long-term fate of produced water chemicals.

PROVAN evolved over several years to become the Dose-related Risk and Exposure Assessment Model (DREAM), a three-dimensional, time-dependent numerical model that computes transport, exposure, dose, and effects in the marine environment (Reed et al. 2001; Chapter 9, this volume). Reed and Rye (Chapter 9, this volume) describe DREAM and its application in modelling the fate and effects of produced water discharges to the marine environment (Fig. 1.4).

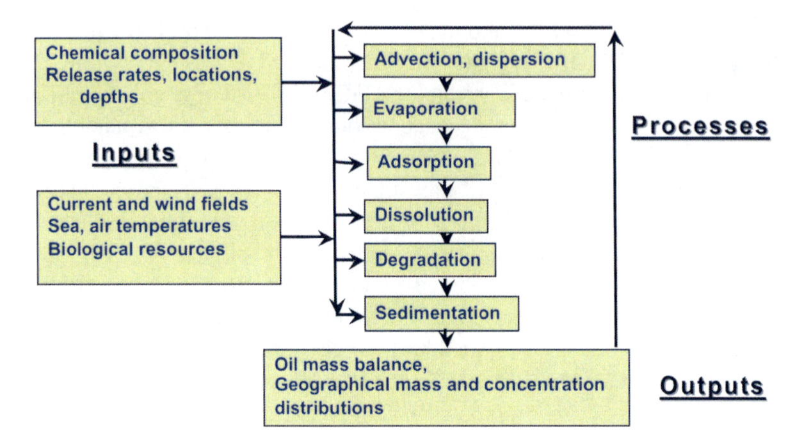

Fig. 1.4 Diagram of the processes governing the physical, chemical, and biological behavior of target chemicals from produced water in DREAM (Chapter 9, this volume)

4.3 Model Validation with Field Measurements

Field measurements are important for both understanding of the fate of produced water and for model validation. Traditionally, this was achieved by either collecting water samples at pre-determined stations or by continuous towing of a fluorometer (Murray-Smith et al. 1996; Smith et al. 1994, 1996, 2004; Terrens and Tait 1996). Traditional ship-based sampling methods often are expensive and time consuming and, therefore, only limited information can be collected. The greater water depth of recent offshore production activities also increases the level of sampling error. Recently, new and innovative means of conducting field measurements using Autonomous Underwater Vehicles (AUV) have been proposed and may be used in future studies of produced water fate and to collect data to validate mathematical fate/transport models (Niu et al. 2007; Chapter 11, this volume).

Field measurements of produced water dilutions usually are highly variable but confirm the predictions of modelling studies, that dilution usually is rapid. The comparison of field measurements of concentrations of hydrocarbons and various other organic produced water components from both fixed sampling stations and continuous towed fluorometers with modeled data showed that measured dilutions were generally much higher than predicted (Murray-Smith et al. 1996). The measured dilutions for alkanes and methylnaphthalenes at 100–1,000 m from the discharge ranged from 5,000 to 50,000-fold, much greater than predicted. The study also found that the measured concentrations of other organics such as phenol and methylphenol, in the water column were below detection limits, probably because of their rapid biodegradation. Similar results were reported for the concentrations of five heavy metals in produced water samples discharged to the Java Sea where a high dilution ratio was observed as predicted in the models (Smith et al. 1996).

Results of the DREAM model have been validated with field measurements in the North Sea. Concentrations of PAH were measured in the water column and in blue mussels (*Mytilus edulis*) deployed at different distances from production platforms in the Norwegian sector off the North Sea (Johnsen et al. 1998; Røe Utvik et al. 1999; Durell et al. 2006; Neff et al. 2006). Direct measurements of PAH in the water gave inconsistent results, because concentrations were too low and variable. However, the mussels did bioaccumulate PAH from the water. PAH concentrations in mussels decreased with distance down-current from platforms. The advantage of using mussels to estimate concentrations of nonpolar organic chemicals in the water column is that they integrate (average) water concentrations over time, whereas discrete water samples or fluorometer surveys can only measure discrete concentrations.

PAH residues in mussel tissues were used to estimate PAH concentrations in surface waters (Neff and Burns 1996). Estimated surface water total PAH concentrations ranged from 0.025 to 0.35 μg/L (parts per billion) within 1 km of platform discharges and reached background levels of 0.004–0.008 μg/L within 5–10 km of the discharge, representing a 100,000-fold dilution of the total PAH concentration in the discharge water (Durell et al. 2006; Neff et al. 2006). Dilution modelling showed that most of the produced water plume was restricted to the upper 15–20 m of the water column. Dilution was very rapid.

Total and individual PAH concentrations in the upper water column, estimated from residues in mussel tissues, were compared to concentrations predicted by the DREAM model (Durell et al. 2006; Neff et al. 2006). There was very good agreement for measured and modeled concentrations of individual and total PAH in the Ekofisk but not the Tampen field. The poor predictions in the Tampen field were caused by lack of accurate physical oceanographic data for the period of the mussel deployment. DREAM tended to overestimate concentrations of naphthalenes and underestimate concentrations of 4- and 5-ring PAH (Fig. 1.5).

Harman et al. (2010) confirmed these results by deploying passive sampling devices, semipermeable membrane devices (SPMDs) and polar organic integrative chemical samplers (POCIS), in the vicinity of an offshore oil production platform discharging produced water to the North Sea. There was a gradient of decreasing concentrations in the receiving waters of low molecular weight PAH and alkylphenols with distance from the platform. However, there was no gradient of concentrations of high molecular weight PAH with distance from the platform.

The DREAM model predicted that the concentrations of PAH and other chemicals in the produced water plume at different distances down-current from the discharge exhibited wide cyclic concentration variations due to tidal and wind-driven current flows (Fig. 1.6). Because of rapid dilution and fluctuating water-column concentrations, the model predicted that potentially toxic concentrations and contact times of PAH would not occur even in the near-field.

Terrens and Tait (1996) measured concentrations of BTEX and several PAH in ambient seawater 20 m from a platform in the Bass Strait off the southeastern Australia discharging produced water at a rate of 11,000-m^3/day (Table 1.15). There

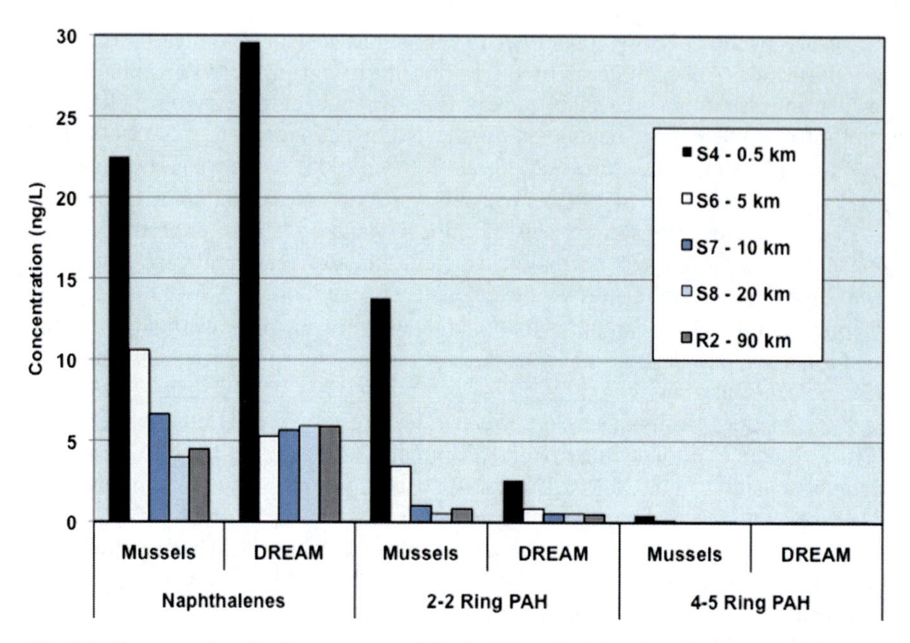

Fig. 1.5 Concentrations (ng/L = parts per trillion) of three PAH fractions in surface waters at five distances down-current from production facilities in the Ekofisk field in the Norwegian North Sea, based on measured concentrations in tissues of deployed mussels and predictions of the DREAM model; concentrations in water at R2 are considered background for the southern North Sea (data from Durell et al. 2006; Neff et al. 2006)

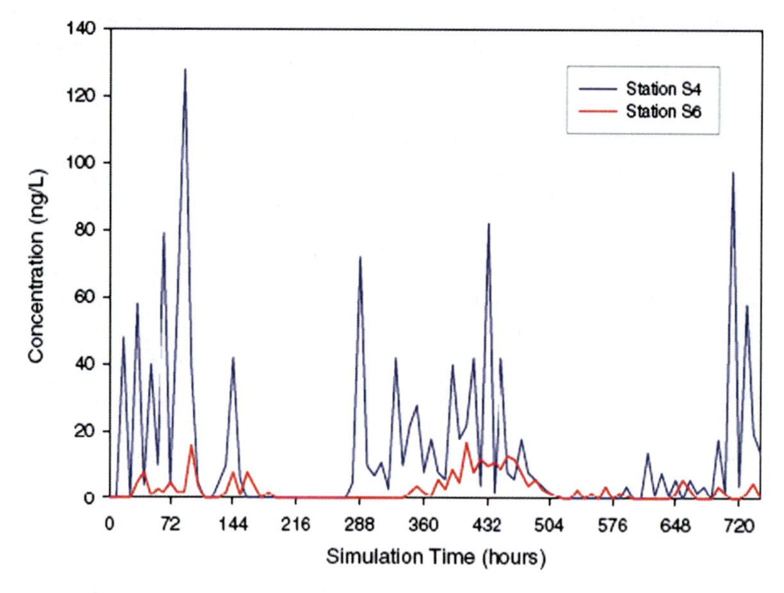

Fig. 1.6 Time series of modeled (DREAM) total PAH (naphthalenes through 5-ring PAH) concentrations at Station S4 and Station S6, 0.5 and 5 km down-current, respectively from a produced water discharge in the Ekofisk Field (from Durell et al. 2006)

Table 1.15 Concentrations (μg/L = parts per billion: ppb) of toluene and three PAH in produced water discharged from the Kingfish B Platform in Bass Strait, Australia, and in near-surface seawater collected 20 m down-current from the discharge (from Terrens and Tait 1996)

Chemical	Kingfish produced water	Ocean 20 m from discharge	Dilution
Toluene	3,000	0.18	16,700
Naphthalene	440	0.04	11,000
Phenanthrene	18	0.003	6,000
Fluoranthene	0.8	0.0002	4,000

was an inverse relationship between molecular weight (and volatility) and the dilution of individual aromatic hydrocarbons. Twenty meters down-current from the produced water discharge containing an average of 6,410 μg/L BTEX, the average BTEX concentration in the plume was 0.43 μg/L, a dilution of 14,900-fold. PAH were diluted by 11,000-fold (naphthalene) to 2,000-fold (pyrene). Concentrations of higher molecular weight PAH were below the detection limit (0.0002 μg/L) in the ambient seawater 20 m from the outfall. The inverse relationship between aromatic hydrocarbon molecular weight and their rates of dilution probably was due in large part to the high temperature (95°C) of the discharged produced water, favoring evaporation of the lighter aromatic hydrocarbons.

Continental Shelf Associates (1993) measured a dilution factor of 426 at 5 m and of 1,065 at 50 m from the discharge for ^{226}Ra in a 1,070 m^3/day produced water discharge plume at a water depth of 18 m in the Gulf of Mexico. The current speed at the time of the measurements was 15 cm/s and there was little vertical stratification of the water column. The OOC (but not the CORMIX) model accurately predicted the dilutions measured in the field. Produced waters from the Gulf of Mexico often contain high concentrations of dissolved barium. It is probable that the radium in the produced water co-precipitated rapidly with barium sulfate in the sulfate-rich receiving waters.

In summary, the majority of plume dispersion and chemical fate/transport models developed to date focus on the process of dispersion and treat produced water as a single conservative contaminant. Only a few models have attempted to include other transformation processes, such as biodegradation, metal speciation, evaporation, and adsorption. Among these models, the most comprehensive appears to be DREAM that is capable of handling a multitude of complex processes and data including discharge volumes, physical, chemical and biological fates of discharged substances, biological uptake and effects (Reed et al. 2001; Chapter 9, this volume). DREAM is capable of predicting the fate of individual chemicals associated with produced water and currently is used extensively by North Sea operators to achieve the regional regulatory goal of "zero harmful discharges". The model can also be used to predict environmental effects by two approaches: the environmental impact factor (EIF) and a body burden related risk assessment model (Chapters 9 and 27, this volume).

5 Environmental Effects of Produced Water Discharges

Based on the concentrations and relative toxicity of chemicals in most produced waters and predicted dispersion and biodegradation/transformation rates in the receiving waters, it is likely that there is only a limited potential for acute toxicity beyond the immediate vicinity of produced water discharges to offshore waters. This hypothesis is supported by sensitive biotests – primarily regulatory acute toxicity assays, and the rapid dispersion and degradation of the produced water plume in the receiving waters (Lee et al. 2005a). However, Holdway (2002) proposed that the chronic impacts associated with long-term exposures must be quantified to fully assess the potential long-term ecological impact of produced water discharges. Continual chronic exposure may cause sub-lethal changes in populations and communities, including decreased community and genetic diversity, lower reproductive success, decreased growth and fecundity, respiratory problems, behavioral and physiological disorders, decreased developmental success and endocrine disruption.

Fisheries and Oceans Canada and other environmental regulatory agencies worldwide are performing ongoing chronic toxicity studies to support the development of cost-effective and sensitive monitoring and environmental assessment protocols for regulatory use.

5.1 Potential for Effects in Water-Column Organisms

Harmful biological effects in water-column biological communities near open-ocean produced water discharges are expected to be minimal and localized, because of the rapid dilution, dispersion, and transformation rates of most produced water chemicals. However, some produced waters contain chemicals that are highly toxic to sensitive marine species, even at low concentrations. When discharge is to shallow, enclosed coastal waters, or when discharge is of a low-density produced water in an area with low water turbulence and current speeds, concentrations of produced water chemicals may remain high for long enough to cause ecological harm (Neff 2002). The chemicals of greatest environmental concern in produced water, because their concentrations may be high enough to cause bioaccumulation and toxicity include aromatic hydrocarbons, some alkylphenols, and a few metals. Highly alkylated phenols (octyl- and nonyl-phenols) are well-known endocrine disruptors, but rarely are detected in produced water at high enough concentrations to cause harm to water column animals following initial dilution (Thomas et al. 2004; Boitsov et al. 2007; Sundt et al. 2009). Most metals and naturally occurring radionuclides are present in produced water in chemically reactive dissolved forms at concentrations similar to or only slightly higher than concentrations in seawater and, therefore, are unlikely to cause adverse effects in the receiving water environment (Neff 2002). Nutrients (nitrate, phosphate, ammonia and organic acids) may stimulate microbial and phytoplankton growth in the receiving waters (Rivkin et al. 2000; Khelifa et al. 2003). Some production treatment chemicals are toxic, and if they are discharged at high concentration in produced water, could cause localized harm (Sverdrup et al.

2002). Inorganic ions (e.g., sodium, potassium, calcium and chloride) are not of concern in produced water discharges to the ocean (Pillard et al. 1996), but are of environmental concern when the treated water is discharged to land or surface fresh or brackish waters.

5.2 Potential for Accumulation and Effects in Sediments

If produced water is discharged to shallow estuarine and marine waters, some metals and higher molecular weight aromatic and saturated hydrocarbons may accumulate in sediments near the produced water discharge (Neff et al. 1989; Means et al. 1990; Rabalais et al. 1991), possibly harming benthic communities. In well-mixed estuarine and offshore waters, elevated concentrations of saturated hydrocarbons and PAH in surficial sediments sometimes are observed out to a few hundred meters from a high-volume produced water discharge. The concentrations of PAH in sediments near offshore produced water discharges are related to the volume and density of produced water discharged, the PAH concentration in it, water depth, and local mixing regimes. PAH in sediments near offshore platforms also may come from drilling discharges, particularly if oil-based drilling muds are used and oily drill cuttings are discharged, a practice no longer allowed in most marine waters (Neff 2005).

Barium, iron, and manganese are the metals most often greatly enriched in produced waters compared to their concentrations in natural seawater. Speciation occurs following the ocean discharge of produced water, in which the metals precipitate rapidly when produced water is discharged to well-oxygenated surface waters containing a high natural sulfate concentration. Trefry and Trocine (Chapter 5, this volume) showed that dissolved barium in produced water precipitates more slowly than predicted when the produced water is discharged to sulfate-rich seawater. However, precipitation of barium and dilution of the resulting barite in the produced water plume are rapid enough that dissolved barium concentrations rarely exceed acutely toxic concentrations. Other alkaline earths, such as strontium, magnesium, and radium, co-precipitate with barium and are rapidly deposited with barium sulfate in bottom sediments (Neff 2002; Chapter 5, this volume).

Dissolved iron and manganese precipitate rapidly as oxyhydroxides when the anoxic produced water plume mixes with oxygen-rich receiving waters. The extremely fine-grained iron and manganese oxides adsorb to or co-precipitate with several other metals from the produced water plume (Lee et al. 2005a; Azetsu-Scott et al. 2007). These particulate metals tend to settle slowly out of the water column and accumulate to slightly elevated concentrations in surficial sediments over a large area around the produced water discharge (Neff 2002; Lee et al. 2005a). In addition, the transport and concentration of inorganic constituents within produced water (e.g., metals) to the surface microlayer may be promoted by the interaction between residual oil droplets and metal precipitates (Burns et al. 1999; Lee et al. 2005a). Toxicity assessment using the Microtox® test, a regulatory bioassay protocol based on inhibition of a primary metabolic function of a bioluminescent bacterium, showed that unfiltered produced water samples containing metal precipitates generally had higher toxicity than filtered samples (Azetsu-Scott et al. 2007). Current results from regulatory environmental effects monitoring programs generally show that natural dispersion processes appear to control the concentrations of toxic metals in the water column and sediments just slightly above natural background concentrations.

5.3 Aquatic Toxicity of Produced Water

Most treated produced water has low to moderate toxicity. A typical distribution of produced water toxicities can be seen in data for produced waters discharged to the Gulf of Mexico off the Louisiana coast (Table 1.16). A small number of produced water samples are moderately toxic to mysids (a small shrimp-like crustacean) and sheepshead minnows, with acute and chronic toxicities less than 0.1% (1,000 mg L^{-1}) produced water. A few produced waters are practically nontoxic with acute and chronic toxicities higher than 35 or 40%. Most produced waters have moderate toxicities, with acute and chronic toxicities between about 2 and 10% for mysids and 5–20% for sheepshead minnows. Based on earlier toxicity studies for produced waters from the Gulf of Mexico, Neff (1987) reported that nearly 52% of all median lethal concentrations (LC_{50}) were greater than 10% produced water, 37% were between 1 and 9.9%, and 11% were less than 1%. These toxicity threshold limits are consistent with those reported for Atlantic Canada. A 1:100 dilution of the produced water, as usually occurs within a few tens of meters of the discharge pipe, would render all but a few of these produced water samples not acutely toxic.

In a comprehensive study on the acute effects of produced water recovered from a Scotian Shelf offshore well on the early life stages of haddock, lobster and sea scallop in terms of survival, growth and fertilization success, Querbach et al. (2005)

Table 1.16 Acute and chronic toxicity of more than 400 produced water (PW) samples from the Gulf of Mexico off Louisiana, USA, to mysids (*Mysidopsis bahia*) and sheepshead minnows (*Cyprinodon variegatus*); exposure concentrations are percent produced water (from Neff 2002)

Test	Number of tests	Mean value (% PW)	Standard deviation	Maximum value
Mysidopsis bahia				
96 h acute toxicity	412	10.8	10.4	86.3
Chronic survival (NOEC)	407	3.4	5.8	50.0
Chronic growth (NOEC)	391	2.4	3.6	42.0
Chronic fecundity (NOEC)	274	2.7	3.2	25.0
Cyprinodon variegates				
96 h Acute toxicity	359	19.2	14.8	>100
Chronic survival (NOEC)	401	6.3	9.0	>100
Chronic growth (NOEC)	395	5.2	8.1	>100

NOEC: no observed effect concentration

noted that fed, stage I lobster larvae were the most sensitive with an observed LC_{50} of 0.9%. Feeding stage haddock larvae and scallop veligers were the least sensitive with LC_{50} values of 20 and 21% respectively. In terms of chronic responses, the average size of scallop veligers was significantly reduced after exposure to produced water concentrations >10%.

There are poorly characterized species differences in the toxicity of produced waters to marine organisms. When bioassays were performed with two or more marine taxa and the same sample of produced water, crustaceans were generally more sensitive than fish (Neff 1987; Louisiana Department of Environmental Conservation 1990; Jacobs and Marquenie 1991; Terrens and Tait 1993). Survival of mummichog (*Fundulus heteroclitus*) embryos was significantly decreased during exposure to 10% or higher concentrations of produced water from the Hibernia gravity base structure (Burridge et al. Chapter 17 this volume). A variety of developmental abnormalities, commonly called blue sac disease, occurred in embryos exposed to 1% produced water during development (initial concentration of total methylated PAH = 7 μg/L). Growth of juvenile cod (*Gadus morhua*) was unaffected by exposure for 45 days to 0.05% of the same produced water. However, no differences in mortality were observed between control and experimental copepods (*Calanus finmarchicus*), a major prey species for fish in the northwest Atlantic, when they were exposed to 5% Hibernia produced water for 48 h (Payne et al. 2001).

Gamble et al. (1987) introduced produced water at a concentration equivalent to a 400–500 fold dilution of produced water into 300 m^3 mesocosm tanks containing natural assemblages of phytoplankton, zooplankton, and larval fish. The produced water concentrations were what could be expected within 0.5–1.0 km of the Auk and Forties platforms in the North Sea. Bacterial biomass increased but phytoplankton production and larval fish survival were unaffected in the produced water-dosed containers. However, early life stages of copepods were sensitive to the produced water and suffered high mortalities. The decrease in zooplankton abundance resulted in an increase in the standing stock of phytoplankton and a reduction in the growth rates of the fish larvae. In other mesocosm studies summarized by Stephenson et al. (1994), larval mollusks and polychaete worms also were adversely affected. These mesocosm studies show that low concentrations of produced water may have subtle effects on marine planktonic communities. However, it should be pointed out that mesocosm studies represent conservative, worst-case exposure scenarios, because produced water chemicals in the mesocosm enclosures do not degrade and disappear as rapidly as they do in well-mixed ocean environments.

5.4 Bioaccumulation and Biomarkers as Evidence of Exposure

Bioaccumulation is the uptake and retention of a bioavailable chemical from one or more possible external sources such as water, food, substrate or air (Neff 2002). Marine animals near a produced water discharge may bioaccumulate metals, phenols, and hydrocarbons from the ambient water, their food, or bottom sediments. An attempt was made to measure bioaccumulation of four metals (arsenic, barium cadmium and mercury), BTEX, phenol, and PAH by two species of bivalve molluscs from platform legs, and five species of fish collected within 100 m of produced water discharging and non-discharging platforms in the Gulf of Mexico (Chapter 24, this volume). There was no difference in concentrations of any of the metals, phenol, or BTEX in tissues of bivalves and fish from discharging and non-discharging platforms. Concentrations of total PAH were low and highly variable in tissues of the two bivalves and five species of fish from all platforms. Total PAH concentrations were higher (usually by an order of magnitude or more) in the mollusc tissues than in the fish tissues, probably because of the high activity of PAH-metabolizing enzymes in fish. Total PAH concentrations were significantly higher in tissues of one or both species of bivalves from the discharging platforms than from the reference platforms at the time of one or both field surveys. Alkyl naphthalenes, phenanthrenes, or dibenzothiophenes, all characteristic of petroleum sources, were the individual PAH that were present most frequently at elevated concentrations in bivalves from discharging platforms. Thus, there was evidence of exposure to and bioaccumulation of PAH, but not metals, phenol, or BTEX from produced water by bivalves associated with the biofouling community on submerged structures of platforms in the Gulf of Mexico discharging production water.

Biomarkers are biochemical, physiological, or histological changes in an organism caused by exposure to and bioaccumulation of specific chemicals in water, food, or sediments (Forbes et al. 2006). Biomarkers usually are not direct indicators of harmful effects caused by the exposure, but can be used as early warnings of possible risk to the exposed organism. The most useful biomarkers respond to a single or small group of chemical contaminants, and so, can be used as evidence of exposure to a particular class of chemicals. For example, any of several measures of the induction (increase in activity) of the enzyme system, cytochrome P450 mixed function oxygenase (CYP1A or MFO), can be used as evidence of exposure to PAH, polychlorinated biphenyls (PCB), and any of several chlorinated hydrocarbon pesticides.

As discussed above, produced water often is a source of PAH in waters and sediments in close proximity to offshore oil and gas production facilities. Exposure to and effects of PAH, including those in produced water, have been the focus of numerous laboratory and field studies using endpoints based on biochemical, histopathological, immunological, genetic, reproductive, and developmental parameters (Neff 2002; Payne et al. 2003).

Børseth and Tollefsen (2004) monitored bioaccumulation and biomarker responses in mussels (*Mytilus edulis*) and Atlantic cod (*Gadus morhua*) held in cages in the vicinity of the Troll B Platform on the Norwegian continental shelf. Cages were deployed for 6 weeks both inside (500 and 1,000 m from the source) and outside the zone of expected influence of the produced water plume. They reported that concentrations of metals and PAH in soft tissues of the caged mussels correlated well with distance from the discharge, with highest body burdens in mussels closest to the platform. The PAH assemblage in mussel tissues was dominated by alkyl homologues of naphthalene, phenanthrene, and dibenzothiophene, suggesting exposure to PAH from the produced water discharge. Biomarker responses in the mussels were weak, providing only equivocal evidence of exposure to produced water chemicals.

Durell et al. (2006) and Neff et al. (2006) confirmed the results of Børseth and Tollefsen (2004) with mussels deployed at different distances from production platforms in the Ekofisk and Tampen fields off Norway. Concentration of total PAH, decalins (decahydronaphthalenes), and heterocycyclic aromatic compounds (dibenzothiophenes) in mussel tissues decreased with distance from production platforms in the Ekofisk field (Table 1.17). Concentrations of total PAH in the water were estimated from tissue residues in the mussels, and were low at all distances from the produced water discharges. The PAH assemblage in the near-field receiving waters (0.5 km) was dominated by alkyl-decalins, and naphthalenes. Decalins were lost rapidly from the water column with distance from the discharges, probably because of their relatively high volatility (Durell et al. 2006). There were only traces of the 4-through 5-ring PAH that are largely responsible for CYP1A induction (Neff 2002). Concentrations in the water of these high molecular weight PAH did not decrease much with distance from the discharge, probably because of deposition of pyrogenic PAH from the atmosphere throughout the area.

Børseth and Tollefsen (2004) found no significant difference in levels of plasma vitellogenin (an indicator of exposure to endocrine-disrupting chemicals) in male cod from exposed and reference sites. No significant differences were detected in ethoxyresorufin-*O*-deethylase (EROD) activity (biomarker of exposure to chemicals, including PAH, that induce the cytochrome P450 mixed function oxygenase enzyme system) in livers of fish from exposed and reference locations, indicating little or no exposure to PAH. Levels of PAH metabolites in cod bile were low, confirming the low level exposure to PAH. Concentrations of naphthalene metabolites in cod bile decreased with distance from the platform, indicating that the low-level exposure to PAH was probably from the platform's produced water discharge. Other biomarkers showed little or no evidence that the cod were exposed to chemicals from the produced water plume. The authors concluded that mussels and cod deployed near a produced water discharge probably were exposed to low concentrations of produced water chemicals, below levels that might represent a health risk to water-column organisms. The low biomarker responses can be explained by the low

Table 1.17 Concentrations of total PAH, decalins, and dibenzothiophenes in tissues of mussels (*Mytilus edulis*) following deployment at different distances down-current from production platforms in the Ekofisk field off Norway, and estimated concentrations of total PAH in the receiving waters based on PAH residues in mussel tissues (from Neff et al. 2006)

Distance from PW discharge (km)	Mussels (ng/g dry wt)	Water (μg/L)
0.5	8,630	0.086
5	2,710	0.025
10	231	0.008
20	100	0.005
90	189	0.006

concentrations of PAH, particularly the higher molecular PAH that are the strongest inducers of CYP1A biomarkers (Durell et al. 2006; Neff et al. 2006).

As part of the Biological Effects of Contaminants in Pelagic Ecosystems (BECPELAG) Program, bioaccumulation and several biomarkers were measured in wild and caged marine animals along a transect away from a Statfjord platform in the North Sea (Hylland et al. 2006). Produced water discharge is 74,100 m^3 per day from three platforms in the Statfjord field (Durell et al. 2006), among the highest discharge rates of any offshore field in the world. Førlin and Hylland (2006) measured hepatic EROD activity and bile metabolites in juvenile cod caged at several distances down-current from one of the discharges. There were no significant trends in EROD activity in male and female cod with distance from the discharge, though there was a trend for EROD activity in female cod to increase with distance from the discharge, a trend opposite the expected one (Fig. 1.7). However, concentrations of alkyl naphthalene metabolites (alkylnaphthalenes are abundant in produced water) in fish bile were highest in cod near the platform and decreased with distance from the platform (Fig. 1.8). There were no distance trends in concentrations of other PAH metabolites in cod bile. The authors concluded that the cod were exposed to low levels of PAH from the produced water discharges, but exposure levels were well below those that would pose a health risk to fish living near the platforms.

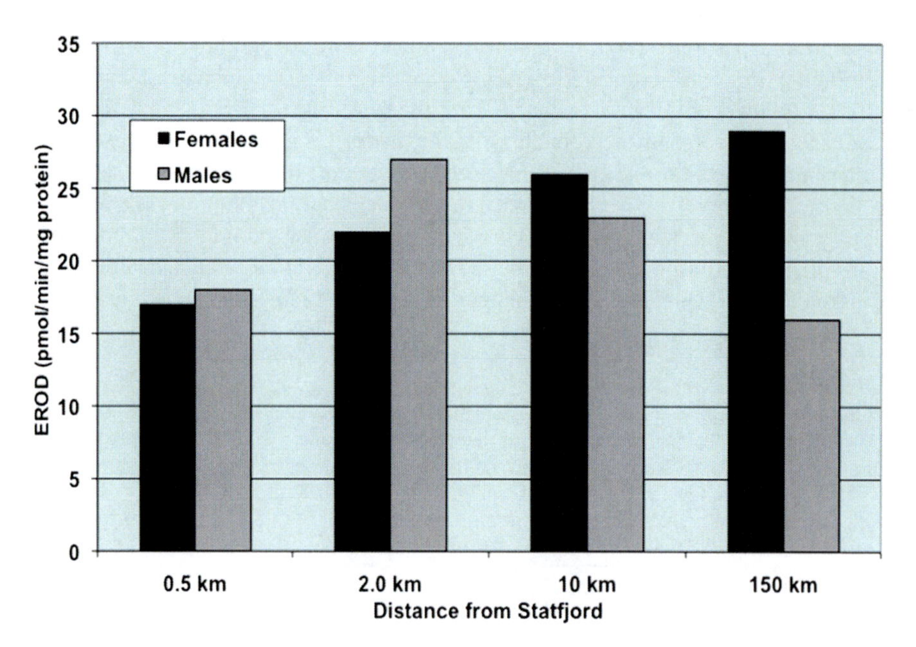

Fig. 1.7 Hepatic EROD activity in juvenile cod after 5–6 weeks deployment in cages at different distances from production platforms in the Statfjord oil field in the North Sea (from Førlin and Hylland 2006)

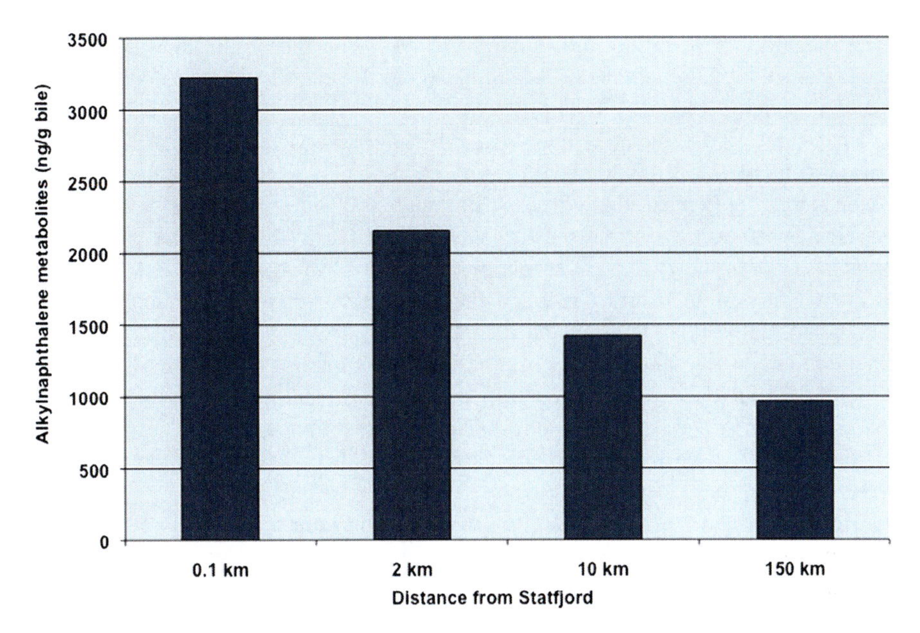

Fig. 1.8 Alkylnaphthalene metabolite concentrations in the bile of cod deployed for 5–6 weeks in cages at different distances from production platforms in Statfjord oil field (from Førlin and Hylland 2006)

Sturve et al. (2006) exposed juvenile Atlantic cod (*Gadus morhua*) to North Sea oil, nonylphenol and a combination of the North Sea oil and an alkylphenol mixture in a flow-through system. Although exposure to North Sea oil resulted in strong induction of CYP1A protein levels and EROD activities, exposure to oil plus nonylphenol resulted in decreased CYP1A levels and EROD activities. Thus, nonylphenol appeared to down-regulate CYP1A expression in Atlantic cod. Meier et al. (2007, 2010) described the effects of produced water and alkylphenols (AP) on early life stages and the reproductive potential of first-time spawning Atlantic cod. Cod were fed with feed paste containing a mixture of four reference alkylphenols, at a range of concentrations for either 1 or 5 weeks. The AP-exposed female fish had impaired oocyte development, reduced estrogen levels, and an estimated delay in the time of spawning of 17–28 days. Male AP-exposed fish had impaired testicular development, with an increase in the amount of spermatogonia and a reduction in the amount of spermatozoa present. Meier et al. (2007) concluded, based on the results of these laboratory studies, that AP associated with a produced water discharge may have a negative influence on the overall reproductive fitness of cod populations.

Abrahamson et al. (2008) exposed juvenile cod for 2 weeks in the laboratory to several concentrations of produced water from the Oseberg C platform in the Norwegian North Sea. There was a dose-related increase in gill EROD activity in the fish. However, when cod were caged for 6 weeks at 0.5–10 km from the Troll B and Statfjord B platforms, gill EROD activity was low in all fish and there was not a clear gradient of decreasing activity with increasing distance from the discharge. Thus, the concentration of higher molecular weight PAH in the produced water plume usually is not high enough at 0.5–1.0 km from North Sea produced water discharges to induce hepatic and gill EROD activity.

Elevated CYP1A enzyme activity has been observed in fish larvae collected downstream of the Hibernia field (Payne et al. 2003). However, induction may be occurring only near the platform site with the induced larvae being transported downstream by currents. Petro-Canada and Husky Energy have performed biomarker studies with American plaice (*Hippoglossoides platessoides*) collected in the vicinity of the Terra Nova and the White Rose offshore developments on the Grand Banks of Newfoundland (DeBlois et al. 2005; Husky Energy 2005; Chapter 20, this volume). These studies showed that the overall health of the American plaice collected in the vicinity of Terra Nova and White Rose was similar to the health of American plaice collected at distant reference sites. American plaice are demersal (bottom-living) fish and may not have been exposed to the produced water plume in the upper water column.

Meier et al. (2010) exposed cod to produced water during the embryonic, early larval (up to 3 months of age), or the early juvenile stages (from 3 to 6 months of age). Alkylphenols bioconcentrated in fish tissue in a manner that was dependent on both dose and developmental stage. However, juveniles appeared able to effectively metabolize the short chain APs. Exposure to produced water had no effect on embryo survival or hatching success. A concentration of 1% produced water, but not 0.01 or 0.1%, interfered with development of normal larval pigmentation. After hatching, most of the larvae exposed to 1% produced water failed to begin feeding and died of starvation. This inability to feed was linked to an increased incidence of jaw deformities in exposed larvae. Cod exposed to 1% produced water, had significantly higher levels of the biomarkers, vitellogenin and CYP1A, in plasma and liver, respectively.

Hamoutene et al. (Chapter 16, this volume) investigated the effects of produced water on cod immunity, feeding, and general metabolism by exposing fish to diluted produced water at concentrations of 0, 100 and 200 ppm for 76 days. No significant differences were observed in weight gain or food intake. Similarly, serum metabolites, whole blood fatty acid percentages, and mRNA expression of a brain appetite-regulating factor (cocaine and amphetamine regulated transcript) remained unchanged between groups. Other than an irritant-induced alteration in gill cells found in treated cod, resting immunity and stress response were not affected by produced water. Catalase and lactate dehydrogenase changes in activities were recorded in livers but not in gills, suggesting an effect on oxidative metabolism subsequent to hepatic detoxification processes. To evaluate potential effects of produced water discharges on cod immunity, fish from the three groups were challenged by injection of *Aeromonas salmonicida* lipopolysaccharides (LPS) at the end of exposure. LPS injection affected respiratory burst activity of head-kidney cells, and circulating white blood cell ratios, and increased serum cortisol in all groups. The most pronounced changes were seen in the group exposed to the highest dose of produced water (200 ppm).

In a followup study, Pérez-Casanova et al. (2010) reported that chronic exposure of juvenile cod to produced water had no significant effects on growth, hepatosomatic index, condition factor or plasma cortisol. The immune response of respiratory burst (RB) of circulating leukocytes was significantly elevated and the RB of head–kidney leukocytes was significantly decreased during exposure to low concentrations of produced water. There also was a significant up-regulation of the mRNA expression of β-2-microglobulin, immunoglobulin-M light chain, and interleukins-1β and 8, and down-regulation of interferon stimulated gene 15 at slightly higher exposure concentrations. However, because toxic effects are directly linked to dosage and exposure time, the ecological significance of laboratory biomarker studies is questionable. Factors such as fish movement and contaminant uptake/elimination are not taken into account and alkylphenol concentrations in seawater near platforms usually are below the limits of detection.

Burridge et al. (Chapter 17, this volume) exposed juvenile mummichogs (*Fundulus heteroclitus*) and cod (*Gadus morhua*) to a range of dilutions of produced water from the Hibernia production facility. The undiluted produced water contained 70 μg/L total alkyl PAH and 1520 μg/L total alkyl phenols (methyl though butyl phenols). EROD activity was not induced in mummichogs following exposure for 24 hours to the produced water plume. There was a significant (~7-fold) induction of hepatic EROD activity in juvenile cod following 48-h exposure to 1.67% produced water, but no induction following exposure to lower concentrations or during and following exposure for 45 days to 0.05% produced water. Plasma vitellogenin concentration (a biomarker of exposure to endocrine disruptors) also was not altered by the 45-day exposure. Payne et al. (2005) reported significant induction of EROD activity and increases in plasma vitellogenin concentrations in juvenile cod during exposure to higher concentrations of Hibernia produced water.

Gagnon (Chapter 15, this volume) and Codi King et al. (Chapter 14, this volume) performed similar studies with tropical fish in the vicinity of offshore production platforms on the Northwest Shelf of Australia. Gagnon (Chapter 15, this volume) collected three species of fish from surface waters near three production platforms or FPSOs discharging large volumes of produced water and measured several physiological parameters and biomarkers in the fish. Condition factor was slightly reduced in fish from one platform and liver somatic index was elevated in fish captured at two of the platforms. EROD activity and incidence of DNA damage were high at one facility discharging high volumes of produced water. Stress proteins (HSP70) were elevated in fish collected at all three facilities. High concentrations of naphthalene and pyrene metabolites were detected in the bile of fish collected at all three facilities. Gagnon (Chapter 15, this volume) concluded that the chemical composition and discharge rate of produced water affected the biological responses observed in resident fish.

Codi King et al. (Chapter 14, this volume) deployed juvenile Spanish flag snapper (*Lutjanus carponotatus*) in cages for 10 days at ~200 m, ~1,000 m, and a distant reference location from a platform on the Northwest Shelf of Australia. The produced water contained 10–14 ppm monocyclic aromatic hydrocarbons, about 2.6 ppm phenol and C_1 and C_2-phenols, and about 1 ppm total PAH, mostly naphthalenes and phenanthrenes. None of these chemicals were detected in bulk water samples collected from the deployment sites; however, low concentrations of several PAH were detected in SPMDs deployed for 10 days with the fish cages. A large number of biomarkers were evaluated for evidence of response to chemicals in the produced water. Bile metabolites, CYP1A, CYP2K-like and CYP2M-like protein, and liver histopathology provided evidence of exposure and effects after 10 days at the two sites near the platform, in comparison to results for fish at the reference site. Hepatosomatic index, cholinesterase, and total cytochrome P450 were not significantly different in fish from the three sites, whereas EROD activity was inconclusive. Principal component analysis (PCA) validated that the most useful diagnostic tools for assessing exposure to, and effects of exposure to produced water in snapper were the CYP proteins.

Payne et al. (Chapter 21, this volume) studied the toxicity to cunner of particulate barite ($BaSO_4$) that forms when produced water containing a high concentration of dissolved barium mixes with seawater, rich in inorganic sulfate (Chapter 5, this volume). Cunner were exposed on a weekly basis for 40 weeks to 200 g 'clouds' of microparticulate barite in a 1,800-L tank (nominal concentration, 111 mg/L $BaSO_4$, compared to a seawater solubility of 0.08 mg/L). Barite that accumulated on the bottom of the tank was not removed. Fish survival and indices of fish health, as assessed by fish and organ condition as well as detailed histological studies on liver, gill and kidney tissue did not differ between control and experimental groups. However, slightly elevated activity of EROD was observed in the exposed fish. The cause of the induction is not known.

A preliminary study was performed with scallops. The scallops were exposed to 2,000 ppm produced water from the Hibernia field every 2 days for a period of ~4 months. No differences in mortality or condition indices were observed between the control and exposed groups. A similar long-term study with mussels found no effect of produced water on mortality (J. Payne, DFO – personal communication).

6 Ecological Risk of Produced Water Discharges

The toxicity and ecological effects of a complex chemical mixture, such as produced water, to marine organisms and communities is a product of the rate of discharge, its chemical composition, environmental fates of each component in the mixture, and the relative toxicities of each component and its degradation products. Increasingly complex and sophisticated fate and effects models, such as DREAM, are being developed to predict the long-term effects and ecological risks of produced water discharges to different marine environments (Chapter 9, this volume). Risk assessments can be performed with the DREAM model by two approaches: (1) a dose-effect risk assessment model, in which the dose is measured as the concentration of each component in the ambient water or tissues of the target marine animals, and (2) the determination of an environmental impact factor (EIF). The assessment is based on the ratio of the predicted environmental concentration (PEC) to a predicted no-effect concentration (PNEC), known as the PEC/PNEC ratio (Karman and

Reerink 1998). This can be followed up with the calculation of an environmental impact factor (EIF) to be used for produced water impact reduction, management and regulation (Chapter 27, this volume). The EIF is a measure of the volume of seawater that contains high enough concentrations of produced water chemicals to exceed a pre-determined risk criterion. The EIF provides a regional-scale, quantitative estimate of the potential ecological risks to marine organisms of produced water discharges (Johnsen et al. 2000).

The Norwegian oil and gas industry advocates ecological risk assessment as the basis for managing produced water discharges to the North Sea. Neff et al. (2006) compared estimates of ecological risks of PAH from produced water to water-column communities based on data on hydrocarbon residues in soft tissues of blue mussels deployed for a month near offshore platforms and based on predictions of the DREAM model. The study was performed near produced water discharges to the Tampen and Ekofisk regions of the Norwegian Sector of the North Sea. Because PAH are considered the most important contributors to the ecological hazard posed by produced water discharges, comparisons focused on this group of compounds.

In the DREAM model, predicted environmental concentrations (PECs) for three PAH fractions were estimated in the three-dimensional area around the produced water discharge. Predicted no effects concentrations (PNECs) for each fraction were based on the chronic toxicity of a representative PAH from each fraction divided by an application factor to account for uncertainty in the chronic toxicity value. The risk characterization ratio (RCR) is the sum of the ratios of PEC estimated by the DREAM to the PNEC for several PAH groups (Fig. 1.9). The hazard index (HI) is the sum of the ratios of the measured concentrations of individual PAH in the water to an equilibrium partitioning/toxicity-based PNEC value for each target PAH. A risk value of 1 or higher indicates a possible risk to the health of marine organisms from the site.

The deployed mussel approach is based on PECs of individual PAH, estimated from PAH residues in mussels that had been deployed at different distances from produced water discharges, and PNECs based on a K_{OW} regression model. The mussel method gave much lower estimates of ecological risk than the DREAM method (Fig. 1.9). The differences are caused by the much lower PNECs used in DREAM than derived from the regression model, and by the lower concentrations of aqueous PAH predicted by DREAM than estimated from PAH residues in mussel tissues. However, the two methods rank stations at different distances from produced water discharges in the same order and both identify two and three-ring PAH as the main contributors to the ecological risk of PAH in produced water discharges. Neither method identifies a significant ecological risk of PAH in the upper water column of the oil fields. The DREAM model may produce an overly conservative estimate of ecological risk of produced water discharges to the North Sea, because of the extremely conservative PNEC values for PAH fractions.

Myhre et al. (2004) have studied the reproductive effects of alkylphenols (APs) on fish stocks in the North Sea using the DREAM model. The fish stock distributions (cod, saithe and haddock, from the international bottom trawl surveys (IBTS) database) and a PNEC for APs of 4 ng/L were used for the

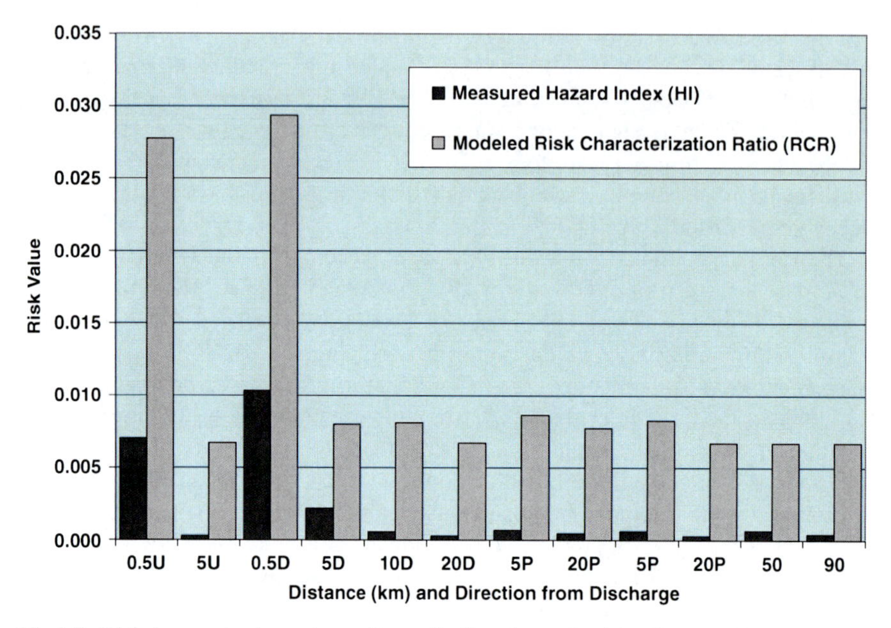

Fig. 1.9 Risk characterization ratios and hazard indices for total polycyclic aromatic hydrocarbons at different distances from produced water discharges in the Ekofisk field in the North Sea; stations are identified in the *x*-axis by their distance (km) and direction (U = up-current; D = down-current; P = parallel to) relative to the nearest produced water discharge (Neff et al. 2006)

calculations of effects from the combined produced water discharges from three major Norwegian oil fields (Tampen, Ekofisk and Sleipner). The total amount of APs >C_4 discharged from all the oil installations was estimated to be 25.6 kg per day, dissolved in 364,300 m^3/day of produced water (~7 μg/L total >C_4-APs). DREAM predicted that none of the fish accumulated APs to concentrations above the critical body burden of 2 μg/kg in any of the simulations. The highest accumulated body burden in any of the fish was 0.09 μg/kg. Myhre et al. (2004) concluded that the overall results of the simulations with DREAM show that there is not a significant ecological risk from >C_4-alkylphenols in produced water discharges to the Norwegian North Sea.

The accuracy of contaminant risk assessment models is dependent on the identification and quantification of the various chemicals that induce the toxic effects. Unfortunately, the causative agents of toxicity in the most toxic produced waters are not known. Toxic responses may be linked to the extremely high concentrations of total dissolved solids (salinity), altered ratios of major seawater ions, and elevated concentrations of ammonia in some Gulf of Mexico produced waters (Moffitt et al. 1992). Salinity and ion ratios quickly return to those in seawater following ocean discharge of produced water, and ammonia evaporates or degrades rapidly. Thus, these contaminants of concern within the produced water discharge stream rarely cause acute toxicity responses in the field.

Bacteria have very short generation times and respond rapidly to environmental changes. Because bacteria are involved in primary production processes including the production of organic carbon, nutrient cycling and the biodegradation and biotransformation of contaminants, their use has been recommended for environmental effects monitoring programs (Lee and Tay 1998; Wells et al. 1998). Studies (Anderson et al. 2000; Chapter 19, this volume) with naturally occurring bacteria have indicated the potential for produced water to both inhibit (short-term exposure at high concentrations) and enhance bacterial growth (lower concentrations over an extended period).

Chemical reactions that occur following the release of hypoxic produced water into well-oxygenated open ocean water alter the toxicity of produced water over time following its discharge (Lee et al. 2005a; Chapter 7, this volume). The significance of this process is clearly illustrated in controlled dose-response experiments using natural microbial populations as the test organisms (Fig. 1.10). A typical toxicity dose-response curve (initial increase in productivity at low concentrations of produced water due to addition of nutrients, followed by inhibition above a threshold value) is observed with fresh produced water. However, additions of produced water that had been aerated for 44 h (to simulate equilibration in the ocean following discharge) over the same concentration gradient elicited a stimulatory response. The difference is attributed to the loss of low molecular weight hydrocarbons, precipitation due to oxidation or reduction of (sulfate-sensitive) metals, such as barium, iron, and manganese that sequester toxic metals, and photo oxidation of some organic produced water chemicals. The results summarized in Fig. 1.10 imply that accurate comparisons of toxicological studies with similar end-points (e.g., LC_{50}) cannot be made unless sample collection, handling, and storage protocols are standardized prior to toxicity testing.

In a modelling study to assess potential perturbations in food web structure and energy flow due to the discharge of produced water, Rivkin et al. (2000) predicted significant increases in productivity and sedimentation fluxes over large spatial domains in response to ammonia and dissolved organic carbon from produced water. However, at current discharge rates, the effects of produced water discharges may be limited. Yeung et al. (Chapter 18, this volume) monitored changes in indigenous microbial community structure in response to produced water discharges from an offshore platform on the Grand Banks of Canada by denaturing gradient gel electrophoresis (DGGE). The DGGE results showed that the produced water did not have a detectable effect on microbial community structure in the surrounding water. Cluster analysis showed a >90% similarity for all near surface water (2 m) samples, ~86% similarity for all the 50 m and near bottom samples, and ~78% similarity for the whole water column from top to bottom across a 50 km range, based on two consecutive yearly sampling events. However, there were clear differences in the composition of the bacterial communities within the produced water compared to seawater near the production platform (~50% similar), indicating that the effect of produced water may be restricted to the region immediately adjacent to the platform. Members of the genus *Thermoanaerobacter* and of the Archaea genera *Thermococcus* and *Archaeoglobus* were identified as significant components of

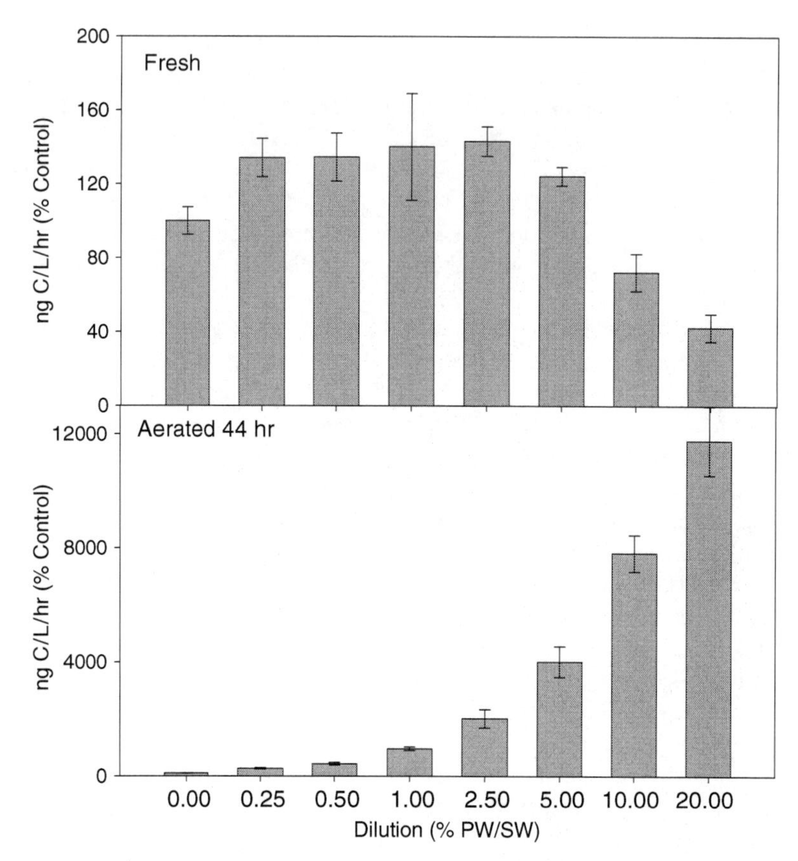

Fig. 1.10 Toxicity dose–response data illustrating the effect of produced water recovered from the Terra Nova FPSO, the Grand Banks, on microbial productivity in natural seawater (measured by ^{3}H-thymidine uptake into DNA as ng C/L/hr, normalized to the controls); identical samples were evaluated under identical conditions, immediately after collection (fresh) and after aeration (aerated) for 44 h

the produced water. These particular signature microorganisms could become useful markers to monitor the dispersion of produced water into the surrounding ocean.

7 Environmental Effects Monitoring and Research Needs

This overview has highlighted our advances in scientific knowledge during the past decade regarding the composition, environmental fate and biological effects of produced water discharges to the ocean.

The principal alternative to ocean discharge of treated produced water is underground injection. However, the feasibility of this practice at offshore installations is dependent on a number of site-specific factors including access to a suitable disposal geologic formation, chemical interactions that may result in precipitates that might plug the receiving formation, and cost. Furthermore, the net environmental benefit of reinjection must be considered. On the basis of energy requirements, it is estimated that 2.6–4.3 g of CO_2 emissions are produced for each litre of produced water reinjected into a sub-surface well (Shaw et al. 1999).

The general consensus of the 2007 International Produced Water Conference (St. John's, Newfoundland, Canada) was that any effects of produced water on individual development sites in the open ocean are likely to be minor. The toxicity threshold limits for acute effects are not likely to occur beyond the immediate vicinity of the discharge pipe due to the effectiveness of natural dispersion processes driven by tides and currents. However, unresolved questions regarding aspects of produced water composition and its fate and potential effects on the ecosystem remain. The chronic effects on important marine communities may become evident only after monitoring several life stages, generations of keystone species, or long-term ecological effects. It is important to acknowledge the consequences of long-term effects from offshore oil and gas facilities that may have a 15-20 year life-cycle. Furthermore, cumulative effects linked to future expansion of production operations must be considered. It is evident that additional information is needed to improve the accuracy of existing risk assessment models for produced water discharge. Multidisciplinary scientific studies are needed under an ecosystem-based management approach to provide information on the environmental fates (dispersion, precipitation, biological and abiotic transformation) and effects of chronic, low-level exposures of the different chemicals in produced water.

Numerical models need to be improved to better predict the fate and effects of chemical constituents in produced water plumes that are rapidly dispersed. There is a need to develop improved sample recovery and analytical techniques to support model validation needs. At present, many of the potential contaminants of concern in produced water cannot be detected in the open ocean environment with standard analytical protocols. The future development of high efficiency, cost-effective produced water treatment technologies is dependent on the identification and monitoring of the primary target constituents of environmental concern (e.g. PAH, phenols, naphthenic acids, metals) in produced water and the produced water plume.

Interpretation of ecological risk from biological effects studies based on biomarker techniques remains a challenge. Biomarkers may be used to indicate that: (1) an organism has been exposed to a specific chemical or group of chemicals, (2) an organism is affected by a contaminant and is responding to it, (3) the organism has been injured. However, as discussed by Gray (2002) in an editorial comment entitled, "Perceived and real risks: Produced Water from oil extraction," the question is, "What is the risk to populations in the field?"

For a comprehensive protection plan, there is a need to support the development of improved monitoring protocols to provide early warning of any potential problems related to sediment and water quality (such as primary productivity), fish quality, and fish health. Development of real-time monitoring systems (contaminant-specific sensors and data transfer technologies) will enhance our capacity to manage the ocean and its living resources. In consideration of natural perturbations currently affecting the ocean (climate change, for instance) and the impacts potentially associated with other users of the oceans (marine transport, fisheries, etc.), an ecosystem-based integrated management approach must be taken to fully evaluate the risks of offshore produced water discharge.

References

Abrahamson A, Brandt I, Brunström B, Sundt RC, Jøgensen EH (2008) Monitoring contaminants from oil production at sea by measuring gill EROD activity in Atlantic cod (Gadus morhua). Environ Pollut 153:169–175

AMEC (2006) Revision of the physical environmental assessment for Deep Panuke production site – oceanographic component: produced water, cooling water, drilling wastes. Report prepared by AMEC for Jacques Whitford (November 2006). 50 pp

Anderson MR, Rivkin RB, Warren P (2000) The influence of produced water on natural populations of marine bacteria. Proceedings of the 27th annual toxicity workshop. Can Tech Rep Fish Aquat Sci 2331:91–98

Ayers RC, Parker M (2001) Produced water waste management. Canadian Association of Petroleum Producers (CAPP). Calgary, AB

Azetsu-Scott K, Yeats P, Wohlgeschaffen G, Dalziel J, Niven S, Lee K (2007) Precipitation of heavy metals in produced water: influence on contaminant transport and toxicity. Mar Environ Res 63:146–167

Barman Skaare B, Wilkes H, Veith A, Rein E, Barth T (2007) Alteration of crude oils from the Troll area by biodegradation: analysis of oil and water samples. Org Geochem 38:1865–1883

Barth T (1991) Organic acids and inorganic ions in waters from petroleum reservoirs, Norwegian continental shelf: a multivariate statistical analysis and comparison with American reservoir formation waters. Appl Geochem 6:1–15

Baumgartner DJ, Frick WE, Roberts PJW (1994) Dilution models for effluent discharges, 3rd edn. EPA/600/R-94/086., U.S. Environmental Protection Agency. Available at http://www.environmental-engineer.com/docs/dos_plumes/rsb_um_plumes.Pdf

Berry JA, Wells PG (2004) Integrated fate modeling for exposure assessment of produced water on the Sable Island Bank (Scotian Shelf, Canada). Environ Toxicol Chem 23(10):2483–2493

Berry JA, Wells PG (2005) Environmental modeling of produced water dispersion with implications for environmental effects monitoring design. In: Armsworthy SL, Cranford PJ, Lee K (eds) Offshore oil and gas environmental effects monitoring: approaches and technologies. Battelle Press, Columbus, OH, pp 111–129

Bierman VJ Jr, Heinz SC, Justic D, Scavia D, Veil JA, Satterlee K, Parker M, Wilson S (2007) Predicted impacts from offshore produced-water discharges on hypoxia in the Gulf of Mexico. SPE 106814. Presented at the 2007 environmental and safety conference, Galveston, TX 5–7 March 2007. Society of Petroleum Engineers, Richardson, TX, 14 pp

Boitsov S, Mjøs SA, Meier S (2007) Identification of estrogen-like alkylphenols in produced water from offshore installations. Mar Environ Res 64:651–665

Borgund AE, Barth T (1994) Generation of short-chain organic acids from crude oil by hydrous pyrolysis. Org Geochem 21:943–952

Børseth JF, Tollefsen K-E (2004) Water column monitoring 2003 – Summary Report RF-Akvaliø Report RF-63319

Bothamley M (2004) Offshore processing options vary widely. Oil Gas J 102.45:47–55

Brandsma MG (2001) Near-Field produced water plume, platform Irene. Report prepared for Arthur. D. Little, Inc, Santa Barbara, CA by Brandsma Engineering, Durango, Colorado, July, 2001

Brandsma MG, Smith JP (1996) Dispersion modeling perspectives on the environmental fate of produced water discharges. In: Reed M, Johnsen S (eds) Produced water 2. Environmental issues and mitigation technologies. Environmental Science Research. Volume 52. Plenum Press, New York, pp 215–224

Burns KA, Codi S (1999) Non-volatile hydrocarbon chemistry studies around a production platform on Australia's northwest shelf. Estuar Cstl Shelf Sci 49:853–876

Burns KA, Codi S, Furnas M, Heggie D, Holdway D, King B, McAllister F (1999) Dispersion and fate of produced formation water constituents in an Australian northwest shelf shallow water ecosystem. Mar Pollut Bull 38:593–603

Chase, D (1994) CDFATE user's manual. Report prepared for US Army Engineer Waterways Experiment Section. Available from: http://el.erdc.usace.army.mil/elmodels/pdf/cdfate.pdf

Cherry MI, Cherry RD, Heyraud M (1987) Polonium-210 and lead-210 in Antarctic marine biota and sea water. Mar Biol 96:441–449

Clark CE, Veil JA (2009) Produced water volume and management practices in the United States. Report ANL/EVS/R-09-1 to the U.S. Dept. of Energy, Office of Fossil Energy, National Technology Laboratory, Washington, DC, 64 pp

Collins A.G (1975) Geochemistry of oilfield waters. Elsevier, New York, 496 pp

Continental Shelf Associates (CSA) (1993) Measurements of naturally occurring radioactive materials at two offshore production platforms in the Northern Gulf of Mexico. Final Report to the American Petroleum Institute, Washington, DC

DeBlois M, Dunbar DS, Hollett C, Taylor DG, Wight FM (2007) Produced water monitoring: use of rhodamine dye to track produced water plumes on the Grand Banks. In: Lee K, Neff J (Chairs) Abstracts of the international produced water conference: environmental risks and advances in mitigation technologies. St. John's Newfoundland, Canada, October 17–18, 2007. National Energy Board of Canada, Environmental Studies Research Fund (ESRF)

DeBlois EM, Leeder C, Penny KC, Murdoch M, Paine MD, Power F, Williams UP (2005) Terra Nova environmental effects monitoring program: from environmental impact statement onward. In: Armworthy SL, Cranford PJ, Lee K (eds) Offshore oil and gas environmental effects monitoring: approaches and technologies. Battelle Press, Columbus, OH, pp 475–491, 631

Doneker RL, Jirka GH (2007) CORMIX user manual: a hydrodynamic mixing zone model and decision support system for pollutant discharges into surface waters, EPA-823-K-07-001, Dec. 2007

Dórea HS, Kennedy JRLB, Aragão AS, Cunha BB, Navickiene S, Alves JPH, Romão LPC, Garcia CAB (2007) Analysis of BTEX, PAHs and metals in the oilfiled produced water in the State of Sergipe, Brazil. Michrochem J 85:234–238

Durell G, Johnsen S, Røe Utvik T, Frost T, Neff J (2006) Oil well produced water discharges to the North Sea. Part I: comparison of deployed mussels (Mytilus edulis), semi- permeable membrane devices, and the DREAM Model predictions to estimate the dispersion of polycyclic aromatic hydrocarbons. Mar Environ Res 62:194–223

Faksness L-G, Grini PG, Daling PS (2004) Partitioning of semi-soluble organic compounds between the water phase and oil droplets in produced water. Mar Pollut Bull 48:731–742

Fisher JB (1987) Distribution and occurrence of aliphatic acid anions in deep subsurface waters. Geochim Cosmochim Acta 51:2459–2468

Flynn SA, Butler EJ, Vance I (1995) Produced water compostion, toxicity, and fate. A review of recent BP North Sea studies. In: Reed M, Johnsen S (eds) Produced water 2. Environmental issues and mitigation technologies. Plenum Press, New York, pp 69–80

Forbes VE, Palmqvist A, Bach L (2006) The use and misuse of biomarkers in ecotoxicology. Environ Toxicol Chem 25:272–280

Førlin L, Hylland K (2006) Hepatic cytochrome P4501A concentration and activity in Atlantic cod caged in two North Sea pollution gradients. In: Hylland K, Lang T, Vethaak D (eds) Biological effects of contaminants in marine pelagic ecosystems. SETAC Press, Pensacola, FL, pp 253–261

Gamble JC, Davies JM, Hay SJ, Dow FK (1987) Mesocosm experiments on the effects of produced water discharges from offshore oil platforms in the northern North Sea. Sarsia 72: 383–386

Garland E (2005) Environmental regulatory framework in Europe: an update. SPE 93796. Paper presented at the 2005 SPE/EPA/DOE exploration and production environmental conference, Galveston, TX 7–9 March, 2005. Society of Petroleum Engineers, Richardson, TX, 10 pp

Getliff JM, James SG (1996) The replacement of alkyl-phenol ethoxylates to improve environmental acceptability of drilling fluid additives. SPE 35982. Proceedings of the international conference on health, safety & environment, New Orleans, LA. Society of Petroleum Engineers, Richardson, TX

Gray JS (2002) Perceived and real risks: produced water from oil extraction. Mar Pollut Bull 44(11):1171–1172

Grewer DM, Young RF, Whittal RM, Fedorak PM (2010) Naphthenic acids and other acid-extractables in water samples from Alberta: what is being measured? Sci Tot Environ 408: 5997–6010

Harman C, Farmen E, Tollefsen KE (2010) Monitoring North Sea oil production discharges using passive sampling devices coupled with in vitro bioassay techniques. J Environ Monit 12: 1699–1708

Hart AD, Graham BD, Gettleson DA (1995) NORM associated with produced water discharges. SPE 29727. 12 pp. In: Proceedings of the SPE/EPA exploration & production environmental conference, Houston, TX. Society of Petroleum Engineers, Inc., Richardson, TX

Henderson SB, Grigson SW, Johnson P, Roddie BD (1999) Potential impact of production chemicals on toxicity of produced water discharges in North Sea oil platforms. Mar Pollut Bull 38(12):1141–1151

Hodgins DO, Hodgins SLM (1998) Distribution of well cuttings and produced water for the Terra Nova development. Report prepared by Seaconsult Marine Research Ltd. for Terra Nova Alliance. August, 1998

Hodgins DO, Hodgins SLM (2000) Modelled predictions of well cuttings deposition and produced water dispersion for the proposed white rose development. Report prepared by Seaconsult Marine Research Ltd. for Husky Oil Operations Limited. June, 2000

Holdway DA (2002) The acute and chronic effects of wastes associated with offshore oil and gas production on temperate and tropical marine ecological processes. Mar Pollut Bull 44:185–203

Holowenko FM, MacKinnnon MD, Fedorak PM (2002) Characterization of naphthenic acids in oil sand waste waters by gas chromatography-mass spectrometry. Water Res 36:2843–285

Husky Energy (2005) White rose environmental effects monitoring program (2005) Vol. 1 and 2. http://www.huskyenergy.ca/dowloads/AreasOfOperations/EastCoast/HSE/EEMP_June2004.pdf

Hylland K, Lang T, Vethaak D (eds) (2006) Biological effects of contaminants in marine pelagic ecosystems. SETAC Press, Pensacola, FL, 474 pp

Jacobs RPWM, Grant ROH, Kwant J, Marquenie JM (1992) The composition of produced water from Shell operated oil and gas production in the North Sea. In: Ray JP, Engelhardt FR (eds) Produced water. Technological/Environmental Issues and Solutions. Plenum Press, New York, pp 13–21

Jacobs RPWM, Marquenie JM (1991) Produced water discharges from gas/condensate platforms: environmental considerations. First International Conference on Health, Safety and Environment Society of Petroleum Engineers, Richardson, TX, SPE 23321, pp 98–96

Johnsen S, Frost TK, Hjelsvold M, Utvik TR (2000) The environmental impact factor – a proposed tool for produced water impact reduction, management and regulation. Society of Professional Engineers. Paper Number 61178

Johnsen S, Røe TI, Durell G, Reed M (1998) Dilution and bioavailability of produced water components in the northern North Sea. A combined modeling and field study. SPE 46578. 1998 SPE international conference on health, safety and environment in oil and gas exploration and production. Society of Petroleum Engineers, Richardson, TX, pp 1–11

Johnsen S, Røe Utvik TI, Garland E, de Vals B, Campbell J (2004) Environmental fate and effects of contaminants in produced water. SPE 86708. Paper presented at the Seventh SPE international conference on health, safety, and environment in oil and gas exploration and production. Society of Petroleum Engineers, Richardson, TX, 9 pp

Jonkers G, Hartog FA, Knaepen AAI, Lancee PFJ (1997) Characterization of NORM in the oil and gas production (E&P) industry. Proceedings of the international symposium on radiological problems with natural radioactivity in the non-nuclear industry (1997, Amsterdam). KEMA, Arnhem

Karman C, Reerink HG (1996) Dynamic assessment of the ecological risk of the discharge of produced water from oil and gas platforms. J Hazard Mater 61:43–51

Khelifa A, Pahlow M, Vezina A, Lee K, Hannah C (2003) Numerical investigation of impact of nutrient inputs from produced water on the marine planktonic community. Proceedings of the 26th Arctic and marine Oilspill Program (AMOP) Technical Seminar, Victoria, BC, June 10–12, 2003, pp 323–334

KLIF (Climate and Pollution Authority) (Formerly SFT: The Norwegian Pollution Control Authority) (2010) Emissions/discharges from the oil and gas industry in 2009. www.Klif.no/no/english/Whats-new/Emissions-and-discharges-in2009/?cid

Kraemer TF, Reid DF (1984) The occurrence and behavior of radium in saline formation water of the U.S. Gulf coast region. Isot Geosci 2:153–174

Krause PR (1995) "Spatial and temporal variability in receiving water toxicity near an oil effluent discharge site." Arch Environ Contam Toxicol 29:523–529

Large, R (1990) Characterization of produced water, phase 1: literature survey. Report to Conoco (UK) Ltd., Aberdeen, Scotland

Lee K, Azetsu-Scott K, Cobanli SE, Dalziel J, Niven S, Wohlgeschaffen G, Yeats P (2005a) Overview of potential impacts of produced water discharges in Atlantic Canada. In: Armworthy SL, Cranford PJ, Lee K (eds) Offshore oil and gas environmental effects monitoring: approaches and technologies. Battelle Press, Columbus, OH, pp 319–342, 631

Lee K, Bain H, Hurley GV (eds) (2005b) Acoustic monitoring and marine mammal surveys in the gully and outer scotian shelf before and during active seismic programs. December 2005. Environmental Studies Research Funds (ESRF) Report No.151 (ISBN 0-921652-62-3), Calgary, 154 p + appendices

Lee K, Tay KL (1998) Measurement of microbial exoenzyme activity in sediments for environmental impact assessment. In: Wells PG, Lee K, Blaise C (eds) Microscale aquatic toxicology: advances techniques and practice. CRC Press, Boca Raton, FL, pp 219–236, Incorporated

Louisiana Dept. of Environmental Quality (1990) Spread sheet containing volumes, salinity, radium isotopes, and toxicity of produced water from coastal Louisiana. Obtained from Stephenson MT, Texaco, Inc, Bellaire, TX

MacGowan DB, Surdam RC (1988) Difunctional carboxylic acid anions in oilfield waters. Org Geochem 12:245–259

Mackay D (1991) Multimedia environmental models: the fugacity approach. Lewis Publishers, Boca Raton, FL

McCormack P, Jones P, Hetheridge MJ, Rowland SJ (2001) "Analysis of oilfield produced water and production chemicals by electrospray ionization multistage massspectrometry [ESI-MSn]." Water Res 35(15):3567–3578

Means JL, Hubbard N (1987) Short-chain aliphatic acid anions in deep subsurface brines: a review of their origin, occurrence, properties, and importance and new data on their distribution and geochemical implications in the Palo Duro Basin, Texas. Org Geochem 11: 177–191

Means JC, McMillin DJ, Milan CS (1989) Characterization of produced water. In: Boesch DF, Rabalais NN (eds) Environmental impact of produced water discharges in Coastal Louisiana. Report to Mid-Continent Oil and Gas Association, New Orleans, LA, pp 97–110

Means JC, Milan CS, McMillin DJ (1990) Hydrocarbon and trace metal concentrations in produced water effluents and proximate sediments. In: St Pé KM (ed) An assessment of produced water impacts to low-energy, brackish water systems in Southeast Louisiana. Report to Louisiana Dept. of Environmental Quality, Water Pollution Control Div., Lockport, LA, pp 94–199

Meier S, Andersen TE, Norberg B, Thorsen A, Taranger GL, Kjesbu OS, Dale R, Morton HC, Klungsoyr J, Svardal A (2007) Effects of alkylphenols on the reproductive system of Atlantic cod (Gadus morhua). Aquat Toxicol 81: 207–218

Meier S, Grøsvik BE, Makhotin V, Geffen A, Boitsov S, Kvestad KA, Bohne-Kjersem A, Goksøyr A, Flkkvord A, Klungsøyr J, Svardal A (2010) Development of Atlantic cod (Gadus morhua) exposed to produced water during early life stages. Effects on embryos, larvae, and juvenile fish. Mar Environ Res 70:383–394

Michel J (1990) Relationship of radium and radon with geological formations. In: Cothern CR, Ribers PA (eds) Uranium in drinking water. Lewis Publishers, Chelsea, MI, pp 83–95

Moffitt CM, Rhea MR, Dorn PB, Hall JF, Bruney JM (1992) Short-term chronic toxicity of produced water and its variability as a function of sample time and discharge rate. In: Ray JP, Engelhardt FR (eds) Produced water. Technological/Environmental Issues and Solutions. Plenum Press, New York, pp 235–244

Mukhtasor HT, Vaith B, Bose N (2004) An ecological risk assessment methodology for screening discharge alternatives of produced water. Hum Ecol Risk Assess 10:505–524

Murray-Smith RJ, Gore D, Flynn SA, Vance I, Stagg R (1996) Development and appraisal of a particle tracking model for the dispersion of produced water discharges from an oil production platform in the North Sea. In: Reed M, Johnsen S (eds) Produced water 2: environmental issues and mitigation technologies. Plenum Press, New York, London, pp 225–245

Myhre LP, Bausant T, Sundt R, Sanni S, Vabø R, Skjoldal HR, Klungsøyr J (2004) Risk assessment of reproductive effects of alkyl phenols in produced water on fish stockes in the North Sea. Report AM-2004/018. Prepared by RF-Akvamiljø and Institute of Marine Research (IMR) for the Norwegian Oil Industry Association's (OLF) Working Group for Discharges to Sea. Available: www.imr.no/filarkiv/2004/08/AM_2004-018_Final_report_AP_risk_03112004_rev_3.pdf

NEB (National Energy Board of Canada) (2002) Offshore waste treatment guidelines. Offshore chemical selection guidelines for drilling & production activities on Frontier Lands. National Petroleum Board, Canada-Nova Scotia Offshore Petroleum Board, Canada-Newfoundland and Labrador Offshore Petroleum Board. Calgary, Alberta, Canada, 23 pp

Nedwed TJ, Smith JP, Brandsma MG (2004) Verification of the OOC mud and produced water discharge model using lab-scale plume behavior experiments. Environ Model Software 19: 655–670

Neff JM (1987) Biological effects of drilling fluids, drill cuttings and produced waters. In: Boesch DF, Rabalais NN (eds) Long-term effects of offshore oil and gas development. Elsevier Applied Science Publishers, London, pp 469–538

Neff JM (1991) Technical review document: process waters in cook Inlet, Alaska. Report to the Alaska Oil and Gas Association, Anchorage, AK

Neff JM (1997) Potential for bioaccumulation of metals and organic chemicals from produced water discharged offshore in the Santa Barbara Channel, California: A Review. Report to the Western States Petroleum Association, Santa Barbara, CA, from Battelle Ocean Sciences, Duxbury, MA, USA, 201 pp

Neff JM (2002) Bioaccumulation in marine organisms. Effects of contaminants from oil well produced water. Elsevier, Amsterdam, 452 pp

Neff JM (2005) Composition, environmental fates, and biological effect of water based drilling muds and cuttings discharged to the marine environment: a synthesis and annotated bibliography. Report prepared for the Petroleum Environment Research Forum (PERF). Available from American Petroleum Institute, Washington, DC, 73 pp

Neff JM, Burns WA (1996) Estimation of polycyclic aromatic hydrocarbon concentrations in the water column based on tissue residues in mussels and salmon: an equilibrium partitioning approach. Environ Toxicol Chem 15:2240–2253

Neff JM, Foster K (1997) Composition, fates, and effects of produced water discharges to offshore waters of the Java Sea, Indonesia. Report to Pertimena/Maxus, Jakarta, Indonesia

Neff JM, Johnsen S, Frost T, Røe Utvik T, Durell G (2006) Oil well produced water discharges to the North Sea. Part II: comparison of deployed mussels (Mytilus edulis) and the DREAM Model to predict ecological risk. Mar Environ Res 62:224–246

Neff JM, Rabalais NN, Boesch DF (1987) Offshore oil and gas development activities potentially causing long-term environmental effects. In: Boesch DF, Rabalais NN (eds) Long-term effects of offshore oil and gas development. Elsevier Applied Science Publishers, London, pp 149–174

Neff JM, Sauer TC, Maciolek N (1989) Fate and effects of produced water discharges in nearshore marine waters. API Publication No. 4472. American Petroleum Institute, Washington, DC, 300 pp

Nelson R (2007) A preliminary investigation into levels of naturally occurring radioactive materials (NORM) from offshore oil and gas production. In: Lee K, Neff J (Chairs) Abstracts of the international produced water conference: environmental risks and advances in mitigation technologies St. John's Newfoundland, Canada, October 17–18, (2007) National Energy Board of Canada, Environmental Studies Research Fund (ESRF)

Niu H, Adams S, Lee K, Husain T, Bose N (2009d) The application of autonomous underwater vehicles in offshore environmental effect monitoring. J Can Petrol Technol 48(5): 12–16

Niu H, Husain T, Veitch B, Bose N, Adams S, He M, Lee K (2007) Ocean outfall mapping using an Autonomous Underwater Vehicles. Proceedings of the MTS/IEEE Oceans 2007 Conference, 29 Sep-04 Oct, 2007, Vancouver, BC, Canada, 4 pages, DOI: 10.1109/OCEANS.2007.4449238

Niu H, Husain T, Veitch B, Bose N, Hawboldt K, Mukhtasor (2009a) Assessing ecological risks of produced water discharge in a wavy marine environment. Adv Sustainable Petrol Eng Sci 1(1):1–11

Niu H, Lee K, Husain T, Veitch B, Bose N (2009b) The PROMISE model for evaluation of the mixing of produced water in marine environment. Proceedings of the 20th IASTED international conference on modelling and simulation, MS 2009, July 6–8, 2009, Banff, AB, Canada

Niu H, Zhan C, Lee K, Veitch B (2009c) Validation of a buoyant jet model (PROMISE) against laboratory data and other models, In: Proceedings of the IASTED International Conference on Modelling, Simulation and Identification, MSI 2009, Oct 12–14, 2009, Beijing, China

Nozaki Y (1991) The systematics and kinetics of U/Th decay series nuclides in ocean water. Rev Aquat Sci 4:75–105

NRPA (Norwegian Radiation Protection Authority) (2004) Natural radioactivity in produced water from the Norwegian oil and gas industry in 2003. Strålevern Rapport 2005:2

OGP (International Association of Oil and Gas Producers) (2004) Environmental Performance in the E&P Industry. 2003 Data. Report No. 359. OGP, London, UK, 32 pp

OGP (International Association of Oil and Gas Producers) (2002) Aromatics in produced water: Occurance, fate and effects and treatments. OGP Publications. Report No. 1.20/324, 24 pp

OOC (Offshore Operators Committee) (1997) Definitive Component Technical Report. Gulf of Mexico Produced Water Bioaccumulation Study. Report to OOC, New Orleans, LA, from Continental Shelf Associates, Inc., Jupiter, FL

Otto GH, Arnold KE (1996) U.S. produced water discharge regulations have tough limits. Oil Gas J 94:54–61

Payne JF, Andrews CD, Guiney JM, Lee K (2005) Production water releases on the Grand Banks: potential for endocrine and pathological effects in fish. Dixon DG, Munro S, Niimi AJ (eds) Proceedings of the 32nd annual toxicity workshop. Can Tech Rep Fish Aquat Sci 2617:24

Payne J, Fancey L, Andrews C, Meade J, Power F, Lee K, Veinott G, Cook A (2001) Laboratory exposures of invertebrate and vertebrate species to concentrations of IA-35 (Petro-Canada) drill mud fluid, production water and Hibernia drill mud cuttings. Can Data Rep Fish Aquat Sci 2560: 27p

Payne JF, Mathieu A, Collier TK (2003) Ecotoxicological studies focusing on marine and freshwater fish. In: Douben PET (ed) Polycyclic aromatic hydrocarbons: an ecotoxicological perspective. Wiley, London, pp 192–224

Pérez-Casanova JC, Hamoutene D, Samuelson S, Burt K, King TL, Lee K (2010) The immune response of juvenile Atlantic cod (Gadus morhua) to chronic exposure to produced water. Mar Environ Res 79:36–34

Pillard DA, Tietge JE, Evans JM (1996) Estimating the acute toxicity of produced waters to marine organisms using predictive toxicity models. In: Reed M, Johnsen S (eds) Produced water 2. Environmental issues and mitigation technologies. Plenum Press, New York, pp 49–60

Querbach K, Maillet G, Cranford PJ, Taggart C, Lee K, Grant J (2005) Potential effects of produced water discharges on the early life stages of three resource species. In: Armsworthy S, Cranford PJ, Lee K (eds) Offshore oil and gas environmental effects monitoring (approaches and technologies). Battelle Press, Columbus, OH, pp 343–372

Rabalais NN (2005) Relative contribution of produced water discharge in the development of hypoxia. OCS Study MMS 2005-044. U.S. Dept. of the Interior, Minerals Management Service, Gulf of Mexico Region, New Orleans, LA, 56 pp

Rabalais NN, McKee BA, Reed DJ, Means JC (1991) Fate and effects of nearshore discharges of OCS produced waters. Vol. 1: executive summary. Vol. 2: Technical Report. Vol. 3. Appendices. OCS Studies MMS 91-004, MMS 91-005, and MMS 91-006. U.S. Dept. of the Interior, Minerals Management Service, Gulf of Mexico OCS Regional Office, New Orleans, LA

Rasmussen LL (2003) Radium isotopes as tracers of coastal circulation pathways in the Mid-Atlantic Bight. PhD Thesis, Massachusetts Institute of Technology, Cambridge, MA, 214 pp

Reed M, Johnsen S, Melbye A, Rye H (1996) PROVANN a model system for assessing potential chronic effects of produced water. In: Reed M, Johnsen S (eds) Produced water 2: environmental issues and mitigation technologies. Plenum Press, New York and London, pp 317–330

Reed M, Rye H, Johansen Ø, Johnsen S, Frost T, Hjelsvold M, Karman C, Smit M, Giacca D, Bufagni M, Gauderbert B, Durrieu J, Utvik TR, Follum OA, Sanni S, Skadsheim A, Bechham R, Bausant T (2001) DREAM: a dose-related exposure assessment model – Technical description of physical-chemical fates components. Proceedings of the 5th international marine environmental modeling seminar, New Orleans, Louisiana, USA, 9–11 October 2001

Reid DF (1983) Radium in formation waters: how much and is it a concern? 4th Annual Gulf of Mexico Information Transfer Meeting, New Orleans, LA. U.S. Dept. of the Interior, Minerals Management Service, Gulf of Mexico OCS Office, New Orleans, LA, pp 187–191

Reuters (2007) Canada Hibernia oil output declines as field ages. Press release, Oct. 22, 2007. Reuters

Rittenhouse G, Fulron RB, III, Grabowski RJ, Bernard JL (1969) Minor elements in oil field waters. Chem Geol 4:189–209

Rivkin RB, Tian R, Anderson MR, Payne JF (2000) "Ecosystem level effects of offshore platform discharges: identification, assessment and modelling." Proceedings of the 27th annual aquatic toxicity workshop, St John's, Newfoundland, Canada, October 01–04, 2000. Can Tech Report Fish Aquat Sci 2331:3–12

Røe Utvik TI (1999) Chemical characterization of produced water from four offshore oil production platforms in the North Sea. Chemosphere 39:2593–2606

Røe Utvik TI, Durell GS, Johnsen S (1999) Determining produced water originating polycyclic aromatic hydrocarbons in North Sea waters: comparison of sampling techniques. Mar Pollut Bull 38:977–989

Santschi PH, Honeyman BD (1989) Radionuclides in aquatic environments. Radiat Phys Chem 34:213–240

Shaw DG, Farrington JW, onner MS, Trippm BW, Schubel JR (1999) Potential environmental consequences of petroleum exploration and development on Grand Banks. New England Aquarium Aquatic Forum Series Report 00-3, Boston. 64 pp

Skåtun HM (1996) A buoyant jet/plume model for subsea releases. In: Reed M, Johnsen S (eds) Produced water 2: environmental Issues and mitigation technologies. Plenum Press, New York and London, pp 247–255

Smith JP, Brandsma MG, Nedwed TJ (2004) Field verification of the Offshore Operators Committee (OOC) mud and produced water discharge model. Environ Model Software 19:739–749

Smith JP, Tyler AO, Rymell MC, Sidharta H (1996) Environmental impacts of produced waters in the Java Sea, Indonesia. SPE 37002, In: 1998 SPE Asia Pacific Oil and Gas Conference, Adelaide, Australia, 28–30 October 1996

Smith JP, Mairs HR, Brandsma MG, Meek RP, Ayers RC (1994) Field validation of the Offshore Operators Committee (OOC) produced water discharge model. SPE paper No. 28350, Society of Petroleum Engineers

SOEP (1996) Sable Offshore Energy Project Environmental Impact Statement. Canada Nova Scotia Offshore Petroleum Board, 481 pp

Somerville HJ, Bennett D, Davenport JN, Holt MS, Lynes A, Mahieu A, McCourt B, Parker JG, Stephenson RR, Watkinson RJ, Wilkinson TG (1987) Environmental effect of produced water from North Sea oil operations. Mar Pollut Bull 18:549–558

Spaulding ML (1994) "MUDMAP": a numerical model to predict drill fluid and produced water dispersion. Offshore, Houston Texas issue

Stephenson MT (1991) Components of produced water: a compilation of results from several industry studies. Proceedings international conference on health, safety and environment, Hague, Netherlands, November 10–14, 1991. Soc Petrol Eng, Pap. No. 23313, pp 25–38

Stephenson MT (1992) A survey of produced water studies. In: Ray JP, Engelhardt FR (eds) Produced water. Technological/environmental issues and solutions. Plenum Press, New York, pp 1–11

Stephenson MT, Ayers RC, Bickford LJ, Caudle DD, Cline JT, Cranmer G, Duff A, Garland E, Herenius TA Jacobs RPWM, Inglesfield C, Norris G, Petersen JD, Read AD (1994) North Sea produced water: fate and effects in the marine environment. Report No. 2.62/204. E&P Forum, London, England, 48 pp

Strømgren T, Sørstrøm SE, Schou L, Kaarstad I, Aunaas T, Brakstad OG, Johansen Ø (1995) Acute toxic effects of produced water in relation to chemical composition and dispersion. Mar Environ Res 40:147–169

Sturve J, Hasselberg L, Falth H, Celander MC, Forlin L (2006) Effects of North Sea oil and alkylphenols on biomarker responses in juvenile Atlantic cod (Gadus morhua). Aquat Toxicol 78:573–578

Sundt RC, Baussant T, Beyer J (2009) Uptake and tissues distribution of C_4-C_7 alkylphenols in Atlantic cod (*Gadus morhua*): relevance for biomonitoring of produced water discharges from oil production. Mar Pollut Bull 58:72–79

Sverdrup LE, Fürst CS, Weideborg M, Vik EA, Stenersen J (2002) Relative sensitivity of one freshwater and two marine acute toxicity tests as determined by testing 30 offshore E&P chemicals. Chemosphere 46:311–318

Terrens GW, Tait RD (1993) Effects on the marine environment of produced formation water discharges from Esso/BHPP's Bass Strait Platforms. Esso Australia Ltd., Melbourne, Australia. 25 pp

Terrens GW, Tait RD (1994) Effects on the marine environment of produced formation water discharges from offshore development in Bass Strait, Australia. SPE Adv Technol Series 4(2):42–50

Terrens GW, Tait RD (1996) Monitoring ocean concentrations of aromatic hydrocarbons from produced formation water discharges to Bass Strait, Australia. SPE 36033. Proceedings of the international conference on health, safety & environment. Society of Petroleum Engineers, Richardson, TX, pp 739–747

Thomas KV, Balaam J, Hurst MR, Thain JE (2004) Identification of in vitro estrogen and androgen receptor agonists in North Sea offshore produced water discharges. Environ Toxicol Chem 23:1156–1163

Thomas KV, Langford K, Petersen K, Smith AJ, Tollefsen KE (2009) Effect-directed identification of naphthenic acids as important in vitro xeno-estrogens and anti-estrogens in North Sea offshore produced water discharges. Environ Sci Technol 43:8066–8071

Tomczyk NA, Winans RW, Shinn JH, Robinson RC (2001) On the nature and origin of acidic species in petroleum. 1. Detailed acid type distribution in a California crude oil. Energy Fuels 15:1498–1504

Turner NB, Tyler AO, Falconer RA, Millward G.E (1995) Modeling contaminant transport in estuaries. Water Res 30(1):63–74

Veil JA (2006) Comparison of two international approaches to controlling risk from produced water discharges. Paper presented at the 70th PERF meeting, Paris, France, March 21–22, 2006

Veil JA, Kimmell TA, Rechner AC (2005) Characteristics of produced water dicharged to the Gulf of Mexico Hypoxic Zone. Report to the U.S. Dept. of Energy, National Energy Technology Laboratory, Argonne National Laboratory, Washington, DC

Veil JA, Puder MG, Elcock D, Redweik RJ Jr (2004) A white paper describing produced water from production of crude oil, natural gas, and coal bed methane. Report to the U.S. Dept. of Energy, National Energy Technology Laboratory. Argonne National Laboratory, Washington, DC, 79 pp

Wells PG, Lee K, Blaise C (eds) (1998) Microscale testing in aquatic toxicology: advances, techniques, and practice. CRC Press, Boca Raton, FL, 679 pp

Witter AE, Jones AD (1999) Chemical characterization of organic constituents from sulfide-rich produced water using gas chromatography/mass spectrometry. Environ Toxicol Chem 18: 1920–1926

Zhang X (1995) Ocean outfall modeling – interfacing near and far field models with particle tracking method. Ph.D. thesis, Massachusetts Institute of Technology

Zhao L, Chen Z, Lee K (2008) A risk assessment model for produced water discharge from offshore petroleum platforms-development and validation. Mar Pollut Bull 56:1890–1897

Part II
Composition/Characterization

Chapter 2
Measurement of Oil in Produced Water

Ming Yang

Abstract The measurement of oil in produced water is important for both process control and reporting to regulatory authorities. The concentration of oil in produced water is a method-dependent parameter, which is traditionally evaluated using reference methods based on infrared (IR) absorption or gravimetric analysis, although Gas Chromatography and Flame Ionisation Detection (GC-FID) have recently become more accepted. This chapter will give a brief overview of hydrocarbon chemistry and discuss the definition of oil in produced water and the requirement for its measurement. Reference and non-reference (bench-top and online monitoring) methods for the measurement of oil in produced water are reviewed. Issues related to sampling, sample handling and calibration, as well as methods of how to accept a non-reference method for the purpose of reporting, are discussed.

1 Introduction

Discharge of produced water from oil and gas productions is generally regulated. Historically one of the key parameters used for compliance monitoring is the oil concentration in produced water, is because it is relatively easy to measure.

Measurement of oil in produced water is required from an operational point of view, because process optimisation is increasingly being implemented by operators so that less oil is discharged, less chemicals are used, process capacity is increased, and oil and gas production is maximised.

Despite the importance of measuring oil in produced water, it is not widely understood what oil in produced water really is, or what the standard methods or field analyses are actually measuring. A very good example of this is the definition of the performance standard under OSPAR (Oslo-Paris Convention). The definition of 40 mg/L, which was reduced to 30 mg/L as of 1 January 2007, dealt with dispersed oil in produced water (OSPAR Recommendation 2001). However, what the old infrared-based OSPAR method measured was not just the dispersed oil. As a result, OSPAR faced a number of challenges in relation to the change in the method of measuring oil in produced water, in order to meet the OSPAR target of 15% reduction of year 2000 total oil by 2006.

M. Yang (✉)
TUV NEL Ltd, Scottish Enterprise Technology Park, East Kilbride, Glasgow G75 0QF, UK

K. Lee, J. Neff (eds.), *Produced Water*, DOI 10.1007/978-1-4614-0046-2_2,

Oil in produced water is a method-dependent parameter. This point cannot be emphasised enough. Without the specification of a method, reported concentrations of oil in produced water can mean little, as there are many techniques and methods available for making this measurement, but not all are suitable in a specific application.

Other issues include sampling, calibration, and the acceptance of a non-reference method for the purpose of reporting. Sampling is often not addressed by an analytical method. However, sampling can lead to significant uncertainty in the final result, which can be important in terms of regulatory compliance and process optimisation.

While uncertainties associated with calibrations are often included in reference method precision data, if the calibration procedures are not strictly followed, they can produce quite different results. A good example showing such an effect was given by a study carried out by Statoil who confirmed an average of 19% reduction by using synthetic oil as a calibration oil compared to using a crude oil in a single wavelength method (Paus et al. 2001).

Also, as a reference method is not always suited for use in the field, alternative methods are used. For the purpose of compliance monitoring and reporting, accept ability of results from a non-reference method is of interest to many. Guidance for such acceptance is given.

2 Basic Hydrocarbon Chemistry

Oil in water is essentially petroleum compounds in water (Total Petroleum Hydrocarbon Criteria Working Group Series 1998). Petroleum compounds can be divided into two main groups: hydrocarbons and heteroatom compounds.

Hydrocarbons are usually measured as Total Petroleum Hydrocarbons (TPHs). These are molecules that only contain carbon and hydrogen. The heteroatom compounds are those that contain not only carbon and hydrogen but also heteroatoms such as sulphur, nitrogen and oxygen.

Hydrocarbons are in general grouped into three categories: saturated, unsaturated and aromatics.

Saturated hydrocarbons are characterised by single C–C bonds with all other remaining bonds saturated by H atoms. This group can be subdivided into aliphatic and alicyclic.

Aliphatic hydrocarbons are straight or branched with a general molecular formula: C_nH_{2n+2}. The common names for these types of compounds are alkanes and isoalkanes, which are often referred to by the petroleum industry as paraffins and isoparaffins respectively.

Alicyclic hydrocarbons are saturated hydrocarbons containing one or more rings with a general molecular formula: C_nH_{2n}. They are also called cycloalkanes or naphthenes or cycloparaffins by the petroleum industry.

Unsaturated hydrocarbons are characterised by two or more bonds (C = C for alkenes or C≡C for alkynes) between two carbon atoms. They are not usually found in crude oils, but are produced in cracking processes (converting large molecular

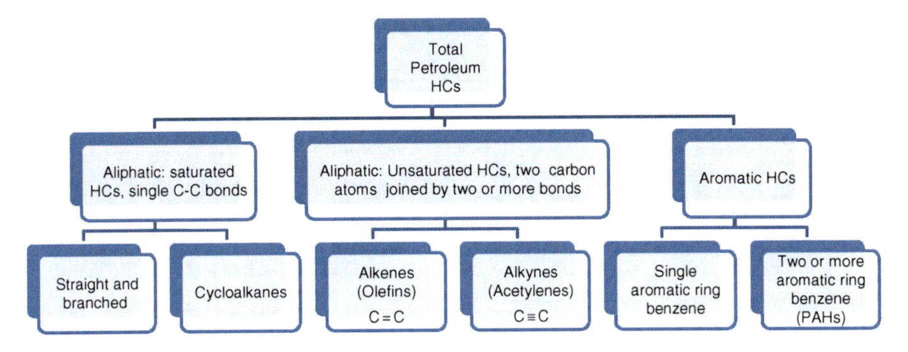

Fig. 2.1 Hydrocarbon chemistry – total petroleum hydrocarbons (TPHs)

hydrocarbons to smaller ones). Unsaturated hydrocarbons can be sub-grouped into alkenes/olefins and alkynes/acetylenes.

Alkenes/olefins are those that contain two carbon bonds with a general molecular formula C_nH_{2n}.

Alkynes/acetylenes are those that contain three carbon bonds with a general molecular formula C_nH_{2n-2}.

Aromatic hydrocarbons are characterised by a benzene ring structure. The benzene ring contains six carbons; each carbon in the ring binds with one hydrogen. Depending on the number of rings that an aromatic hydrocarbon molecule contains, they are often further divided into single ring aromatics and polycyclic aromatics (containing two rings or more).

A summary of the different types of hydrocarbons is given in Fig. 2.1. Having an understanding of the basic hydrocarbon chemistry is useful. It will help with an appreciation of what is meant by oil in water, in particular, when covering topics such as aliphatic, aromatic hydrocarbons, dissolved and dispersed oils, solvent extract clean-up to remove the polar components, etc.

3 Definition of Oil in Produced Water

Oil in produced water is a general term. As shown in Fig. 2.2, it can mean different things to different people. Since the results are method dependent, without specifying the method used to determine the oil concentration, values reported for oil in produced water can be misleading.

Dispersed oil – usually means oil in produced water in the form of small droplets, which may range from sub-microns to hundreds of microns. Dispersed oil will contain both aliphatic and aromatic hydrocarbons.

Dissolved oil – usually means oil in produced water in a soluble form. Aliphatic hydrocarbons in general have very low solubility in water. It is the aromatic hydrocarbons, together with things like organic acid that form the bulk of dissolved oil.

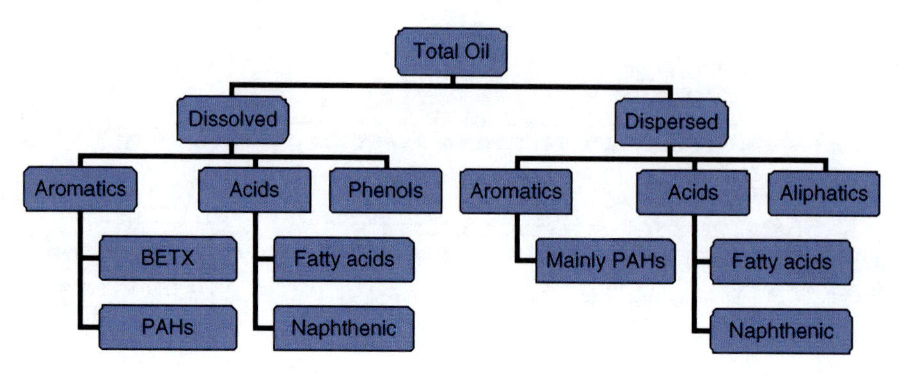

Fig. 2.2 Oil in produced water (Yang 2006)

3.1 Solubility of Hydrocarbons

The solubility of pentane, hexane, heptane, octane, nonane decane in water is given in Table 2.1. As the carbon number increases in a straight-chain saturated hydrocarbon molecule, solubility decreases rapidly. For *n*-octane, it is already around 1 mg/L, which becomes insignificant when compared to a performance standard such as that in the North Sea of 30 mg/L. This is why straight chain aliphatic hydrocarbons are generally ignored when discussing dissolved oil in a produced water sample.

Solubility of some of the single ring aromatic hydrocarbons is given in Table 2.2. It is quite clear that when they are present in a produced water sample, their concentration can be significant in comparison to a performance standard or discharge limit 30 mg/L.

3.2 Definitions

When produced water samples are taken and acidified, many of the organic acids and phenols are converted from water soluble (dissolved) to oil soluble (dispersed), which are then extracted into a solvent. Treatment of the extract using florisil before analysis usually removes them and they are not included in the oil measured.

In the OSPAR makes a distinction between total oil and dispersed oil. Total oil means total hydrocarbons (OSPAR Recommendation 2001). Dispersed oil means

Table 2.1 Solubility of saturated straight chain aliphatic hydrocarbons (http://home.flash.net/~defilip1/solubility.htm)

Compound	*n*-pentane	*n*-hexane	*n*-heptane	*n*-octane	decane
Solubility (mg/L)	39	11	2	1	0

Table 2.2 Solubility of BETX and other single ring aromatic hydrocarbons (Frintrop 2007)

Compound	Molecular formula	Solubility at 25°C (mg/L)
Benzene	C_6H_6	1790
Toluene	C_7H_8	526
Ethylbenzene	C_8H_{10}	169
o-xylene	C_8H_{10}	178
p-xylene	C_8H_{10}	162
m-xylene	C_8H_{10}	161
1,2,3-Trimethylbenzene	C_9H_{12}	75
1,2,4-Trimethylbenzene	C_9H_{12}	57
1,3,5-Trimethylbenzene	C_9H_{12}	48
p-Ethyl toluene	C_9H_{12}	95
o-Ethyl toluene	C_9H_{12}	75
m-Ethyl toluene	C_9H_{12}	40
Iso-propyl benzene	C_9H_{12}	61
n-Propy benzene	C_9H_{12}	52
n-Butyl benzene	$C_{10}H_{14}$	12

the hydrocarbons as determined according to the OSPAR reference method (OSPAR Agreement 2005). Currently, dispersed oil is defined as follows:

> The sum of the concentrations of compounds extractable with *n*-pentane, not adsorbed on florisil and which may be chromatographed with retention times between those of *n*-heptane (C_7H_{16}) and *n*-tetracontane ($C_{40}H_{82}$) excluding the concentrations of the aromatic hydrocarbons toluene, ethylbenzene and the three isomers of xylene.

In the USA, oil in produced water is referred to as "Oil and Grease". It is defined as those materials that are extracted by *n*-hexane not evaporated at 70°C and capable of being weighed (http://water.epa.gov/scitech/methods/cwa/oil/1664.cfm).

Oil in produced water measured by the OSPAR method can be very different from what is quantified in the USA. This difference will also vary for the range of oily waters from the different types of oil and gas installations.

While there is an international standard method now available (ISO 9377-2 GC-FID method), for both historical and economic reasons, there is as yet no unified method for the measurement of oil in produced water.

4 Reference Methods

As oil in produced water is method-dependent, it is critical that a reference method is defined and available so that the data obtained from different installations can be directly compared. There are three main types of reference method available:

- Infrared absorption;
- Gravimetric;
- Gas Chromatography and Frame Ionisation Detection (GC-FID).

4.1 Infrared Absorption

In a typical infrared absorption based method, an oily water sample is first acidified, then extracted by a chlorofluorocarbon (CFC) solvent. Following the separation of the extract from water sample, the extract is then removed, dried and purified by the removal of polar compounds. A portion of the extract is placed into an infrared instrument, where the absorbance is measured. By comparing the absorbance obtained from a sample extract to those that are prepared with known concentrations, the oil concentration in the original sample can be calculated.

The fundamental principle of infrared based measurement is that of the Beer–Lambert law, which is given by the following equation:

$$A = \log I_0/I = ELc \tag{2.1}$$

where

A is the absorbance;
I_0 is the incident light intensity;
I is the transmitted light intensity;
E is a constant;
L is the path length;
c is the hydrocarbon concentration in the sample extract.

There are two types of infrared-based reference method:

- Single wavelength infrared methods, and
- Triple peak or three wavelength methods.

In a *single wavelength* method, quantification (measurement of absorbance) is done by using a single wavelength, usually at around 2930 cm^{-1}, which corresponds to the CH_2 stretch vibration frequency. Several examples of single wavelength infrared reference methods are listed in Table 2.3.

In theory, a single wavelength method quantifies all of the CH_2 contained in a sample extract. These include all of the aliphatic hydrocarbons and those that are contained in aromatic hydrocarbons such as ethylbenzene. The results will depend on how calibration is carried out and whether it uses a synthetic oil or field specific oil. The effect of using different calibration oils will be discussed in Section 7.

For a single wavelength infrared method, both fixed wavelength and scanning infrared instruments can be used. Many of the fixed wavelength instruments available on the market are portable, and easy to use, which offers advantages for offshore applications.

In a *triple peak or three wavelength* infrared method, instead of measuring the absorbance at one fixed wavelength, infrared absorbance at three different wavelengths is recorded. The three wavelengths are respectively related to the stretch vibration frequency of aromatic C–H at 3030 cm^{-1}, methylene CH–H at 2960 cm^{-1} and methyl CH_2–H at 2930 cm^{-1}. Oil content is quantified by using an equation such

Table 2.3 Examples of single wavelength reference methods

Reference method	Wavelength (cm^{-1})	Solvent used	Calibration	Status
Old OSPAR (OSPAR Agreement 1997)	2925	CCl_4; Freon; Tetrochloroethylene	Field specific oil	Superseded
USA EPA 413.2	2930	Freon	*n*-hexadecane + isooctane	Superseded
USA EPA 418.1	2930	Freon	*n*-hexadecane + isooctane	Superseded
ASTM 3921-85	2930	Freon	*n*-hexadecane + isooctane or oil in question	Superseded
ASTM D 7066-04 (2004)	2930	S-316	*n*-hexadecane + isooctane or oil in question	Still in use

as one shown in Eq. (2.2) below, which takes into account the absorbance obtained at the three wavelengths (DECC Guidance Notes 2010). The equation has three coefficients (*X*, *Y*, *Z*) that will have been determined by measuring the absorbance at the same three wavelengths of known concentration calibration standards.

$$C_{\text{total}} = \left\{[X\,(A_{2930})] + [Y\,(A_{2960})] + \left[Z\,(A_{3030} - A_{2930}/F)\right]\right\} 10vD/VL \qquad (2.2)$$

where

C_{total} is the oil-in-water concentration;
X, Y and Z are factors calculated using absorbance obtained with known concentration calibration standards;
v is the volume of extraction solvent;
D is the dilution factor (if the sample is not diluted, $D = 1$);
V is the volume of produced water sample;
A is the absorbance at the specified wavelengths;
F is A_{2930}/A_{3030} for hexadecane standard;
L is the cell path length.

With a three wavelength method, the aromatic part of hydrocarbons in the samples is more properly quantified. Also such a method allows the calculation of aliphatic and aromatic hydrocarbons as given in the Eqs. (2.3) and (2.4).

$$C_{\text{aliphatic}} = [X\,(A_{2930})] + [Y\,(A_{2960})]\,10vD/VL \qquad (2.3)$$

$$C_{\text{aromatic}} = C_{\text{total}} - C_{\text{aliphatic}} \qquad (2.4)$$

Table 2.4 Examples of three wavelength based infrared methods

Reference method	Country	Solvent used	Calibration	Status
DIN Standard (DIN 38409 PT 18 English 1981)	Germany	Freon	No calibration, fixed coefficients	Superseded
Her Majesty's Stationary Office (1983)	UK	Freon	*n*-Hexadecane, pristane and toluene	Superseded
IP 426/98 (Energy Institute)	UK	Tetrachloroethylene	*n*-Hexadecane, pristane and toluene	Still in use
GB/T 17923-1999 (1999)	China	Carbon tetrachloride	*n*-Hexadecane, pristane and toluene	Still in use

For a triple peak method, while a fixed wavelength instrument may be used, in general a scanning infrared instrument is employed. There are several examples of three wavelength infrared methods shown in the Table 2.4.

It is important to realise that although three wavelength infrared methods quantify both aliphatic and aromatic hydrocarbons and allow the calculation of aliphatic, aromatic and total hydrocarbons, it does not mean that the results obtained from a three wavelength method will be higher than those obtained from single wavelength methods.

Extraction solvents play an important part in infrared reference methods. They are used to extract oil from a water sample. Obviously anything that is not extracted will not be included in the analysis.

A good solvent should possess a number of properties in addition to good extraction ability. These may include

- sufficient infrared transmission (infrared transparency)
- environmental friendliness,
- safe to use,
- heavier than water,
- reasonably priced and easily available.

Over the past 20 years, different solvents have been used in infrared based reference methods. These are given in the Table 2.5.

All these solvents are infrared transparent and have been known to be a good extractants for oil in water (Wilks 2001). They are all heavier than water which means that they can be easily separated from the water phase and drained from the extraction funnel.

Carbon tetrachloride was commonly used with infrared methods, but due to its carcinogenicity, it was replaced by Freon in the late 1900s. When Freon was found to be an ozone depleting substance, it was gradually phased out in the late 1900s

Table 2.5 Commonly used solvents for infrared based methods (Yang 2002)

Solvent	Formula	Density at 25°C (g/mL)	Boiling point (°C)	Toxicity	Ozone depletion
Carbon tetrachloride	CCl_4	1.59	77	Carcinogen	Yes
Freon	$C_2F_3Cl_3$	1.56	49	Low	Yes
Tetrachloroethylene	C_2Cl_4	1.623	121	Moderate (suspected carcinogen)	No
S-316	$CClF_2CClF$ $CClFCClF_2$	1.75	134	Low	Not sure

and early 2000s and was replaced by tetrachloroethylene and S-316. However people have been concerned with the fact that tetrachloroethylene is suspected to be carcinogenic and that S-316 is expensive.

In some developing countries, carbon tetrachloride is still used. The United Nations Industrial Development Organization (UNIDO Kortvelyssy 2008) has set up a series of workshops where alternatives to carbon tetrachloride have been discussed.

While infrared methods are excellent for the measurement of oil in water, they are being used less frequently now, mainly due to problems related to solvents.

4.2 Gravimetric

Gravimetric-based methods measure anything extractable by a solvent that is not removed during a solvent evaporation process and is capable of being weighed.

In a typical gravimetric-based method, an oily water sample is extracted by a solvent. After separating the solvent (now containing oil) from the water sample, it is placed into a flask, which has been weighed beforehand. The flask is placed into a temperature controlled water bath, and the solvent is evaporated at a specific temperature, condensed and collected. After the solvent is evaporated, the flask now containing the residual oil, is dried and weighed. Knowing the weight of the empty flask, the amount of residual oil can be calculated. Examples of gravimetric methods are shown in Table 2.6.

Table 2.6 Examples of gravimetric based methods

Reference method	Country	Solvent used	Evaporation temperature (°C)	Status
ASTM D4281-95	USA	Freon		Superseded
Method 5520 B (2001)	USA	*n*-hexane	85	In use
EPA 413.1	USA	Freon	70	Superseded
EPA 1664 A (http://water.epa.gov/scitech/methods/cwa/oil/1664.cfm)	USA	*n*-hexane	85	In use

Before the sample extract is placed into the flask for the evaporation process, it may go through a cleaning step to remove the polar components, e.g. organic acids, phenols.

The most widely used gravimetric method is probably the USA EPA Method 1664 A (http://water.epa.gov/scitech/methods/cwa/oil/1664.cfm). The method was developed following the phase-out of Freon. In the method a litre of water is acidified to pH less than 2 and then extracted using three volumes of *n*-hexane. The hexane extract is then combined and dried before it is distilled at 85°C. Depending upon whether the hexane extract undergoes a silica gel treatment (cleaning) process, the residual oil obtained is called either Hexane Extractable Material (HEM) or Silica Gel Treated – Hexane Extractable Material (SGT-HEM).

In the USA, HEM is synonymous with "Oil and Grease". For regulatory compliance monitoring of the offshore oil and gas industry, it is this "Oil and Grease" as determined by the EPA Method 1664 A that is compared to the discharge limits of a monthly average of 29 mg/L and a daily maximum of 42 mg/L.

4.3 GC-FID

Unlike infrared and gravimetric methods, the use of GC offers the potential for obtaining details of the different types of hydrocarbons in the oil fraction. In general a typical GC-FID instrument will include the following components:

- Carrier gas supplier;
- Injector;
- Chromatographic column;
- Detector;
- Data handling system.

In a typical GC-FID method, an oily water sample is acidified and extracted by a solvent like other reference methods. The extract is then dried and purified before a small amount of the extract is injected into a GC instrument. With the help of a carrier gas and the chromatographic column, different groups of hydrocarbons will then leave the column at different times and be detected. Carrier gas is used to move the components through the column while the column acts to separate the different groups of hydrocarbons so that they leave the column and are detected at different times.

As hydrocarbons leave the column, they are burned and detected by a Flame Ionisation Detector (FID), which responds to virtually all combustible components. For oil-in-water analysis, it is the sum of all the responses within a specific carbon range or retention time that is related to the oil concentration by reference to standards of known concentrations.

Examples of GC-FID reference methods are shown in Table 2.7. The OSPAR GC-FID method, which is currently used as the reference method for measuring oil

Table 2.7 Examples of reference GC-FID methods

Reference method	Country	Solvent used	Hydrocarbon index	Status
ISO 9377-2 (2000)	International	Boiling range (36°C to 69°C)	C_{10}–C_{40}	In use
OSPAR GC-FID (OSPAR Agreement 2005)	OSPAR countries	*n*-pentane	C_7–C_{40} minus TEX[a]	In use
TNRCC Method 1005 (2001)	USA	*n*-pentane	C_6–C_{35}	In use

[a]TEX stands for Toluene, Ethylbenzene and Xylene

Table 2.8 Difference between the ISO 9377-2 and OSPAR GC-FID methods

	ISO 9377	OSPAR GC-FID
Solvent	*n*-Pentane/*n*-hexane	*n*-pentane
Extract concentration step	Yes	No
Hydrocarbon index	C_{10}–C_{40}	C_7–C_{40} minus TEX
GC Column	No need to separate TEX, normal resolution GC	Must separate TEX from the rest, high GC resolution
Precision information	Yes	No

in produced water in North Sea countries, is a modified version of the ISO 9377-2. The key difference between the OSPAR GC-FID method and the ISO 9377-2 is summarised in the Table 2.8. Although the OSPAR GC-FID is the official reference method for the measurement of oil in produced water in the North Sea region, the use of GC-FID offshore for produced water analysis is still limited. To date only operators in the Norwegian sector are using it offshore. Most of the operators in other OSPAR nations either have collected the samples and sent them onshore for GC-FID analysis or have been using an alternative method that is correlated to the reference method.

4.4 Discussion of Reference Methods

Reference methods are important. Without them, discharge limits or performance standards, like 30 mg/L in the North Sea become meaningless. The universal use of reference methods means that one can compare data obtained from various sources, regulators will be able to formulate future legislation based on sound information, and operators can establish appropriate policy towards produced water discharge. However, for historical reasons and the fact that produced water is a waste stream, there has never been an oil-in-water reference method intentionally accepted and adopted by regulators and operators alike.

It should be emphasised that different methods will produce different results; therefore, it is not really possible to compare results obtained from different reference methods. It should also be said that different methods will require different instruments and procedures, which affects costs (capital and operational), training, health and safety.

Generally speaking, infrared-based methods are very well established, commonly used and easy to deploy with portable fixed wavelength instruments. However, due to the use of CFC solvents, and the lack of compositional details, they are becoming less used. Gravimetric methods are simple and relatively cheap, but again they do not provide details of composition. Due to the evaporation procedures involved in gravimetric methods, there is some loss of volatile components. GC-FID methods do not need require CFCs, have no issues with the loss of volatiles, and have the potential to provide detailed information on composition, but they necessitate sophisticated instruments which require skilled operators.

5 Field Measurement Methods

While reference methods are essential for compliance monitoring, comparison of results and the development of future legislation, developments, they are not always user-friendly, and in some cases they may even be impossible to apply. Also, from an operational standpoint, one often requires results quickly. This is particularly true when a production process is being optimised. Analysis using some of the reference methods such as GC-FID and gravimetry, can be time consuming. Instruments and methods for use in the field that are easy, inexpensive and rapid, offer advantages and are often needed.

Field instruments and methods can be grouped into two major categories – laboratory bench-top and online monitors. Bench-top instruments and methods used for routine oil-in-water analysis may be correlated to the reference methods so that results can be submitted in compilance reports. Online monitors are used for process trending and detection of process deviation. They are advantageous in process optimisation.

5.1 Bench-Top

There are a significant number of bench-top instruments on the market for the measurement of oil in produced water. Techniques used include the following:

- Colorimetric;
- Fibre optical chemical sensor;
- Infrared
  - Horizontal Attenuated Total Reflection (HATR);
  - Using S-316 as an extraction solvent;
  - Using Supercritical CO_2 as an extraction solvent;

  - Solventless;
  - Using a non-conventional wavelength for detection;

- UV absorption;
- UV fluorescence.

5.1.1 Colorimetric

In a colorimetric method, oil in water is determined by extracting a sample with a solvent and then directly measuring the colour in the sample extract using a visible spectrophotometer at a wavelength, for example, of 450 nm. It is critically important that the oil in the contaminated water must show colour. A colourless oil is undetectable by this method. As a result for water samples collected from gas and gas condensate installations, such a method may not work properly.

A well-established colorimetric method is provided by Hach (Hach Method 8041), which has the measurement range of 0–80 ppm. But according to a comparison study (Lambert et al. 2001), such a method is best suited for water samples containing dark coloured oil in a concentration range of between 10 and 85 ppm. The colorimetric method is widely used by Petrobras in Brazil.

5.1.2 Fibre Optical Chemical Sensor

A fibre optical chemical sensor is essentially an optical fibre coated with a polymer that can absorb hydrocarbons. Such a sensor comes as a probe. When the probe is inserted into an oily water sample, hydrocarbons are absorbed by the polymer which can change its refractive index. By measuring the amount of transmitted light before and after the absorption of hydrocarbons and through calibration with known concentrations of hydrocarbons, it is possible to measure the oil concentration in an unknown water sample.

In principle it should work; however, measurements will depend on the types of oils and how easily the hydrocarbons are absorbed by the polymer. It is also equilibrium based, and as a result, it can take some time before a reading is taken. The probe must be cleaned and re-zeroed after each measurement. According to the manufacturer (Saini and Virgo 2000; http://www.petrosense.com/products.html), PetroSense PHA-100 WL systems have been used by oil operators in the Gulf of Mexico. It is uncertain how well the technology is able to cope with dispersed oil which can actually form the main phase of oil in produced water derived from oil and gas installations.

5.1.3 Infrared – Horizontal Attenuated Total Reflection (HATR)

In a conventional infrared method, oil concentration is quantified by measuring infrared absorbance after transmitting an infrared light through a cuvette containing sample extract. With HATR, infrared light is reflected at a crystal surface above which a layer of oil is deposited once the solvent has evaporated from the sample aliquot (Fig. 2.3). At each reflection, infrared light is absorbed in a similar way as

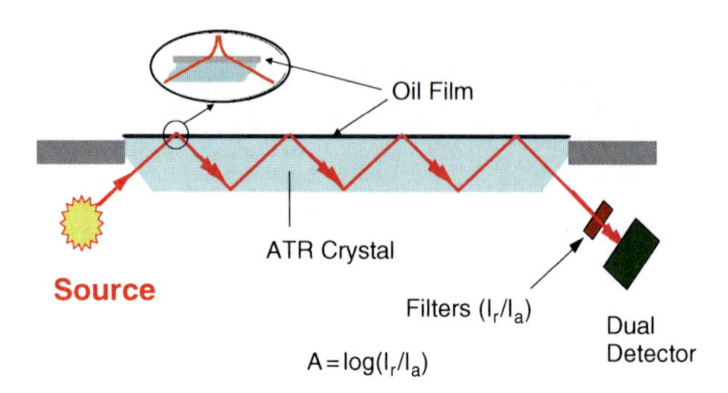

Fig. 2.3 A schematic diagram of the HATR technique (Wilks 1999); ATR = attenuated total reflectance, A = absorbance, I = irradiance, r = reflected, a = attenuated

in the conventional infrared analysis. Therefore by calibration and comparing the absorbance obtained from a sample extract to that obtained from calibration standards with known concentrations, the oil concentration in the sample is determined.

This is a well-established technology with portable instruments available from Wilks Enterprise, Inc. These instruments are believed to be widely used by the offshore oil and gas industry. According to Wilks Enterprise, the measurement range is 4–1000 ppm. The sample extract requires evaporation before measurement. It is inevitable that some of the volatile components in the sample extract will be lost during the evaporation process. Therefore, such a method may be less suited for samples taken from gas and gas condensate installations, which may contain a significant amount of volatile hydrocarbons.

5.1.4 Infrared – Using S-316 and Horiba Instrument

This is a semi-automatic version of a conventional infrared analysis method. Small volumes of water sample and extraction solvent S-316 are injected into the Horiba instrument. Extraction and separation are then carried out automatically, and a portion of the extract is diverted to a measurement cell. Oil concentration is obtained by measuring the infrared absorbance.

The instrument (http://www.qlimited.com/pdf/Horiba-OCMA-310-Q.pdf) is believed to have been well established and widely used, especially in Asian countries. Although the measurement process is essentially the same as a conventional infrared-based reference method, it is not a reference method.

According to the instrument supplier, Horiba, the measurement range is 0–200 ppm. Key issues are the use of S-316, which is a CFC, and more importantly, the expens as discussed in the early sections.

5.1.5 Infrared – Using Supercritical CO_2 Extraction

In this method, one uses supercritical CO_2 as an extraction solvent to extract the oil from water samples. Infrared quantification is then carried out on the supercritical CO_2 that contains the extracted oil just like a conventional infrared method.

Fig. 2.4 A picture of supercritical CO_2 IR technology (Private communication with Ed Ramsey of Critical Solutions Technology)

In theory it is easy to understand. In practice, however, the technology is more complicated as it involves pumping liquid CO_2 into a high pressure metal vessel in which a sample bottle is placed until it reaches a supercritical state. On reaching the supercritical state, a portion of the supercritical CO_2 (now containing oil) is diverted to a high pressure infrared cell for quantification. After analysis is carried out, CO_2 is vented directly to the atmosphere. An example of such an instrument is shown in Fig. 2.4. This interesting technology does not require the use of CFC. Although CO_2 has its health and safety issues, the technology does not produce CO_2, therefore one can claim that it is environmentally neutral and friendly.

The author has tested such an instrument (Yang 2003) and found the results similar to those using other solvents, e.g. tetrachloroethylene. However, for water samples containing heavy oil, it was found that cleaning of the high pressure vessel after extraction can be time consuming.

5.1.6 Infrared – Solventless Approach

Recently a solventless approach has been developed (Martin et al. 2008). A sample of oil in water is filtered through a special membrane which can retain oil (dispersed). The membrane is dried before it is placed into the instrument, and infrared

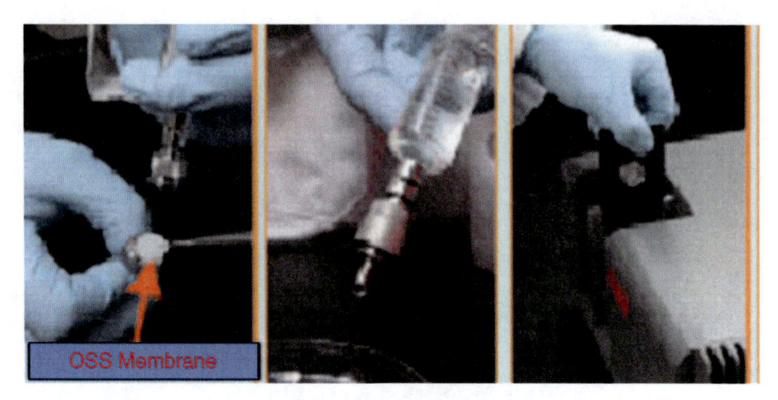

Fig. 2.5 Pictures showing the steps involved in the OSS solventless IR technology (Martin et al. 2008)

quantification is done just like a conventional infrared reference method. The steps involved are illustrated in Fig. 2.5.

The key element is the membrane, which apparently can transmit infrared light. The technology is still being perfected before entering the market. One of the key issues will be the effect of solid particles that are often present in a produced water sample. Other issues are related to sample size, and the possible loss of oil (as oil can stick to the internal surface of an injection syringe). According to the supplier, a detection range from 3 to 200 ppm is covered (Smith and Martin 2008). Also it is believed that the supplier of the technology is working towards a single-laboratory validated method acceptable to ASTM.

5.1.7 Infrared – Using a Non-conventional Wavelength for Detection

This is a new technology based on Quantum Cascade Laser Infrared (QCL-IR) technology in which an oily water sample is extracted by a cyclic hydrocarbon, such as cyclohexane or cyclopentane. Quantification is then carried out by measuring absorbance at a wavelength in the region of 1350–1500 cm^{-1} using mid-infrared spectroscopy that employs a Quantum Cascade Laser as a light source.

To illustrate the measurement principle, Fig. 2.6 shows attenuated total reflection (ATR) absorbance spectra for samples containing crude oil collected in the North Sea in comparison to that obtained from the extraction solvent cyclohexane. While the samples with crude oil show absorption in the region 1350–1400 cm^{-1}, no absorption occurs in this range for the solvent cyclohexane. By exploring the difference and through calibration, one can calculate the oil concentration in a water sample. The technology is new. There is now an instrument available called "Eracheck" which according to the supplier has a measurement range of 0.5–1000 ppm.

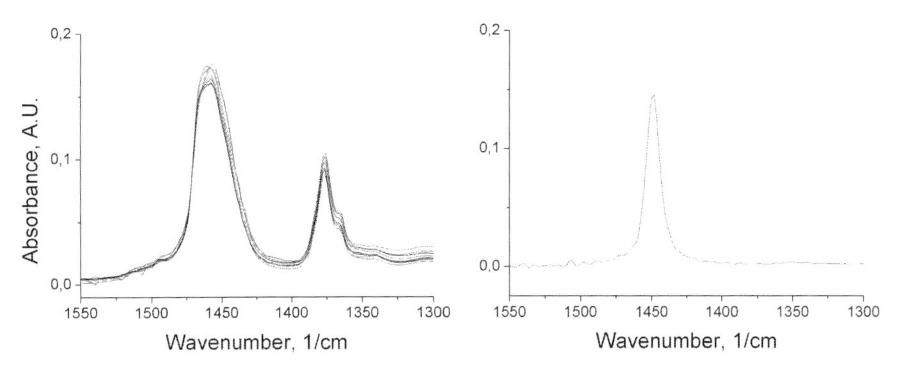

Fig. 2.6 FTIR absorbance spectra of different samples from the North Sea measured with the ATR technique (*left*) and that from the solvent cyclohexane (*right*) (Lendl and Ritter 2008)

5.1.8 UV Absorbance

Like aliphatic hydrocarbons which absorb infrared at certain wavelengths, aromatic hydrocarbons absorb ultraviolet (UV) light. Therefore by measuring the UV absorbance of a sample extract in a similar fashion to the reference infrared method, but using UV spectroscopy, one can quantify aromatic hydrocarbons in an oily water sample. Provided that the ratio of aromatic hydrocarbon content to that of the total hydrocarbon content remains relatively constant, the total hydrocarbon content can be obtained via calibration. The UV absorbance technique has not been widely used for oil in produced water measurement.

5.1.9 UV Fluorescence

When aromatic hydrocarbons absorb UV light, they emit fluorescent light at a longer wavelength. By measuring the intensity of the UV fluorescence light, one can determine the amount of aromatic hydrocarbons, which can be related to the total amount of hydrocarbons providing that the ratio of aromatics to total hydrocarbons remains relatively constant.

UV fluorescence instruments have been widely used by the oil and gas industry for the measurement of oil in produced water. Portable instruments are available on the market, for example, the TD-500 from Turner Designs Hydrocarbon Instruments (Brost 2008) and Fluorocheck from Arjay Engineering (Reeves 2006). Both of these are handheld and easy to use.

The UV fluorescence technique is the second most widely used and accepted after infrared. It is very sensitive and requires no use of CFC solvents, but it is important to remember that should the ratio of aromatic to aliphatic (or aromatic to total hydrocarbons) change due to, for example, the inclusion of new oil streams from different fields, then a new calibration should be established before a measurement is carried out.

5.2 Online

There are many techniques that can be used for online oil in water monitors. A list of these techniques is given below.

- Focused ultrasonic acoustics
- Fibre Optic Chemical Sensors
- Image analysis
- Light scattering and turbidity
- Photoacoustic sensor
- UV fluorescence including Laser-Induced Fluorescence (LIF).

5.2.1 Focused Ultrasonic Acoustics

In this technique, a highly focused acoustic transducer is inserted directly into a produced water stream (Anaensen and Volker 2006). The transducer focuses and a time window determines the measurement volume. Any particles, such as oil droplets, solid particles and gas bubbles, that pass through the measurement volume will produce acoustic echoes. These signals are detected, classified and used to work out particle size and size distribution. Oil concentration is then calculated from the size and size distribution obtained.

A relatively new development of this technique is from TNO TDP. They had licensed the technology to Roxar who have built an online oil-in-water monitor that is being commercialised. According to Roxar, it has a measurement range of 0–1000 ppm.

5.2.2 Fibre Optical Chemical Sensors

This technique has been described in the previous sections (http://www.petrosense.com/products.html). Here a Fibre Optical Chemical Sensor probe is inserted into a produced water by-pass line. Because it is an equilibrium based technique, a back washing mechanism has to be arranged so that the probe can be cleaned and re-zeroed after each measurement. This cleaning and re-zeroing process means that only one measurement can be taken every hour or so. Therefore, although it is an online monitor, it cannot provide data on a minute-by-minute basis as others can.

5.2.3 Image Analysis

Image analysis instruments are based on using a high resolution video microscope to examine the content of a sample stream. Video images are captured in sequence. Particles on each of the images are then counted and analysed with their volumes calculated. To calculate the concentration of the particles, sample volume related to each of the images is determined by multiplying the image area (width by length) by the focal depth.

To distinguish oil droplets from solid particles, a shape factor is often used. For example, the ViPA (Visual Process Analyser) from Jorin (Butler et al. 2001), has the shape factor defined as 4π*Area/Perimeter^2. For a perfect circle (sphere), the shape factor is always 1. As the length of perimeter increases compared to the area enclosed, the shape factor value decreases very quickly. For a particle to be classified as an oil droplet, the shape factor has to be very close to 1. Gas bubbles that are often present in a produced water stream will be spherical too, but its optical property is markedly different from oil, so they can be easily recognised and excluded from oil droplet size and concentration calculation.

Over the past few years, several image analysis systems have been made available. In addition to Jorin, there are other commercial online monitors from J.M. Canty (O'Donoghue and Relihan 2009) and Fluid Imaging (Petersen and Ide 2008). Image analysis has been poplar for produced water re-injection applications where knowledge of the size and size distribution of oil droplets and solid particles in addition to concentration is important. The instruments are increasingly being used for process optimisation (O'Donoghue and Relihan 2009; Iqbal 2009).

5.2.4 Light Scattering

Light scattering was probably the most popular technique for online measurement of oil in water. Ships above a certain size have to be fitted with a type-approved bilge water treatment system that must include an oil content meter to comply with the strict regulations of the International Maritime Organisation (IMO). The vast majority of these oil content meters are based on light scattering.

The technique involves passing visible light through an oily water sample. Due to the presence of particles (oil droplets, solid particles and gas bubbles), some light will be scattered, and a reduced amount will be transmitted. By measuring the amount of transmitted light together with the amount of light that is scattered at different angles, it is possible to distinguish the oil droplets from solid particles and gas bubbles, and determine the oil concentration.

There are many manufacturers who supply online light scattering based oil in water monitors. The best-known suppliers for the oil and gas industry include Deckma (Schmidt 2003) and Rivertrace Engineering (http://www.rivertrace.com/products/industrial_applications).

5.2.5 Photoacoustic Sensor

The principle of the photoacoustic sensor is simple (Whitaker et al. 2001). A pulsed laser light is focused on to a small sample of oily water. Oil (dissolved and dispersed) will absorb the optical energy, which causes sudden local heating. The local heating produces thermal expansion which generates high frequency pressure waves. These pressure waves are detected and correlated with oil concentration.

When the sensor was first made, it was a probe type. Tests indicated that there was a linear response in concentration range of 20–2000 ppm. The development work was focused on subsea separation and produced water re-injection applications. Despite extensive testing and development by AkerKvaerner, a commercial system is yet to emerge.

5.2.6 UV Fluorescence

The principle of UV fluorescence for oil-in-water measurement has been described in section 5.1.9. The key difference between an online monitor and a bench-top analyser is that no solvent extraction is required online. For online monitors, in addition to using a UV lamp as a light source, Laser Induced Fluorescence (LIF) has been developed and applied in recent years. With LIF, a probe type monitor is possible.

There are several manufacturers who supply UV fluorescence online monitors. The best-known ones that use lamps as a light source include Arjay Engineering (http://www.arjayeng.com/arjayeng_oil_and_water.htm), Sigrist (Rechsteiner and Schuldt 2007), and Turner Designs Hydrocarbon Instruments (Bartman 2002). These are well established and have traditionally dominated the market.

The best-known manufacturers, who apply LIF technology, are Advanced Sensors (Thabeth and Gallup 2008), ProAnalysis (Skeidsvoll et al. 2007) and Systektum (Bublitz et al. 2005). These are relative new-comers to the oil-in-water measurement market. Nevertheless, over the last 10 years the development and commercialisation of these monitors has been rapid.

Using LIF, a probe can be easily constructed, which delivers an advantage in that it can be inserted into a produced water stream or fitted inline. Other UV fluorescence monitors that use a lamp as a light source are generally fitted into a produced water by-pass stream. Having a by-pass runs the risk of non-representative sampling. The drawback comes during maintenance or instrument replacement, when an inline monitor can pose greater difficulty than one fitted to a by-pass line.

With the help of developments and commercialisation of the LIF-based monitors, UV fluorescence is now the most widely used technique for online measurement of oil in produced water.

5.3 Field Instrument Selection

As discussed in the previous sections, there are many different techniques and instruments to choose from when it comes to a specific application. For laboratory bench-top instruments, parameters to be considered may include the following:

Purpose of measurement: e.g. for process optimisation and control or for reporting. For process control and optimisation, repeatability is perhaps more important while for reporting, in addition to repeatability, accuracy also becomes very important.

Property and characteristics of produced water: e.g. oil colour, presence of chemicals and solid particles. Colour of oil is extremely important if one is to use a colorimetric based method. Some production chemicals can be extracted together with oil and can affect UV methods. Solids will have a detrimental effect on the filtration infrared method. For methods involving evaporation, light components can be lost.

The use of solvents: if a solvent is used, one needs to check the method must be environmentally sound, and user friendly. Availability and cost must be considered.

Calibration procedures: what is involved in the calibration? What calibration check is required and at what frequency?

Instrument compactness and ease of use: for offshore operations, space is limited and, handheld or portable instruments are more welcome. Most bench-top instruments are easy to use with minimal training, and should not be an issue for a laboratory technician.

Costs: these include Capital Expenditure (Capex) and Operation Expenditure (Opex). Costs of instrument and measurement (operations) vary significantly. Costs associated with Opex may include the purchase of solvent and its disposal. Opex depends on the number of analyses involved.

Maintenance and after-sale service: this not only affects the operating costs, it can also have an impact on production.

For online monitors, most of these items apply. However, certain issues specific to online monitor applications may need to be carefully considered when it comes to instrument selection:

Properties and characteristics: in addition to those mentioned for bench-top methods, online monitoring requires consideration of things such as gas bubbles in the produced water stream.

Issue of inline or online: inline monitors require no sampling by-pass line, but may affect calibration and instrument retrieval when it comes to repair or maintenance. Online using a by-pass can be more easily isolated, and therefore facilitates instrument retrieval, but one has to verify that fluid in the by-pass line has the same property and concentration as in the main flow stream.

Space available for fitting the instrument: for online monitoring, instruments should be fitted downstream of a turbulent region where oil is well mixed with water. If a by-pass line is to be used, then the length of this line should be minimised. Online monitors come in a variety of configurations. Some even come with a sample pre-conditioning system to generate uniform dispersion. The amount of available space might determine the type of online monitor chosen.

Previous applications and field test data: although no two applications will be the same, it will be very useful if the instrument suppliers have already found similar applications and can provide test results.

Others: additional considerations include pressure, temperature, flow rate (minimum and maximum for by-pass line) and pressure rating.

6 Sampling and Sample Handling

Measurement methods are very important for obtaining good results, but if the samples that are analysed are not representative of the flow stream, the results obtained will be of little use regardless of how the samples are analysed. It is important to realise that a measurement method can only give a result as good as the sample can provide!

6.1 Sampling

To obtain a representative sample from a produced water stream, there are a number of aspects that one has to consider. These include:

- location of taking a sample;
- sampling devices;
- iso-kinetic sampling;
- sample bottles.

Sampling location will depend upon what is to be determined from the sample. For regulatory compliance monitoring, samples may have to be taken at specified locations. Without the approval from a regulatory body, one may not change such a location. For process control and optimisation, this may not be an issue, i.e. samples can be taken at any location. Generally speaking, when a produced water sample is taken for the purpose of measuring oil in water, the sample needs to be taken downstream of a turbulent region where oil is expected to have mixed well with the water. In an ideal situation, the sample should also be taken from a vertical pipe. In a horizontal pipe, stratification may take place.

Once a location is chosen, a sampling device is selected that will remove a section of the main flow for subsequent analysis. An example of a good sampling probe is shown in Fig. 2.7 (IMO Resolution MEPC.107(49) 2003). For produced water sampling, a half-inch (12.7 mm) sampling pipe is probably more appropriate than a quarter inch (6.35 mm) pipe.

A side wall sampling point, like the one shown in Fig. 2.8, should be avoided. For a two phase flow like oil in produced water, oil has a different density than water, and due to inertia it can become difficult to change the particle direction by 90° as side wall sampling demands.

For the same reason, it is important that one considers iso-kinetic sampling. Iso-kinetic sampling means that samples are taken such that the velocity of fluid in the sampling pipe is the same as that in the main flow pipe. In general, the flow rate inside the main produced water flow line is measured, and therefore with a known pipe diameter, one can easily calculate the velocity of fluids. To check if sampling conforms to iso-kinetic conditions, one needs to work out the velocity of the fluid in the sampling pipe. This can be done by using a container with known volume and a

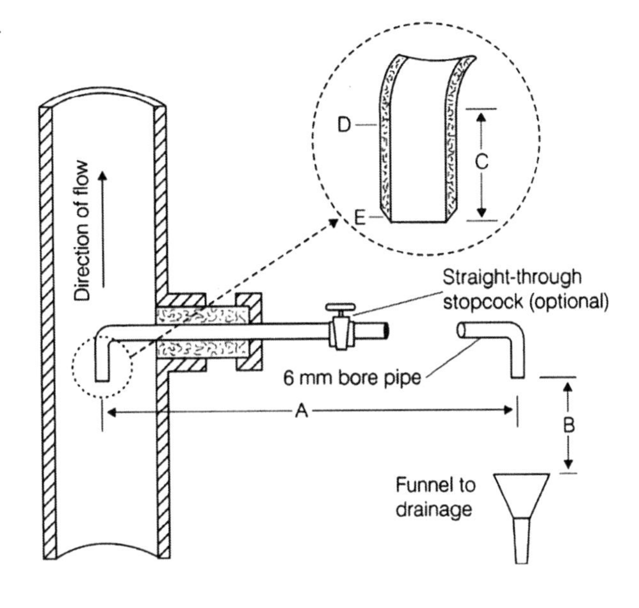

Fig. 2.7 A good example of an oil-in-water sampling approach (IMO Resolution MEPC.107(49) 2003): $A < 400$ mm, B high enough for sample bottles, $C < 60$ mm, $D \leq 2$ mm, E chisel-edged chamfer (30°)

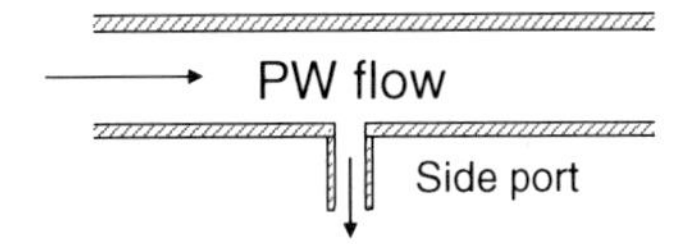

Fig. 2.8 Schematic of side wall sampling point

stopwatch to obtain the flow rate, and knowing the sampling pipe entrance diameter, the fluid velocity can be calculated.

When a sample is procured from a sampling line that is not left to flow continuously, it is important that one allows the sample line to flow for at least 1 min before the sample is taken.

Sample bottles are also important. They must be extremely clean with no oil/hydrocarbon contamination. To achieve this, the sample bottles are often washed with soapy water, rinsed with clean water, then left to dry. The dried bottles are then washed with solvent and dried before use. In general, with oil concentrations in the range of a few ppm to several hundred ppm, 500 mL glass bottles are adequate for taking produced water samples. Plastic bottles should not be used as hydrocarbons may diffuse through these. Similarly for the glass bottles, the caps should be Teflon lined to prevent possible contact of hydrocarbon with plastic.

6.2 Sample Handling

Once a representative sample is obtained, the sample must be properly handled. Sample handling will depend on when, where and how the sample is to be analysed and also on whether the samples are for regulatory compliance or process optimisation. In general, sample handling may include the following aspects:

- Acidification;
- Cooling;
- Transportation;
- Storage.

Acidification serves two purposes: (a) to preserve the samples by killing bacteria which can degrade oil; (b) to dissolve precipitates such as iron oxide and calcium carbonate, which can stabilise an emulsion and therefore prevent a complete separation between solvent extract and water after the extraction process.

For a 500 mL sample, 2.5 mL of diluted HCl is generally sufficient for lowering the pH of the sample to less than 2. Acid can be added to the sample bottle either before or after sample procurement. As a best practice, in particular from a safety point of view, adding acid after filling the sample bottle is probably better.

Produced water can be hot with a temperature as high as 90°C. Before an extraction is carried out, the sample should be left to cool. For the OSPAR GC-FID method, it is actually specified that it should be cooled to 10°C before extraction to prevent the loss of volatile hydrocarbons during the extraction process.

If a produced water sample is to be transported (e.g. sending a sample to an onshore laboratory), in addition to sample preservation by adding acid, the sample should be stored and transported in a suitable sealed container to prevent the ingress of light. Exposure to light may degrade hydrocarbons in the water sample.

Similarly if samples are to be stored for whatever reason, according to ISO5667-3 (2003), in an ideal situation, they should be stored in a refrigerator with a temperature kept between 1°C and 5°C. The ISO standard also states that the maximum recommended preservation time before analysis for an oily water is one month.

7 Calibration

Calibration is a set of operations that establish, under specific conditions, the relationship between the output of a measurement system (i.e. the response of an instrument) and the accepted values of the calibration standards (i.e. the amount of analyte present) (Barwick 2005). For general calibration issues, such as the number of calibration standards, number of replicate samples, plotting the results and performing regression, etc., a best practice guidance entitled "Preparation of calibration curves – a guide to best practice" (Barwick 2005) is available, which is a very good and useful reference. The following discussion will however focus specifically on calibration that is related to oil in produced water measurement.

7.1 Calibration for Lab Methods

Two issues are specifically discussed here.

- Calibration oil
- Method for preparing calibration standards

For a reference method, usually calibration oil is defined. For example, in the ISO9377-2 and OSPAR GC-FID methods, this calibration oil is a 50:50 (in weight) mix of two types of mineral oils, Type A and Type B, both containing no additives. An example of Type A is a diesel fuel; an example of Type B is a lubricant.

Before the OSPAR GC-FID method came into force in the North Sea in January 2007, a reference method based on using a single wavelength infrared was used (OSPAR Agreement 1997). Despite the specification in the method for using a stabilised crude oil from the individual installations as a calibration standard, not everyone used crude oil as a calibration standard. The use of different calibration oils can have a marked effect on the final oil in water results. This is because instruments can respond differently to the different calibration oils.

To demonstrate this effect, Fig. 2.9 shows two calibration curves (not real data) – one is prepared using a synthetic oil (a 50:50 mixture of isooctane and cetane) and one by using crude oil – taking the single wavelength method NS 9803 IR (Norsk Standard NS 9803 1993) as an example. For the calibration standards prepared using synthetic oil, as the oil contains all aliphatic with no aromatic hydrocarbons, they can give higher IR absorbance (in comparison to those prepared using crude oil, which contains aromatic hydrocarbons). As a result, when an oily water sample is collected and analysed, the calibration curve prepared using the synthetic oil would give lower concentrations compared to that from using crude oil as the calibration standard. A Norwegian study carried out in 2000 showed that using synthetic

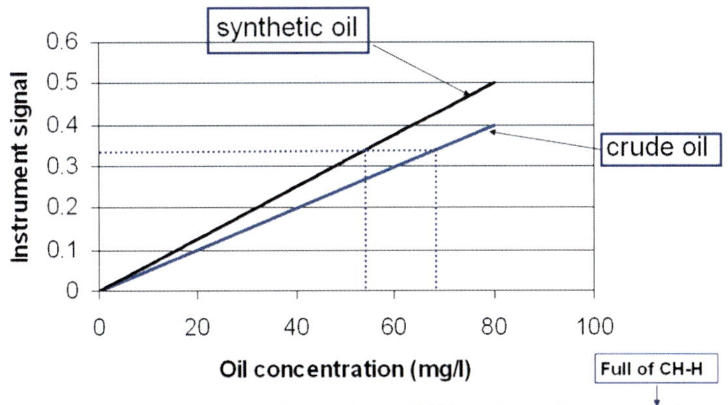

Fig. 2.9 A schematic showing the effect of calibration oil on results (Yang 2009)

oil resulted in an average reduction of 19% as opposed to using crude oil (Paus et al. 2001).

The second issue is related to how standard solutions are prepared. There are two methods that have been used. The first one (*direct dissolving method)* is by weighing a certain amount of the calibration oil and dissolving it directly into a solvent to make up a stock solution, and from this stock solution, subsequent standard solutions are prepared by dilution. A second method (*back-extraction method*) is by weighing a certain amount of calibration oil, spiking it into a salt water solution, and then back extracting the oil to form a stock solution. Subsequent standards are then prepared from the stock solution by dilution. The first one is easy to do, but the sample matrix, i.e. water solution, is not taken into consideration. The second method will have more steps involved (procedure-wise) and will be more time consuming, but it takes the sample matrix into consideration.

Using the same analogy in demonstrating the calibration oil effect, it is not difficult to see that when a calibration curve is generated by using standards that are prepared by directly dissolving oil into a solvent, it will lead to lower concentrations of oil in water compared to the cases where a calibration curve is obtained by using the back-extraction method. The reason is that only the oil that is back extracted will respond to the infrared absorption, and therefore will produce lower absorbance as is shown in the Fig. 2.10.

The foregoing discussion has clearly demonstrates the importance of calibration. By choosing a different calibration oil or a different method for preparing the calibration standard solutions, different oil in water results can be produced. The discussion also shows that if one wants to generate an artificially low oil in water result, one can do so by establishing the calibration curve using synthetic oil, and by preparing the standard solutions using the *direct dissolving method.*

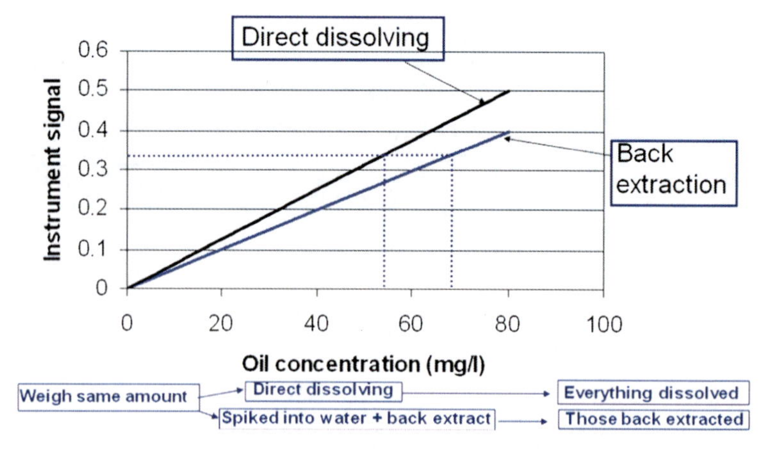

Fig. 2.10 A schematic showing the effect of calibration standard preparation on the results for oil in water (Yang 2009)

7.2 Calibration for Online Monitoring

Calibration for online oil in produced water monitoring is extremely difficult. This is because each produced water stream is different and also the fact that produced water is notoriously difficult to simulate in a laboratory environment, which means that it is very difficult to calibrate online monitors offsite. In the author's view, calibration of online monitoring can only be realistically achieved in the field.

8 Non-reference Method Acceptance

Reference methods cannot always be directly applied in the field. For example, on an FPSO movement of the vessel can prevent the accurate weighing that is required for the gravimetric method. Also there are situations where operators may not feel comfortable with the use of certain instruments for reasons such as the purchase of a new and costly instrument, safety, shortage of skilled personnel, etc. There is often a need to be able to use a non-reference method in the field, but a requirement for use of the results in reporting. This is the crux of non-reference method acceptance.

To be able to accept a non-reference method for the purpose of reporting, one has to demonstrate that the results obtained from the non-reference method are statistically equivalent to those that are produced by the reference method. Two possible ways to overcome this problem that are briefly discussed here are:

(i) Statistical significance tests (F-test and Student's t-test);
(ii) Establishing a valid correlation between a non-reference method and a reference method.

Statistical significance tests allow someone to compare two sets of results (in this case one from the reference method and one from the non-reference method) in an objective and unbiased way. The F-test compares the spread of results from the two methods to check if they can be reasonably considered to have come from the same parent distribution. The Student's t-test is a statistical procedure used to compare the mean values of the two data sets (Burke 1997). By carrying out such tests, one is able to state if the results from one method are statistically equivalent to those from the other method.

Unfortunately, statistical significance tests are limited to comparing two methods with a single concentration or a narrow range of concentrations (same order of magnitude). For oil in produced water, measurement methods often need to cover a concentration range from 0 to 100 mg/L (at least two orders of magnitude). Hence, establishing a correlation between a non-reference method and a reference method is thought to be more appropriate for accepting the non-reference method (Yang and McEwan 2005). The key here is to establish a valid correlation between the two methods and then continue to validate the correlation over time.

To establish this correlation and continue to validate it, two approaches have been suggested by OSPAR. Details can be found in the OSPAR Guidance document (2006).

9 Summary

The measurement of oil in produced water is essential for regulatory compliance monitoring, oil and gas production process control, and optimisation. The measurement is method-dependent. Using a different method will almost certainly result in different values.

There have been many reference methods for the analysis of oil in produced water, which are mainly based on three principles – infrared absorption, gravimetry and gas chromatography. Infrared-based methods have been dominant and popular until recent years, but due to issues associated with the use and availability of the chlorofluorocarbons, they are becoming obsolete. The gravimetric method is particularly popular in the USA where it is used for compliance monitoring. GC-FID is becoming increasingly popular, in particular in the OSPAR nations around the North Sea where it has recently become the reference method for compliance monitoring of produced water discharge.

Reference methods are important, since without them the comparison of results is impossible, and the regulatory framework for compliance monitoring cannot be constructed. Yet reference methods are not always user-friendly and practical in the field. As a result, alternative methods that may be inexpensive, easy to use, and can produce results quickly, are needed in particular for routine measurements. There are many techniques and instruments now available for both laboratory bench-top and online monitoring. For laboratory bench-top types, UV fluorescence and HATR instruments are probably most widely used for oil in produced water measurements. Recently, however, a solventless approach based on membrane filtration and infrared has been developed. Also, a new infrared method that uses a non-traditional wavelength has been made available. For online monitoring, light scattering and UV fluorescence instruments were popular. However with recent the development of Laser Induced Fluorescence, UV fluorescence is now becoming the main technology for online analysis of oil in produced water.

Sampling can lead to a significant amount of uncertainty in the final results. One has to remember that measurement can only provide results as good as the samples can give.

Instrument calibration is an integral part of a measurement method (both reference and non-reference). Calibration procedures are often well defined in a reference method and therefore errors associated with the procedures are often included in the precision data (i.e. repeatability and reproducibility). However experience in the North Sea has shown that the exact calibration procedures are not always adopted and followed. When different calibration procedures are used, results can be significantly different. These have been clearly demonstrated by using different calibration oils and different methodologies in preparing the calibration standards.

To accept an alternative (non-reference) method for the purpose of reporting, two possible approaches are discussed – one based on using statistical significance testing and the other based on establishing a valid correlation. For the determination of oil in produced water, establishing a valid correlation between an alternative method and a reference method is thought to be more appropriate. This is now used in North Sea countries.

Acknowledgement The author thanks TUV NEL (www.tuvnel.com) for permitting the publication of this chapter. TUV NEL is a leading provider of pipeline fluid management services to the global petroleum industry. It offers consulting, training, R&D and laboratory testing services in subject areas that include flow, environment, and measurement.

References

Anaensen G, Volker A (2006) Produced water characterization using ultrasonic oil-in-water monitoring – recent development and trial results. In: A paper presented at NEL's 8th oil-in-water monitoring workshop, 21 September 2006, Aberdeen, UK

ASTM D 3921 – 85 (1985) Standard test method for oil and grease and petroleum hydrocarbons in water

ASTM D 4281 – 95 (1995) Standard test method for oil and grease (fluorocarbon extractable substances) by gravimetric determination

ASTM D 7066-04 (2004) Standard test method for dimer/trimer of chlorotrifluoroethylene (S-316) recoverable oil and grease and nonpolar by Infared determination, 2004

Bartman G (2002) UV fluorescence for monitoring oil and grease in produced water – bench scale and continuous monitors: real data from the fields. In: A paper presented at TUV NEL's 4th oil in water monitoring workshop, 22–23 May 2002, Aberdeen, UK

Barwick V (2005) Best practice guidance for preparing calibration curves. In: A paper presented at TUV NEL's 7th oil in water monitoring workshop, 23–24 November 2005, Aberdeen, UK

Brost DF (2008) Sample preparation methodology for oil in water analysis by UV fluorescence, a presentation made at TUV NEL's 10th Oil in Water Monitoring Club meeting, Aberdeen, UK, 19 September 2008

Bublitz J et al (2005) Adaptation of approved laser-induced time resolved fluorescence spectroscopy in offshore applications: experience of 24 months measurements in produced water. In: A paper presented at TUV NEL's 7th oil in water monitoring workshop, 23–24 November 2005

Burke S (1997) Statistics in context: significance testing. VAM Bull 17, 18–21, Autumn 1997

Butler E, Roth N, Gaskin R (2001) A new approach to online oily water measurement-field experience in BP installations. In: A paper presented at TUV NEL's 3rd oil-in-water monitoring workshop, 23 May 2001, Aberdeen, UK

DECC Guidance Notes for the sampling & analysis of produced water other hydrocarbon discharges – August 2010, version 2.6. https://www.og.decc.gov.uk/environment/opaoppcr_guide.htm

DIN 38409 PT 18 English (1981) German standard methods for the analysis of water, waste water and sludge; summary action and material characteristics parameters (group H); Determination of Hydrocarbons (H18)

EPA Method 413.1 (Issued in 1974, Editorial revision 1978) Standard test method for oil and grease using gravimetric determination

EPA Method 413.2 (Issued in 1974, Editorial revision 1978) Standard test method for Oil and grease analysis using Freon extraction and IR absorbance without the Freon extract being treated by silica gel

EPA Method 418.1 (Issued in 1978) Standard test method for Oil and grease analysis using Freon extraction and IR absorbance with the Freon extract being treated by silica gel
Frintrop P (2007) A change of the oil in produced water analysis reference method – why and its implications. In: A presentation made at TUV NEL's OSPAR GC-FID method implementation session, Aberdeen, 18 September 2007
GB/T17932 (1999) Chinese National Standard – analysis method for oil-bearing waste water from marine petroleum development industry, 1999
Hach Method 8041- Colorimetric extraction method, https://www.hach.com/fmmimghach?/CODE%3AOILINWATER_8041_COLO9189%7C1
http://www.arjayeng.com/arjayeng_oil_and_water.htm
http://water.epa.gov/scitech/methods/cwa/oil/1664.cfm, Method 1664, revision A (1999): N-Hexane extractable material (HEM; Oil and Grease) and silica gel treated n-hexane extractable material (SGT-HEM; Non-polar material) By extraction and gravimetry
http://www.norweco.com/html/lab/test_methods/5520bfp.htm, 5520 Oil and Grease 5520B. Partition-Gravimetric Method, 2001
http://www.petrosense.com/products.html
http://www.qlimited.com/pdf/Horiba-OCMA-310-Q.pdf
http://www.rivertrace.com/products/industrial_applications
IMO Resolution MEPC.107(49): Revised guidelines and specifications for pollution prevention equipment for machinery space bilges of ships, adopted on 18 July 2003, www.imo.org
IP 426/98 (1998) Determination of the oil content of effluent water – extraction and infra-red spectrometric method
Iqbal MF (2009) Produced water optimisation using visual processing technology. In: A paper presented at TUV NEL's 7th produced water workshop, 29–30 April 2009, Aberdeen, UK
ISO – ISO 9377-2:2000 (2000) Water quality – determination of hydrocarbon oil index – Part 2: Method using solvent extraction and gas chromatography
ISO 5667-3: Water quality – sampling – part 3: guidance on the preservation and handling of water samples, 3rd edn, 15/12/2003
Kortvelyssy G Plan for the phase-out of carbon tetrachloride(CTC) in Argentina, a technical report prepared for UNIDO, 6th of January 2008
Lambert P, Fingas M, Goldthorp M (2001) An evaluation of field total petroleum hydrocarbon (TPH) systems. J Hazard Mater 83: 65–81
Lendl B, Ritter W (2008) A new Mid-IR laser based analyser for hydrocarbons in water, a paper presented at TUV NEL's 10th oil in water monitoring workshop, Aberdeen, UK, 18 September 2008
Martin T, Smith D, Schwarz T, Doucette L, Tripp CP (2008) A fast, solventless 'green' device and method for oil in water analysis using infrared spectroscopy. In: A paper presented at TUV NEL's Oil in Water Monitoring Workshop, Aberdeen, UK, 18 September 2008
Norsk Standard NS 9803 (1993) Norwegian standard for detection of oil in water, equivalent to the Swedish Standard Method SS02, 81 45, 1991)
O'Donoghue A, Relihan E (2009) Inline oil in water particle analysis and concentration monitoring for process control and optimisation in produced water plants In: A paper presented at TUV NEL's 7th produced water workshop, 29–30 April 2009, Aberdeen, UK
OSPAR Agreement (1997-16) on Sampling and analysis procedure for the 40 mg/l target standards
OSPAR Agreement (2005-15) Sampling and analysis procedure for the 40 mg/l target standard. www.ospar.org
OSPAR Agreement (2006-6), Oil in produced water analysis – guideline on criteria for alternative method acceptance and general guidelines on sample taking and handling, www.ospar.org
OSPAR Recommendation (2001/1) for the management of produced water from offshore installations. www.ospar.org
Paus J, Grini PG, Flatval K (2001) Comparative study of Freon-IR (NS 9803) and GC-FID (ISO 9377-2) for analysis of oil in produced water on 33 platforms in the Norwegian

sector. In: A paper presented at TUV NEL's 3rd oil in water monitoring workshop, May 2001, Aberdeen, UK

Petersen K, Ide M (2008) Field test results of optical-based oil in water and particle analysis instrumentation technology. In: A paper presented at TUV NEL's 6th produced water workshop, 23–24 April 2008

Private communication with Ed Ramsey of Critical Solutions Technology (2004)

Rechsteiner HR, Schuldt W (2007) UV fluorescence based online oil in water monitoring for produced water applications – advantages and field experiences. In: A paper presented at TUV NEL's Produced Water – Best Management Practices, 28–29 November 2007, Abu Dhabi, UAE

Reeves G (2006) Alternative oil in water measurements using fluorescence and electrochemical nanotechnology. In: A paper presented at TUV NEL's 8th oil in water monitoring workshop, Aberdeen, UK, 21 September 2006

Saini D, Virgo M (2000) An operational perspective of the use of fibre optic techniques for the measurement of oil-in-water offshore. A paper presented at TUV NEL's 2nd oil-in-water monitoring workshop, May 2000, Aberdeen, UK

Schmidt AA (2003) Oil-in-water monitoring using advanced light scattering technology – theory and applications. In: A paper presented at TUV NEL's 5th oil in water monitoring workshop, 21–22 May, 2003, Aberdeen, UK

Skeidsvoll J, Ottoy MH, Vassgard EG, Oa JA (2007) Efficient produced water management through online oil in water monitoring. Case study: StatoilHydro's Snore B. In: A paper presented at TUV NEL's produced water – best management practices, 28–29 November 2007, Abu Dhabi, UAE

Smith D, Martin T (2008) Green solventless infrared oil and grease method, a presentation made at TUV NEL's 10th Oil in Water Monitoring Club meeting, Aberdeen, UK, 19 September 2008

Source: http://home.flash.net/~defilip1/solubility.htm

Texas Natural Resource Conservation Commission Total Petroleum Hydrocarbons TNRCC Method 1005, Revision 03, 1 June 2001

Thabeth K, Gallup DL (2008) A new, self-cleaning continuous online oil in water analyser for the petroleum industry. In: A paper presented at the produced water seminar, 15–18 January 08, Houston, USA

The determination of hydrocarbon oil in waters by solvent extraction and either infrared absorption or gravimetry (1983) Her Majesty's Stationary Office, London, 1983

Total Petroleum Hydrocarbon Criteria Working Group Series (1998) Analysis of petroleum hydrocarbons in environmental media, March 1998 http://www.qros.co.uk/Total%20Petroleum%20Hydrocarbon%20Criteria%20Working%20Group%20Series%20Volume%201%20Analysis%20of%20Petroleum%20Hydrocarbons%20in%20Environmental%20Media.pdf

Whitaker T, Terzoudi V, Butler E (2001) Application of photoacoustic technology for monitoring the oil content in subsea separated produced water. In: A paper presented at TUV NEL's 3rd oil-in-water monitoring workshop, 23 May 2001, Aberdeen, UK

Wilks P (1999) The case for hexane as an alternate solvent for offshore oil-in-water monitoring. A paper presented at TUV NEL's 1st oil-in-water monitoring workshop, April 1999, Aberdeen, UK

Wilks P (2001) Selecting a solvent for TOG/TPH analysis of produced water by Infrared analysis In: A paper presented at TUV NEL's 3rd Oil-in-Water Monitoring Workshop, May 2001, Aberdeen, UK

Yang M (2002) An evaluation of current, pending and alternative oil-in-water monitoring methods. In: A report for the UK Offshore Operator Association and the DTI, TUV NEL Project No. 042/2002

Yang M (2003) Testing and evaluation of the supercritical fluid extraction and infrared (SFE-IR) oil-in-water analysis method, a report for BP, DTI and Statoil, TUV NEL Report No. 2003/223, December 2003

Yang M (2006) SPE 102991, Oil in produced water analysis and monitoring in the North Sea. In: A paper presented at the 2006 SPE technical conference and exhibition, held in San Antonio, Texas, USA, 24-27, September 2006

Yang M (2009) The problem of regulating on the basis of oil in water. In: A presentation made at OGP's Produced Water Discharges: Harm and Risk, 20–21 January 2009, Edinburgh, UK

Yang M, McEwan D (2005) oil-in-water analysis method (OIWAM) JIP, A JIP report for 10 organisations including 8 operators and 2 government bodies, TUV NEL Report No: 2005/96, July 2005

Chapter 3
Evaluation of Produced Water from Brazilian Offshore Platforms

Irene T. Gabardo, Eduardo B. Platte, Antônio S. Araujo, and Fernando H. Pulgatti

Abstract Chemistry and toxicity of produced water (PW) from offshore platforms operated by Petrobras in Brazil were investigated. Three studies – PW monitoring, detailed composition and temporal variability – were conducted during 1996, 2001 and 2006 in the Campos, Santos and Ceara Basins. For approximately 50 samples the median concentrations were ammonia 70 mg L^{-1}, barium 1.3 mg L^{-1}, iron 7.4 mg L^{-1}, BTEX 4.7 mg L^{-1}, PAH 0.53 mg L^{-1}, TPH 28 mg L^{-1}, phenols 1.3 mg L^{-1}, ^{226}Ra 0.15 Bq L^{-1} and ^{228}Ra 0.09 Bq L^{-1}. Acute toxicity median values were $LC50_{96\,h} = 3.57\%$ for *Mysidopsis juniae*, $LC50_{48\,h} = 52.55\%$ for *Artemia sp.*, $EC50_{72\,h} = 8.43\%$ for *Skeletonema costatum* and $EC50_{15\,min} = 16.05\%$ for *Vibrio fischeri*. Median chronic toxicity using *Lytechinus variegatus* showed a NOEC = 1.3%. These results for Brazilian PW are similar to those for the North Sea, Gulf of Mexico, Australia and other regions of the world. Dispersion plumes modelled using CORMIX and CHEMMAP predicted that PW can be diluted rapidly after discharge and that permissible levels for all chemical parameters in seawater cited in the Brazilian Resolution CONAMA 357/05 are attained within 500 m of the discharge point. Over 10 years (1998–2010) of monitoring in the vicinity of the Brazilian platforms did not show alterations in sea water quality, supporting the predictions of the dispersion plume modelling. Despite no observed alteration in seawater quality around oil and gas production platforms, the authors recognize the importance of continuous evaluation of the impact of PW discharges from a risk assessment perspective, and studies of bioaccumulation and the use of biomarkers, among other initiatives currently implemented by Petrobras in areas with large volumes of PW discharge. Up to and including 2011, Petrobras remains the major producer of oil and gas in Brazil and the total discharge of produced water by the country is essentially the volume that is discharged by offshore Petrobras operations. In 2005, the average total volume of PW discharged offshore on the Brazilian coast was 73 million m^3/year, representing less than 3% PW discharged onto other oceans worldwide.

I.T. Gabardo (✉)
Department of Environmental Monitoring and Assessment, Petrobras Research Center/CENPES, Rio Janeiro, RJ, Brazil

K. Lee, J. Neff (eds.), *Produced Water*, DOI 10.1007/978-1-4614-0046-2_3,

1 Introduction

Input of petroleum hydrocarbons to the marine environment can occur from urban outfall, industrial and domestic effluent, navigation, transportation, offshore production and accidental release (Fingas 2001; GESAMP 1993; Bouloubassi et al. 2001; Readman et al. 2002; NRC 2003). Annually 1.3×10^6 tonnes of crude are released from natural seepage, production, transport and consumption, of which offshore oil and gas production activities contribute only 2.9% and >90% of this load is due to produced water (PW) discharge (NRC 2003). Soluble components like aromatics, organic acids and paraffins are degraded by marine bacteria (Stephenson 1992). Other factors that aid hydrocarbon weathering are dispersion, dilution, volatilization, physical—chemical reactions and sedimentation (OGP 2005).

Produced water is the largest volume waste stream in oil and gas production, consisting of natural formation and flood water injected into the formation to maintain well pressure. Initially, PW consists mainly of formation water but as the well matures the proportion of injected seawater or re-injected PW can reach over 10 times the volume of the oil produced for mature fields (OGP 2005; E&P Forum 1994). Oil platforms produce high volumes whereas gas platforms generate less volume but with high concentrations of organic contaminants such as monoaromatics (BTEX), naphthalenes and phenols. Once PW is released into seawater, the distribution of individual compounds into solid (added to particulate matter) or liquid phases depends on their chemical characteristics (Neff et al. 1989; E&P Forum 1994; OLF 2005; OGP 2005, 2002). The major constituents are inorganic salts, dispersed and dissolved hydrocarbons, organic acids and phenols which contribute to its toxicity in addition to the ionic imbalance (Swan et al. 1994) which sometimes cannot be rectified by salinity adjustment (Schiff et al. 1992); hence, toxicity factors are not easily resolved (Fucik 1992). Nevertheless, Rand (1995) considered toxicity testing as a primary approach to evaluate PW effects. Acute and chronic PW toxicity have been documented for Australia (Swan et al. 1994), the North Sea (E&P Forum 1994), Gulf of Mexico and Indonesia (Holdway 2002). There is little evidence of acute PW toxicity beyond the immediate mixing zone (Holdway 2002). Toxicity due to salinity is rapidly reduced by simple dilution in the offshore (Pillard et al. 1996).

Total oil production in Brazil was 596 million bbl in 2005, of which 521 million bbl was produced in the Brazilian offshore area (www.anp.gov.br), constituting 87% of national production. In April 2006 Brazil achieved self-sufficiency, producing 1.8 million bbl per day, implicating an increase in PW discharge. The 107 Brazilian Petrobras offshore facilities are few in comparison with the Gulf of Mexico (~4,000) and North Sea (~500). In 2005 the average annual Brazilian Petrobras offshore facilities PW discharge was 73 million m^3/year (Gabardo 2007), which amounts to less than 3% worldwide (OGP 2005).

World demand for petroleum is expected to increase 47% by the year 2030 as reported by the Energy Information Department (EIA 2006). The supply of crude oil and gas remains an important component of Brazil's current and future energy needs. Most of Brazilian oil and gas production is located in offshore areas in the

States of Rio de Janeiro and Espirito Santo, but in the Santos Basin, Petrobras recently made the biggest oil discovery in Brazil (http://www.petrobras.com.br/minisite/presal/en/questions%2Danswers/).

Prior to 2007 Brazil had no specific criterion for offshore PW discharge. Since then, CONAMA Resolution 393/2007 has established limits for PW oil and grease of 29 mg L^{-1} as a monthly average and a 42 mg L^{-1} daily maximum, besides regulatory monitoring and compliance with the seawater quality criteria of CONAMA Resolution 357/2005 outside a 500 m mixing zone. Also, offshore PW discharge requires detailed monitoring twice a year encompassing several organic and inorganic parameters (Freitas and Mendes 2010).

This chapter deals with surveys conducted in 1996, 2001 and 2005–2006 as well as other samples for PW characterization. The main objectives of this work were (a) to investigate the chemical composition and toxicity of Brazilian PW; (b) to compare Brazilian PW with other regions of the world; (c) to evaluate a single platform for PW temporal variability characteristics; (d) to apply dispersion models to predict environmental effects; (e) to compare modelled predictions in the platform vicinity with field data.

2 Experimental

Figure 3.1 shows the platforms with PW discharge in 2005, and their positions along the Brazilian coast. Table 3.1 presents the detailed platforms data including location, coast distance, water depth, discharge depth and discharge flow. Although there were 107 platforms on the Brazilian coast, only 24 platforms discharged effluents.

2.1 Methods

Study 1 was concerned with monitoring of ammonia (APHA 1995; Standard Method 4500-NH3, colorimetry), Ba, B, Fe, ^{226}Ra and ^{228}Ra (Godoy et al. 1994), BTEX (EPA 5021, headspace, GC-PID), TPH (EPA 1664/8015, GC-FID), PAH (EPA 3510C/3630C/8270D, GC-MS sum of 36 and 38 PAH compounds in 2001 and 2006, respectively), phenols (EPA 3510C/8270D, GC-MS) and toxicity for approximately 55 samples: these data were collected during 1996, 2001 and 2005–2006.

Study 2 consisted of detailed composition profiles from a single sample at each of the 21 platforms in Campos Basin, one in Santos Basin and one in Ceara Basin (Fig. 3.1, Table 3.1), collected from September 2005 to January 2006, and analysed for temperature, salinity (Standard Method 4500, chloride potentiometry), pH, density (densimeter), TSS (Standard Method 2540D, gravimetry), alkalinity (Standard Method 2320, potentiometry), anions (chloride, fluoride, sulphate, nitrate; ion chromatography, electrolytic suppressor/conductivity detector), mercury (UOP 938, auto-analyser NIC SP-3D), metals (Standard Method 3120B, ICP-OES), ^{226}Ra

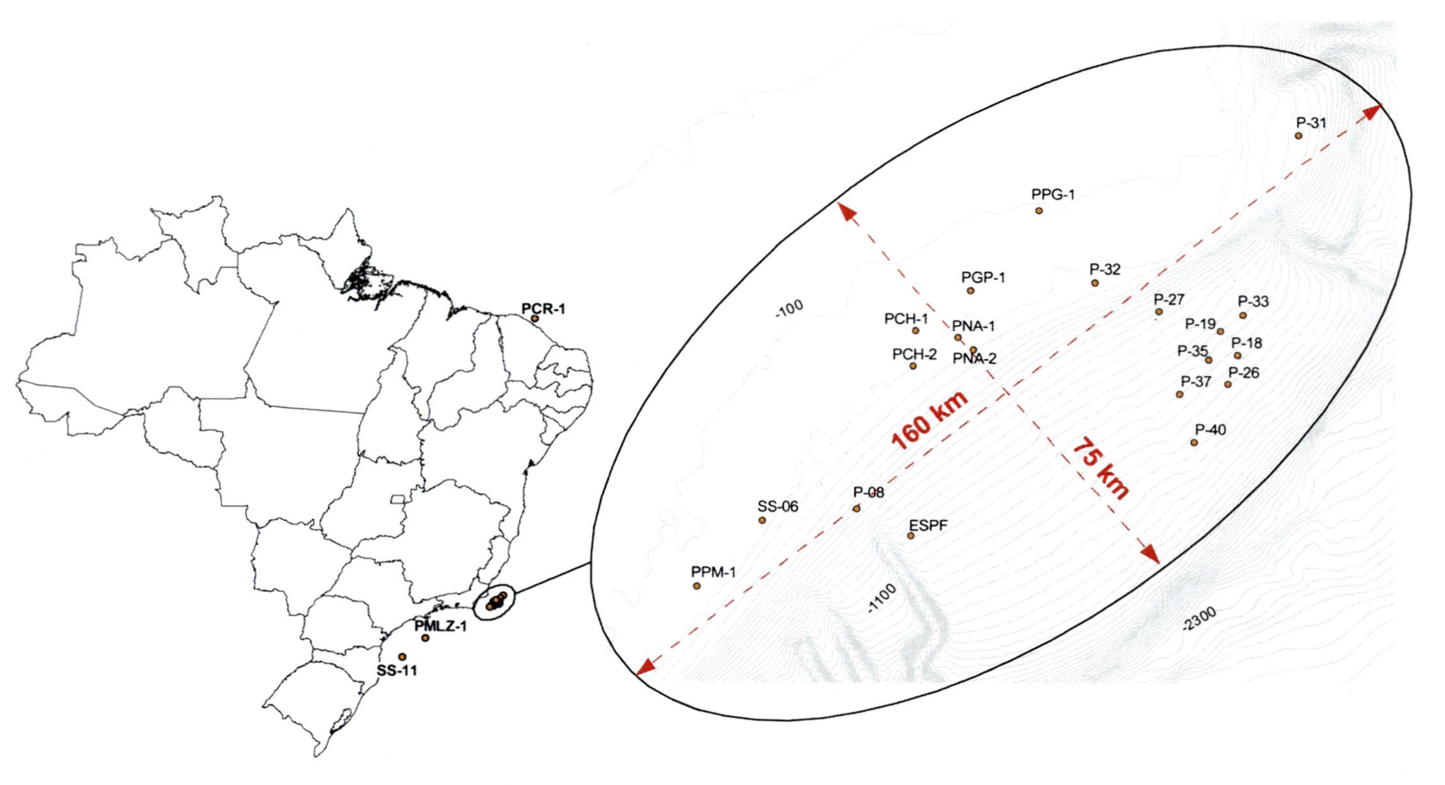

Fig. 3.1 Platforms with produced water discharge along the Brazilian coast (Gabardo 2007, from database 2005)

Table 3.1 Platforms with produced water discharge along the Brazilian coast[a]

Platform	Production	Treatment	Latitude S	Longitude W	Distance (km)	Water depth (m)	Discharge depth (m)	Discharge (m^3/day)
Southeast region – Campos Basin, Rio de Janeiro State, 40 platforms, 21 discharging PW								
PCH-1	Oil and gas	Hydrocyclone, degasser	22°25′56.998″	40°28′48.309″	71	117	28	845
PCH-2	Oil and gas	Hydrocyclone, degasser	22°27′ 56.030″	40°28′ 06.158″	74	142	–	4,124
PNA-1	Oil and gas	Hydrocyclone, degasser	22°26′ 17.507″	40°25′ 26.815″	76	145	49	2,498
PNA-2	Oil and gas	Hydrocyclone degasser	22°27′ 00.000″	40°24′ 41.000″	77	170	30	5,240
PGP-1	Oil and gas	Oil electrostatic treater, oil/water gravitational separator tank	22°22′ 27.330″	40°25′ 01.440″	72	120	32	4,685
PPM-1	Oil and gas	Oil electrostatic treater, hydrocyclone and dissolved gas flotation	22°47′ 51.010″	40°45′ 43.750″	87	115	63	13,276
PPG-1	Oil and gas	Hydrocyclone, degasser	22°15′ 18.180″	40°19′ 46.425″	72	101	40 or 60	16,823
P-08	Oil and gas	Hydrocyclone, degasser	22°40′21.199″	40°32′ 45.618″	87	423	Surface	980
P-18	Oil and gas	Hydrocyclone, degasser	22°25′ 40.887″	40°01′ 41.456″	109	910	Surface	2,650
P-19	Oil and gas	hydrocyclone, degasser	22°23′ 32.058″	40°03′ 15.660″	105	748	Surface	6,162
P-26	Oil and gas	Hydrocyclone, degasser	22°28′05.623″	40°01′ 42.174″	110	990	Surface	6,568
P-27	Oil and gas	Hydrocyclone, degasser	22°22′ 49.047″	40°08′ 44.085″	96	540	Surface	2,039
P-31	Oil and gas	Hydrocyclone, degasser	22°07′ 46.640″	39°57′ 58.510″	105	320	Surface	9,538
P-32	Oil	Oil electrostatic treater, hydrocyclone, slop tank	22°20′ 53.801″	40°14′ 24.184″	85	160	Surface	1,535
P-33	Oil and gas	Oil electrostatic treater, hydrocyclone, dissolved gas flotation, slop tank	22°22′ 18.464″	40°01′ 29.933″	107	780	Surface	1,781
P-35	Oil and gas	Oil electrostatic treater, slop tank	22°26′ 11.668″	40°04′ 03.948″	106	850	Surface	8,092
P-37	Oil and gas	Oil electrostatic treater, slop tank	22°29′05.010″	40°05′ 43.790″	106	905	Surface	9,586
ESPF	Oil and gas	Oil electrostatic treater, hydrocyclone, slop tank	22°42′ 34.441″	40°27′ 33.918″	96	795	Surface	4,166
SS-06	Oil	Oil electrostatic treater, hydrocyclone, dissolved gas flotation	22°42′ 04.790″	40°40′ 36.180″	80	120	11	11,412
P-40	Oil and gas	Oil electrostatic treater, hydrocyclone, induced gas flotation	22°32′ 48.889″	40°04′ 01.511″	113	1,070	Surface	8,300[b]
FPBR[b]	Oil and gas	Hydrocyclone, slop tank	21°56′ 01.480″	39°49′ 00.090″	120	1, 258	Surface	
Northeast Region – 65 platforms in 4 States (SE, AL, CE, RN) but only one discharging PW located in Ceara Basin								
PCR-1	Oil and gas	Oil electrostatic treater	3°05′25.568″	38°47′34.984″	40	45	21	1,664
South region – Santos Basin, São Paulo State, 2 platforms, 2 discharging PW								
SS-11	Oil and gas	Oil electrostatic treater, hydrocyclone, degasser	26°38′ 51.220″	40°52′ 24.810″	160	150	Surface	330
PMLZ-1[c]	Gas and condensate	Gas treatment & compression	25°15′59.304″	45°15′08.552″	145	189	Surface	23[c]

[a]Daily discharge (during the year 2005)
[b]Discharge by batches 5′000–15′000 m^3/batch
[c]Sampled only in 2000 due to very small volume of discharge
Datum SAD69 for the geographical coordinates

and ^{228}Ra (Godoy et al. 1994), TOC (Standard Method 5310B, auto-analyser), organic acids (ion chromatography, electrolytic suppressor/conductivity detector) and toxicity. For the southern area, PMLZ-1 was the only platform producing gas and condensate, and was sampled only in 2000 due to the low volume discharged in 2005. In the northeast area of Brazil, PW is pumped to on-shore treatment facilities and only PCR-1 (Curimã platform) discharges PW: this platform was producing a significant amount of gas.

Study 3 on temporal variability at PCR-1 consisted of nine sequential samples collected in 48 h on December 11–12, 2003, with intervals of 3 h, and four individual samples collected in July 2001, June 2003, December 2004 and January 2006. These were analysed for salinity, pH, TOC, TSS, ammonia, sulphide (Standard Method 4500-S-Sulfide, potentiometry), cyanide (Standard Method 4500-CN, voltammetry), anions, BTEX and PAH.

Several laboratories were involved, depending on the campaigns, including CENPES, the Petrobras Research Center (anions, organic acids, BTEX, PAH, phenols); metals and radioisotopes by PUC-Rio Catholic University; PAH, BTEX and phenols by Analytical Solutions and CTGAS, and toxicity by CENPES (Microtox, *Artemia sp.*), LABTOX and University of Itajai Valley.

Samples were collected by laboratory technicians or trained platform operators from Petrobras, who also performed the on-site physical and chemical analysis. Oil and grease, currently are determined gravimetrically according to CONAMA Resolution 393/2007, but this was established after the samples had been analysed. Instead, we used the TPH data from the GC-FID analysis to quantify total oil content in 22 samples from Study 2. Samples were refrigerated during transportation by helicopter from the platforms. As PW is very saline, samples usually have to be diluted 10–1,000 fold, so metals were often below the detection limit.

Acute toxicity tests were done using *Mysidopsis juniae* (96 h survival, with organisms 1–8 d old, CETESB L05.251/1995, ABNT 15308/2005), *Artemia* sp. (48 h survival, CETESB L05.021/1987), *Skeletonema costatum* (72 h growth, ISO 10.253:1995) and *Vibrio fischeri* (Microtox® System, 15 min, CETESB L05.227/2001, ABNT 15411-3/2006), while chronic tests were performed with the sea-urchin *Lytechinus variegatus* (embryo development 24 h after fertilization, CETESB L05.250/1992, ABNT 15350/2006). Salinity was adjusted to normal seawater when necessary, except for brine shrimp (*Artemia* sp.) that tolerates high salinity since it is normally found in salt lakes and inland brackish waters (Veiga and Vital 2002). All organisms used in this study are indigenous to Brazil. The methods were developed by the São Paulo State Environmental Agency (CETESB), but have been recently adapted by the Brazilian Standardization Association as Brazilian Standard Methods (ABNT 2005, 2006a/b). Results for acute toxicity tests are expressed in Effect Concentration to 50% of exposed organisms (EC_{50}) and Lethal Concentration to 50% of exposed organisms (LC_{50}), for chronic toxicity tests; the results are in Lowest Observed Effect Concentration (LOEC) and No Observed Effect Concentration (NOEC).

2.2 Quality Assurance and Quality Control

Daily checks of calibration curves with a second standard, use of blanks, surrogates, percent recoveries, standard spiking for each batch and repeated analysis of some sample extracts were conducted. For the toxicity tests, a minimum of three dilution replicates were analysed for each sample and quality assurance was based on sensitivity tests, variability of the controls and maximum acceptable effect on controls.

2.3 Statistical Analysis

Statistical analysis was performed using SPSS for Windows 10.0 (SPSS, Inc. 1989–1999) and STATISTICA 6.0 (Statsoft, Inc. 1984–2001). Outliers and extreme values were not excluded (except for ANOVA), so we chose medians instead of means as being more representative results. The interquartile range was used as a measure of spread. An outlier is any value that lies more than 1.5 times the interquartile range from either end of the box. An extreme value is one that lies more than 3 times the interquartile range from either end of the box (Barnett and Lewis 1994).

2.4 Modelling

Applied Science Associates (ASA) South America in Sao Paulo, Brazil, used two models: Cornell Mixing Zone Expert System (CORMIX, developed by Cornell University, USA) and Chemical Discharge Model System (CHEMMAP, ASA Inc., USA). The CORMIX model was used for near-field studies where effluent speed and density dominate in the first few minutes and the principal mechanism is dilution. In the initial dilution or jet phase, the plume rapidly entrains ambient sea water.

The CHEMMAP model was used for the far-field modelling where effects due to site dynamics and passive transportation in the plume happen in hours or days. The model uses individual compound density, vapour pressure, solubility, degradation rate, adsorbed and dissolved partition coefficients, viscosity, surface tension and ambient forcing from wind, currents and seawater density and considers mixtures of products. Far-field dilution is only important during the first hours, after which other concentration reduction mechanisms prevail. The model was used in the stochastic mode to predict the trajectory and biogeochemical transformations (fate) of compounds. Spreading, advection, dispersion, evaporation, volatilization, entrainment, dissolution, partitioning, sedimentation, adsorption and degradation were simulated. The mass of the chemical component is transported by three-dimensional currents as determined by the hydrodynamic model that is forced by tides, wind, oceanic currents, buoyancy and dispersion. The plume was simulated as Lagrangian particles of known mass.

In probabilistic mode, simulations covered January to March (summer) and June to August (winter) for the Campos Basin, and May to July (winter) and October to

December (summer) for the Ceara Basin. Each scenario for plume behaviour was composed of 30 simulations in which there was constant PW release over 24 h and variable meteorological and oceanographic data. To determine the area of influence of the plume, multiple trajectories were used to produce contour curves of maximum expected concentrations for each chemical component at each grid point and the average of the highest concentrations was calculated for the 30 simulations. Data on effluent properties and field conditions were used as input data for the modelling studies' predictions of water column concentrations.

The modelling studies occurred in 2004/5 and the PW composition inputs were the median values of real PW compositional data obtained in 2001. To emphasize plume shape and characteristics, a dilution value equivalent to 10,000 times the CONAMA Resolution 357/2005, Class I regulation criteria for seawater was considered as the threshold value to stop the simulation. Also concentrations of each component at the 500 m limit of the mixing zone (established by CONAMA Resolution 393/2007) were predicted.

3 Results

The results of the PW composition and toxicity obtained for all 3 studies are summarized in Table 3.2. Details can be found in the PhD thesis by Gabardo (2007).

3.1 Results of Study 1

In order to verify the variability between different samples and sampling times, the results for approximately 50 samples were compiled using selected PW parameters (Table 3.3).

3.2 Results of Study 2

For the detailed PW study, 23 samples were collected. pH ranged from 6 to 8.2, temperature from 33 to 90°C (median = 59°C) and TSS from 1.9 to 106 mg L^{-1} (median = 10.6 mg L^{-1}). Median concentrations of anions were sulphates 481 mg L^{-1}, bicarbonates 436 mg L^{-1}, nitrates <0.1 mg L^{-1}, fluorides 2.1 mg L^{-1} and chlorides 45,776 mg L^{-1}. In all samples the concentration of cyanide was below the detection limit (<10 μg L^{-1}). Salinity was 38,182–179,766 mg L^{-1} with a median of 75,434 mg L^{-1}. Ammonia concentration ranged from 22 to 91 mg L^{-1}, with a median of 51.7 mg L^{-1}. Radionuclide activity for ^{226}Ra ranged from 0.02 to 10.9 Bq L^{-1} and for ^{228}Ra from 0.04 to 10.5 Bq L^{-1}.

TOC, measured by TOC Automatic Analyser, ranged from 86 to 971 mg L^{-1} (median = 307 mg L^{-1}).

Table 3.2 Median results of PW from Brazilian platforms; n = number of samples

Parameter	Study 1 monitoring general parameters 1996–2006		Study 2 monitoring detailed composition 2005/6		Study 3 variability at same platform 1996–2006	
	n	mg L^{-1}	n	mg L^{-1}	n	mg L^{-1}
Benzene		1.7		1.6		9.1
Toluene		1.9		2.1		5.2
Ethylbenzene	53	0.2	22	0.2	11	0.4
Xylenes		0.9		0.9		1.1
BTEX		4.7		4.9		15.8
TPH	45	28	22	10.0	10	37.4
Phenols	46	1.3	23	0.73	11	2.0
PAH[a]	45	0.53	23	0.44	10	0.61
Ammonia	47	70	23	51.7	4	47
Ba	55	1.3	23	2.0	12	1.0
B	55	31	23	36.4	12	17.5
Fe	53	7.4	23	1.1	12	17
Ra-226 (Bq L^{-1})	36	0.15	23	0.42	–	–
Ra-228 (Bq L^{-1})	36	0.09	23	0.41	–	–
Toxicity tests						
L. variegatus NOEC %	45	1.3	24	1.97	11	12.5
L. variegatus LOEC %	45	5	24	3.92	11	25
M. juniae LC_{50}%	36	3.57	24	2.95	2	1.1
Artemia sp. LC_{50}%	44	52.55	23	61.3	3	64.7
S. costatum EC_{50}%	–	–	16	8.43	–	–
V. fischeri EC_{15}%	12	16.05		–	–	–

[a]In 1996 PAH were quantified by UV Fluorescence and were not integrated with the 2001 and 2005 results that were obtained by GC-MS

TPH concentrations ranged from 4 to 66 mg L^{-1} (median = 10 mg L^{-1}). From this analysis, GC-FID fingerprints were obtained and in general, oil profiles were typical (a) for some PW samples, but for others the profile showed an atypical pattern (Fig. 3.2b), that was investigated using GC-MS. Cyclic sulphur compounds in PW extracts were probably generated from the reaction H_2S scavenger and identified alkylbenzene peaks from C8 to C10 were from demulsifier solvents.

BTEX concentrations ranged from 1.39 to 20 mg L^{-1} (median = 4.87 mg L^{-1}). Individual concentration ranges were benzene, 0.6–13.46 mg L^{-1}; toluene, 0.4–5.97 mg L^{-1}, ethylbenzene, 0.05–0.77 mg L^{-1} and xylenes, 0.23–3.90 mg L^{-1}. The highest BTEX concentrations were obtained at platforms P-35 (13.2 mg L^{-1}), SS-11 (20 mg L^{-1}), PPG-1 (9.7 mg L^{-1}) and PCR-1 (7.2 mg L^{-1}).

Total phenols (sum of 14 compounds) ranged from 49.7 to 5,735 μg L^{-1} with a median of 730 μg L^{-1}. The compound concentrations were phenol, 4–450 μg L^{-1}; C1-phenols 0.013–0.99 μg L^{-1}; C2-phenols 0.029–3.68 μg L^{-1}, and C3-phenols, 0.003–0.62 μg L^{-1}.

Table 3.3 Summary of chemical analysis and toxicity of PW samples obtained for Study 1

Parameter	Min	Max	Mean	Median	SD	*n*
Ammonia (mg L^{-1})	22.3	800	85.4	70	111	47
Ba (mg L^{-1})	0.2	45	7.1	1.3	10.0	55
B (mg L^{-1})	6	120.4	34.2	31	19.6	55
Fe (mg L^{-1})	0.04	25	7.5	7.4	6.9	53
^{226}Ra (Bq L^{-1})	0.01	10.9	1.24	0.15	2.5	36
^{228}Ra (Bq L^{-1})	<0.02	10.5	1.39	0.09	2.8	36
Benzene (μg L^{-1})	490	13,462	3,324	1,653	3,493	53
Toluene (μg L^{-1})	458	8,639	2,572	1,917	1,957	53
Ethylbenzene (μg L^{-1})	38	770	242	211	162	53
Xylenes (μg L^{-1})	208	3,904	975	859	656	53
BTEX (μg L^{-1})	1,384	21,624	7,115	4,690	5,749	53
TPH (mg L^{-1})	4.0	251	45	28	51.6	45
Phenols (mg L^{-1})	0.05	83.5	3.48	1.42	12.03	47
PAH (μg L^{-1})	42	1,558	595.9	527.2	348.3	45
Toxicity						
L. variegatus $_{NOEC\ \%}$	<0.1	12.5	3.44	1.3	4.2	45
L. variegatus $_{LOEC\ \%}$	≤0.1	25	7.74	5	8.1	45
M. juniae $_{LC_{50}\%}$	<0.6	9.5	3.87	3.57	2.3	36
Artemia sp $_{LC_{50}\%}$	1.6	>100[a]	53.96	52.55	26.0	44

[a]No toxicity observed

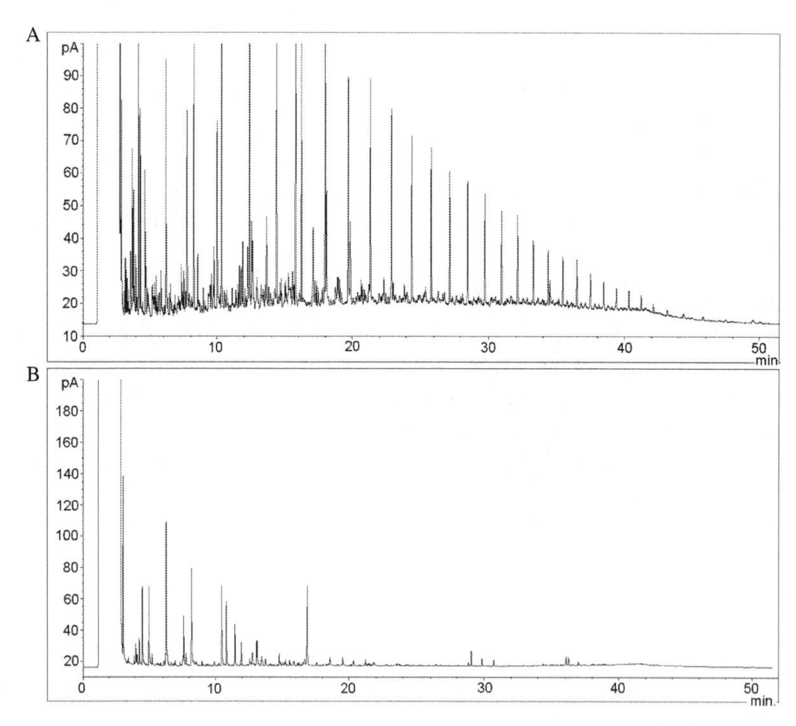

Fig. 3.2 TPH fingerprints obtained by GC-FID for two Brazilian PW with oil signature (*top*) and atypical profile (*bottom*)

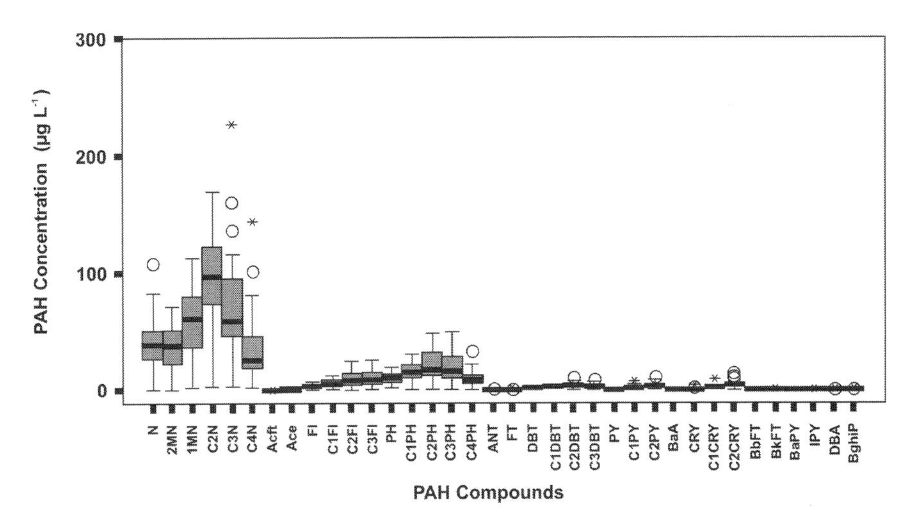

Fig. 3.3 Minimum, maximum and median of the PAH concentrations in Brazilian PW sampled in Study 2: (n = 23 samples) box-and-whisker plot outliers are marked with *open dots*, and extreme values are marked with *asterisks*.
N: Naphthalene; 2MN: 2-MethylNaphthalene; 1MN: 1-MethylNaphthalene; C2N: C_2Naphthalenes; C3N: C_3Naphthalenes; 4N: C_4Naphthalenes; Aceft: Acenaphthylene; Ace: Acenaphtene; FL: Fluorene; C1FL: C_1Fluorenes; C2FL: C_2Fluorenes; C3FL: C_3Fluorenes; PH: Phenanthrene; C1PH: C_1Phenanthrenes; C2PH: C_2Phenanthrenes; C3PH: C_3Phenanthrenes; C4PH: C_4Phenanthrenes; ANT: Anthracene; FT: Fluoranthene; DBT: Dibenzothiophene; C1DBT: C_1Dibenzothiophenes; C2DBT: C_2Dibenzothiophenes; C3DBT: C_3Dibenzothiophenes; PY: Pyrene; C1PY: C_1Pyrenes; C2PY: C_2Pyrenes; BaA: Benz(a)anthracene; CRY: Chrysene; C1CRY: C_1Chrysenes; C2CRY: C_2Chrysenes; BbFT: Benz(b)fluoranthene; BkFT: Benz(k)fluoranthene; BaPY: Benz(a)pyrene; IPY: Indeno(1,2,3-cd)pyrene; DBA: Dibenz(a,h)anthracene; BghiP: Benzo(ghi)perylene

Polycyclic Aromatic Hydrocarbons. Figure 3.3 presents a box-and-whisker plot of the data for Study 2. Outliers are marked with open dots, and the extreme values with asterisks. Based on 38 PAH compounds analyzed the mean was 476.4 $\mu g\ L^{-1}$ and the median 438.5 $\mu g\ L^{-1}$, indicating homogeneity of the data. Ninety-one percent of all PAH were composed by two and three-ring aromatic compounds that include naphthalenes, phenanthrenes and dibenzothiophenes (NPD), and the more condensated PAH (4–6 rings) comprise only 2.71%. This was the pattern for all the Brazilian PW analysed and is common worldwide (Terrens and Tait 1996; Utvik et al. 1999; Neff 2002; Durell et al. 2006; OGP 2005, 2002; E&P Forum 1994).

Toxicity. Figure 3.4 shows the maximum, minimum and the median values obtained in each study using different organisms. Acute toxicity ranged from 0.6 to 9.5% PW for *M. juniae*; from 1.5 to 22.4% for *S. costatum*; 9.2 to 25.6% for *V. fischeri* and 5.3 to >100% for *Artemia* sp. The NOEC for *L. variegatus* ranged from <0.1–5%.

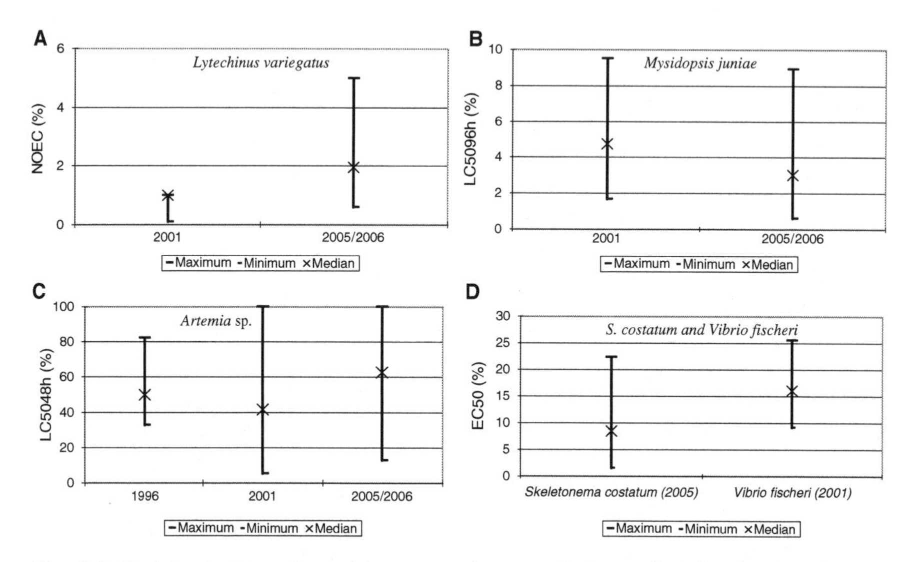

Fig. 3.4 Toxicity tests results: minimum, maximum and the median for the organisms. **a** *Lytechinus variegatus*, **b** *Mysidopsis juniae*, **c** *Artemia* sp. and **d** *S. costatum* and *Vibrio fischeri* for all the surveys

3.3 Results of Study 3

Figure 3.5 presents the results for Study 3 (Variability at the same platform: PCR-1). The only means that were found to be significantly different were the LOEC (23.4% for continuous 24 h sampling vs. 6.3% for individual samples) for *L. variegatus*. ANOVA evaluation was not performed on Ba, B, Fe, TPH, phenols or PAH due to the small sample size and the presence of outliers. Although the trace analyses were performed by different laboratories the results were consistent. When comparing PCR-1 chemistry data with other Brazilian platforms, more elevated BTEX and phenol concentrations were noticed in the PCR-1 produced water, probably, due to its relatively high production of gas. NOEC chronic toxicity values ranged from 0.8 to 12.5%. Fluctuations in toxicity were lower in the continuous 24 h sampling than in the punctual surveys as expected, considering the differences in the PW treatment processes and in the laboratories used (Fig. 3.5).

3.4 Modelling Results

Table 3.4 presents the effluent properties input for the modelling studies and the predicted CORMIX model near-field initial dilution for the platforms studied. For the Campos Basin (southeast region) the dilutions ranged from 96 to 279 times in the summer and 106–348 times in the winter. For PCR-1 in the Ceara Basin (northeast region), the dilutions in near-field were between 713 (winter) and 895 (summer) fold.

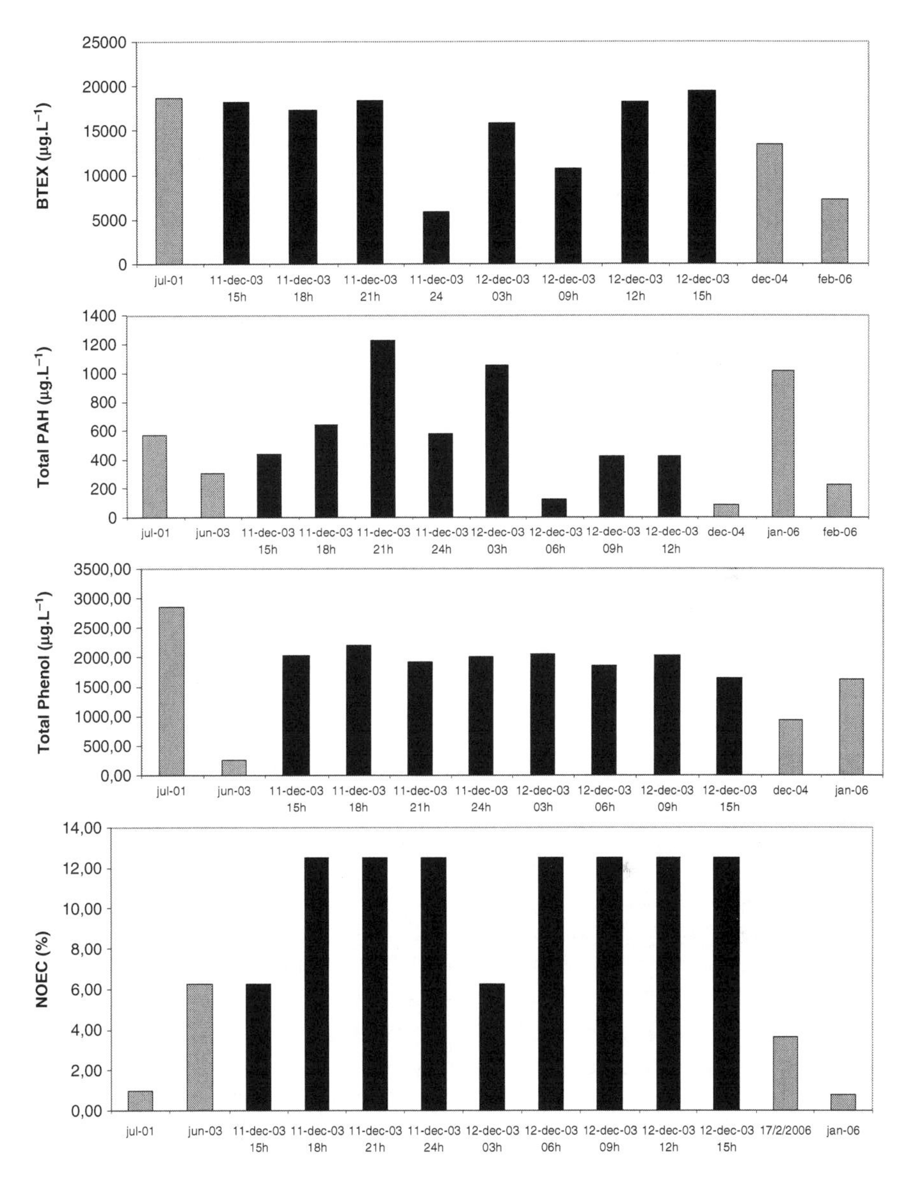

Fig. 3.5 Temporal variability of BTEX, PAH, phenols and chronic toxicity using *Lytechinus variegatus* for the same platform, PCR-1 (*Black bars* = 2 days continuous sampling, *gray bars* = individual samples)

For more than 90% of results, the concentrations of principal analytes, even at near-field dilution, were below the CONAMA Resolution 357/2005 quality criteria for Class I seawater. CONAMA Resolution 393/2007 determines 500 m as the PW mixing zone, and beyond this the seawater quality criteria must be achieved (www.mma.gov.br/port/conama/legiano.cfm?codlegitipo=3), and it was, for the predicted concentrations of all parameters at all the platforms studied (Gabardo 2007).

Table 3.4 Input parameters for plume dispersion studies using CORMIX and CHEMMAP models and the results of dilution and distances from the discharge point in near-field dispersion

	Campos basin – SE				Ceara basin – NE
Parameter	P-32	P-26	PPG-1	SS-06	PCR-1
Pipe diameter	10″	12″	40″	12″	24″
Discharge direction	Vertical	Vertical	Vertical	Vertical	Vertical
Water depth	160 m	990 m	101 m	120 m	45 m
Depth of discharge	Surface	Surface	60 m	8 m	21 m
PW discharge flow m^3/day	4,500	6,400	20,000	12,000	1,000
PW density kg/m^3	1,028	1,028.8	1,054.3	1,060.4	1,056
Benzene (mg L^{-1})	0.917	1.338	1.585		10.18
Toluene (mg L^{-1})	2.265	2.310	0.796	0.349	5.71
Total phenols (mg L^{-1})	1.229	1.577	nr		2.02
Chrysene-PAH (mg L^{-1})	0.0009	0.0012	0.0035		0.0025
Sulphides (mg L^{-1})	6.8	0.05	0.05		1.21
Ammonia (mg L^{-1})	78.0	81.0	90.0		82.0
Barium (mg L^{-1})	1.09	1.21	12.5		1.6
Boron (mg L^{-1})	26.40	21.10	27.2		20.0
Lead (mg L^{-1})				0.71	
Cadmium (mg L^{-1})				0.40	
Nickel (mg L^{-1})				2.60	
Dilution and distance for near-field dispersion (based on Cormix Model)					
Summer dilution (times)	101	156	96	279	895
Summer distances (m)	27	39	111	189	44
Winter dilutions (times)	106	152	110	348	713
Winter distances (m)	29	34	129	221	33

As an example, Fig. 3.6 shows aerial views of the modelled plumes for each parameter analysed at 500 m from the discharge point in the winter for two platforms (PCR-1 and PPG-1), and a comparison with the regulatory parameters for seawater Class I (Conama 357/2005). For each point, the model calculates the average maximum concentration in time and space in the water column; therefore, the predicted results are very conservative from an environmental point of view. For all the platforms studied, all the parameters complied with the Class I seawater legislation criteria within a distance of 500 m from the PW discharge point in summer and winter conditions (Gabardo 2007).

3.5 Environmental Monitoring Data

Environmental monitoring, which included direct measurement of PW constituents as well as seawater chemistry and toxicity, has been required to maintain permits according to Brazilian law (Scofano et al. 2010; Soares and Scofano 2010). Petrobras has been conducting extensive field studies on fate and effects of PW

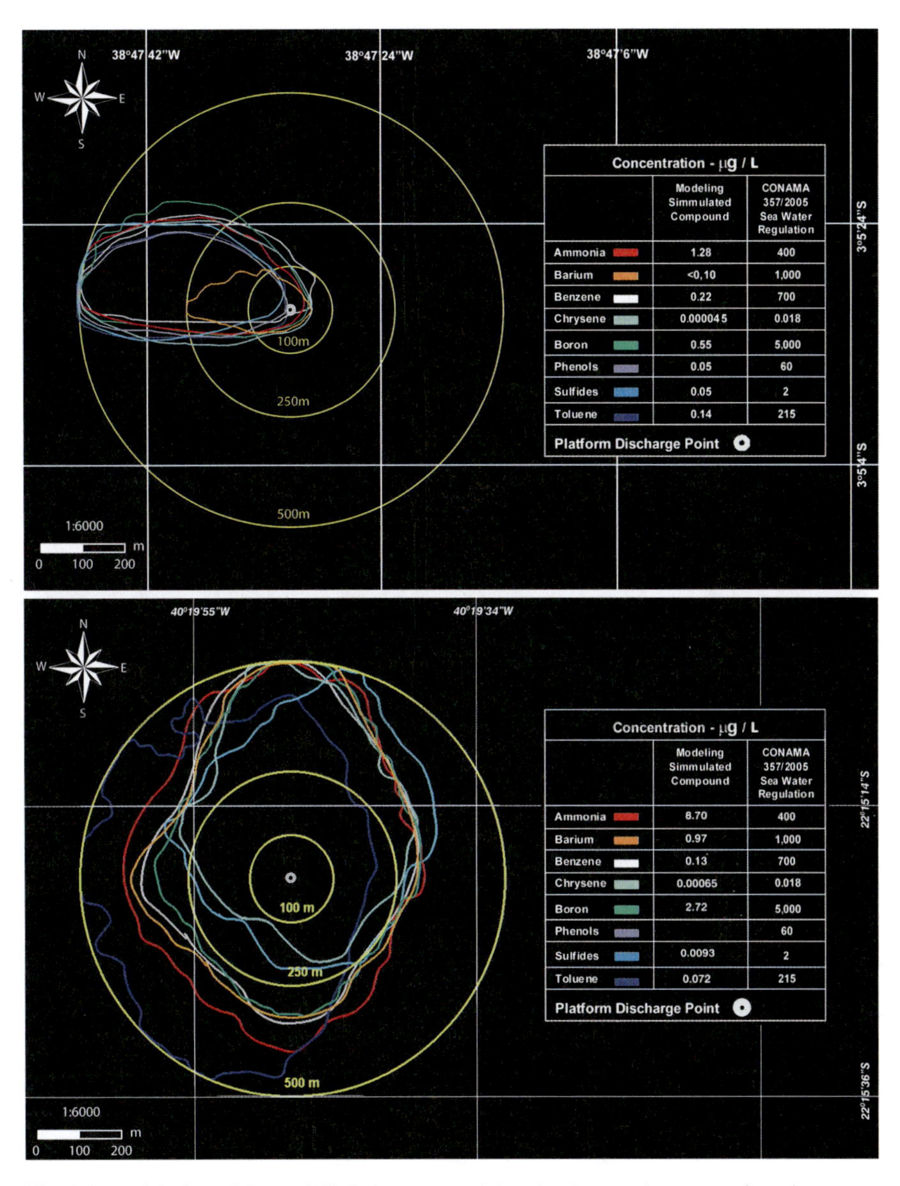

Fig. 3.6 Aerial view of the modelled plumes containing the shape and average of maximum predicted concentration at 500 m distant from the discharge point in winter conditions for PCR-1 (*top*) and PPG-1 (*bottom*) platforms

in the vicinity of its platforms since 1998. Only as an illustrative example, data collected around six platforms are presented in Table 3.5, with about 800 seawater samples collected and more than 12,000 results obtained. Also, there was no observed acute or chronic toxicity for those seawater samples. The data were

Table 3.5 Seawater data in the vicinity of six Brazilian platforms compared to criteria established in CONAMA Resolution 357/2005[a]

Parameter	PPM-1	PPG-1	P-40	FPBR	SS-06	PCR-1	CONAMA 357/2005 seawater class I
Date	Jan–Jul 1998	Jan–Jul 1998	2001–2004	2003–2004	2001–2005	2001–2003	
Number of surveys	2	2	4	2	7	4	
Number of samples/survey	24	24	63	13	41	9	
Total of samples	48	48	252	102	287	36	
Regulatory parameters							
Total phosphorus (mg L^{-1})	A	A	A (99%)	A (95%)	A	–	0.062
Nitrate (mg L^{-1})	A	A	A (92%)	A	A	A	0.4
Nitrite (mg L^{-1})	A	A	A	A	A	A	0.07
Ammonia (mg L^{-1})	A	A	A	A	A	A	0.4
Dissolved oxygen (mL L^{-1})	A	A	A	A	A (98%)	A	> 4.2
pH	A	A	–	A	A	–	6.5–8.5
COT (mg L^{-1})	–	–	–	A	–	–	< 3
Phenols (μg L^{-1})	A	A	–	A	A	A	60
Bz(a)Ant (μg L^{-1})	–	–	A	A	A	A	0.018
Bz(a)Py (μg L^{-1})	–	–	A	A	A	A	0.018
Bz(b)Flu (μg L^{-1})	–	–	A	A	A	A	0.018
Bz(k)Flu (μg L^{-1})	–	–	A	A	A	A	0.018
Chrysene (μg L^{-1})	–	–	A	A	A	A	0.018
Dibz(a,h)Ant (μg L^{-1})	–	–	A	A	A	A	0.018
In(cd) Py (μg L^{-1})	–	–	A	A	A	A	0.018
Non-regulatory parameters							
Chlorophyll a (μg L^{-1})	<0.01–0.79	<0.01– 0.89	<0.02–0.45	<0.02–0.54	0.01–1.87	<0.02–0.25	–
Salinity (g L^{-1})	35.4–37.8	35.0–38.9	34.5–37.9	35.8–37.3	35–37.5		–
Vanadium (μg L^{-1})	2.2–4.9	1.7–2.9	–	–	–		–
Orthophosphate (μg L^{-1})	<0.001–0.0551	0.008– 0.059	0.002–0.210	0.001–0.076	0.01–0.051	<0.1–39	–
TSS (mg L^{-1})	–	–	5.1—12.6	0.27–11.63	3–9.02	3–12.1	–
n-alkanes (μg L^{-1})	–	–	3.6–20.9	0.3–15.1	1.3–6.3	0.03–4.70	–
TPH (μg L^{-1})	–	–	–	0.003–2.89	–	1.45–2.38	–

[a]Conama 357/05: criteria for PAH only for areas with intensive aquaculture for human consumption
A – Result is below the criteria; ( ) – % sample results below the criteria

reported in reports submitted to IBAMA, as part of the documents to obtain environmental permits from those platforms (Petrobras 2001, 2002, 2003, 2004, 2006a, b, c, 2007).

4 Discussion

4.1 Inorganic Constituents

In the present study, total suspended solids (TSS) was in the range of 1.9–106 mg L^{-1} (median = 10.6 mg L^{-1}), while data reported for the North Sea showed concentrations of TSS between 3 and 85 mg L^{-1}, and a study of 10 Louisiana platforms found concentrations ranging from 12 to 840 mg L^{-1} (E&P Forum 1994; OGP 2005).

Metals and radioisotopes are the principal trace inorganic constituents of environmental concern. Brazilian PW metal concentrations ranked closely with the minimum values found for other PW reports (OGP 2005; E&P Forum 1994). For many elements the concentrations were below the detection limit.

Brazilian PW data for Hg was in the range of <0.2–0.63 μg L^{-1}, with a median of <0.2 μg L^{-1} (23 samples). Surprisingly, these values are three orders of magnitude (micrograms/litre) lower than Hg concentrations found in PW from other oil platforms expressed in milligrams per litre (0.02–0.25 mg L^{-1}). The mean concentration for gas platforms was 23 mg L^{-1} (OGP 2005).

Vanadium in PW from the Gulf of Mexico ranged from 6.3 to 22 mg L^{-1} (US-MMS 1992) while for Trinidad and Tobago the literature presents a concentration of 0.011 mg L^{-1} (Maharaj et al. 1996). In this study, the range was <0.002–0.37 mg L^{-1} V, with median of <0.002 mg L^{-1} for 23 samples, which is lower than worldwide documented concentrations.

Other studies have reported maximum Ba concentrations up to 650 mg L^{-1} (Tibbetts et al. 1992) or in the range of 0.2–228 mg L^{-1}, with a median of 87 mg L^{-1}, and those of Fe in the range of 0–15 mg L^{-1} with a median of 4.3 mg L^{-1} (Frost et al. 1998). The concentrations of Ba found in this study (0.2–45 mg L^{-1}; median = 1.3 mg L^{-1}) were less than those mentioned above, while those for Fe in this study were higher (0.04–25 mg L^{-1} with median of 7.4 mg L^{-1}; 55 samples).

Boron occurs frequently in Brazilian PW, ranging from 6–120 mg L^{-1} with a median of 31 mg L^{-1}, but this element has not been found to be toxic (Neff 2002).

For other elements in 23 PW samples, the median results were: As (<0.2 mg L^{-1}), Cd (<0.02 mg L^{-1}) , Pb (<0.1 mg L^{-1}), Cr (<0.005 mg L^{-1}) , Sn (<0.05 mg L^{-1}) Zn (<0.2 mg L^{-1}) , Ni (<0.01 mg L^{-1}), Ag (<0.003 mg L^{-1}) and Ni (<0.01 mg L^{-1}).

Several other elements of little environmental concern were also quantified in Study 2 such as phosphorus (range 0.03–3 mg L^{-1}, mean 0.52 mg L^{-1}, median 0.05 mg L^{-1}), manganese (range 0.04–5.9 mg L^{-1}, mean 0.96 mg L^{-1}, median 0.35 mg L^{-1}), aluminium (range <0.003–0.3 mg L^{-1}, mean 0.09 mg L^{-1}, median 0.1 mg L^{-1}) and selenium (range <0.02–0.4 mg L^{-1}, median <0.02 mg L^{-1}).

Rapid dilution or precipitation of metals with particulate matter when discharged into the ocean has been documented (Hartley 1994; Neff 2002; Neff et al. 1989; Trefry et al. 1995; Trocine and Trefry 1983). Precipitation as metal hydroxides or sulphides is the principal fate of heavy metals in the aquatic environment. The complexation, oxidation and precipitation reactions do not remove the heavy metals from the marine environment, but they do convert them to forms that are not bioavailable (E&P Forum 1994). No acute toxic effects to the sea organisms living around platforms due to metals in PW have been reported as yet (Neff 2002; OGP 2005).

Several naturally occurring radionuclides are present in PW. The most abundant are usually radium-226 and radium-228 (Neff 2002). In this study ^{226}Ra ranged from <0.016 to 10.9 Bq L^{-1} and ^{228}Ra from <0.032 to 10.5 Bq L^{-1} which is similar to a previous study in Campos Basin, where the levels of these radionuclides ranged from 0.012 to 6.0 Bq L^{-1} for ^{226}Ra and from <0.05 to 12.0 Bq L^{-1} for ^{228}Ra (Vegueria et al. 2002). Brazilian PW exhibits the same radioisotope levels as those reported elsewhere (E&P Forum 1994; Utvik 1999; Lysebo and Strand 1998; Guzella et al. 1996).

4.2 Organic Constituents

It is cited in the literature that the most complete non-specific measure of the total amount of organic components is TOC. A North Sea survey conducted in the 1990's showed that TOC concentrations in PW varied from 14 to 552 mg L^{-1}. The range worldwide is 100–700 mg L^{-1}, although one source reported a wider range of 0–1500 mg L^{-1} (OGP 2005). TOC values obtained in this study were in the same order of magnitude.

Total oil content measured as TPH. The range of oil concentrations was 4–66 mg L^{-1}, with a mean of 14.5 mg L^{-1} and median of 10 mg L^{-1}. This is similar to the mean of 17.8 mg L^{-1} in 2006 for PW from the North Sea (OSPAR 2009).

Carboxylic acids. In this study, the carboxylic acids (including acetic and propionic acid) were in the range of 45–928 mg L^{-1}. Reports for produced water in the North Sea showed a range from 81 to 930 mg L^{-1} (E&P Forum 1994), similar to the levels found in Brazilian PW. Carboxylic acid levels are not a cause of environmental concern due to its high biodegradability, but these compounds promote pipeline corrosion.

Aromatic compounds. Volatile aromatic hydrocarbons (BTEX) occur in all PW, but there are significant differences in concentration between oil and gas fields. In PW from PCR-1 there was a high contribution of monoaromatics to the total organic fraction. The same pattern was observed for the other platforms. The main reason for the high BTEX concentrations in PW is due to its solubility. The solubility of the monoaromatics is in range of the hundreds to the thousands of mg L^{-1}, compared with PAH solubility of one to tens of mg L^{-1}, decreasing drastically to very low solubility for the 4–6 ring PAH (Mackey et al. 1992a, b; Neff

2002; Merck Index 2006) like naphthalene (30 mg L^{-1}), phenanthrene (1 mg L^{-1}) and chrysene (0.002 mg L^{-1}). As expected, the levels of monoaromatics (BTEX, 4,690 μg L^{-1}) when compared to the PAH content, represent 92% of the total aromatics, followed by the sum of naphthalenes, phenanthrenes and dibenzothiophenes (NPD 399.2 μg L^{-1}) which was 7.9%. The sum of 14 EPA PAH, excluding naphthalene and phenanthrene which were already counted in the NPD fraction, was 5.9 μg L^{-1} or 0.11% of the total aromatics. Aromatic content of Brazilian PW is in the intermediate to low range of other PW worldwide.

Phenol and alkylated phenols. These compounds occur naturally in oil and will partition into produced water depending on their molecular weight. Data for Brazilian PW are similar to those for the North Sea (Utvik 1999). Phenols with heteroatoms (nitro and chlorophenols) have never been detected in Brazilian PW.

4.3 Toxicity

Acute PW toxicity previously reported in the literature, ranging from 5.2 to 14.5% PW for *Mysidopsis bahia* (Schiff et al. 1992); 4.5–53.5% for *S. costatum* (E&P Forum 1994); 2.4–24.4% for *V. fischeri* (Flynn et al. 1996); and 16–58.8% for *Artemia* sp. (E&P Forum 1994; Holdway 2002), is comparable with the results of this study.

There are few available data for PW chronic toxicity in the literature, but Schiff et al. (1992) published some results for the sea-urchin (*Strongilocentrotus purpuratus*) fertilization test which ranged from 0.74 to 1.73%, similar to the Brazilian NOEC range using *L. variegatus* of <0.1–12.5%. Other studies of PW chronic toxicity reported much lower values, but used different methods and more sensitive endpoints (Krause et al. 1992; Holdway 2002). Chronic toxicity in Study 3 exhibited a narrow range (NOEC 6.25–12.5%) during the 24 h sampling, but a 5-fold range during the annual surveys, indicating that variability is higher in the long-term. Holdway (2002) reported that PW toxicity may fluctuate up to 10-fold, especially due to operational changes in the production process and chemical blends used for oil treatment, but the author did not see significant variation in the intrinsic PW composition.

Considering the differences in sensitivity of the species and methodologies, the overall acute and chronic toxicity results obtained for Brazilian PW can be considered similar to other studies. Gabardo (2007) found no strong correlation between toxicity and chemistry for Brazilian PW as has been also documented in other studies (E&P Forum 1994; Swan et al. 1994).

4.4 Modelling Produced Water Dispersion with Field Data

The impact of marine discharges and its potential environmental effects depend both on the concentrations of the discharged materials and on the capacity of the receiving environment (Smith et al. 2004; Brandsma and Smith 1999). For the majority of the parameters analysed, the seawater quality criteria (Class I, CONAMA Resolution 357/05) were achieved at near-field dilution, except for ammonia (PPG-1, P-32 and P-26 platforms) and sulphides (P-32) (Gabardo 2007). The concentration limits must be achieved at a distance of 500 m from the discharge point, which is considered the limit of the mixing zone by the CONAMA Resolution 393/2007. Based on the modelling of PW dispersion plumes, all the parameters analysed achieved the quality criteria for seawater before the 500 m limit (Gabardo 2007). Adopting Saline Water Class I (CONAMA 357/2005) criteria for water in the vicinity of offshore PW discharge platforms is very conservative but in agreement with environmental preservation.

4.5 Environmental Monitoring

The vast majority of the environmental monitoring results showed very good quality seawater and 99.7% of the parameter concentrations complied with the Brazilian regulatory limits for high quality seawater. The lack of observed acute and chronic toxicity in seawater reinforced the predicted PW dispersion plume obtained with the CORMIX and CHEMMAP models. Same behaviour was previously reported in the literature for environmental monitoring studies around platforms (US MMS 1992; Veil et al. 2005; OGP 2002, 2005). Concerning the Brazilian environment, two additional factors should be borne in mind: a) the low volume of PW discharged in the northeast ($<0.5\ m^3\ s^{-1}$) and southeast ($1.39\ m^3\ s^{-1}$) compared to the input rivers of the same regions that are 3–4 orders of magnitude higher; b) the hot weather and warm seawater of Brazil certainly favour an increased rate of hydrocarbon biodegradation and weathering (Gabardo 2007).

Despite the lack of observed toxicity of seawater adjacent to the platforms, further studies are being implemented to investigate the possible biological effects of PW discharges in the ocean. Recent efforts have been made to evaluate the chronic and sub lethal environmental effects due to the PW discharge including studies with oysters to assess bioaccumulation and biomarkers.

5 Summary

This chapter provides a technical summary of 10 years (1996–2006) of monitoring PW discharge from offshore platforms operated by Petrobras in Brazil. The following conclusions were reached with this study:

- In 2005, the average total volume of PW discharged into Atlantic Ocean by 24 platforms offshore in the Brazilian coast was 73 million m^3/year, representing less than 3% worldwide PW discharged to other oceans;
- Results of organic (TPH, BTEX, PAH, phenols, organic acids) and inorganic (metals, anions, cations) parameters, as well as acute and chronic toxicity obtained for Brazilian PW were similar to the literature;

- Brazilian platforms with greater gas production presented relatively high concentrations of BTEX, phenols and low molecular weight PAH in its PW composition, corroborating the worldwide information previously reported;
- Barium is one metal of significant presence in Brazilian PW and both ^{226}Ra and ^{228}Ra concentrations are highly correlated with this element. In contact with sulphate rich seawater, barium precipitates as barium sulphate, reducing concentrations of Ba and radium isotopes in the water column;
- Modelled PW dispersion plumes showed dilution factors of 100–700 within 200 m of platform discharge points;
- Dilution of several PW chemical components was confirmed by sampling seawater within that distance from the platforms;
- Modelling simulations in winter and summer conditions predicted concentrations for all the parameters compliant with the Brazilian regulatory limits for seawater within 500 m from the discharge point;
- Despite the lack of observed alteration in seawater quality around production platforms, the importance of continuing to evaluate the impact of PW discharges from a risk assessment perspective has been recognized, and studies of bioaccumulation and the use of biomarkers among other initiatives are currently being implemented.

Acknowledgements The authors would like to thank the entire project team without whom this challenge could not have been accomplished. We are grateful to Eduardo Yassuda from ASA (Applied Science Associates), South America, colleagues Angelo Sartori Neto, Renato Parkinson Martins and Jose Antonio Moreira Lima for the modelling studies, and to Ivanil Ribeiro Cruz and Carlos German Massoni for helping with the figures and artwork for the manuscript. We wish to acknowledge Fabiana Dias Costa Gallotta, Priscila Reis da Silva, Maria de Fatima G. Meniconi, Dr. John Veil, José Marcos Godoy and Janaina Medeiros for their constructive comments on the manuscript. Finally, we specially acknowledge the Petrobras E&P Department for their assistance and cooperation.

References

ABNT (2005) Associação Brasileira de Normas Técnicas Aquatic ecotoxicology: acute toxicity: test with misids (Crustacea). ABNT NBR 15308:2005. (In Portuguese)

ABNT (2006a) Associação Brasileira de Normas Técnicas. Aquatic ecotoxicology – Chronic toxicity – Test with sea urchin (Echinodermata: Echinoidea). ABNT NBR 15350:2006. (In Portuguese)

ABNT (2006b) Associação Brasileira de Normas Técnicas (2006) Aquatic ecotoxicology – Determination of the inhibitory effect of water samples on the light emission of *Vibrio fischeri* (Luminescent bacteria test). ABNT NBR 15411-3:2006. (In Portuguese)

APHA (1995) American Public Health Association, American Water Works Association (AWWA), Water Environment Federation – (WEF). Standard Methods for Examination of Water and Waste Water. Eaton AD, Chesceri LS, Greenberg. 1368 p

Agência Nacional do Petróleo, Gás Natural e Biocombustíveis. (In Portuguese). http://www.anp.gov.br. Accessed 28 Aug 2009

Barnett V, Lewis T (1994) Outliers in statistical data, 3rd edn. Willey, Pondicherry, 584 pp

Bouloubassi I, Fillaux J, Saliot A (2001) Hydrocarbons in surface sediments from the Changjiang (Yangtze River) estuary, East China Sea. Mar Pollut Bull 42: 335–1346

Brandsma MG, Smith JP (1999) Offshore operators committee mud and produced water discharge model. Report and User Guide. Exxon Mobil Production Operations Division. EPR.29PR.99, 168 pp

CETESB L5. (021/1987). Companhia de Tecnologia de Saneamento Ambiental. Água do Mar – Teste de Toxicidade Aguda com *Artemia*. Norma Técnica L05.021/1987. (In Portuguese)

CETESB L5. (250/1992). Companhia de Tecnologia de Saneamento Ambiental. Água do mar – Teste de Toxicidade crônico de curta duração com *Lytechinus variegatus* – Lamark, 1816 (Echinodermata, Echinoidea) Norma Técnica L5.250/1992. (In Portuguese)

CETESB L5. (251/1995). Companhia de Tecnologia de Saneamento Ambiental. Água do Mar – Teste de Toxicidade Aguda com *Mysidopsis juniae*. Norma Técnica L05.251/1995. (In Portuguese)

CETESB L5. (227/2001). Companhia de Tecnologia de Saneamento Ambiental. Teste de Toxicidade com Bactéria luminescente *Vibrio fischeri*, Norma Técnica L5. 227/2001. (In Portuguese)

CONAMA. (393/2007, 357/2005) http://www.mma.gov.br/port/conama/legiano.cfm?codlegitipo=3. Accessed Apr 2010

Durell G, Utvik TIR, Johnsen S, Frost T, Neff J (2006) Oil well produced water discharges to the North Sea. Part I: Comparison of deployed mussels (*Mytilus edulis*), semi-permeable membrane devices, and DREAM model predictions to estimate the dispersion of polycyclic aromatic hydrocarbons. Mar Environ Res 62:194–223

EIA (2006) Energy Information Administration. Official Energy Statistics from the U.S. Government. Report DOE/EIA-0484-2006 http://www.eia.doe.gov/oiaf/ieo/oil.html. Accessed Feb 2007

EPA 1664 – Revision A:N-Hexane Extractable Material (HEM; Oiland Grease) and Silica Gel Treated N-Hexane Extractable Material (SGTHEM; Non-polar Material) by Extraction and Gravimetry. Environmental Protection Agency. http://www.epa.gov/waterscience/methods/method/oil/1664guide.pdf. Accessed May 2010

EPA 3510C Separatory Funnel Liquid-Liquid Extraction. Test method for evaluation solid waste physical/chemical methods. Laboratory U.S. Environmental Protection Agency. http://www.epa.gov/waterscience/methods/method/oil/1664guide.pdf. Accessed May 2010

EPA 3630C – Silica gel cleanup. Test method for evaluation solid waste physical/chemical methods. U.S. Environmental Protection Agency. http://www.epa.gov/waterscience/methods/method/oil/1664guide.pdf. Accessed May 2010

EPA 5021 Volatile organic compounds in soils and other solid matrices using equilibrium headspace analysis. U.S. Environmental Protection Agency. http://www.epa.gov/waterscience/methods/method/oil/1664guide.pdf. Accessed May 2010

EPA 8015B Nonhalogenated Organics Using CG/FID. Test method for evaluation solid waste physical/chemical methods. U.S. Environmental Protection Agency. http://www.epa.gov/waterscience/methods/method/oil/1664guide.pdf. Accessed May 2010

EPA 8270D Semivolatile organic compounds by gas. chromatography/mass spectrometry (CG/MS). Test method for evaluation solid waste physical/chemical methods. U.S. Environmental Protection Agency. http://www.epa.gov/waterscience/methods/method/oil/1664guide.pdf. Accessed May 2010

E&P FORUM (1994) Oil industry international exploration and production forum. North Sea produced water: fate and effects in the marine environment. London, Report No. 2.62/204, 50 p

Fingas M (2001) The basics of oil spill cleanup. Edited by Jennifer Charles, 2nd edition, Lewis Publishers, Boca Raton, FL, 233p

Flynn SA, Butler EJ, Vance I (1996) Produced water composition, toxicity, and fate. A review of recent BP North Sea studies. In: Reed M, Johnsen S (eds) Produced water 2: environmental issues and mitigation technologies. Plenum Press, New York, NY, pp 69–79

Freitas ALS, Mendes LCL (2010) Petrobras. Brazilian Regulatory Framework Concerning Produced Water Discharged. SPE 126974. The Tenth SPE International Conference on Health,

Safety, and Environment in Oil and Gas Exploration and Production, Rio de Janeiro, RJ, Brazil, April 12–14
Frost TK, Johensen S, Utvik, TR (1998) Produced water discharges to the North Sea: Fate and effects in the water column. Summary Report. The Norwegian Oil Industry Association. OLF, Hydro, Statoil. 39p. Website http://www.olf.no/ accessed Jan 2007
Fucik KW (1992) Toxicity identification and characteristics of produced water discharges from Colorado and Wyoming. In: Ray JP, Engelhardt FR (eds) Produced water: technological/environmental issues and solutions. Plenum Press, New York, NY, pp 187–198
Gabardo IT (2007) Chemical and toxicological characterization of produced water in oil and gas Brazilian platforms and dispersion behaviour in the ocean. (Natal, RN). PhD Thesis, Federal University of Rio Grande do Norte, 235p. (In Portuguese)
GESAMP (1993) Impact of oil and related chemicals and wastes on the marine environment. Reports and Studies n. 50, 180 pp
Godoy JM, Lauria DC, Godoy ML, Cunha RP (1994) Development of a sequential method for the determination of U-238, U-234, Th-232, Th-228, Ra-226, Ra-228 and Pb-210 in Environmental Samples. J Radioanal Nucl Chem 182:165–169
Guzella L, Bartone C, Ross P, Tartari G, Muntau H (1996) Toxicity identification evaluation of Lake Orta (Northern Italy) Sediments using the Microtox System. Ecotox Environ Saf 35: 231–235
Hartley JP (1994) Environmental monitoring of offshore oil and gas drilling discharges – A caution on the use of Barium as a tracer. Mar Pollut Bull 32:727–733.
Holdway DA (2002) The acute and chronic effects of wastes associated with offshore oil and gas production on temperate and tropical marine ecological processes. Mar Pollut Bull 44: 185–203
ISO 10.253:1995 (E). Water quality – Marine algal growth inhibition test with *Skeletonema costatum* and *Phaeodactylum tricornutum.* 1st ed. 8p
Krause PR, Osenberg CW, Schmitt RJ (1992) Effects of produced water on early life stages of a sea urchin: stage specific responses and delayed expression. In: Ray JP, Engelhardt, FR (eds) Produced water: technological/environmental issues and solutions. Plenum Press, New York, NY, pp 431–444
Lysebo I, Strand T (1998) NORM in oil production: activity level and occupational doses: NORM II, Second International Symposium, 10–13/11–98, Krefeld, RFA
Mackey D, Shiu WY, Ma KC (1992a) Illustrated handbook of physical chemical properties and environmental fate for organic chemicals. Vol. I: monoaromatic hydrocarbons, chlorobenzenes and PCBs, 1st edn. CRC-Press, Boca Ratón, FL, 704 p
Mackey D, Shiu WY, Ma KC (1992b) Illustrated handbook of physical chemical properties and environmental fate for organic chemicals. Vol II: polynuclear aromatic hydrocarbons, polychlorinated dioxins and dibenzofurans. Lewis Publishers, Boca Raton, FL, 608 pp
Maharaj US, Mungal R, Roodalsingh R (1996) Produced water monitoring programme in Petrotin. Society of Petroleum Engineers SPE 36146, pp 675–684
Merck Index (2006) The Merck index. An encyclopedia of chemicals, drugs, and biologicals. Merck & Co., Inc., 14th Whitehouse Station, NJ, EUA. 1756 pp
Neff JM (2002) Bioaccumulation in marine organisms. Effect of contaminants from oil well produced water. Elsevier, London, 1st edn, 452 p
Neff JM, Sauer TC, Macioleck N (1989) Fate and effects of produced water discharge in near shore marine waters. API Publication No. 4472, American Petroleum Institute, Washington, DC, 300 pp
NRC National Resource Council (2003) Oil in the sea – inputs, fates and effects, 2nd. edn. National Academy Press, Washington, DC, 265 pp
OLF, The Norwegian Oil Industry Association (2005) Nilssen I, Johnsen S, Utvik T. Water column monitoring Summary Report, 2005. Discharges Risk Assessment Monitoring, 48p
OGP The International Association of Oil & Gas Producers (2002) Aromatics in produced water: occurrence, fate & effects, and treatment. Report I.20/324. January 2002, 24 pp

OGP The International Association of Oil & Gas Producers (2005). Fate and effects of naturally occurring substances in produced water on the marine environment. Report 364. February 2005, 36 pp

OSPAR Commission (2009) Discharges, spills and emissions from offshore oil and gas installations in 2007. Including assessment of data reported in 2006 and 2007. ISBN 978-1-906840-92-1. Publication Number: 452/2009, 58 pp

PETROBRAS (2001) Monitoramento Ambiental da Atividade de Produção de Petróleo na Bacia de Campos. Relatório Final. Ed. Centro de Pesquisas e Desenvolvimento Leopoldo Miguez de Mello (CENPES), da PETROBRAS, GERÊNCIA de Biotecnologia e Ecossistemas. Rio de Janeiro. 222 pp. (in Portuguese).

PETROBRAS (2002a) Monitoramento Ambiental da Área de Influência do Emissário de Cabiúnas Região de Macaé/RJ – Caracterização pré-operação e monitoramento pós-operação. Relatório Final, Ed. Centro de Pesquisas e Desenvolvimento Leopoldo Miguez de Mello (CENPES), da PETROBRAS, 276 pp

PETROBRAS (2002b) Relatório do Monitoramento pré-operação da Plataforma SS-06, Bacia Campos. Centro de Pesquisas e Desenvolvimento Leopoldo Miguez de Mello (CENPES). PETROBRAS. Universidade do Rio de Janeiro, Instituto de Biologia – Departamento de Zoologia. Rio de Janeiro. Maio. 182 pp

PETROBRAS (2003) Atividade de produção de Óleo e Gás Campo de Roncador, FPSO-Brasil. Relatório da 1.a Campanha de Monitoramento Ambiental. Centro de Pesquisas e Desenvolvimento Leopoldo Miguez de Mello, CENPES. OCEANSAT. Rio de Janeiro. Novembro. 98 pp

PETROBRAS (2004) Relatório do Programa de Monitoramento Ambiental na Bacia de Campos no Campo de Roncador FPSO Brasil. Centro de Pesquisas e Desenvolvimento Leopoldo Miguez de Mello, CENPES. Fundação COPPETEC – UFRJ. Rio de Janeiro. 124 pp

PETROBRAS (2006a) Projeto de avaliação da qualidade da água e efluentes da plataforma SS-06, Bacia de Campos, Rio de Janeiro. Relatório Técnico. Petrobras- Petróleo Brasileiro S/A. – Fundação Bio-Rio – Instituto de Biologia/ UFRJ. 311 pp

PETROBRAS (2006b) Programa de Monitoramento da qualidade da água e efluentes da plataforma Semi-Submersível SS-06, Bacia de Campos, Rio de Janeiro. Relatório de Dados Brutos. Fase II Campanha 7. Petrobras – Petróleo Brasileiro S/A. – Fundação Bio-Rio – Instituto de Biologia/ UFRJ. 138 pp

PETROBRAS (2006c) Programa de Monitoramento da qualidade da água e efluentes da plataforma Semi-Submersível SS-06. 8.a Campanha Oceanográfica-Fase II e 6.a Campanha de Bioincrustação e Ictiofauna. Petrobras – Petróleo Brasileiro S/A. – Concremat. Relatório PT- 3.5.8.023-RT-AMA-005-RO. Outubro

PETROBRAS (2006d) Relatório Consolidado do projeto de Monitoramento Ambiental do Campo de Marlim-Sul – Bacia de Campos, Rio de Janeiro. Petrobras – Petróleo Brasileiro S/A. Fundação BIO-RIO, Universidade Federal do Rio de Janeiro – Instituto de Biologia. 126 pp

PETROBRAS (2007) Relatório Consolidado das Campanhas de Monitoramento Ambiental da Bacia do Ceará. Rio de Janeiro, Petrobras – Petróleo Brasileiro S/A

PRESAL Program (2010) (http://www.petrobras.com.br/minisite/presal/en/questions%2Danswers/). Accessed Nov 10

Pillard DA, Evans JM, Dufresne DL (1996) Acute toxicity of saline produced waters to marine organisms. SPE 35845 of SPE International Conference on Health, Safety & Environment, pp 675–682

Rand GM (1995) Fundamentals of aquatic toxicology – Effects, environmental fate, and risk assessment, 2nd edn. Taylor & Francis, Washington, DC, 1125 pp

Readman JW, Fillmann G, Tolosa I, Bartocci J, Villeneuve JP, Catinni C, Mee LD (2002) Petroleum and PAH contamination of the Black Sea. Mar Pollut Bull 44:48–62

Scofano AM, Xavier AG, Marcon EH, Gabardo IT, Rocha MF, Curbelo-Fernandez MP, Cotta PS, Cavalcanti TBRO (2010) Offshore Regional Environmental Monitoring Model for Campos Basin, Brazil: an innovative proposal. SPE 127040. The Tenth SPE International Conference on Health, Safety, and Environment in Oil and Gas Exploration and Production, Rio de Janeiro, RJ, Brazil, April 12–14

Schiff KC, Reish DJ, Anderson JW, Bay SM (1992) A comparative evaluation of produced water toxicity. In: Ray JP, Engelhardt FR (eds) Produced water: technological/environmental issues and solutions. Plenum Press, New York, NY, pp 199–208

Smith JP, Brandsma MG, Nedwed (2004) Field verification of the Offshore Operators Committee (OOC) mud and produced water discharge model. Environ Model Softw 19:739–749

Soares CRU, Scofano AM (2010) Petrobras environmental permitting offshore Brazil: a new approach for oil and gas activities. SPE 127032. The Tenth SPE International Conference on Health, Safety, and Environment in Oil and Gas Exploration and Production, Rio de Janeiro, RJ, Brazil, April 12–14

Stephenson MT (1992) Components of produced water: a compilation of industry studies. J Petrol Technol 44:548–603

Swan JM, Neff JM, Young PC (1994) Environmental implications of offshore oil and gas development in Australia – the findings of an independent scientific review. Australian Petroleum Exploration Association, Sydney, 696 pp

Terrens GW, Tait RD (1996) Monitoring ocean concentrations of aromatic hydrocarbons from produced formation water discharges to Bass Strait, Australia. SPE 36033. Society of Petroleum Engineers International Conference on Health, Safety & Environment, New Orleans, Louisiana, 9–12 June 1996, pp 739–747

Tibbetts PJC, Buchanan IT, Gawel LJ, Large R (1992) A comprehensive determination of produced water composition. In: Ray, JP, Engelhardt FR (eds) Produced water: technological environmental issues and solutions. Plenum Press, New York, NY, pp 97–112

Trefry JH, Naito KL, Trocine RP, Metz S (1995) Distribution and bioaccumulation of heavy metals from polluted water discharges to the Gulf of Mexico. Water Sci Technol 32:31–36

Trocine RP, Trefry JH (1983) Particulate metal tracers of petroleum drilling mud dispersion in the marine environment. Environ Sci Technol 17:507–512.

UOP-938. Total Mercury and Mercury Species in Liquid Hydrocarbon – Annexes.

U.S. MMS (1992) (U. S. Department of Interior Mineral Management Service). Gulf of Mexico. Sales 142 and 143: Central and Western Planning Areas. Final Environmental Impact Statement. Vol. I and II, New Orleans. (Report MMS 92-0054), 709 pp

Utvik TIR (1999) Chemical characterization of produced water from four offshore oil production platforms in the North Sea. Chemosphere 39:2593–2606

Utvik TIR, Durell G, Johnsen S (1999) Determining produced water originating Polycyclic Aromatic Hydrocarbon in North Sea waters: comparison of sampling techniques. Mar Pollut Bull 38:977–89

Vegueria SFJ, Godoy JM, Miekeley N (2002) Environmental impact studies of barium and radium discharges by produced waters from the "Bacia de Campos" oil-field offshore platforms, Brazil. J Environ Radioact 62:29–38

Veiga LF, Vital NAA (2002) Testes de toxicidade aguda com o microcrustáceo *Artemia* sp. In: Nascimento IA, Sousa ECPM, Nipper M (eds) Métodos em Ecotoxicologia Marinha. Artes Gráficas, Salvador, pp 111–122 (in Portuguese)

Veil JA, Kimmell TA, Rechner A (2005) Characteristics of produced water discharged to the Gulf of Mexico Hypoxic Zone. Report prepared for U.S. Department of Energy., by Argonne National Laboratory. University of Chicago. National Energy Technology Laboratory, August 2005, 76 p. Website: http://www.osti.gov.bridge/ Accessed Jan 2007

Chapter 4
Biodegradation of Crude Oil as Potential Source of Organic Acids in Produced Water

Bent Barman Skaare, Jan Kihle, and Terje Torsvik

Abstract The concentrations of organic acids in produced water are highly variable. Because of the information shown in others chapters of this book about the toxic effects of organic acids, it is of major interest to understand the mechanisms controlling the occurrence of this compound group in produced water. This chapter focuses on in-reservoir biogeochemical processes which may produce organic acids as products or by-products. The biodegradation processes inside reservoirs decrease the hydrocarbon content of petroleum. Additionally, an increase in oil acidity as measured by total acid number (TAN) is frequently observed. Since the formation waters of reservoirs are in close contact with these processes, the production of smaller, more polar petroleum constituents will also have an effect on the composition of organic molecules in the produced water. This manuscript reviews the literature with respect to acids production of petroleum biodegradation and the effects of water re-injection on such processes.

1 The Composition of Crude Oil and Formation Water

Crude oil composition is normally characterized by four groups of compounds: saturated (aliphatic) hydrocarbons, aromatic hydrocarbons, resins and asphaltenes (SARA). The aliphatic hydrocarbons make up between 40 and 97% (w/w) of the crude oil and include *n*-alkanes, branched alkanes and cycloalkanes. Twenty to forty-five per cent (w/w) of the oil is made up of aromatic hydrocarbons. The rest of the oil, 0–40% (w/w), is resins or asphaltenes. These are high molecular weight constituents containing heteroatoms. The asphaltenes are defined as the part of the crude oil which is not dissolved in *n*-pentane or *n*-hexane (Speight 1999). These quantities are, however, based on group-type separation, and other results may be encountered by using, e.g., spectroscopic methods for quantitative analysis. As both resins and asphaltenes contain heteroatoms (e.g. oxygen), their composition may strongly influence the composition of organic constituents in formation water – and hence produced water.

B.B. Skaare (✉)
Department of Environmental Technology, Institute for Energy Technology, 2027 Kjeller, Norway

K. Lee, J. Neff (eds.), *Produced Water*, DOI 10.1007/978-1-4614-0046-2_4,

The composition of polar organic components in formation water will be controlled by the composition of the same group of components in the crude oil.

2 Microbial Processes in Pristine Petroleum Reservoirs

The remaining petroleum resources on earth are dominated by biodegraded crude oil (Roadifer 1987). The world's largest single-petroleum resources are located in the Venezuelan Orinoco belt and the Athabasca tar sands in Canada (Head et al. 2003). Both of these accumulations are saturated with heavy or super-heavy oil, most of which was produced by microorganisms (Head et al. 2003). A comparison between these accumulations (Orinoco 1,200 billion barrels and Athabasca almost 900 billion barrels) and the largest light oil fields in the Middle East (Ghawar – Saudi Arabia and Burghan – Kuwait, approximately 190 billion barrels each) shows the heavy oil accumulations to be huge as compared to the largest non-biodegraded oil accumulations in the world (Roadifer 1987; Head et al. 2003).

The presence of bacteria in oil reservoirs has been known since the 1920s (Bastin et al. 1926), but their role in the reservoir did not receive much attention until the 1960s. At that time biodegradation was considered to be an aerobic process in which bacteria and meteoric water played a crucial role in addition to oxygen (Palmer 1993). However, the organisms described by Bastin et al. (1926) were sulphate reducers, and biodegradation of crude oil from Saskatchewan was attributed to sulphate-reducing bacteria (SRBs) as early as the 1970s (Bailey et al. 1973a, b). It has also been suggested that SRBs are able to grow only on the residue from aerobic biodegradation of crude oil (Jobson et al. 1979). Over the last 20 years, however, a number of anaerobic hydrocarbon degrading bacteria have been isolated and identified (Vogel and Grbic-Galic 1986; Aeckersberg et al. 1991; Zengler et al. 1999). New calculations have shown that the water volumes needed for the necessary oxygen transport into biodegrading reservoirs are unrealistically high (Horstad et al. 1992). Hence, the current opinion is that these in-reservoir biodegradation processes are anaerobic with the most important processes being sulphate reduction (Holba et al. 1996) and methanogenesis (Holba et al. 2004; Jones et al. 2008). In oil fields where nitrate is added to the injection water (in order to inhibit SRB and reduce reservoir souring), there is also evidence that nitrate reducing organisms (NRBs) may take part in in-reservoir biodegradation (Myhr et al. 2002).

Empirical studies of biodegraded oils as well as the information available on successful and unsuccessful microbial incubations indicate that life in petroleum reservoirs reaches an upper temperature at 80°C (Wilhelms et al. 2001).

3 Effects of Crude Oil Biodegradation

The different components making up crude oil show various affinities towards degradation by microorganisms. Hence, the molecular effects of biodegradation on the composition are not easily generalized (Bennett and Larter 2008). However, the bulk effects of such processes are well known, and they may have huge impacts on the quality and value of oil (Wenger and Isaksen 2002). These include decrease in API gravity, gas to oil ratio, gas wetness and the concentration of saturated hydrocarbons. These effects are accompanied by an increase in the sulphur content as well as an increase in the acidity of the oil as measured by the total acid number (TAN; mg KOH/g oil) (Head et al. 2003). The sum of negative effects of in-reservoir biodegradation processes makes biodegradation of crude oil a priority research topic for many oil companies.

A lot of research has been carried out in order to identify the components responsible for the increase in TAN. Most of the acidic species in oil contain oxygen, and generally the abundance of oxygen in oil is rather low – normally less than 2% (w/w) (Speight 1999). Alkyl phenols are low molecular weight oxygen containing constituents occurring widely in crude oil (Ioppolo et al. 1992), and they have been shown to be up to seven times more abundant in high TAN oil from Azerbaijan than the carboxylic acids (Samadova and Guseinova 1993).

The structures of carboxylic acids may be derived from the hydrocarbons found in crude oils (Nascimento et al. 1999). Hence, most of the carboxylic acids with less than eight carbons are aliphatic (Speight 1999). There are also more complex organic acids present in crude oil. Polycyclic saturated acids and complex mixtures of alkyl-substituted acyclic and cycloaliphatic carboxylic acids are usually known as naphthenic acids (Tomzcyk et al. 2001). The monocyclic carboxylic acids found in crude oils occur from C_6 and they dominate from C_{14} (Speight 1999). Furthermore, aromatic acids found in crude oils have been identified as metabolites from hydrocarbon biodegradation (Wilkes et al. 2003; Aitken et al. 2004).

4 Biodegradation of Hydrocarbons Under Anaerobic Conditions

Hydrocarbons are highly reduced organic compounds from which microorganisms may yield large amounts of energy. This is especially true for the *n*-alkanes. But biochemical pathways always occur at functional groups in organic molecules. Therefore, oxidation of the hydrocarbons is necessary before metabolic processes can be undergone (Widdel et al. 2006). These oxidation reactions also make the organic molecules more polar and water soluble, and because the primary zone for biodegradation processes is at or near the oil–water contact (OWC) (Moldowan and Mccaffrey 1995; Horstad and Larter 1997; Larter et al. 2003), they will change the composition of organic constituents in the formation water. A recent review of hydrocarbon metabolic pathways is given in Widdel et al. (2006).

5 Organic Acids in Formation Water

Not many studies describe the detailed composition of organic acids in formation water. The first real attempt to attribute short-chain organic acids to geochemical processes was published in 1978 (Carothers and Kharaka 1978). Analyses of

Californian and Texan oil field waters showed the aliphatic acids to be more or less absent in water samples which were sampled at sites colder than 80°C, but larger concentrations were found in waters sampled at temperatures between 80 and 200°C. In this temperature range, there was a negative correlation between the concentration of aliphatic acids and the temperature. Hence, the conclusion was that the aliphatic acids were produced during petroleum generation and degraded by anaerobic bacteria in reservoirs with temperatures below 80°C. In a similar study based on acids from the Norwegian continental shelf, no such correlation between the short-chain organic acids and the reservoir temperature was found (Barth 1991). These results were not very different from the trends found in a study of the composition of other American oil field waters (Fisher 1987; Barth 1991).

Later, organic constituents in the produced water from four production sites on the Norwegian continental shelf were characterized and compared (Utvik 1999). Of special interest in this study are the results summed up in a principal component analysis (PCA) showing the short-chain organic acids to be most prominent in produced water from one of these fields, the Troll field. Of these fields, Troll is the only field in which crude oil is considerably biodegraded, indicating that the organic acids in this region may be produced during biodegradation rather than petroleum generation. This has previously been shown to be the case in some oil fields in the USA (Carothers and Kharaka 1978). This is supported by metabolic pathways showing acetic acid to be one of the by-products of *n*-alkane biodegradation as it goes through β-oxidation (Wilkes et al. 2002). The composition of aromatic acids in the Troll field has also been shown to be dependent on the ongoing in-reservoir biodegradation processes (Skaare et al. 2007). The study included analyses of oil and water from seven different raisers at the Troll C production unit offshore Norway and showed the composition of organic acids in the waters to vary within the Troll reservoir. Figure 4.1 presents examples of Total Ion Chromatograms (TIC)

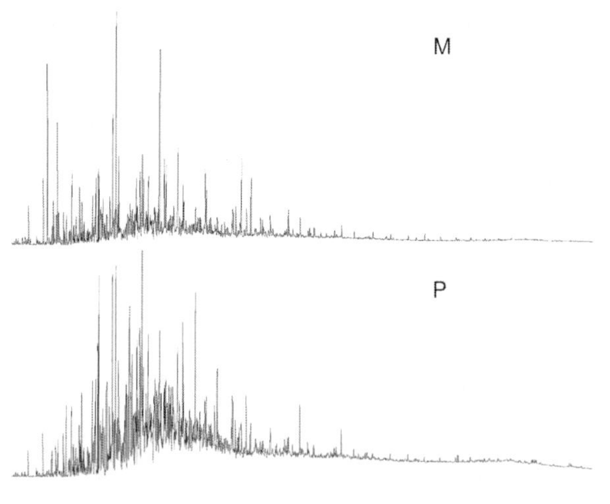

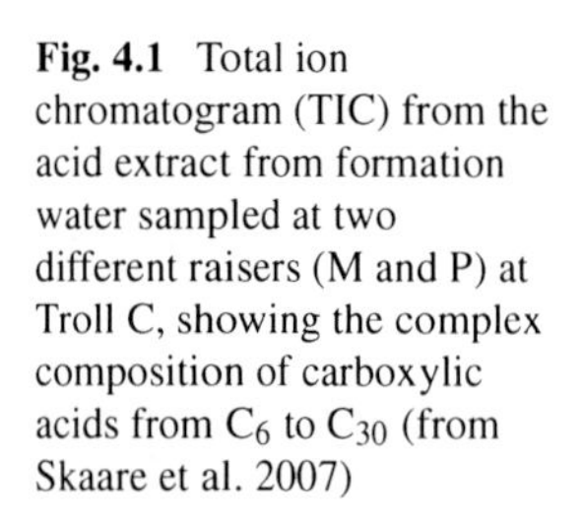

Fig. 4.1 Total ion chromatogram (TIC) from the acid extract from formation water sampled at two different raisers (M and P) at Troll C, showing the complex composition of carboxylic acids from C_6 to C_{30} (from Skaare et al. 2007)

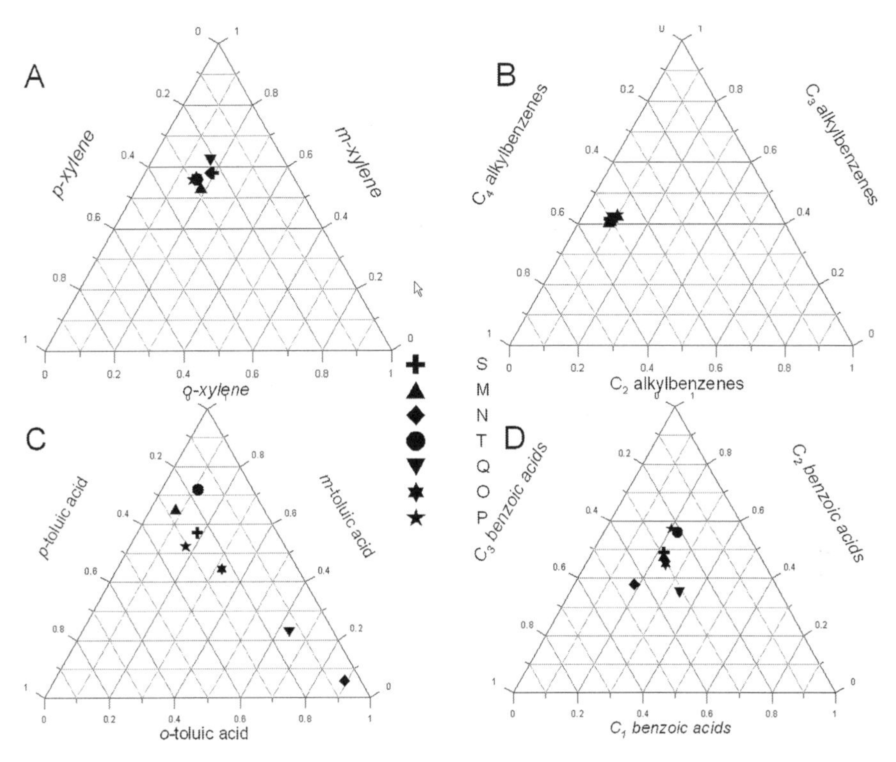

Fig. 4.2 The composition of xylenes **a** and alkyl benzenes **b** in the crude oils produced from seven raiser at the Troll C platform as compared to the patterns of their corresponding acids, the toluic acids **c** and the alkyl benzoic acids **d** in the co-produced water from the same raisers; letters in the legend identify the raiser from which each sample was taken (from Skaare et al. 2007)

showing the composition of the acid extract from two of the seven raisers, indicating the level of variation in the water composition. The composition of xylenes (Fig. 4.2a) and alkyl benzenes (Fig. 4.2b) in the co-produced oils from the seven raisers shows little variation between the samples, but the compositional variation in the acid extracts from the produced water phase can be observed within the group of monoaromatic acids (Fig. 4.2c and d). Thus, it is unlikely that the occurrence of monoaromatic acids in the water samples are due to abiotic oxidation of alkyl benzenes, but these acids are rather products of ongoing biodegradation of hydrocarbons in the reservoir. Additionally, biomarkers previously described as specific markers of anaerobic biodegradation of naphthalene, tetralin and/or 2-naphtoic acid (Annweiler et al. 2002; Aitken et al. 2004) and derivatives of these were found in the formation water/produced water of the Troll field (Skaare et al. 2007). It is likely that this, to a large extent, will be the case for other organic acids in the Troll formation water because of the fact that the metabolic reactions take place at functional groups (Widdel et al. 2006). This means that the bacteria need to invest energy in order to oxidize the hydrocarbons before they can take advantage of the energy yield. To obtain energy from organic acid degradation, no such investment is needed. Hence, it may be expected that organic acids which are produced during petroleum generation (Carothers and Kharaka 1978) are among the first components to be degraded by indigenous bacteria (Skaare et al. 2007). This is in agreement with observations from oil fields which show the composition of organic acids to vary throughout the whole biodegradation process (Peters et al. 2005). The composition of short-chain organic acids in a reservoir model column described by Myhr and Torsvik (2000) also shows large variations in the organic acids of produced water as a consequence of biodegradation processes. The model reservoirs, which are 2 m high sand columns, have sampling points distributed evenly throughout their height. Figure 4.3 shows concentration curves of short-chain organics acids at different heights of the column throughout one such experiment, the concentrations of acids in the water phase decreasing rapidly at the start of the column. Further up in the column, the patterns are more obscure indicating both production and depletion in the measured acid concentrations (Barman Skaare 2007). Altogether, these results clearly indicate that biodegradation of crude oil is one of the main controls of organic acids in co-produced water.

Altogether, these results indicate that biodegradation processes in petroleum reservoirs may both produce and degrade aliphatic and aromatic acids in formation water.

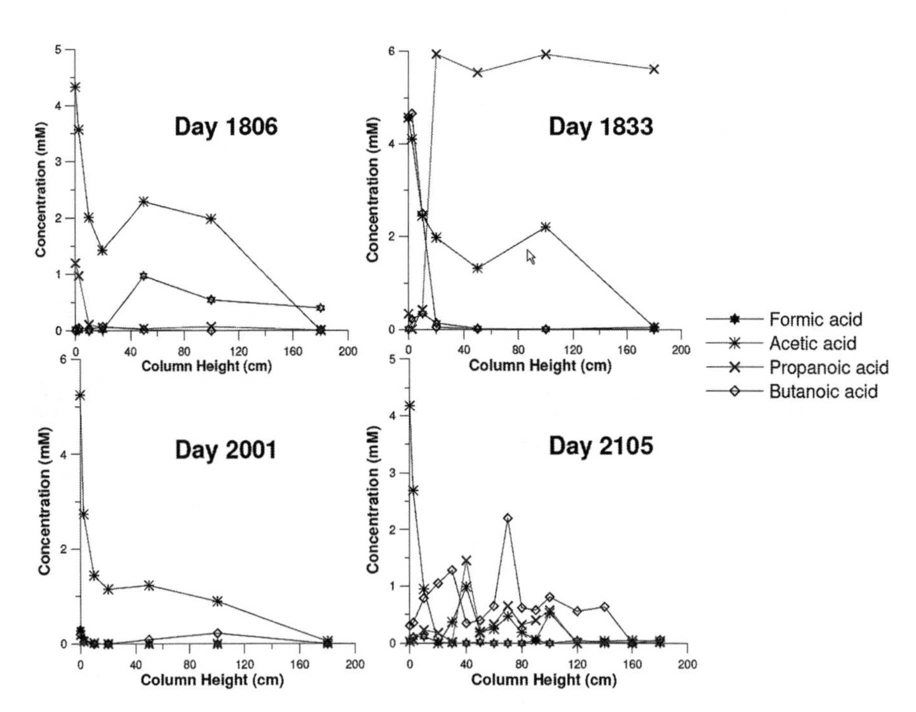

Fig. 4.3 The composition of short-chain organic acids in a model reservoir used to study anaerobic biodegradation during produced water re-injection (from Barman Skaare 2007)

6 Effects on Biodegradation and Produced Water Organic Acids of Seawater and Produced Water Injection

Injection of water or gas to maintain reservoir pressure and enhance oil recovery is part of the secondary recovery process in many reservoirs. In offshore oil fields, the injection water is most often seawater or produced water. The effects of such injection on in-reservoir biodegradation depend on the chemical and physical properties of the injected water which differs between oil fields, but there are also similarities in the effects.

The biodegradation rates in petroleum reservoirs are highest in low temperature petroleum reservoirs, and they decrease with increasing temperature of the reservoir until it reaches 80°C (Wilhelms et al. 2001; Head et al. 2003; Larter et al. 2003). This is considered the upper limit for microbial processes in petroleum reservoirs. This is supported by unsuccessful experiments of growing bacteria indigenous to petroleum reservoirs at 85°C indicating that these microorganisms may not be hyperthermophilic (Grassia et al. 1996; Orphan et al. 2000).

The environment inside the reservoir changes drastically when cold seawater is injected into a reservoir. One effect is reduced reservoir temperature, especially in the area close to the injection well bore, favouring growth of mesophilic microorganisms which are introduced in large quantities during seawater injection (Rozanova et al. 2001). Seawater injection also contributes to an increase in microbial activity as it continuously recharges electron acceptors (i.e. sulphate) and mineral nutrients in the water phase of the reservoir. Additionally, formation water containing toxic soluble oil components and heavy metal ions are diluted (Sunde and Torsvik 2005). All together these factors encourage growth of SRB resulting in increased reservoir souring.

When nitrate is used to inhibit SRB, acting as an alternative electron acceptor to sulphate, NRB are enriched resulting in increased microbial diversity and increased potential for biodegradation of crude oil (Myhr et al. 2002).

PWRI is often implemented at a late stage in the life of an oil field. Now the reservoirs, already harbouring a complex mixture of oil degrading microbes, are fed water saturated with soluble oil components and the degradation products (i.e. organic acids) produced during previous microbial activity in the reservoir. This chain of events may result in production of a complex mixture of organic acids in produced water.

Such effects were observed in a model experiment where the effects of water injection on the composition of short-chain organic acids in the waters from a model oil reservoir were studied (Barman Skaare 2007). The reservoir model system, described by Myhr and Torsvik (2000), was shown to simulate produced water in a realistic manner (Myhr et al. 2002). The study showed that the composition of short-chain carboxylic acids at different sampling points in the model reservoir may differ to large extent (Fig. 4.3). The composition will be a consequence of the organic constituents in the water phase as well as the ongoing processes which degrade the hydrocarbons of the petroleum.

7 How to Get Information About Biodegradation of Oil Fields

Exploration scientists working for oil companies are continuously analysing and re-evaluating their models of the petroleum reservoirs. In this process, detailed analyses of the oil and, in some cases, the formation water are essential ingredients. Hence, these groups will be able to provide information about in-reservoir biodegradation processes at specific oil fields.

Numerous analytical methods seem to provide information about the biodegradation status in a given oil reservoir. In the early part of the exploration, non-destructive fluorescence spectroscopy studies are frequently used to provide information about the petroleum contents of fluid inclusions, cores and cuttings, i.e. Hydrocarbon Core Scanner (HCS^{TM}) and CUSC (Kihle 1996; Munz 2001). These instruments are founded on steady-state fluorescence emission technology measuring emitted fluorescence from sorbed hydrocarbons in cores and cuttings. The HCS^{TM} uses a Peltier-cooled ultra high sensitivity UV-ICCD detector coupled to a 280 mm focal length spectrometer. Excitation source is the isolated 365 nm plasma line from a high pressure Hg plasma source. The CUSC is a miniature instrument using a combined micro-detector and spectrometer and a 373 nm UV laser as the excitation source. Two constraints are routinely measured; I_{max} – detected photons emitted per time unit (HC quantity parameter) and Q_{580} – a fluorescence emission integral quotient (HC quality parameter).

For calibration, reference Berea sandstone wafers were saturated with a collection of 112 petroleums of known compositions and API densities. After 2 weeks of exposure to air, these wafers were measured by both HCS and CUSC. The resultant plots are illustrated in Fig. 4.4.

These methods provide crude information about the amount of petroleum in the cores or cuttings as well as the API gravity, and hence the biodegradation level, of the hydrocarbons in the samples (biodegraded oils have lower API gravity) (Fig. 4.4). Infrared (IR) spectroscopy coupled with multivariate analyses have been shown to be a useful non-destructive method for studying biodegradation of crude oil (Genov et al. 2008).

However, the most commonly used method for analyses of oil composition is gas chromatography (GC). Whole oil GC is a routine analysis in geochemical laboratories and provides good information on the composition of the petroleum. Figure 4.5 shows the distinct differences between non-biodegraded oil (the Norwegian standard oil, NSO-1) and a biodegraded oil from the Troll field offshore Norway (Skaare et al. 2007). The differences may become even clearer than this, with the baseline of the chromatogram increasing into a hump (Peters et al. 2005).

8 Summary

The composition of organic constituents in produced water will, to a large degree, be controlled by the composition of polar organic compounds in the associated petroleum. Hence, in-reservoir processes which alter this compound group in

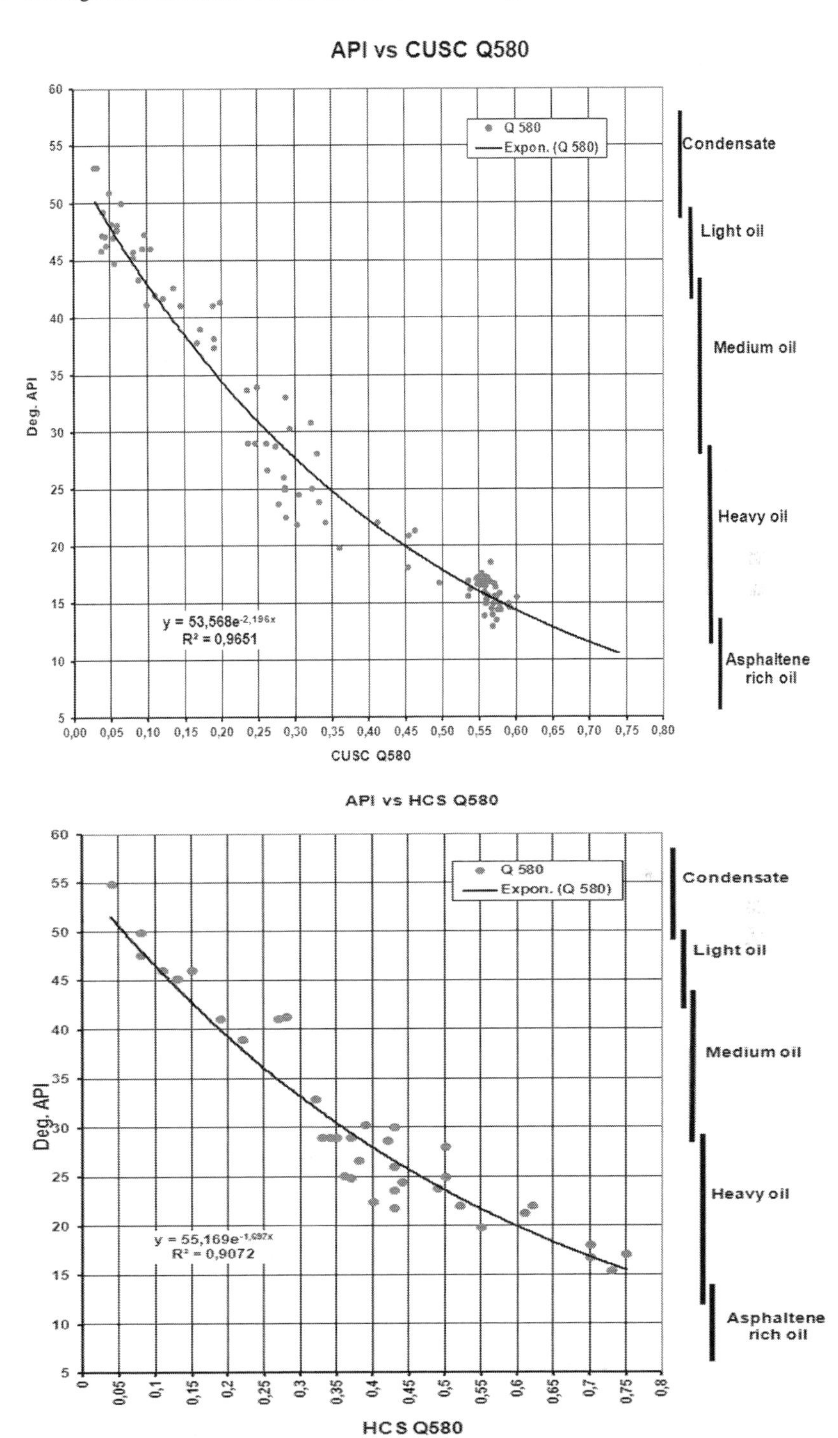

Fig. 4.4 The relationship between fluorescence analyses of cuttings (CUSC and HCS) and the API gravity of 112 reference sorbed petroleums

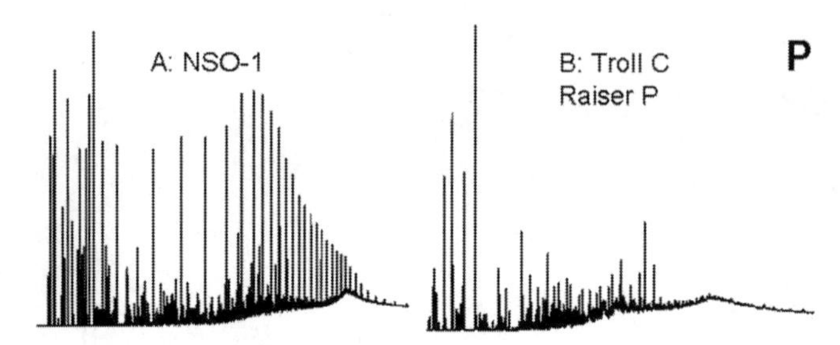

Fig. 4.5 Whole oil chromatogram of (**a**) the Norwegian standard oil NSO-1 which is non-biodegraded and (**b**) a biodegraded oil from the Troll field offshore Norway (taken from Skaare et al. 2007)

petroleum will also affect the composition of organic molecules in the waters. Previously, both petroleum generation and crude oil biodegradation have been shown to produce short-chain organic acids. Biodegradation is of special interest in this context because it occurs at a later time than petroleum generation and because the nature of the microbial processes makes it probable that ongoing processes will both degrade and generate organic acids. Short-chain organic acids have been shown to be both degraded and produced simultaneously in different parts of a model oil reservoir simulating produced water re-injection. Additionally, aromatic acids which could be coupled to in-reservoir biodegradation processes have been identified in the produced water from the Troll oil field offshore Norway. The fact that most of the world's remaining petroleum resources have been altered by microorganisms makes the process important on a global scale. Therefore, there is reason to believe that knowledge of petroleum reservoir biodegradation may give valuable information as to the contents of organic acids in the produced water from single installations worldwide.

It is, however, difficult to give a clear presentation with respect to which organic acids may be encountered. There are thousands of organic compounds in the crude oil matrix, and each one of these has at least one specific biodegradation pathway. Most of these degradation pathways are currently unknown. This explains much of the complexity which can be seen with respect to organic acids in formation water and produced water.

Acknowledgements The authors acknowledge the advices given by Tanja Barth during the work which is presented as a part of this review as well as the help in the microbial laboratory at University of Bergen by Bente-Lise Pollen Lillebø. Thanks to two reviewers who helped improve the manuscript greatly.

References

Aeckersberg F, Bak F et al (1991) Anaerobic oxidation of saturated-hydrocarbons to co2 by a new type of sulfate-reducing bacterium. Arch Microbiol 156:5–14

Aitken CM, Jones DM et al (2004) Anaerobic hydrocarbon biodegradation in deep subsurface oil reservoirs. Nature 431:291–294

Annweiler E, Michaelis W et al (2002) Identical ring cleavage products during anaerobic degradation of naphthalene, 2-methylnaphthalene, and tetralin indicate a new metabolic pathway. Appl Environ Microbiol 68:852–858

Bailey NJL, Jobson AM et al (1973a) Bacterial degradation of crude-oil – comparison of field and experimental-data. Chem Geol 11:203–221

Bailey NJL, Krouse HR et al (1973b) Alteration of crude-oil by waters and bacteria – evidence from geochemical and isotope studies. Aapg Bull 57:1276–1290

Barman Skaare B (2007) Effects of initial anaerobic biodegradation on crude oil and formation water composition. University of Bergen, Bergen. PhD: 193 pp

Barth T (1991) Organic-acids and inorganic-ions in waters from petroleum reservoirs, Norwegian continental-shelf – a multivariate statistical-analysis and comparison with American Reservoir Formation Waters. Appl Geochem 6:1–15

Bastin E, Greer FE et al (1926) The presence of sulphate reducing bacteria in oil field waters. Science 63:21–24

Bennett B, Larter SR (2008) Biodegradation scales: applications and limitations. Org Geochem. doi: 10.1016/j.orggeochem.2008.02.023

Carothers WW, Kharaka YK (1978) Aliphatic acid anions in oil-field waters – implications for origin of natural-gas. Aapg Bull 62:2441–2453

Fisher JB (1987) Distribution and occurrence of aliphatic acid anions in deep subsurface waters. Geochim Cosmochim Acta 51:2459–2468

Genov G, Nodland E et al (2008) Comparison of biodegradation level and gas hydrate plugging potential of crude oils using FT-IR spectroscopy and multi-component analysis. Org Geochem. doi: 10.1016/j.orggeochem.2008.04.006

Grassia GS, McLean KM et al (1996) A systematic survey for thermophilic fermentative bacteria and archaea in high temperature petroleum reservoirs. FEMS Microbiol Rev 21:47–58

Head IM, Jones DM et al (2003) Biological activity in the deep subsurface and the origin of heavy oil. Nature 426:344–352

Holba AG, Dzou LIP et al (1996) Reservoir geochemistry of South Pass 61 Field, Gulf of Mexico: Compositional heterogeneities reflecting filling history and biodegradation. Org Geochem 24:1179–1198

Holba AG, Wright L et al (2004) Effects and impact of early-stage anaerobic biodegradation on Kuparuk River Field, Alaska. In: Cubitt JM, England WA, Larter S (eds) Understanding petroleum reservoirs: towards an integrated reservoir engineering and geochemical approach, 237 Geological Society, London, pp 53–88

Horstad I, Larter SR (1997) Petroleum migration, alteration, and remigration within Troll field, Norwegian North Sea. Aapg Bull 81:222–248

Horstad I, Larter SR et al (1992) A quantitative model of biological petroleum degradation within the Brent group reservoir in the Gullfaks field, Norwegian North-Sea. Org Geochem 19: 107–117

Ioppolo M, Alexander R et al (1992) Identification and analysis of C0-C3 phenols in some Australian crude oils. Org Geochem 18:603–609

Jobson AM, Cook FD et al (1979) Interaction of aerobic and anaerobic-bacteria in petroleum biodegradation. Chem Geol 24:355–365

Jones DM, Head IM et al (2008) Crude-oil biodegradation via methanogenesis in subsurface petroleum reservoirs. Nature 451:176–U176

Kihle J (1996) Adaptation of fluorescence excitation-emission micro-spectroscopy for characterization of single hydrocarbon fluid inclusions. Org Geochem 23:1029–1042

Larter S, Wilhelms A et al (2003) The controls on the composition of biodegraded oils in the deep subsurface – part 1: biodegradation rates in petroleum reservoirs. Org Geochem 34: 601–613

Moldowan JM, Mccaffrey MA (1995) A novel microbial hydrocarbon degradation pathway revealed by hopane demethylation in a petroleum reservoir. Geochim Cosmochim Acta 59:1891–1894

Munz IA (2001) Petroleum inclusions in sedimentary basins: systematics, analytical methods and applications. Lithos 55:195–212

Myhr S, Lillebo BLP et al (2002) Inhibition of microbial H2S production in an oil reservoir model column by nitrate injection. Appl Microbiol Biotechnol 58:400–408

Myhr S, Torsvik T (2000) Denitrovibrio acetiphilus, a novel genus and species of dissimilatory nitrate-reducing bacterium isolated from an oil reservoir model column. Int J Syst Evol Microbiol 50:1611–1619

Nascimento LR, Reboucas LMC et al (1999) Acidic biomarkers from Albacora oils, Campos Basin, Brazil. Org Geochem 30:1175–1191

Orphan VJ, Taylor LT et al (2000) Culture-dependent and culture-independent characterization of microbial assemblages associated with high-temperature petroleum reservoirs. Appl Environ Microbiol 66:700–711

Palmer SE (1993) Effects of biodegradation and waterwashing on crude oil composition. In: Engel MH, Macko SA (eds) Organic Geochemistry, Principles and Applications. Plenum, New York, NY, pp 511–533

Peters KE, Walters CC et al (2005) The BIOMARKER GUIDE. Volume 2. Biomarkers and isotopes in petroleum exploration and earth history. Cambridge University Press, Cambridge

Roadifer RE (1987) Size distribution of the world's largest known oil and tar accumulations. In: Meyer RF (ed) Exploration for heavy crude oil and natural Bitumen. American Association of Petroleum Geologists, Tulsa, pp 3–23.

Rozanova EP, Borzenkov IA et al (2001) Microbiological processes in a high-temperature oil field. Microbiology 70:102–110

Samadova FI, Guseinova BA (1993) Distribution of acids and phenols in Azerbaidzhan crudes. Chem Technol Fuels Oils 28:515–517

Skaare BB, Wilkes H et al (2007) Alteration of crude oils from the Troll area by biodegradation: analysis of oil and water samples. Org Geochem 38:1865–1883

Speight JG (1999) The chemistry and technology of petroleum. Marcel Dekker, Inc., New York, NY

Sunde E, Torsvik T (2005) Microbial control of hydrocarbon sulfide production in oil reservoirs. In: Ollivier B, Magot M (eds) Petroleum microbiology. ASM Press, Washington, DC, pp 201–213

Tomzcyk NA, Winans RE et al (2001) On the nature and origin of acidic species in petroleum. 1. Detailed acid type distribution in a California Crude oil. Energy Fuels 15:1498–1504

Utvik TIR (1999) Chemical characterisation of produced water from four offshore oil production platforms in the North Sea. Chemosphere 39:2593–2606

Vogel TM, Grbic-Galic D (1986) Incorporation of oxygen from water into toluene and benzene during anaerobic fermentative transformation. Appl Environ Microbiol 52:200–202

Wenger LM, Isaksen GH (2002) Control of hydrocarbon seepage intensity on level of biodegradation in sea bottom sediments. Org Geochem 33:1277–1292

Widdel F, Boetius A et al (2006) Anaerobic biodegradation of hydrocarbons including methane. In: Dworkin M, Falkow S, Rosenberg E, Schleifer K-H (eds) The prokaryotes. Volume 2: ecophysiology and biochemistry, 2. Springer, New York, NY, pp 1028–1049

Wilhelms A, Larter SR et al (2001) Biodegradation of oil in uplifted basins prevented by deep-burial sterilization. Nature 411:1034–1037

Wilkes H, Kuhner S et al (2003) Formation of n-alkane- and cycloalkane-derived organic acids during anaerobic growth of a denitrifying bacterium with crude oil. Org Geochem 34: 1313–1323

Wilkes H, Rabus R et al (2002) Anaerobic degradation of n-hexane in a denitrifying bacterium: further degradation of the initial intermediate (1-methylpentyl)succinate via C-skeleton rearrangement. Arch Microbiol 177:235–243

Zengler K, Richnow HH et al (1999) Methane formation from long-chain alkanes by anaerobic microorganisms. Nature 401:266–269

Chapter 5
Chemical Forms and Reactions of Barium in Mixtures of Produced Water with Seawater

John H. Trefry and Robert P. Trocine

Abstract High concentrations of Ba in produced water have been linked to adverse biological impacts during toxicity testing. This observation, along with interest in modelling chemical processes during produced water discharges to the ocean, led to the present study. Essentially all the Ba in produced water collected for this study was dissolved and passed through 0.4-μm pore size filters and 1-KDa ultrafilters. Using 1:99 and 1:199 mixtures of produced water with seawater, initial amounts of dissolved Ba of 180–600 μg/L did not change over 72 h and were oversaturated with respect to $BaSO_4$ by factors of ~5–16. For 1:9 mixtures of produced water with seawater, precipitation was observed to either start immediately or be delayed for as long as 12 h; however, the dissolved Ba that remained after 24–72 h in these mixtures ranged from 340–1,200 μg/L, or ~10 to >30 times above apparent saturation. Field sampling of a discharge in the Gulf of Mexico showed that >98% of the Ba added to seawater via produced water was still dissolved 5 min after discharge at a distance of 20 m from the source and at >1,600-fold dilution. Overall, the results show the importance of considering both dilution and precipitation when interpreting the behavior or impact of Ba based on laboratory and field studies.

1 Introduction

Concentrations of Ba in produced water typically range from 1,000 to >500,000 μg/L and thus are about 100 to >50,000 times greater than seawater values of 10–20 μg/L (Trefry et al. 1996; Neff 2002). When discharged into the ocean, produced water is generally diluted by factors of a few hundred to >5,000 within 100 m of the discharge site (Brandsma and Smith 1996). Despite such rapid dilution, concern has been raised about adverse effects to sea urchin eggs (Higashi et al. 1992) and mussel embryos (Spangenberg and Cherr 1996) from exposure to Ba at concentrations of 200–900 μg/L over a period of 48 h. These Ba values that are reported to induce adverse effects are well above the solubility of barite ($BaSO_4$) in seawater at ~35 μg Ba/L (Church and Wolgemuth 1972; Monnin et al. 1999), but well within

J.H. Trefry (✉)
Department of Marine & Environmental Systems, Florida Institute of Technology, Melbourne, FL 32901, USA

K. Lee, J. Neff (eds.), *Produced Water*, DOI 10.1007/978-1-4614-0046-2_5,

the range of possible dilutions for some produced water at <100 m from the point of discharge. Toxicity testing is carried out at a variety of dilutions of produced water with seawater where concentrations of Ba at the start of the experiment can be much greater than 200 μg/L.

To better understand proposed scenarios whereby organisms inhabiting areas around produced water discharges could be exposed to high concentrations of dissolved Ba, more information is needed regarding the concentrations and chemical forms of Ba in produced water and in mixtures of produced water with seawater. Previous studies of the kinetics of barite precipitation have focused on scale formation in the down-hole reservoir or during oil recovery. For example, Granbakken et al. (1991) showed that >95% of the Ba in a 9:1 mixture of produced water with seawater (i.e., 90% produced water) precipitated as barite within 10–15 min at 91.4°C and a pressure of 312 atm. However, they also showed that the precipitation process extended from 2 to >4 h at 25°C and 1 atm of pressure. Fernandez-Diaz et al. (1990) demonstrated that nucleation of barite crystals, and thus the onset of precipitation (i.e., time from when oversaturation was created until a critical nucleus was formed), occurred over periods of about 1 min at 283-fold oversaturation relative to >5 min as the degree of oversaturation approached 100. However, Fernandez-Diaz et al. (1990) also showed that the induction of nucleation was significantly impeded (as much as tenfold) when 15 mg/L of polyacrylic acid inhibitor were added, except at the greatest degree of oversaturation (283-fold).

As part of their toxicity studies, Spangenberg and Cherr (1996) emphasized the importance of the acetate anion as a complexing agent by noting that concentrations of dissolved Ba could reach 200–900 μg Ba/L and cause adverse biological effects only when Ba was added as barium acetate. At dissolved Ba values >900 μg/L (about 25 times greater than apparent saturation), Spangenberg and Cherr (1996) observed precipitation of barite and no incidences of toxicity.

Knowledge about the concentrations and rates of precipitation of barite when produced water is mixed with seawater can be used to help explain and predict concentrations of dissolved Ba in toxicity experiments and also to help augment models for the dispersion and dilution of constituents in produced water so that they can include a reaction term. The specific objectives of this study were as follows: (1) to determine the concentrations and degree of saturation of Ba over time in various laboratory mixtures of produced water with seawater, (2) to determine how long it takes for barium sulfate to precipitate when produced water is mixed with seawater, and (3) to determine the behavior of Ba when produced water is discharged from an offshore platform to the ocean.

2 Methods

2.1 Sample Collection

Eight different samples of produced water were collected from six different locations in the Gulf of Mexico for the mixing experiments (Table 5.1). Two of the

Table 5.1 Background information for samples of produced water and mixtures of produced water (PW) with seawater (SW)

Source of produced water	Sample provider	Date sampled	24-h screening	Mixing times for PW:SW mixtures (min)
1. Grand Isle	Exxon-Mobil	5/19/97	Yes	12, 30, 60, 120, 480, 1440
2. Fourcheon	Chevron	1/23/00	Yes	–
3a. West Delta 73(#1)	Exxon-Mobil	1/17/00	Yes	–
3b. West Delta 73(#2)	Exxon-Mobil	9/26/00	No	0.5, 1, 2, 3, 4, 5, 6, 7, 8, 9, 10, 60, 120, 720, 1440, 4320
4a. Eugene Island 314A(#1)	Exxon-Mobil	1/20/00	Yes	–
4b. Eugene Island 314A(#2)	Exxon-Mobil	6/29/00	No	0.5, 1, 2, 3, 4, 5, 6, 7, 8, 9, 10, 60, 120, 720, 1440, 4320
5. Bay Marchland	Chevron	1/23/00	Yes	–
6. West Delta 109A	Texaco	2/08/00	Yes	–

samples (Grand Isle and Fourchon) were collected from onshore treatment plants and the remaining samples were collected from holding tanks on offshore platforms. The sample from Grand Isle had been treated with chemicals to prevent scaling and corrosion during deep-well injection. Onshore treatment can be more rigorous than that carried out for produced water discharged at sea. Immediately following collection, the produced water sample from Grand Isle was returned to the laboratories of the Louisiana Universities Marine Consortium in Chauvin, LA, for filtration and to begin the mixing experiments. Samples from the other sites were collected by personnel working at each site in bottles prepared by Florida Institute of Technology. The samples were filled to the top of the container to minimize oxidation and shipped via overnight delivery.

Seawater for the mixing experiments was collected on the day that a particular test was initiated. About 75 L of seawater were collected from the Gulf of Mexico at a water depth of 10 m from a location 10 km due south of Grand Isle, Louisiana, for use in the mixing experiments with the produced water from Grand Isle. The seawater was collected by pumping directly into four carboys through acid-washed tubing using a peristaltic pump. Samples of seawater from the Atlantic Ocean at Indialantic, Florida, were collected directly in 19-L carboys. All carboys of seawater were stored at 4°C in the dark until used to minimize biological growth.

2.2 Mixing Experiments

An initial screening of Ba precipitation in the samples was carried out by mixing produced water with seawater in a proportion of 1:9 (produced water:seawater), shaking on a wrist-action shaker for 24 h and analyzing the mixtures for dissolved and particulate Ba. Subsequently, the sample of produced water from Grand Isle and a second set of samples from West Delta 73(#2) and Eugene Island 314A(#2) were used for more detailed study at mixing times that ranged from 0.5–4,320 min (72 h) for both 1:9 and 1:99 (or 1:199 for the Grand Isle sample) mixtures of produced water with seawater (Table 5.1). Each reaction was terminated by filtering the sample. One portion of each mixture was filtered through a 0.4-μm (pore size) polycarbonate membrane filter. The filter was rinsed with two, 20-mL rinses of distilled–deionized water and saved in a plastic petri dish for later analysis. This filter contained the particulate fraction (>0.4 μm). The filtrate from the <0.4-μm pore size filter was acidified to a pH <2 with concentrated Ultrex II nitric acid and saved for analysis and operationally defined as the dissolved (<0.4 μm) fraction.

Some mixtures of the produced water sample from Grand Isle also were processed using Amicon ultrafilters with nominal molecular cutoffs of 10,000 (10K) and 1,000 (1K) Daltons (Da). To meet the time requirements of the experimental design, separate samples were prepared for each of the three possible filtrations (0.4 μm, 10K and 1K). Thus, no sample was processed in a sequential mode (i.e., no samples were passed through more than one filter). Ultrafiltration was carried out under N_2 pressure (~3.4 atm) in a sealed plastic vessel. The solutions that passed through the 10 KDa and 1 KDa ultrafilters were defined as dissolved (<10K) and dissolved (<1K), respectively.

2.3 Sample Analysis

The original unfiltered samples, as well as solutions processed by filtration (0.4 μm) and ultrafiltration, were analyzed for salinity using a Reichert-Jung Model 1040 Optical Refractometer, pH using an Orion probe, and DOC using a Shimadzu TOC-500 total organic carbon analyzer. The manufacturer's specifications and methods were followed in each case. Concentrations of sulfate were determined according to standard methods (Clesceri et al. 1989). Data for each of these supporting parameters were used to help calculate the solubility of Ba and explain observed changes in metal concentrations.

The original produced water and the various filtered fractions were analyzed for Ba by the method of standard additions using a Perkin-Elmer ELAN-5000 inductively coupled plasma-mass spectrometry (ICP-MS). Concentrations of Fe in the dissolved fractions for the experiments using the Grand Isle sample were determined directly following dilution with reagent water by graphite furnace atomic absorption spectrometry using a Perkin-Elmer model 5100 instrument with Zeeman background correction. In addition to the samples, the following standard reference materials were analyzed: Riverine Certified Reference Material SLRS-1 from the National Research Council of Canada and Trace Metals in Water Standard Reference Material (SRM) #1643d from the U.S. National Institute of Standards and Technology (NIST).

Suspended particles were collected on acid-washed polycarbonate filters (47 mm diameter, 0.4-μm pore size). Prior to use, the filters were acid washed in 5N, trace metal grade HNO_3 and rinsed three times in 18-megohm resistivity reagent water. Filtration of the samples and all handling of the filters were carried out in a laminar-flow hood using acid-washed glassware and Teflon forceps. After filtration, the filters were rinsed with pH 8 reagent water, returned to their labeled petri dishes and stored in plastic bags.

Digestion of the particulate samples was carried out in stoppered Teflon test tubes following the methods of Trefry and Trocine (1991) using ultra-high purity HNO_3, HF, and HCl. The digested samples and reagent water rinses of the Teflon digestion tubes were then transferred to acid-washed, 15-mL, polyethylene bottles for analysis. Each set of digested particulate samples also included reagent blanks, filter blanks, and milligram quantities of the SRM Buffalo River Sediment (#2704) obtained from the NIST. Particulate Ba concentrations were determined by ICP-MS using a Perkin-Elmer ELAN-5000 instrument.

3 Results and Discussion

3.1 Characterization of Undiluted Samples of Produced Water

Salinities for produced water collected during this study ranged from 76 to 124‰ (Table 5.2). Salinities of 50–180‰ are typical of produced water from the Gulf of Mexico and other locations (Trefry et al. 1996; Neff 2002). The produced water used in the mixing experiments was about 2–4 times saltier than the coastal seawater (at 32.5–35‰) with which it was mixed (Table 5.2).

Table 5.2 Data for produced water and seawater used during this study

Sample identification	Salinity (‰)	Ba (μg/L)	Sulfate (g/L)	DOC (mg/L)	Comments
Produced water					
1. Grand Isle	124	33,000	< 0.03	371	Fe[a], amber color
2. Fourchon	113	24,000	< 0.03	259	Fe, dark yellow
3a. West Delta 73(#1)	81	52,000	< 0.03	52	No color or Fe
3b. West Delta 73(#2)	76	42,400	< 0.03	64	Fe, No color
4a. Eugene Island 314A(#1)	88	67,000	< 0.03	151	Fe, amber
4b. Eugene Island 314A(#2)	99	59,400	< 0.03	348	Fe, amber
5. Bay Marchland	121	88,000	< 0.03	12	Fe, amber
6. West Delta 109A	101	133,000	< 0.03	432	Fe, amber
Seawater					
1. Gulf of Mexico	32.5	18.7	2.5	4	No color
2. Atlantic Ocean	35	8.4	2.7	2.4	No color
3. Atlantic Ocean	35	7.7	2.7	2.9	No color

[a]Fe denotes samples that contained iron oxide precipitates

The unfiltered samples of produced water contained concentrations of Ba that ranged from 24,000 to 133,000 μg/L relative to 7.7–18.7 μg/L in the seawater samples (Table 5.2). Thus, the enrichment factor for the produced water relative to the ambient seawater for Ba ranged from about 3,000 to 17,000. Barium concentrations in the produced water collected during this study were generally lower than values of 81,000–342,000 μg/L reported in previous studies in the Gulf of Mexico (e.g., Trefry et al. 1996). Dissolved sulfate in the produced water was non-detectable at <0.03 g/L, relative to concentrations of 2.5–2.7 g/L in ambient seawater (Table 5.2). Such low sulfate concentrations are common to anoxic, highly reducing produced water (Barth 1991) and play a major role in facilitating high concentrations of dissolved Ba. No simple relationship was observed for Ba concentrations versus salinity in the undiluted samples of produced water from this study.

More than 97% of the total Ba in the original, undiluted sample of produced water from Grand Isle passed through a 0.4 μm filter (Table 5.3). Furthermore, concentrations of Ba in the solutions that passed through 10K and 1K ultrafilters were not significantly different (t-test, ($\alpha = 0.05$, $p < 0.01$)) from values obtained for Ba in solutions that passed through the 0.4-μm filter. Thus, essentially all the Ba in the original produced water was present in what sometimes is referred to as the "truly dissolved" fraction (dissolved at <1 KDa).

In seawater from the Gulf of Mexico, concentrations of Ba averaged 18.7 ± 0.7 μg/L for the 0.4-μm-filtered portion and 19.3 ± 0.3 and 18.3 ± 0.3 μg/L for the <10K and <1K samples, respectively (Table 5.2). These results show that >95% of the Ba in the seawater from the Gulf of Mexico also was dissolved (<1 KDa).

Produced water used in this study had concentrations of DOC that ranged from 12 to 432 mg/L (Table 5.2). No simple relationship was observed between concentrations of DOC and Ba in samples of produced water from this study. In a manner similar to that observed for Ba, >95% of the DOC in the produced water sample from Grand Isle that passed through a 1-μm glass fiber filter also passed though 10K and 1K ultrafilters (Table 5.3). Thus, ~95% of the DOC in the undiluted produced water from Grand Isle was made up of molecules with a molecular weight <1,000 Da. The molecular weights of some organic components in the produced

Table 5.3 Concentrations of barium (Ba) dissolved organic carbon (DOC) and iron (Fe) in various fractions of the produced water from Grand Isle and the seawater sample from the Gulf of Mexico

	Total (unfiltered)	<0.4 μm	<10K	<1K
Produced water	(Grand Isle)			
Ba (μg/L)	33,000 ± 600	32,000 ± 1,100	32,900 ± 500	31,500 ± 800
Organic C (mg/L)	–	371 ± 12[a]	361 ± 1	353 ± 12
Fe (μg/L)	31,800 ± 1,300	10,000 ± 7,300	8,800 ± 6,200	4,600 ± 3,700
Seawater	(Gulf of Mexico)			
Ba (μg/L)	–	18.7 ± 0.7	19.3 ± 0.3	18.3 ± 0.3
Organic C (mg/L)	–	4[a]	4	1

[a]Organic C filtered through glass fiber filter with a pore size of 1 μm

water may have been even lower, based on the contribution that simple organic acids (e.g., acetic acid with MW = 60 Da) often make to the DOC of produced water (Barth 1991).

The concentration of DOC in seawater from the coastal Gulf of Mexico was 4 mg/L for both the 0.4-μm filtered and 10K ultrafiltered samples, but <1 mg/L for the 1K ultrafiltered sample (Table 5.3). Therefore, the dissolved organic matter in the seawater sample from the Gulf of Mexico was composed of predominantly higher molecular weight substances (1 to >10K Da) that were distinctly different in molecular size from those in the produced water. Concentrations of DOC in the seawater sample from the Atlantic Ocean were 2.4–2.9 mg/L (Table 5.2).

Rapid oxidative precipitation of Fe in slightly aerated samples of undiluted produced water was observed in most cases (Table 5.2). Iron concentrations in the produced water sample from Grand Isle were investigated to determine the possible influence of the formation of Fe oxides on precipitation of Ba. The results for Fe in the undiluted produced water from Grand Isle showed that only ~30% of the total Fe passed through a 0.4-μm filter (Table 5.3), yet all the Ba was dissolved, suggesting that precipitation of iron oxides did not impact concentrations of Ba.

3.2 *Preliminary Mixing Experiments*

An initial screening of the produced water was carried out for 24 h by mixing produced water with seawater in a proportion of 1:9. After 24 h, 49–97% of the dissolved Ba had precipitated in the six samples investigated (Fig. 5.1). In produced water from the four offshore samples (West Delta 73, Eugene Island 314A, Bay Marchland and West Delta 109A), 93 ± 3% of the dissolved Ba precipitated in 24 h. In contrast, 74% and 49% of the dissolved Ba precipitated over 24 h for samples from the two onshore facilities, Grand Isle and Fourchon, respectively. The difference in behavior for the produced water from the onshore, deep-well injection facilities, relative to samples from the offshore platforms, is most likely

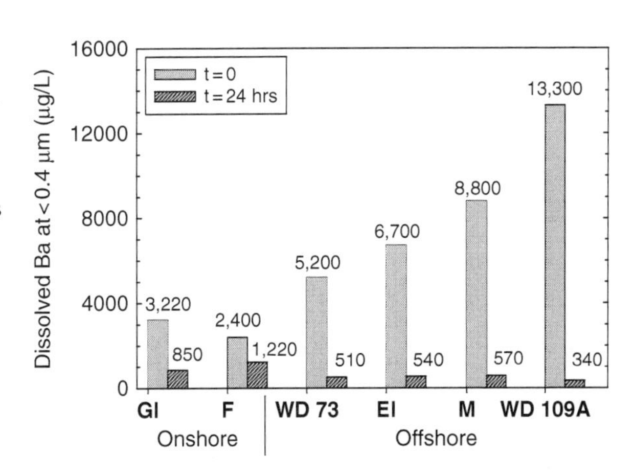

Fig. 5.1 Concentrations of dissolved barium (Ba) at <0.4 μm in mixtures of produced water (PW) with seawater (SW) (1 PW:9 SW) at time (t) = 0 and at t = 24 h; values for Ba at t = 0 were calculated based on the dilution of the original PW with SW (GI = Grand Isle, F = Fourchon, WD 73 = West Delta 73, EI = Eugene Island 314A, M = Bay Marchland, WD 109A = West Delta 109A)

due to the presence of greater amounts of additives that inhibit barite precipitation at the onshore sites. However, such distinction cannot be inferred from the DOC data (Table 5.2) because only small amounts (10–20 mg/L) of chemical are needed to inhibit Ba precipitation (Fernandez-Diaz et al. 1990).

Initial concentrations of Ba in the six different mixtures of 1:9 (produced water: seawater) ranged from 2,400–13,300 μg/L (Fig. 5.1). Thus, the amounts of Ba that precipitated in each case showed an equally wide range of 1,180–12,960 μg/L. In the four offshore samples, with widely varying initial concentrations of Ba (Fig. 5.1), concentrations of dissolved Ba after 24 h averaged a relatively uniform 490 ± 100 μg/L. In the two samples from onshore treatment facilities, the residual amounts of Ba at 24 h were 850 and 1,220 μg/L for Grand Isle and Fourchon, respectively. As will be detailed below, all concentrations of Ba after 24 h were >8 times oversaturated with respect to $BaSO_4$. Results from the 24-h mixing experiments, along with the convenience of obtaining samples, led to the decision to use fresh samples from Grand Isle, WD 73(#2) and EI 314A(#2) in the more comprehensive mixing experiments.

3.3 Mixing Experiments with 1:199 and 1:99 Proportions of Produced Water: Seawater

3.3.1 Grand Isle

Initial concentrations of total Ba in the 1:199 mixture of produced water from Grand Isle with seawater water from the Gulf of Mexico were calculated to be 184 μg/L $\left[((1 \times 33,000) + (199 \times 18.7))/200\right]$, and thus ~5.1 times oversaturated with respect to barite where $\left[Ba^{2+}\right]_{Sat} = 36$ μg/L in the 199:1 mixtures based on calculations by Church and Wolgemuth (1972), Falkner et al. (1993) and Monnin et al. (1999).

$$\left[Ba^{2+}\right]_{sat} = K_d / \left(\Gamma_{Ba^{2+}} \times \left[SO_4^{2-}\right]_{measured} \times \Gamma_{SO_4^{2-}}\right)$$

where $K_d = 1.1 \times 10^{-10}$ @1 atm, 25°C

$\Gamma_{Ba^{2+}} = \gamma_{Ba^{2+}}$ (simple activity coefficient) $\times f_{Ba^{2+}}$ (fraction of free ions)

$= (0.24)\,(0.93) = 0.22$

$\Gamma_{SO_4^{2-}} = \gamma_{SO_4^{2-}} f_{SO_4^{2-}} = (0.17)\,(0.39) = 0.066$

By substituting a molal sulfate value of 28.1×10^{-3} m for the 1:199 mixture of produced water with Gulf of Mexico seawater, a saturation value for Ba was calculated to be 36 μg/L (2.7×10^{-7} m). Thus, the calculated Ba concentration of 184 μg/L was 5.1 times oversaturated.

No changes in dissolved or particulate Ba concentrations were observed throughout the 24-h mixing period for the Grand Isle sample (Fig. 5.2 and Table 5.4) as

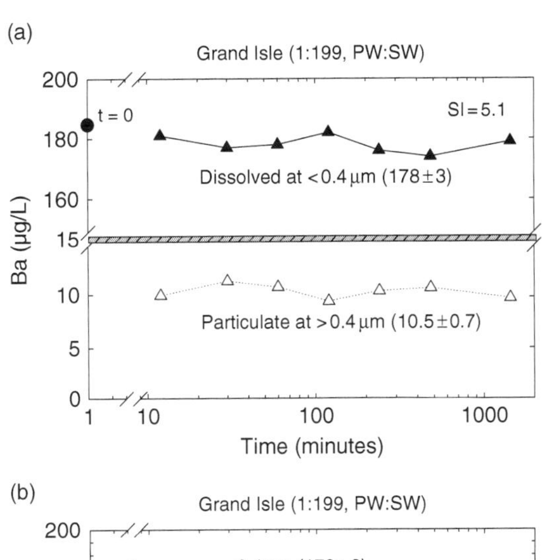

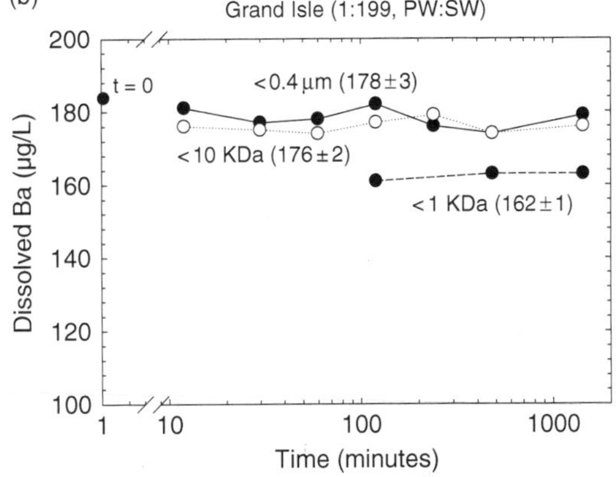

Fig. 5.2 Concentrations of barium (Ba) in 1:199 mixtures of produced water (PW) from Grand Isle, Louisiana, with seawater (SW) as a function of time (t) after the two solutions were combined: **a** dissolved at <0.4 μm and particulate at >0.4 μm (*shaded* bar shows scale break on y-axis) and **b** dissolved at <0.4 μm, <10 KDa and <1 KDa; the value for Ba at a $t = 0$ min was calculated based on the dilution of the original PW with SW; Saturation Index (SI) = (Ion Activity Product/K_{sp})

concentrations of dissolved Ba (<0.4 μm) averaged 178 ± 3 μg/L during the entire 1,440 min of the experiment. Concentrations of particulate Ba (>0.4 μm) averaged 10.5 ± 0.7 μg/L (Table 5.4) for all samples over the entire 24 h period and were within 20% of those determined for ambient seawater (12.3 ± 0.5 μg/L). The concentration of dissolved plus particulate Ba at the end of the experiment (189 μg/L) was in reasonably good agreement with the Ba value of 184 μg/L calculated for $t = 0$ h (Table 5.4). Thus, no precipitation of barium was observed in the 1:199 mixtures that initially were ~5.1 times oversaturated with respect to barite. This observation is consistent with the work of Fernandez-Diaz et al. (1990) who noted that the kinetics of barite precipitation was significantly slower when solutions were < 8 times oversaturated with respect to barite.

Concentrations of Ba in the filtrate from the 10K ultrafiltration after 24 h averaged 176 ± 4 μg/L ($n = 3$) and were not statistically different from the value of 179 ± 2 μg/L in the <0.4 μm fraction (Fig. 5.2b). The ~9% lower values for Ba of 162 ± 1 μg/L in the <1K fraction (Fig. 5.2b) relative to the <0.4 μm and <10K fractions may indicate the slow onset of barite nucleation and particle growth.

Table 5.4 Concentrations of dissolved and particulate barium (Ba) for 1:199 mixtures of produced water (PW) from Grande Isle with seawater (SW) from the Gulf of Mexico

Time (min)	Filter type	Dissolved Ba (μg/L)	Measured particulate Ba (μg/L)	Total Ba[a] (μg/L)
0	–	(184)[b]	–	–
12	0.4 μm	181	10.2	191
	10K	176	–	–
30	0.4 μm	177	11.6	188
	10K	175	–	–
60 (#1)	0.4 μm	176	10.6	187
	10K	174	–	–
60 (#2)	0.4 μm	179	11.3	190
	10K	174	–	–
120	0.4 μm	182	9.6	192
	10K	177	–	–
	1K	161	–	–
240	0.4 μm	176	10.6	187
	10K	179	–	–
480	0.4 μm	174	10.9	185
	10K	174	–	–
	1K	163	–	–
1,440 (#1)	0.4 μm	178	9.6	188
	10K	180	–	–
1,440 (#2)	0.4 μm	181	9.5	191
	10K	173	–	–
1,440 (#3)	0.4 μm	178	10.6	189
	10K	174	–	–
	1K	163	–	–

[a]Total Ba = (measured dissolved Ba) + (measured particulate Ba)
[b]Dissolved Ba value at a time = 0 was calculated based on the dilution of the original PW with SW

However, the final concentration of dissolved Ba (<1 KDa) in the 1:199 mixtures, (163 μg/L), was still oversaturated with respect to barite by a factor of ~4.5.

3.3.2 West Delta 73(#2) and Eugene Island 314A(#2)

Produced water from the platforms at West Delta 73 (WD 73) and Eugene Island 314A (EI 314A) was mixed with seawater in the ratio of 1:99 (produced water to parts seawater). The degrees of Ba oversaturation in the 1:99 mixtures at $t = 0$ were 11.8 and 16.5 for the WD 73(#2) and EI 314A(#2), respectively. During the 72-h mixing periods, concentrations of dissolved Ba were uniform at 453 ± 5 μg/L and 616 ± 10 μg/L for the WD 73 and EI 314A samples, respectively (Fig. 5.3 and Table 5.5). These mean values were each within 5% of the calculated values for $t = 0$ (Table 5.5). Concentrations of particulate Ba were relatively uniform at 1–2 μg/L from 0.5–720 min in the WD 73 sample, and the final value of 7 μg/L for particulate Ba at 72 h (4,320 min) may indicate the onset of barite precipitation. The sum of dissolved + particulate Ba for the WD 73 sample after 72 h was 105 ± 1% of the calculated total of 432 μg/L (at $t = 0$). Concentrations of particulate Ba for the

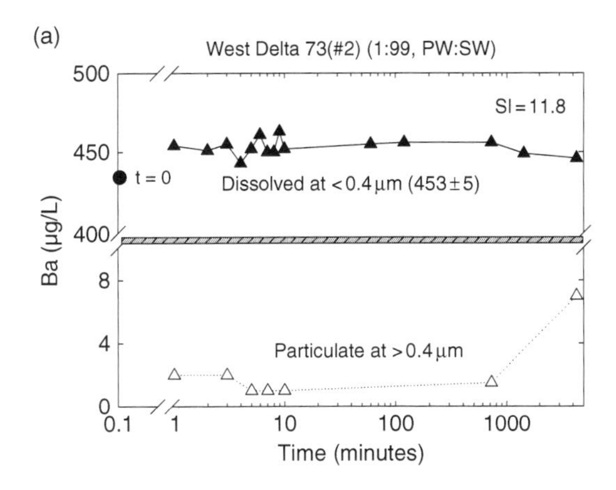

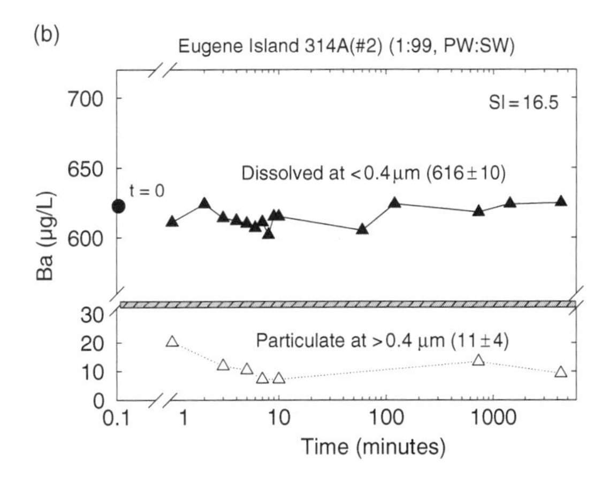

Fig. 5.3 Concentrations of barium (Ba) in 1:99 mixtures of produced water (PW) with seawater (SW) as a function of time (t) after the two solutions were combined: graphs show dissolved (<0.4 μm) and particulate (>0.4 μm) for samples from **a** West Delta 73(#2) and **b** Eugene Island 314(#2); the value for Ba at a $t = 0$ min was calculated based on the dilution of the original PW with SW; Saturation Index (SI) = (Ion Activity Product/K_{sp}); shaded bar shows scale break on y-axis

sample from EI 314A averaged 11 ± 5 μg/L during the 72-h time interval (Fig. 5.3). The concentration of dissolved + particulate Ba for the sample from EI 314A was 104 ± 1% over 72 h relative to the calculated total at $t = 0$. The results for both samples show that even at oversaturation values of 11.8 and 16.5, the kinetics of any precipitation of barite was too slow to observe over 72 h.

3.4 Mixing Experiments with 1:9 Proportions of Produced Water and Seawater

3.4.1 Grand Isle

The first detailed mixing experiment in this series was carried out using the sample from Grand Isle with six mixing-time intervals between 12 and 1,440 min (Table 5.1). Just 12 min after preparing the 1:9 mixture of produced water with seawater from Grand Isle, the concentration of dissolved Ba (<0.4 μm) decreased

Table 5.5 Concentrations of dissolved and particulate barium (Ba) for 1:99 mixtures of produced water (PW) from West Delta 73(#2) and Eugene Island 314A(#2) with seawater (SW)

	West Delta 73(#2)			Eugene Island 314A(#2)		
Time (min)	Dissolved Ba (μg/L)	Measured particulate Ba (μg/L)	Total Ba[a] (μg/L)	Dissolved Ba (μg/L)	Measured particulate Ba (μg/L)	Total Ba[a] (μg/L)
0	(432)[b]	–	–	(602)[b]	–	–
0.5	460	–	–	641	–	–
1	454	2	456	611	20	631
2	451	–	–	625	–	–
3	455	2	457	615	12	627
4	443	–	–	612	–	–
5	452	1	453	610	10	620
6	461	–	–	607	–	–
7	450	1	451	611	7	618
8	450	–	–	602	–	–
9	463	–	–	615	–	–
10 (#1)	450	1	451	617	7	624
10 (#2)	455	1	456	613	7	620
60	455	–	–	605	–	–
120	456	–	–	624	–	–
720 (#1)	452	2	454	613	13	626
720 (#2)	460	1	461	623	14	637
1,440	449	–	–	624	–	–
4,320	446	7	453	625	9	634

[a]Total Ba = (measured dissolved Ba) + (measured particulate Ba)
[b]Dissolved Ba value at a time = 0 was calculated based on the dilution of the original PW with SW

by 83% to 542 μg/L (Fig. 5.4a and Table 5.6). Such a decrease in dissolved Ba is consistent with previously referenced studies (e.g., Templeton 1960; Granbakken et al. 1991). Concentrations of dissolved Ba (<0.4 μm) reached minimum values of 494 ± 34 μg/L between 60 and 240 min. Then, concentrations of dissolved Ba (<0.4 μm) increased to 700 μg/L at 480 min and to 850 ± 30 μg/L at 1,440 min (Table 5.6). The concentration of dissolved Ba (<1 KDa) after 24 h was 654 μg/L (Table 5.6).

In sharp contrast with results for dissolved Ba for the Grand Isle sample, concentrations of particulate Ba increased slowly during the first 60 min with larger increases at 240 and 480 min (Fig. 5.4a and Table 5.6). At the end of the experiment (1,440 min), the average sum of concentrations of dissolved plus particulate Ba was 3,310 ± 60 μg/L relative to the original calculated total Ba content of 3,220 μg/L (Table 5.6). However, at shorter time periods (e.g., 30, 60, and 120 min), concentrations of total Ba were only 15–20% of values at the beginning and end of the experiment. Thus, values for calculated particulate Ba were more than double measured concentrations after the first 4 h of the experiment (Table 5.6). This observation is believed to be due to early formation of very fine-grained solid barite that

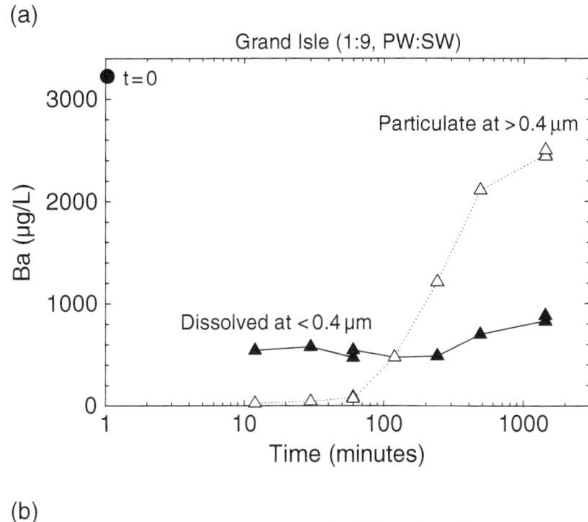

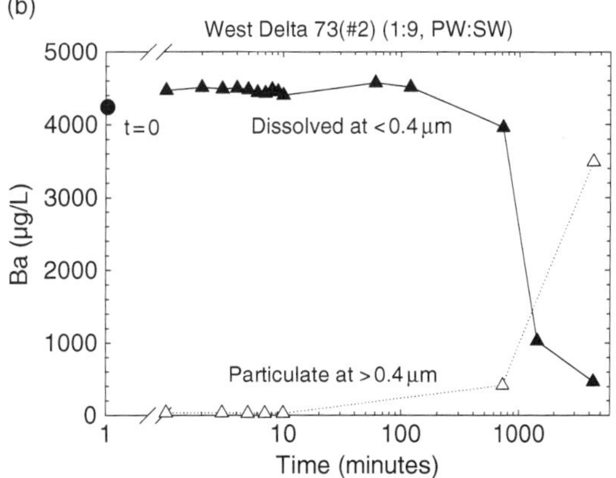

Fig. 5.4 Concentrations of barium (Ba) in 1:9 mixtures of produced water (PW) with seawater (SW) as a function of time (t) after the two solutions were combined: graphs show dissolved (<0.4 μm) and particulate (>0.4 μm) for samples from **a** Grand Isle and **b** West Delta 73(#2); the value for Ba at a $t = 0$ min was calculated based on the dilution of the original PW with SW

passed through the 0.4-μm filter and, as a finely dispersed solid, was not analyzed as part of the dissolved fraction. For example, the size of the barite critical nucleus has been reported to be only 0.001–0.005 μm (Benton et al. 1993). At 480 and 1,440 min, the particles were large enough to be trapped on the 0.4-μm filter and thereby yield the expected amounts of dissolved plus particulate Ba (Table 5.6). The initial degree of oversaturation at ~77 decreased to ~16 after 12 min and then increased to ~24 after 1,440 min.

Results from previous work (Rollheim et al. 1993; Christy and Putnis 1993) suggest that the rate of precipitation should not have slowed so dramatically (i.e., no significant change from 0.2 to 24 h) when the degree of oversaturation with respect to barite was still as high as 16. However, similar degrees of oversaturation were found for the 1:99 mixtures.

Concentrations of DOC varied by <5% (RSD) during the 24-h experiment with lower molecular weight organic matter (<1 KDa) as predominant in the produced water. The toxicity studies of Spangenberg and Cherr (1996) showed that adding Ba acetate, but not $BaCl_2$, could yield solutions with Ba concentrations in the range of

Table 5.6 Concentrations of dissolved and particulate barium (Ba) for 1:9 mixtures of produced water (PW) from Grand Isle with seawater (SW)

Time (min)	Filter type	Dissolved Ba (μg/L)	Measured particulate Ba (μg/L)	Calculated particulate Ba (μg/L)[a]
0	–	(3,220)[b]	–	–
12	0.4 μm	542	24.7	2,680
	10K	677	–	2,540
30	0.4 μm	576	42.5	2,640
	10K	411	–	2,810
60 (#1)	0.4 μm	472	84.3	2,750
	10K	392	–	2,830
60 (#2)	0.4 μm	544	72.8	2,680
	10K	371	–	2,850
120	0.4 μm	474	474 ± 54[c]	2,750
	10K	461	–	2,760
	1K	700	–	2,520
240	0.4 μm	486	1,210	2,730
	10K	567	–	2,650
480	0.4 μm	700	2,110	2,520
	10K	937	–	2,280
	1K	1,140	–	2,080
1,440 (#1)	0.4 μm	829	2,450	2,390
	10K	810	–	2,410
1,440 (#2)	0.4 μm	827	2,440	2,390
	10K	810	–	2,410
1,440 (#3)	0.4 μm	882	2,500	2,340
	10K	832	–	2,390
	1K	654	–	2,570

[a]Total Ba = (measured dissolved Ba) + (measured particulate Ba)
[b]Dissolved Ba value at a time = 0 was calculated based on the dilution of the original PW with SW
[c]Concentrations of dissolved and particulate Ba at 120 min were the same

200–900 μg/L. Studies with various additives (e.g., Prieto et al. 1990) show that the rate of nucleation of barite crystals is slowed in the presence of polyacrylic acid or polymalaic acid at concentrations as low as 5 mg/L. The trends for concentrations of particulate and dissolved Ba in this study support a slow second stage of crystal growth following nucleation with likely inhibition due to complexation of Ba by organic molecules.

3.4.2 West Delta 73(#2)

The degree of oversaturation in the 1:9 mixture of produced water from West Delta 73(#2) at $t = 0$ was ~106. During the first 120 min, concentrations of dissolved Ba were uniform with an average of 4,480 ± 40 μg/L (Fig. 5.4b and Table 5.7). This mean value is only ~5% higher than the calculated value of 4,250 μg/L at $t = 0$. Concentrations of particulate Ba also were relatively uniform at 29 ± 4 μg/L for the first 120 min (Table 5.7).

Table 5.7 Concentrations of dissolved and particulate barium (Ba) for 1:9 mixtures of produced water (PW) from West Delta 73(#2) with seawater

Time (min)	Dissolved Ba (μg/L)	Measured particulate Ba (μg/L)	Total Ba[a] (μg/L)
0	(4,250)[b]	–	–
0.5	4,490	–	–
1	4,470	34	4,500
2	4,510	–	–
3	4,490	33	4,520
4	4,500	–	–
5	4,480	27	4,510
6	4,440	–	–
7	4,430	26	4,460
8	4,480	–	–
9	4,450	–	–
10 (#1)	4,380	28	4,410
10 (#2)	4,430	24	4,450
60	4,570	–	–
120	4,510	–	–
720 (#1)	3,990	402	4,390
720 (#2)	3,920	425	4,340
1,440	1,030	–	–
4,320	467	3,490	3,960

[a]Total Ba = (measured dissolved Ba) + (measured particulate Ba)
[b]Dissolved Ba value at a time = 0 was calculated based on the dilution of the original PW with SW

After 720 min, concentrations of dissolved Ba (<0.4 μm) in duplicate samples decreased by 520 μg/L (~12%) relative to the mean for the first 120 min (Table 5.7). Concentrations of particulate Ba at 720 min increased to 402 and 425 μg/L in the duplicate samples in response to precipitation of $BaSO_4$. Concentrations of dissolved Ba further decreased to 1,030 μg/L after 1,440 min to 467 μg/L after 4,320 min (Fig. 5.4b). The total Ba (dissolved + particulate Ba) at 72 h was 3,960 μg/L or about 88% of the value for total Ba for the 0.5–120-min period.

Precipitation of barite in the 1:9 mixture of produced water with seawater from the WD 73 platform occurred at a much slower rate than observed for the sample from Grand Isle. No change in concentrations of dissolved Ba was observed for 12 h in the West Delta 73(#2) sample relative to 12 min for the Grand Isle sample.

3.4.3 Eugene Island 314A(#2)

The 1:9 mixture of produced water with seawater from the platform at EI 314A(#2) had an initial degree of Ba oversaturation of ~148. During the first 120 min, concentrations of dissolved Ba (<0.4 μm) averaged 1,940 ± 120 μg/L, then decreased to 1,530 and 517 μg/L after 720 and 4,320 min, respectively. Concentrations of total Ba throughout the experiment were only 26–36% of the calculated total Ba at $t = 0$. As described previously, this apparent discrepancy may have resulted from early formation of colloidal barite that passed through the 0.4-μm filter or adsorbed on the plastic sample bottle and was not analyzed as part of the dissolved Ba.

Concentrations of particulate Ba averaged 32 ± 12 μg/L for the first 120 min. At 720 min, the concentrations of particulate Ba in duplicate samples increased to 459 and 478 μg/L and were in balance with the decrease in dissolved Ba. Concentrations of particulate Ba further increased to 1,020 μg/L after 4,320 min, again in balance with the decrease in dissolved Ba. The mixing experiment for the Eugene Island sample was repeated with the same trends.

3.5 Field Discharge of Produced Water

To complement the laboratory studies, samples of water were collected from a produced water plume in the Gulf of Mexico during May 2002. The primary goal of that effort was to determine how field results for reactions of Ba discharged with produced water compared with laboratory results. The experiment was carried out around the platform at Main Pass 288 (29°14.38′ N, 88°24.57′ W). Discharge of produced water from the platform was via a pipe at a depth reported to be ~6 m below the sea surface. The produced water was held in a 5,000-L holding tank and partial discharges from the tank occurred periodically based on the incoming flow to the tank.

Prior to initiating the tracking experiment, a sample of the produced water with a salinity of 108‰ and a DOC of 39.7 mg/L was collected directly from the tank. After collecting a sample of produced water, two persons remained on the platform to add Rhodamine WT dye to the holding tank and to monitor concentrations of the dye in that tank over time. About 500 mL of Rhodamine WT dye were added at $t =$ 0 and again at $t = 20$ min. Periodic discharges and refilling of the holding tank made continuous assessment of the dye content of the produced water being discharged more difficult. Once the dye was added and a discharge event occurred, samples were collected from a Zodiac positioned adjacent to the platform and equipped with a field fluorometer and a peristaltic pump attached to Tygon tubing to obtain samples. The dye was visible at the sea surface within ~3 min after discharge. Samples were collected in 1-L polyethylene bottles for subsequent laboratory determinations of concentrations of dye, dissolved and particulate Ba, and DOC.

In the upper 20 m of the water column, the salinity was 33–34‰ and the temperature was 26–28°C. The salinity increased to 36.5‰ at 40 m and was uniform to the bottom at ~400 m. Concentrations of Rhodamine WT dye in the field samples collected adjacent to the platform ranged from <1 to 33 μg/L. The dilution factors based on the data for the dye in the reservoir and a measured 3-min lag time, ranged from 1,600–5,600 (Table 5.8).

To augment the dilution factors calculated based on dye concentrations, similar factors were calculated using data for dissolved Ba (Table 5.8). Concentrations of dissolved Ba in the discharge plume ranged from 12.0–16.7 μg/L relative to 10.4 ± 0.1 μg/L in the seawater adjacent to the platform prior to the discharge. Values for excess dissolved Ba were calculated by subtracting the ambient seawater

Table 5.8 Concentrations of Rhodamine WT dye, dissolved and particulate barium, and corresponding dilution factors from sampling a produced water discharge plume in the Gulf of Mexico

Sample ID	Rhodamine dye (μg/L)	Rhodamine dilution factor[a]	Dissolved Ba (μg/L)	Excess Ba dilution factor[b]	Particulate Ba (μg/L)
1028	0	–	10.4 ± 0.2	–	0.076 ± 5
1033	9	4,800	15.7	3,130	0.075
1035	19	1,600	14.2	4,370	0.142
1121	28	2,700	16.2	2,860	0.080
1123	33	3,200	16.7	2,630	0.097
1125	13	2,000	15.7	3,130	0.061
1137	9	3,400	15.1	3,530	0.106
1150	3	5,600	15.3	3,390	0.061
1156	4	3,500	14.5	4,050	0.109
Seawater	0	–	10.4 ± 0.2	–	0.076 ± 5
Produced water	6,000–33,000	1	16,600	–	–

[a]Rhodamine dilution factor = (Rhodamine in platform tank)/(Rhodamine in seawater)
[b]Excess Ba dilution factor = (Ba in platform tank)/(dissolved Ba in seawater – 10.4 μg/L) where Ba in platform tank = 16,600 μg/L

concentration of 10.4 μg/L from the observed concentration of dissolved Ba in the discharge plume (Table 5.8). The Ba content of the undiluted produced water (16,600 μg/L) was then divided by the concentration of excess dissolved Ba to obtain dilution factors based on excess Ba (Table 5.8). The dilution factors based on excess Ba ranged from 2,600 to 4,400 and averaged 1.0 ± 0.4 times those based on the dye (Table 5.8).

The observed degrees of dilution are consistent with calculated results for a distance of ~20 m from the discharge point using model calculations made by Joe Smith of Exxon-Mobil using the dispersion model developed for the Offshore Operators Committee (Brandsma and Smith 1996). The model for dispersion of produced water was run to simulate a 500 bbl/day (7.9 × 10^4 L/day) discharge from a 15-cm diameter pipe located 6 m below the sea surface with a current velocity of 0.1 m/s at all depths, produced water was assumed to have a temperature of 25°C and a salinity of 100 g/kg. A 100-fold dilution was calculated within the first 5 m, a 1,000-fold dilution was calculated at 20 m and a 30,000-fold dilution at 300 m.

Concentrations of excess particulate Ba were calculated by subtracting the concentration of particulate Ba in the sample collected adjacent to the platform prior to the discharge (0.076 μg/L) from the values for particulate Ba in the discharge plume (Table 5.8). The excess particulate Ba made up only 1.2 ± 0.7% of the total excess Ba. In other words, 98.8 ± 0.7% of the Ba introduced to the waters adjacent to the platform was still dissolved when the mixture of produced water and seawater was sampled within ~5 min after discharge.

4 Conclusions

Results from the laboratory mixing experiments and field sampling exercise have been used to make the following conclusions:

- Essentially all the Ba in the produced water collected was dissolved and passed through 0.4-μm pore size filters and 1-KDa ultrafilters.
- In mixing experiments with 1:99 and 1:199 proportions of produced water with seawater, initial concentrations of dissolved Ba in three different samples of 184, 432, and 602 μg/L were essentially unchanged over 72 h, even though they were oversaturated by factors of 5.1, 11.8, and 16.5, respectively.
- After 24 h, at a proportion of 1:9 (produced water: seawater), 93 $\pm$ 3% of the dissolved Ba precipitated from samples from four offshore platforms, and 74% and 49% of the dissolved Ba precipitated from samples from two onshore facilities, with the difference most likely due to greater amounts of additives to inhibit barite precipitation in the onshore samples. However, different time intervals were found for the onset of precipitation that ranged from $<$12 min to $>$120 min.
- Based on field sampling of a discharge in the Gulf of Mexico, $>$98% of the Ba added to seawater via produced water was still dissolved 5 min after discharge at a distance of 20 m from the source and at a 2,000-fold dilution.
- Concentrations of dissolved Ba in the range of 340–1,200 μg/L can exist for at least 24 h in a static mixture of produced water with seawater. Such behavior is believed to be due to a slow rate of nucleation as well as a slow second stage of precipitation following nucleation. Each of these processes can be further inhibited by the presence of various natural and added organic substances in the produced water.
- Discharge of produced water at sea is a dynamic process wherein dilution seems to be more important to the concentrations of dissolved Ba than precipitation.

Acknowledgments The authors thank the American Petroleum Institute for the opportunity to carry out this study and Tom Purcell for his support and guidance. Thanks to Joe Smith of Exxon-Mobil for his keen interest and intellectual support for this project as well as logistical assistance for obtaining samples and arranging the discharge study in the Gulf of Mexico. Field assistance during the discharge experiment from Steve Viada and Frank Johnson of Continental Shelf Associates, as well as Steven Rabke of MI and Michelle McElvaine of Florida Institute of Technology, is greatly appreciated.

References

Barth T (1991) Organic acids and inorganic ions in waters from petroleum reservoirs, Norwegian continental shelf: a multivariate statistical analysis and comparison with American reservoir formation waters. Appl Geochem 6:1–15

Benton WJ, Colllins IR et al (1993) Nucleation, growth and inhibition of barium sulfate-controlled modification with organic and inorganic additives. Faraday Discuss 95:281–297

Brandsma MG, Smith JP (1996) Dispersion modeling perspectives on the environment fate of produced water discharges. In: Reed M and Johnsen S (eds) Produced water 2: environmental issues and mitigation technologies. Plenum, New York

Christy AG, Putnis A (1993) The kinetics of barite dissolution and precipitation in water and sodium chloride brines at 44–85°C. Geochim Cosmochim Acta 57:2161–2168

Church TM, Wolgemuth K (1972) Marine barite saturation. Earth Planet Sci Lett 15:35–44

Clesceri LS, Greenberg AE et al (eds) (1989) Standard methods for the examination of water and wastewater. Am Publ Health Assoc, Washington, DC

Falkner KK, Klinkhammer GP et al (1993) The behavior of barium in anoxic marine waters. Geochim Cosmochim Acta 57:537–554

Fernandez-Diaz L, Putnis A, Cumberbatch TJ (1990) Barite nucleation kinetics and the effect of additives. Eur J Mineral 2:495–501

Granbakken D, Haarberg T et al (1991) Scale formation in reservoir and production equipment during oil recovery. III. A kinetic model for the precipitation/dissolution reactions. Acta Chem Scan 45:892–901

Higashi RM, Cherr GN et al (1992) An approach to toxicant isolation from a produced water source in the Santa Barbara Channel. In: Ray JP, Englehardt FR (eds) Produced water: technological/environmental issues and solutions. Plenum, New York

Monnin C, Jeandel C et al (1999) The marine barite saturation state of the world's oceans. Mar Chem 65:253–261

Neff JM (2002) Bioaccumulation in marine organism: effects of contaminants from oil well produced water. Elsevier, Amsterdam

Prieto M, Putnis A, Fernandez-Diaz L (1990) Factors controlling the kinetics of crystallization: oversaturation evolution in a porous medium. Geol Mag 127:485–495

Rollheim M, Shamsili RG, et al (1993) Scale formation in reservoir and production equipment during oil recovery IV. Experimental study of $BaSO_4$ and $SrSO_4$ scaling in steel tubings. Acta Chem Scan 47:338–367

Spangenberg JV, Cherr GN (1996) Developmental effects of barium exposure in a marine bivalve (Mytilus californianus). Environ Toxicol Chem 15:1769–1774

Templeton CC (1960) Solubility of barium sulfate in sodium chloride solutions from 25° to 95° C. J Chem Eng Data 5:514–516

Trefry JH, Trocine RP (1991) Collection and analysis of marine particles for trace elements. In: Hurd DC and Spencer DW (eds) Marine particles: analysis and characterization, Geophys Mono 63. Am Geophys Union, Washington, DC

Trefry JH, Trocine RP et al (1996) Assessing the potential for enhanced bioaccumulation of heavy metals from produced water discharges to the Gulf of Mexico. In: Reed M, Johnsen S (eds) Produced water 2: environmental issues and mitigation technologies. Plenum, New York

Chapter 6
The Distribution of Dissolved and Particulate Metals and Nutrients in the Vicinity of the Hibernia Offshore Oil and Gas Platform

Philip A. Yeats, B.A. Law, and T.G. Milligan

Abstract Water column samples for trace metal and nutrient analysis were collected in the vicinity of the Hibernia offshore oil and gas platform on CCGS Hudson cruises in July 2005 and June 2006 as part of an investigation of chemical tracers of produced water discharges. Measurements of metals and nutrients in two produced water samples from the Hibernia platform show that produced water concentrations for SiO_2, NH_3, Ba, Fe and Mn are >100 times those in seawater. The concentrations of dissolved Fe and Mn increase with depth in the waters outside the platform's exclusion zone, with highest concentrations in deep water samples immediately south and west of the platform. Particulate Ba, Fe and Mn are elevated compared to Al in a number of the deep and bottom water samples located in the vicinity of the platform in 2005 and 2006. Elevated concentrations of SiO_2, NH_3, Fe and Mn were also found within 150 m of the platform at 10 m depth in 2005.

1 Introduction

Produced waters discharged from offshore oil and gas wells are frequently anoxic and have salinities and temperatures that are much higher than those of the receiving waters. They can also have elevated concentrations of nutrients and heavy metals. In the Canadian regulatory environment, only the concentrations of hydrocarbons in the discharges are controlled. Dilution (which will be rapid) is relied upon to deal with any environmental effects of anoxic conditions, temperature and salinity anomalies, or potential toxic effects of heavy metals or other chemicals in the discharges.

Although acute toxicity of produced water (PW) may be adequately mitigated by treatments to remove hydrocarbons and dilution of the plumes, concerns remain over potential for chronic and cumulative effects of the discharges and bioaccumulation of contaminants. In order to understand these more subtle effects (or conversely to

P.A. Yeats (✉)
Department of Fisheries and Oceans, Bedford Institute of Oceanography, Dartmouth, NS, Canada B2Y 4A2

K. Lee, J. Neff (eds.), *Produced Water*, DOI 10.1007/978-1-4614-0046-2_6,

show that there are no long-term effects), we need a better understanding of how chemicals in the PW discharge are transported and diluted. We have previously conducted some studies to investigate chemical reactivity of PW on dilution with seawater in the laboratory (Azetsu-Scott et al. 2007), and others have conducted tracer studies in which dyes have been introduced into produced water discharges (e.g. DeBlois et al. 2007). It is never certain that dyes become properly integrated with the PW and faithfully follow the plume, and rather more certain that they will not reflect the transport of contaminants that react chemically and separate from the plume. Hydrodynamic models have been used quite extensively to predict dispersion of the plumes (e.g. Reed and Johnsen 1995, chapters 19–23; Niu et al. 2007), but these also do not account for chemical reactivity of the discharge. Several chemicals occur in PW at concentrations that can be >1000 times those in the ambient seawater and thus provide natural tracers for the discharge components that would, depending on the choice of tracer, inherently account for the reactivity.

In the summer of 2005 and 2006, we participated in a study designed to investigate potential biological effects of PW discharges from the Hibernia offshore oil and gas platform on the Grand Banks of Newfoundland. In our component of this program, we collected water samples for nutrients and dissolved and particulate metals at various locations in the vicinity of the platform. Our purpose was (1) to provide some information on the levels of inorganic contaminants in the vicinity of the platform as a component of field investigation of biological effects and (2) to investigate whether or not we could find any evidence of PW plumes away from the immediate vicinity of the PW discharges. In this chapter we will focus on the second goal, documenting observations of elevated concentrations of Ba, Fe and Mn in bottom waters between 500 and 4500 m from the platform and Fe, Mn, SiO_2 and NH_3 in the upper water column closer to the platform.

2 Methods

Between 30 June and 3 July 2005 and 16 and 23 June 2006, water samples were collected (CCGS Hudson cruises 2005-028 and 2006-022) at a grid of stations radiating mostly N, S, E and W from the edge of the 500 m exclusion zone to 6.5 km from the Hibernia platform (46°45′N, 48°47′W). The station numbers indicate both the direction from the platform and the distance from the edge of the exclusion zone. A CTD Rosette was deployed at each station. Water samples for dissolved and particulate metals were collected at three depths (10, 35 and 60 m in 2005; 10 35 and 70 m in 2006) at each station with hydrowire casts using General Oceanics Lever Action Niskins (LAN) that were modified and cleaned for contamination-free trace metal sampling and deployed on a stainless steel hydrowire. After initially drawing off unfiltered samples for Hg (preserved with 0.2 N BrCl), salinity and nutrient (preserved by freezing) analyses, the remainder of the water in the sampler was pressure filtered through prewashed and tared 0.4 mm pore size Nuclepore polycarbonate filters. Two litres of the filtered water was retained (acidified with 2 ml of Seastar HNO_3) for dissolved metal analysis and the filters were rinsed with Milli-Q water to remove any residual seawater and dried for later analysis of suspended particulate matter (SPM) concentrations and particulate metals. These techniques have been tested over a number of years and found to give good control of contamination. Nutrient samples were also collected from the sampling bottles on the rosette.

Near-bottom samples for nutrients and particulate metals were collected with a modified Benthic Organic Seston Sampler (BOB, Muschenheim et al. 1995) with self-triggering 2 L syringes at 5, 15, 25, 35 and 45 cm above the seabed. When the sampler is lowered to the seabed, a 45 s time delay before syringe triggering allows for resettlement of material suspended by the sampler. During this study two pooled samples (5-15-25 cm and 35–45 cm) were collected for trace metal and nutrient analysis.

On the 2005 cruise CTD data were collected between July 1 and July 4 at 22 sampling stations within the platform's 500 m exclusion zone to the south of the platform using a portable CTD (Seabird model #25) deployed by hand from one of the ship's launches. Unfiltered water samples (LAN) for chemical analyses were collected at 10 m depth at 12 of these stations.

Samples were analysed for total Hg and dissolved Cd, Cu, Fe, Mn, Ni, Pb and Zn in our clean lab using well-proven analytical techniques (Landing et al. 1995; Yeats and Dalziel 2008). Total Hg samples were concentrated by $SnCl_2$ reduction and trapping on Au columns and analysed by atomic fluorescence spectrophotometry. Dissolved Cd, Cu, Fe, Ni, Pb and Zn samples were extracted with an APDC/Freon extraction procedure and back extraction into HNO_3 and analysed by grahite furnace atomic absorption spectrophotometry (GFAAS). For Mn, the extraction was with oxine/MIBK followed by GFAAS. SPM concentrations were determined by gravimetric analysis of the filters using a microbalance. Particulate metal concentrations were determined by digesting the filters with HNO_3/HF, taking the solutions to dryness and redissolving them in dilute HNO_3, and analysing the samples using inductively coupled plasma emission spectroscopy and mass spectrometry. The digestions of the particulate metal samples were conducted in the clean lab at BIO and the analyses by a commercial laboratory (RPC, Fredericton NB). The nutrient samples were analysed by Technicon autoanalyser using standard techniques for seawater analysis. Unfiltered samples from the LAN sampler deployed from the ship's launch were analysed using the techniques described above for filtered samples. Sources of contamination were not as well controlled in the handling of samples on the launch and some contamination of both Pb and Zn is likely, so results for these elements have not been reported for the sub-set of samples collected from the ship's launch.

Ten litre acid washed carboys for produced water samples were flown to the platform in advance of our cruises. The carboys were filled to capacity by Rig staff and returned to CCGS Hudson via launch once the samples were collected. Onboard ship, samples were filtered and dissolved and particulate samples preserved for subsequent chemical analysis following protocols used for seawater samples. Nutrient samples were analysed by autoanalyser and metal samples (dissolved and particulate) by ICPMS. Unavoidable delays between collection and processing of the samples mean that some precipitation will have occurred before filtration.

3 Results

3.1 Produced Water

If we are looking for evidence of PW plumes, we need to know which components are likely to be elevated and thus potential tracers of the plumes. Analysis of the two PW samples from the Hibernia platform (Table 6.1) and assessment of data in the literature suggests that aqueous Fe, Mn, Ba, SiO_2 and NH_3 concentrations have the best potential as tracers as described below. More extensive measurements of major ion composition of PW from Hibernia (Ayers and Parker 2001) give a range of salinities of 46–196 and Fe between an unspecified detection limit and 89 mg/L. Results for our two samples are at the low end of these ranges. Our analyses show that PW salinities are higher than seawater, concentrations of Fe are ~10,000 times ambient seawater, based on our seawater observations in this study (the maximum Fe content reported by Ayers and Parker is 300,000 times seawater concentrations), Mn, SiO_2 and NH_3 are >1000 times surface seawater concentrations, and Ba, Hg, Pb and Zn are all in the 50–100 times range. Dissolved Ba concentrations in Hibernia PW discharge appear to be considerably lower than in discharges from other areas (Neff 2002). Ayers and Parker commented on the role of high sulphate ion concentrations in the Hibernia PW in the precipitation of Ba. SPM concentrations are elevated in our PW samples but particulate metals (on weight of metal per weight of SPM basis) are not.

To focus the search for PW tracers in the environment, we also need to know about reactivity and transport pathways for the PW. Our earlier laboratory studies of mixing with seawater (Azetsu-Scott et al. 2007) indicated three different pathways: components that stayed in solution would sink or rise and dilute along with PW plume; a second group of chemicals would oxidize/precipitate to form insoluble inorganic compounds that would be heavier than seawater and sink; and a third

Table 6.1 Nutrient and metal concentrations in produced water samples from Hibernia

sal	Si µM	P µM	NO_3 µM	NH_3 µM	Hg_t ng/L	Ba_d µg/L	Cd_d µg/L	Cu_d µg/L	Fe_d mg/L	Pb_d µg/L	Mn_d µg/L	Ni_d µg/L	Zn_d µg/L
45.6	814	8.6	0.34	641	7.1	354	0.2	<2	3.43	0.34	563	1.7	24
45.5	890	15.5	0.37	642	6.9	343	<0.1	<2	3.44	0.62	565	1.1	25
	SPM mg/L	Al_p µg/g	Ba_p µg/g	Cd_p µg/g	Cr_p µg/g	Cu_p µg/g	Fe_p mg/g	Pb_p µg/g	Mn_p µg/g	Ni_p µg/g	V_p µg/g	Zn_p µg/g	
	75.9	121	217	0.11	6.3	5.0	68.0	4.8	93	3.7	2.4	19	
	75.8	100	214	0.05	6.9	5.5	68.5	6.4	94	3.2	2.6	17	

group of chemicals would associate with oil droplets that are lighter than seawater and rise to the surface. Results of our lab experiments suggested that Cu, Pb and Zn are more likely to be associated with the buoyant oily particles and Ni, Fe, Mn and Al with the denser settling ones. Dye studies conducted at the Terra Nova platform on the Grand Banks (Deblois et al. 2007) indicated another important pathway that would not have been seen in the laboratory experiments. In this study, gas entrained in the discharge (or degassing from the PW) gave buoyancy to the discharge plume, rapidly transporting water and dye to the surface.

3.2 Observations in the Immediate Vicinity of the Platform

Produced water is discharged from the Hibernia platform at ~40 m water depth. The PW discharge point is on the south side of the platform and mean currents are weak and not very directional (Petrie and Warnell 1988) with perhaps some residual deep water movement in northwesterly or southwesterly directions (Tedford et al. 2003). We would expect from available data on temperature and salinity of the PW discharge that at the time of our surveys it would have been substantially warmer, saltier and denser than the ambient waters. Temperature and salinity characteristics would suggest that the plume should sink on discharge to the seabed, but visual observations of upwelling water adjacent to the platform at the time of our 2005 sampling indicated that water was being transported to the surface.

CTD data from within the exclusion zone show that near-surface waters within about 150 m of the platform are saltier and colder (and have greater density) than water from farther away (Fig. 6.1). If PW were being entrained with gas bubbles and brought to the surface, we would expect to see high temperature anomalies at the surface, not low temperature ones. The observed temperature/salinity anomalies in the top few metres have characteristics that are more indicative of entrainment of water from the water column as the bubbles move from the discharge pipe to the surface. If large quantities of PW were being rapidly transported to the surface, we would also expect to see visible colouring of the water as a result of precipitation of the very high concentrations of Fe that are found in the PW. The metal and nutrient analyses for samples collected at 10 m adjacent to the platform, however, would suggest that PW may be contributing to the TS anomalies near the surface. As shown in Table 6.2, the samples from closest to the rig have higher concentrations of Cu, Fe, Mn, SiO_2 and NH_3 than those collected at greater distances. The differences (as well of those for salinity) are small and only those for Fe, SiO_2 and NH_3 are significant ($P < 0.05$). An estimate of the amount of PW in the 10 m samples at 50–150 m from the platform based on concentrations in Table 6.1 suggests that they could contain about ~0.03% PW (even less if the PW salinity and Fe concentrations are closer to those in Ayers and Parker 2001). Not enough Fe, perhaps, to generate a visible plume at the surface.

A calculation of contributions of metals and nutrients from three different sources (dissolved metals and nutrients from ambient water at 10 m water depth based on

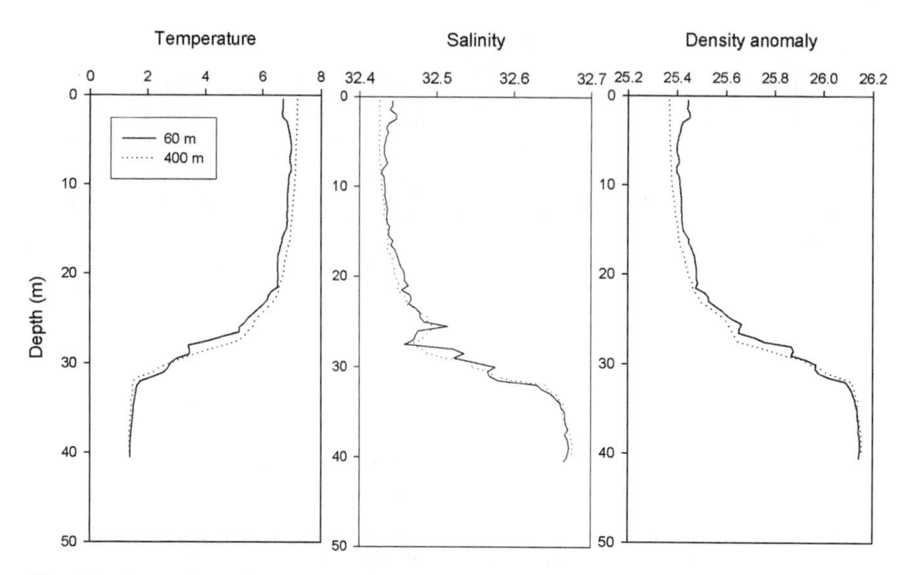

Fig. 6.1 Comparison of temperature, salinity and density profiles for a station 60 m from the platform with profiles for one at 400 m

Table 6.2 Dissolved metal and nutrient concentrations in samples collected in 2005 at 10 m within the platform's 500 m exclusion zone using the ship's launch, and outside the exclusion zone from CCGS Hudson

Distance		sal	Cd	Cu	Fe	Mn	Ni	Si	P	NO_3	NH_3
			µg/L	µg/L	µg/L	µg/L	µg/L	µM	µM	µM	µM
50–150 m	*x*	32.452	0.020	0.24	1.36*	0.21	0.25	0.61*	0.37	0.54	0.61*
$n = 7$	SD	0.038	0.003	0.09	0.59	0.04	0.05	0.15	0.02	0.08	0.17
150–500 m	*x*	32.437	0.022	0.22	0.84	0.19	0.24	0.56**	0.34	0.51	0.45**
$n = 5$	SD	0.006	0.003	0.05	0.65	0.06	0.04	0.08	0.03	0.05	0.12
500–1000 m	*x*	32.397	0.020	0.21	0.26	0.18	0.27	0.35	0.36	0.52	0.29
$n = 4$	SD	0.046	0.004	0.07	0.26	0.05	0.01	0.15	0.03	0.05	0.05

*Concentrations at 50–150 m significantly ($P < 0.05$) greater than those at 500–1000 m
**Concentrations at 150–500 m significantly ($P < 0.05$) greater than those at 500–1000 m

hydrowire samples collected from the Hudson in the vicinity of the platform, particulate metals from the same samples (because the metal measurements in the ambient water are dissolved but those from the launch are unfiltered), and PW based on our PW measurements in Table 6.1 and estimated 0.03% PW contribution) is shown in Fig. 6.2. This plot shows that particulate material only makes a contribution to the Fe concentrations and that PW could be making a contribution to the overall observed concentrations for Fe, Mn, SiO_2 and NH_3. For Fe, SiO_2 and NH_3, the proportions observed in our PW samples are quite consistent with the proportions needed to explain the observations, but Mn is at too high a concentration in our PW samples.

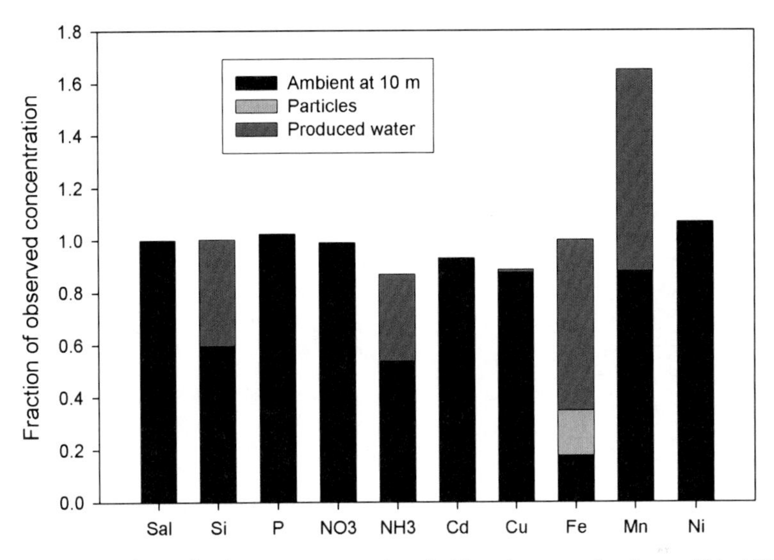

Fig. 6.2 Computed contributions to concentrations in 10 m deep samples from within 150 m of the platform

3.3 Observations from CCGS Hudson

3.3.1 Temperature and Salinity

Predictions for the fate of PW discharged from Hibernia based on considerations of the plume density and hydrodynamic models of the plume dispersion indicate that the discharge should sink to a trapping layer at or near the bottom (Ayers and Parker 2001). CTD observations from 2005 at the stations at the edge of the exclusion zone show no indication of high salinity and high temperature anomalies between 40 m and the bottom. The CTD profiles in 2005 (Fig. 6.3) do show some differences for depths > 40 m between profiles for stations close to the platform (and predominantly to the south and west) and those from farther away. At depths > 40 m the near-platform stations have lower salinities and higher temperatures. These observations could be generated by downward entrainment of water from shallower depths as the PW plume exits the platform and sinks to the sea floor.

3.3.2 Particulate Metal Concentrations

Precipitation reactions as the plume mixes with seawater will contribute Fe and Mn oxide/hydroxide as well as Ba sulphate particles that would add a net downward settling velocity to components of the produced water plume that are already sinking because of plume density. As a result of these two processes, we might anticipate accumulation of particulate Fe, Mn and Ba in the bottom boundary layer 'downstream' of the platform.

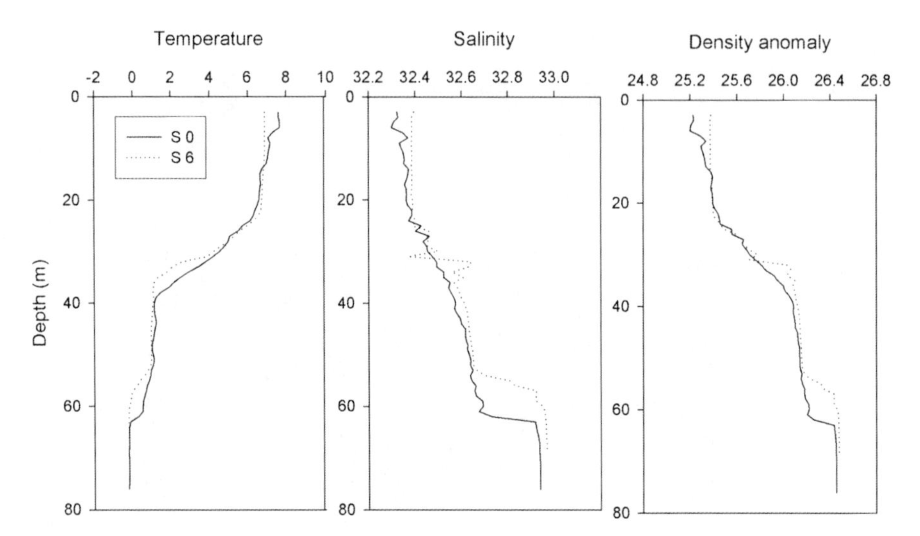

Fig. 6.3 Comparison of temperature, salinity and density profiles for station S0 and station S6 in 2005

The SPM concentrations measured in these near-bottom samples were very similar in the 2 years and not very high (0.3–0.8 mg/L compared to 0.1–0.2 mg/L in the LAN samples collected 10–20 m above the bottom in 2005, and 0.2–0.3 mg/L in 2006) indicating that a strong nepheloid layer was not established in these waters at the times of our sampling. The Al content of the near-bottom particles collected in both years and the 60 m deep water column particles in 2005 were in the 3000–12,000 mg/kg range, indicating that the inorganic (clay) content of these particles is rather low, only 10–20% of the total. Organic matter undoubtedly contributes much of the remainder, especially in the water column in 2006 when SPM concentrations were higher but particulate Al concentrations lower. Authigenic precipitation of metal oxides and salts from PW could also contribute SPM to the deep and near-bottom samples.

The strongest indication of near-bottom enrichment of particulate metals is generated by the particulate Ba results. Ba is a well-known PW component that precipitates in seawater to form barium sulphate ($BaSO_4$). Elevated near-bottom particulate Ba concentrations are indicated by the plot of particulate Ba vs. particulate Al (Fig. 6.4). If concentrations were in the range normally expected for sediments on the Grand Banks, they should be in the area of this plot bounded by the dashed lines. All but a very few of the near-bottom samples are clearly in excess of this background concentration, as are most of the deep water LAN samples. Station numbers here and elsewhere in the text refer to the distance (km) and direction from the edge of the Hibernia platform's exclusion zone not from the centre of the platform. The spatial pattern of particulate Ba concentrations in the near-bottom samples (Fig. 6.5) clearly shows a pattern of high concentrations in the immediate vicinity of the platform extending out at least 2 km to the west and south. The

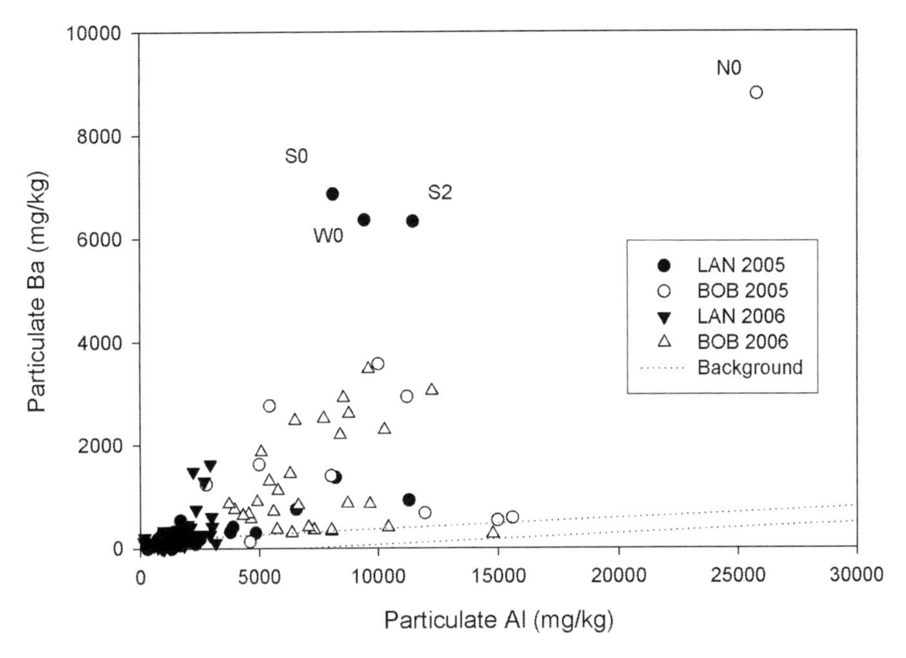

Fig. 6.4 Plot of particulate Ba vs. particulate Al for all samples collected from the Hudson in 2005 and 2006; the *dotted lines* show anticipated range for concentrations in uncontaminated shelf sediments

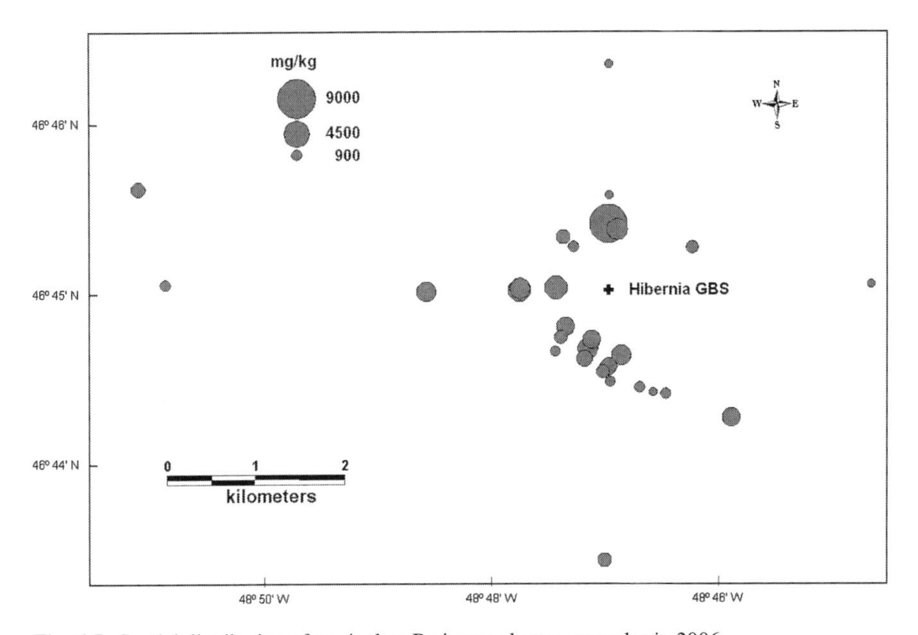

Fig. 6.5 Spatial distribution of particulate Ba in near-bottom samples in 2006

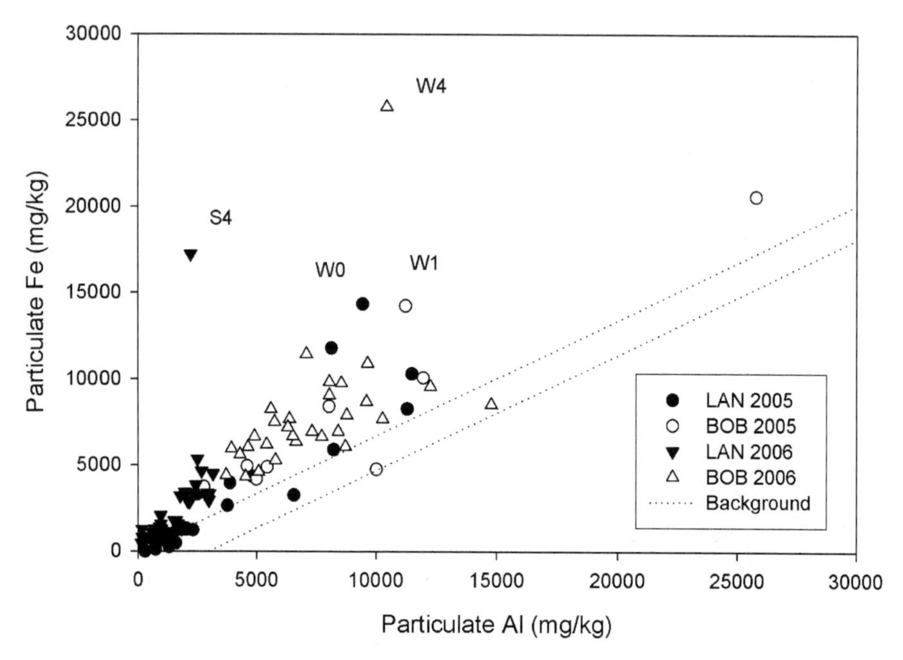

Fig. 6.6 Plot of particulate Fe vs. particulate Al for all samples collected from the Hudson in 2005 and 2006; *dotted lines* show anticipated range for concentrations in uncontaminated shelf sediments

picture for particulate Fe (Fig. 6.6) is similar but deviations from the expected background concentrations are approximately half as large as those for Ba. As with Ba, anomalously high concentrations are focused to the south and west of the platform. Particulate Mn, Cu, Pb and Zn are similar to particulate Fe, but with even fewer anomalously high concentrations, all in the near-bottom samples. Mn had a single anomaly at W4 in 2006, Cu two anomalies, W1 and S0, both in 2005, Pb, S0 in 2006 and Zn, W1 in 2005.

3.3.3 Dissolved Metal Concentrations

No discernible horizontal gradients in the concentrations of dissolved Cd, Cu, Ni, Pb, Zn and total Hg were observed in the study area. The vertical distributions of Cd, Cu, Hg, Ni, Pb and Zn (Table 6.3) share three common features: first, concentrations in 2005 and 2006 are very similar; second, trends with depth, salinity and nutrients are as expected from observations elsewhere; and third, there are generally one or two anomalies. Cd shows increasing concentrations with depth and strong positive correlation with phosphate ($r > 0.9$ in both years), as expected given the known oceanic covariance of Cd and P. Cu shows the expected decreasing concentrations with depth and salinity but has two outliers in 2005 – both from surface waters close to platform (NE0 and W2). Hg, Ni, Pb and Zn are similar, concentrations and trends

Table 6.3 Average dissolved metal concentrations for all samples collected from CCGS Hudson between 500 and 6500 m from the Hibernia platform

Yr/depth		sal	Hg_t	Cd_d	Cu_d	Fe_d	Mn_d	Ni_d	Pb_d	Zn_d
			ng/L	ng/L	μg/L	μg/L	μg/L	μg/L	ng/L	μg/L
05/10 m	*x*	32.409	0.14	19	0.21	0.27	0.18	0.26	10	0.64
n = 8	SD	0.035	0.03	3	0.07	0.24	0.03	0.09	4	0.43
05/35 m	*x*	32.560	0.09	16	0.15	0.25	0.18	0.25	9	0.40
n = 8	SD	0.078	0.03	4	0.02	0.18	0.03	0.01	7	0.21
05/60 m	*x*	32.983	0.10	40	0.15	0.73	0.27	0.30	12	0.46
n = 8	SD	0.028	0.03	6	0.01	0.70	0.08	0.02	6	0.23
06/10 m	*x*	32.805	0.18	20	0.19	0.10	0.16	0.28	10	0.32
n = 16	SD	0.029	0.07	3	0.02	0.02	0.03	0.02	7	0.10
06/35 m	*x*	32.954	0.11	21	0.18	0.07	0.16	0.28	7	0.28
n = 16	SD	0.101	0.05	3	0.02	0.03	0.02	0.02	3	0.09
06/70 m	*x*	33.350	0.14	51	0.17	0.47	0.31	0.29	8	0.37
n = 16	SD	0.097	0.05	5	0.02	0.58	0.06	0.02	3	0.15

are as expected based on oceanic observations, with no more than one or two outliers and no strong association of outliers with the platform. There is an anomalously high total Hg sample at 10 m at N5 in 2006, a high dissolved Zn sample at 10 m at S5 and particulate Zn at 35 m at W2, both in 2005.

The picture for Mn and Fe is definitely different. Concentrations of dissolved Mn (Fig. 6.7) are in the range expected for open coastal waters but do not show the expected pattern of decreasing concentrations with increasing depth and salinity.

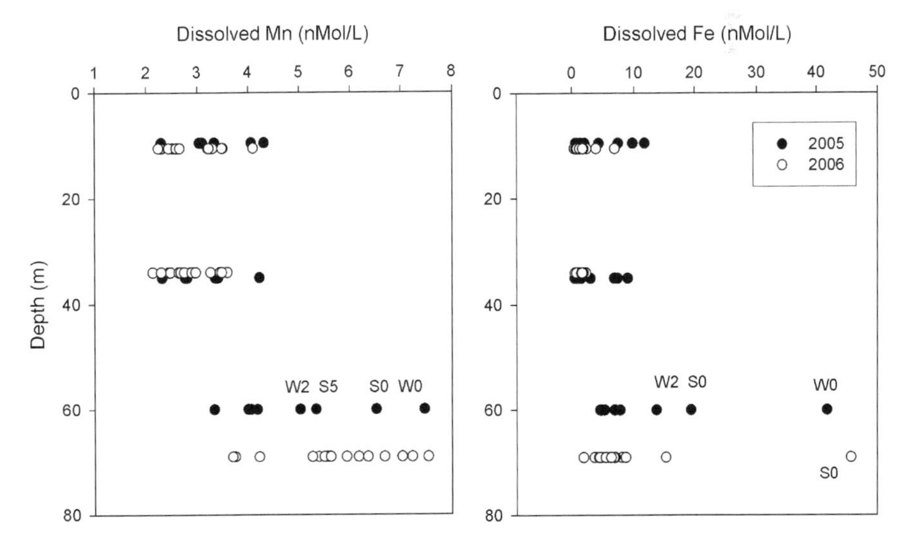

Fig. 6.7 Plots of dissolved Mn and Fe concentrations vs. depth

Correlation coefficients for the linear regressions of Mn on salinity are $r = 0.64$ in 2005 and $r = 0.75$ in 2006. In 2005, elevated deep water (60 m) concentrations are found at W0, S0, W2 and S5. In 2006, the high deep water concentrations are more broadly distributed, but most are located close to the platform (0–2 km) or at greater distances to the west (W4 and W5). For dissolved Fe (Fig. 6.7), anomalously high deep water concentrations are seen at S0, W0 and S2 in 2005, and at S0 in 2006. The spatial patterns for high dissolved Mn and Fe concentrations are quite similar to those described above for particulate Ba and Fe.

3.3.4 Nutrients

Samples for silicate, phosphate, nitrate, nitrite and ammonia were collected from all the LAN samples, from water samplers on the rosette used for CTD measurements and collection of water samples for other chemical and biological studies, as well as from the near-bottom (BOB) samples collected in 2006.

Nitrate, phosphate and silicate concentrations show expected concentration ranges and expected vertical profiles, i.e. depletion of the nutrients in the biologically active surface mixed layer and higher concentrations beneath the photic zone. Concentrations are consistent with observations made in the Atlantic Zonal Monitoring Program (AZMP) sampling on the Grand Banks (AZMP database, www.meds-sdmm.dfo-mpo.gc.ca). Ammonia shows similar vertical profiles with concentrations that range from <1 μM at the surface to >5 μM near the bottom. On the Scotian Shelf we see no marked vertical ammonia gradient and seldom see concentrations >2 μM except in relatively stagnant deep basin waters above organically enriched anaerobic sediments. The AZMP Flemish Cap line ammonia data (G. Maillet, personal communication) show distributions that are like those on the Scotian Shelf in April, but concentrations and gradients that are very similar to the ones we see near Hibernia in August. In our Hibernia data, there are no discernable horizontal gradients in any of the nutrients, and concentrations near the platform are essentially equivalent to those at the reference stations 50 and 100 km away. The nutrient concentrations (including NH_3 and NO_2) are all highly correlated. Distributions are evidently controlled by biogeochemical cycles of uptake and regeneration of the nutrients, and any PW signals for ammonia or silicate are lost in the natural variability.

4 Discussion

The strongest indication of a PW tracer signal is the particulate Ba distribution. The Ba signal shows a pattern of contamination to both the south and west of the platform that is evident in both the deepest LAN samples and the BOB samples. The association of this signal with PW is strengthened by observations of elevated dissolved and particulate Mn and Fe concentrations in several of the same samples. Lee et al. (2007) reported elevated productivity and bacteria numbers at 50 m depth at stations 500 m and 1 km to the south of the platform. The CTD profiles are also consistent with a PW plume that is sinking as it disperses from the discharge pipe predominantly in southward and westward directions, the expected direction of mean current drift (Tedford et al. 2003).

Metals such as Fe, Mn and Ba precipitate, flocculate and settle to the seabed. Flocculation theory shows that aggregates sink at ~1 mm/s and therefore would take ~10 h to settle from the 40 m water depth of the discharge pipe. But the plume will be dense and expected to sink rapidly, so actual settling times will be <10 h. But, flocculation theory also tells us that flocs generally break up at over 0.1 Pa which is equivalent to the shear stress generated by a flow velocity of >25 cm/s. Although conditions are energetic at the Hibernia oil field, shear stress near the bottom can remain below 0.1 Pa for extended periods of time, thus facilitating settling of material to the bottom.

Bottom core samples were collected in 2005 and 2006 at some of the water sampling stations outside of the platforms exclusion zone using a slo-corer. Slo-corer is a hydraulically damped gravity corer that uses 750 lbs of weight to slowly push an acrylic core barrel into the seabed. This corer is capable of preserving the sediment–water interface for all bottom types, including sands, and is essential for measuring particle dynamics and contaminants at the sediment surface. Elevated Mn concentrations were found at or near the surface in sediments at stations N1 and S1 in 2005, and at NE0 in 2006. Fe was elevated at N1 and S1 in 2005, and at W0 and NE0 in 2006. Ba was elevated at stations N0, N1, S0, S1 and W0 in 2005 and at N0, S0, S1, NW0 and NE0 in 2006 (B. Law, unpublished data).

An alternative explanation for the Ba anomalies would be the resuspension of sedimentary Ba that originated with drilling muds. Drilling mud Ba, however, will be dense (4650 kg/m^3) and unlikely to be resuspended as much as 25 m above the seabed. In addition, resuspension of drilling muds would not explain the observations of high Fe and Mn concentrations, nor could it produce the observed biological anomalies.

The other location where we are perhaps seeing some evidence of PW in the environment is in the 10 m samples collected from the ship's launch within about 150 m of the platform. High concentrations of Fe, SiO_2, NH_3 and Mn in these samples could be an indication of a detectable PW component. The proportions of Fe, SiO_2 and NH_3 lend some credence to the suggestion that PW is contributing to observed concentrations in these samples, but a better idea of variability in PW concentrations would be required to really address this question. The lack of a visible plume caused by precipitation of Fe from PW in surface waters is one possible argument against rapid transport of PW plumes to the surface. However, the apparent dilution indicated by our calculations of the potential PW content of the 10 m samples may be reducing particulate Fe concentrations sufficiently to minimize flocculation and generation of visible precipitates.

There are several problems with association of these anomalies with PW. First, our produced water samples have insufficient Ba to explain the concentrations found in the deep and bottom water samples. In addition, the Fe:Ba ratio in our produced water is ~20 compared to ~5 in the deep and bottom water samples. The data for PW in Ayers and Parker (2001) would suggest higher Ba concentrations and a ratio that is closer to 5, so perhaps our two samples are not representative of the average situation. Second, the NH_3 and SiO_2 concentrations in PW should have been high enough to generate anomalies in the samples that show anomalies in dissolved Fe and/or Mn. However, we see no anomalies for these nutrients, nor any deviations from linearity in the relationships between SiO_2 (or NH_3) and NO_3, a nutrient that is not enriched in PW. Lee et al. (2007) do see, however, evidence of increased biological activity at some stations within a kilometre or two of the platform. Can biological activity reduce SiO_2 and NH_3 concentrations to background in the short time it would take PW to reach these stations?

5 Conclusion

Sampling in the vicinity of the Hibernia platform in 2005 and 2006 suggests that there is some evidence for PW in both the surface water in the immediate vicinity of the discharge and in a deep water plume that is transported predominantly to the south and west. In the surface water (10 m depth) within 150 m of the platform, elevated concentrations of SiO_2, NH_3, Fe and Mn are observed. Entrainment with air rising to the surface from the discharge is the likely mechanism for transport of the initially rather dense PW to the surface.

Elevated Fe, Mn and Ba concentrations are also found in bottom waters and surficial sediments. Initial downward transport of the dense PW plume combined with precipitation, flocculation and settling of particle reactive metals will be the mechanisms that transport these metals to the near-bottom waters and the sediments. Observations of elevated near-bottom concentrations are mostly restricted to 1–2 km from the rig which is consistent with expectations for flocculation and settling.

None of the observations described in this chapter are indicative of toxicity or other direct biological effects. They do, however, provide some indications for transport pathways for more hazardous chemicals contained in the produced water discharge. Further sampling at the Hibernia platform is needed to extend the results of these surveys and better define the transport pathways for contaminants. Water column and surficial sediment sampling should be coordinated in order to investigate processes occurring at the sediment water interface.

References

Ayers RC, Parker M (2001) Offshore produced water waste management. Canadian Association of Petroleum Producers, Technical Report 2001-0030

Azetsu-Scott K, Yeats P, Wohlgeschaffen G et al (2007) Precipitation of heavy metals in produced water: influence on contaminant transport and toxicity. Mar Environ Res 63: 146–167

DeBlois EM, Dunbar DS, Hollett C, Taylor DG, Wight FM (2007) Produced water monitoring: use of rhodamine dye to track produced water plumes on the Grand Banks. International Produced Water Conference, St. John's Nfld

Landing WM, Cutter GA, Dalziel JA et al (1995) Analytical intercomparison results from the 1990 Intergovernmental Oceanographic Commission open-ocean baseline study for trace metals: Atlantic Ocean. Mar Chem 49:253–265

Lee K, Cobanli SE, King T et al (2007) Application of microbiological methods to assess the potential impact of produced water discharges. International Produced Water Conference, St. John's Nfld

Muschenheim DK, Milligan TG, Gordon DC, Jr (1995) New technology and suggested methodologies for monitoring particulate wastes discharged from offshore oil and gas drilling platforms and their effects on the benthic boundary layer environment. Canadian Data Report of Fisheries and Aquatic Sciences 2049:x + 55 pp

Neff JM (2002) Bioaccumulation in marine organisms. Effects of contaminants from oil well produced waters. Elsevier, Amsterdam, 452 pp

Niu H, Husain T, Veitch B et al (2007) Experimental and modeling studies on the mixing behavior of offshore discharged produced waters. International Produced Water Conference, St. John's Nfld

Petrie B, Warnell D (1988) Oceanographic and meteorological observations from the Hibernia Region of Newfoundland Grand Banks. Can Data Rep Hydrog Ocean Sci #69:270 pp

Reed M, Johnsen S (1995) Produced water 2: environmental issues and mitigation technologies. Plennum Press, New York

Tedford T, Drozdowski A, Hannah CG (2003) Suspended sediment drift and dispersion at Hibernia. Can Tech Rep Hydrog Ocean Sci #227:57 pp

Yeats PA, Dalziel JA (2008) Heavy metal distributions in the waters of Sydney Harbour. Proc N.S. Inst Sci 44:171–186

Chapter 7
The Effect of Storage Conditions on Produced Water Chemistry and Toxicity

Monique T. Binet, Jennifer L. Stauber, and Trevor Winton

Abstract It is widely accepted that toxicity tests on environmental samples should commence as soon as possible after sample collection. However, constraints involved with sampling and transporting produced water (PW) from offshore oil and gas facilities can cause lengthy delays, during which time some of the toxic constituents may degrade. This can lead to an underestimation of PW toxicity. The objective of this study was to determine whether storage conditions (time and temperature) affected the toxicity and chemical constituents in undiluted PW over a 4-day period from the time of sampling. In addition, the toxicity and chemical composition of PW diluted in seawater, when stored under natural day/night conditions in open and closed test containers, was also assessed. Toxicity was determined after 0, 4, 15, 24, 48, 72 and 96-h storage, using the Microtox® bacterial bioassay. When undiluted PW was stored in the dark for 96 h, refrigeration was not required to prevent changes in PW toxicity indicating that storage temperature was not important for reducing chemical degradation of the PW. For PW that was diluted in seawater, many measured PW constituents were readily degraded (by up to 90%) due to volatilization (BTEX and TPHs C6–C9) and/or photodegradation (PAHs, TPHs C10–C28, phenols). Despite this degradation, there was only a small decrease in toxicity of PW for both open and closed tests (i.e. the EC_{50} increased from 2 to 4% PW for both tests) over the 96-h period, indicating that some toxicant(s) persisted. While it was beyond the scope of this project to identify the cause of toxicity in the PW, it was unlikely that BTEX, naphthalene or ammonia were contributing to toxicity. Phenols, TPHs (in the C10–C14 fraction) and production chemicals were possible toxicants.

1 Introduction

Produced water (PW) is the waste water generated during oil and gas production. PW from oil wells generally consists of formation water (water associated with the oil in the reservoir) and injection water (water that is injected into the reservoir

M.T. Binet (✉)
Centre for Environmental Contaminants Research, CSIRO Land and Water, Kirrawee, NSW 2232, Australia

K. Lee, J. Neff (eds.), *Produced Water*, DOI 10.1007/978-1-4614-0046-2_7,

to force oil to the surface), which have been separated from the oil (Ekins et al. 2007). For gas and condensate wells, the PW is predominantly saturation water that has condensed during the pressure drop between the reservoir and the surface (Veil et al. 2004). In addition, chemicals are often added to the PW during oil and gas production to aid with extraction processes and to protect the system from biofouling and corrosion.

The chemical composition of PWs varies substantially between production platforms, depending on the geological formation of the reservoir, the age of the production well, and the petroleum product being extracted. However, in general, PWs contain inorganic compounds (trace metals), volatile aromatic compounds (benzene, toluene, ethylbenzene, xylenes), polycyclic aromatic hydrocarbons (PAHs) (e.g. naphthalene), phenols, organic acids and additives (Manfra et al. 2007).

Several treatment technologies are available to remove hydrocarbons and other possible toxicants from PW (e.g. Sadiq et al. 2005; Meijer-Akzo and Kuijvenhoven 2001; Tellez et al. 2002); however, they are seldom used on offshore facilities due to space, weight and vulnerability limitations. In addition, the technologies often require additional chemicals (e.g. flocculants, coagulants and de-oilers) and energy (heat and/or pressure that would need to be generated), which pose significant environmental issues (OGP 2002; Veil et al. 2004). Therefore, PW is usually discharged directly into the ocean from offshore platforms. In the North Sea, it is estimated that 400 million cubic metres of PW was discharged in 2003 (Durell et al. 2006), while in 2000, discharges from the UK sector were over 244 million tonnes (Ekins et al. 2007). In Australia, estimates of annual PW discharges have not been published; however, reported discharge from a single platform (Harriet A) was 6600 m^3/day (Jones and Heyward 2003), while for other platforms in the same region it has been known to range from 200 to 8000 m^3/day (IRCE 2005).

There is now increasing global concern over the fate and environmental effects of PW on the marine environment. Environmental studies have included (but are not limited to) composition, fate and dispersion of PW (Burns et al. 1999; Cianelli et al. 2008), biomarker studies (Burns and Codi 1999), acute and chronic toxicity assessments (Holdway 2002; Manfra et al. 2007; Jones and Heyward 2003; Azetsu-Scott et al. 2007), Toxicity Identification Evaluations (Sauer et al. 1997), bioaccumulation studies (Trefry et al. 1995) and environmental risk and effect modelling (Durell et al. 2006; Neff et al. 2006).

In Australia, offshore oil and gas production activity is limited to the Northwest Shelf (NWS) in Western Australia, the Bass Strait (Tasmania) and Timor Sea (Northern Territory) (Jones and Heyward 2003). Currently, the majority of this activity is on the NWS, which is situated off the continental margins between the northwest Cape and Dampier (King et al. 2005). The offshore platforms in this tropical region are remotely situated up to 200 km away from the mainland of Australia. One of the major difficulties researchers face when sampling from these remote tropical locations is weather, as cyclones and rough seas can cause lengthy delays between sampling PW and the receipt of samples in testing laboratories.

Prior to the commencement of a direct toxicity assessment (DTA) of six PWs from the NWS and Timor Sea, determination of appropriate storage and handling conditions for PWs was desirable, to minimize any changes to PW chemistry and toxicity during potential delays in sample transport. The objective of the current study was to determine whether storage conditions affect PW toxicity and chemical composition over a 4-day period from the time of sampling. Storage conditions considered were time (length of storage period) and temperature (refrigeration versus ambient room temperature). In addition, the toxicity and chemical composition of PW diluted in seawater, when stored under natural day/night conditions in open and closed test containers, was also assessed. This was to determine whether any changes in toxicity over time were due to degradation in light and volatilization after dilution in seawater.

Toxicity of PW was determined using the Microtox® test, which measures the light output of the luminescent marine bacterium *Vibrio fischeri* before and after exposure to PW. The Microtox® test was chosen as it was one of the most sensitive tests to PW samples collected from NWS platforms in a previous study (IRCE 2005). In addition, it is a rapid 15-min test, which can easily be used to trace the toxicity of PW over time, and the standard reagent solutions and desktop instrumentation required could be easily transported to nearby Karratha, WA, where toxicity tests were carried out within 3 h of PW collection. *Vibrio fischeri* is commercially supplied as a freeze-dried concentrate that is reconstituted prior to testing. Unlike other test species (e.g. fish, algae), the bacteria are kept in a freezer, and these do not have specific culture requirements (e.g. holding tanks, culture cabinets).

2 Methods

2.1 PW Collection

The site chosen for this study was a gas/condensate platform on Australia's NWS. PW from this platform was previously the most toxic out of six PWs from the same region tested for toxicity to a range of marine biota (IRCE 2005).

The PW samples were collected in 4-L amber glass bottles in May, 2006 by personnel on the platform. Bottles were completely filled to ensure that there was no headspace. All bottles had been pre-cleaned by soaking overnight in 10% nitric acid, followed by five rinses with demineralized water and five rinses with Milli-Q® water. The samples were immediately airlifted to Karratha, WA.

Because PW needed to be tested for toxicity as soon as possible after sampling in order to accurately track toxicity over time, a temporary ecotoxicology laboratory was established in Karratha. From this laboratory, toxicity tests and chemical sampling were carried out within 3 h of PW collection.

2.2 Chemical Analyses

Chemical analyses of PW sub-samples for ammonia, total petroleum hydrocarbons (TPHs), polycyclic aromatic hydrocarbons (PAHs), total organic carbon (TOC), total phenols and the monocyclic aromatic hydrocarbons benzene, toluene,

Table 7.1 Analytical methods used for measurement of PW constituents

Analysis	Preservation	Method	Limit of reporting (μg/L)
BTEX	4°C, pH < 2 (HCl)	Gas chromatography/mass spectrometry (GC/MS) with purge and trap (P&T) (USEPA 8021A)	0.5 (benzene, toluene, ethylbenzene) 2-(*m*- and *p*-xylene) 1-(*o*-xylene)
PAHs (16)	4°C	GC/MS (modified USEPA 8270C)	1–2
Ammonia-N	4°C, pH < 2 (H_2SO_4)	Colorimetric method (APHA 4500)	100
TOC	4°C, pH < 2 (H_2SO_4)	Combustion-infrared (APHA 5310-B)	1000
Total phenols	4°C, pH < 2 (H_2SO_4)	Distillation with flame ionization detector (FID) (APHA 5530-D)	10
TPH (C6–C9)	4°C, pH < 2 (HCl)	GC/MS with FID and P&T (USEPA 8015B)	20
TPH (C10–C14)	4°C	GC/MS with FID and P&T (USEPA 8015B)	20
TPH (C15–C36)	4°C	GC/MS with FID and P&T (USEPA 8015B)	100

ethylbenzene and xylenes (BTEX) were carried out by Amdel Ltd in Sydney. Analytical methods used are listed in Table 7.1.

Chemical analyses of PWs collected previously from the same platform showed that metal concentrations were well below that known to cause toxicity to Microtox®, therefore metal analyses were not included in the current study.

All sample bottles for each PW sub-sample were pre-prepared by Amdel according to the preservation needs for each analyte (Table 7.1). All sub-sample bottles were completely filled (no headspace) and refrigerated before being sent to Amdel for analysis.

Sub-samples were collected for chemical analyses immediately upon arrival at the laboratory in Karratha, and then again at several time points throughout the toxicity tests (Fig. 7.1). Physico-chemical characteristics of the PW were also measured on receipt at Karratha and throughout the 4-day storage period. Measurements were made of salinity and conductivity using a WTW LF 320 m with a TetraCon 325 electrode, of pH using a WTW 320 m with a SenTix 41 sensor, and of dissolved oxygen using a WTW Oxi 330 m with a CellOx 325 probe. All metres were calibrated daily.

2.3 *Microtox®*

The Microtox® test system measures the light output of the luminescent marine bacterium *V. fischeri* before and after exposure to PW and compares this to the light output of controls (no PW), to determine the toxic effect of the PW on the bacteria.

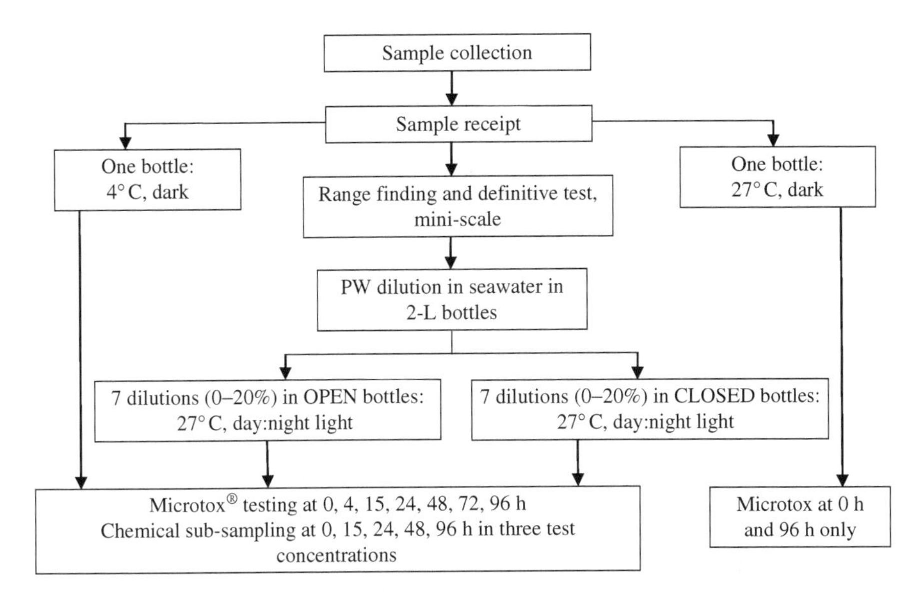

Fig. 7.1 Experimental design of PW chemistry and toxicity testing regime

Microtox® tests were carried out using the 90% Protocol (Azur Environmental 1998) with filtered (0.45 μm) seawater as the diluent. For all tests conducted, two controls and seven to eight dilutions of PW were tested in duplicate. Prior to addition of diluted PW, the initial light output of the bacteria was measured in the photometer (Microtox Model 500 Analyser®). The PW was then added to each vial containing bacteria, and the samples were again measured to determine light output after a 5- and 15-min exposure. Two exposure times were used as per the standard Microtox® test, in case some contaminants increased in toxicity over the exposure duration. For each batch of bacteria reconstituted, the reference toxicant phenol (four concentrations) was also tested for quality assurance purposes.

2.4 Storage Conditions

The PW was tested for toxicity and sub-sampled for chemical analyses several times over a 4-day period under three separate storage/test systems. Due to the time taken for sample receipt, preparation, dilution and preliminary toxicity testing, storage/test times of 0, 4, 15, 24, 48, 72 and 96 h discussed herein correspond to times since PW collection of 8.5, 13, 23, 33, 57, 81 and 105 h (Fig. 7.1).

To simulate immediate and continued storage at 4°C, one bottle of PW was placed in the refrigerator immediately upon arrival at the laboratory. This bottle was sub-sampled after 0, 4, 15, 24, 48, 72 and 96-h storage for toxicity testing and at 0 and 96 h for chemistry.

Another bottle of PW was stored in an unopened bottle at 27°C in the dark and was tested for toxicity at the end of the test program, to determine whether refrigeration (i.e. storage at 4°C) was necessary to avoid degradation in this PW sample.

An initial range finding and subsequent definitive toxicity tests were carried out on the PW with Microtox® immediately on sample receipt. Based on the results of these tests, a concentration range of 0.02–18% PW was chosen for all subsequent toxicity tests.

Seven dilutions of PW (0.027–20%) and a control (0.45 μm filtered seawater) were then prepared in 2-L amber glass bottles. Each dilution was evenly divided into clear 1-L glass Schott bottles. The lids on half of all the Schott bottles in each dilution were loosened, to become the "Open" test system, while the lids on the remaining Schott bottles were tightened to become the "Closed" test system. All Schott bottles were then randomly placed on the laboratory bench and stored under ambient tropical day/night light (approximately 12:12 h light:dark) and temperature (20–28°C) conditions.

At each time point (0, 4, 15, 24, 48, 72 and 96-h storage), the same volume was removed from each Schott bottle and used for chemical sub-sampling and/or Microtox® testing. Chemical sub-sampling was only carried out at 0, 15, 48 and 96 h from three concentrations of PW: 0.008% (the NOEC value), 2.2% (close to the EC_{50} value) and 20% (the highest test dilution prepared). The temperature in the laboratory was digitally logged over the entire test period using multi-trip temperature loggers (TempRecord II, with Multitrip Temprecord Bent Probe, Cole Parmer, IL, USA) with the temperature probe submerged in seawater.

The Microtox® test system dilutes each prepared test dilution by a further 10% during the addition of bacteria to the test vials. For example, although 20% was the highest dilution of PW prepared, this was further diluted to 18% during the toxicity test. Therefore, the concentration of each constituent in test solutions of 18, 2.0, and 0.074% PW for the open and closed tests was extrapolated from the measured value in each prepared test solution (20, 2.2, and 0.074%), i.e. extrapolated value = measured value × 0.9.

The Microtox® test system can test only one sample every 30 min; therefore, the order in which the PW samples were sub-sampled for toxicity testing and chemistry at each time-point was randomized.

All glassware used throughout the study were pre-cleaned by soaking overnight in 10% nitric acid, followed by five rinses with demineralized water and five rinses with Milli-Q® water.

2.5 Statistical Analyses

Duplicate values for each test treatment were combined prior to conducting statistical analyses. Toxicity was expressed as the concentration of PW that caused a 50% reduction in the light output of the bacteria, i.e. EC_{50}. The lower the EC_{50}, the more toxic the sample. EC_{50} values were determined using linear interpolation. The lowest observable effect concentration (LOEC) and no observable effect concentration (NOEC) were calculated using Bonferroni's *t*-test. All statistical calculations were done using ToxCalc Version 5.0.23 (TidePool Software).

For comparison of EC_{50} values, the 95% confidence limits around each value were assessed. Where confidence limits overlapped, the values were deemed not to be significantly different. Where confidence limits did not overlap, the EC_{50} values were considered to be significantly different.

3 Results

3.1 PW Composition

The composition of the PW is summarised in Table 7.2. As received, the PW had a pH of 4.7 and a salinity of 0.1‰. The dissolved oxygen content was low (51% saturation). Total phenols, which are known toxicants to Microtox®, were detected

Table 7.2 Chemical composition of PW

Chemical	As received[a]	Test end[b]
pH	4.7	4.9
Salinity (‰)	0.1	0.1
Conductivity (μS)	140	150
Dissolved oxygen (% saturation)	51	75
Ammonia (NH_4-N mg/L)	11.4	11.6
TOC (mg/L)	95	110
Total phenols (mg/L)	32	33
PAHs (μg/L) [c]:		
Naphthalene	200	170
Acenaphthene	1	<1
Fluorene	4	4
Phenanthrene	2	2
BTEX (mg/L):		
Benzene	5.1	5.1
Toluene	7.4	6.8
Ethylbenzene	0.26	0.21
m- and *p*-Xylene	2.2	1.8
o-Xylene	0.74	0.62
Xylenes	2.9	2.4
TPHs (mg/L):		
C6–C9	16	15
C10–C14	32	28
C15–C28	1.7	2.1
C29–C36	0.26	<0.1

[a]Sampled immediately upon sample receipt, 2.5 h from the time of collection
[b]Sampled after 96-h storage at 4°C, 105 h from the time of collection
[c]Only PAHs that were above the detection limit (1 μg/L) are tabulated

in the PW at a concentration of 32 mg/L. Naphthalene (200 μg/L) accounted for 97% of all the PAHs detected in the PW, with concentrations of other PAHs ranging from 1 to 4 μg/L. The concentration of BTEX (benzene, toluene, ethylbenzene and xylene) compounds ranged from 0.26 mg/L for ethylbenzene to 7.4 mg/L for toluene. Long-chained TPHs (C15–C36) were detected at low concentrations of 0.26–1.7 mg/L. The majority of the TPHs detected were within the C10–C14 fraction (32 mg/L) and the C6–C9 fraction (16 mg/L). There was little degradation of the measured PW constituents when stored at 4°C in the dark over 96 h (Table 7.2).

For the open and closed tests, there were substantial changes in PW composition over the 4-day study. The concentration of BTEX chemicals decreased for each test dilution sampled in the open test system; however, in the closed test system, the concentrations remained relatively stable (Fig. 7.2). Similarly, the concentration of the C6–C9 fraction of TPHs decreased more rapidly in the open test system than the closed (Fig. 7.3).

The concentrations of ammonia (not shown), long-chain TPHs (C10–C14 and C15–C28) and the most dominant PAH, naphthalene, appeared to decrease at a similar rate over the 4-day study in both the open and closed test systems (Fig. 7.3). There were no compounds within the C29–C36 fraction of TPHs detected in either open or closed test solutions.

Unlike any other group of chemicals detected in the PW, the phenols behaved differently in lower and higher PW test concentrations (Fig. 7.4). While the concentration of total phenols remained relatively stable in 18% PW over the 4 days for both the open and closed tests, there was a 90% decrease in total phenols in the 2% solutions for the same time period. This may be due to the lower pH and salinity at higher PW concentrations, which may have altered the stability of PW constituents. In 20% PW, the pH ranged from 6.3 to 7.4, and the salinity was 28–29‰ throughout the study period, whereas in 2.2% PW the pH was closer to 7.9, while the salinity was 34–35‰.

3.2 PW Toxicity

3.2.1 Quality Assurance

The phenol standard gave nominal 5-min EC_{50} values of 21–26 mg/L, within the range of 13–26 mg/L (Azur Environmental 1998) indicating test acceptability. The temperatures recorded in seawater ranged from 20 to 28°C over the duration of the 4-day study, with a mean of 24°C. This was similar to the air temperature in Karratha over the same period (19–31°C) (Bureau of Meteorology 2006).

The physico-chemical properties of each test solution were measured throughout the 4-day study. The pH (6.3–8.1) and salinity (29–35‰) of the test solutions were within acceptable limits for Microtox® (pH 5–9; salinity 20–35‰). However, the pH in higher concentrations of PW (6.7 and 20%) was lower (pH 7.3 and 6.3, respectively) than other test solutions (pH of around 8.1), which may have in turn affected the bioavailability of other toxicants within these solutions.

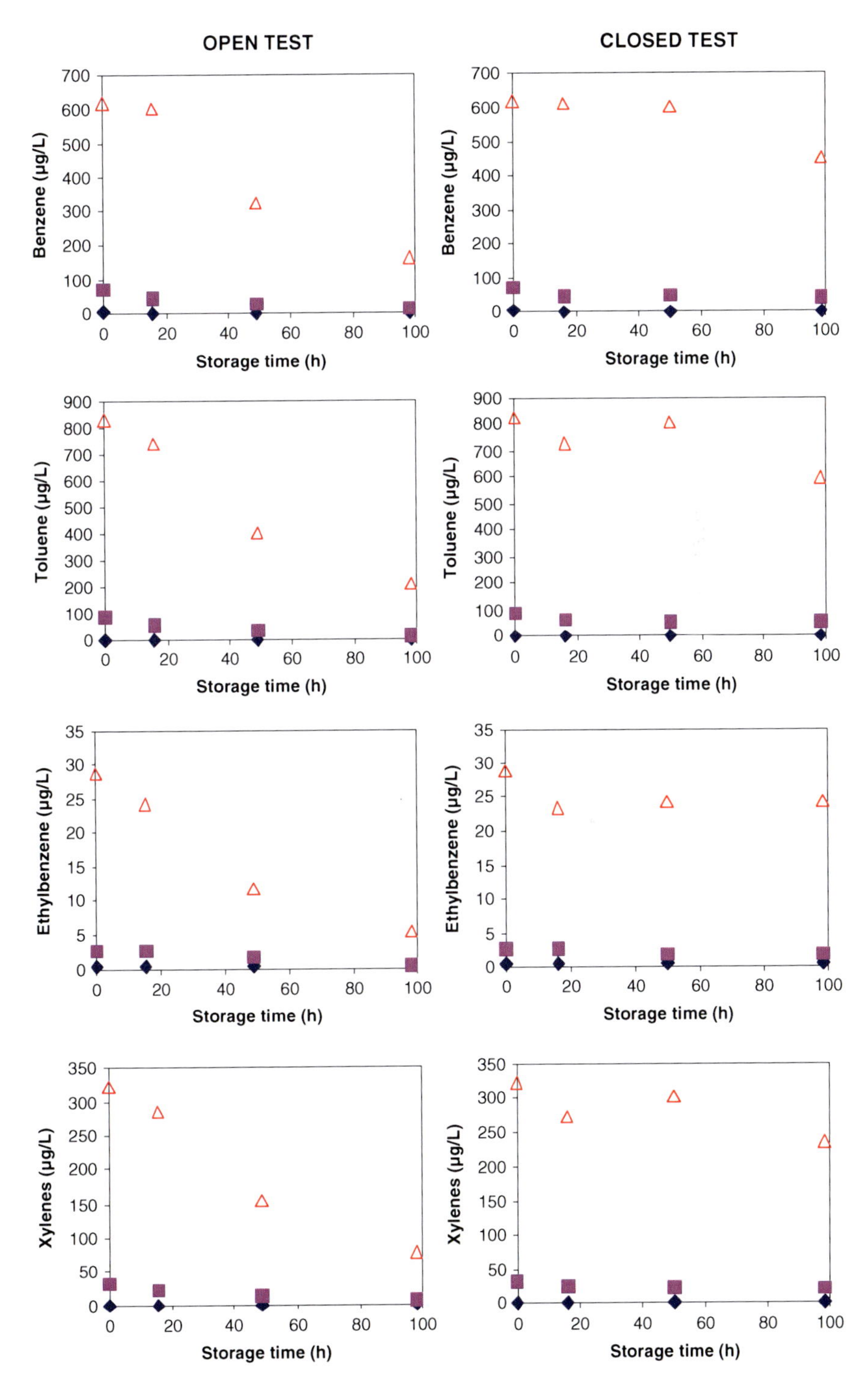

Fig. 7.2 Concentrations of BTEX compounds in dilutions of PW for open and closed test systems over 4 days. Measurements below the detection limit were plotted as half the detection limit value

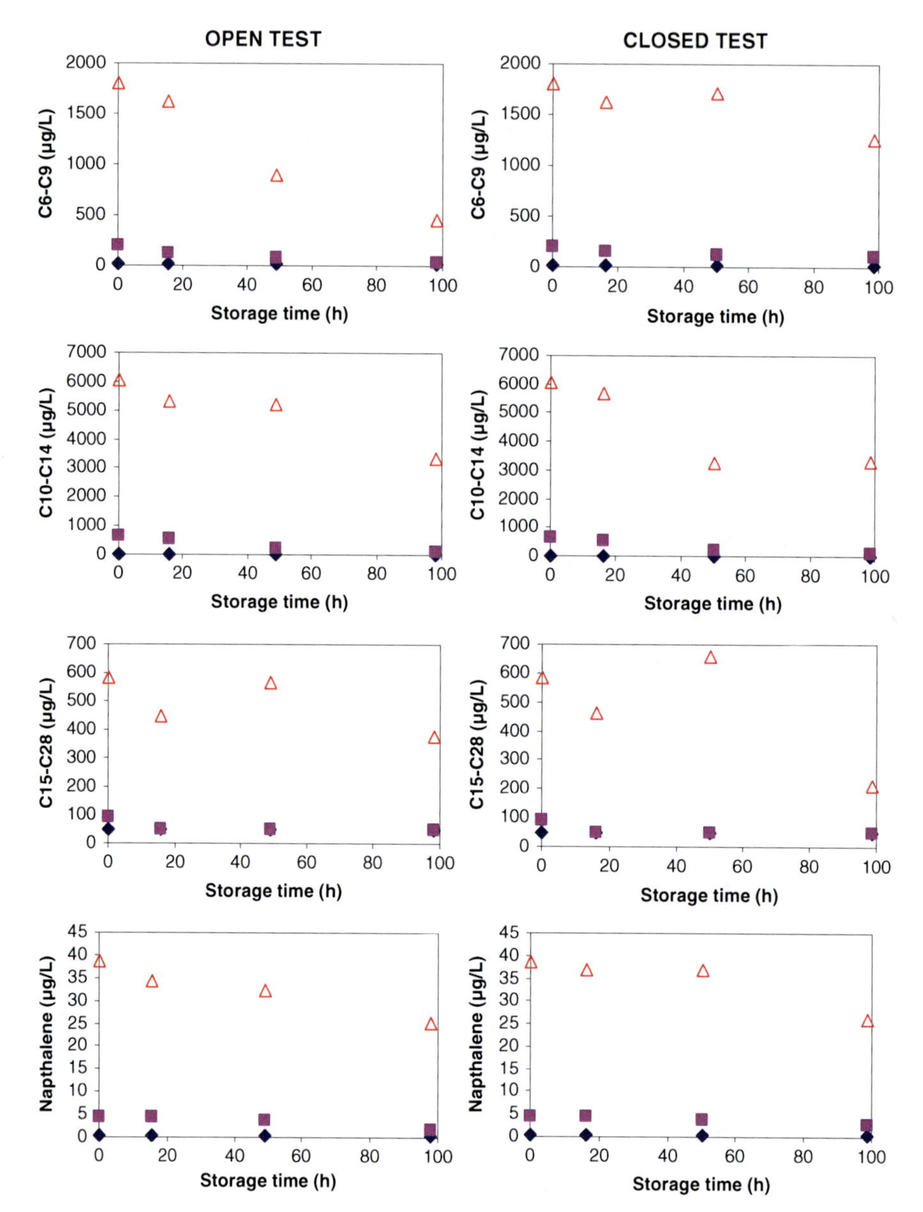

Fig. 7.3 Concentrations of naphthalene and TPHs in dilutions of PW for open and closed test systems over 4 days. Measurements below the detection limit were plotted as half the detection limit value

There was no change in salinity of PW over the 4-day study. Similarly for dilutions that were made from the 4°C-stored PW, there was no change in pH or dissolved oxygen (DO). However, in the highest test dilution of the open and closed tests (20% PW), the pH increased by 0.5 (closed) and 1 (open) pH unit, while the

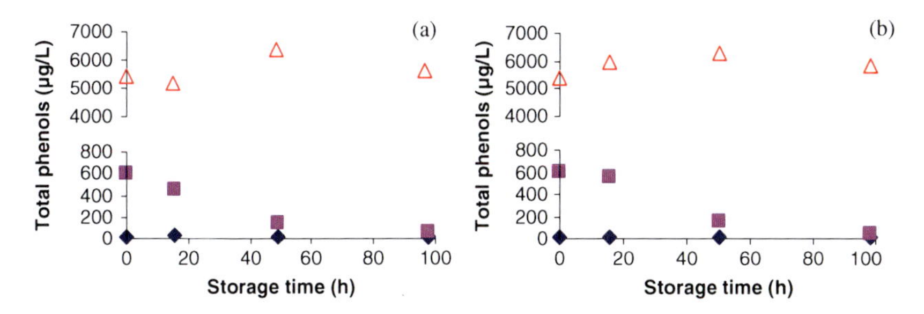

Fig. 7.4 Concentrations of total phenols in dilutions of PW for **a** open and **b** closed test systems over 4 days. Measurements which were below the detection limit were plotted as half the detection limit value

dissolved oxygen decreased from around 90 to 70% saturation for both open and closed test solutions.

3.2.2 Toxicity Testing

The initial range-finding test found the PW to be very toxic to Microtox®, with a 5-min EC_{50} of 2.1% (where 100% is undiluted PW). This result was confirmed in the subsequent definitive test which gave a 5-min EC_{50} value of 1.8%. Significant inhibition of luminescence was observed at concentrations of PW as low as 0.074% (i.e. the LOEC was 0.074% and the NOEC was 0.025%).

The toxicity of the 4°C-stored PW did not change over the 4-day test (Fig. 7.5a), with no significant differences between the EC_{50} values for each time point. This is in agreement with the chemical analyses that found little difference between PW sampled as received and after 96-h storage at 4°C (Table 7.2). In addition, the PW that had been stored unopened at room temperature in the dark had identical toxicity to that of the 4°C-stored PW.

The toxicity of diluted PW solutions in the open test system decreased significantly over the 4 days (Fig. 7.5b), with 5- and 15-min EC_{50} values of 4.1 and 3.9% (respectively) on day 4, compared to 2.3 and 2.4% (respectively) on day 0. The first significant decrease in PW toxicity for the open test after 15 min exposure was evident after 72-h storage, i.e. the EC_{50} values for days 3 and 4 were significantly less than those on day 0.

The toxicity of diluted PW solutions in the closed test system decreased at a similar rate to that observed in the open test system over the 4-day study (Fig. 7.5b), with 5- and 15-min EC_{50} values on day 4 (4.2 and 4.1%, respectively) significantly higher than that found on day 0 (2.3 and 2.4%, respectively). For the 5-min endpoint, after 48-h storage, there was a significant decrease in toxicity of the diluted PW. The 15-min endpoint was slightly less sensitive, with no significant decrease in toxicity until day 3 (72-h storage).

These findings indicate that one (or more) of the toxicants causing toxicity to Microtox® was degrading over time at a similar rate in both the open and closed test solutions.

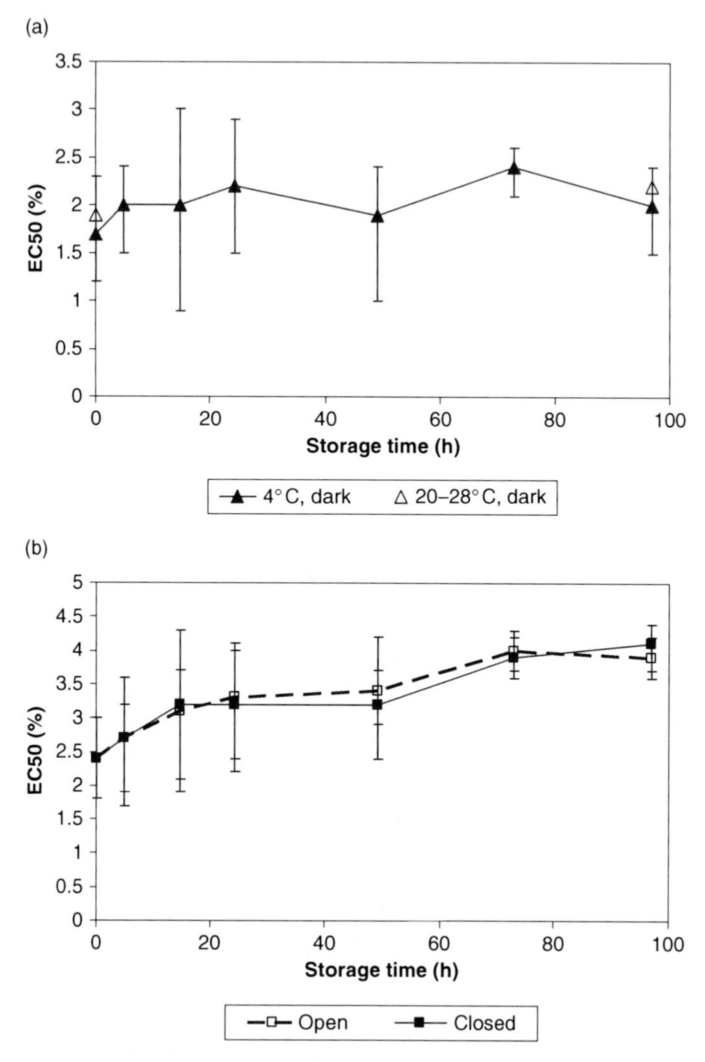

Fig. 7.5 Toxicity of aging PW to Microtox® after 15-min exposure for **a** undiluted PW stored at 4°C and room temperature (20–28°C) in the dark and **b** diluted PW stored in open and closed test containers at room temperature (20–28°C) with natural day:night light cycles

4 Discussion

The PW tested was toxic to Microtox®, with an EC_{50} of 2% (v/v). This is among the most toxic PWs tested to Microtox®, with other reported EC_{50} values ranging from 2.4 to 37% (Grigson et al. 2006; Manfra et al. 2007; Flynn et al. 1996). This is not surprising as PWs from gas platforms have been shown to be more toxic than those from oil platforms (e.g. Flynn et al. 1996; Jacobs et al. 1992). This is most likely because PWs from gas platforms have higher concentrations of hydrocarbons (Veil et al. 2004).

Undiluted PW that was stored at 4°C showed no change in toxicity and very little change in chemical composition over the 4-day storage period. In addition, undiluted PW that was stored at ambient room temperature (20–28°C) in the dark showed no change in toxicity, indicating that refrigeration was not necessary to avoid degradation in this PW sample. This is particularly useful information for large-scale toxicity studies, where cooling large volumes of PW required for toxicity testing and chemical analyses (up to 40 L) can be difficult, particularly since PW can be close to boiling at the point of discharge. However, for prolonged storage, cooling is still desirable to avoid the growth of bacteria (and subsequent biodegradation of PW constituents) in the sample.

When diluted in seawater and exposed to natural day/night light cycles for 4 days, significant degradation of the PW occurred. In both open and closed tests, the concentrations of TPHs (C10–C14), phenols and naphthalene decreased by up to 90%. In addition, in open test solutions, BTEX and TPH (C6–C9, C15–C28) concentrations also decreased by up to 80%. These results indicate that many of the components within the PW are easily degraded once diluted in seawater as a result of photodegradation and volatilization, and possibly biodegradation, although this was not examined in the current study.

Despite the loss of these components from the diluted PW, there was only a small decrease in toxicity observed after 4-day storage (increase in EC_{50} of 2–4%), indicating that the major compound(s) responsible for toxicity was relatively persistent in the PW, even after dilution in seawater. The decrease in toxicity was similar in both the open and closed tests suggesting that volatile compounds were unlikely to be major contributors to toxicity. Flynn et al. (1996) reported similar results when, after sparging with nitrogen to remove volatile compounds from a PW, its toxicity to Microtox® was only reduced by half, leading the authors to conclude that volatile compounds were not necessarily the most important compounds in determining overall PW toxicity.

The concentrations of ammonia, BTEX and naphthalene in the PW were well below those known to cause toxicity to Microtox® (Adams et al. 2008; Grigson et al. 2006; Kaiser and Devillers 1994). In addition, although the concentration of metals was not measured in the current PW, previous studies on PW from the same platform have shown that since 1993, concentrations of most metals were low and often below the metal detection limits (IRCE 2005). The two most abundant metals previously detected in the PW were copper (0.076 mg/L) and zinc (1.9 mg/L), both of which were detected at concentrations well below that known to cause toxicity to Microtox® (EC_{50} values of 0.78 mg Cu/L and 5.7 mg Zn/L) (Grigson et al. 2006). Therefore, it is unlikely that these toxicants were contributing to the observed toxicity.

Phenols have been shown to be important contributors to PW toxicity to Microtox® (Flynn et al. 1996; Johnsen et al. 1996). In the current study, the role of phenols is unclear. Unlike any other group of chemicals measured, phenols behaved differently in lower and higher dilutions of PW over the 4-day storage period. For both the open and closed tests, while the concentration of total phenols remained relatively stable in 18% PW over the 4 days, there was a 90% decrease in total phenols in the 2% solutions for the same time period. This may be due to the lower pH and salinity at higher PW concentrations, which may have altered the stability of PW constituents. It is unclear whether this decrease in total phenols at the EC_{50} concentration is contributing to the observed decrease in toxicity. If total phenols were a major contributor to toxicity, it is unlikely that a 90% decrease in concentration near the EC_{50} dilution (2%) would only cause the toxicity of the PW to decrease by half (i.e. EC_{50} increased from around 2 to 4%). However, it is possible that one particular phenol degraded over time, causing the decrease in toxicity. Individual phenol compounds were not analysed in the current study.

The majority of TPHs detected in the undiluted PW were in the C10–C14 fraction (32 mg/L), followed by the C6–C9 fraction (16 mg/L). The concentrations of higher molecular weight TPHs in the C15–C28 and the C29–C36 fractions were around 10 times lower. This is probably because the solubility of TPHs decreases with increasing molecular weight (Veil et al. 2004), and most of the heavier TPHs are removed with the oil during oil separation techniques. The concentration of the C6–C9 fraction of TPHs decreased more rapidly in the open test system than the closed. This is similar to that found for BTEX compounds, which reside within the C6–C9 fraction of TPHs. The concentrations of higher molecular weight TPHs, (C10–C14 and C15–C28) appeared to decrease at a similar rate over the 4-day study in both the open and closed test systems, indicating that compounds within these fractions may be contributing to toxicity. In particular, the C10–C14 fraction was detected at higher concentrations than any other group measured, and it is this fraction of TPHs that contains many of the phenols in PWs (IRCE 2005), further suggesting that phenols were contributing to toxicity.

In addition to the chemicals measured, production chemicals that can increase the toxicity of a PW by 10-fold (Holdway 2002) may also be contributing to the observed toxicity. Cortron IRN100 and sodium hypochlorite are routinely added to the PW tested in this study. Based on known dosage and discharge rates, the concentrations of these chemicals in the PW tested are estimated to be 130 mg/L Cortron and 0.09 mg/L sodium hypochlorite. Cortron IRN100 is a corrosion inhibitor, with an alkyl pyridine derivative as its active component. According to its material safety data sheet, the 15-min Microtox® EC_{50} is 7 mg/L, which is well below the estimated concentration of Cortron in the PW (130 mg/L). For sodium hypochlorite, the reported Microtox® EC_{50} value is 0.1 mg/L (Kaiser and Devillers 1994). Therefore, if present in the PW, both these water-soluble chemicals are likely to contribute to toxicity.

It is interesting to note, however, that when Henderson et al. (1999) added process chemicals to a PW its toxicity did not increase, despite the fact that each of the 11 chemicals tested were determined to be toxic to Microtox® when tested individually. The chemicals used in the study were not named, however, based on the description of their active components, neither Cortron IRN100 nor sodium hypochlorite were among the 11 chemicals tested. Therefore, further studies would be required to confirm the contribution of these two process chemicals to PW toxicity.

Microtox® was chosen for use in this study due to its sensitivity to organics, its portability and the speed with which tests can be carried out, which made it ideal for use in tracking the toxicity of PW over time. However, results from this study cannot be used to infer toxicity of PWs to other species. Other species are likely to be sensitive to contaminants within the PW to which Microtox® was not (e.g. metals and ammonia).

For the subsequent direct toxicity assessment (DTA), it was recommended that amber bottles be completely filled (no headspace) when sampling PW, and that they be transported in sealed insulated containers (Eskies) ensuring dark conditions to minimize PW degradation.

5 Conclusions

Delays in testing PWs can cause significant changes in PW composition and toxicity if the PW is not stored correctly. For delays of up to 96 h, refrigeration was not required to avoid degradation of constituents or changes in toxicity of undiluted PW to Microtox®. This is a particularly useful information for large-scale toxicity studies, where cooling large volumes of PW required for toxicity testing and chemical analyses can be difficult.

When diluted in seawater and exposed to normal day/night light conditions for 96 h, many of the measured constituents of PWs were readily degraded by volatilization and photodegradation; however, there was only a small decrease in toxicity of the diluted PW, indicating most of the toxicant(s) persisted. This will be important when considering the fate of PW once discharged.

References

Adams MS, Stauber JL, Binet MT, Molloy R, Gregory D (2008) Toxicity assessment of a secondary-treated sewage effluent to marine biota in Bass Strait, Australia: development of action trigger values for a toxicity monitoring program. Mar Poll Bull 57:587–598

Azetsu-Scott K, Yeats P, Wohlgeschaffen G, Dalziel J, Niven S, Lee K (2007) Precipitation of heavy metals in produced water: influence on contamination transport and toxicity. Mar Environ Res 63:146–167

Azur Environmental (1998) Microtox® acute toxicity test. Azur Environmental, Carlsbad, CA

Bureau of Meteorology (2006) Karratha, Western Australia, May 2006 Daily Weather Observations. http://www.bom.gov.au/climate/dwo/200805/html/IDCJDW6064.200805.shtml. Accessed 9 Feb 2009

Burns KA, Codi S (1999) Non-volatile hydrocarbon chemistry studies around a production platform on Australia's Northwest shelf. Estuar Coast Shelf Sci 49:853–876

Burns KA, Codi S, Furnas M, Heggie D, Holdway D, King B, McAllister F (1999) Dispersion and fate of produced water constituents in an Australian Northwest Shelf shallow water ecosystem. Mar Poll Bull 38:593–603

Cianelli D, Manfra L, Zambianchi E, Maggi C, Cappiello A, Famiglini G, Mannozzi M, Cicero AM (2008) Near-field dispersion of produced formation water (PFW) in the Adriatic Sea: an integrated numerical-chemical approach. Mar Environ Res 65:325–337

Durell G, Roe Utvik T, Johnsen S, Frost T, Neff J (2006) Oil well produced water discharges to the North Sea. Part I: Comparison of deployed mussels (*Mytilus edulis*), semi-permeable membrane devices, and the DREAM model predictions to estimate the dispersion of polycyclic aromatic hydrocarbons. Mar Environ Res 62:194–223

Ekins P, Vanner R, Firebrace J (2007) Zero emissions of oil in water from offshore oil and gas installations: economic and environmental implications. J Cleaner Prod 1302–1315

Flynn A, Butler E, Vance I (1996) Produced water composition, toxicity and fate: a review of recent BP North Sea studies. In: Reed M, Johnsen S (eds) Produced water 2: Environmental issues and mitigation technologies. Plenum Press, New York

Grigson S, Cheong C, Way E (2006) Studies of produced water toxicity using luminescent marine bacteria. Environ Toxicol 10:111–121

Henderson SB, Grigson SJW, Johnsen P, Roddie BD (1999) Potential impact of production chemicals on the toxicity of produced water discharges from the North Sea oil platforms. Mar Poll Bull 38:1141–1151

Holdway D (2002) The acute and chronic effects of associated offshore oil and gas production on temperate and tropical marine ecological processes. Mar Poll Bull 44:185–203

IRCE (2005) Produced formation water assessment. International Risk Consultants Environment Document No. ENV-REP-02-078-NRA REV 2. IRC Environment, Perth

Jacobs RPWM, Grant ROH, Kwant J, Marqueine JM, Mentzer E (1992) The composition of produced water from Shell operated oil and gas production in the North Sea. In: Ray JP, Englehart FR (eds) Produced Water. Plenum Press, New York

Johnsen S, Smith AT, Brendehaug J (1996) Identification of acute toxicity sources in produced water. Second international conference on health, safety and environment in oil and gas exploration and production, Jakarta, Indonesia. Society of Petroleum Engineers, pp 383–390

Jones RJ, Heyward AJ (2003) The effects of produced formation water (PFW) on coral and isolated symbiotic dinoflagellates of coral. Mar Freshwater Res 54:153–162

Kaiser KLE, Devillers J (1994). Ecotoxicity of chemicals to *Photobacterium phophoreum*, Volume 2. Gordon and Breach Science Publishers, Switzerland

King SC, Johnsen JE, Haasch ML, Ryan DAJ, Ahokas JT, Burns KA (2005) Summary results from a pilot study conducted around an oil production platform on the Northwest Shelf of Australia. Mar Poll Bull 50:1163–1172

Manfra L, Moltedo G, Virno Lamberti C, Maggi C, Finoia G, Giuliani S, Onorati F, Gabellini M, Di Mento R, Cierco AM (2007) Metal content and toxicity of produced formation water (PFW): Study of the possible effects of the discharge on marine environment. Arch Environ Contam Toxicol 53:183–190

Meijer-Akzo DT, Kuijvenhoven CAT (2001) Field proven removal of dissolved hydrocarbons from offshore produced water by the Macro Porous Polymer-Extraction technology. Offshore Technology Conference, Houston, Texas

Neff JM, Johnsen S, Frost TK, Roe Utvik TI, Durell GS (2006) Oil well produced water discharges to the North Sea. Part II: Comparison of deployed mussels (*Mytilus edulis*) and the DREAM model to predict ecological risk. Mar Environ Res 62:224–246

OGP (2002) Aromatics in produced water: occurrence, fate and effects, and treatment. International Association of Oil and Gas Producers Report No. 1.20/324. http://www.ogp.org.uk/pubs/324.pdf. Accessed 9 Feb 2009

Sadiq R, Khan FI, Veitch B (2005) Evaluating offshore technologies for produced water management using *GreenPro-1* – a risk-based life cycle analysis for green and clean process selection and design. Comput Chem Eng 29:1023–1039

Sauer TC, Costa HJ, Brown JS, Ward TJ (1997) Toxicity identification evaluations of produced-water effluents. Environ Toxicol Chem 16:2020–2028

Tellez GT, Nirmalakhandan N, Gardea-Torresdey JL (2002) Performance evaluation of an activated sludge system for removing petroleum hydrocarbons from oilfield produced water. Adv Environ Res 6:455–470

Trefry JH, Naito KL, Trocine RP, Metz S (1995) Distribution and bioaccumulation of heavy metals from produced water discharges to the Gulf of Mexico. Water Sci Technol 32:31–36

Veil JA, Puder MG, Elcock D, Redwick R Jr (2004) A white paper describing produced water from production of crude oil, natural gas and coal bed methane. Argonne National Laboratory. http://s3.amazonaws.com/propublica/assets/natural_gas/doe_produced_water_2004.pdf. Accessed 9 Feb 2009

Chapter 8
Centrifugal Flotation Technology Evaluation for Dissolved Organics Removal from Produced Water

Marcel V. Melo, O.A. Pereira Jr, A. Jacinto Jr, and L.A. dos Santos

Abstract The aim of this chapter is to introduce an compact system for produced water treatment that is under development, the CFS (Centrifugal Flotation System) which combines the classic flotation and centrifugal separation processes. In the CFS, the bubbles generation, the flocculation and the droplet-bubble contact pass through a pneumatic flocculator, while the separation of the gas phase and the phase rich in oil is achieved using a cylindrical hydrocyclone. In the quest for new produced water treatment concepts, this chapter aims to evaluate the preliminary performance of a Centrifugal Flotation prototype with a feed flow rate of 10 m^3/h, designed and built at Petrobras Research and Development Center (CENPES) to remove dispersed oil, sulphide and soluble compounds such as benzene and toluene, using hydrogen peroxide as oxidant. During the field tests, it was possible to achieve a sulphide removal above 95% and considerable removal values for benzene and toluene from the produced water using a hydrogen peroxide concentration of 100 ppm and a gas flow rate of 10 Nm^3/h.

1 Introduction

The three combined areas—unit installation size, oil–water separation efficiency and the handling of huge volumes of produced water—are currently a global challenge for the oil industry. Improvements in all three areas are fundamental to meet the required produced water disposal limit which has shown a marked tendency to continue to decrease into the future, especially for dissolved organics (Melo et al. 2006). In addition, optimization of offshore Platform space is related directly to the increase of oil production, and to the efficiency of related operational systems and addresses current issues and concerns associated with environmental sustainability (Knudsen et al. 2004). During the past two decades, the pre-disposal treatment of waters contaminated by dispersed oil has been classically performed by oil/water hydrocyclones and by conventional flotation units which occupy what may now be

M.V. Melo (✉)
Petrobras Research and Development Center (CENPES/PDEDS/TTRA), Cidade Universitária, ZIP 21941-598, Ilha do Fundão, Rio de Janeiro, Brazil

K. Lee, J. Neff (eds.), *Produced Water*, DOI 10.1007/978-1-4614-0046-2_8,

regarded as large spaces. The development of a new and alternative system could result in significant reductions of installation sizes of the treatment units. This chapter presents experimental results obtained from the Centrifugal Flotation System (CFS) using a prototype developed at the Petrobras Research and Development Center (CENPES), and preliminarily tested on an offshore Platform to evaluate its performance for dissolved organics removal from produced water.

2 CFS Prototype Description

In Figs. 8.1 and 8.2 the description of the CFS set-up used in this work is presented. The system basically consists of a serpentine flocculator (MS-20) in which compressed natural gas (flow rate of 10 Nm^3/h) and a chemical oxidant are continuously injected and mixed with the produced water. This device works as a turbulent flocculator with a plug-flow pattern (with a hydraulic retention time of 2 s) in order to generate "aerated" oily flocs which can be easily separated because of the trapped gas. In addition, due to the high turbulence and mixing inside the device, the soluble organics can be rapidly oxidized. Alternatively, the system can work with air or nitrogen, however, at offshore platforms it is common to use natural gas, due to the fact that it is already produced onboard. Two pressure indicators (PI) are used to measure the pressure drop along the serpentine. Two cylindrical hydrocyclones (both with a hydraulic retention time of 3 s) were designed at CENPES for promoting the gas/liquid/liquid separation, whereby the gas and the rich oil phase are removed at the hydrocyclone overflow (Q_O) while the rich water phase is withdrawn

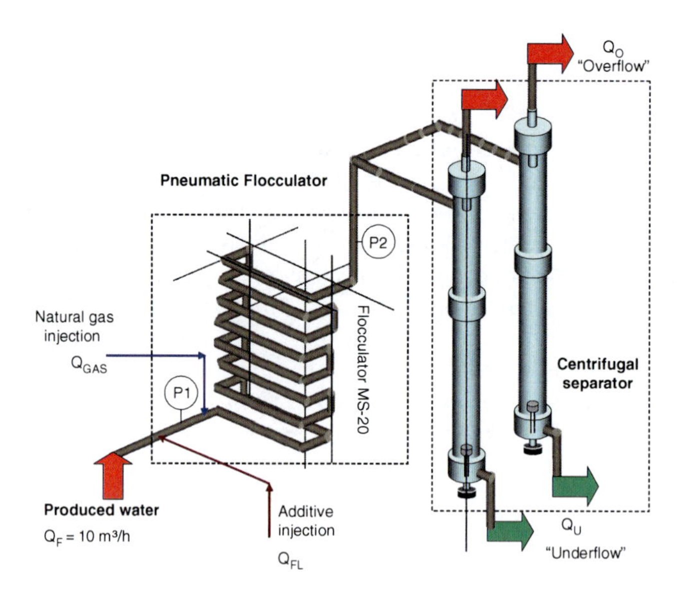

Fig. 8.1 Schematic of the CFS prototype

Fig. 8.2 CFS prototype installation

at the underflow discharge (Q_U). The cylindrical shape of the hidrocyclones is recommended for this case in order to avoid the break of the "aerated" flocs. During the normal and stable operational conditions, the liquid overflow flow rate (Q_O) is adjusted to be around 5% of the inlet flow rate (Q_F).

The excess of gas and the liquid phase rich in oil removed at the overflow are recovered and recycled to the process, after its storage in a specific atmospheric vessel, depending on the installation site. The centrifugal acceleration (around 200 g) is not so high in comparison with the conventional liquid/liquid hydrocyclones and it is promoted by a tangential inlet at the entrance of the hydrocyclone. The prototype presented has an oily water feed flow rate (Q_F) of 10 m^3/h.

This new system including the association of the pneumatic flocculation, cylindrical hydrocyclones and advanced oxidation was not applied until now for the treatment of produced water at offshore installations. This treatment concept was idealized, developed and evaluated previously in a small scale apparatus (feed flow rate of 4 m^3/h) using synthetic produced water, with different inlet oil concentrations, at the Petrobras Research and Development Center (CENPES). According to the results and experience achieved during these initial experiments, it was possible to design and assemble a larger prototype in order to evaluate the technology under real field conditions.

3 Produced Water Used for the System Evaluation

The offshore Platform chosen for the installation of the CFS does not have an onboard produced water treatment system, consequently, all the water production is being offloaded to a relief ship in order to undergo treatment before discharging.

The produced water used in the test is rich in iron sulphide colloids and dissolved organics such as benzene and toulene. Due to operational problems, the dispersed oil concentration in the produced water was relatively low (below 30 ppm) during the tests. As consequence, at this stage of the experimental investigation, it was only possible to evaluate the system performance in terms of dissolved organic compounds removal. Due to the characteristics of the CFS, it is not necessary to pre-treat for solids in the produced water. The iron sulphide is oxidized by the hydrogen peroxide (generating sulphate) and the others colloids are flocculated by the pneumatic flocculator as well as dispersed oil droplets. As a result, the "aerated" oily flocs are removed by the cylindrical hydrocyclones.

4 Chemical Additive and Experimental Measurements

For the oxidation of dissolved organics, hydrogen peroxide was injected in-line into the system before the pneumatic flocculator. It is not possible to inject compressed oxygen as an alternative oxidant into the system, due to safety concerns. Hydrogen peroxide injection was performed using a dosage pump with flow rate control in order to vary the concentration from 50 to 100 ppm.

The dispersed oil, sulphide, benzene, ethyl-benzene, toluene and xylene concentrations were measured in the feed and in the treated produced water streams in order to evaluate the CFS performance.

5 Results, Observations and Conclusions

The CFS prototype (10 m^3/h) was installed in a Petrobras offshore platform and operated for three weeks using the operational conditions listed in Table 8.1.

Initially, in order to evaluate which oxidant was best for removing iron sulphide, the first screening tests were carried out using sodium hypochlorite as oxidant. This additive showed good results when high hydraulic retention times are available, however, for the proposed compact system, the sodium hypochlorite did not show good results. Alternatively, in a second step of chemical selection for the CFS hydrogen peroxide (a stronger oxidant) was used which showed better destabilization of emulsified oil in water. Figure 8.3 compares the raw produced water (feed) with samples taken from the CFS treated water stream using a hydrogen peroxide concentration of 100 ppm.

Table 8.1 Operational conditions used in the CFS during the tests

Q_F (m^3/h)	Q_U (m^3/h)	P1 (kgf/cm)	P2 (kgf/cm^2)	P_{GAS} (kgf/cm^2)	ΔP (kgf/cm^2)
11	8	2.2	0.5	2.8	1.7

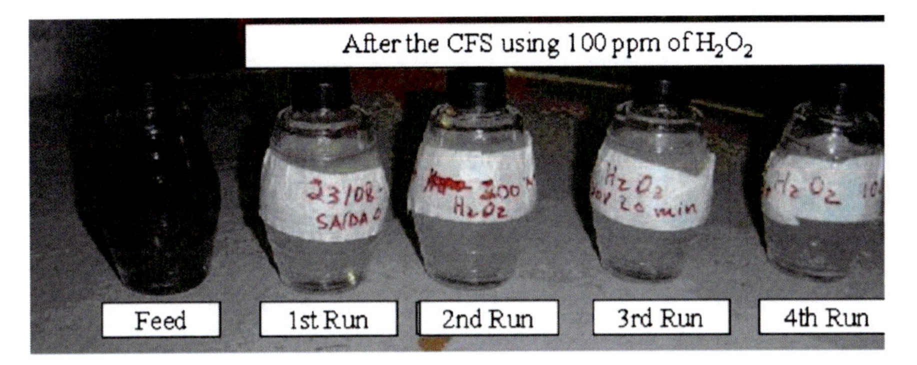

Fig. 8.3 Produced water samples before and after treatment using the CFS

Table 8.2 CFS performance working with 50 and 100 ppm of H_2O_2 injection

Component	Feed 1	Outlet 50 ppm H_2O_2	Removal (%)	Feed 2	Outlet 100 ppm H_2O_2	Removal (%)
Dispersed oil	29 mg/L	0 mg/L	100	11 mg/L	0 mg/L	100
Sulphide	13 mg/L	0.9 mg/L	93	31 mg/L	0.13 mg/L	99.6
Benzene	1340 μg/L	191 μg/L	86	1110 μg/L	116 μg/L	90
Toluene	278 μg/L	68.8 μg/L	75	254 μg/L	91.6 μg/L	64
Ethyl-benzene	12.4 μg/L	6.7 μg/L	46	12.5 μg/L	3.7 μg/L	70
m+*p*-Xylenes	28.8 μg/L	7.2 μg/L	75	23.9 μg/L	16.9 μg/L	30
o-Xylene	17.5 μg/L	3.6 μg/L	79	14.3 μg/L	9.4 μg/L	34

During the field tests carried out in optimized conditions using a concentration of 50 ppm of hydrogen peroxide, it was possible to achieve a sulphide removal rate above 95% and considerable removal rates for benzene and toluene from the produced water as shown in Table 8.2. In addition, using a higher oxidant concentration it was possible to achieve even better results. A cost–benefit study should be done in order to evaluate what is the better choice in terms of oxidant concentration ranging from 50 to 100 ppm. Oxidant concentrations below 50 ppm showed similar results in comparison to using sodium hypochlorite as the chemical oxidant.

Finally, the Centrifugal Flotation technology seems to have some advantages over conventional processes, namely, a low area requirement, absence of mobile parts, simple design and low mechanical and electrical energy requirements. This Centrifugal Flotation prototype is currently being adapted in another offshore platform in order to evaluate this technology using a produced water feed with a higher oil concentration. In conclusion, the presented technology seems to have potential; however, further experiments should be carried out as well a cost–benefit study in order to thoroughly compare this technology against conventional methods.

In addition, for the specific case of adding a chemical oxidant for the oil in water emulsion destabilization, further environmental studies have to be done in order to evaluate the potential toxicity of discharging the treated produced water. The possible consequences of remnant oxidative capacity in the discharged stream (not yet evaluated at this preliminary study) should be investigated in further studies.

References

Knudsen BL et al (2004) Meeting the zero discharge challenge for produced water. 7th SPE International Conference, SPE 86671

Melo MV et al (2006) Advances in non-conventional flotation for oily water treatment. Filtration 6(1):31–34

Part III
Modelling, Fate and Transport

Chapter 9
The DREAM Model and the Environmental Impact Factor: Decision Support for Environmental Risk Management

Mark Reed and Henrik Rye

Abstract DREAM (Dose-related Risk and Effect Assessment Model) is a three-dimensional, time-dependent numerical model that computes transport, exposure, dose, and effects in the marine environment. The model can simulate complex mixtures of chemicals. Each chemical component in an effluent mixture is described by a set of physical, chemical, and toxicological parameters. Because petroleum hydrocarbons constitute a significant fraction of many industrial releases, DREAM incorporates a complete surface slick model, in addition to the processes governing contaminant behavior and fates in the water column. The model can also calculate exposure, uptake, depuration, and effects for fish and zooplankton simultaneously with physical–chemical transport and fates. The Environmental Impact Factor (EIF), first developed for the water column, has been extended to include ecological stresses in the benthic community. The EIF is a standardized method for marine environmental risk assessment that does not require explicit information on the local biological resources. This makes the methodology relatively easy to apply to new geographical areas, and it has been used in northern and southern European, as well as in North American, South American, and African waters. This chapter gives an overview of the model system and a summary of ongoing developments.

1 Introduction

The numerical model DREAM (Dose-related Risk and Effect Assessment Model) has been developed at SINTEF with support from StatoilHydro, ENI, Total, ExxonMobil, Petrobras, ConocoPhillips, Shell, and British Petroleum. The model is a decision support tool for management of operational discharges to the marine environment. DREAM is integrated with the oil spill model OSCAR within a graphical user interface called the Marine Environmental Modelling Workbench. The system has been in continuous development for the past 15 years. A drilling discharge capability has recently been added to the system.

M. Reed (✉)
Division for Marine Environmental Technology, SINTEF Materials and Chemistry, Trondheim, Norway

K. Lee, J. Neff (eds.), *Produced Water*, DOI 10.1007/978-1-4614-0046-2_9,

DREAM is a three-dimensional, time-dependent, multiple-chemical transport, exposure, dose, and effects assessment model. DREAM can account simultaneously for up to 200 chemical components, with different release profiles for 50 or more different sources (Reed et al. 2001). Each chemical component in the effluent mixture is described by a set of physical, chemical, and toxicological parameters. Because petroleum hydrocarbons constitute a significant fraction of many industrial releases, DREAM incorporates a complete surface slick model, in addition to the processes governing contaminant behavior and fates in the water column. The model can also calculate exposure, uptake, depuration, and effects for fish and zooplankton simultaneously with physical–chemical transport and fates.

This software tool and the associated risk assessment methodology were designed and developed to support rational management of environmental risks associated with operational discharges of complex mixtures. The petroleum operators on the Norwegian Continental Shelf, the Norwegian Petroleum Directorate, and the Norwegian Pollution Control Authority agreed in the year 2000 to work toward a reduction of the potential environmental risks associated with operational offshore releases, primarily focused on produced water and drilling activities. The goal was to reach a level of "zero harmful discharge." To more clearly define this goal, the Environmental Impact Factor (EIF) was developed as an indicator of the potential impacts from produced water and drilling releases. A reduction in the EIF is used as a measure of the benefit achieved when alternate measures are considered for reducing potential environmental risks and is therefore also a measure of the degree of movement toward the goal of zero harmful discharge.

2 General Model Description

DREAM is a software tool designed to support rational management of environmental risks associated with operational discharges of complex mixtures. The model has evolved over a number of years (Reed 1989; Reed et al. 1996, 2002; Johnsen et al. 1998; Rye et al. 1998, 2008). Governing physical–chemical processes are accounted for separately for each chemical in the mixture, including

- vertical and horizontal dilution and transport,
- dissolution from droplet form,
- volatilization from the dissolved or surface phase,
- particulate adsorption/desorption and settling,
- biodegradation, and
- sedimentation to the sea floor.

The algorithms used in the computations, and verification tests of the resulting code, are presented in Reed et al. (2002). The model has also been verified against field measurements (Neff et al. 2006; Durell et al. 2006).

Chemical concentrations in the water column are computed from the time and space-variable distribution of pseudo-Lagrangian particles. These particles are of two types: those representing dissolved substances and those representing droplets composed of less soluble chemical components or solid particulate matter in the release. These latter particles are pseudo-Lagrangian in that they do not necessarily move strictly with the currents, but may rise or settle according to their physical characteristics.

Each mathematical particle represents conceptually a Gaussian cloud (or 'puff') of dissolved chemicals, droplets, or sinking particles, as described for example by Csanady (1973). Concentration fields are constructed by the model from the superposition of all of these clouds of contaminants. Each cloud consists of an ellipsoid with a particle at its center and semi-axes, a function of the time-history of the particle. (Ellipsoids encountering boundaries are truncated, with mass being conserved through reflection from the boundary, sorption to the boundary, or some combination of the two.)

Particles representing dissolved substances carry with them the following attributes:

- x, y, and z spatial coordinates,
- mass of each chemical constituent represented by the particle,
- distance to and identity of the nearest neighbor particle,
- time since release, and
- spatial standard deviations in x, y, and z.

Particles representing non-dissolved substances, such as oil droplets, drill muds, or cuttings, carry two additional attributes:

- mean droplet diameter, and
- droplet density.

Concentrations are computed within one of three user-specified three-dimensional grid systems. The first is a translating, expanding grid that follows the evolution of a release, thus providing higher resolution during the early stages and lower resolution as time progresses. The second is a fixed grid, with resolution defined by the user. The third is a grid with fixed horizontal resolution, but time-variable vertical resolution. This latter grid is useful, for example, in resolving surface releases of oil, in which the near-surface vertical evolution may be of particular interest.

As mentioned earlier, the position of each particle locates the center of a moving, spreading ellipsoidal cloud, with axes a function of the time-history of the particle. The theoretical distribution of mass within the ellipsoid is assumed Gaussian. Each such ellipsoid will typically contribute mass to many cells in the concentration field, and neighboring ellipsoids will typically overlap spatially. Thus a given cell in the concentration field will in general contain a concentration resulting from the presence of multiple nearby particle clouds. This hybrid numerical-analytic scheme removes much of the dependence of the computed concentration field on both the number of particles and the resolution of the physical three-dimensional grid.

The model is driven by winds and currents either produced by other numerical models or measured as time series in the region of interest. Global data sets of bathymetry and coastlines are supplied with the system and can be augmented by the user with standard GIS and/or ASCII formats.

Processes governing the behavior of contaminants in DREAM are presented in Fig. 9.1. DREAM employs surface oil spill model algorithms to simulate the behavior and fates of surface slicks. Such slicks can occur in the model as the result of rising oil droplets, or if oil is released at the air–water interface. In the water column, horizontal and vertical advection and dispersion of entrained and dissolved hydrocarbons are simulated by random walk procedures. Vertical turbulence is a function of wind speed (wave height) and depth; horizontal turbulence is a function of the age of a contaminant 'cloud.' Contaminants near the sea surface may evaporate to the atmosphere. Partitioning between particulate adsorbed and dissolved states is calculated based on linear equilibrium theory. The contaminant fraction that is adsorbed to suspended particulate matter settles with ambient particles. Contaminants at the bottom are mixed into the underlying sediments and may dissolve back into the water. Degradation in water and sediments is represented as a first-order decay process, with the possibility of producing intermediate metabolites. Results of model simulations are stored at discrete time-steps in data files for subsequent viewing and analysis.

For spilled oil, processes such as advection, spreading, entrainment, and vertical mixing in the water column are not directly dependent on oil composition, although all tend to be linked through macro-characteristics such as viscosity and density. Other processes, such as evaporation, dissolution, and degradation, are directly dependent on oil composition.

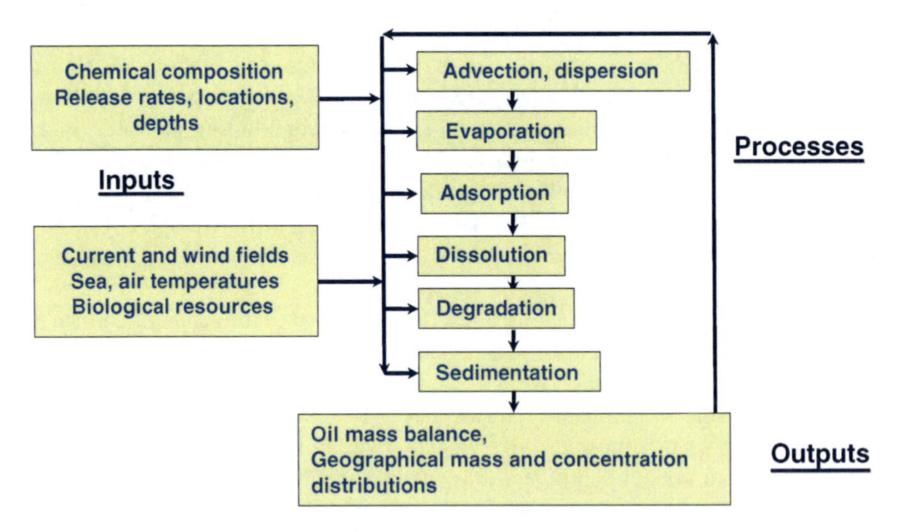

Fig. 9.1 General schematic of the DREAM model

DREAM focuses primarily on underwater releases, such that surface phenomena are of secondary interest. Oil droplets contained in produced water, for example, may rise to the surface and form a surface slick, such that related processes must also be represented in the model. DREAM uses the same algorithms for these processes as used in the oil spill contingency and response model OSCAR. These algorithms are described in detail in Reed et al. (2001).

The DeepBlow model (Johnsen 2000), developed in response to the interest in petroleum exploration in deep waters, has been generalized and serves as the three-dimensional dynamic near-field module for DREAM as well as for the oil spill model OSCAR. DeepBlow is a Lagrangian element model, the plume being represented by a sequence of elements. Each element, which can be thought of as a conical cylindrical section of a bent cone, is characterized by its mass, location, width (radius), length (thickness), average velocity, contaminant concentrations, temperature, and salinity. These parameters will change as the element moves along the trajectory, i.e., the element increases in mass due to shear-induced and forced entrainment, while rising or sinking according to buoyancy and becoming sheared over by the cross flow. This modified version, call Plume-3D, functions as a near-field module for produced water and drilling discharges, as well as other releases of complex mixtures in an aquatic environment. This module is activated automatically whenever a release is specified to originate under water. Depending on depth and other input parameters, the module automatically computes the near-field plume, and the release of dissolved, solid, and droplet-related contaminants from the plume and into the far field.

Figure 9.2 is a snapshot from an animation of simulation results in DREAM. Both dissolved concentrations of chemicals and distribution of drilling muds and cuttings are depicted, as is the distribution of previously deposited materials on the seafloor and the instantaneous mass balance. An EIF can be computed for each of the water column and the sediment compartments, and for each moment in time, such that the maximum and time-averaged values can be reported. To date, the maximum value has been used in the Norwegian management of produced water.

3 The Environmental Impact Factor

Development of the Environmental Impact Factor has been guided by the principle that areas of uncertainty should be resolved in favor of protecting the environment (i.e., conservative environmental assumptions are invoked). The methodology is therefore conservative in the sense of over-protecting rather than under-protecting the environment. On the other hand, the method is not designed to serve as an estimator of impact. It is rather a measure of environmental risk that is intended to be used to quantify the comparative benefit to the environment of alternate management strategies.

The EIF method is based on the ratio of an exposure concentration to a no-effect concentration, such that the concentration for each compound discharged into the recipient is compared to a concentration threshold for that compound. When the

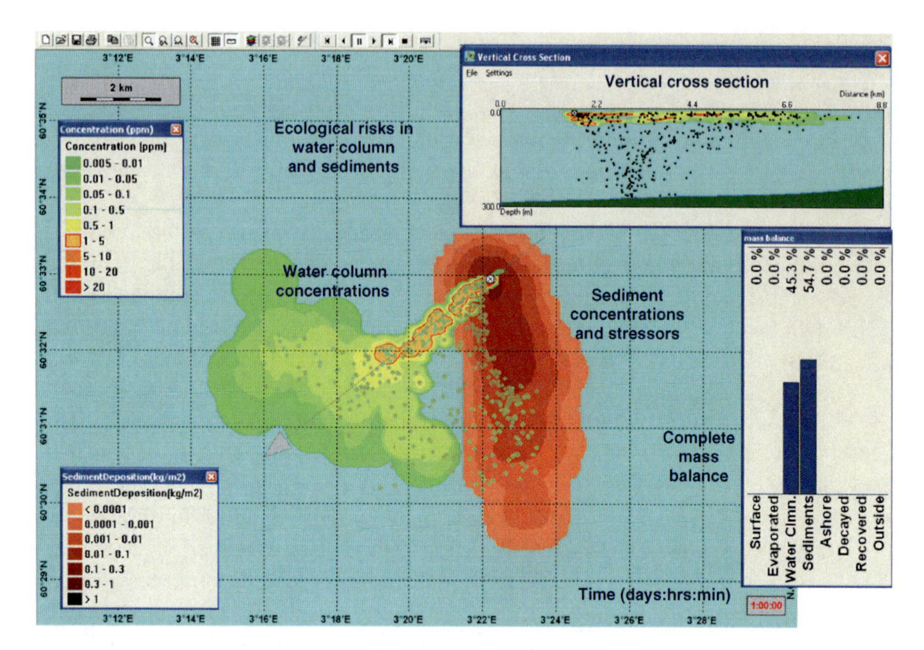

Fig. 9.2 Snapshot of output from the DREAM model, showing both the bird's-eye view and the vertical section: the release includes inorganic particles that sink to the seafloor, as well as chemicals that are carried in the water column, either in dissolved or in droplet form

predicted (modeled) environmental concentration (PEC) is larger than the probable no-effect concentration (PNEC), there may be a risk for ecological injury. When the PEC is lower than the PNEC threshold, the risk for injury from that single substance is considered to be acceptably near zero. By computing the environmental risk due to each component, and adding the risks as independent probabilities, the total risk at any given spatial point at any given time can be calculated (Johnsen et al. 2000).

The approach has the advantage that it provides a quantitative measure of the environmental risks involved when produced water is released into the sea and is thus able to form a basis for reduction of impacts in a systematic and a quantitative manner. It is important to note that the EIF reflects a level of environmental risk, but does not in any direct way measure impacts or effects. It is conservative in terms of risk assessment, but should only be used to compare the relative benefits of alternative mitigation actions, not as an impact assessment tool.

3.1 The Predicted Environmental Concentration (PEC)

The PEC is the three-dimensional and time variable concentration in the recipient caused by the discharge of the produced water. The PEC is calculated by the DREAM model for all compounds that are assumed to represent a potential for harmful impact on the biota. The calculations use the numerical DREAM model.

This model is fully three-dimensional and time variable. It calculates the fate in the recipient of each compound considered under the influence of

- currents (tidal, residual, meteorological forcing),
- turbulent mixing (horizontal and vertical),
- evaporation at the sea surface, and
- reduction of concentration due to biodegradation.

Figure 9.3 shows an example of a concentration field calculated with the DREAM model. The ocean current field used in the calculations is based on a hydrodynamic model operated by the Norwegian Meteorological Institute in Oslo. The output from the hydrodynamic model is used as input to DREAM. Further details are presented in Reed et al. (2001).

3.2 The Predicted No-Effect Concentration (PNEC)

The PNEC is the estimated lower limit for effects on the biota in the recipient for a single chemical component or component group. A PNEC level is given for each component in the produced water. It is derived from laboratory testing of toxicity for each component (or chemical product) in question. The PNEC value is derived from EC_{50}, LC_{50}, or NOEC values from laboratory testing, where the EC_{50}, LC_{50}, or NOEC value determined is divided by some assessment factor in order to arrive at the expected PNEC.

A major data collection work has been performed in order to obtain data of sufficient reliability to be selected for determination of PNEC values. Different procedures have been selected for determination of the PNEC values for natural constituents in produced water and for added chemicals. Table 9.1 shows the actual PNEC values used for natural compounds (or component groups) in produced water. For added chemicals, the PNEC values are usually based on the information found in the Harmonized Offshore Chemical Notification Format scheme. Further details can be found in Johnsen et al. (2000) and Smit et al. (2004).

3.3 EIF Produced Water

The following assumptions or model process specifications have been made with this guidance in mind.

- Simulations should be carried out during times when local biological resources are most vulnerable, either because of sensitivity of life stages or because of low turbulent mixing and resultant possibility of higher levels of exposure. In Norway, in the North and Norwegian Seas, the month of May has been identified as the month meeting both these criteria, by agreement between the authorities, research organizations, and the industry.

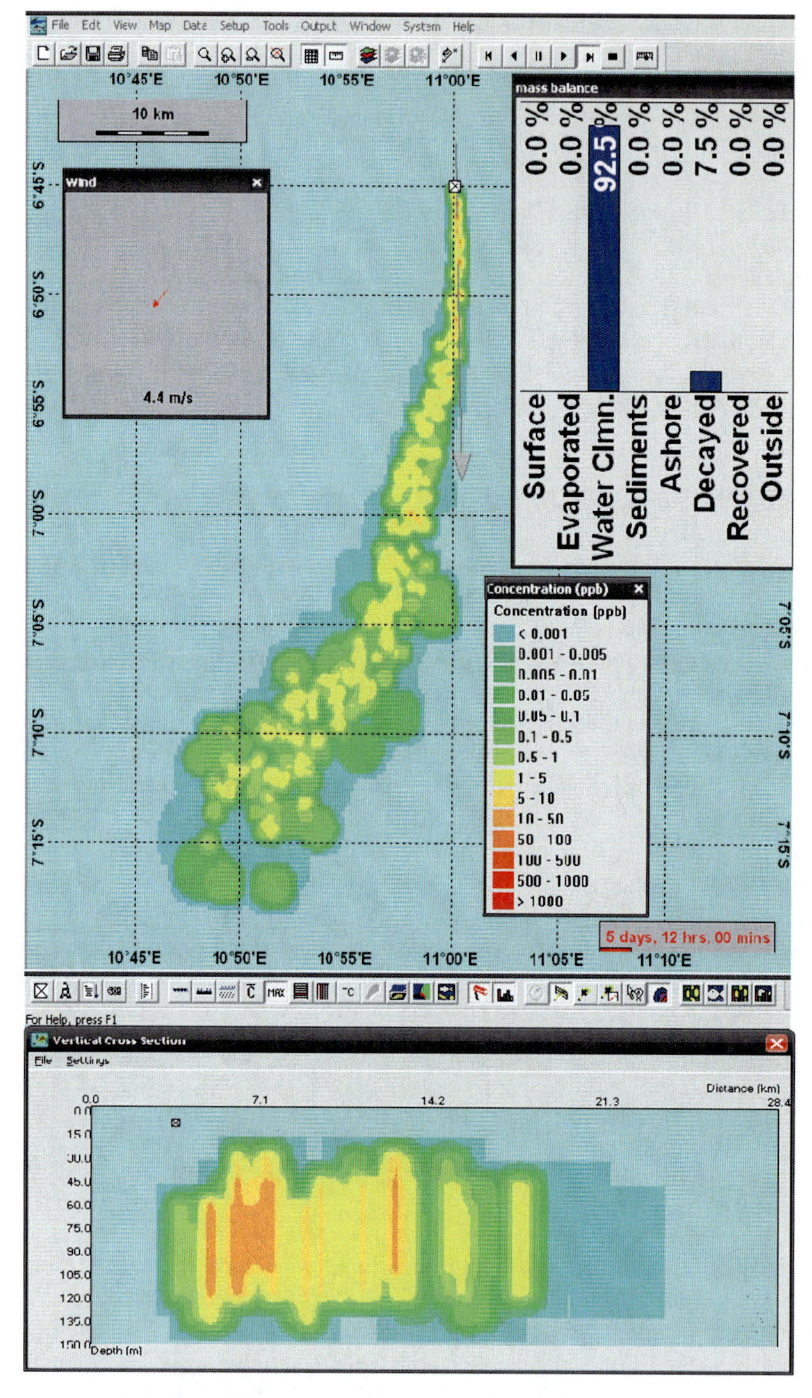

Fig. 9.3 Example concentration field for discharge of produced water simulated in DREAM, including a vertical section south through the discharge point at 10 m depth; the concentration field shown is the total concentration including all substances in the release; snapshot of the concentration field after 5.5 days of discharge

Table 9.1 PNEC values in parts per billion (ppb) for natural constituents in produced water applied to EIF calculations and presently applied in the DREAM EIF simulations for produced water; benzenes, toluenes, ethylbenzenes, xylenes (BTEX); polycyclic aromatic hydrocarbon (PAH)

Natural compounds	PNEC ppb
Dispersed oil	40.4
BTEX mono-aromatics	17
Naphthalenes	2.1
PAH 2–3 ring (excl. naphthalenes)	0.15
PAH 4–6 ring	0.05
Phenols C0–C3	10
Phenols C4–C5	0.36
Phenols C6 +	0.04
Zinc (Zn)	0.46
Copper (Cu)	0.02
Nickel (Ni)	1.22
Cadmium (Cd)	0.028
Lead (Pb)	0.182
Mercury (Hg)	0.008

- The simulations should account for potential effects of hydrocarbons, process chemicals, and any released solids, whether in dissolved, droplet, or solid form.
- When applied to the water column, both evaporation of released chemicals from the upper water column to the atmosphere and sequestering of contaminants through adsorption and settling are de-activated during the computational process. This was decided during initial trials of the system in response to the difficulty of getting the necessary chemical parameters for the many industrial chemical products in use during produced water processing offshore, while still adhering to the conservative principle.
- The EIF is defined as the *maximum* value over a 30-day simulation (the month of May in the Norwegian applications).

3.4 EIF for Drilling Discharges

The EIF for drilling discharges is necessarily more complex than that for produced water, since both the water column and the seafloor are involved. There are also additional stress factors beyond toxicity that need to be taken into account, including effects of particles in the water column, changes in particle size on the seafloor, burial, and potential oxygen depletion in the sediments due to biodegradation of organic components in the discharge. These issues are only mentioned here and are reviewed in more detail in Rye et al. (2006, 2008), Singsaas et al. (2008), and Smit et al. (2008).

3.5 Environmental Risk and the EIF

The EIF for a single component or component group is related to the recipient water volume where the ratio PEC/PNEC exceeds unity. The ratio PEC/PNEC is related to the probability of biological injury according to a method developed by Karman

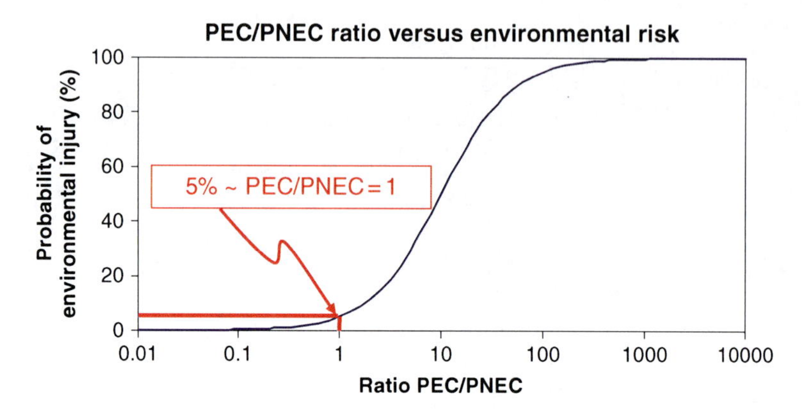

Fig. 9.4 Relation between the PEC/PNEC level and the risk level (in %) for injury to biota (based on Karman 1994); PEC/PNEC = 1 corresponds to a level at which there exists a possibility of injury to the 5% most sensitive species

(1994) (and also published in Karman and Reerink 1997). When PEC/PNEC = 1, this corresponds to a level at which there exists a possibility of injury to the 5% most sensitive species. Figure 9.4 shows the relation between the PEC/PNEC ratio and the probability of injury.

The EIF method has the advantage over other risk assessment methods in that it can calculate risk contributions from a sum of chemicals and/or natural compounds in the recipient. The total risk for a mixture of chemicals is calculated using formulas for the sum of independent probabilities. For two components A and B, the total risk is the sum of the individual probabilities minus their intersection:

$$P\,(A + B) = P\,(A) + P\,(B) - P\,(A) \times P\,(B) \tag{9.1}$$

where $P(A)$ and $P(B)$ are the probability of environmental risk from each compound at a particular time and spatial location. For small risks (that is, $P(A)$ and $P(B)$ are both small), or risks from chemicals which are toxicologically similar in their activity, the risks can be considered to be additive. The method does not account for interactions among chemicals. For n components, the generalized formula is

$$P_{\text{total}} = 1.0 - \Pi_i^n\,(1.0 - P_i) \tag{9.2}$$

The total risk resulting from all components in a release is calculated by the DREAM model in space and time within the model domain. The resultant three-dimensional risk fields can then be viewed as a time series, a snapshot example being shown in Fig. 9.5. Results can also be presented as risk in percent. The water volume indicated by red then indicates the water volume where the nominal PEC/PNEC is larger than one. Note that the PEC/PNEC ratios for all individual components in the release may be less than unity, but the cumulative risk from all components may exceed 5%, such that the nominal PEC/PNEC ratio produced by the

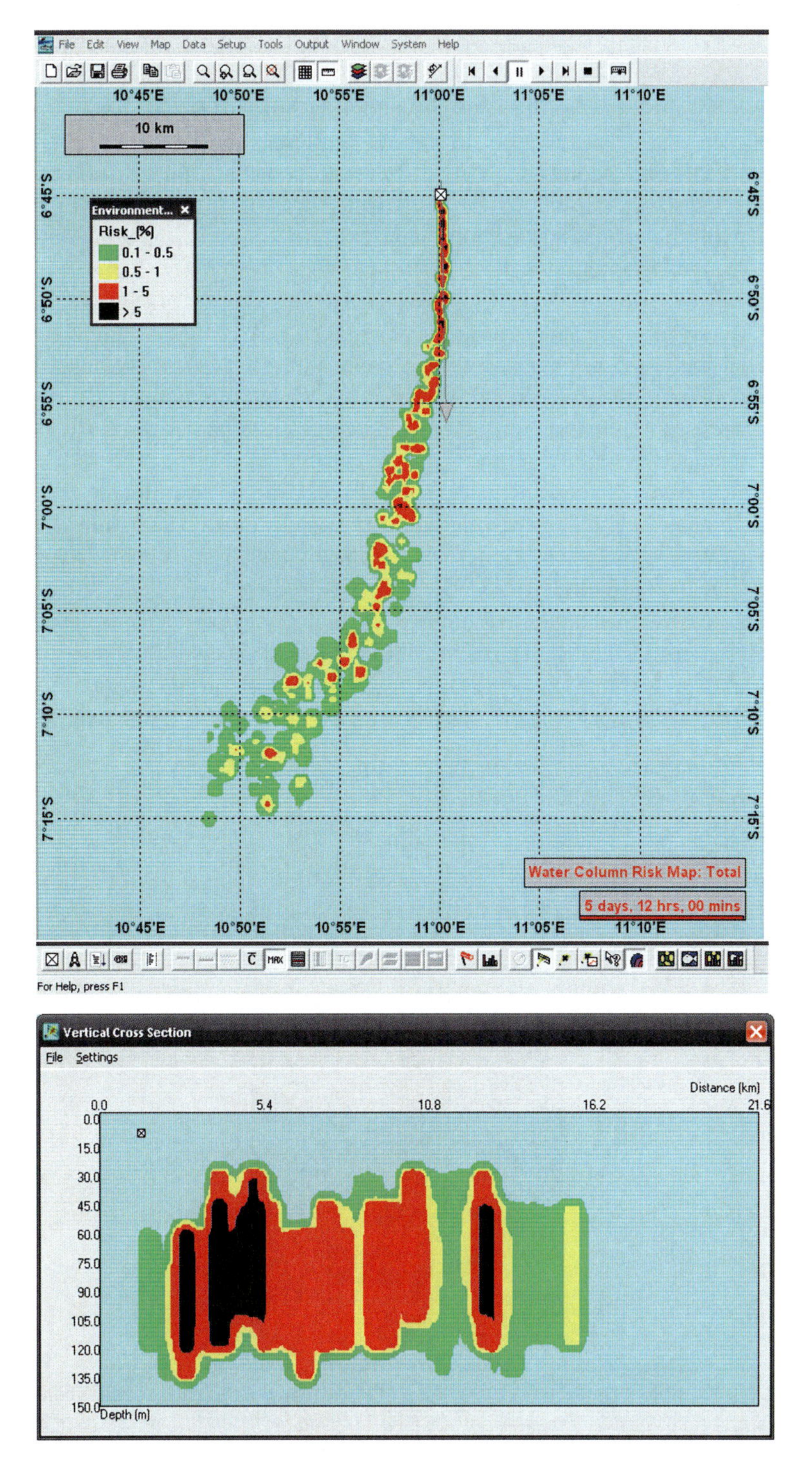

Fig. 9.5 Snapshot of risk field and vertical section, corresponding to the concentration field shown in Fig. 9.3 at 5.5 days into the simulation

procedure described above, and representing a conglomerate value for the release, exceeds unity.

An EIF of unity is defined as a water volume 100 m × 100 m × 10 m $\left(10^5\ \mathrm{m}^3\right)$ in which there is a risk of injury to the 5% most sensitive species. For a single component, this corresponds to a PEC/PNEC ratio exceeding unity. In addition, the EIF water volume is adjusted upward by a factor of two for those compounds that have a small biodegradation factor combined with a large bioaccumulation factor. Details are given in Johnsen et al. (2000).

An attractive feature of the EIF approach is that the method is able to discriminate among the various contributors to environmental risk. An example of the distribution of contribution to risk among components in a release at a specific point in time is shown in Fig. 9.6. This capability provides useful information when comparing alternative proposed methodologies for reducing environmental risks associated with a discharge.

Thus it is possible to separate a chemical product into its constituents and calculate the EIF contribution from each of them. The results of the calculations can then be used to improve the product in terms of replacing the constituents in the product with the largest contribution to the EIF.

As a standard for calculation of the EIF for the North Sea area, an ocean current field database for the North Sea is applied (the OLF ocean current data base). This current database 5 years (1990–1994) for the North Sea area. Time resolution is 2 h

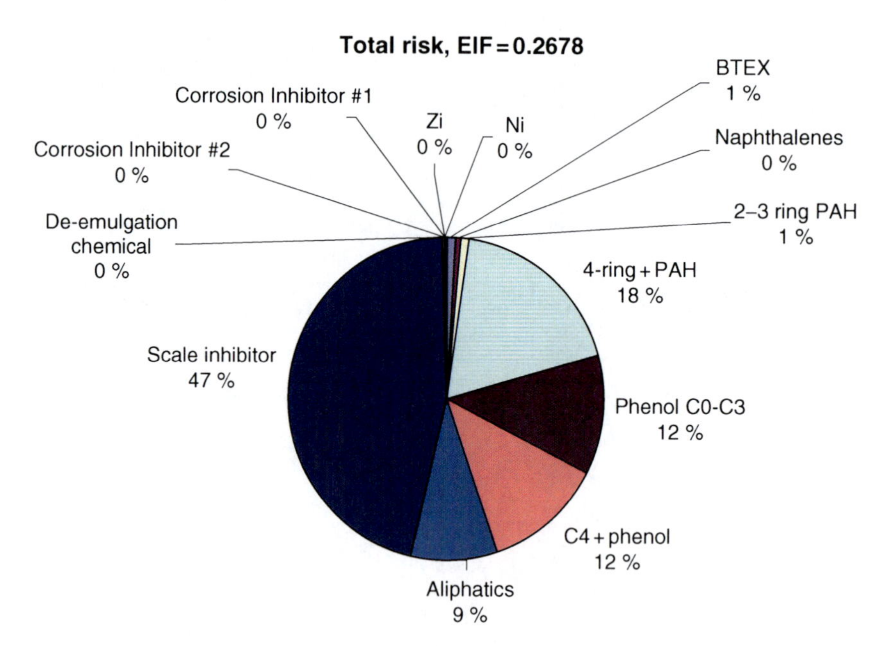

Fig. 9.6 Distribution of contribution to risk for an EIF calculation; the scale inhibitor contributes 47% of the total risk; standard procedure in Norway is to report the maximum EIF recorded over the entire simulation time as a conservative measure of risk

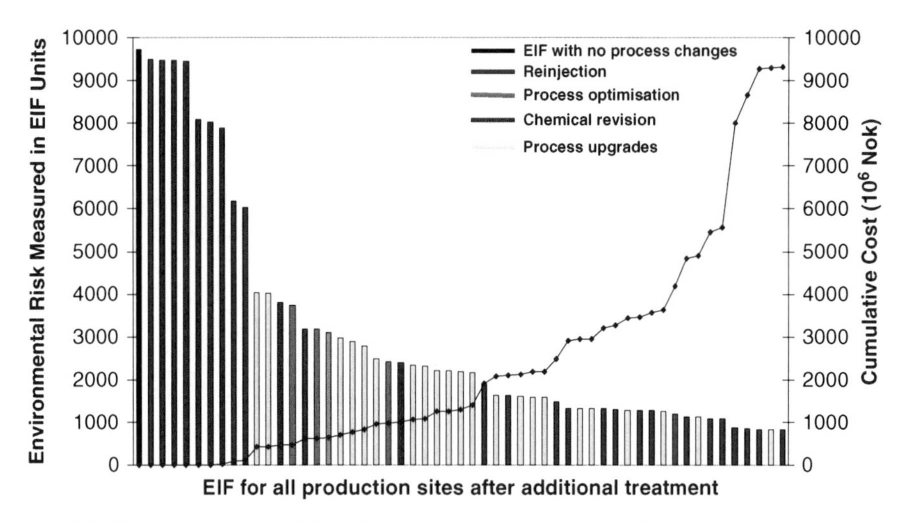

Fig. 9.7 Reduction in potential environmental risk, measured in EIF units, as a function of investment in water treatment technologies (based on SFT 2004)

and grid size is 20 km in the horizontal. A number of layers are included in order to account for vertical variations in the currents. The month of May 1990 is used as a standard for the EIF calculations for the North Sea area. The standard procedure in Norway is to report the maximum EIF recorded over the entire simulation time, as a conservative measure of risk. The time-averaged value is also available from DREAM.

The EIF is calculated on a yearly basis. The total EIF for the discharge is then summed over the expected lifetime of the platform or field. When a risk-reduction measure is considered, the revised EIF is calculated for each year when the measure is implemented. The delta EIF is then also summed over the years in question. The cost efficiency of the measure can then be calculated. Figure 9.7 depicts the cumulative reduction in the potential environmental risk, as reflected in the EIF measure, for progressively costly treatment strategies, as applied to a constellation of production platforms in the Norwegian sectors of the North and Norwegian Seas. Thus the EIF can provide a quantitative approach to support environmental management within a cost-benefit framework.

4 Conclusions and Future Directions

The Marine Environmental Modelling Workbench, of which DREAM is a central part, is a work in progress. There is always room for improvement by inclusion of new capabilities, and in algorithms, parameter estimates, and coding practices. By working in the direction of increased modularity in the structure of the underlying code, we will assist present and future developers in maintaining the goal of constant improvement of the tools.

DREAM was originally developed with an eye toward offshore releases of produced water, such that input variables and internal process representations could be on scales of kilometers for many calculations. Present work is focused on implementing algorithms that will function well in nearshore areas, such that discharges from land-based industry to marine and freshwater environments can be addressed. Additional capabilities now being implemented and tested include dilution of cooling water discharges, potential effects of particulate organic matter in an effluent, and potential oxygen deficits in the receiving water due to biodegradation of the effluent. Associated with each characteristic of a release is an environmental risk curve of the type shown in Fig. 9.4, such that the total environmental risk may be composed of several stressors besides toxicity. This approach allows a relatively wide variety of stress factors to be integrated into a single risk measure.

The EIF is not a measure of environmental impact. It is a reflection of potential environmental risk and is conservative in that it assumes the most sensitive species always to be present. It is therefore a relative measure of potential risk, not an absolute measure in that no actual impact is necessarily implied. The EIF provides an objective quantitative measure of risk that has proven to be a very useful decision support tool for environmental management.

Aknowledgments Development of the DREAM model has been supported by the oil companies ConocoPhillips, Eni, ExxonMobil, Hydro, Petrobras, Shell, Statoil, and Total. The companies are acknowledged for financial support as well as scientific input during development. SINTEF, a non-profit research foundation, has also contributed to the development through internal funding procedures.

References

Csanady GT (1973) Turbulent diffusion in the environment. D. Reidel Publishing Company, Dordrecht, Holland, 246 pp

Durell G, Utvik TR, Johnsen S, Frost T, Neff J (2006) Oil well produced water discharges to the North Sea. Part I: comparison of deployed mussels (Mytilus edulis), semi-permeable membrane devices, and the DREAM model predictions to estimate the dispersion of polycyclic aromatic hydrocarbons. Mar Environ Res 62(2006):194–223

Johansen Ø (2000) DeepBlow, a Lagrangian plume model for deep water blowouts. Spill Sci Technol Bull 6(2):103–111

Johnsen S, Frost TK, Hjelsvold M, Utvik TR (2000) "The environmental impact factor – a proposed tool for produced water impact reduction, management and regulation". SPE paper 61178 presented at the SPE international conference on health, safety and environment in oil and gas exploration and production held in Stavanger, Norway, 26–28 June 2000

Johnsen S, Røe TI, Durell GS, Reed M (1998) Dilution and bioavailability of produced water compounds in the northern North Sea. A combined modeling and field study. SPE paper 46269. 1998 SPE international conference on HSE in oil and gas E&P, Caracas, Venezuela, 7–10 June 1998, 11 pp

Karman C (1994) "Ecotoxicological risk of produced water from oil production platforms in the Statfjord and Gullfax fields". TNO Environmental Sciences. Laboratory for Applied Marine Research, den Helder, The Netherlands. Report TNO-ES, February 1994

Karman C, Reerink HG (1997) Dynamic assessment of the ecological risk of the discharge of produced water from oil and gas producing platforms. Paper presented at the SPE conference in 1997, Dallas, USA. SPE paper No. SPE 37905

Neff J, Johnsen S, Frost TK, ROE Utvik TI, Durell GS (2006) Oil well produced water discharges to the North Sea. Part II: comparison of deployed mussels (Mytilus edulis) and the DREAM model to predict ecological risk. Mar Environ Res 62(2006):224–246

Reed M (ed) (1989) Oil and chemical pollution, special issue: resource damage assessment in the marine environment. Elsevier Appl Sci 5(2 and 3):85 pp

Reed M, Hetland B (2002) DREAM: a dose-related exposure assessment model technical description of physical-chemical fates components. SPE 73856

Reed M, Johnsen S, Rye H, Melbye A (1996) PROVANN: a model system for assessing exposure and bioaccumulation of produced water components in marine pelagic fish, eggs, and larvae. In: Reed M, Johnsen S (eds) Environmental aspects of produced water. Proceedings of the 1995 international produced water seminar, Trondheim, Norway. Plenum Press, NY, pp 317–332

Reed M et al (2001) DREAM: a dose-related exposure assessment model. Technical description of physical-chemical fates components. Proceedings of the 5th international marine environmental modelling seminar, New Orleans, USA, Oct. 9–11 2001

Rye H, Reed M, Durgut I, Ditlevsen MK (2006) The use of the diagentic equations to predict impact on sediment due to discharges of drill cuttings and mud. Paper presented at the 9th international marine environmental modelling seminar, Rio de Janeiro Brazil, October 9–11, 2006

Rye H, Reed M, Ekrol N, Johnsen S, Frost T (1998) Accumulated concentration fields in the North Sea for different toxic compounds in produced water. SPE 46621

Rye H, Reed M, Frost TK, Smit MGD, Durgut I, Johansen Ø, Ditlevsen MK (2008) Development of a numerical model for calculation of exposure to toxic and non-toxic stressors in the water column and sediment from drilling discharges. SETAC J Int Environ Assess Manage 4:194–203

SFT (2004) Offshore petroleum operators' work to reach the goal of zero harmful releases (in Norwegian). www.sft.no/publikasjoner/vann/1996/ta1996.pdf

Singsaas I, Rye H, Frost TK, Smit MGD, Garpestad E, Skare I, Bakke K, Veiga LF, Buffagni M, Follum OA, Johnsen S, Moltu UE, Reed M (2008) Development of a risk-based environmental management tool for drilling discharges. Summary of a Four-Year project. SETAC J Int Environ Assess Manage 4:171–176

Smit MGD, Holthaus KIE, Tamis JE, Karman CC (2004) From PEC_PNEC ratio to quantitative risk level using Species Sensitivity Distributions; Methodology applied in the Environmental Impact Factor, TNO-report R 2004

Smit MGD, Jak RG, Rye H, Frost TK, Singsaas I, Karman CC (2008) Assessment of environmental risks from toxic and nontoxic stressors: a proposed concept for a risk-based management tool for offshore drilling discharges. SETAC J Int Environ Assess Manage 4:177–183

Chapter 10
Diffuser Hydraulics, Heat Loss, and Application to Vertical Spiral Diffuser

Maynard G. Brandsma

Abstract The ready availability of plume models allows analysts to predict plume behavior and performance of wastewater outfalls. Such models need inputs describing the flow rate of effluent; the number, diameter, and depth of discharge ports; and fluid properties of the effluent and receiving water and the water depth. Unfortunately, there is a crucial step in outfall design that is oftentimes neglected—hydraulic analysis of the outfall. Hydraulic analysis predicts the flows from each port of a diffuser and the hydraulic head needed to drive the diffuser for a required total effluent discharge rate. Hydraulic analysis is especially important to a relatively new class of wastewater outfall suitable for offshore oil and gas facilities, the vertical spiral diffuser. This chapter describes the rules of thumb for outfall design, the influence of the major design parameters on vertical diffuser behavior, the methodology for calculating the hydraulic performance of vertical diffusers, and an illustration of vertical spiral diffuser performance. The reconciliation of current regulatory practices in US federal waters with correct hydraulic analysis is also discussed.

1 Introduction

Produced water is the principal waste stream of offshore oil and gas production. The water is typically heated and has a salinity that differs from that of the receiving water body. Offshore oil and gas installations treat their produced water to remove as much of the hydrocarbons as possible before discharging this water overboard. The wastewater is discharged to an outfall that may be a single pipe or a multiple-port diffuser.

In his consulting practice, the author has repeatedly come across diffuser installations that do not provide the dilution performance that the installation designers and owners think exists. The author has also observed that regulatory practices of the US Environmental Protection Agency foster deficient designs of vertical diffusers. The

Brandsma: deceased.

K. Lee (✉)
Centre for Offshore Oil, Gas and Energy Research (COOGER), Fisheries and Oceans Canada, Bedford Institute of Oceanography, Dartmouth, NS, Canada

K. Lee, J. Neff (eds.), *Produced Water*, DOI 10.1007/978-1-4614-0046-2_10,

deficiencies arise from apparent ignorance of hydraulic and heat loss principles that must be accounted for in diffuser designs, especially ones with vertically distributed ports. This chapter attempts to correct this situation by assembling information on design rules of thumb and calculation methods for hydraulic performance and heat loss from heated effluents. While the information is available in literature dating from the 1950s, 1960s, and 1970s, it is difficult to find and apparently not well known.

The goal of any diffuser design should be to improve initial mixing by taking maximum advantage of the laws of nature. To do this, the diffuser must distribute the flow evenly along its length. The designer has control of initial mixing (in the near-field) only. Following initial mixing, waste plumes are said to exist in the far-field and their behavior is controlled by currents and ambient turbulence. A diffuser works by spreading a wastewater stream over a much larger cross-section of the ambient water flux than could be achieved if the wastewater were discharged from a single pipe.

Plume models and tables of dilution issued by regulatory authorities are frequently used to design multiple-port wastewater diffusers. The diffusers may or may not work as designed, depending on whether or not hydraulic and thermal issues have been considered.

This chapter attempts to collect in one place the analysis techniques required to correctly determine the thermal and hydraulic behavior of a diffuser so that plume models and tables of dilution can be correctly applied.

2 Consequences of Neglecting Hydraulics

Region 6 (which includes US water in the western portion of the Gulf of Mexico) of the US Environmental Protection Agency issued a general permit under the National Pollutant Discharge Elimination System (NPDES). The general permit provides a set of critical dilution concentration tables showing precalculated effluent concentrations at the 100 m regulatory mixing zone boundary for effluent discharges from single vertical pipes based on three parameters, the distance between the bottom of the pipe and the seabed, the discharge rate and the diameter of the discharge pipe (http://www.epa.gov/Region6/6wq/npdes/genpermt/gmg29000finalpermit2007.pdf). Other important parameters used in calculating the tables are current speed = 0.1 m/s, effluent density = 1070 kg/m^3, the ambient density gradient = 0.15 kg/m^3/m, and the ambient density at the mouth of the discharge pipe = 1017 kg/m^3. Operators are allowed to increase mixing by using a horizontal diffuser that must be analyzed with CORMIX (Doneker and Jirka 1990) or by using multiple vertically spaced ports separated by prescribed distances based on a port flow rate. The general permit states that "The critical dilution [concentration] value shall be based on the port flow rate (*total flow rate divided by the number of discharge ports*) and based on the diameter of the discharge ports (or smallest discharge port if they are of different styles)." The portion of the permit language in italics ignores the role hydraulics plays in determining the flows from each port. For multiple vertically aligned ports, the allowable critical dilution concentration is provided by the permit's tables for single port discharges, assuming sufficient vertical separation of ports. The tables of critical dilution concentrations have gone through several iterations, and it has been advantageous for operators to provide large diameter discharge ports that yielded lower critical dilution concentrations from multiple vertically spaced ports. Thus operators operating in deep enough water would divide their total effluent discharge rate by a number of ports sufficient to provide a port flow rate included in the range of the tables and select the port diameter predicted to provide the best dilution for the port flow rate. Following this process inspired one operator to have an outfall consisting of two 0.5 m (20 in.) diameter ports spaced 6.7 m (21.7 ft) one above the other, connected by a 0.25 m (10 in.) diameter downpipe, discharging 0.046 m^3/s (25,000 bbl/day). Based on the requirements of the general permit, the facility operator imagined that the plume issuing from the two ports looked like Fig. 10.1a and provided a high dilution ratio (low critical dilution concentration). A hydraulic analysis showed that, in reality, all the produced water was issued from the lower port (Fig. 10.1b), and that ambient water was actually sucked into the upper port. The consequence was that the actual dilution of the produced water was significantly less than predicted. The Offshore Operators Committee (OOC) discharge model (Brandsma et al. 1992; Brandsma and Smith 1999) was used to independently calculate the concentration distributions shown in Fig. 10.1 and to predict dilutions of 128:1 and 65:1 for the plumes in Fig. 10.1a and b, respectively.

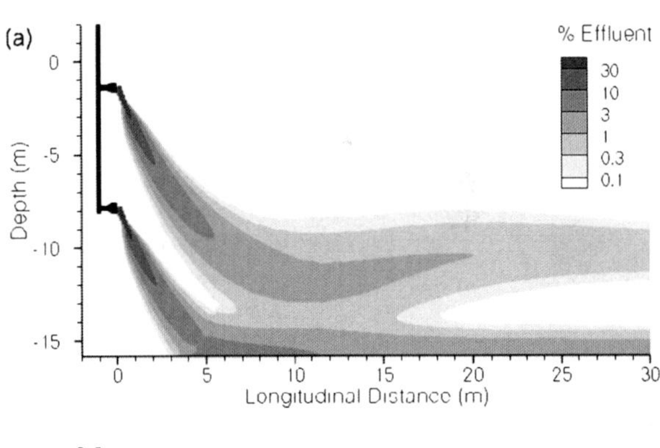

(b)
Depth (m)
0
-5
-10
-15
0
5
10
15
20
25
30
Longitudinal Distance (m)

Fig. 10.1 Behavior of plumes issuing from two-port vertical diffuser with incorrectly sized ports: (**a**) as implicit in the US EPA Region 6 general permit and (**b**) as calculated using hydraulic principles (*arrows* indicate ambient water flowing into the upper port)

3 Diffuser Design Considerations

A produced water effluent is typically heated well above the temperature of the receiving water and may have a salinity ranging from fresh water to very salty (100 parts per thousand or more). The design goals for a horizontal diffuser to increase the initial mixing of such an effluent include (1) an even flow distribution, (2) no seawater intrusion into the diffuser (all ports are filled with effluent), and (3) minimum hydraulic head (only low pressure needed to drive the diffuser). For offshore oil and gas installations, horizontal diffuser lengths are typically limited to the dimensions of the structure available to support the diffuser. In shallow water, longer diffusers supported on pilings may be used.

When a vertical diffuser is used instead of a horizontal one, some additional considerations arise, namely (1) the vertical diffuser is much more sensitive to the differences in density between the effluent and the ambient receiving water; (2) port sizes must be precisely varied to govern port flows; (3) port sizes that are too large will allow effluent to escape from the diffuser in an uncontrolled way.

4 Influence of Major Design Parameters on Vertical Diffuser Performance

Consider the third point in the previous paragraph in relation to an effluent being discharged from process equipment into a three-port diffuser with too-large ports. The effluent surface inside the diffuser is exposed to atmospheric pressure. If the effluent density is higher than the ambient density, the effluent will seek the deepest possible port to exit the vertical diffuser, and the level of the top of the effluent in the diffuser will lie below the mean sea level outside the diffuser. If the effluent density is lower than the ambient density, the effluent will seek the shallowest possible port to exit the vertical diffuser, and the top surface of the effluent will lie above the mean sea level. If the diffuser ports are large enough that the flow distribution between ports is not controlled, some or all of a dense effluent will escape from the deepest port(s). Similarly some or all of an effluent that is less dense than the surrounding effluent will escape from the shallowest port(s). These behaviors are illustrated in Fig. 10.2.

5 Design Rules of Thumb

The problems illustrated in Figs. 10.1 and 10.2 arose because the ports of the two- and three-port vertical diffusers were too large to regulate the flows issuing from them. In Fig. 10.1b, there was no constraint on wastewater flow, and the wastewater took the path of least resistance by escaping from the lower port. In order to force a more even distribution of flow, port diameters small enough to pressurize the wastewater within the diffuser are required. The job of designing an appropriate

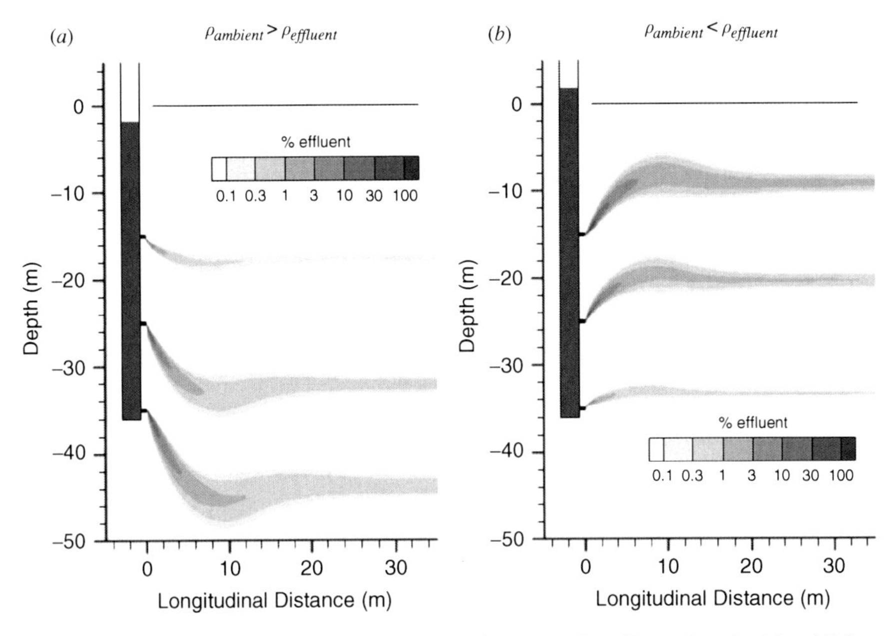

Fig. 10.2 Illustration of unbalanced flows from too-large ports for effluents heavier (**a**) and lighter (**b**) than the ambient fluid

diffuser is aided by observing certain rules of thumb, such as these suggested by Fischer et al. (1979).

1. The port diameters should be relatively small. The optimal ratio of the total area of ports downstream of any point in the diffuser to the diffuser pipe cross-sectional area at that point should be between 1/3 and 2/3. This allows reasonably even flow distribution between ports.
2. Each port should flow full. The densimetric Froude number of the flow from each port should be at least 1 and preferably 2 or more. (The densimetric Froude number is the ratio of kinetic to potential energy of the effluent leaving a port and is $F = \frac{u}{\sqrt{g\frac{\Delta\rho}{\rho}d}}$, where u is the port exit velocity, g is the acceleration of gravity, ρ is density, $\Delta\rho$ is the density difference between effluent and ambient, and d is the port diameter.)
3. It is essential that the distal end(s) of the diffuser be capped with (a) removable plate(s) in order to force effluent out the ports. The plate(s) should be removable to aid maintenance.
4. The hydraulic head to drive the diffuser may have to be minimized consistent with all the ports flowing full (being fully occupied by exiting wastewater).

In cases where flows through diffusers vary significantly, the first rule may have to be bent by allowing ratios of less than 1/3. Ratios of greater than 2/3 should not be used.

6 Other Important Issues

The type of diffuser selected will naturally depend on the characteristics of the site. In shallow water, a horizontal diffuser (ports aligned with a horizontal line) will be the natural choice. At deep-water sites, a vertical or near vertical diffuser with its ports aligned with a platform leg may be more appropriate. A vertical diffuser is much more sensitive to density differences between the wastewater issuing from the outfall and the receiving ambient water, especially if the ports are too numerous or too large. With a vertical diffuser, port diameters of a diffuser providing reasonably uniform flow from all ports can be expected to vary significantly. With a horizontal diffuser, a uniform port diameter is likely to work well. Figure 10.2 illustrated the behavior of wastewater effluents that are respectively heavier than and lighter than the average density of a stratified receiving water body. Notice the surface of a heavy effluent inside the vertical diffuser is below the receiving water surface and the surface of a light effluent is above the receiving water surface. The heavy plumes sink and the lighter plumes rise before becoming trapped by the stratification.

Cooling of heated produced water reduces buoyancy as the produced water effluent approaches the distal end of the diffuser. When a plume model is used, especially for a vertical diffuser, more accurate results are likely if the correct effluent temperatures from each port are used. In the case of a long outfall with a multiport diffuser at the end, the average temperature at the midpoint of the diffuser will be sufficient. In any case, wastewater temperature changes due to heat loss can be an important consideration.

7 Overview of Manifold Calculations for Diffuser Hydraulics

A satisfactory outfall design (vertical or horizontal) must properly account for hydraulics to establish effluent flows from each port of the outfall. When the temperature of an effluent entering the diffuser differs significantly from the ambient temperature, the heat lost from the effluent as it passes through the diffuser must also be accounted for. The dilution of effluent discharged from each port is strongly dependent on the effluent's initial momentum and buoyancy at the port exit. The momentum is dictated by the port flow rate. The buoyancy of produced water is dictated by the salinity and temperature of the effluent. The effluent buoyancy is a critical factor in the design of a vertical multiport outfall because a buoyant fluid forced into a denser fluid will try to escape the outfall as quickly as possible. If the ports are too large, all the effluent will flow from the upper ports and seawater will flow into the lower ports. If the ports are too small, hydraulic head requirements will be unduly large. The design process is thus an interactive one in which repeated trials are made while varying the following:

- Layout of ports (port depths and diameters)
- Selection of manifold (downpipe) diameter(s)
- Hydraulic calculations to establish port flows

- Heat loss calculations to establish effluent temperature at each port
- Dilution modelling
- Checks for non-interference of plumes from adjacent ports.

8 Manifold Hydraulics

The hydraulic calculations follow the methods outlined by Rawn et al. (1961) and summarized more recently by Fischer et al. (1979), with some additional refinements by the present author.

As will be evident shortly, it is not possible to specify a particular total flow rate and explicitly calculate flows from individual ports. Instead, one must begin at the port furthest from the effluent inlet (port #1) and estimate a hydraulic head for that port. Hydraulic head is the total energy per unit weight, expressed as a distance above some datum. The port flow is calculated from the head. Then the difference in hydraulic head from port #1 up to port #2 is calculated based on friction losses and effluent and ambient densities. This determines the hydraulic head at port #2 and in turn the flow rate at port #2. This process continues up the diffuser until all port flows are calculated. The additional hydraulic head up to the sea surface is also calculated. The results are the total flow consisting of the sum of the individual port flows and the total hydraulic head at the sea surface (which will be the height of effluent inside the manifold (downpipe) above the sea surface). The following development assumes that each port is at the end of a riser assembly consisting of a tee fitting in the diffuser pipe, a riser consisting of a straight section, a bend, and a tapered nozzle ending in the port diameter (Fig. 10.3). (If the length of taper is zero, the nozzle will be an orifice plate.)

Begin with the following variables using any consistent system of units:

n	port number, beginning with 1 at the far end
D	diameter of diffuser (manifold or downpipe)
D_r	diameter of riser pipe
d_n	diameter of the nth port
a_n	area of the nth port
V_n	the mean velocity in the diffuser between the nth and $(n+1)$th port
ΔV_n	$V_n - V_{n-1}$, the increment of velocity in the diffuser due to discharge from the nth port
h_n	$\Delta P_n/\gamma$ = the difference in pressure head between the inside and outside of the diffuser pipe just upstream of the nth port
E_n	$h_n + \frac{V^2}{2g}$ = the total head at the nth port
C_D	discharge coefficient for ports (varies by port)
Q_n	discharge from the nth port
h_{fn}	friction loss between $(n+1)$th and nth ports
L_n	the distance between $(n+1)$th and nth ports
L_r	total length of straight sections of riser pipe
L_c	length of contraction section of nozzle ending in port diameter

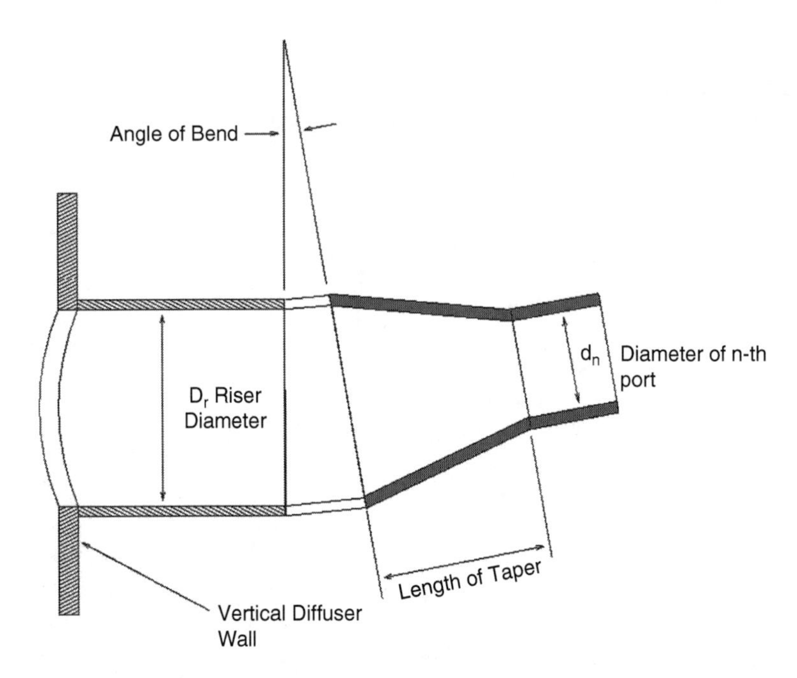

Fig. 10.3 Cross-section of vertical diffuser with riser port

f	the Darcy–Weisbach friction factor
Δz_n	the change in elevation between the $(n+1)$th and the nth port (positive when the $(n+1)$th port is shallower than the nth port)
$\Delta s/s$	the relative difference of specific gravities between the effluent and ambient fluid
g	acceleration of gravity.

To begin, we must select a value for E_1. Then the flow, q_1, for the first port is

$$q_1 = C_\mathrm{D} a_1 \sqrt{2\,gE_1} = C_\mathrm{D} \frac{\pi}{4} d_1^2 \sqrt{2\,gE_1} \tag{10.1}$$

Next determine the velocity in the diffuser pipe:

$$V_1 = \Delta V_1 = \frac{q_1}{\frac{\pi}{4} D^2} \tag{10.2}$$

Moving up the diffuser to port #2, determine the hydraulic head there as

$$E_2 = E_1 + h_{\mathrm{f}1} + \frac{\Delta s}{s} \Delta z_1 \tag{10.3}$$

The ratio $\left(V_1^2/2\,g\right)/E_2$is calculated to determine the coefficient of discharge for the port and port riser assembly, using

$$C_D = \frac{-r^2\left(\frac{V}{\sqrt{2gE}}\right) + \sqrt{X\left(1 - \frac{V^2}{2gE}\right) + r^4}}{X + r^4} \tag{10.4}$$

where r is the ratio of port diameter to diffuser diameter (d_n/D), and

$$X = \frac{1}{C_c^2} + \left(x_{en} + f\frac{L_r}{D_r} + x_{bend} + x_c\right)\left(\frac{d_n}{D_r}\right)^4 \tag{10.5}$$

where x_{en}, x_{bend}, and x_c are the head loss coefficients for the entrance from the diffuser pipe to riser, elbow, and nozzle contraction, respectively. (There is a typographical error in the Fischer et al. (1979) version of Eq. (10.4); the radical sign over the second term of the numerator is missing.) The contraction coefficient for the tapered section of the nozzle, C_c, can be calculated using free streamline theory and is a function of d_n/D_r and the angle $\alpha = \tan^{-1}\left(\frac{D_r - d_n}{2L_c}\right)$. Fischer et al. (1979) give a table of values for C_c.

With the discharge coefficient, C_D, for port #2 determined, carry on to find the port flow:

$$q_2 = C_D\frac{\pi}{4}d_2^2\sqrt{2gE_2} \tag{10.6}$$

Next, find the velocity in the diffuser pipe upstream of port #2:

$$V_2 = V_1 + \frac{q_2}{\frac{\pi}{4}D^2} \tag{10.7}$$

The calculations are completed by repeating the above steps for each diffuser port in turn, moving up the riser toward the sea surface, using the following generalized equations:

$$C_D = \text{function of}\left(\frac{\frac{V_{n-1}^2}{2g}}{E_n}\right) \tag{10.8}$$

$$q_n = C_D\frac{\pi}{4}d_n^2\sqrt{2gE_n} \tag{10.9}$$

$$\Delta V_n = \frac{q_n}{\frac{\pi}{4}D^2} \tag{10.10}$$

$$V_n = V_{n-1} + \Delta V_n \tag{10.11}$$

$$h_{fn} = f\frac{L_n}{D}\frac{{V_n}^2}{2g} \tag{10.12}$$

$$E_{n+1} = E_n + h_{fn} + \frac{\Delta s}{s}\Delta z_n \tag{10.13}$$

After the shallowest port is reached, Eq. (10.13) is applied again to the diffuser pipe from the shallowest port up to the surface.

The above procedure can be incorporated in a worksheet that takes as input a set of parameters for each diffuser port: the port number, port depth, port diameter, riser diameter, port nozzle taper length, riser length, riser bend angle, and diffuser pipe diameter upstream of the port. Determining the total flow for a specific vertical diffuser is thus a short process of trial and error until the starting head, E_1, results in the total flow from all ports equaling the desired total flow. Note that if $E_1 = 0$ yields more than the desired flow, the port diameters are much too big. Conversely, if the total head is too large, the port diameters are too small.

Typical coefficient values are $f = 0.02$; $x_{en} = 0.406$; x_{bend} values are obtainable from any hydraulics reference (e.g., King and Brater 1963); $x_c = 0.02$.

9 Heat Loss Calculation Method

Produced water is typically heated during treatment. As mentioned previously, temperature is a key determinant of produced water density and thus it is important to know the produced water temperature as it exits each port of a diffuser. Heated effluent flowing through the vertical diffuser pipe will experience a heat flux to the surrounding seawater, and this heat flux will gradually reduce the effluent temperature.

The heat flux can be calculated by methods in Byrd et al. (2007). The heat flux is determined by three components:

1. heat transfer through the boundary layer on the inside wall of the pipe,
2. conduction of heat through the pipe wall and surface coating (if any), and
3. forced convection acting on the outside wall of the pipe.

The effects of all three components acting together can be determined by finding a suitable heat transfer coefficient for each and then combining these in series.

Consider a vertical pipe segment (Fig. 10.4) and define the following variables:

x	longitudinal distance along diffuser pipe ($x = 0$ at sea surface)
dq	heat flux in a pipe length of dx
D	pipe diameter
$T_e(x)$	bulk (average) temperature of effluent inside pipe at distance x from sea surface, $T_e(0) = T_i$ in Fig. 10.4
T_a	ambient sea water temperature
dT_e	change of effluent temperature over distance dx
C_p	specific heat constant for water
M	mass flow rate of effluent through pipe
U	overall heat transfer coefficient
u_a	ambient current speed
w	pipe wall thickness
c	pipe coating thickness.

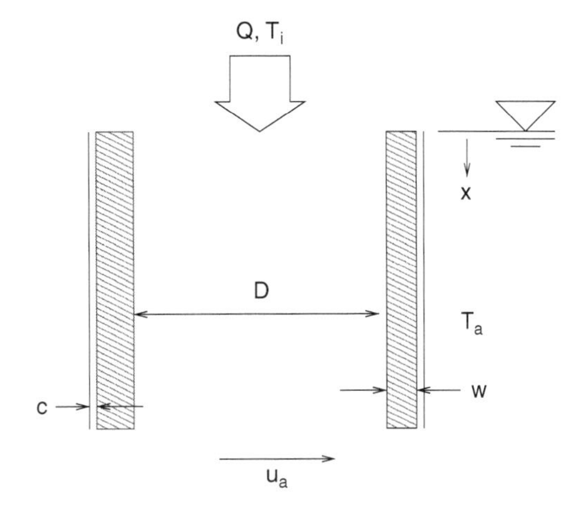

Fig. 10.4 Heat loss (T_i) from fluid flowing (Q) in vertical pipe

The steady-state heat flux from a pipe element of length d_x can be written as

$$dq = \pi D \left(T_e\left(x\right) - T_a\right) \, U \, dx \tag{10.14}$$

and as

$$dq = -dT_e \, C_p \, M \tag{10.15}$$

Equation (10.14) governs the overall heat flux, from the effluent out to the seawater flowing outside the diffuser. Equation (10.15) represents the heat contained in the effluent inside the diffuser. Equations (10.14) and (10.15) may be equated and integrated with the overall heat transfer coefficient, U, held constant to yield

$$T_e\left(x + \Delta x\right) = T_a + \left(T_e\left(x\right) - T_a\right) \, \exp\left(-\frac{\pi D U \, \Delta x}{C_p M}\right) \tag{10.16}$$

The overall heat transfer coefficient accounts for the effects acting in series of the heat transfer coefficient at the inside boundary of the pipe, the heat conduction through the pipe wall and pipe coating, and the heat transfer coefficient at the outside wall of the coated pipe, thus

$$U = \left(\frac{1}{h_i} + \frac{w}{K_p} + \frac{c}{K_c} + \frac{1}{h_o}\right)^{-1} \tag{10.17}$$

where

h_i heat transfer coefficient at the inside boundary of the pipe (Btu hr^{-1} ft^{-1} $°F^{-1}$)
K_p thermal conductivity of the pipe wall material (Btu hr^{-1} ft^{-1} $°F^{-1}$)
K_c thermal conductivity of the coating material (Btu hr^{-1} ft^{-1} $°F^{-1}$)
h_o heat transfer coefficient at the outside wall of the coated pipe (Btu hr^{-1} ft^{-1} $°F^{-1}$).

For a steel pipe, $K_p = 26.2$ Btu hr^{-1} ft^{-1} °F^{-1} (ASHRAE 1977) and for an amine adduct coal tar epoxy coating, $K_c = 0.8$ Btu hr^{-1} ft^{-1} °F^{-1} (Perry and Chilton 1973).

The heat transfer coefficient at the inner surface of the pipe, for turbulent flow, is

$$h_i = 0.026\frac{K_w}{D}(\mathrm{Re}_i)^{0.8}(\mathrm{Pr}_i)^{0.333} \tag{10.18}$$

where K_w is the thermal conductivity of water (0.348 Btu hr^{-1} ft^{-1} °F^{-1}; ASHRAE 1977), Re_i is the Reynolds number inside the pipe (VD/ν), V is the average effluent velocity inside the pipe, ν is the kinematic viscosity of water, Pr_i is the Prantdl number inside the pipe $\left(C_p\mu/K_w\right)$, and μ is the dynamic viscosity of water which varies with temperature. Eckert (1963) provides a table of Prandtl numbers for water as a function of temperature (°F) and

$$\mathrm{Prandtl}(T_F) = 13.6/(1 + 0.019(T_F - 32)^{1.11}) \tag{10.19}$$

is a good approximation of this data.

The heat transfer coefficient for turbulent flow at the outer surface of the coated pipe is

$$h_o = \frac{0.016\,\mathrm{Re}_o^{-0.352}K_w}{D}\mathrm{Re}_o\,\mathrm{Pr}^{0.333} \tag{10.20}$$

where Re_o is the Reynolds number on the outside boundary of the pipe (u_aD/ν).

The calculations of this section can be implemented in a worksheet program for each segment of the vertical diffuser, segments being delineated by the ports of the diffuser. Thus heat loss is calculated from the water surface down to the shallowest port, using the inlet temperature of effluent entering the diffuser. This determines the temperature of the water leaving the shallowest port. Then heat loss is calculated from the shallowest to the next shallowest port, using the exit temperature from the shallowest port as an initial condition. This provides the exit temperature from the second shallowest port. This procedure is repeated moving down the diffuser port by port. The input to the worksheet defines the port number, port discharge flow rate, diffuser pipe diameter, pipe wall thickness, and length of diffuser pipe upstream to the next port or to the water surface. The worksheet can incorporate the effects of temperature variations on dynamic and kinematic viscosities.

10 Spiral Port Arrangement Allows Shorter Vertical Diffuser

As flows increase, vertical diffusers with ports all facing in one direction must get longer and longer to provide the required initial dilution. When the produced water density differs significantly from the ambient density, the port must be more and more constricted to control the distribution of flow, and this requires a higher and higher hydraulic head to drive the diffuser. An effective way to address this problem is by arranging the diffuser ports in a spiral pattern along the diffuser. This directs the individual plumes from each port in different directions during the initial mixing, prevents interference between the plumes until after most of the initial mixing has occurred, and allows the spacing between ports along the diffuser to be reduced. It is most effective to provide an odd number of ports for each complete circuit of ports around the perimeter of the diffuser. With an odd number of ports, a current directed in opposition to one port will not carry the plume from one port directly toward the plume from another port. An angular interval between ports of 72° (5 ports per circuit of the diffuser) works well. It is also possible to angle each port above or below horizontal by the use of pipe elbows and short risers. An orifice plate at the end of each riser is the easiest way to control flows from the riser ports. An orifice plate is simply a flat plate bolted to the end of the riser with a hole of the required diameter bored through the center of it.

11 Spiral Diffuser Example

Consider a vertical diffuser needed to dispose of 0.11 m^3/s (60,000 bbl/day) of produced water. The produced water has a salinity of 100 ppt and temperature of 21.0°C. Ambient current speed was 0.1 m/s. A linear ambient density stratification of 0.15 $kg/m^3/m$ was the design condition. With the aid of the methods described previously, a 15-port vertical diffuser was developed with an angular increment between ports of 72° (Fig. 10.5) and ports separated by a longitudinal distance along the diffuser of 2 m (6.5 ft). All risers were 6.35 cm in diameter, 7.62 cm

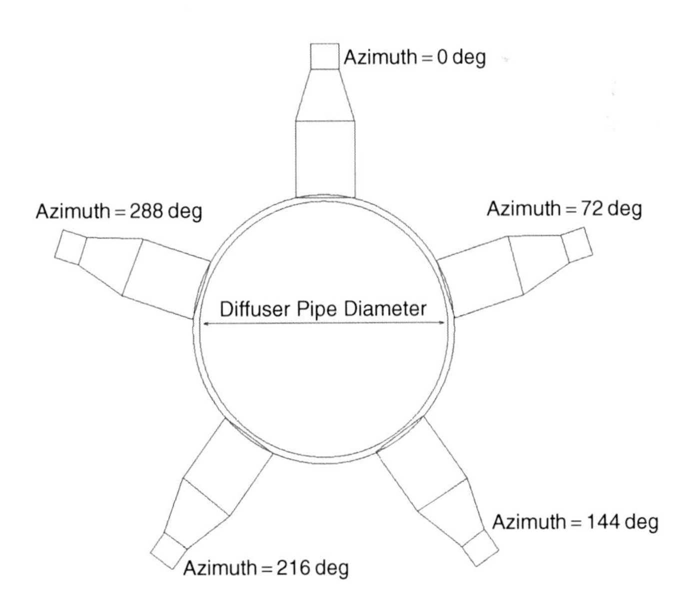

Fig. 10.5 Top view of spiral diffuser riser ports

Table 10.1 Ports and flows for example spiral diffuser

Riser #	Riser centerline depth from MSL (m)	Riser Azimuth (degrees)	Riser vertical angle from horizontal (degrees)	Port flow (m^3/s)
15	30	288	0	0.006265
14	32	216	0	0.006426
13	34	144	0	0.006587
12	36	72	0	0.006748
11	38	0	0	0.006909
10	40	288	0	0.007069
9	42	216	0	0.007228
8	44	144	0	0.007384
7	46	72	0	0.007538
6	48	0	0	0.007689
5	50	288	0	0.007837
4	52	216	0	0.007980
3	54	144	0	0.008118
2	56	72	0	0.008252
1	58	0	0	0.008380

long, with a taper over 5.08–3.81 cm diameter ports. Other riser properties and the calculated port flows are shown in Table 10.1. The Offshore Operators Committee (OOC) model (Brandsma et al. 1992; Brandsma and Smith 1999) was used to simulate the plumes issuing from each port. The surfaces of the plumes predicted by the model were colored by the average concentration within the plume using a grayscale mapping. Figure 10.6 shows a view of the 15 plumes originating from the spiral

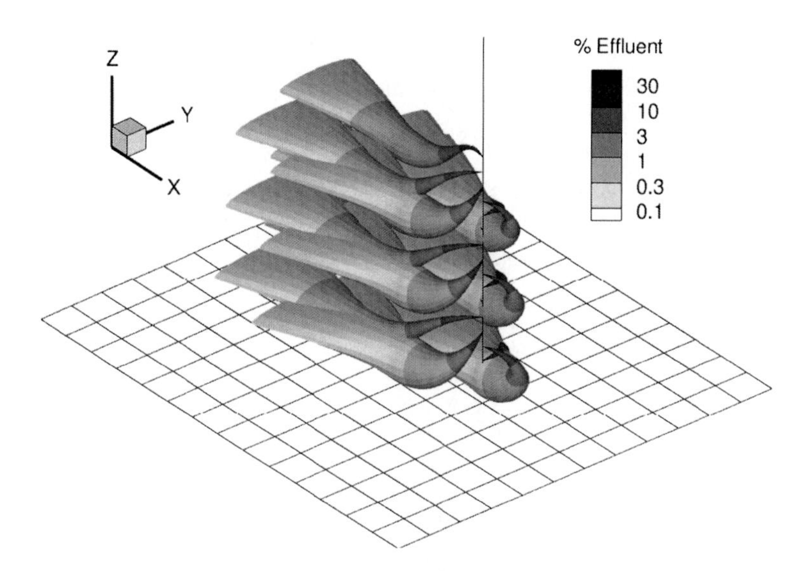

Fig. 10.6 View of plumes issuing from spiral diffuser

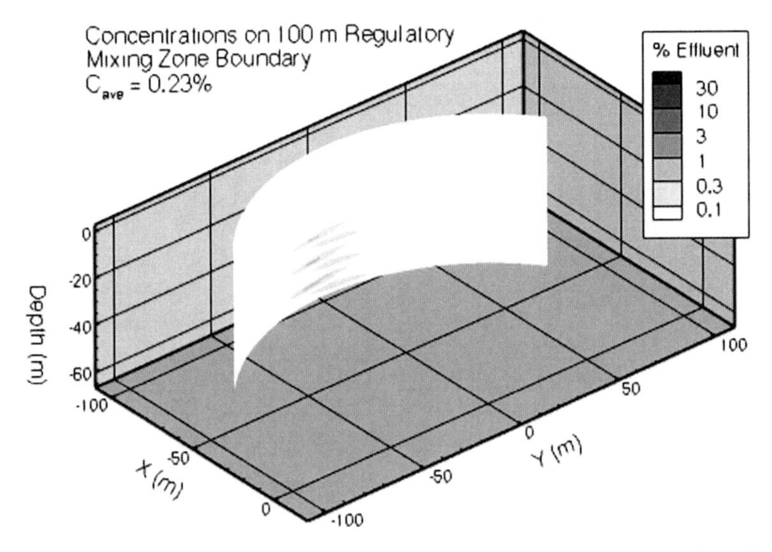

Fig. 10.7 Concentrations of combined 15 plumes on 100 m regulatory mixing zone boundary

diffuser (the vertical line). It is evident that the plumes remain clear of one another until dilution factors of 100 or more are achieved. Looking further down current, to a 100 m regulatory mixing zone boundary, the average produced water concentration at the boundary was found to be 0.23% (Fig. 10.7) corresponding to a dilution factor of 435.

12 Conclusions

This chapter was written because reliance on tables of dilution like those promulgated by Region 6 of the US Environmental Protection Agency and reliance on computer models like CORMIX (http://www.mixzon.com) and Visual Plumes (http://www.epa.gov/ceampubl/swater/vplume/) are not sufficient to ensure a properly performing diffuser, especially a vertically oriented one. Analysis techniques originally developed for horizontal diffusers can be applied to vertical diffusers as well. While ignoring the techniques might still allow a reasonable performance from a horizontal diffuser, a vertical diffuser is almost sure to fail if the techniques are ignored. The problem is that the tables and models deal only with dilution. It is left to the designer to ensure that the port discharge rates used with the tables and model will in fact occur. By the use of the analysis techniques summarized here, the designer can do just that.

It is possible to summarize a recommended design and modelling procedure for a vertical diffuser by the following steps.

1. Establish the effluent discharge conditions (total flow rate, produced water salinity, and temperature).
2. Establish the required concentration or dilution to be achieved at the edge of the mixing zone.
3. Estimate the number of ports likely to be needed to reach the required dilution. One can divide the total flow by the number of ports and use plume models to find the dilution for flow from a single port.
4. A candidate port layout for a vertical diffuser will define how the ports are spaced longitudinally and directionally, along with the depths of the deepest and shallowest ports. In addition, candidate port diameters, riser arrangements, and vertical port angles must be set.
5. Next one must guess an initial head with which to begin the hydraulic analysis. A recommended starting value at port #1 (the deepest one) is 0.1. The procedure in Section 8 can be applied to each port in turn to calculate port flows. If the total flow exceeds the desired value, the initially selected port diameters are probably too large.
6. Depending on the properties of the effluent, precise variation of the port diameters may be needed. One method of making a set of trial port diameters is to vary areas of adjacent ports on the diffuser by a ratio whose value, starting from the deepest port, is slightly greater than 1.0. Thus port diameters will increase as a function of distance up the diffuser.
7. Once a reasonable set of port flows has been established, heat losses can be calculated using the methods described in Section 9. In contrast to the hydraulics calculations, which begin at the deepest port and work upward, the heat loss calculations begin from the sea surface and work downward, finding exit temperatures at each port in succession.
8. Only when the port properties are defined and the flow rate and effluent temperature for each port is it possible to conduct definitive plume simulations to calculate dilutions.

While uniform port sizes work well for a horizontal diffuser, vertical diffusers may require precise variation of port sizes to control the flow distribution.

References

ASHRAE (American Soc. Heating, Refrigerating and Air-Conditioning Engineers) (1977) Fundamentals. ASHRAE, New York, NY

Brandsma MG, Smith JP (1999) Offshore Operators Committee Mud and produced water discharge model – Report and user guide. EPR.29PR.99. Production Operations Division. Exxon Production Research Company, Houston, TX

Brandsma MG, Smith JP, O'Reilly JE, Ayers RC Jr, Holmquist AL (1992) Modeling offshore discharges of produced water. In: Ray JP, Engelhardt FR (eds) Produced water, pp. 59–71, Plenum Press, New York, NY

Byrd RB, Stewart WE, Lightfoot EN (2007) Transport phenomena, 2nd edn. Wiley, New York, NY

Doneker RL, Jirka GH (1990) Expert system for hydrodynamic mixing zone analysis of conventional and toxic submerged single port discharges (CORMIX1). U.S. EPA Report EPA/600/3-90/012 (PB90-187196), February 1990

Eckert ERG (1963) Introduction to heat and mass transfer. McGraw-Hill, New York, NY

Fischer HB, List EJ, Koh RCY, Imberger J, Brooks NH (1979) Mixing in Inland and Coastal waters. Academic, New York, NY

King HW, Brater EF (1963) Handbook of hydraulics, 5th edn. McGraw-Hill, New York, NY

Perry RH, Chilton CH (eds) (1973) Chemical engineers handbook. McGraw-Hill, New York, NY

Rawn AM, Bowerman FR, Brooks NH (1961) Diffusers for disposal of sewage in sea water. Trans. American Society of Civil Engineers. vol 126, Part III. pp 344–388, American Society of Civil Engineers, Reston, VA

Chapter 11
Experimental and Modelling Studies on the Mixing Behavior of Offshore Discharged Produced Water

Haibo Niu, Kenneth Lee, Tahir Husain, Brian Veitch, and Neil Bose

Abstract A probabilistic based steady state model, PROMISE, was developed to predict the mixing behaviors of produced water in the marine environment. The model was also coupled with a MIKE3 model to study the dispersion in non-steady state conditions. Laboratory experiments were conducted in a 58 m towing tank to calibrate the near field model. Field experiments using an Autonomous Underwater Vehicle (AUV) were also performed to test the ability of an AUV in produced water plume mapping.

1 Introduction

The recent increase in offshore oil and gas development off the east coast of Canada has created concern regarding the capacity of the marine environment to act as an intermediate buffer zone for receiving produced water discharge, and the subsequent mixing of the waste materials with offshore water. To assess the potential environmental risks of produced water to the marine ecosystem and to manage produced water in the marine environment in an environmentally safe and cost-effective manner, a research project entitled "Offshore Produced Water Management – An Integrated Approach" is being conducted by the Faculty of Engineering & Applied Science, Memorial University of Newfoundland. The study of the mixing behaviors of produced water in the marine environment has been a key focus during the development of a decision support system for produced water management (DISSPROWM) under the project. As a result, a PROduced-water Mixing In Steady-state Environment (PROMISE) model has been developed and is being integrated with the DISSPROWM.

The PROMISE model can be used to predict the steady state mixing processes such as initial mixing, buoyant spreading, and turbulent diffusion, in both deterministic and probabilistic forms. The PROMISE model has also been coupled with

H. Niu (✉)
Centre for Offshore Oil, Gas and Energy Research (COOGER), Fisheries and Oceans Canada, Bedford Institute of Oceanography, Dartmouth, NS, Canada, B2Y 4A2

K. Lee, J. Neff (eds.), *Produced Water*, DOI 10.1007/978-1-4614-0046-2_11,

a three-dimensional non-steady state model MIKE3 to study the non-steady state far field dispersion of produced water. To calibrate the model coefficients, laboratory experiments have been performed in a towing tank to investigate the near field mixing behavior of vertically discharged buoyant jets, and various discharge and ambient conditions have been studied. Mapping produced water plumes is important for planning Environmental Effects Monitoring (EEM) programs and for validating mathematical models; a traditional ship-based mapping methods are generally expensive, alternative methods are being explored. The potential of using a new generation of Autonomous Underwater Vehicle for produced water plume mapping was tested as part of the research work. Two plume mapping missions were conducted in August 31 and September 7, 2006 using the MUN Explorer AUV in Holyrood Bay, Newfoundland. The objective of this chapter is to give an overview of the studies described above.

2 Development of PROMISE Model

2.1 Origination

Over the past few decades, a number of computer models have been developed to study the fate of produced water in the marine environment. A detailed review of available models can be found in Niu et al. (2011). Both deterministic and probabilistic approaches may be applied in produced water modelling. The advantage of a probabilistic model is that it gives a probability distribution instead of a single value description of the predicted concentration. The use of probabilistic models in place of deterministic models is an accepted practice for the assessment of effluent discharges (Huang et al. 1996).

A number of models using the probabilistic method have been developed to study the dispersion of produced water (Mukhtasor 2001; Niu et al. 2004, 2009). The limitation of these models is that they can only be used for limited discharge and ambient conditions and cannot be widely used. For example, these models can be used only in uniform ambient environment and do not take into account the stratification of seawater. The near field component of these models are two-dimensional, and thus, cannot be used for horizontal discharges for which the plume trajectories are often three dimensional. There is a need to develop a more general model which can be applied for various combinations of discharge and ambient conditions.

The effects of ocean waves on the initial mixing process has long been a concern within the modelling community. The model developed by Tate (2002) considered the effects of ocean internal waves, and, although there have been a number of experimental studies (Shuto and Ti 1974; Ger 1979; Sharp 1986; Chin 1987; Koole and Swan, 1994; Kuang and Hsu 1999; Chyan et al. 2002) on the effects of ocean surface waves on initial dilution, the findings have not been incorporated into existing models. Another objective of this study is to integrate both effects into the proposed PROMISE model.

2.2 Approaches

The proposed model has four sub-components: (1) PROMISE1 – a near field model which simulates the initial mixing behavior before boundary interaction occurs; (2) PROMISE2 – a wave effect model which accounts for the effects of both internal and surface waves; (3) PROMISE3 – a boundary interaction model which may include an upstream intrusion and a downstream control model depending on the impinging angle; (4) PROMISE4 – a far field dispersion model which models the buoyant spreading and turbulent diffusion process.

An integral type of near field model was adopted in PROMISE1 which is based on the Lagrangian formulation of Tate (2002). The system of equations was formulated by the conservation of mass, momentum, and buoyancy and was closed by an entrainment function which is the sum of shear and vortex entrainments. A modification of the formulation was made to compute the maximum centerline concentration as well as the averaged concentration.

The wave effect model, PROMISE2, uses the same internal wave effect model as Tate (2002). The surface wave model was developed based on the analysis of available experimental data, and an empirical equation was derived based on dimensional analysis. Unlike the similar equations reported in the literature, the newly developed equations consider the effects of discharge characteristics by the addition of the densimetric Froude number, *Fr*. The effect of water depth was also taken into account by using different empirical constants.

After the plume reaches a boundary (surface, seabed, or internal density layer), impingement may take place and additional mixing/dilution occurs. PROMISE3 calculates the dilution in this region using a formulation similar to Mukhtasor (2001) and Huang et al. (1996). While the impingement angle was estimated in the formulation by Mukhtasor (2001) and Huang et al. (1996), PROMISE3 uses an improved approach to compute the angle using the advanced integral model PROMISE1.

PROMISE4 uses a unified buoyant spreading and turbulent diffusion model. The model is a modification of the Huang and Fergen (1997) formulation by expanding the model to stratified conditions.

In a traditional deterministic based modelling application, the uncertainties associated with model parameters are not considered. A single (in most cases, the average) value from a range of possible values will be used to give a single value prediction of pollutant concentration. For example, for an entrainment coefficient that may range from 0.3 to 0.7, an average value of 0.5 can be used in the deterministic based application and a concentration of 35 mg/L may be predicted for the location 100 m east of the discharge point. In a probabilistic based application, the probability distribution of the model parameters can be conducted first, and the Monte Carlo Simulation (MCS) method will be used. In an MCS analysis, one value for each model parameter can be randomly generated from their probability distributions, and these values can be used by the model to give a single value prediction. This process is the same as the deterministic application; however, it is repeated several times and therefore several values for each parameter from their probability distributions are used and result in several predictions for a given point. For example, for a 5 times MCS analysis, the entrainment coefficient used may be [0.3, 0.4, 0.5, 0.6, 0.7] and the predicted concentration at 100 m east of the discharge may be [25, 30, 35, 40, 45] mg/L. From the MCS analysis, one may conclude that the average concentration is 35 mg/L while the concentration may range from 25 mg/L minimum to 45 mg/L maximum. With the increasing number of MCS, more detailed statistical information can be obtained for any point of prediction. Therefore, the probabilistic approach is superior to the deterministic approach as it can provide more comprehensive information.

In the case of non-steady state environments, the concentration field must be computed by coupling PROMISE with ocean circulation models, such as DHI MIKE3. In such a simulation, the hydrodynamic information is first obtained by running the MIKE3 model. The simulated currents and ambient density profile are then used by PROMISE to determine the minimum grid size and time step for coupling. The grid size of MIKE3 is then refined and the HD module is executed again to generate the velocity and density information. The PROMISE is executed to predict the near field concentration and the size of plume. This information is used to create the initial source term for the MIKE3 Advection–Diffusion module (AD). After the concentration field for this time step is calculated, the MIKE3 model moves forward to the next time step. If t is less than t_c, the model will continue running until t_c is reached. At this point, the predicted field will be used by PROMISE as the accumulated background concentration to consider the re-entrainment of far field returned pollutants. To maintain a mass balance, a sink term in the MIKE3 will be created to remove the same amount of pollutant re-entrained by PROMISE. This process is repeated several times until the end of simulation ($t = t_N$).

2.3 Results

Figure 11.1 shows a sample predicting near field centerline concentration by PROMISE1. The predictions from the CORMIX and VISJET are also plotted. It can be seen from the figure that PROMISE agrees well with these two models, especially the VISJET. This may be because the formulation of PROMISE is more similar to VISJET. Both PROMISE and VISJET use the Lagrangian formulation, but CORMIX uses the Eulerian formulation.

The probabilistic based modelling results from a hypothetical study are shown in Fig. 11.2. At about 1,500 m from the discharge, the predicted mean concentration is about 2% of the initial effluent concentration, while the predicted 95%-tile concentration is about 16.5% of the initial effluent concentration, and it is much higher than the mean concentration. Similarly, the model also predicts the minimum and maximum concentration profile.

A sample prediction from the coupled PROMISE/MIKE3 model is shown in Fig. 11.3. At 100 m downstream, the predicted concentration is about 4.6%, which is close to the average value of 6% predicted by the PROMISE model alone (Fig. 11.2). Although these two cases are not directly comparable due to the different currents, concentrations at this location from both models are in the same order of magnitude.

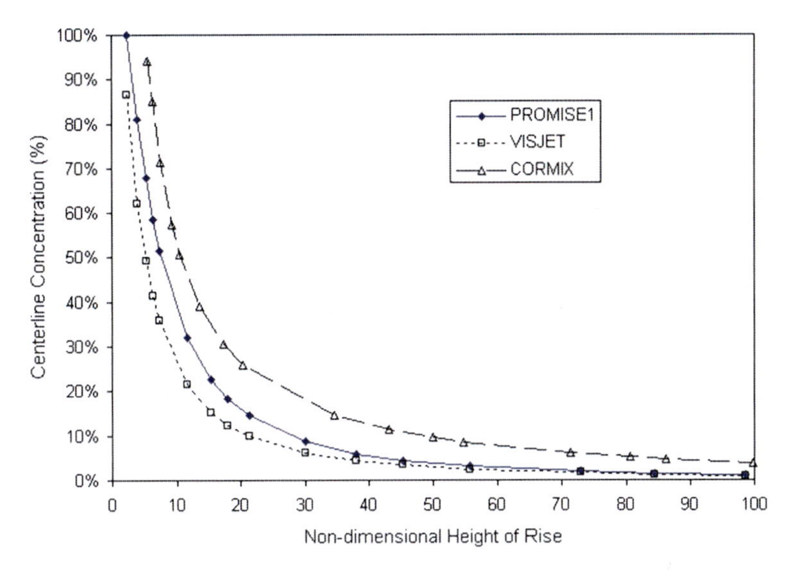

Fig. 11.1 Comparison of PROMISE with CORMIX and VISJET

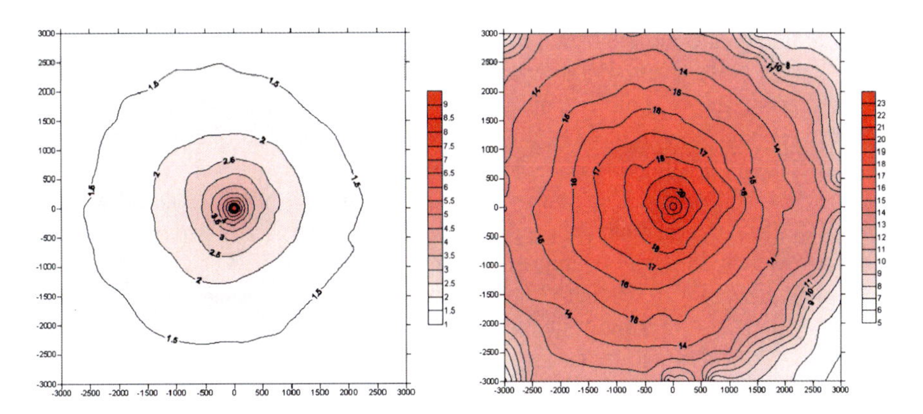

Fig. 11.2 Mean (*left*) and 95%-tile (*right*) concentration profile

3 Laboratory Experiments

3.1 Origination

To study the mixing behaviors of produced water or other types of buoyant jet discharges, it is important to conduct laboratory experiments for both mathematical model development and validation. A successful laboratory model must have four features: (1) the densimetric Froude number of the laboratory model and prototype must be equal. The equality of Froude number means that the ratio of the jet

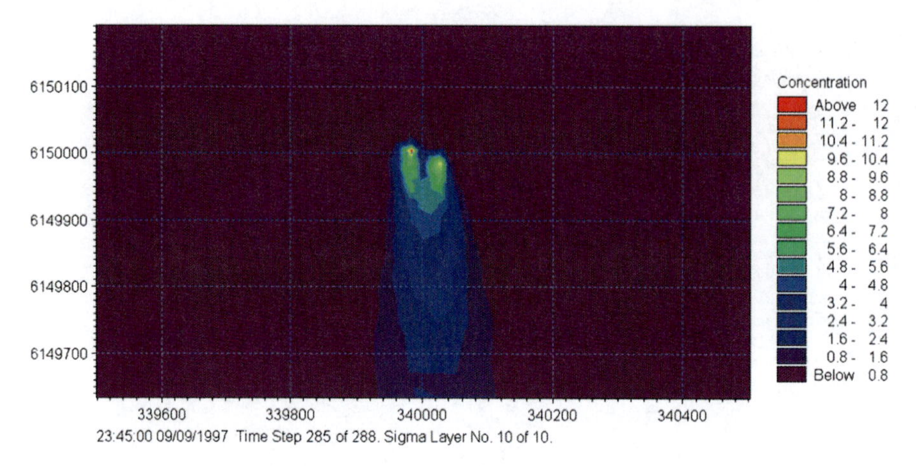

Fig. 11.3 Predicted concentration from the coupled PROMISE/MIKE3 model

momentum flux to jet buoyancy flux will be correct, and therefore the jet entrainment will be correctly simulated; (2) the ratio of jet velocity to current velocity must be equal in laboratory model and prototype; (3) the jet discharge angle must be identical in laboratory model and prototype; and (4) the laboratory model Reynolds number must be large enough to ensure that the laboratory jet is fully turbulent so that the jet mixing is similar at model and full scale. The laboratory experiments should be conducted at a scale as close to the prototype as possible to minimize the scale effects and obtain more realistic results.

A review of 19 sets of laboratory experiments on buoyant jets has shown that almost all the experiments were conducted at small scales. Considering a prototype discharge pipe of 35.6 cm in diameter (the diameter used on the FPSO for the White Rose site, east coast of Canada), the scales of these models range from 32.4:1 to 197.8:1. Therefore, it is the objective of this study to conduct a relatively larger scale (14.2:1) experiment by taking advantage of a long towing tank facility at the Ocean Engineering research Centre, Memorial University.

3.2 Methods

To simulate the mixing behavior of produced water, salt water with densities ranging from 1,022.03 to 1,046.37 kg/m^3 was discharged into the towing tank (58 m long, 4.5 m wide, 2.2 m water depth) filled with freshwater with a density ranging from 999.02 to 999.58 kg/m^3. The discharge system was mounted on the towing carriage which was running at speeds from 0.1 to 0.2 m/s. This configuration gives densimetric Froude numbers ranging from 9.81 to 17.24 and velocity ratios ranging from 5 to 14.58. A MicroCTD sensor was used to measure the conductivities at several cross-sections downstream. The experimental setup is shown in Fig. 11.4.

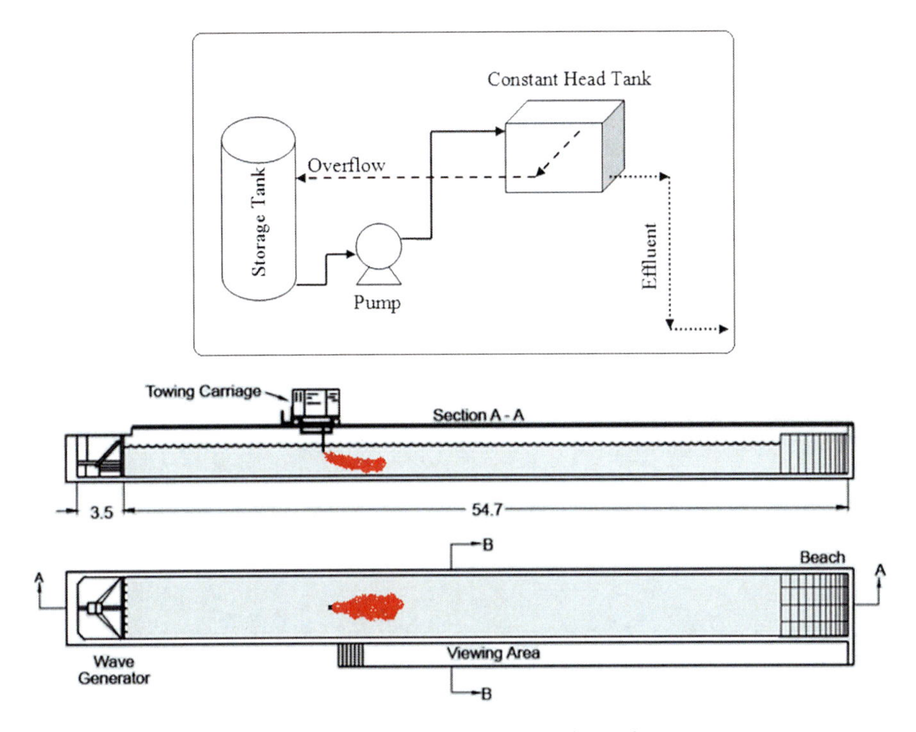

Fig. 11.4 The experimental setup: discharge system and towing tank

3.3 Results

A total of 54 sets of experiments were conducted, and among these there were 2 sets of experiments that failed to log data. To check the validity of the data, they were compared with the CORMIX model and good agreement was found. The data was then used to calibrate the PROMISE1 model coefficient, and a new entrainment coefficient equation was derived. The calibrated model was executed to simulate the 52 sets of experiments. Both trajectory and centerline concentration were predicted.

Figure 11.5 shows the comparison of measured dilution and the predicted dilution. The predictions from the CORMIX model are also plotted. It can be seen from the figure that both models agree well with the experimental data. The CORMIX model prediction is slightly better than that of the PROMISE1 model.

4 Field Experiments Using AUV

4.1 Origination

Field experiments are important for both the environmental effects monitoring and numerical model validation. Very few field tests for model validation have been

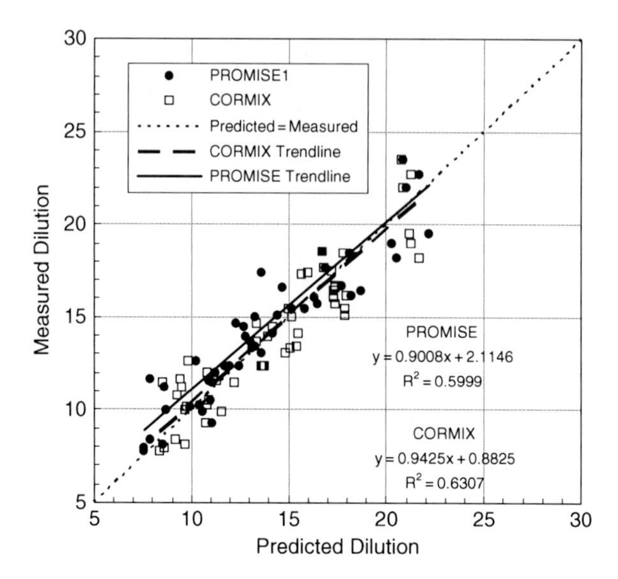

Fig. 11.5 Comparison of experimental data with PROMISE1

reported to date. This is mainly due to the difficulty of collecting data in harsh, remote environments and the high cost associated with these experiments. The data collected using traditional towing tests are often incomplete and provide very limited information. For deepwater monitoring, the increasing water depth also increases the level of sampling error due to the drift of surface vessel platforms and prolonged sampling times. To map a produced water plume more effectively and accurately, new and innovative means of acquiring data need to be used. One solution is the use of a new generation of autonomous oceanographic platform – Autonomous Underwater Vehicles (AUVs) – that are capable of tracking water masses, recording chemical/physical/biological properties, and transmitting data without tether to either the seafloor or a vessel. AUVs are able to provide a detailed four-dimensional view of the dynamic ocean.

Memorial University of Newfoundland has recently acquired a new Explorer class AUV built by International Submarine Engineering Ltd. The AUV is designed as a 4.5 m ocean-going instrumentation platform with a 3,000 m depth capability. The strength of the MUN Explorer AUV is its ability to carry 150 kg of scientific payload (instruments), with a power requirement in the hundreds of Watts, on missions of up to 12 h duration or 100 km. To study the ability of plume mapping using the MUN Explorer AUV, a field test that tracked Rhodamine WT dye was conducted on August 31, 2006 and again on September 7, 2006.

4.2 Methods

The field tests were performed in the south arm of Holyrood Bay at the head of Conception Bay. Holyrood Bay is located about 40 km southwest of St. John's, Newfoundland. There is no existing outfall of this type in the study area, therefore a

Fig. 11.6 MUN Explorer AUV with CTD and fluorometer sensors

temporary artificial outfall was built on a wharf. Freshwater mixed with Rhodamine WT was discharged into the sea through a 2 in. diameter pipe submerged at about 3 m below the sea surface. The distance of the discharge pipe to the wharf was about 4 m. During the experiment, the discharge was started at least 2 h earlier than the launch of the AUV to give enough time for the plume to disperse. Using an example mean current speed of 5.86 cm/s allows the plume to travel up to 421 m downstream.

Before the AUV was put in the water, the missions were planned on the Mission Planning Workstation using the FleetManager software and uploaded into the Vehicle Control Computer (VCC). Once in water, the vehicle followed the pre-planned routes and depths and collected data using a Cyclops-7 fluorometer and a MicroCTD sensor (Fig. 11.6). After the mission was completed, the vehicle returned to the pre-programmed location.

For both tests, the vehicle speed was 1.5 m/s. The vehicle data, including the position, heading, and speeds, were logged to the vehicle computer at a sampling rate of 0.1 s. The sampling interval of the CTD/fluorometer was 0.2 s (5 data points per second). This setup yields a horizontal resolution of 30 cm along the AUV trajectory.

4.3 Results

Example results of the September 7, 2006 mission are presented in Figs. 11.7 and 11.8. The mission time for this test was about 1 h and 10 min. The area surveyed was about 170 m $\times$ 240 m. The vehicle trajectories over the course of the test are shown in Fig. 11.7.

It can be seen from the contour plot (Fig. 11.8) that the plume spread toward the east. The highest concentration was measured at the discharge point and decreased

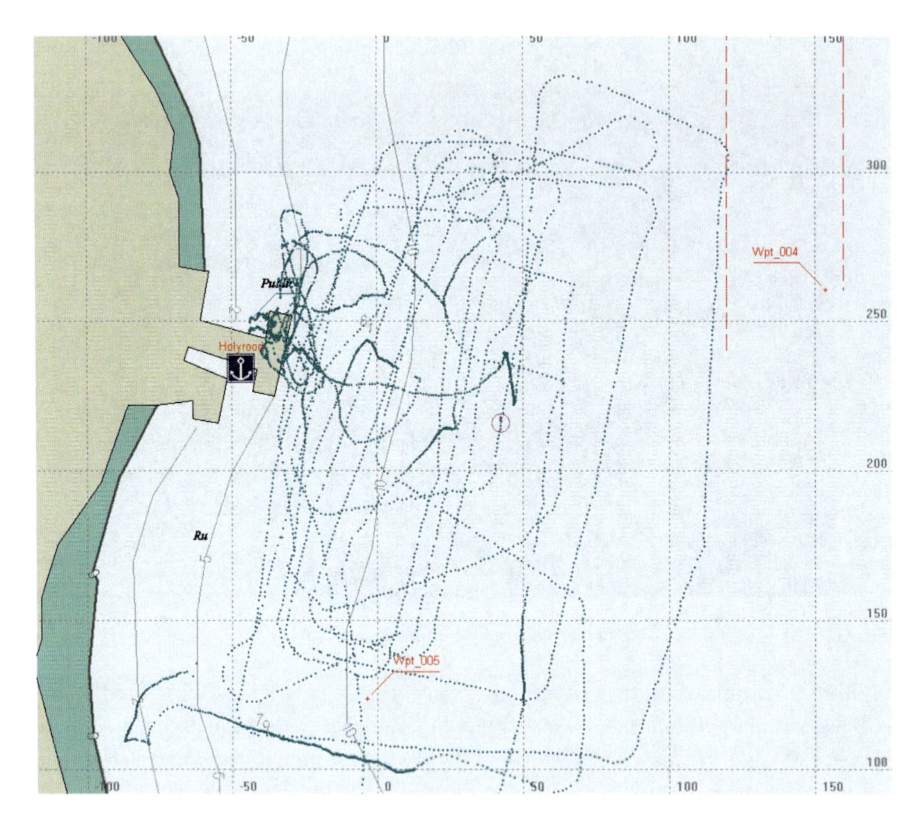

Fig. 11.7 Trajectory of the AUV

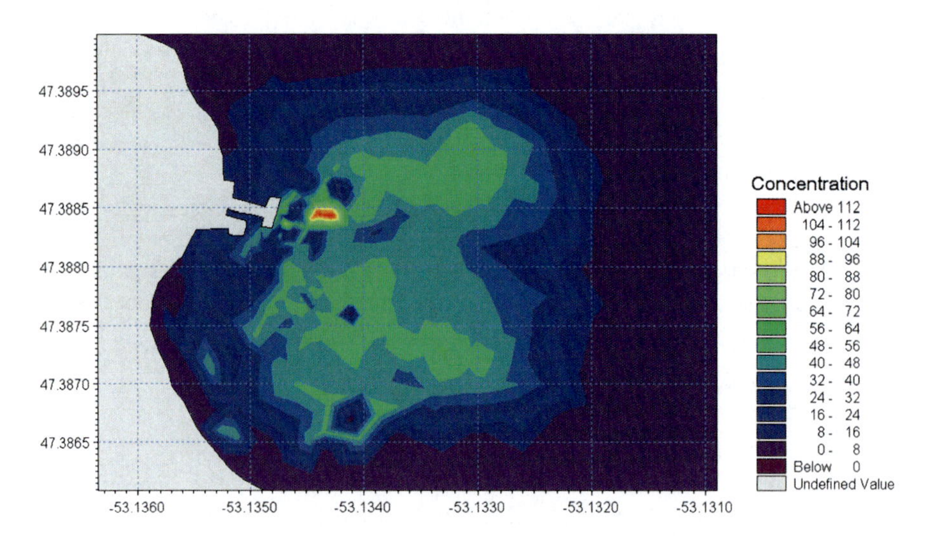

Fig. 11.8 Contour plot of the Rhodamine WT distribution

toward the plume edge. The plume mapped by the AUV for this case is relatively smooth, but still shows patchiness. An interesting pattern shown by the contour is that the plume separated into two centers downstream.

5 Conclusions and Future Research

A steady state model, PROMISE, has been developed, and it can be used in both deterministic and probabilistic analyses. The model considers the effects of ocean waves on the initial dilution process. Compared with other probabilistic based models, PROMISE can be used in a wide range of discharge and ambient conditions.

A set of laboratory experiments were performed to further calibrate the PROMISE model and refine the model coefficient. By comparison with the CORMIX model, the measured data are shown to be of good quality. The comparison of PROMISE predictions with laboratory experiments shows good agreement.

A field mission using the MUN Explorer AUV for plume mapping has been conducted and the results show that the horizontal plume profile can be successfully mapped by an AUV. An AUV is suitable for offshore produced water plume mapping.

Although PROMISE1 has been validated against extensive laboratory data, the overall performance of PROMISE has not been validated. This is mainly due to the lack of sufficient field data. Future work on using an AUV to map a real produced water outfall and validate the PROMISE model is recommended.

Acknowledgments Financial support from the Natural Sciences and Engineering Research Council of Canada and Petroleum Research Atlantic Canada through a Collaborative Research and Development grant (NSERC/PRAC CRD) and the Panel for Energy Research and Development (PERD) is gratefully acknowledged.

References

Chin DA (1987) Influence of surface waves on outfall dilution. J Hydraul Eng ASCE 113: 1006–1018

Chyan JM, Hwung HH, Chang CY, Chen IP (2002) Effects of discharge angels on dilution of Buoyant jets in wave motions. In: Proceedings of the 5th international conference on hydrodynamics, Tainan, Taiwan, Oct 31–Nov 2 2002, pp 485–490

Ger AM (1979) Wave effects on submerged buoyant jets. In: Proceedings of the 18th IAHR Congress, Cagliari, Italy, 10–14 Sept 1979, pp 295–300

Huang H, Fergen RE (1997) A model for surface plume dispersion in an ocean current. In: Proceedings of the 27th IAHR Congress, San Francisco, CA, 10–15 Aug 1997

Huang H, Fergen RE, Proni JR et al (1996) Probabilistic analysis of ocean outfall mixing zones. J Enviro Eng ASCE 122:359–367

Koole R, Swan C (1994) Dispersion of pollution in a wave environment. In: Proceedings of the 24th International on Coastal Engineering Conference, Kobe, Japan, 23–28 Oct 1994, pp 3071–3085

Kuang J, Hsu CT (1999) Experiments on 2-D submerged vertical jets with progressive water surface waves. In: Proceedings of the 2nd international symposium on environmental hydraulics, Hong Kong, pp 155–160

Mukhtasor (2001) Hydrodynamic modeling and ecological risk-based design of produced water discharge from an offshore platform. Ph.D. Thesis, Memorial University of Newfoundland, St. John's, NL, Canada

Niu H, Husain T, Veitch B et al (2004) Probabilistic modeling of produced water outfall. In: Lee JHW, Lam KM (eds) Environmental hydraulics and sustainable water management. Taylor & Francis Group, London, pp 467–472

Niu H, Husain T, Veitch B et al (2009) Assessing ecological risks of produced water discharge in a wavy environment. Advances in Sustainable Petroleum Engineering and Science, Int J Risk Assess Manag. 1, pp 91–102

Niu H, Lee K, Husain T et al (2011) A review of the state-of-the-art of produced water fate/transport models. International Journal of Environment and Waste Management Environ Model Softw. Submitted

Sharp JJ (1986) The effect of waves on buoyant jets. Proc Inst Civil Eng, Part 2 81:471–475

Shuto N, Ti LH (1974) Wave effects on buoyant plumes. In: Proceedings of the 14th Coastal Engineering Conference, Copenhagen, Denmark, 24–28 June 1974, pp 2199–2208

Tate PM (2002) The rise and dilution of buoyant jets and their behavior in an internal wave field. Ph.D. Thesis, School of Mathematics, University of New South Wales, Australia

Chapter 12
A Coupled Model for Simulating the Dispersion of Produced Water in the Marine Environment

Haibo Niu, Kenneth Lee, Tahir Husain, Brian Veitch, and Neil Bose

Abstract This chapter describes a method for simulating the dispersion of produced water in the marine environment. A near field model, PROMISE, has been coupled with the MIKE3 model using the passive offline coupling method. A hypothetical case study has been conducted to evaluate the method and it has been shown that model coupling is critical in accurately simulating the dispersion. A minimum grid size must be maintained in the far field model to introduce the source term from the near field model correctly.

1 Introduction

Produced water is the largest waste stream associated with offshore oil and gas production, it contains various naturally occurring and production chemicals which may pose adverse impacts in the water column and sediments. In order to study the environmental impacts of produced water in the marine environment, it is important to understand the dispersion processes of produced water following its release into the ocean.

Once discharged, the produced water plume will descend or ascend depending on its density relative to the ambient seawater and it will bend in the direction of the ambient current until it encounters the seafloor or reaches the water surface. In the case of a stratified environment, the plume will usually be trapped at a neutrally buoyant level before it encounters the seafloor or reaches the water surface. This phase, named the near field, ends within minutes and within a few meters from the discharge source with a corresponding dilution range of 10–1,000:1. After the plume reaches the boundary (surface/seabed), it spreads as a thin layer and mixing is dominated by two mechanisms: buoyant spreading and oceanic turbulent diffusion. Buoyant spreading is a self-driven dispersion process because the horizontal transverse spreading and vertical collapse of the plume are due to the residual buoyancy contained in the plume. Buoyant spreading is particularly important for a plume that is poorly diluted during the initial mixing process. The far field mixing process starts from the turbulent diffusion region. The turbulent diffusion is a passive dispersion

H. Niu (✉)
Centre for Offshore Oil, Gas and Energy Research (COOGER), Fisheries and Oceans Canada, Bedford Institute of Oceanography, Dartmouth, NS, Canada, B2Y 4A2

K. Lee, J. Neff (eds.), *Produced Water*, DOI 10.1007/978-1-4614-0046-2_12,

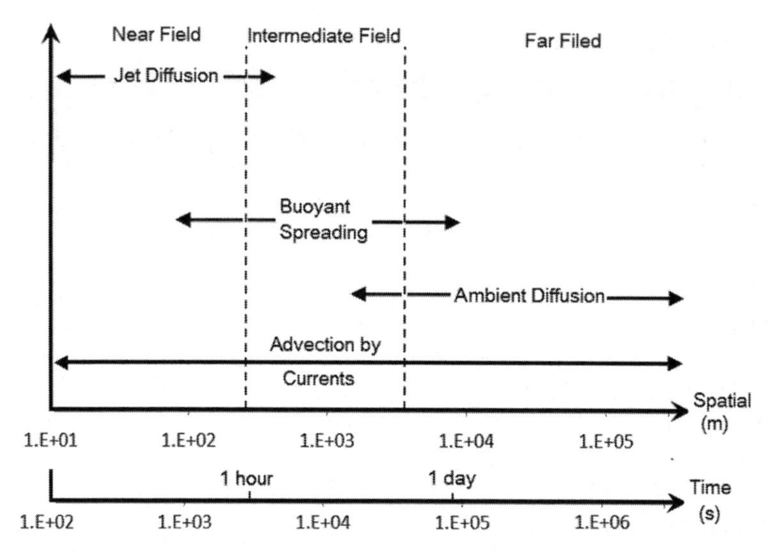

Fig. 12.1 Physical processes and length and time scales of discharged outfalls in marine environments (after Jirka et al. 1975)

process resulting from oceanic turbulence or eddies. Both buoyant spreading and turbulent diffusion could be important over a distance from the discharge point, but the buoyancy effect decreases while the turbulence effect increases as a plume travels downstream. The length and time scales for the mixing process are illustrated in Fig. 12.1.

Mathematic models may be used to describe the above mixing processes. In the immediate vicinity of the discharge, the mixing behaviors are mainly dominated by the source momentum flux, buoyancy flux, outfall geometry, ambient velocity, and stratification. These near field processes can be modeled satisfactorily by many near field models, such as CORMIX (Doneker and Jirka 1990), VISJET (Lee and Chu 2003), Visual PLUMES (Baumgartner et al. 1994), and PROMISE (Niu 2008). The influence of the source characteristics decrease as the plume progresses away from the discharge point. In the far field region, the plume is passively transported and further diluted by ambient currents. Although some models, like CORMIX and PROMISE, include a far field module for these processes, their predictions in this region are more intended for the design goal of minimizing the possible environmental impacts, rather than as an operational or monitoring tool. This is due to the fact that their far field modules are steady state models and cannot account for the effects of time varying current speeds and directions. To model these motions, the nonsteady-state models, such as DELFT3D (Delft 2005), ECOM-sed (Hydroqual 2002), EFDC (Hamrick 1992), and MIKE21/3 (DHI 2007) should be used. While the far field models focus on the three-dimensional motions of the natural water body rather than focus on the jet, plume, or waste field driven motions, they are unable to resolve the detailed near field motions, especially those depth averaged models which are sufficient for large-scale flows, but not for discharge assessment.

To correctly simulate the near field and far field motions, these two types of models need to be coupled together. As stated by Bleninger and Jirka (2006), coupling models means introducing flow quantities, such as momentum or mass, from one model into the other. The flow quantities may be introduced by specifying the model boundary conditions and thus have direct effects on the whole flow, or by modifying the existing flow by adding source terms. A number of coupling studies for general environmental discharges have been conducted (Zhang and Adams 1999; Roberts 1999; Li and Hodgins 2004; Kim et al. 2002; Choi and Lee 2005; Bleninger et al. 2006). For produced water discharges, the DREAM model coupled a near field Plume3D model with a particle tracking model (Reed et al. 2001); the MUDMAP model also coupled a near field model similar to the OOC model (Smith et al. 2004) with a particle tracking far field model (Burns et al. 1999; King and McAllister 1998); the PROTEUS model uses a unified particle tracking model for both near field and far field dispersion (Sabeur et al. 2000; Sabeur and Tyler 2004).

Most of the previous studies use the one-way coupling approach in which the reentrainment of far field plume back into the near field plume was not considered. This may affect the accuracy of prediction, especially in tidal environments. To improve the model accuracy, a two-way coupling method, which considers both the input of near field source term to far field and the feedback of far field plume to near field, should be used. This chapter describes a two-coupling method which uses a Lagrangian based near field model and a finite volume based far field model.

2 Methodology

2.1 Near and Far Field Models

In this study, the near field model for coupling is the PROMISE model (Niu 2008). PROMISE is a Lagrangian model based on the conservation of mass, momentum, buoyancy and closed with an entrainment function. The model has been validated with both laboratory data and other similar models.

There are a number of far field models available that may be coupled with PROMISE. To date, more than 30 circulation models have been developed (TAMU 2007). Among those models, the most cited models are POM, ECOM-si, Delft 3D, Telemac 3D (Electricité de France and Wallingford), MIKE3, and EFDC.

In this study, the MIKE3 was adopted as the far field model to couple with the PROMISE. The main reason for the selection of MIKE3 is that this model has not been used in this type of study before. In addition to this, the MIKE3 is easier to use than other models, especially among non-commercial models. The setup of non-commercial models tend to be extremely complicated and time consuming.

MIKE3 is a professional engineering software package developed by the Danish Hydraulics Institute (DHI). MIKE3 is a general non-hydrostatic numerical modelling system for a wide range of applications in areas such as oceans, coastal regions, estuaries, and lakes. MIKE3 includes several modules. The hydrodynamic module HD is the basic flow module. It simulates unsteady three-dimensional flows, taking into account density variations, bathymetry, and external forcings such as meteorology, tidal elevations, currents, and other hydrographic conditions. The advection-diffusion module can be applied to a wide range of hydraulic and related phenomena. The advantage of MIKE3 is its ability to use flexible mesh which is more efficient and adaptable.

The first step of a MIKE3 simulation is the setting up of a modelling domain. The horizontal grid of MIKE3 is unstructured while the vertical coordinate is structured sigma-coordinate mesh. The simulation period is then specified with a proper time interval controlled by a Courant number. To avoid stability problems, the maximum Courant number must be less than 0.5. The pollutant can be introduced into MIKE3 as source or sink term at given locations and depths.

2.2 Coupling Approaches

The coupling method used in this study is a passive offline coupling. As defined by Bleninger et al. (2006), a passive coupling assumes that the source-induced flow does not change the flow characteristics of the far field, and this is the case for most environmental discharges, such as produced water or sewage outfalls. In a passive coupling approach, only passive flow quantities need to be linked at the location and time that source induced motion are negligible. If the discharge is high enough to affect the ambient flow even in the far field, the passive coupling should not be used because the coupling of flow quantities has to be accomplished as well.

The objective of the present study is to couple a steady-state model with a nonsteady-state time-dependent model. The temporal aspects must be considered. In other words, the time intervals for introducing source terms need to be determined. If a very short period (Δt = order of minutes) is used, this may result in an unrealistic change of near field source location. However, if the periods are too long, too much information will be lost and the effects of ambient flow on near field mixing cannot be correctly represented. Bleninger et al. (2006) have suggested that the period may be estimated by

$$\Delta t_{\mathrm{c}} = (1 \text{ to } 3) \max(t_{\mathrm{M}}, t_{\mathrm{m}}) \tag{12.1}$$

where Δt_{c} is the coupling time step, t_{M} is the jet/plume time scale, t_{m} is the jet/crossflow time scale. Equation (12.1) gives a time step of approximately 1 h for typical wastewater discharges.

The locations of coupling are determined by the near field model predictions. If the buoyant spreading is not important, the coupling location can be defined at the end of the near field. If the buoyant spreading is important, the coupling locations are defined at the end of intermediate mixing. In the latter case, the near field predicted plume traveling time may need to be compared with the coupling time step to verify the assumption of steady state.

To perform the coupling simulation, a minimum far field grid resolution is required to correctly distribute the scalar quantities. More than one grid cell may be required in some cases. Bleninger et al. (2006) have recommended that the size of a domain can be estimated by

$$\mathrm{Size}_{\mathrm{NF}} = (1 \text{ to } 3)\min\left(l_{\mathrm{M}}, l_{\mathrm{m}}, L_{\mathrm{D}}\right) \tag{12.2}$$

where $\mathrm{Size}_{\mathrm{NF}}$ is the minimum grid size, l_{M} is the slot jet/plume transition length scale, l_{m} is slot crossflow length scale, and L_{D} is the length of diffuser. For a typical produced water discharge, the L_{D} is excluded from Eq. (12.2) as no diffuser is used.

The coupling algorithm used in this study is illustrated in Fig. 12.2. First, the boundary and initial conditions must be obtained to run the MIKE3 hydrodynamic module (HD) based on any reasonable (Courant number < 0.5) grid resolution.

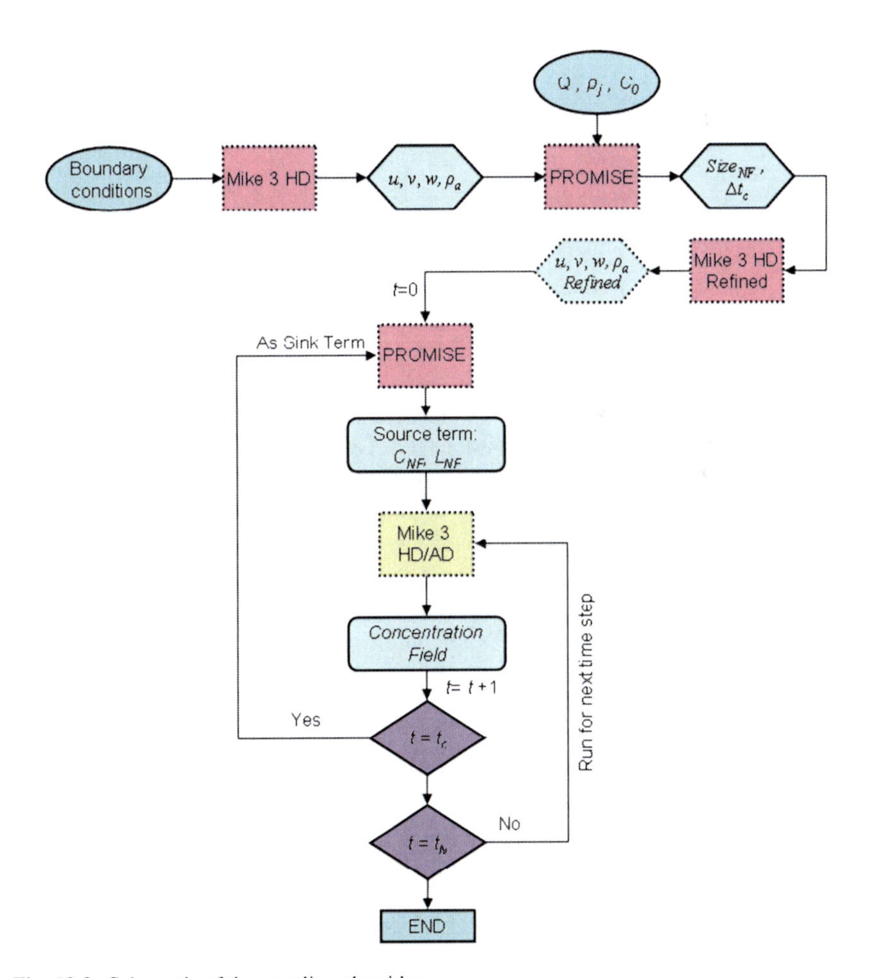

Fig. 12.2 Schematic of the coupling algorithm

The outputs of MIKE3, for example the velocity field and ambient density profile, together with discharge characteristics are used by PROMISE to determine the minimum grid size and the time step of coupling.

The grid size of MIKE3 is then refined and the HD module is executed again to generate the velocity and density information. The PROMISE is executed to predict the near field concentration and the size of plume. This information is used to create the initial source term for the MIKE3 advection-diffusion module (AD).

After the concentration field for this time step is calculated, the MIKE3 model moves forward to the next time step. If t is less than t_c, the model will continue running until t_c is reached. At this point, the predicted field will be used by PROMISE as the accumulated background concentration to consider the re-entrainment of far field returned pollutants. To maintain a mass balance, a sink term in the MIKE3 is created to remove the same amount of pollutant re-entrained by PROMISE. This process is repeated several times until the end of simulation ($t = t_N$).

3 Case Studies

To test the coupling algorithm, a hypothetical case study was performed which simulated an outfall in Oresund, Denmark. The reason for selecting this location is simply because of availability of data. All required data for this case have been provided with the MIKE3 software. It is assumed that an outfall is located at the point (340000, 6150000) at –12.5 m depth. The flow rate of the discharge is assumed to be 0.35 m^3/s via a 0.345 m pipe oriented vertically upward. Three test cases were studied and are described below.

In test case 1, only the far field model MIKE3 was used. The purpose of this case was to examine the ability of the far field model to simulate buoyancy effects. The pollutant with a density of 988 kg/m^3 was introduced at –10 m depth. This discharge density is much smaller than the ambient density (1,013–1,015 kg/m^3) and the plume is expected to rise toward the surface once discharged. In case 1, a coarse grid was used. The advantage of this coarse grid is that the simulation time can be significantly reduced because a longer time step can be used to give a Courant number less than 0.5. The Courant number for any grid must not exceed 0.5 to ensure the stability of the model. A 24 h simulation was performed with a time step of 7.2 s (this gives a Courant number of 0.385).

In test case 2, both PROMISE and the far field model MIKE3 were used. However, only a simple coupling was used in this case. Before the simulation, the PROMISE was used to calculate the dilution, and this concentration was introduced at the terminal level (in this case, the surface). The purpose of this case is to study the effects of coupling under a coarse grid. This case also served as a base case to compare with case 3 to study the effects of grid resolution. The computation time for this case is the same as case 1 because of the same grid resolution.

The purpose of test 3 is to fully evaluate the coupling algorithm described in Fig. 12.2. Unlike case 2, the PROMISE model was used after each coupling step

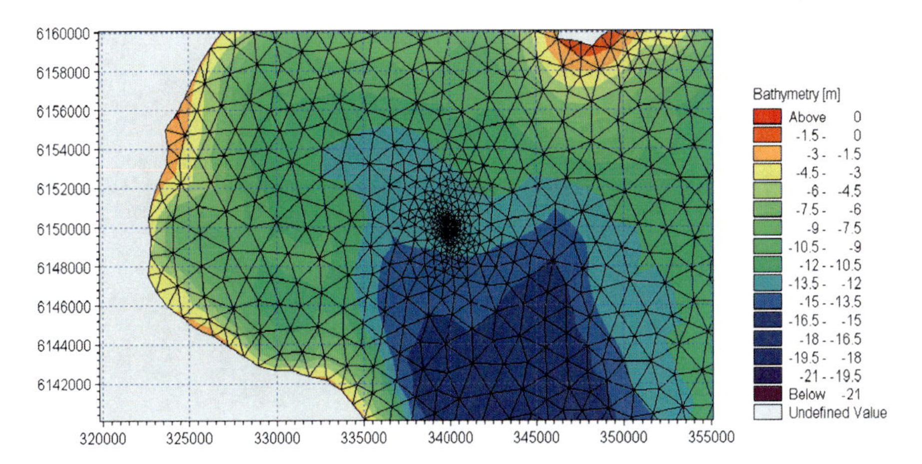

Fig. 12.3 Fine grid used for the simulation of case 3

to calculate the location and concentration for the source term. Further, to correctly introduce the near field term, the computation grid was refined based on the criteria for minimum grid resolution. The grid size at the discharge is only 0.017% that of the coarse grid (Fig. 12.3). The time step of 0.3 s was used in this case, and this gives a Courant number of 0.433 to ensure the stability. Due to the reduced time step, there was a significant increase in computation time. For the same 24 h simulation time, the computation time was 33 h using the same computer excluding the time used in human interaction between each coupling time step ($\Delta t_c = 1$ h). This time is 30 times that of cases 1 and 2.

4 Results and Discussion

It is shown in Fig. 12.4 that the plume center concentration is only 0.26, which is much smaller than the expected value. The near field model PROMISE has predicted a near field concentration of about 70 ppm. The reason for the low concentration is the coarse grid sizes, which cause the source to be dispersed rapidly over the entire grid and results in an unreasonably high dilution. From the near field prediction, the plume is expected to rise until it impinges the surface. However, the far field model using only a coarse grid failed to reproduce the near field plume dynamic processes, and the plume remains in the bottom layer (Fig. 12.5). To correctly predict the mixing and incorporate the near field dilution, the far field model needs to be coupled with a near field model.

To account for the near field mixing, the PROMISE model was executed, and the diluted source was introduced into MIKE3 at the surface layer in test case 2. It can be seen that the plumes predicted by case 2 are wider than that by case 1 (Fig. 12.6). Also, the directions of plumes in these two cases are different. The reason for this difference is the surface current, which is stronger than the bottom currents, and the

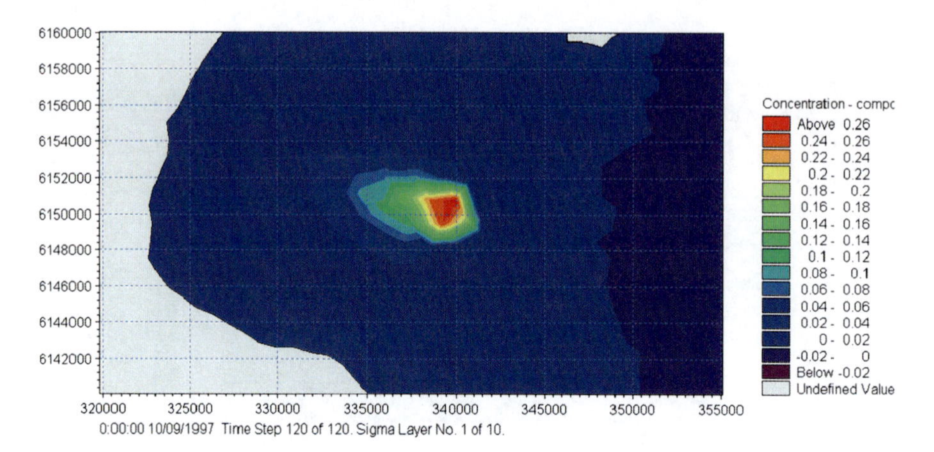

Fig. 12.4 Case 1 horizontal profile, time = 24 h

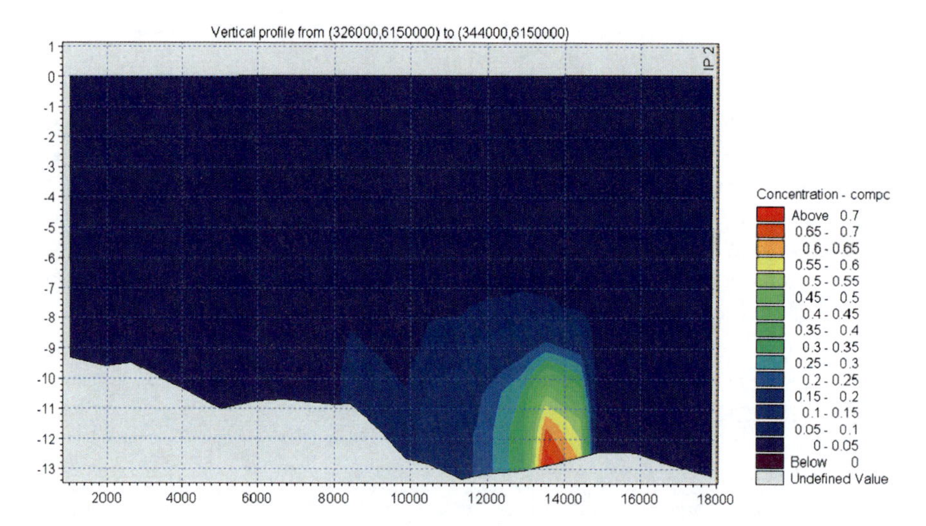

Fig. 12.5 Case 1 vertical profile, time = 24 h

two currents have different directions. In test case 2, the plume is mainly affected by the surface current while the plume in case 1 is mainly advected by the bottom currents.

To correctly predict the concentration, full implementation of the coupling algorithm was conducted in case 3, and the results are shown in Fig. 12.7. It can be seen from the figures that a much smaller plume was predicted in case 3 than the other two cases. This is the result of a finer grid. As the model did not force the pollutants to disperse over a large grid in this case, the introduced near field concentration can be correctly incorporated. As a result of this fine grid and the small plume size, the predicted concentrations in this case are much higher than cases 1 and 2.

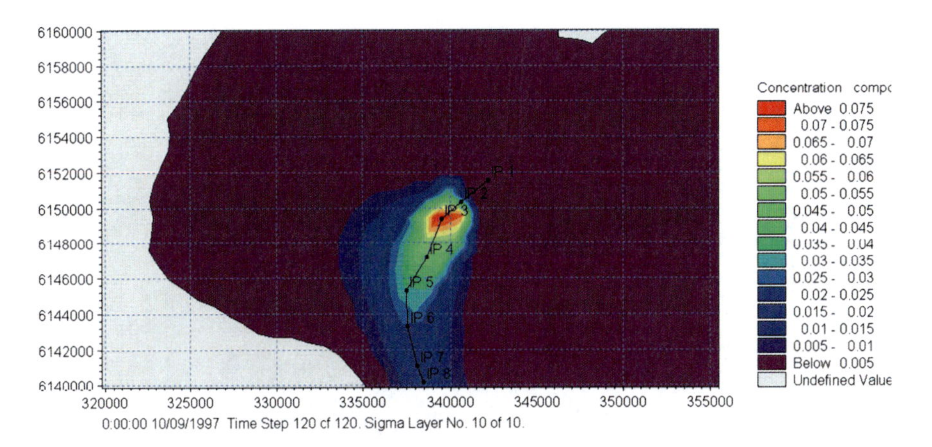

Fig. 12.6 Case 2 horizontal profile, time = 24 h

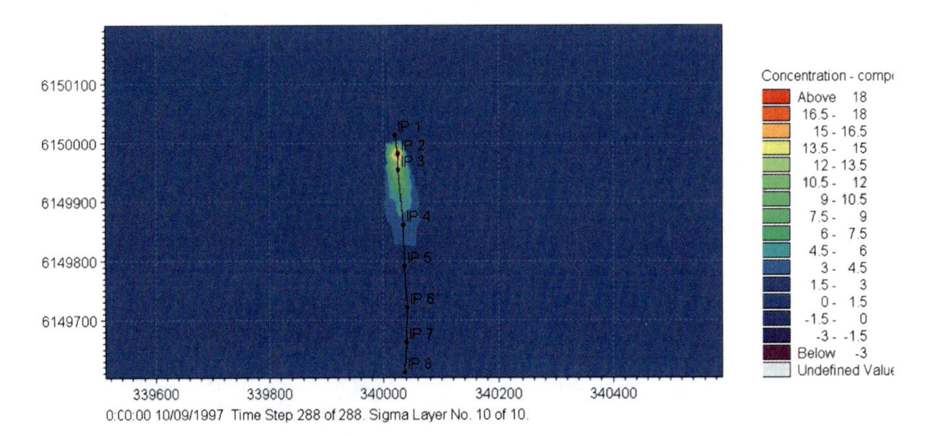

Fig. 12.7 Case 3 horizontal profile, time = 24 h

To show the vertical plume profile longitudinally, the results for the two coupled cases are presented in Figs. 12.8 and 12.9. It can be seen from the longitudinal vertical plots that case 2 gave a bigger plume dimension and therefore a low concentration due to the coarse grid size. Case 3 gave a smaller plume dimension and higher concentration due to the fine grid size. The prediction of case 3 is closer to the near field predictions.

To give a more quantitative description of the predictions, point outputs were generated for the discharge point (340000, 6150000) at two depths, –9 and –1 m. The results for –1 m depth are given in Fig. 12.10. Case 1 is not shown in the figure and only the results for cases 2 and 3 are plotted. To check the model performance, the near field prediction is also shown. It can be clearly seen that the case 2 failed to correctly predict the concentration due to the coarse grid size which forced higher

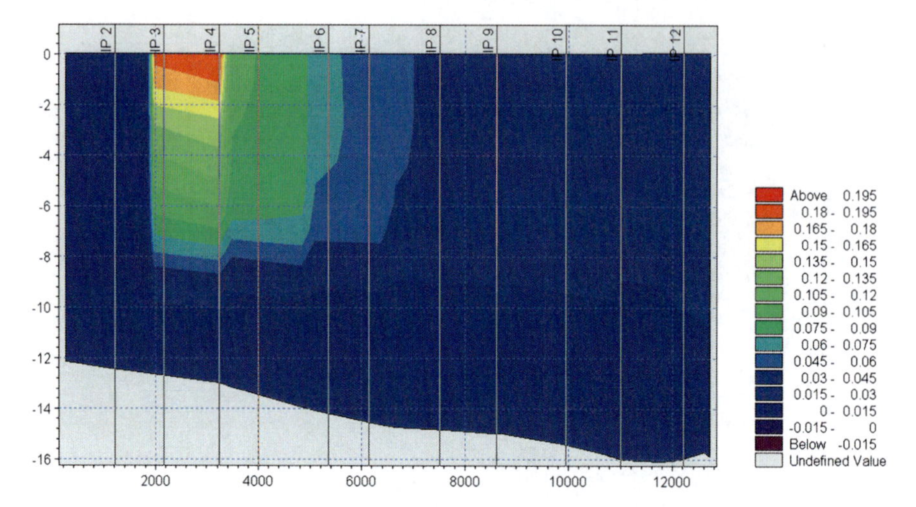

Fig. 12.8 Case 2 longitudinal vertical profile after 24 h

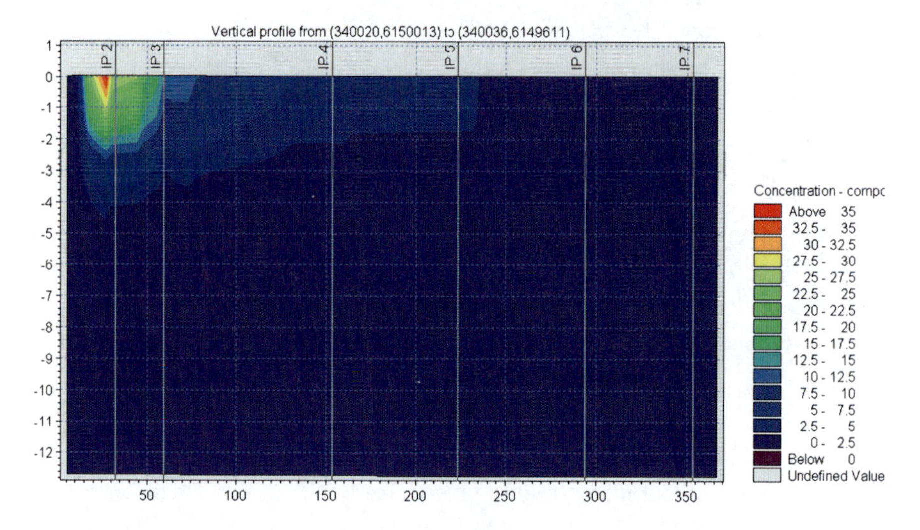

Fig. 12.9 Case 3 longitudinal vertical profile after 24 h

order of dilution of the source over a large grid cell. On the contrary, case 3 predicted the concentration reasonably well from 3 to 16 h after discharge. The result (35 ppm) is close to the values (70 ppm) predicted by the near field model. The small difference between the peak of case 3 and the near field model prediction is mainly because the near field model prediction is for locations at the end of the intermediate field, but case 3 is for locations at the center of the discharge point. There is generally 10–40 m distance between these two locations in the current simulation.

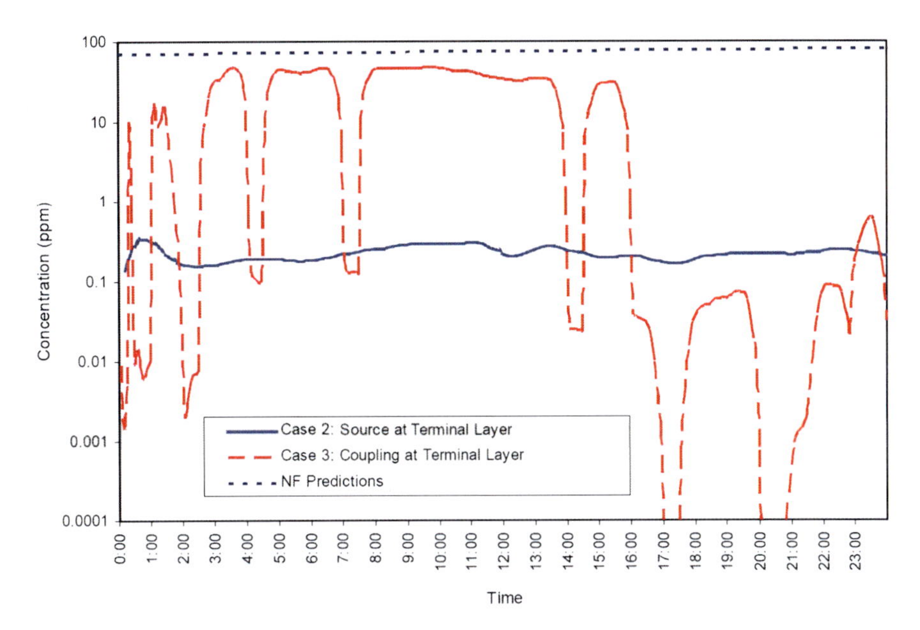

Fig. 12.10 Concentration at the discharge point, depth = –1 m

The near field prediction is the center and has the highest concentration. The concentration decreases as the plume progresses farther away. Therefore, the predicted lower concentration in case 3 after 16 h of discharge compared to the near field is expected and reasonable.

5 Conclusions

In this chapter, the steady-state model, PROMISE, was coupled with a nonsteady-state model, MIKE3. The coupling algorithm used was a two-way passive offline coupling. Three test cases were studied. Case 1 used only the nonsteady-state model with a coarse grid. Case 2 used a simple coupling method with a coarse grid. Case 3 used a fine grid and fully adopted the coupling algorithm.

It can be seen from the case study that a far field model alone fails to resolve the near field dynamics and fails to predict the dilution correctly. The plume was over diluted artificially and remained in the bottom layer rather than being buoyed to the surface. The introduction of a source term into the surface layer in case 2 resulted in an expected surface plume, however, the concentration was also underestimated due to the artificially introduced dilution by the coarse grid. The use of a fine grid in case 3 gives a much smaller plume and a concentration close to the order predicted by the near field model.

It can be concluded that a coupling is necessary and a minimum grid size must be maintained in order to introduce the source term correctly.

Acknowledgments Financial support from the Natural Sciences and Engineering Research Council of Canada and Petroleum Research Atlantic Canada through a Collaborative Research and Development grant (NSERC/PRAC CRD) and the Panel for Energy Research and Development (PERD) is gratefully acknowledged.

References

Baumgartner DJ, Frick WE, Roberts PJW (1994) Dilution models for effluent discharges, 3rd edn. U.S. Environmental Protection Agency, EPA/600/R-94/086, June 1994

Bleninger T, Jirka G (2006) Coupling hydrodynamic models for multi-port diffusers: design and control techniques for submarine outfalls. In: Proceedings of the MWWD-IEMES 2006 conference, Antalya, Turkey, 6–10 Nov, 2006

Burns KA, Codi S, Furnas M, Heggie D, Holdway D, King B, McAllister F (1999) Dispersion and fate of produced formation water constituents in an Australian Northwest Shelf Shallow Water Ecosystem. Mar Pollut Bull 38(7):593–603

Choi KW, Lee JHW (2005) A new approach to effluent plume modelling in the intermediate filed. In: XXXI IAHR congress, Seoul, Korea, 11–16 Sept 2005, pp 4303–4311

Danish Hydraulic Institute (2007) MIKE 21/MIKE 3 flow model FM: hydrodynamic and transport module scientific documentation. Publication of Danish Hydraulic Institute, Horsholm, Denmark

Delft Hydraulics (2005) Delft3D-Flow user's manual: simulation of multi-dimensional hydrodynamic flows and transport phenomena, including sediments. WL Delft Hydraulics, Netherland

Doneker RL, Jirka GH (1990) Expert system for hydrodynamic mixing zone analysis of conventional and toxic submerged single port discharges (CORMIX1). U.S. Environmental Protection Agency, EPA/600/3-90/012, February 1990

Hamrick JM (1992) A three-dimensional environmental fluid dynamics code: theoretical and computational aspects. Special Report No. 317 in Applied Marine Science and Ocean Engineering, Virginia Institute of Marine Science, Gloucester Point, VA

HydroQual (2002) A primer for ECOMSED version 1.3: user's manual. HydroQual Inc., Mahwah, NJ

Jirka HH, Abraham G, Harleman DRF (1975) An assessment of techniques for hydrothermal prediction. Massachusetts Institute of Technology for US Nuclear Regulatory Commission

Kim YD, Seo IW, Kang SW, Oh BC (2002) Jet integral-particle tracking hybrid model for single Buoyant jets. J Hydraul Eng ASCE 128(1):753–760.

King BA, McAllister FA (1998) Modelling the dispersion of produced water discharges. APPEA J 38:681–691

Lee JHW, Chu VH (2003) Turbulent jets and plumes: a Lagrangian approach. Kluwer, Boston

Li S, Hodgins DO (2004) A dynamically coupled outfall plume-circulation model for effluent dispersion in Burrard inlet, British Columbia. J Environ Eng Sci 3:4333–449

Niu H (2008) Dispersion of offshore discharged produced water in the marine environment: hydrodynamic modeling and experimental study. Ph.D. Thesis, Memorial University of Newfoundland, St. John's, Canada

Reed M, Johnsen S, Karman C, Giacca D, Gaudebert B, Utvik TR, Sanni S (2001) DREAM: a dose-related exposure assessment model, technical description of physical-chemical fates components. In: Proceedings 5th international marine environmental modeling seminar, New Orleans, LA, 2001, pp 445–478

Roberts PJW (1999) Modeling Mamala Bay outfall plumes. II: Far field. J Hydraul Eng ASCE 125(6):574–583

Sabeur ZA, Tyler AO (2004) Validation and application of the PROTEUS model for the physical dispersion, geochemistry and biological impacts of produced waters. Environ Model Softw 19(7–8):717–726

Sabeur ZA, Tyler AO, Hockley MC (2000) Development of a new generation modeling system for the prediction of the behaviour and impact of offshore discharges to the marine environment. SPE paper 61263

Smith JP, Brandsma MG, Nedwed TJ (2004) Field verification of the Offshore Operators Committee (OOC) Mud and Produced Water Discharge Model. Environ Model Softw 19(7–): 739–749

TAMU (2007) Ocean/atmosphere circulation modeling projects. Available from: http://stommel.tamu.edu/~baum/ocean_models.html

Zhang XY, Adams EE (1999) Prediction of near field plume characteristics using far field circulation model. J Hydraul Eng ASCE 125(3):233–241

Chapter 13
A New Approach to Tracing Particulates from Produced Water

Barry R. Ruddick and Christopher T. Taggart

Abstract It has been recently discovered that precipitates form when produced water is diluted with seawater. These precipitates are more toxic than the dissolved fraction, can flocculate and sink to the bottom, or can coalesce onto microscopic oil droplets and rise to the surface. This represents new pathways that could bring potentially toxic substances to the surface or bottom where marine life is concentrated. The surface or bottom concentration of toxins and their slow two-dimensional dispersion present potential biohazards that should be evaluated. We present a novel, proven, patent-pending technology specifically designed to trace near-surface or near-bottom particulates to greater distances and dilutions than other measurement technologies can achieve, and present preliminary results demonstrating the effectiveness of the system. We propose a method to experimentally determine the pathways and dilution factors of particulates from oil production platform produced water discharges.

Abbreviation

MAP magnetically attractive particle

1 Introduction

A mixture of oil, gas, and water comes from petroleum reservoirs and is separated in early stages of production. The water, termed "produced water," contains high concentrations of dissolved minerals and emulsified oil droplets and is diluted with seawater before being released from ocean production platforms. The plume is diluted with ambient seawater prior to discharge into the surrounding sea, where it is further diluted by turbulence and ambient currents. Canadian regulatory standards specify only that hydrocarbon concentration at the discharge point not exceed 30 ppm, which can be met by pre-discharge dilution, and do not regulate total quantities of released hydrocarbon or minerals. Post-release dilution and pathways of produced water are rarely measured or quantified. The approach proposed here can accurately track the fate of particulates from produced water, giving the relative

B.R. Ruddick (✉)
Dalhousie University, Halifax, NS, Canada, B3H 4J1

K. Lee, J. Neff (eds.), *Produced Water*, DOI 10.1007/978-1-4614-0046-2_13,

probability of transport to various locations and distances from release. Knowledge of the produced water plume properties can be incorporated to give quantitative estimates of transport, and knowledge of the ambient currents will then give quantitative estimates of concentration.

Azetsu-Scott et al. (2007) found in laboratory experiments that produced water undergoes changes in its physical chemistry, including precipitation of heavy metals, after being mixed with ambient seawater. The particulate fraction was generally more toxic than the dissolved fraction, showing a sustained toxic response for more than a week following the oxidation of freshly discharged produced water that initially elicited little or no toxic response in the Microtox® test. The precipitates tended to aggregate and sink, and there was production of buoyant particles comprised of heavy metal precipitates sequestered onto oil droplets that were transported to the surface. The combination of

- increased toxicity following dilution-induced oxidation,
- rapid transport to both the surface and the bottom where living organisms concentrate,
- slower dilution due to the produced water being near the surface or bottom, and
- slower dispersion in two dimensions in comparison with three

suggest that the heavy metals in produced water may be more toxic than originally believed and have physical pathways that allow them to become concentrated in regions of high environmental sensitivity. These new findings raise the possibility that dilution may not be a sufficient method to deal with produced water from offshore oil platforms. In addition to studies of the changes in physical chemistry and potential for toxicity that occur when produced water is mixed with seawater, it is important to be able to trace the particulate products, both surface and benthic, from existing produced water plumes to assess their fate and dilution factors. Both approaches will be required to accurately assess the potential impact on aquatic life.

Conventional tracing technologies (drifters, current meters, dyes and chemical tracers, survey vessels, and numerical models) are expensive and suffer many time–space limitations, yet models of a dispersing system need appropriate observations sufficient for testing and validation. Dye studies are prohibitively expensive in terms of vessel time. Current meters provide information about currents at a few locations, but do not provide sufficient spatial resolution to deduce Lagrangian paths. Drifters provide Lagrangian paths, but it is the path appropriate to large drifting bodies, not small particles, and their expense precludes their use in large numbers. Dye can be followed in the near field, (DeBlois et al. 2007) but cannot be detected at long distance or times from the source because dilution rapidly reduces the concentration to below detection threshold (Wells, Bailey and Ruddick, 2011). While dye can be detected with purpose-built moored instruments, their expense precludes their application in even moderate numbers. The default approach, sampling from a moving vessel, suffers greatly from the "moving target" problem: the plume tends to move and disperse more rapidly than it can be mapped, and synoptic surveys are prohibitive. Dye is moreover inappropriate for tracing particulates, since it remains dissolved in the water column, dispersing and moving differently from surface or near-bottom particulates. The approach we propose below does not have these deficiencies, has been shown to trace fluid dispersion over distances of ~10 km, and with appropriate accompanying observations, promises to yield quantitative results.

2 A Novel Tracing Technology

A new, patent-pending (Ruddick and Taggart 2006) technology for aquatic environmental tracing is based on small, design-size, design-buoyancy, environmentally benign particles that can be released in either an instantaneous or a time-release fashion from one or more spatial locations. They drift and disperse in the surface (or near-bottom) layer of the ocean, to be collected (sampled) using inexpensive autonomous moored magnetic collectors. The concepts, particles, and collectors form an inexpensive system for directly determining particle dispersal at a range of temporal (minutes to months) and spatial scales (meters to 1000s km^2) in virtually any liquid.

The magnetically attractive particles (MAPs; Fig. 13.1) combine glass microspheres for floatation with fine magnetite plus a binding agent and pigment. Proportions can be adjusted to achieve the desired particle-specific gravity, ranging from positive buoyancy to trace floating particulates to negative buoyancy to trace sediments. The size can be adjusted, but we currently employ the 300–500 μm range, which allows for easy counting of collected particles.

The collectors (Fig. 13.2) consist of plastic flow-through tubes with rare-earth magnets strategically placed such that any MAP will be caught and retained with

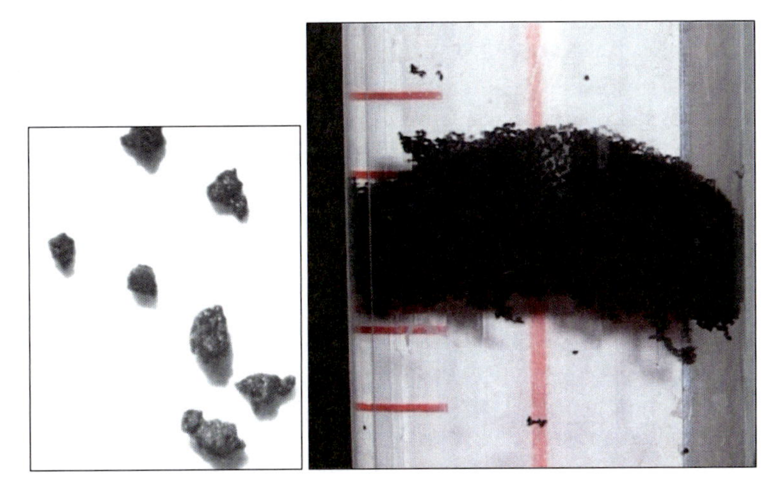

Fig. 13.1 Magnetically attractive particles (MAPs) in the 300–500 μm size range (*left panel*) and suspended at a specific gravity near 1.02 (precise values not known for this sample) in a density gradient settling column (*right panel*); size and specific gravity can be adjusted during manufacture

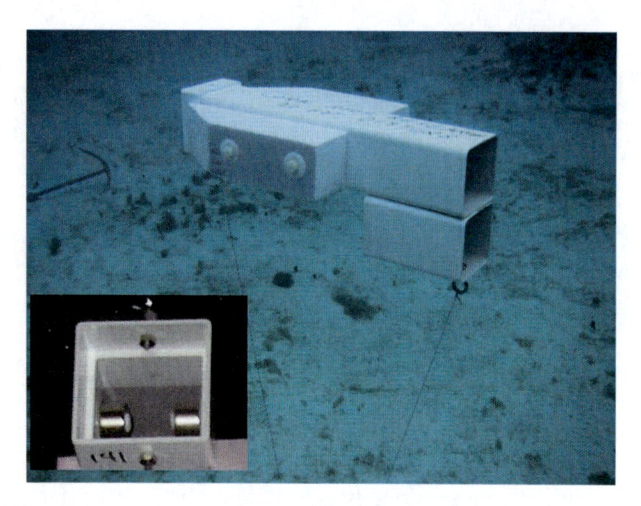

Fig. 13.2 MAP collector in situ on the Belize Barrier Reef, September 2007, with inset showing detail of the collector mouth and magnets

greater than 90% probability in 1 m/s currents. Polyurethane extruded foam is attached to provide floatation, allowing the collectors to be moored prior to particle release. The collectors are autonomous, inexpensive, require no power supply or technical expertise to deploy and use, and they automatically vane into the flow.

In operation, collectors are moored at locations of interest prior to a release. This can be done in good weather, and their simplicity and robustness ensures their likely survival in the event of poor weather. Their low cost means that the number of sample locations is mainly limited by the time required to set and recover the moorings. At the start of an experiment, particles are released in one or more locations, either in an instantaneous release or over a time chosen to sample a variety of flow conditions (e.g., over one or more tidal cycles). During the time of dispersal and particle collection, typically several days to a week, it is possible (but not necessary) to sample the plume using conventional net tows guided by drifters. We emphasize that this is expensive in terms of vessel time and provides limited ancillary information. The collectors are then retrieved and the numbers of particles from each release location are counted in the laboratory.

The system has several advantages over conventional tracing technologies:

It is exceedingly economic. In environmental field applications, the collectors negate the use of expensive instrumentations and their spatial limitations and/or ship time for surveying an ever-increasing area with an ever-diminishing signal from conventional tracers. The low cost of the collectors allows a large number of mooring locations for sampling. A single particle and collector experiment is far cheaper than developing a numerical flow model and making the necessary field observations required to validate such models.

It can validate numerical models. A significant limitation in the plethora of numerical flow models used for open ocean systems is the paucity of model validation, particularly with respect to Lagrangian paths. The technology we propose can function across a large range of scales for the validation and improvement of the models.

It offers direct dispersion measurement. This technology offers direct observation of sub-grid scale dispersion; a poorly known quantity that must be accurately parameterized if numerical models are to work well. Typical current measurements (flow sensor arrays, conventional drifters, etc.) provide information at scales too large to provide this information.

It is resistant to weather. The moored collectors continue to work over day to year periods (no power requirements) in all weather and in environments when or where vessels or other technology cannot. The simple and sturdy collectors are similar to lobster pot marker buoys and are dragged beneath the surface and its wave activity, when currents become extreme. In two coral-tracking experiments, collectors survived the passage of hurricanes with minimal losses.

It can accommodate discrimination of multiple sources and multiple sinks. Variations in color and/or size and/or design density of the particles can allow different batches of particles to be released at identical and/or different times and/or locations, or as a series, in liquids or water masses of different densities. Thus, assessing advection and dispersion of particulates to, from, and among a variety of flows, locations, and water masses can be achieved within a single study.

It is suitable for use in difficult or inhospitable environments. Many environments are too difficult and/or expensive to access, to instrument, and to model easily. Inclement weather can obliterate a dye study making the expensive release and initial observations useless.

It is significantly more immune to dilution than dye. Because we are detecting particles rather than dye, particles are either detected or not, are concentrated by the magnetic collectors, and can be detected at larger distances and greater dilution factors than dyes.

It is appropriate technology for tracing particulates. The design buoyancy and small size of the tracer particles makes them float (or sink, in the case of sediments or flocs), drift, and disperse in the same fashion as the particulates they mimic. This cannot be said of any other tracing technology.

The collectors integrate the particle concentration over time. This accumulation allows detection to lower concentration levels so the resulting data are more statistically robust, spatially smooth, and more representative of average dispersion and dilution.

3 Proof of Concept: The Key Largo Experiment

The particle/collector technology was initially inspired by and developed to answer questions about coral reef connectivity – the numerous factors that determine propagation distance from spawning to settlement in corals and the animals that live on the reefs. A prototype experiment in the reef system off Key Largo, Florida was

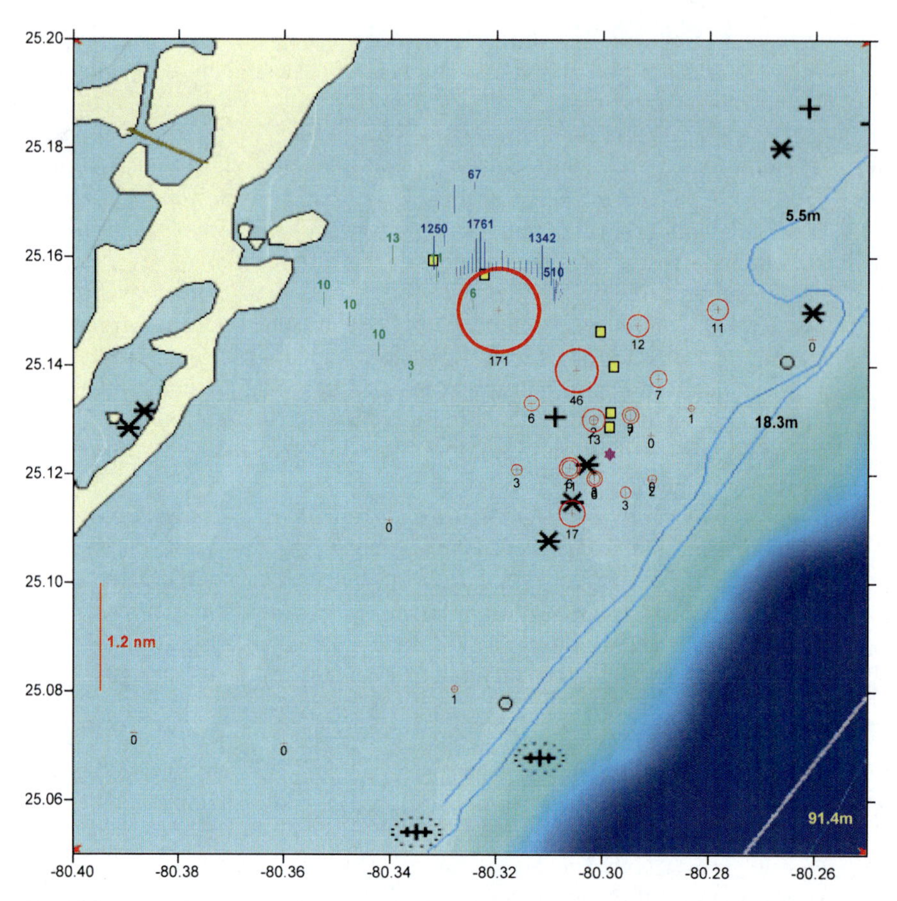

Fig. 13.3 Results from the Key Largo proof of concept tracing experiment; MAPs and drifting buoys were released at the location marked by the *magenta star*; path of one drifter is shown as *yellow squares*; *vertical bars* indicate particle counts (square-root scale) from conventional net tows in the 8 h post-release (*blue*) and ~20 h post-release (*green*); *red circles* indicate numbers (area proportional to count, circle subscript) of particles found in collectors; collectors with zero counts are indicated by a *red* +

planned to use a 40-collector mooring array surrounding a known coral spawning area, where a release of MAPs was planned to coincide with coral spawning. Three hurricanes and 2 mandatory evacuations forced a reduction to an array of 25 collectors and conventional dispersion survey monitoring for <24 h. The severe weather proved the value of the collector technology in that the collectors worked flawlessly when virtually every other aspect of the program was compromised. One collector, moored in <10 m of water, survived and functioned through winds exceeding 45 knots. We believe that the small size and appropriate buoyancy of the collector allowed it to sink beneath the surface during high current events, protecting it from wave damage.

Figure 13.3 shows the release location (magenta star) and the track (yellow squares) of one of four drifters released with the particles that helped to guide the location of conventional net tows. Particle counts from the net tows during the first 8 h post-release (blue vertical bars, square root scale) and longer tows at about 20 h post-release (green) are also illustrated. The small-scale variability on day 1 is due to the released patch being distorted into thin streaks, as confirmed by visual observation. This "streakiness" is a common feature of a tracer patch during initial dispersion (Garrett 1983), makes conventional survey methods more difficult, and makes point sampling of a field so variable as to be almost useless. The particle counts from the moored collectors (red circles, area proportional to particle number) are spatially smooth, almost certainly a result of the advection of individual streaks being swept past collectors and averaged. Notice that significant numbers of particles were collected at locations that were not surveyed by net tows. This was because the survey vessel was mapping the plume as defined by drifter tracks and visual observation, leaving no time for a complete spatial survey – the plume was moving and dispersing faster than the speed at which the vessel could map it.

The comparison of tow collections with moored particle collector counts illustrates the advantage of the MAP/collector technology over conventional dye survey observations. The collectors take observations and average over the whole experiment, even during inclement weather, resulting in a statistically robust, spatially smooth, objective, and synoptic survey and can sample a dispersing plume over longer distances and at lower concentrations than dye.

Prior to deployment, it was discovered that a significant proportion of the particles sank (MAP formulation has since changed), so several near-bottom collectors were deployed. These were found upon recovery to have captured particles from the near-bottom region surrounding the release site, confirming the potential for the MAP/collector system to trace sinking and or sediment-like particles.

4 A Proposed Experiment to Trace Particulates from Produced Water

Produced water can form precipitates rich in heavy metals upon dilution with seawater and consequent oxidation, and the particulate form is more toxic than the dissolved form. These particulates can become concentrated near the surface or bottom where life is most plentiful. The MAP/collector system is an appropriate technology for tracing the path, dispersion, and eventual fate of these potentially toxic particulates; we outline a tracing experiment that could prove its value as a monitoring technology.

An inexpensive but feasible tracing experiment would begin with mooring 25–40 collectors at surface and (optionally) near-bottom at several locations surrounding a production rig out to approximately 10 km distance. Particles would then be released at a known rate into the produced water effluent stream over a minimum of one tidal cycle, so as to effectively sample a variety of flow conditions. Both positively and (optionally) negatively buoyant MAPs could be released, allowing both surface and near-bottom particulates to be traced. After several days, the collector array would be retrieved and the numbers of captured particles counted.

Knowledge of the rate of produced water discharge, rate of discharge of MAPs, and concentration of both MAPs and precipitates in the near field of the discharge would allow the MAP and particulate concentrations to be related, so that particle capture numbers could be related to precipitate concentrations. Furthermore, knowledge of the ambient currents past the collectors would allow particle capture numbers to be related to fluxes and to average concentrations.

How well can we expect this system to perform in such an experiment? The rate of discharge of produced water varies during the lifetime of a well ("Managing Produced Water," Canadian Association of Petroleum Producers information Pamphlet), but a typical value may be 1000 m^3/h, or about 0.3 m^3/s (Zhao et al. 2008). We imagine releasing into a 0.3 m^3/s effluent stream 14 kg ($\sim 10^9$) of MAPs over 12 h duration. The stream will therefore have $10^9/0.3 \times 12 \times 3600\ m^3 =$ 77000 particles/m^3 at inflow. With a dilution factor of 10,000, ten times more dilute than the normal limit for dye detection, there will be 7.7 particles/m^3. In an ambient current of 20 cm/s (typical for the Nova Scotia and Newfoundland shelves in the absence of storm conditions), a collector with a mount area of 25 cm^2 samples $\sim$50 m^3/day, and so would be expected to catch 380 particles in a day; a substantial number. This suggests that the plume of particles should be detectable at significantly greater distances and dilutions than achievable by dyes.

Another approach would be to estimate the dimensions of the released patch of particles after approximately 1 week dispersion. If the eddy diffusivity in the vicinity of the production rig were $K = 100\ m^2/s$, then at a time of 1 week post release, the dispersed patch would occupy an area of $K_t = (100\ m^2/s) \times (1\ \text{week})$ $= 6 \times 10^7\ m^2$, or 60 km^2. If the particles occupy the upper water column (keeping in mind that they float) then the concentration at this late stage of dispersion would be 16.5 particles/m^3. A collector would then be expected to capture 830 particles per day in a 20 cm/s current. The MAP/collector system is therefore expected to track the fate of particulates to dilutions and distances well beyond what can be achieved with dyes or chemical tracers, and at a reasonable cost. Recent (Wells, Bailey and Ruddick, 2011) experiments to trace ballast water released from vessels in Goderich Harbor (ON) used Rhodamine dye to track ballast water to distances of several hundred meters, and particles to track it to distances nearing 10 km. The estimates above and the successful applications of this tracing technology in field experiments suggest that using the MAP/collector system to trace produced water will allow tracing to greater ranges than previously possible, and with appropriate ancillary measurements, would allow fluxes and concentrations of PW constituents to be quantitatively estimated.

5 Summary

We have described a potential environmental problem with produced water (metallic particulate formation and concentration near surface and bottom), and a novel, patent-applied, field-proven technology capable of tracing the particulates to greater distances and dilutions than any other means. This technology is robust, simple, and relatively inexpensive. It can be combined with direct observations of particulates in the effluent stream and near field to result in "calibrated" results. The observations are extremely well suited for testing and/or validating numerical flow/dispersion models used to forecast the fate of dissolved and particulate production rig products.

Acknowledgments We are grateful to Kumiko Azetso-Scott for pointing out the possibility of particulate formation when produced water is diluted with seawater, and lending us her invaluable understanding of the physical chemistry of produced water/seawater mixtures. We thank Ken Lee for his kind invitation to present this concept at IPWC St. John's. Our work is supported by the Natural Science and Engineering Research Council of Canada and a number of other sources.

References

Azetsu-Scott et al (2007) Precipitation of heavy metals in produced water: influence on contaminant transport and toxicity. Mar Environ Res 63:146–167

DeBlois EM, Dunbar DS, Hollett C, Taylor DG, Wight FM (2007) Produced water monitoring: use of rhodamine dye to track produced water plumes on the Grand Banks. Presented to the international produced water conference, NFLD, St. Johns, 17–18 Oct 2007

Garrett C (1983) On the initial streakiness of a dispersing tracer in two- and three-dimensional turbulence. Dyn Atmos Oceans 7:265–277

Ruddick BR, Taggart CT (2006) Apparatus, system and method for evaluating fluid systems. USA and Canada Patent Pending. Appl.No.11/458,287, Filed 07/18/2006. Conf. #8539, with benefit of 60/699,845 filed 07/18/2005

Wells M, Bailey S, Ruddick B (2011) The dilution and dispersion of ballast water discharged into Goderich Harbor. Mar Pollut Bull, 25 Mar 2011. doi: 10.1016/j.marpolbul.2011.03.005

Zhao L, Chen Z, Lee K (2008) A risk assessment model for produced water discharge from offshore petroleum platforms-development and validation. Mar Pollut Bull 56(11):1890–1897. doi:10.1016/j.marpolbul.2008.07.013

Part IV
Biological Effects

Chapter 14
Field Evaluation of a Suite of Biomarkers in an Australian Tropical Reef Species, Stripey Seaperch (*Lutjanus carponotatus*): Assessment of Produced Water from the Harriet A Platform

Susan Codi King, Claire Conwell, Mary Haasch, Julie Mondon, Jochen Müeller, Shiqian Zhu, and Libby Howitt

Abstract There is paucity of data regarding hydrocarbon exposure of tropical fish species inhabiting the waters near oil and gas platforms on the Northwest Shelf of Australia. A comprehensive field study assessed the exposure and potential effects associated with the produced water (PW) plume from the Harriet A production platform on the northwest shelf in a local reef species, Stripey seaperch (*Lutjanus carponotatus*). This field study was a continuation of an earlier pilot study which concluded that there were "warning signs" of potential biological effects on fish populations exposed to PW. A 10-day field caging study was conducted deploying 15 individual fish into 6 separate steel cages set 1-m subsurface at 3 stations in a concentration gradient moving away from the platform. A battery of biomarkers were evaluated including hepatosomatic index (HSI), total cytochrome P450, bile metabolites, CYP1A-, CYP2K- and CYP2M-like proteins, cholinesterase (ChE) activity, and histopathology of liver and gill tissues. Water column and PW effluent samples was also collected. Results confirmed that PAH metabolites in bile, CYP1A-, CYP2K-, and CYP2M-like proteins and liver histopathology provided evidence of significant exposure and effects after 10 days at the near-field site (~200 m off the Harriet A platform). Hepatosomatic index, total cytochrome P450, and ChE did not provide site-specific differences by day 10 of exposure to PW. CYP proteins were shown by principal component analysis (PCA) to be the best diagnostic tool for determining exposure and associated biological effects of PW on *L. carponotatus*. Using a suite of biomarkers has been widely advocated as a vital component in environmental risk assessments worldwide. This study demonstrates the usefulness of biomarkers for assessing the Harriet A PW discharge into Australian waters with broader applications for other PW discharges. This approach has merit as a valuable

S. Codi King (✉)
Australian Institute of Marine Science, Townsville, QLD, Australia

K. Lee, J. Neff (eds.), *Produced Water*, DOI 10.1007/978-1-4614-0046-2_14,

addition to environmental management strategies for protecting Australia's tropical environment and its rich biodiversity.

1 Introduction

The Northwest Shelf of Western Australia is one of the main regions for Australia's offshore oil and gas production activity. Produced water (PW) is a major waste product of offshore petroleum and gas production activities and can be discharged directly into the ocean from offshore processing facilities following separation from the product component. Its treatment and subsequent disposal has been highlighted as potentially the most significant problem associated with offshore production activities in Australia (Swan et al. 1994). In Western Australia, PW discharges are regulated under Australian offshore petroleum legislation which states that maximum allowable PW discharges (as petroleum in water) cannot exceed 50 mg L^{-1} (instantaneous) or 30 mg L^{-1} (over a 24-h period) (Cobby 2002). Harriet A is an oil and gas production platform which lies ~120 km due west of Dampier off the Northwest Shelf of Australia (Fig. 14.1) and discharges directly into the marine environment. Harriet A PW has been the focus of numerous studies which have examined the physical, biological, and geochemical processes affecting the fate of this PW discharge. Results from these studies demonstrate that the Harriet A PW plume is buoyant and forms a fine surface slick in a NNW direction on the ebb tide and SSE direction on the flood tide and can be detected up to distances of 10 km (Burns and Codi 1999; Holdway and Heggie 1998). A zone of potential biological impact has been identified within approximately 1000 m from the Harriet A platform in the direction of predominant tidal flows (Furnas and Mitchell 1999; Holdway and Heggie 1998; Burns et al. 1999; Burns and Codi 1999; King and McAllister 1998). Also identified from these studies were a number of Harriet A PW constituents that are of environmental concern including petroleum hydrocarbons, polycyclic aromatic hydrocarbons (PAHs), and volatile organics consisting of benzene, toluene, ethylbenzene, and xylene—BTEX compounds.

In 1999, the regulation of PW in Western Australian waters had changed under the introduction of the Commonwealth Petroleum (Submerged Lands; Management of Environment) Regulations, which stated that operators of facilities such as Harriet A had to assess the environmental effects and risks associated with the PW discharge. To address this issue, a pilot study evaluated the use of sub-lethal stress indicators in resident fish as potential biomarkers for hydrocarbon exposure from the Harriet A PW discharge. Biomarkers measure the biological response of an organism at the molecular, biochemical, cellular, and physiological levels of biological organization which can be related to exposure to or toxic effects of environmental chemicals (Peakall 1994). This study investigated several biomarkers that have been established for use in PAH assessments in temperate regions (Stagg and McIntosh 1998); however, were being applied in tropical Australian waters for the first time. Ethoxyresorufin-*O*-deethylase (EROD) activity, bile metabolites, and CYP1A mRNA levels were evaluated in two tropical reef species, Gold-Spotted

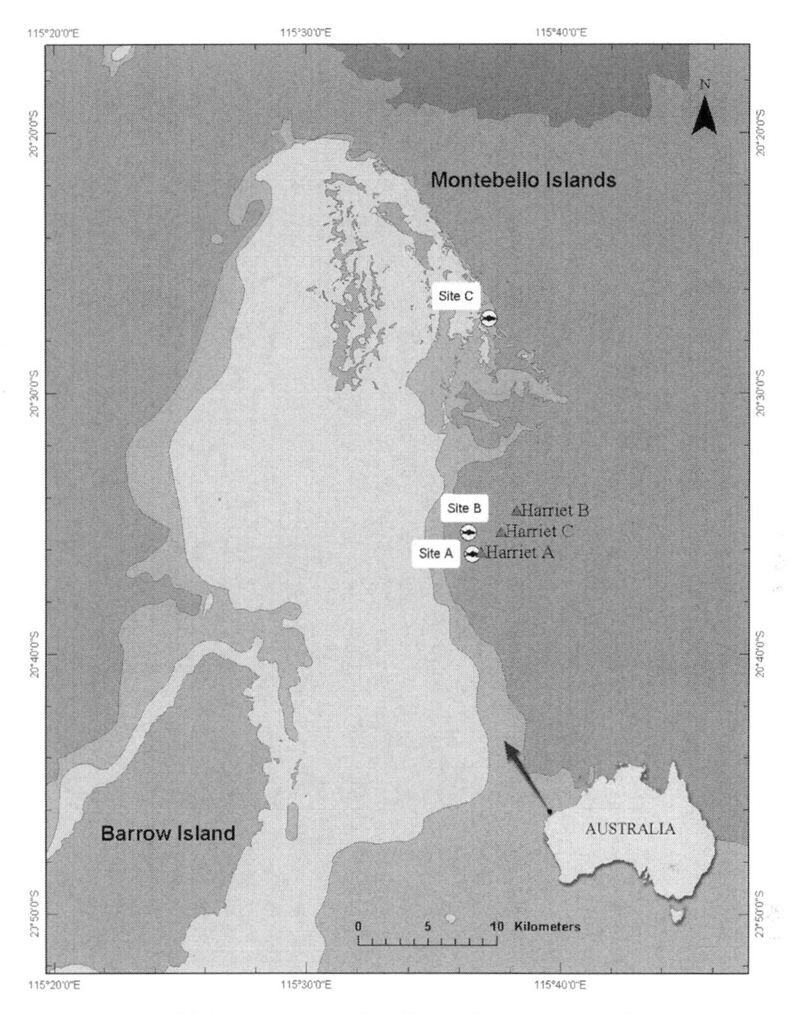

Fig. 14.1 Locations of fish cages deployed at Site A (Harriet A, near-field, ~200 m), Site B (Harriet A, far-field, ~1000 m), and Site C (reference site, off the Montebello Islands) relative to the Harriet A platform

Trevally (*Carangoides fulvoguttatus*) and Bar-Cheeked Coral Trout (*Plectropomus maculatus*) captured from two sites impacted by PW discharge, Harriet A (near-field) and Harriet C (far-field), and these responses were compared to a reference site located outside the zone of PW influence (Codi King et al. 2005). Comparison of hydrocarbon levels and CYP1A induction with other field studies showed that the pilot study data most closely resemble sites with low-level PAH contamination (McDonald et al. 1996; Collier et al. 1995). Both bile metabolites and CYP1A mRNA level provided clear warning signs that there was potential for biological effects in fish populations exposed to PW around the Harriet A production platform

Table 14.1 Suite of biomarker tests conducted in *L. carponotatus* exposed to Harriet A PW

Biomarker	Ecological significance	Type of biomarker
Hepatic somatic index (HSI)	This is a physiological condition index. Different stresses may cause the HSI to fluctuate above or below a normal range. Studies indicate that exposure to pollutants causes the HSI to increase as the liver increases in size to allow greater detoxification of pollutants.	Effects
Cytochrome P450	Cytochrome P450 (CYP1A) comprises an extensive super family of heme proteins that are found in all living organisms. Both hepatic (liver) and extra hepatic tissues (e.g. gills, kidney, intestine, and brain) have detectable enzymatic activity, but the liver is the organ in vertebrates known to have the highest activity. Enzymatic induction by planar molecules can interfere with the normal functioning of an organism.	Exposure
CYP1A-, CYP2K1-, and CYP2M1-like proteins	The induction of cytochrome P450 1A (CYP1A) is considered the most sensitive sub-lethal indicator of PAH exposure. In general, very little is known about the inducibility of CYP1A in tropical fish species, although multiple tropical fish species from Bermuda (Stegeman et al. 1997) and butterfly fish from Florida and Belize (Vrolijk et al. 1994) have been investigated. Recent work in Australia has demonstrated that several species of tropical reef fish have significantly increased induction in CYP1A and CYP2K1 proteins in response to exposure to petroleum hydrocarbons and PAHs from a production platform on the NW Shelf of Australia (Zhu et al. 2006).	Exposure
Bile metabolites (also known as fluorescent aromatic compounds–FACs)	PAHs are readily absorbed by benthic and water-column species, where they are extensively biotransformed from parent compounds to more water-soluble metabolites, which is part of the function of the cytochrome P450 enzymes. Metabolites are concentrated in the bile and due to their water solubility are easily sampled and analysed directly through HPLC using fluorescence detection. This technique has been successfully employed in numerous fish species as the presence of fluorescent aromatic compounds (FACs) can provide a clearer estimation of the biological availability of the PAHs (Krahn et al. 1984; Altenburger et al. 2003).	Exposure

Table 14.1 (continued)

Biomarker	Ecological significance	Type of biomarker
Cholinesterase activity (ChE)	Cholinesterase (ChE) includes acetylcholinesterase (AChE), found mainly in the synapses and blood, and butyrylcholinesterase (BChE), found mainly in the liver. Cholinesterase catalyses the hydrolysis of acetylcholine, an important function for normal synaptic activity in both invertebrates and vertebrates (Ellman et al. 1961). Inhibition of ChE activity is a specific biomarker neurotoxic action by organophosphate and carbamate pesticides. New evidence shows some metallic ions (Guilhermino et al. 2000; Elumalai et al. 2002); hydrocarbons (Galgani et al. 1992), surfactants, and detergents (Guilhermino et al. 2000; Guilhermino et al. 1998); and wood chip extracts (Payne et al. 1996) may also inhibit ChE activity.	Neurological effects
Histopathology	Histopathology helps in identifying target organs of toxicity and mechanisms of action (Johnson and Bergman 1984; Wester et al. 2002). Histopathology may help to detect minute effects (and early changes) which are not discernable on gross visual inspection, but may be highly relevant to the long-term well-being of the fish species concerned. Critical adverse biological effects, such as reproductive abnormalities, may well go undetected if histopathology is not employed (Ibid.). Additionally, the potential toxicity at the cellular and tissue level will be visible at lower dosages (or shorter exposure duration) compared to measuring toxicological endpoints such as mortality or behavioural changes (Ibid.).	Cellular effects

(Codi et al. 2001; Codi King et al. 2005; Zhu et al. 2008) and confirmed earlier findings that within 1000 m (the mixing zone) of the Harriet A platform is the area most impacted and warrants further investigation.

In May 2003, a second study was conducted by the Australian Institute of Marine Science (AIMS) which involved a more detailed biological and chemical assessment of Harriet A PW discharge. A 10-day field caging study was conducted deploying 15 *L. carponotatus* into 6 separate steel cages set 1-m subsurface at three sites: Site A (~200 m, near-field), Site B (~1000 m, far-field), and Site C (>8 km, reference site) in a NNW line from the Harriet A platform. A suite of biomarkers of PAH exposure were evaluated (Table 14.1) along with chemical-based monitoring of PW effluent and water-column samples using grab and time-integrated samples. The main focus was to assess the region in close proximity to the Harriet A PW discharge by comparing biological responses in fish exposed within 1000 m of the Harriet A Platform (at two sites) to responses in unexposed fish. The use of biomarkers provides an important measure of the total external load that is biologically available that can be related to actual or "real-world" exposures as well as provide some insight into the potential mechanisms of biological effects caused by the contaminants (Holdway et al. 1995; Altenburger et al. 2003). Secondly, the extent of the PW plume and its impact both near-field and far-field was assessed to determine whether a proposed Marine Protected Area (MPA), located due west ~7 km of the platform, would be at risk from the Harriet A PW discharge.

2 Materials and Methods

2.1 Fish Collection in Dampier Harbour, Australia

Stripey seaperch (*L. carponotatus*) is an important recreational fishery in Dampier Harbour (Craig Skepper, Western Australian (WA) Fisheries, personal communication), and preliminary biomarker work (Codi et al. 2001) suggested that it would be a good candidate species for this study. Fish were collected using galvanized steel fish traps deployed in Dampier Harbour. A total of 130 *L. carponotatus* were collected over 3 days and maintained in a 1000-L holding tank with running seawater, aerated and transported live to the Harriet A Platform aboard the *RV Cape Ferguson*.

2.2 Site Selection and Fish Cage Deployment

The Harriet A production platform is the only oil production platform in this region and discharges PW at two points, pipe Z2 and C8 (subsurface release into <25 m of water). From January 1998 to September 2003, the average daily total hydrocarbon concentration in PW discharge for Harriet A was 18.4 ± 3.41 mg L^{-1} and 20.0 ± 4.32 mg L^{-1}. Site A (~200 m, near-field) and site B (~1000 m, far-field) were positioned in a NNW line from the Harriet A PW discharge (Fig. 14.1). The hydrocarbon chemistry around this platform is well documented with sea-surface microlayer total petroleum hydrocarbons concentrations detectable to a distance of 1 km. Near-field concentrations of hydrocarbons (within ~300 m of platform) range from 3 to 9 $\mu g\ L^{-1}$ and far-field concentrations (~1 km) range from 0.3 to 1.3 $\mu g\ L^{-1}$ total oil in water (Burns and Codi 1999). The main components of the PW discharge identified in the dissolved phase were naphthalene and its homologues whereas phenanthrene and its homologues mainly associate with particulate matter. The mixing zone was defined from 0.9 to 2 km of the Harriet A platform, where most particulates settle out of the water column (Burns et al. 1999). Site C (>8 km) was chosen to represent a site well outside the zone of PW influence; our reference site against which caging effects and PW effects were assessed (Fig. 14.1).

Fifteen ($n = 15$) *L. carponotatus* were deployed into each cage, and there were two cages deployed per station to assess caging effects. All fish cages were deployed on the same day (with cage numbers randomly assigned) using the mooring design presented in Fig. 14.2.

- Site A (near-field site), fish cages 1 (FC1) and 8 (FC8) deployed at 1 m below the surface; water depth 22 m.
- Site B (far-field site), fish cages 6 (FC6) and 7 (FC7) deployed at 1 m below the surface; water depth 25 m.
- Site C (reference site), fish cages 2 (FC2) and 3 (FC3) deployed 1 m below the surface; water depth 20 m.

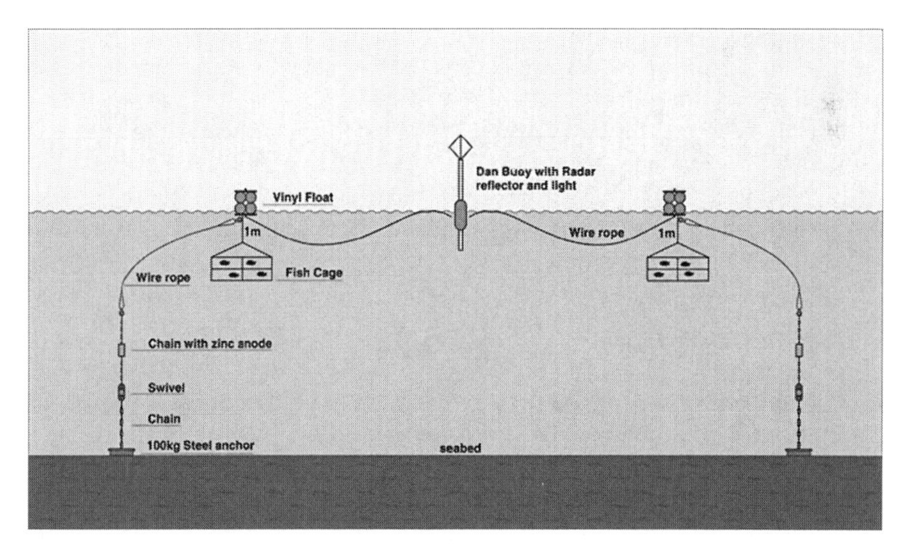

Fig. 14.2 Schematic diagram of mooring deployment for fish cages as designed by Cary McLean (AIMS)

A total of 90 fish (15 individuals × 6 cages) were deployed into fish cages plus 14 individuals were sacrificed from the holding tank to represent exposure at $T = 0$, for a total of 104 fish.

All fish cages were deployed with canisters containing high-quality (IQF) pilchards (Western Australia Bait Supply, Fremantle) collected from clean waters off Albany, WA. In addition, each fish cage contained one semipermeable membrane device (SPMD) containing three membranes to estimate the concentrations of hydrophobic PAHs for the duration of the caging study. Bait canisters were replenished every 2 days to remove starvation as a physiological variable.

2.3 Stripey Seaperch (*L. carponotatus*) Tissue Collection

Five fish per cage (10 per site) were to be sampled on Days 3, 7, and 10 of exposure. However +35 knots winds on Days 6 and 7 prevented the collection of fish for Day 7, so all fish remaining in each cage (~10) were sacrificed on Day 10. Fish were retrieved using small boats and were transported live to the vessel and sacrificed. The liver was excised immediately and 2 g sub-samples were collected for total cytochrome P450 content and immunodetection of CYP1A-, CYP2M1-, and CYP2K1-like proteins. Both liver and gill samples were collected and preserved in 10% formalin for histopathology. Bile fluid was extracted with a syringe and frozen in liquid nitrogen (N_2) for analysis of PAH metabolites. Muscle tissue was collected for assessment of AChE activity. The sex of the fish was determined by examination of the gonads. The sagittal otoliths were removed for age determination. Processing time was kept to a maximum of 15 min. All samples were frozen in liquid N_2, transported back to AIMS, and stored at –80°C until analysed. Total fork length (mm), total and gutted fish weight (g), and gut and liver weight (g) were recorded for each fish.

Of the 90 *L. carponotatus* deployed in fish cages, only a total of 76 fish were recovered. Fourteen individual fish ($n = 14$) were lost from the cages during the study; 7 fish from Site C, 1 fish from Site B, and 6 fish from Site A. Due to this loss, fish sample numbers for Day 10 for all three sites were unequal (<20 fish per site).

2.4 Age Determinations

Otolith samples were processed at James Cook University (Townsville, Queensland). Left sagittal otoliths were weighed to four decimal places using a Sartorius Model BP61 balance. Age estimates were obtained from counts of opaque increments of transverse sections of otoliths as viewed at ×40 magnification using a dissecting microscope. Three independent sets of age readings were done to quantify the precision of the age estimate data. The calculated precision (index of average percent error = 0.87%) demonstrated exceptionally high precision of the data set.

2.5 Biomarker Assessment

2.5.1 Hepatosomatic Index (HSI)

Hepatosomatic index (HSI) was calculated according to Slooff et al. (1983) as follows:

$$HSI = (L_W/S_W) \times 100 \quad (14.1)$$

where L_W represents the total liver weight in grams and S_W represents the somatic weight (total weight of the fish in grams, less gonad, and intestines).

2.5.2 CYP1A-, CYP2M1-, and CYP2K1-Like Proteins

Microsomes were prepared from liver samples and all preparative steps were conducted on ice using a modified Lech method as described in Haasch (2002). The final step involved the quantitative transfer of aliquots of microsomes into labelled cryogenic tubes for determination of total cytochrome P450 content and for immunodetection of CYP1A-, CYP2M1-, and CYP2K1-like proteins. All cryogenic vials were frozen in liquid N_2 and then placed in a –80°C freezer until analysed. Microsomes were processed for CYP1A-, CYP2M1-, and CYP2K1-like proteins by immunodetection using Western blot at The University of Mississippi (University, Mississippi, USA). A detailed description of the methodology is provided in Zhu et al. (2006, 2008). The results were reported as optical density units which refer to the width of the band that is proportional to the amount of the specific protein determined using chemiluminescence.

2.5.3 Fluorescent Aromatic Compounds (FACs) in Bile

Bile samples were analysed as described previously in Codi King et al. (2005). All results were converted to naphthalene equivalents (wet weight) using a calibration curve for naphthalene, and then normalized to protein concentration expressed as $\mu g\ g^{-1}$.

2.5.4 Cholinesterase (ChE) Inhibition Assay

Cholinesterase activity was determined using a spectrophotometric method as described in Humphrey et al. (2007). Enzyme activity was reported as μmole acetylthiocholine iodide (ACTC) min^{-1} mg $protein^{-1}$.

The term cholinesterase (ChE) activity will be used since we did not distinguish between acetylcholinesterase (AChE) and butyrylcholinesterase (BuChE), which are the two major forms of cholinesterase present in marine fish muscle (Sturm et al. 2000).

2.5.5 Protein Assay

All biomarker concentrations were normalized to protein concentration. Protein was determined using a BioRad Protein Assay Kit with bovine serum albumin as a standard, following the method of Lowry et al. (1951).

2.5.6 Histopathological Assessment of Gills and Liver

Gill and liver samples were processed according to the methods described in Schlacher et al. (2007). Histopathological abnormalities detected in the gill and liver were recorded as presence or absence per fish and expressed as percentage of fish affected (prevalence) per exposure treatment. Histopathological alterations in hematoxylin and eosin (H and E) stained liver sections were recorded as present or absent. Only one section from each block was examined to ensure independence of data.

2.6 Chemical Analyses

2.6.1 Effluent and Water-Column Samples

A total of 5 Harriet A PW effluent samples (100% effluent collected directly from the process stream) and 18 water samples were collected 1 m below the surface in duplicate on Days 0, 3, and 10 of exposure (3 sites $\times$ 2 cages $\times$ 3 days $=$ 18 samples). Samples were analysed for the following components: (1) semi-volatile organic compounds (SVOCs) including phenols, polycyclic aromatic hydrocarbons (PAHs), and chlorinated hydrocarbons; (2) volatile organic compounds (VOCs) including monocyclic aromatic hydrocarbons and sulphonated compounds; (3) total petroleum hydrocarbons (TPH); (4) speciated extractable petroleum hydrocarbon; (5) total organic carbon (TOC); and (6) total suspended solids (TSS). After collection, all samples were immediately shipped on ice to Australian Laboratory Services, which is a National Association of Testing Authorities certified laboratory, located in Perth, WA. Samples were analysed according to standard USEPA methods. Only the results for the components measured above the detection limit are reported.

2.6.2 Semi-permeable Membrane Devices (SPMDs)

Semi-permeable membrane devices (SPMDs) consist of a thin film of highly refined lipid sealed inside a thin-walled polyethylene tube (surface area of about 460 cm^2, 90–95 μm thickness, 2.5 cm width containing 1 mL triolein) and are known to sequester hydrophobic compounds such as PAHs. SPMDs were prepared by the National Research Centre for Environmental Toxicology (ENTOX) as outlined in Shaw and Müller (2005) and transported frozen in airtight containers to the field. Field blanks were also transported along with sample SPMDs so contamination associated with handling could be determined. Methods employed in the clean-up and analyses of SPMDs were based upon US Geological Survey protocols (Huckins et al. 2000) with modifications for use in Australian waters (Shaw and Müller 2005). Detection limits were defined as 3× standard deviation of the field blank values. The concentrations of PAHs determined in the SPMDs were converted into estimates of predicted water concentrations using sampling rates (L d^{-1}) determined in the laboratory by Huckins et al. (2000) and corrected for field conditions using deuterated PAHs that acted as performance reference compounds (Shaw and Müller 2005). These compounds were loaded into the SPMD membrane in the laboratory, and their loss in the field was used to calibrate the uptake rates for the PAHs evaluated (Booij et al. 1998). The 16 parent PAHs measured included naphthalene, acenaphthylene, acenaphthene, fluorene, phenanthrene, anthracene, fluoranthene, pyrene, benzo(a)anthracene, chrysene, benzo(b&k)fluoranthene, benzo(a)pyrene, indeno(1,2,3-cd)pyrene, dibenzo(a,h)anthracene, and benzo(g,h,i)perylene.

2.6.3 Conductivity/Temperature/Depth (CTD) Profiles

An SBE 25 Sealogger conductivity, temperature, and depth (CTD) (Sea-Bird Electronics, Bellevue, Washington, USA) instrument was calibrated prior to deployment at each fish cage location on Days 0, 3, and 10 of exposure to record temperature and salinity profiles. In addition, CTD surface profiles were conducted between each cage at ~1 m below the surface to monitor water temperature and salinity.

2.7 Quality Assurance/Quality Control

The biomarkers applied in this study were based on established, published methods. In Australia, there are no standard reference materials available for any Australian tropical marine species. Notwithstanding, the following quality assurance and control procedures were incorporated into each bioassay. All total cytochrome P450 and ChE samples were calculated from duplicate measurements, and one sample was run on consecutive days to check instrument performance and sample integrity. For all CYP proteins, molecular weight standards were processed with each batch to ensure that the detected protein band (by the specific antibody) is within the known molecular weight range of CYP450s. For FACs, calibration curves for naphthalene (5 concentrations $R^2 = 0.99$) were processed with each batch of 10 samples, instrument blanks and a composite bile sample for quality control was processed with every batch (%CV < 10) to check instrument performance.

2.8 Statistics

2.8.1 Univariate Analyses

All data are presented as mean ± standard error unless otherwise stated. Levene's test was used to test for homogeneity of variances. All biochemical data and physiological data were log transformed ($\log_{10}$) prior to analysis, and percentage data were arcsine square root transformed prior to analysis. Normality of data was tested using Shapiro–Wilk W normality test with $\alpha = 0.05$.

Formal analyses were tested for both the effects of cages and for the effects of covariates (length, body weight, and age). For each measure, the significance of caging effects was tested using a 3 factor nested analysis of variance (ANOVA) (Table 14.2). The model tested the effects of site (3 sites), with cage as a nested factor in site (2 cages per site), and for effects of time of sampling. All factors were fixed factors. Caging had no effect on the data, so it was removed from all subsequent analyses. All univariate models were re-run testing for only the effects of site and time of sampling.

Analysis of covariance (ANCOVA) was performed to test for the effects of fish parameters (fish length, weight, and age) on each biomarker. There was only one significant interaction between sampling day, sites, and total weight (DAY × SITE × TOTALWEIGHT) and this was for FACs (Table 14.3; ANCOVA, $p < 0.046$). Therefore, an ANCOVA model was run for FACs with total weight as a co-variant. No other interactions were noted, so covariates were removed from the model and the model was re-run. ANOVA were run using GraphPad Prism 4 software. ANCOVA was run using SYSTAT™ Version 11. If significant differences were detected, a Tukey multiple comparison post hoc test was used ($p < 0.05$).

Table 14.2 Full model ANOVA (fixed factors)

Site
Day
Cage (site)
Site × day
Site × cage (day)

Table 14.3 An analysis of covariance (ANCOVA) using fish weight as the co-variant

Dep Var: FACS_AD, *N*: 90, Multiple *R*: 0.655, Squared multiple *R*: 0.430

Analysis of variance

Source	Sum-of-squares	df	Mean-square	*F-ratio*	*p*
SITE$	1.896	2	0.948	2.932	0.059
DAY$	8.969	2	4.485	13.866	0
TOTALWEIGHT	1.052	1	1.052	3.254	0.075
DAY$ × SITE$	3.411	4	0.853	2.637	0.04
DAY$ × SITE$ × TOTALWEIGHT	3.29	4	0.822	2.543	0.046
Error	24.579	76	0.323		

All but one interaction (DAY$ × SITE$ × TOTALWEIGHT) for FACs indicate the assumption of ANCOVA was met (i.e. homogeneity of slopes)

Histological alterations quantified in gills and liver of Stripey seaperch were analysed by analysis of variance (ANOVA) using the statistical package JMP® (SAS Institute Inc.). Shapiro–Wilk W and Cochran's test were used to test for normality and homogeneity of variances, respectively. Where data were heterogeneous after transformation, the data were tested untransformed but interpreted with caution (Underwood 1981). If significant differences were detected, the Tukey–Kramer comparison of means test was used.

2.8.2 Multivariate Analyses

Principal Components Analysis

Principal components analysis (PCA) was used to examine the contribution of each biomarker and the trends in biomarker response that could not be described by univariate analysis alone. Several PCAs were performed to include various combinations of biomarkers, with the aim of describing the match or grouping of biomarker responses that best described variation in the original data set.

Pearson's Chi-Square Statistic

The histopathological data were reported as presence (1) or absence (0) of specific alterations in holding tank fish ($T = 0$) compared to fish collected at Day 10 at Sites A, B, and C. Due to unequal sample size, the data were converted to percentage frequency. A comparison was made between the Pearson's chi-square values for the frequency at which histopathology markers were observed in all fish at Day 10, to the values observed in all fish from the holding tank (T = 0), which represented the expected frequency (or apparent baseline frequency) of histopathological markers. Data included measurements from holding tank fish ($T = 0$) and Day 10, and were assessed assuming that the expected frequency of marker response in Day 10 was equal to the observed frequency at $T = 0$ (i.e. observed frequency = Day 10 measurements, expected frequency = $T\,0$ measurements).

3 Results

3.1 Fish Biometric Data

The 90 fish used in this exposure study, were comparatively uniform in fork length (265 $\pm$ 2.7 mm, range from 215 to 332 mm, $n = 90$), total weight (300 $\pm$ 9.5 g, range from 65 to 590, $n = 90$), and age (5 $\pm$ 3 yr; range from 2 to 15 year, $n = 89$). There was no significant difference in total length, total weight, and age across all sampling sites (ANOVA, $p > 0.05$) and among the sampling times (3 and 10 d, ANOVA, $p > 0.05$) with the majority of fish collected being males (73%), and this ratio was consistent at all sites (data not shown).

3.2 Biomarker Responses

3.2.1 Hepatosomatic Index (HSI)

For HSI, there was a trend in the data showing an increase by Day 10 at Sites A and B in comparison to the reference site, Site C. However, this difference was not statistically significant from one another (ANOVA, $p > 0.05$), nor in comparison to $T = 0$ fish (ANOVA, $p > 0.05$) (Fig. 14.3a).

3.2.2 Total Cytochrome P450

Total cytochrome P450 showed an increase in concentrations at Sites A and B with increasing exposure time (Day 10), but these changes were not significant (ANOVA, $p > 0.05$), due to higher intra-site variability (Fig. 14.3b).

3.2.3 Immunodetection of CYP1A1-, CYP1M1-, and CYP2K1-Like Proteins

The results of the immunodetection of CYP1A-like proteins indicated that proteins of the appropriate molecular weight range (56.1 kDa) were detectable in *L. carponotatus* at all three sites and over time of the exposure, suggesting the potential of CYP1A immunodetection use as an indicator of PAH exposure (Fig. 14.3c). By Day 10, fish collected at Site A had significantly higher induction of CYP1A-like proteins as compared to the fish exposed at Sites B and C (ANOVA, $p < 0.05$) and as compared to fish sacrificed at $T = 0$ (ANOVA, $p < 0.05$).

Two distinct molecular weight bands were observed in *L. carponotatus* for CYP2M1-like proteins (52.8 and 49.0 kDa) and CYP2K1-like proteins (52.5 and 48.2 kDa) and both bands were quantified. A similar pattern of induction occurred for both CYP2M1-like protein bands with Site A fish having significantly induced responses at Days 3 and 10 (ANOVA, $p < 0.05$) compared to Site B and C fish, and as compared to the holding tank fish, $T = 0$ (ANOVA, $p < 0.05$). (Fig. 14.3d, CYP2M1-like protein 1 only).

For CYP2K1-like proteins, the optical densities for the two proteins were dissimilar. CYP2K1-like Protein 1 was significantly elevated in *L. carponotatus* from Site A Day 3 as compared to Sites B and C on Day 3 (ANOVA, $p < 0.05$, data not presented). No other site-specific comparison differences were observed (ANOVA, $p > 0.05$). Yet for CYP2K1-like Protein 2, the induction pattern was similar

Fig. 14.3 The log transformed data (mean ± SE) for **a** hepatosomatic index (expressed as %), **b** total cytochrome P450 (expressed as nmol mg protein^{-1}), **c** CYP1A-like proteins, **d** CYP2M1-like Protein 1, **e** CYP2K1-like Protein 2, **f** concentration of FACs (expressed as naphthalene equivalents in μg mg protein^{-1}), and **g** concentration of ChE activity (expressed as μmol ACTC min^{-1} equivalents mg protein^{-1}) for *L. carponotatus* collected at Day 0 (holding tank), Days 3 and 10 of exposure at Sites C, B and A. Different letters denote significant differences ($p < 0.05$). Numbers inside each bar represent the individuals (n) sampled at each time point

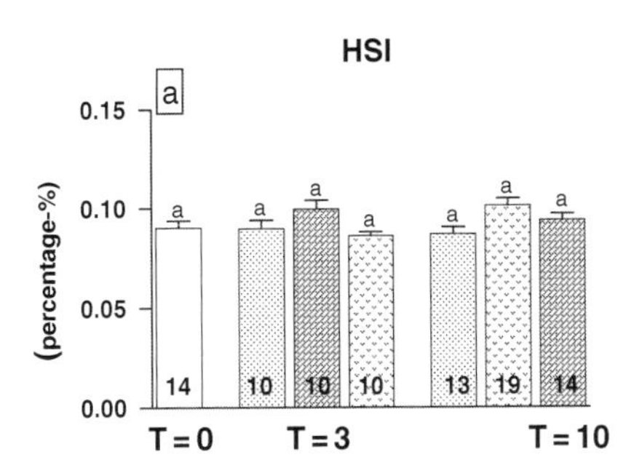

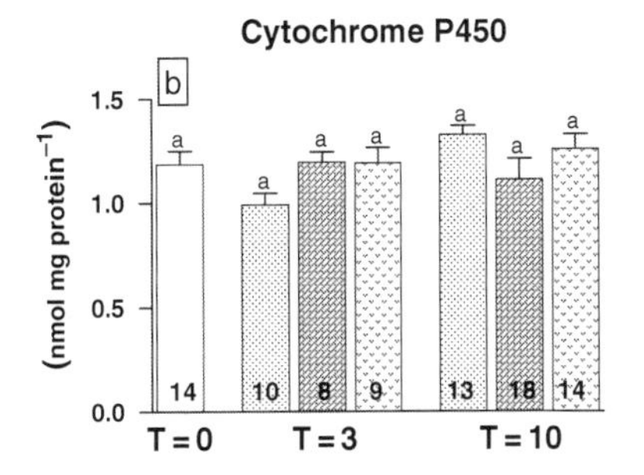

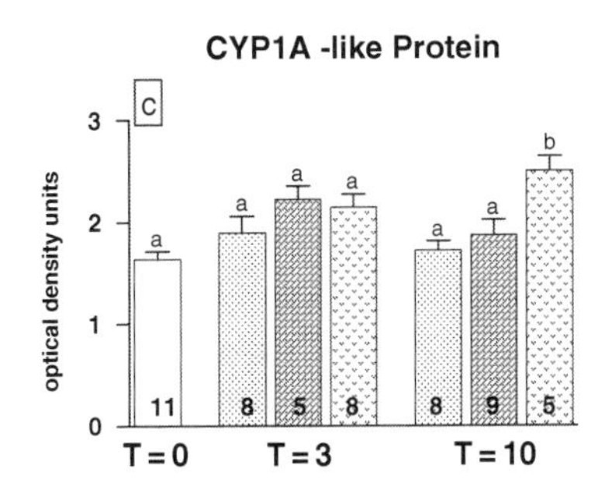

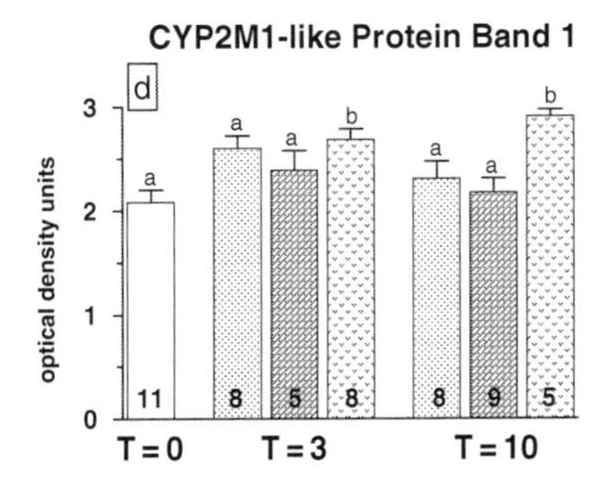

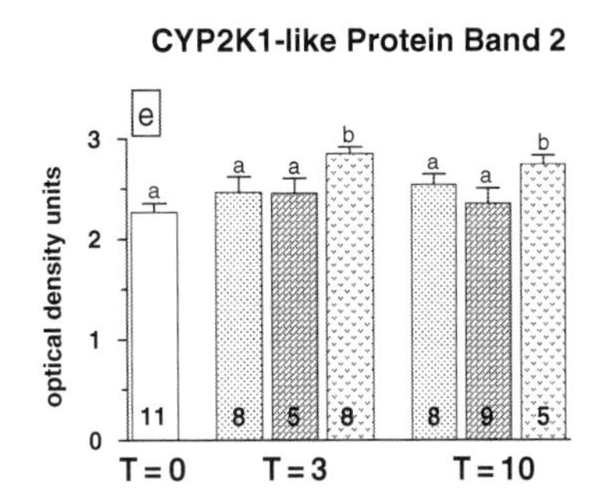

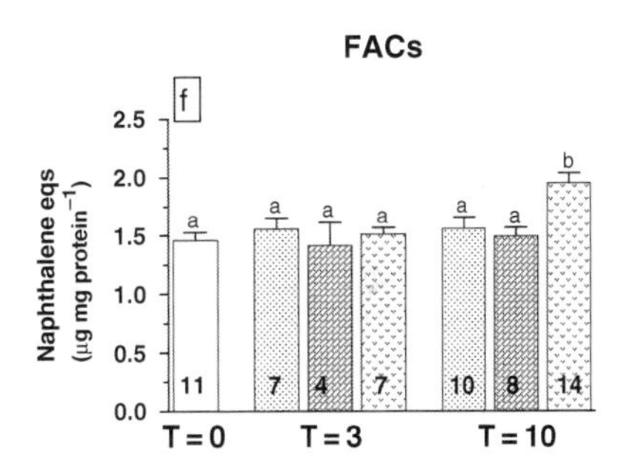

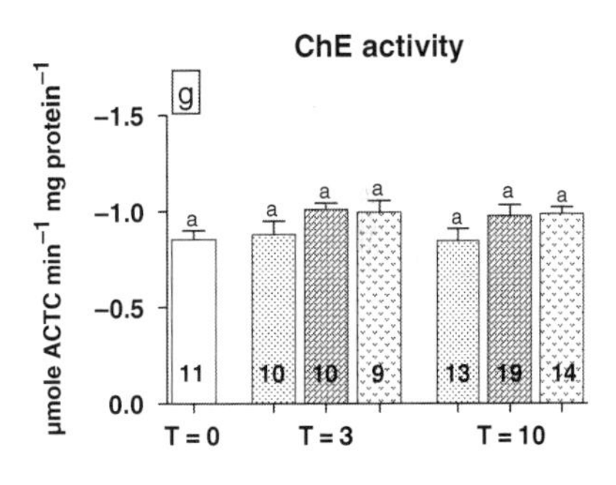

Fig. 14.3 (continued)

to that observed for CYP1A and both proteins of CYP2M1-like proteins in that by Day 10 Site A fish demonstrated the highest mean level of induction (ANOVA, $p < 0.05$) (Fig. 14.3e). Site-specific comparisons demonstrated that Site A fish after 3 and 10 days of exposure had significantly induced levels of CYP2K1-like Protein 2 as compared to $T = 0$ fish (ANOVA, $p < 0.05$). The levels of Protein 2 CYP2K1-like proteins were also orders of magnitude higher in density than the Protein 1 proteins which may relate to the sensitivity of one protein over the other to different components in the PW or to differential primary antibody affinities of the two proteins.

3.2.4 Fluorescent Aromatic Compounds (FACs) in Bile

The levels of FACs measured in fish bile showed that Site A, Day 10 fish had significantly increased levels of FACs (measured as naphthalene equivalents) as compared to Site B fish (ANCOVA, $p < 0.05$) and Site C fish (ANCOVA, $p < 0.05$) (Fig. 14.3f). Specific-site comparisons showed that fish at Site A Day 10 had significantly elevated levels of FACs compared to fish at all three sites at Day 3 (ANCOVA, $p < 0.05$) and $T = 0$ fish (ANCOVA, $p < 0.05$).

3.2.5 Cholinesterase (ChE) Activity

Cholinesterase (ChE) showed a trend in decreasing activity at both sites A and B, with increasing time of exposure (Fig. 14.3g). These results were consistent for Days 3 and 10 exposure, although the decrease in ChE activity was not significantly different from the reference site fish, Site C (ANOVA, $p > 0.05$).

3.2.6 Histopathology

The results of the histopathological alterations measured in the liver of Stripey seaperch indicated a strong response to exposure to the Harriet A PW (Table 14.4). Histopathological markers in Day 10 fish were, with one exception, significantly higher than the expected frequency observed in fish from the holding tank ($T = 0$). The large calculated chi-square statistics were mainly contributed by the increase of histological markers observed in fish from Site A.

A second series of comparisons were made to assess site-specific differences by Day 10 of exposure looking at the prevalence of liver pathology again using holding tank fish ($T = 0$) as our expected frequency (Table 14.5). By Day 10, Site A fish consistently showed significantly higher prevalence of all pathologies. Multifocal necrosis and fatty liver had significantly higher prevalence at Site A ($p < 0.001$) compared to fish from Site B and fish from Site C. A final comparison was conducted using Pearson's chi-square distributions. The results demonstrated that the frequency at which histology markers were observed after pooling the data from Sites A and B was significantly higher than the frequency at which they were observed at Site C (reference site), after 10 days of exposure (Table 14.6).

Table 14.4 Prevalence of histopathological responses observed in liver of *L. carponotatus* from holding tank ($T = 0$), Sites A, B, and C after 10 days of exposure

Pathology	Holding tank ($T = 0$), $n = 14$	Site A, $n = 14$ (near-field site)	Site B, $n = 19$ (far-field site)	Site C, $n = 13$ (reference site)	X^2, p
Multifocal necrosis	21.4 ± 11.7	14 ± 11.7	36.8 ± 10.0	23.1 ± 12.2	29.0, $p < 0.001$
Fatty liver (lacy liver)	21.4 ± 11.9	71.4 ± 11.9	26.3 ± 10.2	23.1 ± 12.4	65.5, $p < 0.001$
Granuloma/cyst (liver)	7.1 ± 10.7	35.7 ± 10.7	21.1 ± 9.2	15.4 ± 11.1	11.2, $p < 0.01$

Data presented as percentages, means ± SE. Pearson's chi-squared values for histopathology markers from fish sampled at Day 10 were calculated based on the expected frequency observed in holding tank fish ($T = 0$), with df = 2

Table 14.5 Individual site comparisons using calculated Pearson's chi-square values to assess site-specific differences in pathology observed in the liver of *L. carponotatus*

Pathology	Site A vs Site B	Site A vs Site C	Site B vs Site C
Multifocal necrosis	19.89***	28.17***	9.97**
Fatty liver (lacy liver)	64***	65.34***	1.23 (ns)
Granuloma/cyst (liver)	9.53**	11.23***	1.71 (ns)

Chi-squared values for histology markers from fish sampled at Day 10 Sites A, B, and C were calculated using holding tank fish ($T = 0$) as the expected frequency with 1 degree of freedom
***$p < 0.001$, **$p < 0.01$, (ns) not significant

Table 14.6 Summary of calculated Pearson's chi-square values for Sites A and B after 10 days of exposure based upon on the expected frequency observed at Site C in *L. carponotatus*

Pathology	χ^2	p
Multifocal necrosis	11.56	$p < 0.001$
Fatty liver (lacy liver)	101.76	$p < 0.001$
Granuloma/cyst (liver)	28.95	$p < 0.001$

Chi-squared values of histology markers from fish sampled at Day 10 Sites A and B were calculated based on the expected frequency observed at Site C with 1 degree of freedom

The results of the gill histopathology measured as percent of gill filaments affected demonstrated that the highest gill pathologies measured were uniformly greatest in holding tank fish ($T = 0$) and although, lamellar fusion, hyperplasia, and cell proliferation after 10 days of exposure were higher at Sites A and B as compared to Site C, these differences were not significant (data not shown). These results strongly advocate that the method of trapping fish had a severe impact on the gills prior to deployment in the field. These results demonstrate that confounding factors can interfere with biomarker responses (e.g. gill histopathological markers) especially in a field exposure study and need to be considered when interpreting results.

3.3 Biomarker Responses: Multivariate Analyses

Principal component analysis (PCA) was conducted on nine biomarker responses including HSI, CYP1A, CYP2M1 Protein 1, CYP2M1 Protein 2, CYP2K1 Protein 1, CYP2K1 Protein 2, total CYT P450, FAC (adjusted values; FAC_AD), and ChE (adjusted values; ChE_AD). Table 14.7 presents the contribution of each principal component, its eigenvalue, and percentage of variance explained by each of the biomarker responses and their respective eigenvectors. There were three principal components (PC) that explained 68.72% of the variation. PC1 was driven largely by the response of CYP proteins (CYP1A, CYP2M1 Protein 1, CYP2M1 Protein 2, CYP2K1 Protein 1, and CYP2K1 Protein 2), which contributed significantly to the loading on PC1, explained 42% of the variance, and provided equal contribution with similar eigenvectors (range from 0.383 to 0.463). Both Day ($p = 0.006$) and Site ($p < 0.001$) had a significant effect on the factor scores contributing to the PC1 (Fig. 14.4). Although PC2 and PC3 both accounted for over 26% of the total variation, there was only an apparent effect from both Site and Day (interaction, $p < 0.001$) for PC3, which was contributed largely by loading from the HSI. This indicated that shifts in ChE (PC2) and HSI (PC3) contributed in different components, and that these contributions varied across time and space for PC3.

Table 14.7 Contribution to each principal component from respective eigenvalues, the percentage of variance explained, and contribution loadings from each biomarker HSI, CYP1A, CYP2M1-P1, CYP2M1-P2, CYP2K1-P1, CYP2K1-P2, total CYTP450, FAC_AD, and AChE_AD to each PC and their respective eigenvectors + (SE)

Principal component	PC1	PC2	PC3
Eigenvalues	3.764	1.250	1.221
Percent of total variance explained	41.827	13.886	13.566
Eigenvectors (+SE) for each component			
HSI	–0.178 + 0.096	0.398 + 3.325	0.548 + 2.422
CYP1A	0.383 + 0.061	–0.171 + 1.727	0.284 + 1.044
CYP2M1-P1	0.463 + 0.036	0.069 + 0.601	0.098 + 0.430
CYP2M1-P2	0.414 + 0.053	0.218 + 0.209	0.024 + 1.333
CYP2K1-P1	0.438 + 0.044	–0.056 + 0.652	–0.106 + 0.361
CYP2K1-P2	0.443 + 0.044	0.183 + 0.459	–0.074 + 1.113
CYTP450	–0.043 + 0.104	–0.762 + 0.398	0.061 + 4.620
FAC_AD	0.135 + 0.097	–0.116 + 3.352	–0.551 + 0.772
AChE_AD	0.167 + 0.096	0.359 + 3.242	–0.534 + 2.187

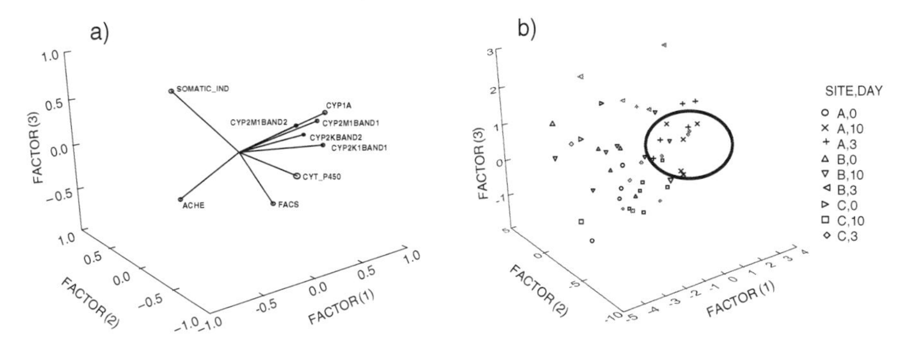

Fig. 14.4 **a** Factors loading plot and **b** variable loading plot for all nine biomarker responses measured in *L. carponotatus* in holding tank ($T = 0$) fish, Sites A, B, and C for Days 3 and 10 of exposure. *Circled* area highlights clustering of Day 10, Site A fish

3.4 Water Chemistry

3.4.1 Water-Column Samples and SPMDs

The results for all water-column samples collected 1-m subsurface at each caging site on Days 0, 3, or 10 of exposure demonstrated no detectable concentrations of VOCs, SVOCs, TPH, or speciated extractable petroleum hydrocarbons.

Only the time-integrated samplers (i.e. SPMDs) deployed in each fish cage for the full exposure period (10 days) demonstrated site-specific differences in hydrophobic concentration of the 16 parent PAHs analysed by ENTOX. Total PAH concentration (field blank corrected) for Site A was 4.66 μg L^{-1}, which was slightly higher than the individual membrane PAH concentrations for Site A of 2.87, 3.27, and 3.57 μg L^{-1}. The main PAHs detected were acenaphthene, fluorene, phenanthrene, anthracene, fluoranthene, pyrene, and chrysene (ranging from 0.1 to 2.1 μg L^{-1}). Total PAH concentrations determined at Site B FC6 and FC7 were 5.26 μg L^{-1} and 2.87 μg L^{-1}, respectively, and were mainly comprised of acenaphthene, fluorene, phenanthrene, and chrysene (ranging from 0.1 to 2.6 μg L^{-1}). Total PAH results for Site C (field blank corrected) for all three individual membranes were below detection limits.

For Sites A and C, individual membranes were analysed to assess field variation in the hydrophobic concentration of PAHs measured. Individual membranes analysed for Site A demonstrated some intra-site variability with a coefficient of variation (%CV) of 10.8%. For Site C, all three membranes were below detection limits for PAHs once corrected for field blanks.

3.4.2 PW Effluent Samples

The volatile or BTEX components represented the highest proportion (~90%) of the monocyclic aromatic hydrocarbons quantified in the Harriet A PW, with mean concentrations of 12.98, 9.57, 11.28, and 8.71 mg L^{-1} (Table 14.8). Toluene was the most abundant BTEX component in the PW samples which is consistent with previous studies that have used toluene as a chemical tracer for the surface PW plume (Holdway and Heggie 1998).

The SVOC concentrations for Harriet A PW comprised of two major components, the phenols and PAHs (Table 14.8). Phenols which included phenol, 2-methylphenol, 3,4-methyphenol, and 2,4-dimethyphenol represented on average 68.3% ($n = 5$) of the total SVOCs components determined in Harriet A PW. These were the four major compounds found in Harriet A PW at mean concentration of 632, 672, 714, and 581 μg L^{-1}, with total average concentrations of all four components of 2.6 mg L^{-1} ($n = 5$).

Table 14.8 Mean (μg L^{-1}±SD) concentration of monocyclic aromatic hydrocarbons, semi-volatile organic compounds (SVOCs) including phenols, total petroleum hydrocarbons (TPH), and speciated extractable petroleum hydrocarbons detected in Harriet A PW samples collected during 10-day exposure study

Components analysed	Limit of reporting (LOR)	Samples (*n*)	PW concentration mean (±SD)	Range (min–max)
Monocyclic aromatic hydrocarbons				
Benzene	1	4	3060 ± 617	2560–3920
Toluene	1	4	4480 ± 634	3770–5160
Ethylbenzene	1	4	229 ± 81	151–336
meta- and *para*-Xylene	1	4	2085 ± 477	1510–2600
Styrene	1	1	118	
ortho-Xylene	1	4	752 ± 113	613–845
Isopropylbenzene	1	4	23 ± 7	17–33
n-Propylbenzene	1	4	33 ± 13	22–52
1,3,5-Trimethylbenzene	1	4	152 ± 44	112–215
1,2,4-Trimethylbenzene	1	4	417 ± 64	352–504
p-Isopropyltoluene	1	4	7 ± 4	4–13
Sulphonated compounds				
Carbon disulfide	1	4	39 ± 22	<LOR–3
Phenols				
Phenol	2	5	632 ± 243	436–1030
2-Methylphenol	2	5	672 ± 242	482–1080
3- and 4-Methylphenol	2	5	714 ± 256	506–1130
2.4-Dimethylphenol	2	5	581 ± 198	425–904
Polycyclic aromatic hydrocarbons				
Naphthalene	2	5	579 ± 197	400–906
2-Methylnaphthalene	2	5	553 ± 199	384–888
Acenaphthene	2	5	4 ± 1	3–5
Fluorene	2	5	23 ± 3	18–27
Phenanthrene	2	5	23 ± 4	16–28
Chlorinated hydrocarbons				
Hexachlorocyclopentadiene	10	5	10 ± 0	10

Table 14.8 (continued)

Components analysed	Limit of reporting (LOR)	Samples (n)	PW concentration mean (±SD)	Range (min–max)
Anilines and benzidines				
Dibenzofuran	2	5	6 ± 1	5–7
Carbazole	2	5	7 ± 1	6–8
Total petroleum hydrocarbons				
C6–C9 fraction	20	4	13,275 ± 2412	9700–14,900
C10–C14 fraction	50	4	11,185 ± 1937	9480–13,700
C15–C28 fraction	100	4	8548 ± 1692	6680–10,300
C29–C36 fraction	50	4	1069 ± 242	822–1310
Total TPH fraction		4	34,077 ± 4532	29,905–39,740
Total speciated petroleum hydrocarbons				
Aliphatic hydrocarbons- $>C_{16}–C_{21}$	50	4	2393 ± 788	1682–3250
Aliphatic hydrocarbons- $>C_{21}–C_{35}$	100	4	1206 ± 75	100–265
Aromatic hydrocarbons- $>C_{16}–C_{21}$	50	4	1822 ± 543	1228–2503
Aromatic hydrocarbons- $>C_{21}–C_{35}$	100	4	228 ± 37	196–269
Total speciated fraction (aliphatic and aromatic)		4	4649 ± 496	4114–5311

The major PAHs detected were naphthalene, 2-methylnaphthalene, acenaphthene, fluorene, and phenanthrene at mean individual concentrations of 580, 550, 4, 23, and 23 $\mu g\ L^{-1}$, respectively (Table 14.8). Naphthalene and 2-methylnaphthalene comprised the greatest proportion of the total in all five of the Harriet A PW samples.

Total petroleum hydrocarbons (TPH) and speciated extractable petroleum hydrocarbons data demonstrated that the lower molecular weight aliphatic and aromatic hydrocarbons are the main constituents of the Harriet A PW (Table 14.8). This point is illustrated in the total ion chromatogram for the Harriet A PW analysed by gas chromatography/mass spectrometry (GC/MS) (Fig. 14.5). The mean concentration of TPH and speciated aliphatic and aromatic fractions for Harriet A PW was 3.4, 2.6, and 2.1 mg L^{-1}, respectively.

3.4.3 Total Suspended Solids (TSS) and Total Organic Carbon (TOC)

Mean total suspended solids (TSS) concentration was higher in Harriet A PW samples (10 mg L^{-1}, $n = 4$) as compared to the water-column samples from Site A (3.3 mg L^{-1}, $n = 6$), Site B (5.7 mg L^{-1}, $n = 6$), and Site C (3.1 mg L^{-1}, $n = 6$). Mean total organic carbon (TOC) was 70 mg L^{-1} ($n = 4$) in the Harriet A PW

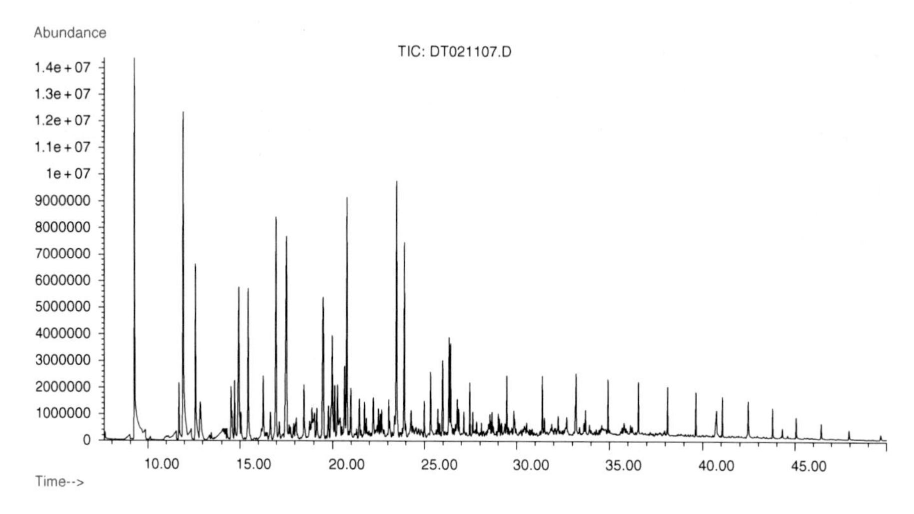

Fig. 14.5 The total ion chromatogram for a Harriet A PW effluent sample analysed by GC/MS

samples. Whereas, TOC concentrations were an order of magnitude lower at Site A (2.5 mg L^{-1}, $n = 6$), Site B (4.2 mg L^{-1}, $n = 6$), and Site C (1.3 mg L^{-1}, $n = 6$).

3.4.4 CTD Profiles

The CTD profiles showed that salinity (35.4–35.6 ppt) and temperature (25.0–27.5°C) were similar at the three caging sites during the 10-day exposure study. Temperatures decreased slightly by Day 10 from the high winds (>45 knots) experienced on Day 7, which produced a homogenous water column due to mixing of this shallow water marine system (<25 m).

4 Discussion

This study evaluated exposure to, and potential effects of, Harriet A PW using a multi-biomarker approach on a tropical reef species, *L. carponotatus,* in a 10-day field caging study. A combination of chemical and biological assessments was conducted with the aim of providing a strong correlation between biomarker responses and the chemical constituents determined in situ at all three of the caging sites—Site A near-field (~200 m from the platform), Site B far-field (~1000 m from the platform), and Site C reference site (no influence of PW). Time-integrated samplers (SPMDs) were the only data useful in determining any site-specific differences in chemical constituents. PAHs are one of several components of the Harriet A PW that are of greatest environmental concern in PW due to both persistence in the marine environment and their associated toxicity (Neff 1987). Biomarkers chosen were aimed at measuring or detecting changes at the molecular, cellular, physiological, and whole organism level and evaluated to determine which ones were most predictive for the Harriet A PW and its constituents. Several of the biomarker responses provided statistically significant evidence that *L. carponotatus* by Day 10 at Site A was responding to the Harriet A PW and they included CYP1A, CY2M1 Protein 1, CY2M1 Protein 2, CY2K1 Protein 1, CY2K1 Protein 2, FACs, and liver histopathology.

4.1 Biomarker Responses

The induction of cytochromes P450 by environmental contaminants has been used with great success as a biomarker of exposure (Goksøyr et al. 1992). Fish represent a very useful organism for monitoring environmental contaminants because of their position in the food web, their lifestyle, their relative abundance, and their adaptable physiology. The induction of CYP1A in fish after 10 days of exposure at the near-field site (Site A, Fig. 14.3c) indicates that even though PW discharge limits (<30 mg L^{-1}) are being met there is enough low-level exposure to CYP1A inducing chemicals within the diluted PW to induce a significant response. It is the continuous low-level exposure to constituents in the PW such as PAHs that can have a significant impact on fish and can produce long-term, adverse health and reproductive effects (Stegeman et al. 2001; Incardona et al. 2004). CYP1A is induced by PAHs, polychlorinated biphenyls, polychlorinated dibenzo-p-dioxins and dibenzofurans, and is probably the most sensitive marker for exposure and effect of these types of contaminants (Myers et al. 1993; Haasch 1996; Haasch et al. 1998a, 1998b). Many of these compounds bioaccumulate and are persistent, and may produce physiological, reproductive, and histopathological adverse effects (Anderson et al. 1996; Haasch et al. 1993). CYP1A biomonitoring has become an important tool in the analysis of exposure of fish and other animals to xenobiotics such as petroleum (Haasch et al. 1993; Payne et al. 1987) and our results validate its usefulness for assessment of the Harriet A PW in Australian waters.

The similar induction pattern of CYP1A1 and CYP2K1-like Protein 2 and CYP2M1-like Proteins 1 and 2 may indicate that all of these isozymes are induced by the same chemical class and that the inducing chemical or chemicals is one or more of the petroleum hydrocarbons or the multi-ringed PAHs (Fig. 14.3d, e). The reason for the induction of the CYP2K1-like Protein 1 at Site A Day 3 and inhibition at Site A Day 10 is unclear at this stage but warrants further investigation. CYP2M1-like and CYP2K1-like proteins have been thoroughly characterized in rainbow trout (*Oncorhynchus mykiss*). CYP2M1-like proteins have been shown to catalyse the ω-6 hydroxylation of lauric acid, whereas CYP2K1-like proteins have been shown to catalyse ω-1 and ω-2 hydroxylation of lauric acid and some longer chain fatty acids, and both are involved in the hydroxylation of estradiol, testosterone, and progesterone (Buhler and Wang-Buhler 1998). It follows that induction of CYP2M1-like and CYP2K1-like proteins in fish may impact on steroid and fatty acid homeostasis (Zhu et al. 2008). These results highlight that these two CYP2 proteins may prove very useful for future monitoring of the biological impacts of the Harriet A PW.

The FACs detected in bile of *L. carponotatus* at Site A, Day 10 demonstrate clear exposure over time to naphthalene and its metabolites, one of the major components of the Harriet A PW (Fig. 14.3f). The use of FACs in fish has become a powerful diagnostic tool to demonstrate direct exposure of different species of fish to PAHs, petroleum hydrocarbons, and acid resins from paper and pulp effluents (Krahn et al. 1987, 1992, 1993; Brumley et al. 1996, 1998). The presence of PAH metabolites in bile are well correlated with levels of exposure (Collier and Varanasi 1991), and there have been numerous laboratory and field studies that have demonstrated its ability as a biological monitoring tool (full review provided in van der Oost et al. 2003). Levels of FACs are certainly sensitive biomarkers to assess recent exposure to PAHs (Altenburger et al. 2003). Our results corroborate earlier findings (Codi King et al. 2005) and substantiate this biomarker as a good diagnostic tool for monitoring PAH exposure to the Harriet A PW in Australian waters.

Cytological changes in liver and gills are useful indicators of possible environmental toxicity. The presence of pollutants in the environment can contribute to deformities and structural abnormalities resulting in alteration of the normal structure and functioning of cells (Vethaak and Rhenalt 1992). There are a number of environmental contaminants including petroleum compounds and metals that are known to cause pathological changes in finfish tissues (for a full review, see Myers and Hendricks 1985; Au 2004). The liver of fish, in particular, is susceptible to damage from a variety of toxicants (Gingerich 1982), due to the role of the fish liver in mediating biotransformation and elimination of xenobiotic compounds (Hinton et al. 2001). Additionally, fish gills are the major site of uptake, being constantly exposed to the external environment, and are usually the first organ affected by aqueous exposure to toxicants (Wood 2001; Nowak et al. 1992). Both the liver and gills have long been recognized as reliable indicators of sub-lethal effects of toxicants in finfish (Hylland et al. 2003; Mallatt 1985). Our results showed that the method of trapping had a severe impact on fish gills prior to deployment in the field. In the future if gill histopathology is to be evaluated as a biomarker of effects, different methods of trapping large numbers of fish may be needed to avoid having such a significant impact.

Fatty infiltration in fish liver has been linked to exposure to petroleum compounds (McCain et al. 1978). The liver is a major storage site of lipids in fish; consequently liver metabolism is a potential target for toxic action of chemicals (Hinton et al. 2001). The presence of abundant fat within liver cells of *L. carponotatus* at Site A (Fig. 14.6a and b) may signify an alteration in lipid metabolism (Hinton and Lauren 1990; DiMichelle and Taylor 1978). Although the exact mechanism of fatty liver is yet to be determined (Hinton et al. 2001), fish exposed to crude oil are known to exhibit abnormally enlarged mitochondria (Hawkes 1977). Oxidation of fats in the liver cells may be inhibited by a lack of mitochondrial activity possibly due to defective enzyme activity (Heath 1995), resulting in an accumulation of lipid unable to be synthesized. Specific alteration at the cellular level in liver tissue was also evident with higher accumulation of fat in the liver (fatty liver) of Site A fish

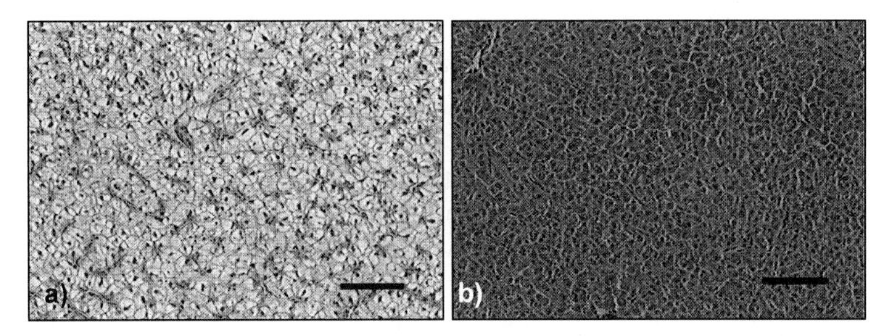

Fig. 14.6 Liver of *L. carponotatus* exhibiting **a** extensive fat accumulation (×200) and **b** liver with normal histology (×200). H and E stain

as compared to fish at Sites B and C, even though the reference fish exhibited elevated histological alterations. Histopathological markers provide information on the responses at lower biological organization, on the individual fish, but this information may be used to extrapolate to population/community effects to assess long-term impacts (Au 2004). These changes provide information on the potential for cumulative effects associated with Harriet A PW and other organisms living in close proximity to the platform itself.

Several other biomarkers applied were not predictive for this study and they included HSI, total cytochrome P450, and ChE. Measurement of HSI and total cytochrome P450 did not provide strong statistical evidence by Day 10 of higher induction for cytochrome P450 or any site-specific differences in the case of ChE activity. PAH contamination can cause an increase in HSI in exposed fish (Fletcher et al. 1982; Baumann et al. 1991); however, our results suggest that a longer exposure period may be necessary in order to demonstrate site-specific difference and the usefulness of this biomarker for PW assessments. Results in both laboratory and field studies using cytochrome P450 have observed >50% of a positive response in fish exposed to contaminants versus non-exposed fish; yet, its value as a predictive and valuable biomarker in PW is limited, since the responses of individual isozymes (notably CYP1A) are more specific and sensitive (Stegeman and Hahn 1994; Altenburger et al. 2003). ChE has been used extensively as a direct measure of toxic action of organophosphate and carbamate pesticides, though studies in Europe have suggested that ChE may be a useful biomarker for exposure to hydrocarbons. Work by Galgani et al. (1992) off a drilling platform in the North Sea showed levels of AChE and BChE decreased at the site closest to the platform compared to sites >5 and >15 km away. There are numerous other chemicals added into a PW production stream including biocides, corrosion inhibitors, oxygen scavengers, scale inhibitors, and chemicals for gas treatment (Neff 2002) that may be contributing to the decrease in ChE activity including metals (Frasco et al. 2005; Brown et al. 2004) and surfactants and detergents (Guilhermino et al. 1998, 2000). More research is necessary to investigate ChE as a potential biomarker for use in PW assessments in Australian water.

4.2 Chemical Assessment

The chemical assessment confirms that the Harriet A PW is complex and heterogeneous in nature. The major organic components quantified were BTEX, phenol, alkylphenols (APs), and PAHs. The volatile organics or BTEX compounds dominate in the Harriet A PW samples as follows: toluene > benzene > total xylenes > ethylbenzene. In contrast, BTEX composition in Indonesian and Gulf of Mexico PW shows benzene as the dominant component (84–2300 and 440–2800 $\mu g\ L^{-1}$, respectively), followed by toluene (89–800 and 340–1700 $\mu g\ L^{-1}$, respectively), then total xylene (13–480 and 160–720 $\mu g\ L^{-1}$, respectively), and finally, ethyl benzene (26–56 and 26–100 $\mu g\ L^{-1}$, respectively) (Dept. of Energy 1997; Neff and Foster 1997; Neff 2002). BTEX are soluble in seawater and both toluene and benzene are excellent chemical tracers for the Harriet A PW plume. They have been detected up to distances of 10 km from the discharge point (Holdway and Heggie 1998). Typically, BTEX are more acutely toxic (Neff 2002); yet, their impact on marine organisms residing around Harriet A is relatively unexplored. Holdway (2002) highlighted the need for further impact studies of PW exposure on early developmental stages of marine invertebrates including the process of settling behaviour and metamorphosis and research is underway to address these issues (Conwell, unpublished data).

Phenol and alklyphenols (2-methylphenol, 3,4-methyphenol, and 2,4-dimethyphenol) were measured in the Harriet A PW at mean concentrations of 632, 672, 714, and 581 $\mu g\ L^{-1}$, ($n = 5$) respectively. Typical concentrations of total phenols in PW can range from 600 to 2300 $\mu g\ L^{-1}$ (Neff 2002). Alklyphenols (APs) have emerged as chemicals of concern due to their estrogenic properties especially the potent endocrine disrupting chemicals (EDCs) octyl- and nonylphenol (Nimrod and Benson, 1996). In particular, 4-*tert*-octyl phenol has been shown to inhibit both CYP2M1 and CYP2K1 proteins in rainbow trout (Katchamart et al. 2002). Alklyphenols in PW from the North Sea demonstrate both estrogenic and anti-androgenic properties using in vitro bioassays with rainbow trout (Tollefsen et al. 2007) and zebrafish, *Danio rerio* (Holth et al. 2008). At the time of this study, EDC research received little attention in Australian marine waters. The North Sea PW studies highlight the importance of these components and their potential to have significant effects on marine organism. As the oil and gas industry grows in Australia, evaluation of APs in PW effluents should be undertaken to assess both estrogenic and anti-androgenic components and their environmental effects.

PAHs tend to be a minor component of the Harriet A PW, and are less soluble but more persistent (Swan et al. 1994). Toxicity in marine organisms in previous studies has been mainly due to benzenes and naphthalene components (Brand et al. 1989). Jones and Heyward (2003) demonstrated minor toxicity of the PW discharged from Harriet A oil platform to surrounding corals. A MicrotoxR® Basic test performed on the Harriet A raw effluent highlighted particulate PAHs as the main component for acute toxicity. The MicrotoxR® test uses the luminescent bacterium, *Vibrio fischeri*, which is sensitive to a wide range of chemicals. The bacteria emit light as a natural product of their metabolic processes; exposure to toxic samples inhibits those processes, and the light output drops in proportion to the toxicity of the sample. A simplified toxicity identification evaluation (TIE) test on the raw Harriet A PW gave an EC_{50} value of 10.4 % (range 5.1–21.2%). Aeration appeared to reduce the response of the bacteria (EC_{50} value 13.6%; range 8.7–21.1%) compared with that of the raw solution, but did not significantly alter toxicity. Filtration of the effluent appeared to have the greatest impact on toxicity of the PW, which was significantly decreased to an EC_{50} value of 36.7% (range 27.4–34.4%) after passage through a C_{18} cartridge (Codi King, unpublished data). The TIE results were consistent with Burns and Codi (1999), which established that most of the total PAHs in Harriet A PW was in the particulate phase (89–93%) and the region in close proximity to the Harriet A is the area most at risk. Other studies have demonstrated effects on the benthic structures in close proximity to oil platforms in the Northern Hemisphere (Olsgard and Gray 1995; Osenberg et al. 1992).

Unfortunately, the lack of detectable concentrations of organic constituents in any of the water-column samples limits our ability to try and discriminate which specific chemicals found in the PW contribute the most to the biomarker responses observed. Although no visible surface slick was noted during site sampling, the modelling data (Global Environmental Modelling Systems, GCOM3D, validated for the North West Shelf) (Libby Howitt, personal communication) clearly demonstrated that sampling at Sites A and B was done when the current was heading in a north westerly direction, just not at times of peak current speed. Conventional grab sampling is one dimensional in space and time. In addition, water concentrations of PW components are usually low (ng to $\mu g\ L^{-1}$) and many laboratories have difficulty attaining such low detection limits with confidence. The use of time-integrated samplers can overcome some of these difficulties as they can be deployed for up to a month and provide an average concentration of hydrophobic PAHs over time, which is a more realistic measure.

Our SPMD data did confirm detection of hydrophobic parent PAHs; however, the field exposure time for this study was relatively short, 10 days. Durell et al. (2006) have demonstrated that depending upon the log of the octanol-water partitioning coefficient (log K_{OW}) of the PAH (determines solubility in water), deployment times can range from 10 to 25 days before PAHs equilibrate from the water into the membrane. Clearly our results underestimate the bioavailable fraction of the 16 PAHs and need to be interpreted with caution. In addition, field blanks confirmed contamination of the membranes during shipping and storage prior to delivery to ENTOX. Since that time, better preparative and handling practices have been implemented to decrease blank contamination to below detection limits with success (Codi King, unpublished data). The use of time-integrated samples is a relativity new tool being applied in Australia to assess in situ hydrophobic PAHs concentrations. Their main implementation has been for PAH assessment along coastal regions of north Queensland (Shaw and Müller 2005; Shaw et al. 2009) and not for oil and gas production. Only the 16 priority pollutant PAHs were reported and this is based upon their toxicity to mammals and aquatic organisms (USEPA, 1987). For accurate assessment of PW discharges using SPMDs, alkyl homologues must be evaluated in future studies because they are a major component of most PW discharges (Durell et al. 2006) including the Harriet A PW.

In summary, several of the biomarkers measured provided statistically significant evidence of exposure and potential impact of PW and these include CYP1A, CY2M1 Protein 1, CY2M1 Protein 2, CY2K1 Protein 1, CY2K1 Protein 2, FACs, and liver histopathology. Although not statistically significant, there were several biomarkers that provided biologically important information and they were HSI, cytochrome P450, and ChE activity. This study evaluated chemical analyses, along with a suite of biomarkers, to help address the need for more pragmatic environmental assessment techniques attempting to link environmental degradation with its causes (Galloway et al. 2004). However, the limitation of grab sampling and the need for better analytical detection limits were highlighted from this study, especially with determining low-level chronic exposure to PW. The results did confirm previous findings in other tropical reef species that there are issues of environmental degradation in water quality close to Harriet A (Codi King et al. 2005).

Understanding the spatial and temporal scale of the PW discharge is essential information for an environmental monitoring program. Although this study was inconclusive in determining far-field impacts of the Harriet A PW discharge at Site B, the data does suggest that PW is not likely to pose a risk on the Marine Protected Area, 7 km away. This was a relatively "short-term" study and the biomarker responses at Site B were not definitive. It is possible that a longer exposure time is needed for any far-field effects to manifest themselves due to distance from the PW and the complex water movement around the Harriet A platform.

One of the main advantages of biomarkers is for use as early warning tools. They have the potential to help focus future research targeted at providing a method of assessment that assists compliance with regulatory needs. For assessment of Harriet A PW, there are two main research areas on which to focus: acute toxicity and chronic toxicity. Acute toxicity is an important consideration of PW and one that can be addressed through laboratory toxicity studies. However, we believe the primary focus should be on cumulative and chronic impacts of Harriet A PW discharge through continued assessment using biomarkers and chemical-based monitoring. This study has demonstrated the suitability of a suite of biomarkers on one important species. Biomarkers have been widely advocated as a vital addition to the risk assessment procedure (Galloway et al. 2004) because their focus is as "early warning" signs before effects manifest themselves at the population and ecosystem level. It is well documented that biomarkers should not be applied singly, but as a suite of techniques ranging over different trophic levels of biological organization (Stebbing and Dethlefsen 1992; Galloway et al. 2004). Fish have been well established as good biomonitoring organisms and this study confirms that approach. However, different organisms may be exposed to contaminants based upon distribution in the water column during different stages of their life cycle (Stebbing and Dethlefsen 1992) and this issue should be addressed. We recommend that additional trophic levels should be included to represent a more holistic approach to environmental monitoring. This will identify if certain species and specific biomarkers are more sensitive and hence, more useful to assess Harriet A PW exposure and effects. This is a weight-of-evidence approach, which can be vital for environmental management to develop better methods for dealing holistically with impacted biota and predicting likely adverse effects of anthropogenic pollutants or activities on ecosystems.

This study demonstrates the usefulness of biomarkers for assessing the Harriet A PW discharge into Australian waters. The approach has merit as a valuable addition to environmental management strategies for protecting Australia's tropical environment and its rich biodiversity.

Acknowledgements This study was funded by the Australian Institute of Marine Science and Apache Energy Limited. The authors are indebted to Cary McLean and Michael K King for their invaluable expertise in mooring design and deployment; Dr. Stephen Newman and Mr. Craig Skepper from Department of Fisheries, Western Australia for all their assistance with field design and fish capture; Craig Humphrey for his laboratory analysis of muscle samples for ChE activity; and finally, the entire crew of the *RV Cape Ferguson* for their valuable assistance in the field.

References

Altenburger R, Segner H, van der Oost R (2003) Biomarkers and PAHs – prospects for the assessment of exposure and effects in aquatic systems. In: Douben PET (ed) PAHs: an ecotoxicological perspective. Wiley, London, pp 297–328

Anderson JW, Bothner K, Vu T, Tukey RH (1996) Using a biomarker (P-450RGS) test method on environmental samples. In: Ostrander GK (ed) Techniques in aquatic toxicology. CRC Press, Boca Raton, FL, pp 277–286

Au DWT (2004) The application of histo-cytopathological biomarkers in marine pollution monitoring: a review. Mar Poll Bull 48:817–834

Baumann PC, Mac MJ, Smith SB, Harshbarger (1991) Tumor frequencies in walleye (*Stizostedion vitreum*) and brown bullhead (*Ameiurus nebulosus*) and sediment contaminants in tributaries of the Laurentian Great Lakes. Can J Fisheries Aquat Sci 48:1804–1810

Booij K, Sleiderink H, Smedes F (1998) Calibrating the uptake kinetics of semipermeable membrane devices using exposure standards. Environ Toxic Chem 17:1236–1245

Brand GW, Gibbs CF, Monahan CA, Palmer DH, Murray AJ, Fabris GJ, Chamberlain T, Nicholson GJ (1989) Produced waters from the bass strait oil and gas field. Chemical characterisation and toxicity to marine organisms. Department of Conservation, Forests and Lands, Fisheries Division, Final Report, 22p

Brown RJ, Galloway TS, Lowe D, Browne MA, Dissanayke A, Jones MB, Depledge MH (2004) Differential sensitivity of three marine invertebrates to copper assessed using multiple biomarkers. Aquat Toxicol 66:267–278

Brumley CM, Haritos VS, Ahokas JT, Holdway DA (1996) Metabolites of chlorinated syringaldehydes in fish bile as biomarkers of exposure to bleached Eucalypt pulp effluents. Ecotoxicol Environ Saf 33:253–260

Brumley CM, Haritos VS, Ahokas JT, Holdway DA (1998) The effects of exposure duration and feeding status on fish bile metabolites: implications for biomonitoring. Ecotoxicol Environ Saf 39:147–153

Buhler DR, Wang-Buhler J (1998) Rainbow trout cytochrome P-450's: purification, molecular aspects, metabolic activity, induction and role in environmental monitoring. Comp Biochem Phys 121C:107–137

Burns KA, Codi S (1999) Non-volatile hydrocarbon chemistry studies around a production platform on Australia's Northwest Shelf. Estuarine, Coastal Shelf Sci 49:853–876

Burns KA, Codi S, Furnas, M, Heggie D, Holdway D, King B, McAllister F (1999) Dispersion and fate of produced formation water constituents in an Australian Northwest Shelf shallow water ecosystem. Mar Poll Bull 38:593–603

Cobby, GL (2002) Changes to the environmental management of produced formation water offshore Australia. APPEA J 677–682

Codi S, Burns KA, Johnson JE, Ramsay M, Ryan DJA, Haasch ML (2001) A pilot study conducted around Harriet A petroleum production on the Northwest Shelf of Australia: assessment of sub-lethal stress indicators in fish as potential indicators of petroleum hydrocarbon exposure. Final report to Apache Energy Pty Ltd., 73p

Codi King S, Johnson JE, Haasch ML, Ryan DAJ, Ahokas JT Burns KA (2005) Summary results from a pilot study conducted around and oil production platform on the Northwest Shelf of Australia. Mar Poll Bull 51:1163–1172

Collier TK, Varanasi U (1991) Hepatic activities of xenobiotic metabolising enzymes and biliary levels of xenobiotics in English sole (*Parophrys vetulus*) exposed to environmental contaminants. Arch Environ Toxicol Chem 20:462–473

Collier TK, Anulacion BF, Stein JE, Goksøyr A, Varanasi U (1995) A field evaluation of cytochrome P4501A as a biomarker of contaminant exposure in three species of flatfish. Environ Toxicol Chem 14:154–162

Department of Energy (DOE) (1997) Radionuclides, metals, and hydrocarbons in oil and gas operational discharges and environmental samples associated with offshore production facilities on the Texas/Louisiana continental shelf and environmental assessment of metals and hydrocarbons. Report to the U.S. Department of Energy, Bartlesville, OK

DiMichelle L, Taylor MH (1978) Histopathological and physiological responses of *Fundulus heteroclitus* to naphthalene exposure. J Fish Res Board Canada 35:1060–1066

Durell G, Utvik TR, Johnsen S, Frost T, Neff JM (2006) Oil well produced water discharges to the North Sea. Part I: comparison of deployed mussels (*Mytilus edulis*), semi-permeable membrane devices, and the DREAM model predictions to estimate the dispersion of polycyclic aromatic hydrocarbons. Mar Environ Res 62:194–223

Ellman GL, Courtney DK, Andres Jr V, Featherstone RM (1961) A new and rapid colorimetric determination of acetylcholinesterase activity. Biochem Pharmacol 7:88–95

Elumalai M, Antunes C, Guilhermino L (2002) Enzymatic biomarkers in the crab *Carcinus maenas* from the Minho River estuary (NW Portugal) exposed to zinc and mercury. Chemosphere 66:1249–1255

Fletcher GL, King MJ, Kiceniuk JW, Addison RF (1982) Liver hypertrophy in winter flounder following exposure to experimentally oiled sediments. Comp Biochem Physiol Part C: Pharmacol Toxicol Endocrinol 73:457–462

Frasco, MF, Fournier D, Carvalho F, Guilhermino L (2005) Do metals inhibit acetylcholinesterase (AChE)? Implementation of assay conditions for the use of AChE activity as a biomarker of metal toxicity. Biomarkers 10:360–375

Furnas MJ, Mitchell AW (1999) Wintertime carbon and nitrogen fluxes on Australia's North West Shelf. Estuarine, Coastal Shelf Sci 49:165–175

Galgani F, Bocquené G, Cadiou Y (1992) Evidence of variation in cholinesterase activity in fish along a pollution gradient in the North Sea. Mar Ecol Prog Ser 91:77–82

Galloway T, Brown R, Browne M, Dissanayake A, Lower D, Jones M, Depledge M (2004) Ecosystem management bioindicators: the ECOMAN project – a multi-biomarker approach to ecosystem management. Mar Environ Res 58:233–237

Gingerich WH (1982) Hepatic toxicology of fishes. In: Weber LG (ed) Aquatic toxicology. Raven Press, New York, NY, pp 55–106

Goksøyr A, Larsen H, Blom S, Forlin L (1992) Detection of cytochrome P450 1A1 in North Sea dab liver and kidney. Mar Ecol Prog Ser 91:83–88

Guilhermino L, Barros P, Silva MC, Soares AMVM (1998) Should the use of inhibition of cholinesterases as a specific biomarker for organophosphate and carbamate pesticides be questioned? Biomarkers 3:157–163

Guilhermino L, Lacerda MN, Nogueira AJA, Soares AMVM (2000) In vitro and in vivo inhibition of Daphnia magna acetylcholinesterase by surfactant agents: possible implications for contamination biomonitoring. Sci Total Environ 247(2–3):137–141

Haasch ML (1996) Induction of anti-trout lauric acid hydroxylase immunoreactive proteins by peroxisome proliferators in bluegill and catfish. Mar Environ Res 42:287–291

Haasch ML (2002) Effects of vehicle, diet and gender on the expression of PMP70- and CYP2K1/2M1-like proteins in the mummichog. Mar Environ Res 54:297–301

Haasch ML, Prince R, Wejksnora PJ, Cooper KR, Lech JJ (1993) Caged and wild fish: induction of P-450 (CYP1A1) as an environmental biomonitor. Environ Toxicol Chem 12:885–895

Haasch ML, Henderson MC, Buhler DR (1998a) Induction of CYP2M1 and CYP2K1 lauric acid hydroxylase activities by peroxisome proliferating agents in certain fish species: possible implications. Mar Environ Res 46:37–40

Haasch ML, Henderson Mc, Buhler Dr (1998b) Induction of lauric acid hydroxylase activity in catfish and bluegill by peroxisome proliferating agents. Comp Biochem Physiol Part C: Pharmacol Toxicol Endocrinol 121:297–303

Hawkes JW (1977) The effects of petroleum hydrocarbon exposure on the structure of fish tissues. In: Wolfe DA, Anderson JW (eds) Fate and effects of petroleum hydrocarbons in marine ecosystems and organisms: proceedings of a symposium, November 10–12, 1976, Olympic Hotel, Seattle, Washington Pergamon Press, NY, pp 115–128

Heath AG (1995) Water pollution and fish physiology. CRC Press, Boca Raton, FL

Hinton DE, Lauren DJ (1990) Liver structural alterations accompanying chronic toxicity in fishes: potential biomarkers of exposure. In: McCarthy JF, Shugart LR (eds) Biomarkers of environmental contamination. CRC Press, Boca Raton, FL, pp 17–57

Hinton DE, Segner H, Braunbeck T (2001) Toxic responses of the liver. In: Schlenk D, Benson WH (eds) Target organ toxicity in marine and freshwater teleosts. Taylor & Francis, London, pp 224–268

Holdway DA (2002) The acute and chronic effects of wastes associated with offshore oil and gas production on temperate and tropical marine ecological processes. Mar Poll Bull 44:185–203

Holdway DA, Brennan SE, Ahokas JT (1995) Short review of selected fish biomarkers of xenobiotic exposure with an example using fish hepatic mixed-function oxidase. Aust J Ecol 20:34–44

Holdway D, Heggie D (1998) Tracking produced formation water discharge from a petroleum production platform to the northwest shelf. APPEA J 38:665–680

Holth TF, Nourizadeh-Lillabadic R, Blaesbjergd M, Grunga M, Holbeche H, Petersenf GI, Aleströmc P, Hylland K (2008) Differential gene expression and biomarkers in zebrafish (*Danio rerio*) following exposure to produced water components. Aquat Toxicol 90: 277–291

Huckins JN, Petty JD, Prest HF, Clark RC, Alvarez DA, Orazio CE, Lebo JA, Cranor WL, Johnson BT (2000) A guide for the use of semi permeable membrane devices (SPMDS) as samplers of waterborne hydrophobic organic contaminants. US Geological Survey, Columbia Environmental Research Center, Columbia, MO

Humphrey CA, Codi King S, Klumpp DW (2007) A multi-biomarker approach in barramundi (Lates calcarifer) to measure exposure to contaminants in estuaries of tropical North Queensland . Mar Poll Bull 54:1569–1581

Hylland K, Feist S, Thain J, Forlin L (2003) Molecular/cellular processes and the health of the individual. In: Lawrence AJ, Hemingway KL (eds) Effects of pollution on fish: molecular effects and population responses. Blackwell Science, Oxford, pp 83–133

Incardona P, Collier TK, Scholz NL (2004) Defects in cardiac function precede morphological abnormalities in fish embryos exposed to polycyclic aromatic hydrocarbons. Toxicol Appl Pharmacol 196(2):191–205

Johnson RD, Bergman HL (1984) Use of histopathology in aquatic toxicology: a critique. In: Cairns VW, Hodson PV, Nriagu JO (eds) Contaminant effects of fisheries. Wiley, New York, NY, pp 19–39

Jones RJ, Heyward AJ (2003) The effects of produced formation water (PFW) on coral and isolated symbiotic dinoflagellates of coral. Mar Fresh Res 54:153–162

Katchamart S, Miranda CL, Henderson MC, Pereira CB, Buhler DR (2002) Effect of xenoestrogen exposure on the expression of cytochrome P450 isoforms in rainbow trout liver, Environ Toxicol Chem 21:2445–2451

King BA, McAllister FA (1998) Modelling the dispersion of produced water discharges. APPEA J 681–691

Krahn MM, Myers MS, Burrows D, Malins DC (1984) Determination of metabolites of xenobiotics in the bile of fish from polluted waterways. Xenobiotica 14:633–646

Krahn MM, Burrows DG, MacLeod WD Jr, Malins DC (1987) Determination of individual metabolites of aromatic compounds in hydrolyzed bile of English sole (*Parophrys vetulus*) from polluted sites in Puget Sound, Washington. Arch Environ Contam Toxicol 16: 511–522

Krahn MM, Burrows DG, Ylitalo GM, Brown DW, Wigren CA, Collier TK, Chan SL, Varanasi U (1992) Mass spectrometric analysis for aromatic compounds in bile of fish sampled after the Exxon Valdez oil spill. Environ Sci Technol 26:116–126

Krahn MM, Ylitalo GM, Buzitis J, Bolton JL (1993) Analyses for petroleum-related contaminants in marine fish and sediments following the Gulf oil spill. Mar Poll Bull 27:285–292

Lowry OH, Rosebrough AL, Farr AL, Randall RJ (1951) Protein measurements with the Folin phenol reagent. J Biol Chem 193:265–275

Mallatt J (1985) Fish gill structural changes induced by toxicants and other irritants: a statistical review. Can J Fish Aquat Sci 42:630–648

McCain BB, Hodgins HO, Gronlund WD, Hawkes JW, Brown DW, Myers MM, Vandermeulen JH (1978) Bioavailability of crude oil from experimentally oiled sediments to English sole (*Parophyrs vetulus*) and pathological consequences. J Fish Res Board Canada 35:657–664

McDonald SJ, Willett KL, Thomsen J, Beatty KB, Connor K, Narasimhan TR, Erickson CM, Safe SH (1996) Sub lethal detoxification responses to contaminant exposure associated with offshore production platforms. Can J Fish Aquat Sci 53:2606–2617

Myers TR, Hendricks JD (1985) Histopathology. In: Rand GM, Petrocelli SR (eds) Fundamentals of aquatic toxicology. Hemisphere Publishing Corporation, London, pp 283–331

Myers CR, Sutherland LA, Haasch ML, Lech JJ (1993) Antibodies to a synthetic peptide that react specifically with rainbow trout hepatic cytochrome P-450 1A1. Environ Toxicol Chem 12:1619–1626

Neff JM (1987) Biological effects of drilling fluids, drill cuttings and produced waters. In: Boesch DF, Rabalais NN (eds) Long-term environmental effects of offshore oil and gas development. Elsevier Applied Science Publishers, London, pp 469–538

Neff JM (2002) Bioaccumulation in marine organisms: effects of contaminants from oil well produced water. Elsevier, London

Neff JM, Foster K (1997) Composition, fates and effects of produced water discharges to offshore waters of the Java Sea, Indonesia. Report to Pertimina/Maxus Southeast Sumatra, Jakarta, Indonesia, 103p

Nimrod AC, Benson WH (1996) Estrogenic responses to xenobiotics in channel catfish (*Ictalwus punctatus*). Mar Environ Res 42:155–160

Nowak BF, Deavin JG, Sarjito, Munday BL (1992) Scanning electron microscopy in aquatic toxicology. J Comput Assist Microsc 4:241–246

Olsgard F, Gray JS (1995) A comprehensive analysis of the effects of offshore oil and gas exploration and production on the benthic communities of the Norwegian continental shelf Mar Ecol Prog Ser 122:277–306

Osenberg CW, Schmitt RJ, Holbrook SJ, Canestro D (1992) Spatial scale of ecological effects associated with an open coast discharge of produced water. In: Ray JP, Engelhardt FR (eds) Produced water: technological/environmental issues and solutions. Plenum Press, New York, NY, pp 387–402

Payne JF, Fancey LL, Rahimtula AD, Porter EL (1987) Review and perspective on the use of mixed-function oxygenase enzymes in biological monitoring. Comp Biochem Physiol Part C: Pharmacol Toxicol Endocrinol 86:233–245

Payne JF, Mathieu A, Melvin W, Fancey LL (1996) Acetycholinesterase, an old biomarker with a new future? Field trials in association with two urban rivers and a paper mill in Newfoundland. Mar Poll Bull 32:225–231

Peakall DB (1994) The role of biomarkers in environmental assessment (1). Introduction. Ecotoxicology 3:157–160

Schlacher TA, Mondon JA, Connolly RM (2007) Estuarine fish health assessment: evidence of wastewater impacts based on nitrogen isotopes and histopathology. Mar Poll Bull 54: 1762–1776

Shaw M, Müller JF (2005) Preliminary evaluation of the occurrence of herbicides and PAHs in the Wet Tropics region of the Great Barrier Reef, Australia, using passive samplers. Mar Poll Bull 51:876–881

Shaw M, Eaglesham G, Müller JF (2009) Uptake and release of polar compounds in SDB-RPS Empore^{TM} disks; implications for their use as passive samplers. Chemosphere 75:1–7

Slooff W, van Kreijl CF, Baars AJ (1983) Relative liver weights and xenobiotic-metabolizing enzymes of fish from polluted surface waters in the Netherlands. Aquat Toxicol 4:1–14

Stagg RM, McIntosh A (1998) Biological effects of contaminants: determination of CYP1A-dependent mono-oxygenase activity in dab by fluorimetric measurement of EROD activity. ICES Techniques in Marine Environmental Sciences, No. 23, International Council for the Exploration of the Sea, Copenhagen, Denmark, pp 1–16

Stebbing ARD, Dethlefsen V (1992) Introduction to the Bremerhaven workshop on biological effects of contaminants. Mar Ecol Prog Ser 91:1–8

Stegeman JJ, Hahn ME (1994) Biochemistry and molecular biology of monooxygenases: current perspectives on forms, functions, and regulation of cytochrome P450 in aquatic species. In: Malins DC, Ostrander GK (eds) Aquatic toxicology: molecular, biochemical and cellular perspectives. CRC Press, Boca Raton, FL, pp 87–206

Stegeman JJ, Woodin BR, Singh H, Oleksiak MF, Celander M (1997). Cyctochrome P450 (CYP) in tropical fishes, catalytic activities, expression of multiple CYP protein and high levels of microsomal P450 in liver of fishes from Bermuda. Comp Biochem Physiol Part C: Pharmacol Toxicol Endocrinol 116(1):61–75

Stegeman JJ, Schlezinger JJ, Craddock JE, Tillitt DE (2001) Cytochrome P450 1A expression in midwater fishes: Potential effects of chemical contaminants in remote oceanic zones. Environ Sci Technol 35:54–62

Sturm A, Wogram J, Segner H, Liess M (2000) Different sensitivity to organophosphates of acetylcholinesterase and butyrylcholinesterase from three-spined stickleback (*Gasterosteus aculeatus*): application in biomonitoring. Environ Toxicol Chem 19:1607–1615

Swan JM, Neff JM, Young PC (1994) Environmental implications of offshore oil and gas development in Australia – the findings of an independent scientific review. APPEA, Western Australia

Tollefsen KE, Harman C, Smith A, Thomas KV (2007) Estrogen receptor (ER) agonists and androgen receptor (AR) antagonists in effluents from Norwegian North Sea oil production platforms. Mar Poll Bull 54:277–283

Underwood AJ (1981) Techniques of analysis of variance in experimental marine biology and ecology. Ocean Mar Biol Ann Rev 19:513–605

USEPA (1987) Quality criteria for water 1986 EPA 440/5-86-001. United States Environmental Protection Agency, Washington, DC

van der Oost R, Beyer J, Vermeulen NPE (2003) Fish bioaccumulation and biomarkers in environmental risk assessment: a review. Environ Toxicol Pharm 13:57–149

Vethaak AD, Rhenalt T (1992) Fish disease as a monitor of marine pollution: the case of the North Sea. Rev Fish Biol Fisheries 2:1–32

Vrolijk N, Targett N, Woodin B, Stegeman J (1994) Toxicological and ecological implications of biotransformation enzymes in the tropical teleost *Chaetodon capistratus*. Mar Biol 119: 151–158

Wester PW, van der Ven LTM, Vethaak AD, Grinwis GCM, Vos JG (2002) Aquatic toxicology: opportunities for enhancement through histopathology. Environ Toxicol Pharm 11:289–295

Wood CM (2001) Toxic responses of the gill. In: Schlenk D, Benson WH (eds) Target organ toxicity in marine and freshwater teleosts, Taylor & Francis, London, pp 1–89

Zhu S, Codi King S, Haasch ML (2006) Environmental induction of CYP1A-, CYP2M1- and CYP2K1-like proteins in tropical fish species by produced formation water on the northwest shelf of Australia (short communication). Mar Environ Res 62:S322–S326

Zhu S, Codi King S, Haasch ML (2008) Biomarker induction in tropical fish species on the Northwest Shelf of Australia by produced formation water. Mar Environ Res 65:315–324

Chapter 15
Evidence of Exposure of Fish to Produced Water at Three Offshore Facilities, North West Shelf, Australia

Marthe Monique Gagnon

Abstract In Western Australia, the discharge of produced water (PW) by the offshore petroleum production facilities is acceptable under specific conditions. Little is known on the effects of PW discharge on the health of marine organisms attracted to the submerged structures. Three offshore facilities have been selected for studying the impact of exposure to PW discharge on fish health, as measured by a suite of biomarkers of fish health. Physiological indices (liver somatic index, condition factor) as well as biochemical markers of exposure (EROD activity, biliary metabolites) and of effect (DNA damage, stress proteins) have been assessed on three different fish species captured in the vicinity of the facilities. Condition factor was slightly reduced at one site only, but liver somatic index was elevated in fish captured at two of the three locations. EROD activity and DNA damage levels were high only at one facility discharging high volumes of PW. Naphthalene and pyrene biliary metabolites were detected at significant levels at all locations. Stress proteins HSP70 were also elevated at all locations. The results suggest that while the chemical characteristics of PW are important, consideration of the loading (concentration × volume) of PW is crucial in assessing environmental effects and risks of PW discharge.

1 Introduction

In Australia, the majority of offshore petroleum production facilities discharge produced water (PW) directly to the ocean. Prior to 2001, PW regulation was limited to achieving compliance with specific maximum concentrations of petroleum in PW, these concentrations being a maximum of 50 mg/L at any one time, and the average petroleum content being less than 30 mg/L PW over 24 h. Since 2001, however, the regulation of PW has to include an assessment of

---

M.M. Gagnon (✉)
Department of Environment and Agriculture, Curtin University, Perth, WA 6845, Australia

K. Lee, J. Neff (eds.), *Produced Water*, DOI 10.1007/978-1-4614-0046-2_15,

the environmental effects and risks of the whole PW discharge, which provides a more meaningful approach than only controlling oil in water concentrations (Cobby 2002). This approach also integrates the various chemicals present in the PW and the loading (contaminant concentration $\times$ volume) of effluent discharged.

PW is usually the largest single aqueous discharge from offshore production platforms. The buoyancy of PW discharges in seawater appears to be a characteristic associated with the high discharge temperatures of PW and low specific gravity (Holdway and Heggie 1998). Of the many contaminants present in PW, the two- and three-ring polycyclic aromatic hydrocarbons (PAHs) are considered the most important contributors to the ecological hazard posed by produced water discharges (Neff et al. 2006). Although studies have been performed overseas on the toxicity of PW, they may be of little relevance to the Australia's North West Shelf environment where oil composition, water temperatures and ocean currents are quite different.

Fish biomarkers have been suggested as a tool to evaluate the impact of PW discharge on fish health (Borseth et al. 2000; Cobby 2002). Biomarkers of exposure and effects in fish are now widely used in environmental monitoring in North America and in the North Sea (George et al. 1995) and have proven to be relevant and useful in the biomonitoring of oil and gas activities in the Norwegian Sea (Borseth et al. 2000). In a relatively uniform water column as found in offshore environments, submerged structures such as platforms, well jackets and subsea connectors may act as artificial reefs. These structures provide substrates for encrusting epi-biota and consequently, fish are attracted as a more complex food chain develops (Black et al. 1994). It is therefore relevant to collect fish in the vicinity of an offshore platform, as these fish will most likely be residents of this environment. Fish will also be mobile within this environment resulting in a variable exposure to petroleum hydrocarbons, not only increasing variation in biomarker measurements but also increasing ecological relevance of the effect assessment as fish movement is part of a normal functioning of ecosystems. As PW is relatively buoyant, field studies would preferably target surface fish as bioindicator species. In addition, the boindicator organism should be collected within 1 km of the platform discharge as PAHs have been shown to be at the highest concentration within this distance from a discharging offshore facility (Durell et al. 2006).

Investigative studies have been performed at three PW discharging facilities located in the North West Shelf of Australia. Surface fish species present at each location have been captured in the close vicinity of the discharging operations, and biopsies were collected in order to quantify biomarkers of exposure and of effects. The concept of biomarkers in field-collected animals is based on the recognized fact that an adverse effect will become apparent at the subcellular level before the effect appears at higher levels of biological organization such as population and community levels (Stein et al. 1998). For this reason, biomarkers are used as an early warning system so that appropriate management action can be implemented before irreversible ecological changes occur.

2 Sampling Sites and Collection of Biopsies

Three petroleum production operations located in the North West Shelf of Australia have been selected for the study: the Wandoo B, the Four Vanguard and the Ocean Legend (Fig. 15.1).

The Wandoo B facility is a concrete gravity structure in 54 m of water producing high viscosity crude of 19°API and, at the time of the study, was discharging 132,000 barrels PW per day (89% water cut). Since this facility was operating for 6 years prior to the study, a temporal reference was not possible and consequently, the Wandoo A monopod located 5-km west of Wandoo B was selected as the reference. Wandoo A is a small monopod platform supported by a single column. It is an unmanned platform but has helicopter landing and boat mooring facilities. No effluents are released from Wandoo A monopod. Sampling of fish in close proximity of the Wandoo A and B structures was performed in October 2002.

The Four Vanguard is a floating production storage and off-loading unit (FPSO) located in 100-m water and producing only 25,000 barrels PW per day (50% water cut). The light sweet crude produced by the Four Vanguard FPSO has a 49°API. This FPSO has commenced operations late 2003, and sampling of reference fish at this facility was conducted prior to the start of the PW discharge in November 2003, while the sampling of PW-exposed fish was conducted after 15 months of operations in February 2005.

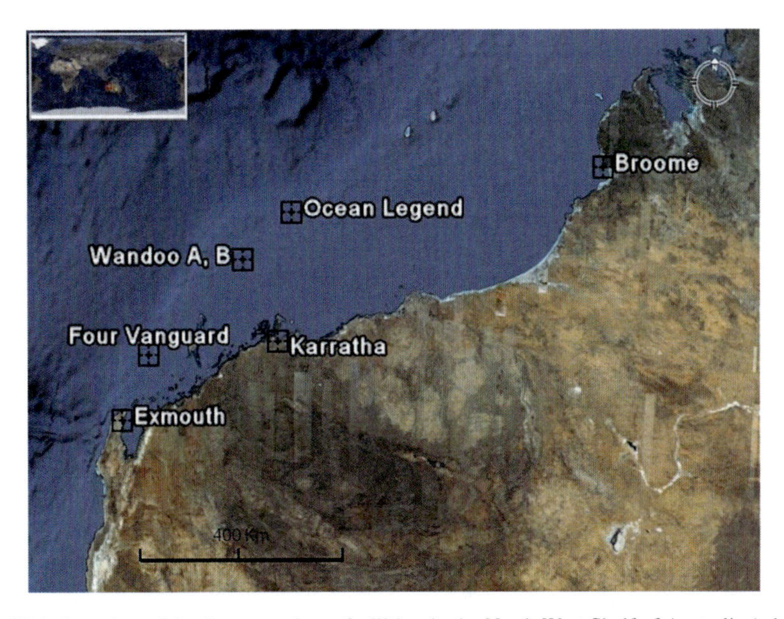

Fig. 15.1 Location of the three petroleum facilities, in the North West Shelf of Australia (adapted from Google Earth 2009)

The third petroleum production facility to be investigated, the Ocean Legend, is operating in 50-m water, discharged 37,000 barrel PW per day and produced a light sweet crude of 43°API. The Ocean Legend commenced operating in the Legendre oil fields late 2001, allowing reference fish to be collected in August 2001, and collection of PW-exposed fish occurred 20 months later in March 2003. This facility retains the PW in the FPSO's retention tank for 4 days, allowing a limited biodegradation to occur before discharge at sea (S. Capper, HSE manager, personal communication).

In all cases, the current pattern is highly influenced by tides, and leads to rapid PW dilution. However, the strong tidal influence which moves water masses back and forth might result in accumulation of petroleum hydrocarbons in the near field. The detailed chemistry of individual PW discharge is not available.

Surface fishing from the lower deck of the facilities or from a fishing boat was performed within 100 m from the PW discharge point with a rod or a snapper winch equipped with 200 lbs line. The target species were species that remain near to the surface, or spend a significant time at the surface of the PW receiving waters to feed or to perform other activities. Typically, rainbow runner (*Elegatis bipunnulata*) or trevally species (*Caranx* sp.) were targeted, but when these were not available in large numbers, resident fish from deeper waters were collected, e.g. gold-banded snapper (*Pristipomoides multidens*). A minimum number of 12 fish per site per sampling was targeted (Hodson et al. 1993). Because of the highly variable weather conditions in the North West Shelf of Australia, the fishing was limited to 6 days per site per sampling which in some cases did not allow capture of the targeted number of fish.

Upon capture of a fish, external pathology examination was conducted with any abnormalities and/or parasites detailed and photographed. The fish were then killed by the method known as *Iki jime,* according to Curtin University Animal Ethics approval number N40-01. Total weight, and total, fork and standard lengths were recorded. A sample of gill was collected and immediately frozen in liquid nitrogen for stress protein determination. The abdominal cavity of the fish was opened, and the bile was collected from the gall bladder for PAH biliary metabolites quantification. The liver was weighed and two 1-g liver samples were immediately frozen in liquid nitrogen for EROD activity measurement and DNA damage determinations. Upon arrival to the laboratory, all samples were transferred to a –80°C freezer until analysis.

3 Biochemical Methodologies

3.1 Liver Detoxification Enzymes (EROD Activity)

The liver detoxification activity of the mixed function oxygenase P450 system was measured by assessing EROD activity. EROD activity was measured using a modified method of Gagnon and Holdway (2002). Each liver sample was thawed on ice, then homogenized in HEPES pH 7.5 using a Heidolph DIAX 900 homogenizer. The homogenate was centrifuged (Jouan CR3i centrifuge) at 9000×g for 20 min at 4°C and the S9 post-mitochondrial supernatant (PMS) collected for immediate use. The reaction mixture contained HEPES pH 7.8, $MgSO_4$, BSA, NADPH solution and PMS. The reaction was initiated by adding ethoxyresorufin, incubated at room temperature for 2 min, and the reaction terminated by adding HPLC grade methanol. Resorufin standards (0.000–0.085 M) and samples were centrifuged to precipitate proteins, and the fluorescence of the supernatant was immediately read on a Perkin-Elmer LS-45 Luminescence Spectrometer at excitation/emission wavelengths of 535/585 nm (slit 10 ex/10 em). Protein content of the PMS was determined according to Lowry et al. (1951). EROD activity was expressed as picomoles of resorufin produced, per mg of total protein, per minute (pmol R/mg Pr/min).

3.2 PAH Bile Metabolites

The biliary metabolite determination was performed by fixed fluorescence (FF) measurement (Lin et al. 1996). The method is semi-quantitative and reports metabolized chemicals fluorescing at the naphthalene, pyrene or benzo(a)pyrene [B(a)P] specific excitation/emission wavelengths. Fluorescent readings were performed at the naphthalene excitation/emission 290/335 nm using 1-naphthol (Sigma) as a reference standard. Metabolites fluorescing at the pyrene and B(a)P wavelengths were measured using 1-hydroxypyrene as a reference standard at 340/380 nm and 380/430 nm for pyrene and B(a)P wavelengths, respectively. Metabolites fluorescing at the naphthalene wavelength are reported in μg of 1-naphthol fluorescence units equivalent per mg biliary protein, and those fluorescing at the pyrene and B(a)P wavelengths are reported in μg of 1-OH pyrene fluorescence units equivalent per mg biliary protein. Therefore, the biliary metabolite levels measured represent fluorescence-equivalents of PAH metabolites used as standards.

Bile samples were thawed on ice and diluted to 1:2000 in 50% HPLC grade methanol/H_2O for determination of metabolites fluorescing at the pyrene and B(a)P wavelengths. The bile was further diluted to 1:5000 for the determination of metabolites fluorescing at the naphthalene wavelength. Collier and Varanasi (1991) have shown that the normalization for protein concentration in the bile can, to a large extent, account for changes in the level of biliary metabolites due to differences in the feeding status of some fish. Bile was diluted in 19 volumes of double-distilled H_2O (bile:water 1:20) and the protein content determined using the method of Lowry et al. (1951). Biliary metabolites are reported on the basis of biliary protein (metabolite/mg protein).

3.3 DNA Damage

The measurement of DNA damage was performed by the alkaline unwinding assay using liver tissue (Shuggart 1996). Briefly, the tissue was hand homogenized with DNAzol® and centrifuged at 8000×g for 10 min at 5°C. Ethanol was added to the isolated supernatant to precipitate the DNA. The isolated DNA was cleaned using a Tris–EDTA buffer, and the double-stranded (DS), single-stranded (SS) and partially unwounded (DSS) DNA were obtained by treating samples with NaOH and incubating at various temperatures according to Shuggart (1996). The fluorescence of each sample was measured using a Perkin-Elmer LS-45 Luminescence Spectrometer at excitation/emission wavelengths of 350/453 nm (slit 5 ex/10 em), and the F-value (representing DNA integrity) was calculated as $F = (\text{DSS–SS})/\text{DS–SS}$.

3.4 Stress Proteins

Stress protein (HSP-70) response was measured using standard electrophoresis protocols as described in Webb and Gagnon (2009). Gill tissue was weighed and homogenized with Tris–PMFS buffer using a Heidolph DIAX 900 homogenizer. The homogenate was centrifuged at 12,000$\times g$ for 98 min at 4°C. Proteins were determined in the supernatant according to Lowry et al. (1951). Supernatant containing 40 μg proteins was mixed with sample buffer (Bio-Rad Laboratories, NSW, Australia) at a ratio of 1:2 supernatant /buffer, then placed in a waterbath at 95°C for 4 min. Samples were loaded in duplicate into 12% Tris–Glycine iGels (Nu-Sep, NSW, Australia) wells with heat shock standardized controls loaded into the two outermost wells. The gels were run at 225 V, 120 mA (60 mA per gel) for 40 min in a mini-PROTEAN 3 Electrophoresis Cell (Bio-Rad Laboratories Pty Ltd, NSW, Australia).

Proteins were transferred from iGels to 0.2 μm supported nitrocellulose membranes in a mini Trans-Blot electrode module (Bio-Rad) at 100 V, 250 mA for 1 h. Following Western Transfer, the blots were blocked in 5% skim milk powder dissolved in Tween-phosphate buffered saline on a shaker for 1 h. The blots were probed overnight at 4°C with monoclonal (mouse) anti-heat shock protein 70 antibody (IgG1, Bio-Scientific, Gymea, Australia), diluted 1:5000 in Tris-buffered saline (TBS), then the secondary antibody (goat anti-mouse IgG peroxidase conjugated, Progen Bioscience, Archerville, Australia) was applied, diluted 1:30,000 in Tween-TBS (TTBS) and allowed to incubate for 2 h. Between each step, the blots were washed three times with TTBS, then finally washed in TBS to remove the Tween.

A working solution of chemiluminescent substrate was prepared using the Super Signal® West Pico Chemiluminescent Substrate kit (Progen Bioscience). Under dark conditions, each blot was wetted with the chemiluminescent solution, then exposed to a radiographic film (CL-Xposure™ X-Ray film, Progen Bioscience), which was immediately developed. The films were scanned and the pixel density of the images analysed using NIH image program. The bands on different blots were calibrated to known standards to enable comparison between blots. HSP-70 levels are reported as pixels per μg total protein (μg HSP-70/μg pr^{-1}).

3.5 Statistical Analyses

For each variable, descriptive statistics including average and standard errors of the mean have been generated using the computer program SPSS, version 17.0. Because each fish species exhibits different physiological and biochemical parameters, each fish species has been treated separately for statistical purposes.

Physiological parameters total weight, fork length, condition factor (CF) and liver somatic index (LSI) as well as biomarker levels were compared between reference and PW-exposed conditions using a Student's *t*-test after verification of normal data distribution (Levene test) and homoscedasticity (distribution of residuals). For all statistical testing, a significance level alpha (α) of 0.05 was applied.

4 Results

4.1 Morphological Parameters

Different species of fish were collected at the different facilities; however, all fish were captured at the surface of the ocean. The species captured, the number of individuals and related physiological details are listed in Table 15.1. All fish were captured outside of their reproductive season and appear to be in good condition and healthy.

For each species, the physiological parameters total weight, fork length, CF and LSI have been tested for differences between sexes however, no difference was

Table 15.1 Fish species, number of individuals and physiological characteristics (mean ±SE) of fish collected at each site during the study, with significant differences, relative to their respective reference group, indicated with* ($p < 0.05$)

Parameter/ facility	Wandoo A (reference)	Wandoo B (exposed)	Four Vanguard (Pre-PW)	Four Vanguard (post-PW)	Ocean Legend (pre-PW)	Ocean Legend (post-PW)
Species collected	Rainbow runner *Elegatis bipunnulata*		Gold-banded snapper *Pristipomoides multidens*		Giant trevally *Caranx ignobilis*	
Number male/female	3 M; 2 F	4 M; 1 F	8 M; 11 F	6 M; 10 F	7 M; 2 F	4 M; 2 F
Total weight (g)	4125 ± 696	5625 ± 447	1816 ± 102	2053 ± 138	6511 ± 1070	7950 ± 200
Fork length (mm)	793 ± 27	867 ± 23	503 ± 7.8	551 ± 9.0*	761 ± 31	823 ± 31
CF[a]	8.46 ± 0.70	8.38 ± 0.25	1.40 ± 0.03	1.23 ± 0.03*	1.61 ± 0.10	1.47 ± 0.01
LSI[b]	0.55 ± 0.06	0.75 ± 0.03*	1.11 ± 0.08	1.50 ± 0.05*	0.85 ± 0.07	0.97 ± 0.09

[a]Due to the varied body shapes of the different species, the CF was calculated as CF = (total weight/fork length^3) $\times$ (10^5 for giant trevally)/(10^2 for gold-banded snapper)/(10^6 for rainbow runner)

[b]LSI = (liver weight/total weight) $\times$ 100

identified between sexes. Therefore the statistical testing has been performed on reference versus exposed groups of fish, on a species by species basis.

For all three species, the reference fish were of similar size (total weight or fork length) to the PW-exposed fish except for the gold-banded snapper at the Four Vanguard which were longer ($p < 0.001$) in the PW-exposed group. This might be due to the fact that that the new Four Vanguard facility, which commenced operations in 2003, was colonized by young individual fish who gained in length during the 20 months separating the pre-PW from the post-PW sampling. The increase in length but not in weight for the gold-banded snapper influenced the CF, with this fish species having a significantly ($p = 0.001$) lower CF in PW-exposed fish while the rainbow runner and giant trevally had similar CF pre- or post-PW exposure. The liver, a very plastic organ known to increase in size when organisms are exposed to chemical pollutants (van der Oost et al. 2003), was larger in the PW-exposed rainbow runner ($p = 0.016$) and gold-banded snapper ($p = 0.001$) relative to the pre-exposed fish of each species. In the giant trevally exposed to the Ocean Legend discharge, however, the liver size remained the same ($p = 0.070$) before and after chronic exposure to PW.

Morphological indicators (weight, length and indices) can not only inform on potential pollution impacts but also be influenced by non-pollutant factors such as season or reproduction. However, these gross indicators are reliable biomarkers of chronic exposure to pollution and represent an initial screening tool to indicate exposure or effects and to provide information on energetic reserves of the organisms (Mayer et al. 1992). Our data indicate that only the gold-banded snapper collected at the Four Vanguard facility has a lower CF, the result of increased length of this fish species at this new facility. The PW-exposed gold-banded snapper appeared to be in good condition and healthy. The LSI of the rainbow runner and of the gold-banded snapper were higher following exposure, similar to other field studies where fish species were exposed chronically to organic pollution (van der Oost et al. 2003). None of the fish species were actively developing gonads when sampling occurred and consequently, the liver enlargement cannot be related to gonad maturation (Glubokov and Orlov, 2008) or to seasonal differences, but rather appear to be related to chronic exposure to PW. A higher LSI does not represent a pathological condition in itself but suggests that an increased metabolic demand is imposed on the liver by the chemicals contained in PW.

4.2 Biochemical Markers

Biochemical markers, or biomarkers, are quantifiable measurements of body fluids, cells or tissues that reflect a biological response of a bioindicator organism to contaminants. Because many species of fish metabolize and depurate hydrocarbons very efficiently (Varanasi et al. 1989), quantification of tissue concentrations generally does not provide a useful indicator of exposure of fish to petroleum hydrocarbons. In addition, chemical concentrations alone are usually not related to effects in organisms or populations (Collier and Varanasi, 1991). For these reasons, biomarkers of exposure and effects in fish are more relevant and do complement each others in (1) confirming exposure and uptake of contaminants and (2) assessing if these contaminants have a biologically significant effect in exposed organisms.

A suite of biomarkers of exposure and of effect has been measured in the fish captured in the vicinity of the offshore petroleum production facilities. PAHs biliary metabolites inform on exposure, hepatic EROD activity suggests uptake and metabolism of contaminants, while DNA damage and stress proteins HSP70 represent effects related to exposure.

EROD activity was elevated at Wandoo B and at the Four Vanguard facilities, but not in fish captured at the Ocean Legend facility (Fig. 15.2).

Wandoo B produces a crude oil with a low 19°API which is expected to contain a significant quantity of high molecular PAHs, which are potential EROD activity inducers. In addition, this facility discharges a high volume of PW in the ocean, resulting in an exposure to PAH sufficiently high to induce EROD activity in rainbow runners. Comparatively, the Four Vanguard facility does discharge a small volume of PW issued from the processing of a 49°API light crude oil, an oil that typically contains a much reduced level of heavy molecular PAHs. However, the PW-exposed gold-banded snapper collected at the Four Vanguard site also had significantly higher EROD after 20 months of exposure, showing that exposure to PAHs was sufficient at this site to induce the detoxification system. EROD activity was not significantly elevated at the Ocean Legend facility, which discharged small volumes of PW associated with an oil of 43°API. The observed levels of EROD activity in the liver of the exposed fish demonstrates that the induction of this biomarker cannot easily be predicted from the characteristics of the PW alone, and that each situation has to be addressed on a case-by-case basis. A second

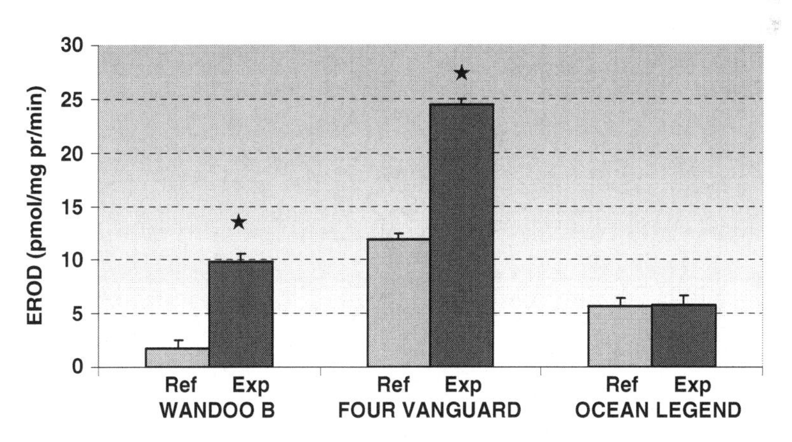

Fig. 15.2 EROD activity (mean ± SE) in fish collected around three offshore petroleum facilities in the North West Shelf of Australia; Ref = fish unexposed to PW, Exp = exposed to PW for at least 20 months; star indicates that the exposed fish are significantly different ($p < 0.05$) to the reference fish

factor that might be considered is the different induction potential of various species of fish. While some fish species as the rainbow trout (*Oncorhynchus mykiss*) might be induced at 200-fold the control fish (Jönsson et al. 2006), the fish present at the offshore platforms might not be biochemically responsive to EROD inducers, limiting the value of this biomarker as an indicator of exposure and uptake of PAHs. Typically, EROD induction in field-caught animals is not induced at levels as high as could be observed under laboratory conditions, e.g. EROD inductions measured in field-caught fish in Australia report two- to sixfold induction only (Cavanagh et al. 2000; Gagnon and Holdway 2002; Rawson et al. 2009). The value of EROD induction in each species would need to be tested under laboratory conditions to confirm the EROD induction potential of specific fish species.

Many species of fish take up and depurate hydrocarbons very efficiently (Varanasi et al. 1989; van der Oost et al. 2003), with PAHs metabolites being directed to the bile for elimination via the intestinal route. Biliary metabolite measurement offers a rapid and inexpensive method of validating exposure to, and uptake of, PAHs. Elevated amounts of biliary PAH metabolites of the naphthalene, pyrene and B(a)P-type of metabolites have been measured in the three fish species collected at all locations (Fig. 15.3). The expression 'type of metabolites' refers to the various conjugated metabolites originating from a common compound, in this case, naphthalene, pyrene or B(a)P (Krahn et al. 1992; Vermeulen et al. 1992). The naphthalene metabolites were high in the rainbow runners collected at the Wandoo B facility, relative to the reference fish. Wandoo B was discharging the largest amount of PW and was producing a heavy oil resulting in a significant exposure of fish to the PAHs contained in the PW. The Four Vanguard and the Ocean legend facilities were both producing a lighter oil. Typical light oil is known to have few PAHs; however, naphthalene usually represents the highest proportion of PAHs in light crude oils. The gold-banded snapper captured at the Four Vanguard facility accumulated a comparatively low level of naphthalene metabolites. This might be the result of the 4-day treatment of PW that was ongoing on the Four Vanguard FPSO before discharge of the PW at sea (S. Capper, HSE manager, personal communication). The giant trevally collected at the Ocean Legend had the highest naphthalene biliary metabolites which might be related to a higher exposure level or to a more efficient PAH uptake and/or metabolism in this fish species. Most importantly, biliary metabolites of naphthalene-type were elevated in all PW-exposed fish at all locations, making this biomarker a sensitive biomarker of exposure to PW.

Biliary metabolites of pyrene and B(a)P were also elevated at all locations, relative to each species' respective reference group (Fig. 15.3b, c). The rainbow runners at the Wandoo A reference site had a noticeably high background level of biliary metabolites, which might be due to a high naturally occurring bile fluorescence in this fish species, or to the presence of trace amounts of PAHs at this site. There is a small possibility that some PAHs from the Wandoo B PW discharge do reach the Wandoo A monopod located 5 km away; however, this would be unexpected as the ocean currents take the PW plume away from Wandoo A. In addition, assessments of hydrocarbon dispersion in the North West Shelf have shown that at distances greater than 1 km, the hydrocarbons from the PW discharge are well mixed throughout the water column (Holdway and Heggie, 1998). However, it is possible that the ocean

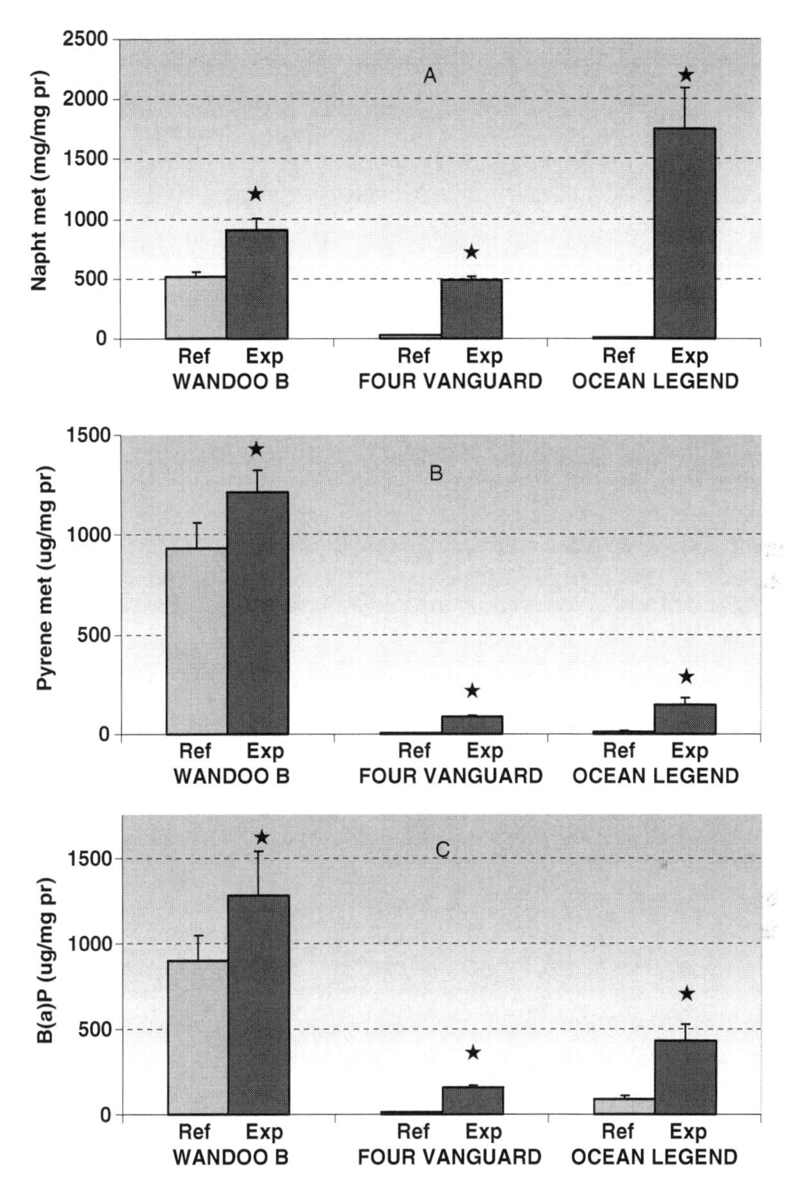

Fig. 15.3 PAH biliary metabolites (mean ± SE) of the (**a**) naphthalene-type, (**b**) pyrene-type, (**c**) B(a)P-type measured in rainbow runners (Wandoo B), in gold-banded snapper (Four Vanguard) and in giant trevally (Ocean Legend) facilities; star indicates that the exposed fish are significantly different ($p < 0.05$) to the reference fish

currents at this location are highly influenced by strong tidal movements causing the same water mass to be relocated in the near field of the Wandoo A platform. This would result in the higher than expected biliary metabolite measurements in the Wandoo A fish.

It has been suggested that in environments where fish are exposed to high PAH levels, a greater P450 liver detoxification activity could contribute to higher levels of DNA-reactive PAH metabolites (van der Oost et al. 2003). Maccubbin (1994) showed that in fish, there is a good correlation between the P450-generated PAH metabolites and carcinogenic potency. It is therefore relevant to measure the integrity of the DNA molecule in fish chronically exposed to PAHs discharged from the offshore petroleum production facilities.

The DNA integrity was reduced in the Wandoo B rainbow runners as well as in the gold-banded snapper collected at the Four Vanguard facility (Fig. 15.4). Coincidently these two offshore facilities where DNA integrity was reduced also had elevated EROD induction levels although no significant correlations were found between these two biomarkers.

DNA damage can be caused not only by pollutants but also by ionizing radiation or heat shock (Shuggart 2000). In order to cope with frequently occurring DNA damage, animals have evolved DNA repair mechanisms involving enzymatic processes that are rapidly activated following a genotoxic damage (Shuggart 2000). Organisms' ability to repair DNA damage might vary in each species, which creates difficulties in correlating DNA damage with other biological responses. However, the observation of higher DNA damage in PW-exposed fish indicates that the anthropogenically enhanced and possibly, naturally occurring genotoxicant exposure levels encountered surpasses the ability of these fish to repair DNA damage in the long term.

Stress proteins are inducible proteins involved in the protection and repair of the cell in response to stress and harmful conditions such as high or low temperatures, anoxia and xenobiotics (van der Oost et al. 2003). Correlations have been established between DNA damage and HSP70 levels in fish (Schröder et al. 2000); however, multiple factors that can affect HSP70 levels (van der Oost et al.

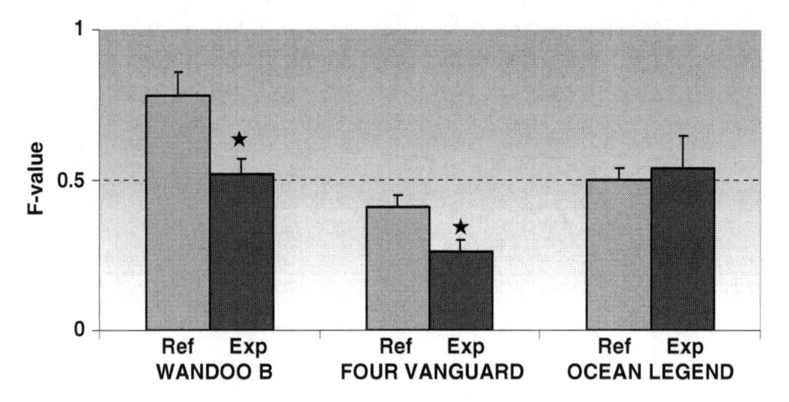

Fig. 15.4 DNA integrity (as calculated by the *F*-value, mean ± SE) measured in the liver of rainbow runners collected at Wandoo B, gold-banded snapper from the Four Vanguard FPSO, and giant trevally from the Ocean Legend facility; star indicates that the exposed fish are significantly different ($p < 0.05$) to the reference fish

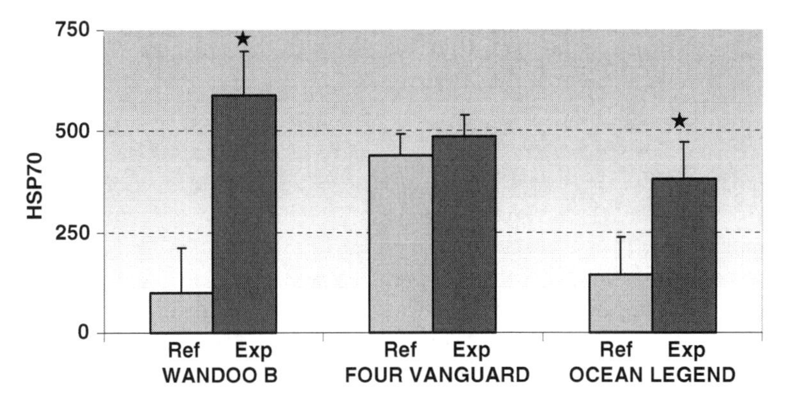

Fig. 15.5 HSP70 levels (in pixels, mean ± SE) measured in fish at three offshore facilities in the North West Shelf of Australia; star indicates that the exposed fish are significantly different ($p < 0.05$) to the reference fish

2003) preclude the conclusion of a direct link between these two biomarkers. In the present study, no significant correlations were established between DNA damage and HSP70 levels, although HSP70 levels were strongly induced at the Wandoo B as well as at the Ocean Legend facilities (Fig. 15.5).

The gold-banded snapper collected at the Four Vanguard FPSO prior to the discharge of PW appeared to have naturally high levels of HSP70, which might be normal for this species, or might be related to the sensitivity of this species to non-contaminant parameters such as nutrition or water temperature. Combined to the high inter-individual variability in HSP70 levels, the influence of confounding factors makes the interpretation of HSP70 levels difficult (Webb and Gagnon, 2009); however, HSP70 levels provide evidence of a metabolic perturbation occurring in exposed fish at two of the locations sampled.

5 Conclusions

Various physiological parameters and biomarkers of fish health have been measured in three species of fish collected in the vicinity of offshore petroleum production platforms located in the North West Shelf of Australia. The induction of these biomarkers varied according to each location, with PAHs biliary metabolites of the naphthalene-, pyrene- and B(a)P-types being the most consistent biomarkers of exposure to PW discharges. Not only the PAH biliary metabolites provided the best discriminatory power between reference and exposed fish, they also had the highest x-fold induction, relative to reference fish, of all biomarkers measured. EROD activity showed similar trends to DNA damage both temporally and geographically, for all fish species. Stress proteins HSP70 did not follow other biomarker trends, maybe due to the influence of confounding factors inducing stress proteins independently of xenobiotics exposure.

The fish collected at the facility with the highest volume of PW discharge and producing the heaviest crude oil had the largest number of induced biomarkers, with all but CF being different between reference and exposed site. Fish collected at the two other facilities with similar PW volumes and crude oil type had varied biomarker response. The variation in biomarker responses at different sites highlights the need to evaluate the impact of an offshore production facility on a case-by-case basis. The results of the present study suggest that larger contaminant load (PW volume $\times$ petroleum hydrocarbon concentration) results in a clearer biomarker response in fish; however, the biochemical responsiveness of individual fish species might also influence the conclusions of an environmental effect assessment. It is therefore recommended not only to measure a suite of biomarkers of fish health but where possible, to measure these biomarkers in two species of fish collected simultaneously from the same location in order to offset the sensitivity, or the lack of biochemical responsiveness, of individual fish species.

Acknowledgements The author thanks the Environment Managers of the three facilities involved in this study: Namek Jivan for the Wandoo B facility (Vermilion Oil & Gas Australia), Sue Capper for the Four Vanguard FPSO (ENI Australia) and Myles Hyams for the Ocean Legend offshore platform (Woodside Energy). Their support with the study has been much appreciated. Thanks are also directed to staff who provided their time and efforts in fishing and collecting biopsies, especially Dr Diane Webb and Dr Peter Sheppard.

References

Black KP, Brand GW, Grynberg H, Gwyther D, Hammond LS, Mourtikas S, Richardson JB, Wardrop JA (1994) Production facilities. In: Swan JM, Neff, JM, and Young PC (eds) Environmental implications of offshore oil and gas development in Australia – the findings of an independent review, Australian Petroleum Production Association, Sydney, pp 209–407

Borseth JF, Grosvik BE, Camus L, le Floch S, Gaudebert B (2000) Biomarkers: a new approach to assess environmental impact of E&P activities. SPE 61202. In: SPE international conference on health, safety and environment, oil and gas exploration and production, Stavanger, Norway, June 2000

Cavanagh JE, Burns KA, Brunskill GJ, Coventry RJ, Ryan D, Ahokas JT (2000) Induction of hepatic CYP 1A in Pikey Bream (*Acanthopagrus berda*) collected from agricultural and urban catchments in far north Queensland. Mar Pollut Bull 41:377–384

Cobby G (2002) Changes to the environmental management of produced formation water, offshore Australia. APPEA J 40:677–682

Collier TK, Varanasi U (1991) Hepatic activities of xenobiotics metabolising enzymes and biliary levels of xenobiotics in English sole (*Parophrys vetulus*) exposed to environmental contaminants. Arch Environ Contam Toxicol 20:462–473

Durell G, Utvik TR, Johnsen S, Frost T, Neff J (2006) Oil well produced water discharge to the North Sea. Part I. Comparison of deployed mussels (*Mytilus edulis*), semi-permeable membranes devices, and the DREAM model predicitons to estimate the dispersion of polycyclic aromatic hydrocarbons. Mar Environ Res 62:194–223

Gagnon MM, Holdway DA (2002) EROD Activity, serum SDH and PAH biliary metabolites in sand flathead (*Platycephalus bassensis*) collected in Port Phillip Bay, Victoria. Mar Pollut Bull 44:230–237

George SG, Wright J, Conroy J (1995) Temporal studies of the impact of the Braer oil spill on inshore feral fish from Shetland, Scotland. Arch Environ Contam Toxicol 29:530–534

Glubokov AI, Orlov AM (2008) Data on distribution and biology of poachers agonidae from the northwestern part of the Bering Sea. J Ichthyol 48:426–442

Hodson PV, Dodson JJ, Bussières D, Gagnon MM (1993) Fish population monitoring – how many fish are enough? In: Proceedings of the 20th aquatic toxicity workshop, October 21, 1993, Canada, pp 43–47

Holdway D, Heggie DT (1998) Tracking produced formation water discharge from a petroleum production platform to the North West Shelf. APPEA J 38: 665–679

Jönsson EM, Abrahamson A, Brunström B, Brandt I (2006) Cytochrome P4501A induction in rainbow trout gills and liver following exposure to waterborne indigo, benzo[*a*]pyrene and 3,3',4,4',5-pentachlorobiphenyl. Aquat Toxicol 79:226–232

Krahn MM, Burrows DG, Ylitalo GM, Brown DW, Wigren CA, Collier TK, Chan S-L, Varanasi U (1992) Mass spectrometric analysis for aromatic compounds in bile of fish sampled after the Exxon Valdez oil spill. Environ Sci Technol 26:116–126

Lowry OH, Rosebrough NJ, Farr AL, Randall RJ (1951) Protein assessment with Folin-phenol reagent. J Biol Chem 193:265-275

Lin ELC, Cormier SM, Torsella JA (1996) Fish biliary polycyclic aromatic hydrocarbon metabolites estimated by fixed-wavelength fluorescence: comparison with HPLC-fluorescent detection. Ecotoxicol Environ Saf 35:16–23

Maccubbin AE (1994) DNA adduct analysis in fish: laboratory and field studies. In: Malins DC, Ostrander GK (eds) Aquatic toxicology; molecular, biochemical and cellular perspectives. Lewis Publishers, CRC press, Boca Raton, FL, pp 267–294

Mayer FL, Versteeg DJ, McKee MJ, Folmar LC, Graney RL, McCume DC, Rattner BA (1992) Metabolic products as biomarkers. In: Huggett RJ, Kimerly RA, Mehrle PM Jr, Bergman HL (eds) Biomarkers: biochemical, physiological and histological markers of anthropogenic stress. Lewis Publishers, Chelsea, MI, pp 5–86

Neff JM, Johnsen S, Frost TK, Utvik TIR, Durell GS (2006) Oil well produced water discharges to the North Sea. Part II: comparison of deployed mussels (*Mytilus edulis*) and the DREAM model to predict ecological risk. Mar Environ Res 62:224–246

Rawson CA, Tremblay LA, St.J.Warne M, Ying G, Kookana R, Laginestra E, Chapman JC, Lim RP (2009) Bioactivity of POPs and their effects on mosquitofish in Sydney Olympic Park, Australia. Sci Tot Environ 407:3721–3730

Schröder HC, Batel R, Hassanein HMA, Lauenroth S, Jenke H-St, Simat T, Steinhart H, Müller WEG (2000) Correlation between the level of the potential biomarker, heat-shock protein, and the occurrence of DNA damage in the dab, *Limanda limanda*: a field study in the North Sea and the English Channel. Mar Environ Res 49:201–215

Shuggart LR (1996) Application of the Alkaline unwinding assay to detect DNA strand breaks in aquatic species. In: Ostrander GK (ed) Techniques in aquatic toxicology. CRC press, Boca Raton, FL, pp 205–218

Shuggart LR (2000) DNA damage as a biomarker of exposure. Ecotoxicology 9:329–340

Stein JE, Percic P, Gnassia-Barelli M, Romeo M, Lafaurie M (1998) Evaluation of biomarkers in caged fish and mussels to assess the quality of waters in a bay of the N.W. Mediterranean Sea. Environ Health Perspect 90:101–109

Van der Oost R, Beyer J, Vermeulen NPE (2003) Fish bioaccumulation and biomarkers in environmental risk assessment: a review. Environ Toxicol Pharmacol 13:57–149

Varanasi U, Stein JE, Nishimoto M (1989) Biotransformation and disposition of polycyclic aromatic hydrocarbons (PAH) in fish. In: Varanasi U (ed) Metabolism of polycyclic aromatic hydrocarbons in the environment. CRC Press, Boca Raton, FL, pp 93–150

Vermeulen NPE, Donne-Óp den Kelder G, Commandeur JNM (1992) Formation of and protection against toxic reactive intermediates In: Testa B, Kyburz E, Fuhrer W, Giger R (eds) Perspectives in medicinal chemistry. Verlag Helvetica Chimica Acta, Basel, pp 573–593

Webb D, Gagnon MM (2009) The value of stress protein 70 as an environmental biomarker of fish health under field conditions. Environ Toxicol 24:287–295

Chapter 16
Effect of Produced Water on Innate Immunity, Feeding and Antioxidant Metabolism in Atlantic Cod (*Gadus morhua*)

Dounia Hamoutene, H. Volkoff, C. Parrish, S. Samuelson, G. Mabrouk, A. Mansour, Ann Mathieu, Thomas King, and Kenneth Lee

Abstract Emerging evidence from North Sea investigations indicates that the discharge of produced water (PW) may impact biota over greater distances from operational offshore platforms than originally predicted. We have investigated the effects of PW on cod immunity, feeding and general metabolism by exposing fish to diluted PW at concentrations of 0, 100 and 200 ppm for 76 days. No significant differences were observed in weight gain or food intake. Similarly, serum metabolites, whole blood fatty acid percentages and mRNA expression of a brain appetite-regulating factor (cocaine- and amphetamine-regulated transcript) remained unchanged between groups. Other than an irritant-induced alteration in gill cells found in treated cod, resting immunity and stress response were not affected by PW. Catalase and lactate dehydrogenase changes in activities were recorded in livers but not in gills, suggesting an effect on oxidative metabolism subsequent to hepatic detoxification processes. At the end of the exposure, fish from the three groups were challenged by injection of *Aeromonas salmonicida* lipopolysaccharides (LPS). LPS injection affected respiratory burst activity of head-kidney cells, and circulating white blood cells ratios, and increased serum cortisol in all groups. The most pronounced changes were seen in the group exposed to the highest PW dose (200 ppm). Our results indicate an effect of PW on cod immunity after immune challenge with LPS as well as an impact on the liver oxidative metabolism.

1 Introduction

Operational discharges from offshore oil and gas production sites include produced water (PW), drilling fluids, mud and cuttings, and oils. PW is the largest volume waste stream in the exploration and production processes. It contains a complex mixture of aliphatic, aromatic and polar compounds. Dispersion modelling of PW

D. Hamoutene (✉)
Science Branch, Northwest Atlantic Fisheries Institute, Fisheries and Oceans Canada, St John's, NL, Canada

K. Lee, J. Neff (eds.), *Produced Water*, DOI 10.1007/978-1-4614-0046-2_16,

discharged into surface waters has indicated that rapid dilution by at least 240× occurs within 50–100 m and up to 9000 × 20 km from the point of discharge (Somerville et al. 1987; Murray-Smith et al. 1996).

The co-existence of the oil industry activities with fish, fish habitat and commercial fishing is dependent on our understanding of harmful cumulative effects of PW discharges on marine organisms around production sites and the development of improved mitigation technologies. Numerous studies have reported that the overall toxicity of PW to marine organisms is low, and acute effects would likely only be seen within the immediate mixing zone (Aas and Klungsoyr 1998; Stephens et al. 2000). Few data are available regarding the chronic effects of PW on long-lived marine organisms. Atlantic cod is a species of considerable ecological and economic importance for Canada. Examining the chronic effects of PW on cod growth, feeding and disease resistance will shed light on the potential long-term impact of oil and gas offshore developments on stocks.

The importance of immune responses in fish disease susceptibility and its potential effect on population decline are well known (e.g. Gulland 1995). Chronic exposure of aquatic animals to low concentrations of toxicants can predispose them to debilitating diseases (Weeks and Warinner 1984), which would affect their growth and reproductive capacity (Kiceniuk and Khan 1987). In particular, a recent study by Bohne-Kjersem et al. (2009) on the effect of a surrogate produced water (crude oil spiked with alkyl phenols and polycyclic aromatic hydrocarbons) on cod shows that recorded changes in protein expression profiles are linked to the complement system, the immune system, as well as increased oxidative stress, impaired cell mobility and fertility-linked mechanisms. To better appreciate the significance of PW dilution and the potential vertical and horizontal migratory behaviour by cod (Turner et al. 2002) in contact with PW discharges in the open ocean, we conducted an experiment that incorporated pulse exposures to low levels of PW. At the end of the exposures, cod fish were injected with lipopolysaccharides from *A. salmonicida*. Injection of fish with an inactivated form of the fish-specific pathogen *A. salmonicida* activated their immune system to enable investigation of the effects of effluents on reactivity against threatening diseases without causing death of the test animals by direct exposure to pathogens (Hoeger et al. 2004).

Metabolic capacities, feeding and digestive physiology of fish are influenced by environmental parameters. Chronic exposure to PW might potentially affect feeding, which in turn might have drastic consequences on growth/health status of fish populations. Food intake and nutritional status of cod can be assessed by using body morphometrics as well as measurements of nonesterified fatty acids (NEFAs), total lipids, triacylglycerol and phospholipids according to Alkanani et al. (2005). Plasma lipids reflect the energetic/nutritional status of gadoids and have previously been used to monitor responses to changing environmental conditions in wild cod. Feeding in fish is regulated by a number of appetite-controlling hormones secreted by both brain and peripheral organs (Volkoff et al. 2005). Little is known about the effects of environmental changes on these factors, although it appears that certain toxins affect their gene expression (Volkoff and Peter 2004). This study assesses the effects of PW on fish feeding physiology by examining possible changes in the brain gene expression of appetite-related hormones. Overall, the current study therefore brings information on potential chronic effects of PW on immune parameters, feeding, antioxidant metabolism and general stress responses in cod.

2 Material and Methods

2.1 Fish Exposure

Atlantic Cod (*Gadus morhua*) were obtained from the Cod genome research group (Joe Brown Aquatic Research Building, Ocean Sciences Centre, Newfoundland). Fish were transferred to three tanks (820 l tanks, 37 fish per tank, average weight of 774 g) and acclimatized for 2 months. Approximately 10 fish per tank were tagged with pit tags in order to assess growth rates. After a month of acclimatization, baseline data including weight, length (tail and fork length) and liver weight were recorded and blood and gill tissues were collected from 5 fish per tank, for a total of 15 fish. Enzymatic data, as well as general blood parameters (see below), were similar for all fish. This baseline data allowed us to verify that there were no tank effects prior to test-exposure treatments. By the end of the second month of acclimatization, intermittent cod fish exposure was initiated three times a week to produced water (representative sample from the discharge line obtained directly from the Hibernia offshore platform on the Grand Banks of Newfoundland) diluted to concentrations of 100 ppm (10,000× dilution; Ea) and 200 ppm (5000× dilution; Eb) for a total period of 76 days. PW was sampled in August 2006 and kept in Nalgene high-density polyethylene jerricans provided to the Hibernia platform (Newfoundland, Canada) staff for collection. Instructions for collection were to fill the jerricans to overflowing to eliminate any headspace. Collection of PW was coordinated so that the samples were returned to scientific staff on board the oceanographic vessel CCGS Hudson (Fisheries and Oceans Canada) as soon as logistically possible. PW was conserved in the fridge and transferred to the Northwest Atlantic Fisheries Center within a week and used immediately upon arrival (at 4°) to initiate cod exposures. All samples were thoroughly mixed before pulse exposures. Based on the flow rates of the tanks, fish would be exposed to PW for no more than 5 h per time (pulse). The consumption of commercial feed pellets was monitored throughout the experiment. At the end of the experiment, 12 cod were sacrificed and investigations pertaining to feeding, immunity, general stress levels and gill enzyme and pathology were conducted. At the end of the 76 days exposure to PW, 10 fish were kept from each experimental tank and injected with lipopolysaccharides (LPS) from atypical *A. salmonicida*. Fish were sacrificed 24 h post-injection and similar investigations as listed for the first group of fish were completed. LPS from *A. salmonicida* was obtained from Dr Joe Banoub (Fisheries and Oceans Canada, St John's, NL) prepared as described in Banoub et al. (2004). Fish were injected intraperitoneally according to Seternes et al. (2001) with 0.5 μg/kg of LPS (200 μl of total injection volume).

To assess the effects of the LPS injection without PW exposure on fish metabolism, another experiment was conducted. A total of 36 fish (average weight of 52 g) were divided into three groups of 12 fish each (a control untreated group, a saline-injected group and a group injected with LPS). Investigations on immune and stress responses were completed on these fish as described below.

2.2 Feeding and Appetite Monitoring

Feeding of fish was monitored daily by weighing pellets consumed at every feeding time. Food intake was calculated by dividing the weight of the pellets eaten by the average weight of the fish and the number of fish per tank. Individual growth ($n = 10$ per treatment) was monitored by calculating growth rates (GR) using weights and fork lengths: $GR = \left[(\ln X2 - \ln X1)/t\right] \times 100$, where $X1$ and $X2$ are the weight/length of the fish (e.g. Bjornsson et al. 2007) at the start and at the end of the experiment, respectively, and t the duration of the experiment (i.e. 76 days). As part of the general assessment of the potential effect of PW on appetite, lipid profiles of blood samples were performed ($n = 10$ per treatment). Lipid samples were extracted using liquid–liquid extraction and concentrated using a flash-evaporator. Lipid class composition was determined using an Iatroscan Mark V TLC-FID, after separation on silica-coated Chromarods according to a three-step development method (Parrish 1987). For all samples, lipid extracts were transesterified using 14% BF3/MeOH for 1.5 h at 85°C. This method has been shown to derivatize over 90% of the acyl lipids. Fatty acid composition was determined by GC-FID.

For evaluation of cocaine and amphetamine regulated transcript (CART; a brain appetite-regulating factor) expression in fish brains, fish were sacrificed with a blow on the head ($n = 6$–8 per treatment). Whole brains were dissected and stored at –20°C in RNAlater (Qiagen) until RNA isolations were performed. Total RNA was isolated from the forebrain using a Trizol/chloroform extraction method with Tri-Reagent (BioShop, Burlington, Ontario, Canada) following the manufacturers protocol. Final RNA concentrations were determined by spectrophotometric readings at 260 nm. Gene expression of CART was performed by semi-quantitative RT-PCR relative to a reference gene, elongation factor, EF1α. PCR primers were designed based on cDNA sequences available for Atlantic cod CART cod peptides (GenBank accession numbers AY822596 and DQ256082, Kehoe and Volkoff 2007) and EF1α (GenBank accession number CO541952) peptides.

2.3 Immunity

Prior to fish sacrifice, blood samples were collected using previously heparinized 5 cc syringes, diluted in isolation buffer (MCHBSS + Alsever's buffer) (Crippen et al. 2001), and kept for respiratory burst (RB) response evaluation. Dissociated head-kidney leucocytes were obtained by pressing through a nylon screen (50–60 μm) in the presence of isolation buffer. The cell suspension was layered on a Percoll gradient 33/51% in PBS and NaCl 1.8% centrifuged at $400\times g$ for

30 min according to Stenvik et al. (2004). The cells at the interface were recovered and kept on ice or at 4°C until use for RB. Oxidative burst was quantified using a flow cytometer as a measure of intracellular hydrogen peroxide production following activation with PMA (1 μg/ml). Flow cytometric assessment of the RB was based on the technique described by Bass et al. (1983). The assay depends upon the cell incorporating 2′–7′ dichlorofluorescin diacetate (DCFH-DA), a nonfluorescent molecule which is oxidized to fluorescent DCF. Briefly, isolated circulating leukocytes (or whole blood) were incubated with DCFH-DA (5 μM) at room temperature and fluorescence measurements were done to acquire baseline fluorescence levels. PMA was added and fluorescence was measured immediately ("time 0") and 10 min after cell stimulation. Between 10,000 and 30,000, events were included in the analysis of every blood sample. For each sample, a stimulation index was determined as the ratio of fluorescence of PMA stimulated cells (10 min) to that of cells at "time 0". Cellular debris was excluded from the analysis by raising the forward-scatter threshold only minimally. The number of samples processed were $n = 12$ for exposures (Cont, 100 ppm and 200 ppm) and $n = 10$ for fish injected with LPS (Cont+LPS, 100 ppm + LPS, 200 ppm + LPS).

Measures of proteins and cortisol were also performed on serum samples of control fish and fish exposed to PW/LPS or saline/PW + LPS. Protein and lactic acid analyses were carried out on a Beckman LX automated analyzer at the haematology/biochemistry laboratory of the General Hospital, St John's, Newfoundland. Cortisol was measured using a Unicel Dxl800 Access® immunoassay analyzer with a Cortisol immunoassay kit (Beckman).

Blood smears (Wright-Giemsa) were prepared and counts were performed on 200 white blood cells to evaluate percentages of lymphocytes, thrombocytes, neutrophils, granulocytes and monocytes/macrophages (n = 8–12 fish per case).

2.4 Enzymatic Analysis

Tissue (liver and gill) samples were homogenized in 10 volumes of Tris- Buffer (0.05 M Tris, pH 7.2) ($n = 12$ per treatment). Homogenates were centrifuged at $9000 \times g$ for 20 min. The supernatant (S9) was retained and stored at –80°C for further analyses. Catalase (CAT) activity was measured according to the procedure originally devised by Chance and Herbert (1950). Lactate dehydrogenase (LDH) activity was evaluated as described by Mitchell et al. (1980). Protein levels were subsequently measured according to Lowry's assay (Lowry et al. 1951).

2.5 Gill Histology

Gill tissues fixed in SAFEFIX II were processed for histological analysis (Lynch et al. 1969) using a Tissue-Tek® VIP Processor. Sections were stained with Mayers Haematoxylin and Eosin (H&E) and Periodic Acid-Schiff (PAS) (Luna 1968). Gill sections were examined by transmission light microscopy and analysed for epithelial lifting, telangiectasis, lamellar hyperplasia, fusion and oedema as well as any other abnormalities. Sections ($n = 12$ per treatment) were also stained with PAS and examined for the presence of mucous cells.

2.6 Chemical Analysis

Produced water samples ($n = 2$) were analysed at the Bedford Institute by GC-MSD for phenols and PAHs. PW was sampled in August 2006 and analysed within a few days to avoid degradation.

2.7 Statistical Analysis

Comparisons between groups were completed by ANOVAs (one way) followed by a multiple comparison test: Holm-Sidak or *t*-tests with a $p < 0.05$ for significance. To ensure that sex or sampling day had no effects on parameters examined, two-way ANOVAs were used with two factors: sex or sampling day and treatment.

3 Results

3.1 Feeding and Appetite Monitoring

Fish used for the PW exposure attained sexual maturity during the course of the experiment (naïve spawners). To ensure sex had no effect on the experiment, two-way ANOVAs were performed on all parameters investigated. No effect of sex was found other than a difference in hepatosomatic indices (HSI) between males and females. HSI data were divided into data for males and females, but no effect of PW was found when comparing HSI data of controls (males or females) with exposed fish (males or females). Similarly, PW exposure did not affect the other morphometric parameters investigated (total body weight, fork length, condition factor; data not shown), growth or food intake (Fig. 16.1a and b). When investigating gene expression of CART, an appetite regulating hormone, no significant differences were observed between control and exposed fish (Fig. 16.2). Lipid analysis (% main lipid classes as well as % major fatty acids) revealed no differences between fish from control or exposed groups. Total lipid concentrations were in the range of 3–4 mg/ml with major lipid classes comprising sterol, phospholipid, triacylglycerol and ethyl ketone, and major fatty acids comprising 22:6ω3, 20:5ω3, 18:1ω9 and 16:0 in accord with Alkanani et al. (2005).

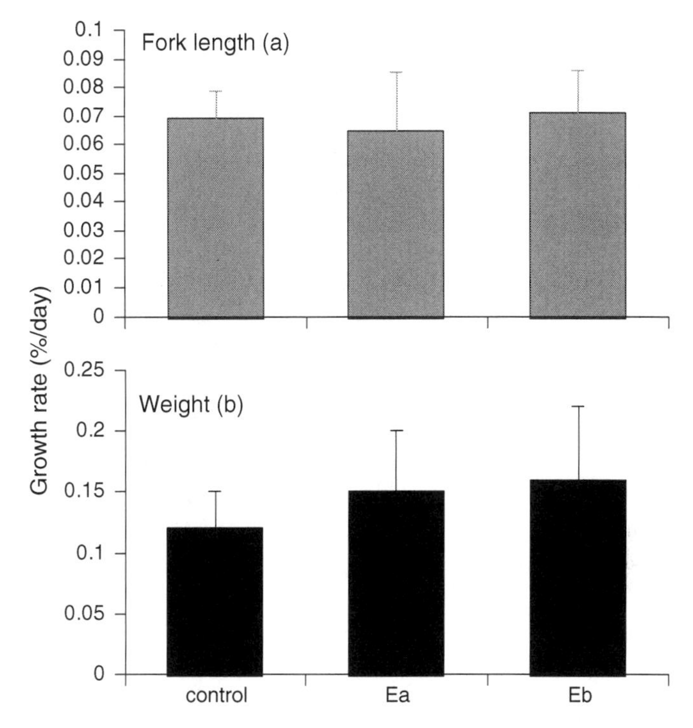

Fig. 16.1 Growth rates (GR) in terms of fork length (**a**) and weight (**b**) [GR $= \left[(\ln X2 - \ln X1)/t\right] \times 100$, where $X1$ and $X2$ are the weight/length of the fish at the start and at the end of the experiment, respectively, and t is the duration of the experiment (76 days), $n = 25$], Ea = 100 ppm, Eb = 200 ppm

3.2 Immunity

3.2.1 Effect of Saline and LPS Injection

To ensure proper understanding of LPS injection on cod metabolism, an experiment was set up to inject 12 fish with saline or LPS and compare immune responses between injected and non-injected fish. Fish used for this trial were significantly smaller (50–100 g) than the ones used for the PW exposure and were all sexually immature. Considering the number of analyses to be conducted, fish sacrifices had to be completed in 2 days with six fish sacrificed per day. As stated in Section 2, a two-way ANOVA was performed to ensure that differences were not masked by effect of "sampling day". No effect of the sampling day was found allowing us to group results of the 2 days (Table 16.1).

Results show that LPS injections induce a significant decrease in lactate and a trend towards increased cortisol concentrations. RB of whole blood is unaffected while a significant decrease in RB HK can be observed after LPS injection.

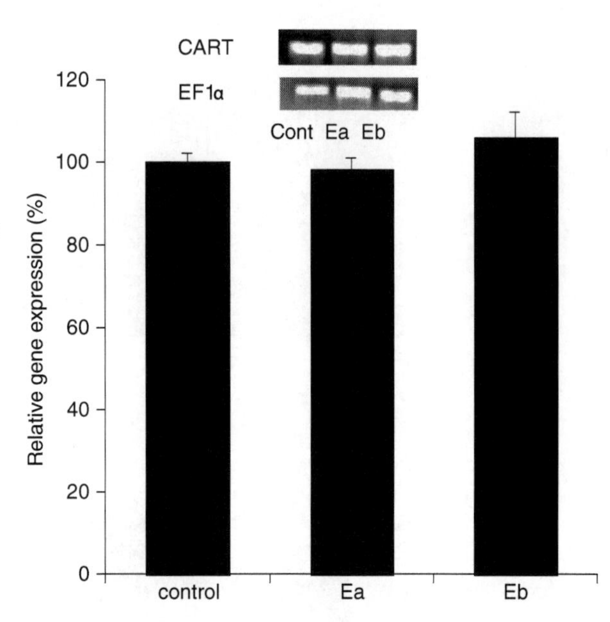

Fig. 16.2 Changes of CART mRNA expression within the forebrain; data are presented as mean ± SEM ($n = 6$–8); insets show representative gels (*upper panel*: CART; *lower panel*: EF1α); Ea = 100 ppm, Eb = 200 ppm

Table 16.1 Blood parameters investigated after fish injection with LPS or saline solutions

Parameter investigated (serum, blood, HK)	Control ($n = 12$)	Saline injected ($n = 12$)	LPS injected ($n = 12$)
Lactate (mmol/l)	1.60 ± 0.26	1.32 ± 0.46	0.95 ± 0.13*
Cortisol (nmol/l)	66.67 ± 27.00	134.00 ± 70.17	124.33 ± 75.94
RB whole blood	5.71 ± 1.86	5.16 ± 0.85	5.01 ± 1.25
RB HK	3.72 ± 1.03	4.93 ± 3.25	2.96 ± 0.77*

RB respiratory burst expressed as ratios of stimulation; *HK* head-kidney
*Statistically significant difference when compared to controls (ANOVA, Holm-Sidak, $p < 0.05$)

3.2.2 Effect of PW and LPS

Similar to investigations carried out on feeding and nutrition, all data collected for immunity were tested to ensure no effect of sex. Two-way ANOVAs were performed on all parameters investigated after fish sacrifice; no effect of sex was found. In addition, two-way ANOVAs were completed to test the effect of sampling day (fish were sacrificed during the course of 2 days); similarly, no effect of sampling day was observed.

Cell counts revealed an effect of LPS injection or a combined effect of PW and LPS injection on percentages of thrombocytes (Fig. 16.3a), monocyte/macrophages (Fig. 16.3b) and lymphocytes (Fig. 16.3c). No effect was seen on neutrophils or granulocytes. Data were compared in two ways, control fish vs. fish exposed to

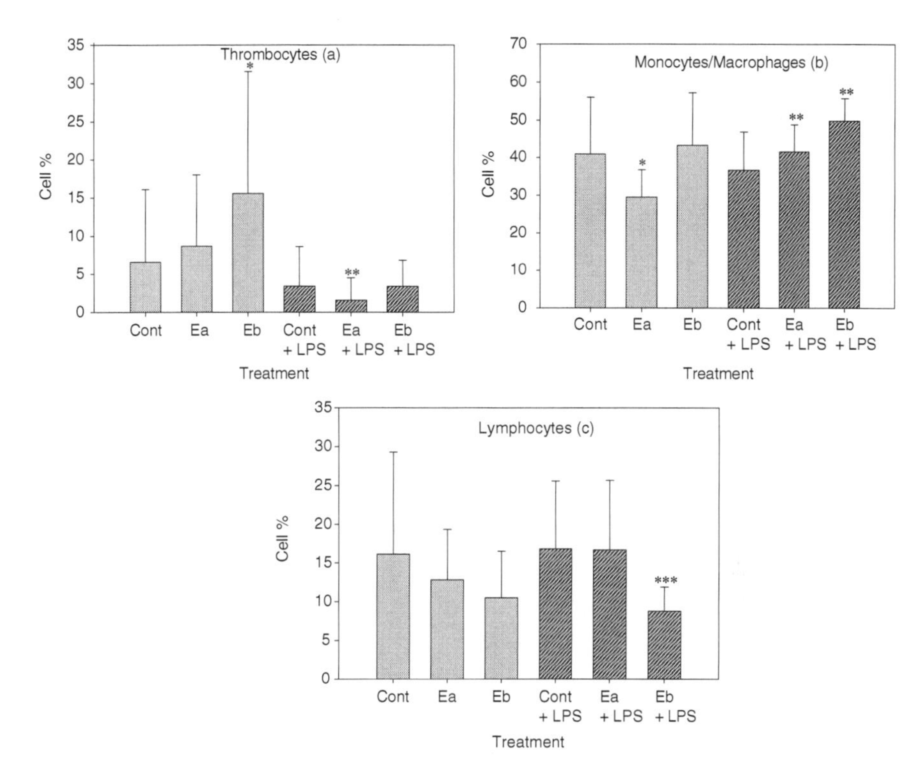

Fig. 16.3 White cell counts (**a** thrombocytes, **b** monocytes/macrophages, **c** lymphocytes) expressed as percentages sampled from fish kept in seawater and fish exposed to PW (Controls, Ea = 100 ppm, Eb = 200 ppm), as well fish injected with LPS from *A. salmonicida* (Cont + LPS, Ea + LPS, Eb + LPS); *statistically significant difference (ANOVA and Holm-Sidak, $p < 0.05$) when compared to non-exposed fish (Cont vs. Ea or Eb, Cont + LPS vs. Ea + LPS or Eb + LPS), **statistically significant difference (*t*-test, $p<0.05$) when compared to non-injected fish (Cont vs. Cont +LPS, Ea vs. Ea + LPS, Eb vs. Eb + LPS), ***statistically significant difference (*t*-test, $p<0.05$) following exposure to two concentrations of produced water (Ea, Eb)

100 ppm PW, or 200 ppm PW, as well as fish non-injected (controls, 100, or 200 ppm) vs. fish injected with LPS. An effect of PW was found with the highest dose in thrombocytes, and with 100 ppm in monocytes. Thrombocyte percentages showed a decrease due to the LPS injection while lymphocyte percentages were influenced by the combination of LPS and highest dose of PW.

Serum analyses revealed an effect of the LPS injection on cortisol and lactic acid (no effect on blood proteins) as seen in Fig. 16.4a and b respectively. It is important to note that the increase in lactic acid and cortisol was greater for the highest dose of PW, suggesting a potential effect of PW. RB ratios (Fig. 16.5a, b) show no significant effect of PW and/or LPS injection on whole blood (white and red blood cells) immune responses. On the other hand, a significant decrease in RB of head-kidney cells can be seen after LPS injection with a stronger decrease in the group exposed to the highest dose of PW.

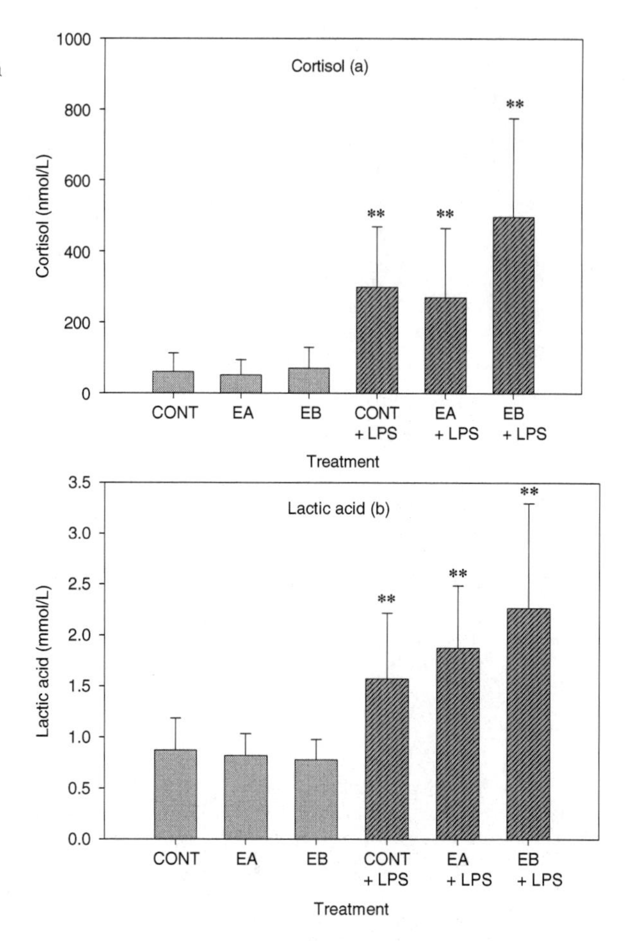

Fig. 16.4 Blood serum data (**a** cortisol, **b** lactic acid) from fish kept in seawater and fish exposed to PW (Controls, Ea = 100 ppm, Eb = 200 ppm), as well fish injected with LPS from *A. salmonicida* (Cont + LPS, Ea + LPS, Eb + LPS); **statistically significant difference (*t*-test, $p < 0.05$) when compared to non-injected fish (Cont vs. Cont +LPS, Ea vs. Ea + LPS, Eb vs. Eb + LPS)

3.3 Enzymatic Analysis

Results of catalase (CAT) and lactate dehydrogenase (LDH) activities in liver and gills show a significant decrease in CAT and an increase in LDH in liver samples (Table 16.2) after exposure to PW.

3.4 Gill Histology

The primary and secondary gill lamellae appeared normal in all fish and no cases of gill lesions associated with chemical toxicity such as epithelial lifting, telangiectasis, severe lamellar hyperplasia and extensive fusion and oedema, were observed. Mucous cells stain fuchsia in colour with the PAS method. Since the number of mucous cells was variable from one gill sample to another, sections were ranked

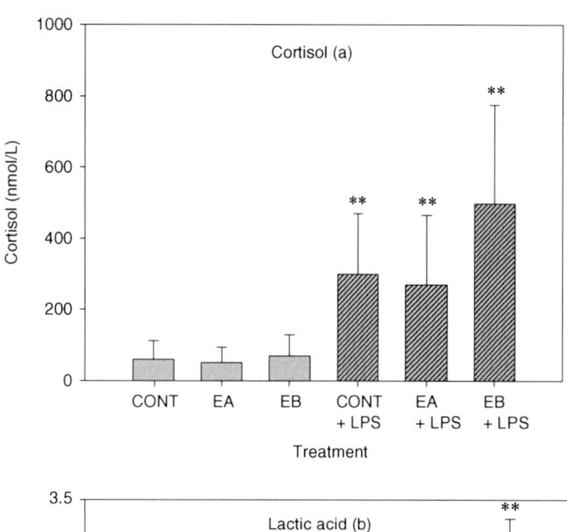

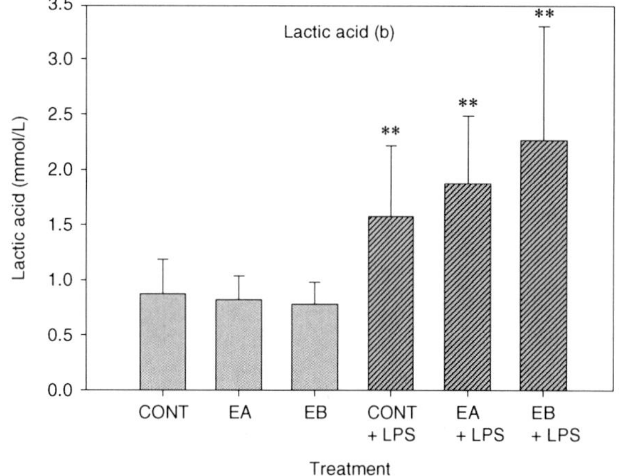

Fig. 16.5 Respiratory burst (RB) ratios (a-in whole blood, b-in head-kidney white blood cells) from fish kept in seawater and fish exposed to PW (controls, Ea = 100 ppm, Eb = 200 ppm), as well fish injected with LPS from *A. salmonicida* (Cont + LPS, Ea + LPS, Eb + LPS); **statistically significant difference (*t*-test, $p < 0.05$) when compared to non-injected fish (Cont vs. Cont +LPS, Ea vs. Ea + LPS, Eb vs. Eb + LPS)

Table 16.2 Enzymatic activities of catalase and lactate dehydrogenase, as well as protein amounts in liver and gill extracts (S9)

	Control ($n = 12$)	100 ppm ($n = 12$)	200 ppm ($n = 12$)
Liver CAT (units/mg prot)	930.0 ± 160.0	731.0 ± 145.0*	767.0 ± 196.0*
Gill CAT (units/mg prot)	334.0 ± 94.0	351.0 ± 74.0	336.0 ± 40.0
Liver LDH (nmol/min/mg prot)	1.7 ± 0.6	3.0 ± 1.4*	3.1 ± 1.7*
Gill LDH (nmol/min/mg prot)	49.9 ± 10.4	47.5 ± 11.8	49.7 ± 9.1
Liver PROT (mg/ml)	12.3 ± 5.1	11.6 ± 4.3	19.2 ± 5.7
Gill PROT (mg/ml)	5.7 ± 2.8	4.5 ± 1.1	3.8 ± 0.8

CAT catalase, *LDH* lactate dehydrogenase, *PROT* proteins
*Statistically significant difference (ANOVA, Holm-Sidak, $p < 0.05$) when compared to controls

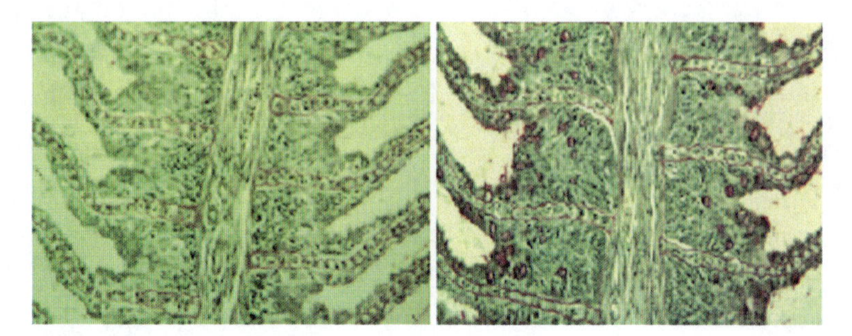

Fig. 16.6 Pictures of cod gill lamellae after PAS colouration observed under ocular microscopy (×250), (*left*) Mucous cells ranked 0, (*right*) Mucous cells ranked 3

Table 16.3 Mucous cell ranks in cod gills' lamellae

Comparison	Control ($n = 12$)	100 ppm ($n = 12$)	200 ppm ($n = 12$)	p value
All groups	1.5 ± 0.8	2.2 ± 0.9	2.2 ± 0.7	0.116
Exposed groups are pooled	1.5 ± 0.8	2.2 ± 0.8*		0.040

*Statistically significant difference when compared to controls (Mann-Whitney Rank Sum test, $p < 0.05$)

for the occurrence of these cells on a relative scale from 0 to 3, with 0 being the least (Fig. 16.6a) and three being the most (Fig. 16.6b). Despite a trend in rank increase, mean ranks did not differ significantly between the 3 groups (ANOVA on Ranks; $p = 0.116$). However, significant differences in the ranks were noted when comparison was carried out between the control and the two exposed groups pooled (Mann-Whitney Rank Sum test; $p = 0.040$: Table 16.3).

3.5 *Chemical Analysis of Produced Water Samples*

Chemical analysis of PW samples is summarized in Fig. 16.7. The detailed analyses of individual PAHs and alkanes determined in PW are withheld to maintain reasonable content. Compositional classes of components in PW were selected to provide clarity in the interpretation of toxic effects.

4 Discussion

Produced water consists of water trapped with oil and gas in the reservoirs together with water and process treatment chemicals injected during drilling to maintain reservoir pressure and stability of the recovery system. In addition to small amounts of dispersed oil, PW contains compounds of environmental concern such

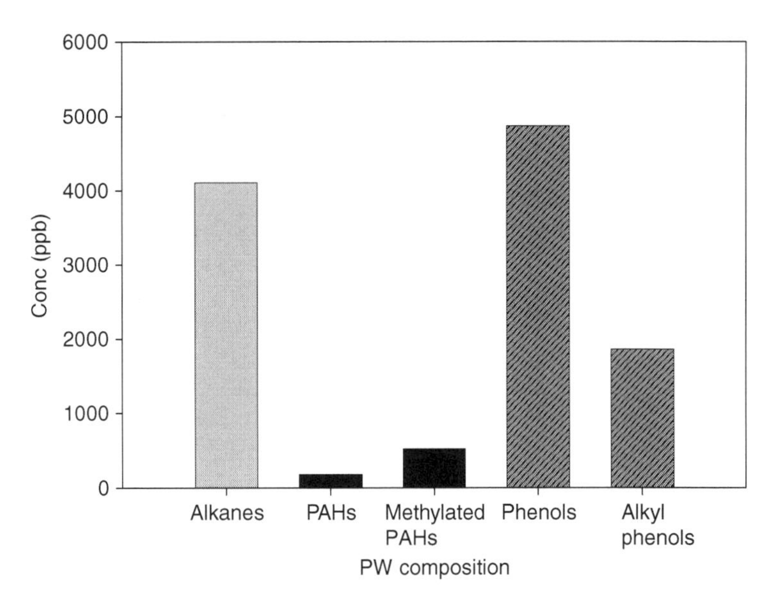

Fig. 16.7 Composition of two samples of PW sampled in August 2006 from Hibernia offshore platform, analysis is done by GC-MSD

as aromatic and aliphatic hydrocarbons, phenols (in particular alkylphenols), metals and traces of chemicals added during production (Roe 1998). It is important to note the fact that PW samples acquired from the oil platforms were kept at 4°C during the duration of the exposure and therefore underwent a degradation process. The handling of PW, which may involve depressurization, aeration, transport, freezing, storage, thawing and more, causes the volatile compounds to escape from solution and some of the more labile compounds to be degraded (Sundt et al. 2009). Sundt et al. (2009) have demonstrated that the use of freezing at –20°C has improved the stability of the more prevalent PAHs (e.g. naphthalenes) when diluting PW for cod exposures. Similarly, Stephens et al. (2000) reported that when using PW kept at 4°C, hydrocarbon content in the exposed fish showed a general decline as the exposure elapsed. It is probably the case in our study too; nonetheless, further research is needed to ensure that freezing/thawing episodes do not impact the other components of PW. Moreover, the weathering process that will take place in the natural environment (platform/field dependant) happens as soon as PW is released; therefore, a certain degree of degradation may be more environmentally realistic than the use of a fresh batch of PW. In addition, it is not unreasonable to assume that pelagic fish may avoid the immediate vicinity of the point of release of PW therefore being exposed to a more degraded (and diluted) PW effluent. More field observations are needed to complete this picture.

This study shows that intermittent exposure of cod fish to weathered PW at environmentally relevant concentrations provoked mild effects on fish. There was no mortality, no effects on feeding and nutrition were observed, and no significant impairment of innate immunity was recorded. Nonetheless, an indication of increased number of mucous cells (or mucous cell hyperplasia) was seen in gills of the treated cod. Mucous cell hyperplasia has been shown to be a common chronic response to injury by a number of toxicants, irritants and infectious agents (Myers and Fournie 2002) and has also been reported in association with crude oil toxicity (e.g., Khan and Kiceniuk 1984). In addition, a non-significant dose-dependent decrease in circulating lymphocytes, accompanied by an increase in thrombocytes, can be observed. Lymphocytes are immunocompetent cells with a well-known antibody synthesizing capacity (e.g. Schreck 1996). Cortisol directly induces apoptosis in lymphocytes so fish stressed by disturbance will show reduced numbers of these cells (Weyts et al. 1998). No significant changes in cortisol were found after PW exposure suggesting no strong stress response at this stage. These results imply a limited effect of PW on cod resting immunity other than an irritant-induced alteration in gill cells. A significant decrease in liver CAT and an increase in liver LDH are observed after exposure with both doses of PW. The fact that no changes of CAT and LDH activities are recorded in gills suggests that the observed effect on oxidative metabolism is subsequent to potential detoxification processes taking place in the liver. For example, CYP1A-mediated metabolism of PAH can lead to the formation of reactive oxygen species (ROS) through the formation of labile metabolites (Sturve et al. 2006) and therefore lead to a state of oxidative stress (Livingstone 2001). Similarly, alkylphenols have been suggested to cause oxidative stress in fish (Hasselberg et al. 2004). The increase in lactate dehydrogenase observed in this study suggests an increased anaerobic metabolism. The oxidative stress provoked by the detoxification of PW seems to have led to an increased LDH; this increase in activity reflects the amplified necessity of using lactate as an oxidative fuel (Phillips et al. 2000). A recent study by Bohne-Kjersem et al. (2009) on the effect of a surrogate produced water on cod shows that, among the protein expression profiles that undergo a change, a few are linked with increased oxidative metabolism.

To ensure a proper understanding of the effect of PW on immune responses of cod, fish were submitted to an immune challenge by injection with LPS of *A. salmonicida*. Researchers using immune indicators in ecotoxicology often advocate that a battery of tests be applied to quantify effects (e.g. Dunier and Siwicki 1993). Ideally, measures of humoral and cell-mediated immunity, and stress would all be examined; however, measures of numerous stress and immune parameters vary temporally and are logistically difficult to assess (Van Den Heuvel et al. 2005). Moreover, attempts to induce specific humoral antibody response in cod have been unsuccessful in spite of giving protection against infection (e.g. Espelid et al. 1991) suggesting that the non-specific immune factors, as well as the cellular immune system, are very important in securing immune defence. Therefore, in this study, efforts were concentrated on investigating the immune cell responses and how both PW and LPS affected them.

When setting up an experiment to test the effect of the LPS injection per se, data showed that lactic acid and RB response of HK cells were significantly decreased. Cortisol increase seemed to be linked with the stress of the injection per se as the saline, and LPS injected fish experience the same increase. The decrease in RB shows that LPS has a significant impact on immune cell reactivity. In contrast to antibody production, non-specific immune parameters, such as oxidative burst, are known to respond rapidly to antigens (Hoeger et al. 2004). In an in vitro study on cod head-kidney macrophages, Sorensen et al. (1997) demonstrated that the production of reactive oxygen species was significantly enhanced by the stimulation of the cells with LPS of *Vibrio anguillarum*. It is therefore suggested that acute stress stimulates the natural immune system rather than suppresses it in order to protect fish against possible trauma (Demers and Bayne 1997). Therefore, stimulation of immune responses has been considered to be an indication of lesser toxicity than an inhibition of these same responses (Pipe et al. 1999). The inhibition seen in this study can be interpreted as a sign of an overwhelming effect of LPS on cell innate reactions in cod. It is important to note that the fish considered for the injection trial only were significantly smaller than the ones exposed to the PW (and injected with LPS only). Using the data obtained with these fish to extrapolate to bigger fish should be done with caution. Magnadottir et al. (1999) have shown that humoral immune parameters of cod are clearly influenced by its size. Nonetheless, other parameters such as haemolytic activities of serum showed a seasonal pattern stronger than a size-related pattern (Magnadottir et al.1999). Therefore, the data collected on smaller fish are used solely to help differentiate the specific effect of *A. salmonicida* LPS from the general stress effect of the injection process.

Results showed a combined effect of LPS injection and PW on white blood cell counts with a greater increase in macrophages and a decrease in lymphocytes in injected fish exposed to the highest dose of PW. As stated earlier, cortisol directly induces apoptosis in lymphocytes so fish stressed by disturbance will show reduced numbers of these cells (Weyts et al. 1998). The cortisol increase seen with LPS injection is not solely associated with the effect of LPS but seems to be a sign of general stress due to the injection process, as shown by the trial on smaller fish. Nonetheless, the increase is higher in the group exposed to 200 ppm of PW suggesting that the cortisol increase as well as the effects on blood cell percentages result from a combination of both PW and injection. This is confirmed by the increase in lactic acid that shows an effect of the LPS (the saline injection had no effect on lactic acid) with a dose-dependent augmentation in fish exposed to PW. Lactate increase has been suggested as a part (in terms of carbohydrate metabolism) of a common pattern of stress response in fish besides concomitant increases of plasma glucose and cortisol (Hontela et al. 1996; Pacheco and Santos 2001). Increases in basal levels of lactate have been already noticed in fish exposed to PAHs (Tintos et al. 2008), as well as in fish exposed to water-soluble fraction of diesel oil (Pacheco and Santos 2001). As clearly seen in the results, the injection affected fish metabolism; nonetheless, the patterns seen when comparing injected fish maintained in seawater (CONT + LPS) with injected fish submitted to PW demonstrate an added effect of PW. This is confirmed by the significant decrease of RB responses of HK in injected fish exposed to the highest dose of PW.

The results indicate that PW affects resting immunity of cod in a limited way but that the combination of LPS injection and PW impairs cod innate immune response. Alterations in the reactivity of selected immune parameters cannot be rated as indicative for the resulting overall immune competence without specific evidence of increased probability of infection (Hoeger et al. 2004). The LPS injection provides us with an insight on how cod innate immune system may react if exposed to a disease and PW at the same time and indicates that release of PW in the proximity of cod habitat may create environmental risks for stocks. Exposure to pulses (5 h exposures 3 times per week) of diluted PW (5000× and 10,000× dilution) were used in this study to mimic the rapid dilution and potential weathering processes that occur following discharge into the environment. Dispersion modelling of PW discharged into surface waters has indicated that rapid dilution by at least 240 times occurs within 50–100 m, up to 9000 times by 20 km from the discharge (Somerville et al. 1987; Murray-Smith et al. 1996). Abrahamson et al. (2008) showed that results of increased gill EROD activity obtained in the laboratory (dilution of 1000×) were not confirmed by caging experiments demonstrating therefore that setting up continuous exposures (vs. intermittent exposures) in tanks does not always capture properly what is happening in the field. Amounts and composition of PW vary between different fields and over time. Furthermore, water masses are stratified, particle concentrations differ over time and water currents fluctuate (Abrahamson et al. 2008). Moreover, in addition to cod seasonal migrations between spawning, nursery and feeding grounds (McKeown 1984), changes in the range and extent of vertical movement are significant because it is linked to migration, feeding and predator avoidance (e.g. Turner et al. 2002). Therefore, the intermittent exposure to low levels of PW completed in this study provides valuable information for use in risk analysis to determine how low levels of PW can affect cod immunity. Degradation of PW samples has certainly occurred during the exposure set-up of this study; therefore, it may be reasonable to assume that risk posed by the PW batch tested is significant. Other studies are being completed to ensure proper understanding of PW effect on resting immunity of smaller fish and in particular on the gene expression of immune factors.

References

Aas E, Klungsoyr J (1998) PAH metabolites in bile and EROD activity in North Sea fish. Mar Environ Res 46:229–232

Abrahamson A, Brandt I, Brunstorm B et al (2008) Monitoring contaminants from oil production at sea by measuring gill EROD activity in Atlantic cod (*Gadus morhua*). Environ Poll 153: 169–175

Alkanani T, Parrish CC, Rodnick KJ et al (2005) Lipid classes and nonesterified fatty acid profiles in plasma of North Atlantic cod (*Gadus morhua*). Can J Fish Aquat Sci 52:2509–2518

Banoub J, Cohen A, El Aneed A et al (2004) Structural reinvestigation of the core oligosaccharide of a mutant form of *Aeromonas salmonicida* lipopolysaccharide containing an O-4 phosphorylated and O-5 substituted Kdo reducing end group using electrospray QqTOF-MS/MS". Eur J Mass Spectrom 10:541–554

Bass DA, Parce WJ, Dechatelet LR et al (1983) Flow cytometric studies of oxidative product formation by neutrophils: A graded response to membrane stimulation. J Immunol 130(4):1910–1917

Bjornsson B, Steinarsson A, Arnason T (2007) Growth model for Atlantic cod (*Gadus morhua*): Effects of temperature and body weight on growth rate. Aquaculture 271:216–226

Bohne-Kjersem A, Skadsheim A, Goksoyr A, Grosvik BE (2009) Candidate biomarker discovery in plasma of juvenile cod (*Gadus morhua*) exposed to crude North Sea oil, alkyl phenols and polycyclic aromatic hydrocarbons (PAHs). Mar Envir Res 68(5):268–277

Chance B, Herbert D (1950) The enzyme-substrate compounds of bacterial catalase and peroxide. Biochem J 46:402–414

Crippen TL, Bootland LM, Leong JA et al (2001) Analysis of salmonid leucocytes by hypotonic lysis of erythrocytes. J Aquat Anim Health 13:234–245

Demers NE, Bayne CJ (1997) The immediate effects of stress on hormones and plasma lysosyme in rainbow trout. Dev Comp Immunol 21:363–373

Dunier M, Siwicki AK (1993) Effects of pesticides and other organic pollutants in the aquatic environment on immunity of fish: a review. Fish Shellfish Immunol 3:423–438

Espelid S, Rodseth OM, Jorgensen TO (1991) Vaccination experiments and studies of the humoral immune response in cod, *Gadus morhua* L., to four strains of monoclonal-defined *Vibrio anguillarum*. J Fish Dis 14:185–197

Gulland FM (1995) The impact of infectious diseases on wild animal populations- a review. In: Grenfell BT, Dobson AP (eds) Ecology of infectious diseases in natural populations. Cambridge University Press, Cambridge, pp 20–51

Hasselberg L, Meier S, Svardal A (2004) Effects of alkylphenols on redox status in first spawning Atlantic cod (*Gadus morhua*). Aquat Toxicol 69:95–105

Hoeger B, Van Den Heuvel MR, Hitzfeld BC et al (2004) Effects of treated sewage effluent on immune function in rainbow trout (*Oncorhynchus mykiss*). Aquat Toxicol 70: 345–355

Hontela A, Daniel C, Ricard AC (1996) Effects of acute and subacute exposures to cadmium on the interregnal and thyroid function in rainbow trout, *Oncorhynchus mykiss*. Aquat Toxicol 35:171–182

Kehoe A, Volkoff H (2007) Cloning and characterization of neuropeptide Y (NPY) and cocaine and amphetamine regulated transcript (CART) in Atlantic cod (*Gadus morhua*). Comp Biochem Physiol A 146(3):451–461

Khan RA, Kiceniuk J (1984) Histopathological effects of crude oil on Atlantic cod following chronic exposure. Can J Zool 62:2038–2043

Kiceniuk JW, Khan RA (1987) Effect of petroleum hydrocarbons on Atlantic cod, *Gadus morhua*, following chronic exposure. Can J Zool 65(3):490–494

Livingstone DR (2001) Contaminant-stimulated reactive oxygen species production and oxidative damage in aquatic organisms. Mar Poll Bull 42:656–666

Lowry OH, Rosebrough NJ, Lewis Farr A et al (1951) Protein measurement with the folin phenol reagent. J Biol Chem 193:265–275

Luna LG (1968) Manual of Histological Staining Methods of the Armed Forces Institute of Pathology. McGraw-Hill, New York, NY

Lynch M, Raphael S, Mellor L et al (1969) Medical laboratory technology and clinical pathology. Saunders Company, Newberg, OR

Magnadottir B, Jonsdottir H, Helgason S et al (1999) Humoral immune parameters in Atlantic cod (*Gadus morhua* L.) II. The effects of size and gender under different environmental conditions. Comp Biochem Physiol B 122:181–188

McKeown BA (1984) Fish migration. Croom Helm, London and Sydney, 224 pp

Mitchell DB, Santone KS, Acosta D (1980) Evaluation of cytotoxicity in cultured cells by enzyme leakage. J Tissue Culture Methods 6:113–16

Murray-Smith R, Gore D, Flynn SA et al (1996) Development and appraisal of a particle tracking model for the dispersion of produced water discharged from an oil production platform in the North Sea. In: Reed M, Johnsen S (eds) Environmental science research 52: produced water2. Environmental issues and mitigation technologies. Plenum Press, New York, NY, pp 225–245

Myers MS, Fournie JW (2002) Histopathological biomarkers as integrators of anthropogenic and environmental stressors. In: Adams M (ed) Biological indicators of aquatic ecosystem stress. American Fisheries Society Publication, Bethesda, MD, pp 221–287

Pacheco M, Santos MA (2001) Biotransformation, endocrine, and genetic responses of *Anguilla anguilla* L. to petroleum distillate products and environmentally contaminated waters. Ecotoxicol Environ Saf 49:64–75

Parrish CC (1987) Separation of aquatic lipid classes by Chromarod thin-layer chromatography with measurement by Iatroscan flame ionization detection. Can J Fish Aquat Sci 44(4):722–731

Phillips MCL, Moyes CD, Tufts BL (2000) The effects of cell ageing on metabolism in rainbow trout (*Oncorhynchus mykiss*) red blood cells. J Exp Biology 203:1039–1045

Pipe R, Coles JA, Carissan FMM et al (1999) Copper induced immunomodulation in the marine mussel *M. edulis*. Aquat Toxicol 46:43–54

Roe TI (1998) Produced water discharges to the North Sea: a study of bioavailability of organic produced water compounds to marine organisms. PhD thesis, Norwegian University of Science and Technology, Trondheim, Norway

Schreck CB (1996) Immunomodulation: endogenous factors. In: Iwana G, Nakanishi T (eds) The fish immune system, organism, pathogen, and environment, Academic, New York, NY, pp 339–366

Seternes T, Dalmo RA, Hoffman J et al (2001) Scavenger-receptor-mediated endocytosis of lipopolysaccharide in Atlantic cod (*Gadus morhua* L.). The J Exp Biol 204:4055–4064

Somerville HJ, Bennett D, Davenport JN et al (1987) Environmental effects of produced water from North Sea oil operations. Mar Poll Bull 18:549–553

Sorensen KK, Sveinbjornsson B, Dalmo RA et al (1997) Isolation, cultivation and characterization of head kidney macrophages from Atlantic cod, *Gadus morhua* L. J Fish Dis 20:93–107

Stenvik J, Solstad T, Strand C et al (2004) Cloning and analyses of a BPI/LBP cDNA of the Atlantic cod (*Gadus morhua* L.). Dev Comp Immunol 28:307–323

Stephens SM, Frankling SC, Stagg RM et al (2000) Sub-lethal effects of exposure of juvenile turbot to oil produced water. Mar Poll Bull 40(11):928–937

Sturve J, Hasselber L, Falth H et al (2006) Effects of North Sea oil and alkylphenols on biomarker responses in juvenile Atlantic cod (*Gadus morhua*). Aquat Toxicol 78S:73–78

Sundt RC, Meier S, Jonsson G et al (2009) Development of a laboratory exposure system using marine fish to carry out realistic effect studies with produced water discharged from offshore oil production. Mar Poll Bull 58:1382–1388

Tintos A, Gesto M, Miguez JM et al (2008) ß-Naphthoflavone and benzo(a)pyrene treatment affect liver intermediary metabolism and plasma cortisol levels in rainbow trout *Oncorhynchus mykiss*. Ecotoxicol Environ Saf 69:180–186

Turner K, Righton D, Metcalfe JD (2002) The dispersal patterns and behaviour of North Sea cod (*Gadus morhua*) studied using electronic data storage tags. Hydrobiol 483:201–208

Van Den Heuvel MR, O'Halloran K, Ellis RJ et al (2005) Measures of resting immune function and related physiology in juvenile rainbow trout exposed to a pulp mill effluent. Arch Environ Contam Toxicol 48:520–529

Volkoff H, Peter RE (2004) Effects of lipopolysaccharide treatment on feeding of goldfish: role of appetite-regulating peptides. Brain Res 998(2):139–147

Volkoff H, Canosa LF, Unniappan S et al (2005) Neuropeptides and the control of food intake in fish. Gen Comp Endocrinol 142(1-2):3–19

Weeks BA, Warinner JE (1984) Effects of toxic chemicals on macrophage phagocytosis in two estuarine fishes. Mar Environ Res 14:327–335

Weyts FAA, Flik G, Rombout JHWM et al (1998) Cortisol induces apoptosis in activated B cells, not in other lymphoid cells of the common carp. Dev Comp Endocrinol 22:551–562

Chapter 17
Effects of Hibernia Production Water on the Survival, Growth and Biochemistry of Juvenile Atlantic Cod (*Gadus morhua*) and Northern Mummichog (*Fundulus heteroclitus macrolepidotus*)

Les Burridge, Monica Boudreau, Monica Lyons, Simon Courtenay, and Kenneth Lee

Abstract Juvenile Atlantic cod and mummichog were exposed to a range of concentrations of produced water (PW) collected from the Hibernia oil production platform in 2005 and 2006. PAH exposure was measured by induction of cytochrome P-450 (CYP1A) as indicated by ethoxyresorufin *O*-deethylase (EROD) activity. In short-term exposures mummichog exposed to PW collected near the produced water discharge of the platform showed no change in EROD response compared to controls. EROD activity in livers collected from juvenile cod was significantly elevated in response to exposure to dilutions of PW but only at concentrations greater than 1.67% by volume. When juvenile cod were exposed to 0.05% PW (by volume) for 45 days, there was no significant change in EROD activity, growth or plasma vitellogenin compared to unexposed fish. Embryo growth and heart rate in mummichog were slowed by exposure to dilutions of raw PW as low as 1%. However, mortality and developmental abnormalities were only observed at high concentrations (10 and 66%). PW from the Hibernia platform poses a low risk to cod and mummichog for the endpoints tested in this study.

1 Introduction

Produced water (PW) is the waste usually generated in largest volume during production of oil and gas from offshore oil and gas wells (Neff et al. 2006). The Canada-Newfoundland and Labrador Offshore Petroleum Board (CNLOPB) report that 5.72×10^6 and 7.18×10^6 tonnes of PW were released from the Hibernia platform in 2007 and 2008, respectively (CNLOPB 2009). Most PW is fossil water (formation water) that has accumulated over millions of years with fossil fuels in

---

L. Burridge (✉)
Fisheries and Oceans Canada, St. Andrews Biological Station, St. Andrews, NB, Canada, E5B 2L9

K. Lee, J. Neff (eds.), *Produced Water*, DOI 10.1007/978-1-4614-0046-2_17,

geologic formations deep in the earth. PW also may contain some surface water that has been injected into the formation for enhanced oil recovery. PW reaches the surface from natural oil seeps worldwide and during production of oil and gas from a well (Neff et al. 2006). The chemical characteristics of PW are different for each production platform or formation from which the oil is extracted. It is typically highly saline and contains elevated levels of heavy metals, hydrocarbons (including polycyclic aromatic hydrocarbons (PAHs)), alkylphenols, ammonia and radionuclides compared to the receiving environment (Lee et al. 2005; Sturve et al. 2006).

With anticipated increases in the number of new offshore platforms, PW discharge has been identified as an issue of concern by both regulators and environmental groups (Zhao et al. 2008). While both monitoring and preliminary modelling efforts have been reported over the past decades, the possible long-term adverse effects of PW discharge into localized marine environments require further study.

The environmental impact of PW has been assessed by monitoring the chemical characteristics of freshly discharged PW and by ecotoxicological measurements (Azetsu-Scott et al. 2007). Because the chemical make-up of PW varies with its source, the results of these studies are variable.

One of the most toxic groups of compounds found in PW, PAHs, has been shown to have cytotoxic, immunotoxic and mutagenic or carcinogenic effects in marine organisms (Sturve et al. 2006). PAH exposure can be estimated by a standardized laboratory assay of cytochrome P-450 (CYP1A) induction (Hodson et al. 1996). The terminal oxidase enzyme of the mixed-function oxygenase (MFO) system is the iron-containing hemoprotein CYP1A. Ethoxyresorufin *O*-deethylase (EROD) is part of the family of cytochrome enzymes and is induced in the presence of certain xenobiotic compounds including PAHs (Hodson et al. 1991). The CYP1A enzyme catalyzes the hydroxylation of PAH to a more soluble and excretable form in the bile, therefore assays of liver CYP1A activity provide a good biomarker of PAH exposure in fish (McCarty et al. 2002).

PAHs and alkylphenols are oestrogenic compounds (Kime 1998). Therefore, PW exposure may adversely affect the hormonal system of fish. It is accepted that exposure to oestrogen-like compounds in the aquatic environment may result in elevation of the yolk protein, vitellogenin (VTG) in male fish (see for example Allen et al. 1999). Scott et al. (2006) have developed an enzyme linked immunosorbent assay (ELISA) for VTG in cod and have shown it to be sensitive to environmentally relevant concentrations of contaminants.

Atlantic cod (*Gadus morhua*) is an important commercial species of groundfish. They are found throughout the North Atlantic including the Grand Banks of Newfoundland. This area is also the site of an expanding oil exploration industry. Mummichog (*Fundulus heteroclitus*) is one of the most common fish species in shallow coastal and estuarine waters of the Northwest Atlantic. Considerable work has been done on their responses to anthropogenic contaminants, and they have been proposed as a model species for detecting physiological responses to environmental change (Burnett et al. 2007). In this chapter we describe the results of lab-based studies wherein Atlantic cod, mummichog and early life stages of mummichog were exposed to PW from the Hibernia production platform. The endpoints studied include biomarkers of exposure (EROD) and of effects (growth and developmental abnormalities).

2 Materials and Methods

2.1 Collection of Hibernia Produced Water

PW is discharged 40 m below the water surface at the Hibernia platform. Sea water samples were collected at the surface of the PW plume approximately 20 m from the platform on 2 July 2005. Approximately 70 L of sea water was collected in acid-washed (1 M HCl) Nalgene jerricans, transported to the Gulf Fisheries Centre (GFC), Moncton, NB, transferred to hexane/acetone-washed 20 L glass bottles, sealed and refrigerated at 4–10°C until use in mummichog experiments on July 19th.

In addition, raw PW was collected on the Hibernia platform 25 June 2006 in Teflon-sealed, acid-washed Nalgene jerricans. The containers were held at 4–10°C, transported to the GFC or the St. Andrews Biological Station (SABS) and stored at 4°C until used in bioassays. Mummichog embryo bioassays were conducted with 2006 PW in July, 2006. Growth and EROD induction experiments with cod were conducted with 2006 PW at SABS from December 2006 to March 2007.

2.2 Fish

Mature northern mummichog (*Fundulus heteroclitus macrolepidotus*) were collected by beach seine in July 2005 from the Kouchibouguac River estuary in Grand Barachois, New Brunswick. Kouchibouguac estuary has limited development (a few cottages along the shores and some recreational boating) and was used as a reference site in a recent study of anthropogenic effects on local populations of mummichog and Atlantic silverside (*Menidia menidia*; Thériault et al. 2007).

Mummichog were held for 1 week at the GFC aquarium facility prior to the EROD assay. Fish were held at densities of less than 2 g/L in recirculated artificial sea water (Kent Sea Salt in RO-treated municipal (Moncton, NB) water) at 15 parts per thousand, 22–23°C and a photoperiod of 16 h light: 8 h dark. Fish were fed twice daily with Aquatox flake food (Zeigler Bros., Gardners, PA, USA) supplemented once a week with frozen, vitamin-enriched bloodworms (chironomid larvae; Bio-pure, Hikari Sales, Hayward, CA, USA).

Atlantic cod (*Gadus morhua*) were obtained from the Genome Canada, Cod Genome Project hatchery at SABS. The fish were held in filtered sea water with simulated natural photoperiod and ambient temperature for 7 months. Oxygen and temperature were recorded daily. The fish were hand-fed once daily with a dry pellet mixture of Gemma starter feed and Europa, 2.0 mm feed from Skretting North

America (Bayside, New Brunswick Canada) and a cod marine, 1.5 mm feed from EWOS Canada, Ltd. (Surrey, British Columbia Canada).

2.3 Exposures and Endpoints

2.3.1 Mummichog Embryonic Development Experiments

Mummichog broodstock were beach seined from the Kouchibouguac estuary (early July, 2006) and held in 1200-L recirculating aquaria (22–23°C, 15 ppt) at the GFC until the assay began later the same month. Eggs were artificially fertilized after being collected from females by ventral pressure. Males were sacrificed by spinal severance to obtain the gonads because it was impossible to draw enough milt by ventral pressure. The testes were ground then stirred with the eggs and left to sit for approximately 10 min to allow fertilization to occur (Boudreau et al. 2004a, b, 2005). Eggs were then rinsed with artificial sea water and placed in covered Pyrex Petri dishes (100 mm diameter; 20 eggs/dish) with 50 mL of test solution. Eggs were examined for development after 24 h, and unfertilized eggs were removed leaving 10–18 live embryos per dish. There were six replicates (Petri dishes) per treatment group. Treatments were 1, 10 and 66% PW collected from the Hibernia platform in June 2006. As the salinity of raw Hibernia PW is ~48 ppt, dilution to 66% is required to reach a salinity of 30–34 ppt. All dilutions were prepared with artificial sea water. Test solutions were renewed three times per week with freshly mixed solutions. Experiments were conducted at 25°C with no aeration. Photoperiod was 16 h light: 8 h dark.

Mortalities were monitored daily and dead animals were removed. Live embryos were observed by microscope to detect morphological abnormalities commonly found in toxicological studies (von Westernhagen 1988; Boudreau et al. 2004a, b, 2005). Specifically, embryos were monitored for abnormalities related to blue sac disease associated with PAH exposure (Brinkworth et al. 2003). Heart rates were measured from five randomly selected embryos per dish on day 7 post-fertilization. Time-to-hatch was noted and larvae were measured at hatch (nearest 0.01 mm). Observations and measurements were made with a computer-based image analysis system (Matrox Inspector, version 3.0, Matrox Imaging, Dorval, QC) linked to a microscope (Leitz, Wild Photomakroskop M400, Leica Microsystems, Inc., Willowdale, ON) (16x–90x) through a video camera (Hitachi, HV-D25, Fisher Scientific Ltd., Nepean, ON).

2.3.2 Mummichog EROD Induction Experiments

Adult mummichog were held in 1200-L recirculating aquaria (22–23°C, 15 ppt) at the GFC until 48 h prior to initiation of bioassays. They were then acclimated to full sea water over a 48-h period. Mummichog were exposed for 24 h to decimal dilutions (0.01–100%) of receiving environment sea water collected off the Hibernia platform in July 2005. In order to test effects of 100% receiving water, all treatment groups including the negative control were tested at a salinity of full sea water (30–34ppt salinity). Also included in the bioassay was a positive control of 10 $\mu g \cdot L^{-1}$ beta-naphthoflavone (BNF; dissolved in 100% methanol). In order to achieve the target concentration of BNF, 1 mL of the stock solution (in methanol) was added to 10 L water resulting in a methanol concentration of 0.01%. There were two replicates per treatment group, each with 10 L of aerated test solution and five fish (mummichog: 5.52 g $\pm$ 0.46) Temperature was maintained at 22–23°C, the photoperiod was 16 h light: 8 h dark.

At the end of the exposure period fish were sacrificed by anaesthetic overdose [320 mg/L of tricaine methanesulfonate (Sigma-Aldrich, Oakville, ON, Canada)] and livers were extracted, weighed, placed in cryovials (Fisher Scientific Company, Ottawa, ON, Canada), frozen in liquid nitrogen and stored in a –80°C freezer until analysis. After removal of the fish from their exposure containers, salinity, pH, dissolved oxygen and ammonia concentration were noted for both replicates.

2.3.3 Juvenile Cod EROD Induction Experiments

Juvenile cod (mean weight = 24.8 g; mean length = 14.5 cm) were exposed for 48 h to one of five concentrations (0.06, 0.19, 0.56, 1.67 and 5%) of PW, a sea water control or a positive control, 10 $\mu g \cdot L^{-1}$ of beta-naphthoflavone (BNF, prepared in methanol). In order to achieve the target concentration of BNF, 10 mL of the stock solution (in methanol) was added to 25 L water resulting in a methanol concentration of 0.04%. Five fish per concentration were exposed at ambient light and water temperature of 10°C $\pm$ 1°C in 10 or 25 L glass aquaria fitted with clear plexiglass lids. Loading rates were 5–10 g of fish$\cdot L^{-1}$ test solution. Half of the exposure water was replaced with fresh solution of PW after 24 h. The 48-h exposure was repeated two times for a total of three bioassays. Dissolved oxygen (DO) was monitored throughout the test and in the first 48-h exposure was maintained at 70–80% except for two measurements made at $T = 24$ prior to the water change when the DO level dropped to 65–70%. There was no treatment-related trend in the DO.

A 500-mL water sample was taken from the aquarium spiked with 0.56% PW at 0, 24 and 48 h and preserved with 0.5 mL 6 N HCl. The water samples were held at 4°C for chemical analysis. At the end of 48 h, all the fish were sacrificed by a blow to the head, blood was collected from the caudal vein followed by cervical transection. Liver and muscle samples were collected. Length, weight and gender were recorded. Tissue samples were quick frozen in liquid nitrogen. Blood samples were centrifuged at 6000 rpm for 10 min and the plasma collected. Blood and tissue were frozen at –80°C.

2.3.4 Juvenile Cod Growth Experiments

One hundred and forty juvenile cod (mean weight = 23.0 g; mean length = 13.7 cm) were tagged with Passive Integrated Transponders (PIT) and allowed to recover for several weeks in ambient (SABS) sea water. Seventy PIT tagged juvenile cod were exposed continuously for 45 days to 0.05% PW and 70 fish were held in running sea water and served as controls. Full-strength PW was delivered by controlled, gravity flow using a mariotte bottle. The flow rate was maintained at

1 mL·min^{-1} and was mixed with flowing sea water (2.5 L·min^{-1}) prior to entering the 400 L exposure tank. The fish were held in a simulated natural photoperiod and 10°C± 2°C sea water. Water temperature was maintained with the addition of hot salt water into a header tank. DO and temperature were monitored throughout the bioassay. The fish were hand-fed once a day a dry pellet of EWOS marine sinking cod diet. Five hundred millilitres of water samples were collected periodically, preserved with 0.5 mL 6 N HCl and stored at 4°C.

At days 3, 7, 14, 28 and 45 of the exposure and days 14 and 60 post-exposure, eight cod were sacrificed by a quick blow to the head, blood was collected from the caudal vein followed by cervical transection. Liver, gill and muscle were collected for biochemical analysis. In addition, gender, length and weight data were recorded. Specific growth rates were determined using changes in fish weights according to Busacker et al. (1990). Tissue samples were quick frozen in liquid nitrogen.

2.4 Chemical and Biochemical Analyses

2.4.1 EROD Assay

Frozen livers were thawed, a sub-sample taken, put in a tared micro-centrifuge tube and weight recorded. Sub-samples of liver were homogenized and diluted with HEPES-KCl buffer (pH 7.5, 0.15 M KCl, 0.02 M HEPES) using a hand-held motor driven Kontes pestle. All steps were carried out on ice. The homogenates were centrifuged at 9850 rpm for 20 min at 2°C. The post-mitochondrial supernatant (S-9 fraction) was collected with a Pasteur pipette, taking care to avoid the pellet and the floating lipid layer (Hodson et al.1991). S-9 fractions were frozen at –80°C in micro-centrifuge tubes until analysis.

Liver samples collected from mummichog were analysed at Queen's University, ON, Canada for ethoxyresorufin *O*-deethylase (EROD) activity. Samples of cod liver were analysed at SABS. The kinetic microplate assay used is described in Hodson et al. (1996). EROD activity in cod liver S-9 fractions was measured using a BioTek FLx 800 spectrofluorometer using excitation and emission wavelengths of 530 and 590 nm respectively. The protein levels in the same S-9 samples were quantified with a Bio-Rad assay (Bradford) dye reagent using a BioTek Powerwave XS UV/Vis spectrophotometer with absorbance set at 600 nm. BioTek Gen 5 software was used to view triplicate results for each sample.

EROD activity was calculated from the slope of the curve over the selected time. Activity was converted to picomoles resorufin per minute per milligram protein in the S-9 fraction.

2.4.2 Plasma Vitellogenin

Blood samples were centrifuged at 6000 rpm for 10 min and the plasma collected. Blood and tissue were frozen at –80°C. Plasma samples were analysed for plasma vitellogenin (VTG) using Biosense cod VTG kit (Biosense Laboratories, Bergin,

Norway) by an enzyme linked immunosorbent assay (ELISA). The sensitivity of the kit was verified by Dr. Robert Roy (Fisheries and Oceans Canada, Mont Joli, QC) using plasma collected from Atlantic cod that had received oestrogen injections at SABS.

2.4.3 Water Analysis

PW collected at the Hibernia platform was analysed at the Bedford Institute of Oceanography. Water samples were extracted and analysed for organic compounds according to King and Lee (2004) and for inorganic compounds according to Azetsu-Scott et al. (2007). The water samples collected during cod bioassays were analysed for organic compounds according to King and Lee (2004).

2.5 Data Analysis

Effects of treatments on mummichog embryo survival and incidence of morphological abnormalities were tested by probit regression. EROD activity, heart rates, time-to-hatch and size-at-hatch were tested for normality (probability plot) and homoscedasticity (F_{max} test) and transformed where necessary. Treatment effects were tested by parametric analysis of variance (ANOVA) models followed by Tukey's multiple range test, or by the non-parametric Kruskal–Wallis test followed by Nemenyi or Noether non-parametric multiple range tests if transformations did not correct non-normal distributions. Nested ANOVA removed the variability added by replicates to test for influence of treatment only. Because size-at-hatch can be affected by the duration of the embryo stage (i.e., longer incubation producing longer larvae; von Westernhagen 1988), analysis of covariance (ANCOVA) was used to remove effects of incubation duration from comparisons of size-at-hatch. All analyses were performed with Systat version 11.0 (SPSS Inc., Chicago, IL, USA). The level of significance was $P < 0.05$. Back-transformed means are accompanied by their 95% confidence interval.

Specific EROD activities were log-transformed to establish normal distribution and homogenous variances. Log-transformed data were treated by analysis of variance (level of significance was $P < 0.05$) to determine if differences were present between treated and untreated fish. Where significant differences were indicated, log means were compared by Tukey's multiple comparison test.

3 Results and Discussion

3.1 Mummichog Embryonic Development

Survival of mummichog embryos exposed to Hibernia PW was reduced by the two highest concentrations tested, 10 and 66% ($P < 0.05$, Table 17.1). Morphological

Table 17.1 Responses of mummichog embryos to Hibernia PW; arrow indicates direction of effect; threshold concentration indicates lowest concentration at which a significantly different response was seen from water control

Endpoint	Direction	Threshold concentration (% v/v)
Survival	↓	10%
Normal development	↓	66%
Heart rate	↓	1% (not significant at 10%)
Time-to-hatch	↑	1%
Size-at-hatch	↓	1%

development of embryos was also significantly affected by PW but only at the highest concentration of 66% ($P<0.05$, Table 17.1). Common morphological abnormalities observed during the study included cardio-vascular lesions such as pericardial oedema (accumulation of fluids in the pericardial sac; Fig. 17.1b), haemorrhaging (accumulation of blood outside the blood vessel; Fig. 17.1b) and haemostasis most commonly along the tail (accumulation of blood inside the blood vessel; Fig. 17.1c) and general retarded development of the embryo (Fig. 17.1d).

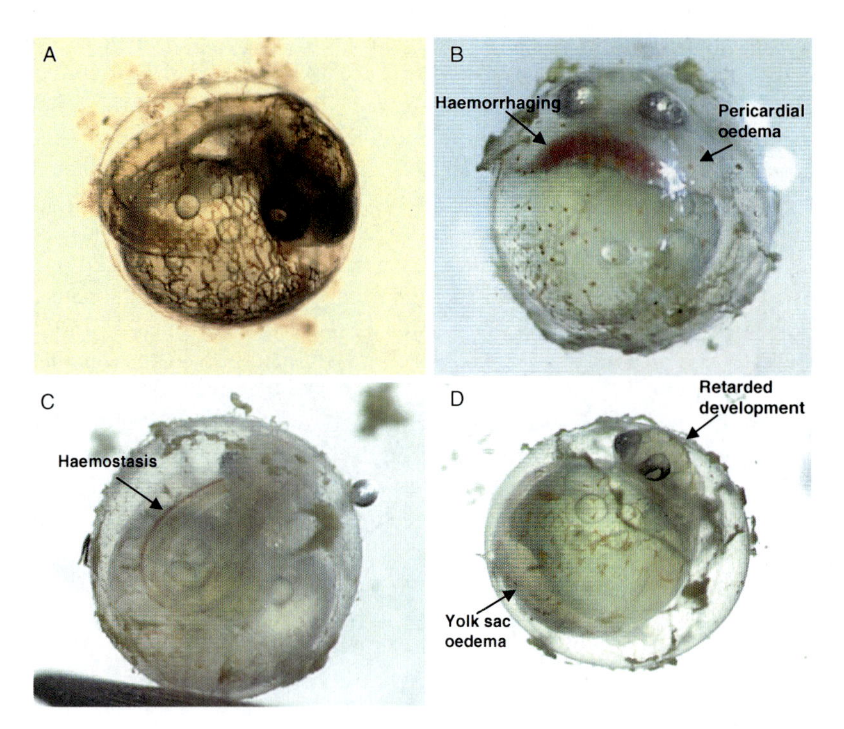

Fig. 17.1 Most common abnormalities observed in mummichog embryos exposed to graded doses of pure Hibernia PW from fertilization: normal embryo (**a**), haemorrhaging and pericardial oedema (**b**; *left* and *right* arrows respectively), haemostasis along the tail (**c**) and retarded development and yolk sac oedema (**d**; *upper* and *lower arrows* respectively); egg diameter is approximately 2 mm

Other endpoints tested proved more sensitive to PW and were all significantly impacted by the lowest concentration tested of 1% PW (Table 17.1). Heart rates were significantly reduced by treatment with 1 and 66% but not 10% ($P<0.05$). Embryos exposed to 1 and 10% PW took longer to hatch than controls ($P<0.05$) and none of the embryos hatched in the highest concentration of 66%. At hatch, embryos in the 1 and 10% treatment groups were smaller than control fish ($P<0.05$). This suite of abnormalities and reduced growth was described originally in fish exposed to dioxins and was referred to as blue sac disease (Helder 1980; Walker et al. 1991). This condition has also been reported in fish exposed to PAHs (Billiard et al. 1999; Brinkworth et al. 2003). PAHs are among the most toxic compounds in oils and are believed to be the causative agents for blue sac disease in fish exposed to oils. In mummichog, these lesions have been reported following exposure to Alaska North Slope crude oil and Mesa light crude oil (Couillard 2002) and we have recently reported blue sac disease in mummichog exposed to Orimulsion (Boudreau et al. 2009). The results of the present study show that a high concentration of Hibernia PW is also capable of producing morphological abnormalities, categorized as blue sac disease.

3.2 EROD Induction

3.2.1 Mummichog

Twenty-four hour exposure to dilutions of sea water taken in the plume of PW from the Hibernia platform did not significantly elevate EROD activity in mummichog, though fish exposed to the positive control (10 $\mu g \cdot L^{-1}$ BNF) did show a significant increase in EROD activity ($P<0.05$; Fig. 17.2a).

Consistently high basal levels of EROD have been observed in previous studies with mummichog (Whyte et al. 2000). Mummichog collected in three other New-Brunswick estuaries (Horton's Creek tributary of the Miramichi River, Shediac Bridge and Saint-Louis) have also shown high basal levels of EROD activity (S. Ramachandran, Queen's University, Kingston, ON, personal communication). Nevertheless, even though basal levels were high in the present assay, significant induction was seen in response to the BNF positive control consistent with Whyte et al. (2000). This suggests that the high basal levels did not preclude response to a strong inducer but they may have rendered the assay insensitive to a weak inducer.

These results suggest that seawater samples taken in the plume of PW from the Hibernia platform contain biologically insignificant concentrations of specific Ah-R active compounds, or the threshold for activation of the CYP1A system in this species of fish is not achieved under our experimental conditions. Hodson et al. (2008) have shown that, in rainbow trout (*Oncorhynchus mykiss*), the alkylphenanthrenes, fluorenes and napthobenzothiophenes are the main components of oil that induce CYP1A. Analysis of PW from Hibernia (2005) showed an average concentration of fluorenes in the dissolved phase of 2205 $ng \cdot L^{-1}$ and methylated phenanthrenes at average concentrations of 277, 501 and 1182 $ng \cdot L^{-1}$ for

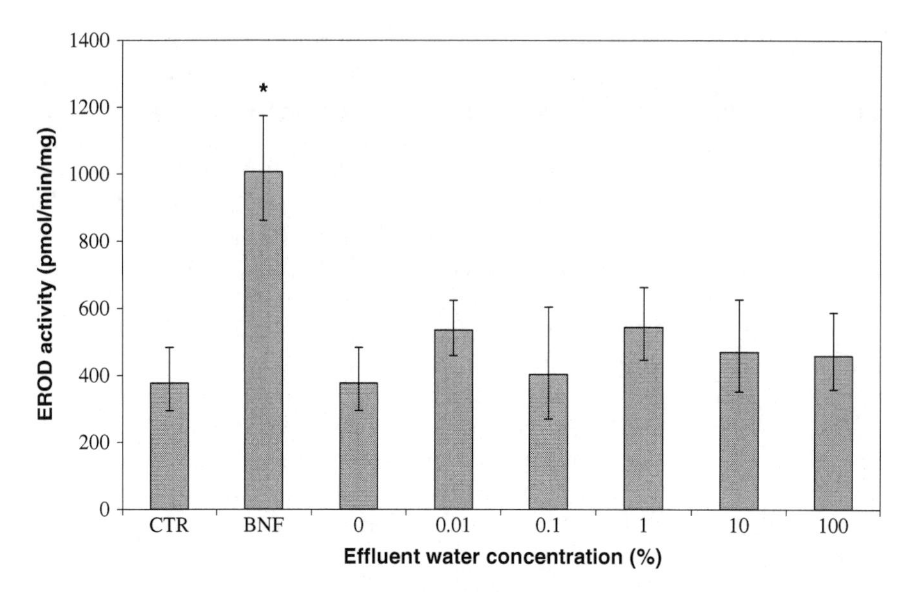

Fig. 17.2 Mean EROD activity (+/– 95% confidence intervals) in mummichog (*Fundulus heteroclitus*) after a 24-h exposure to field collected water samples in the receiving environment of the Hibernia oil platform; asterisk indicates significant difference from water control (Nested ANOVA, replicates nested within treatments followed by Tukey's multiple range test); each treatment group had two replicates of five fish each

2-methlyphenanthrene, dimethylphenanthrene and methylphenanthrene respectively. However, it is not known if these same chemicals produce CYP1A induction in mummichog, and if so at what concentrations. Furthermore, the concentrations to which mummichog were exposed in the seawater dilutions are unknown but would be a small fraction of the concentrations in the PW itself.

3.2.2 Juvenile Cod Dose–Response

The level of EROD activity in livers collected from cod exposed to a range of PW concentrations ranged from 0.2 to 12.6 $pmol \cdot min^{-1} \cdot mg^{-1}$. A comparison of treatment effects is shown in Fig. 17.3. Only in PW concentrations of 1.67 and 5% was EROD activity significantly ($P < 0.05$) elevated compared to controls. BNF exposure resulted in elevated EROD activity in cod livers but the increase was not significant compared to controls. EROD activity in BNF-exposed cod was significantly higher than that observed in cod exposed to 0.06, 0.19 or 0.56% PW. The compounds identified by Hodson et al. (2008) as having CYP1A inducing qualities in trout were detectable in raw PW from Hibernia collected in 2006 and in the PW used in the cod bioassays. Raw PW contained an average of 743, 260 and 507 $ng \cdot L^{-1}$ of methylphenanthrene, 2-methylphenanthrene and dimethylphenanthrene respectively in the dissolved phase (Table 17.2). Of the four PW water samples analysed, one had significantly higher concentrations of the methylated phenanthrenes than the other three thereby increasing the average. Fluorene was present at an average

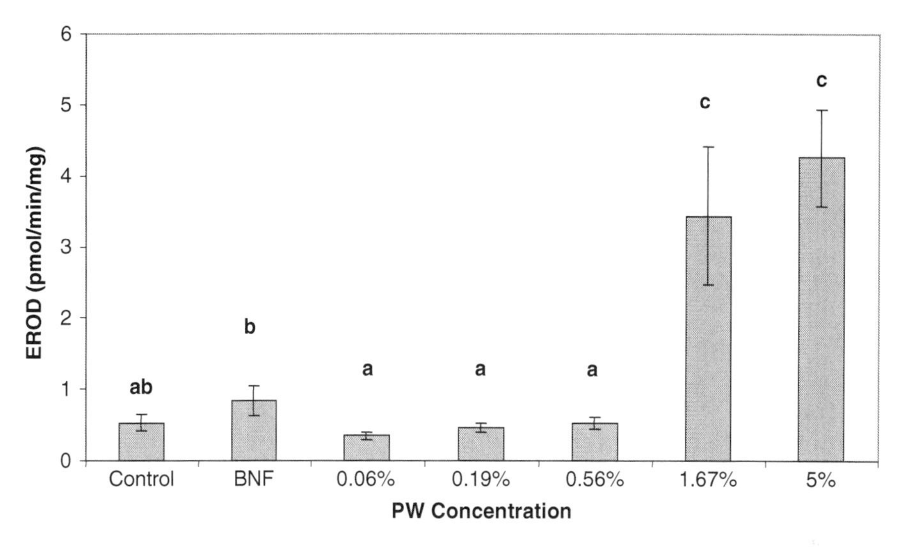

Fig. 17.3 EROD activity in livers collected from Atlantic cod exposed for 48 h to produced water from the Hibernia platform (2006); each bar represents the mean value for a sample of five fish; bars identified with the same letter are not significantly different from each other (ANOVA, ($P<0.05$))

concentration of 1902 $ng \cdot L^{-1}$ and napthobenzothiophene was not detectable in the dissolved phase (Table 17.2). In the acute (48 h) exposures water samples were collected from the 0.56% exposure tanks. Only methylphenanthrene was detected in the water samples and then only in one test (Table 17.2). Although it is not known if these compounds have the same effect in cod as they do in trout, the lack of EROD

Table 17.2 Concentration ($ng \cdot L^{-1}$) of alkylphenanthrenes, fluorenes, napthobenzothiophenes and (in $mg \cdot L^{-1}$) total phenols and alkylphenols in PW collected from the Hibernia platform in 2006 and in water samples collected from bioassays with Atlantic cod conducted with dilutions of the same PW

	Source of water sample(s)		
Analyte (dissolved)	Raw PW Hibernia 2006	48-h dose–response (0.56%) $T=0$	Chronic exposure (0.05% PW)
Total methylated PAH	70,090 ($n=4$)	266 ($n=3$)	ND
Methylphenanthrene	743 ($n=4$)[a]	100[b]	ND
2-Methylphenanthrene	260 ($n=4$)[a]	ND	ND
Dimethylphenanthrene	507 ($n=4$)[a]	ND	ND
Fluorene	1902 ($n=4$)	ND	ND
Napthobenzothiophene	ND	ND	ND
Σ Phenols	3.962 $mg \cdot L^{-1}$	0.012 $mg \cdot L^{-1}$	0.009 $mg \cdot L^{-1}$
Σ Alkylphenols (methyl to butyl phenols, no nonylphenols present)	1.519 $mg \cdot L^{-1}$	0.008 $mg \cdot L^{-1}$	0.004 $mg \cdot L^{-1}$

[a]One PW sample with significantly higher concentrations compared to the other three
[b]Compound only detectable in one sample from three tests

induction at 0.56% PW is consistent with no measurable alkylated phenanthrenes, fluorines or napthobenzothiophenes. It is clear from the EROD induction experiment that PW from Hibernia will activate the CYP1A system in cod at higher concentrations (above 1.67%, Fig. 17.3).

3.2.3 Juvenile Cod Chronic Exposure

During the 45-day exposure of juvenile cod to 0.05% PW from Hibernia, there was no significant difference ($P < 0.05$) in EROD activity between treated and untreated fish (Fig. 17.4). EROD activity was low throughout the study. The maximum measured value was approximately 5 pmol·min^{-1}·mg^{-1}. The EROD activities reported are lower than those reported (in this study) for cod exposed to higher concentrations in an acute exposure and values reported for mummichog. Hepatic EROD induction is also lower than values reported for Atlantic cod collected from areas of known contamination (Aas and Klungsøyr 1998, Lee and Anderson 2005, Schnell et al. 2008). Abrahamson et al. (2008) have reported that there is no difference between gill and liver EROD activity in cod caged near PW outflows and cod from reference sites. During the chronic exposure experiment, the compounds identified by Hodson et al. (2008) as being responsible for EROD induction in trout (methylphenanthrene, 2-methylphenanthrene, dimethylphenanthrene and fluorine) were not detected in samples collected from the exposure tank. Holth et al. (2009) exposed cod to low molecular weight PAHs and alkylphenols for 44 weeks. They found that EROD induction peaked after 4 weeks but was back to control levels at 32 weeks.

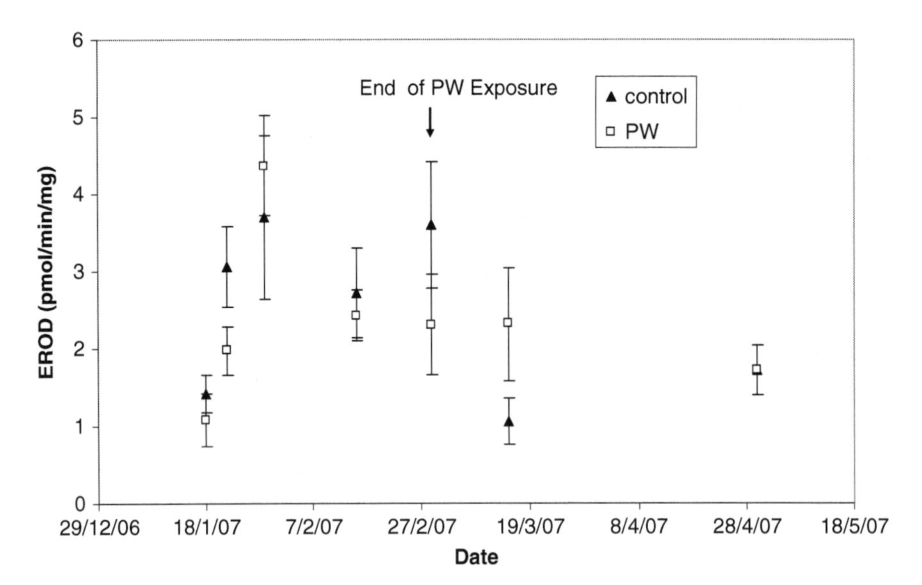

Fig. 17.4 EROD activity in livers collected from Atlantic cod chronically exposed for 45 days to 0.05% PW from Hibernia (2006), then transferred to untreated water for 60 days; each data point represents the mean value for a sample of eight fish

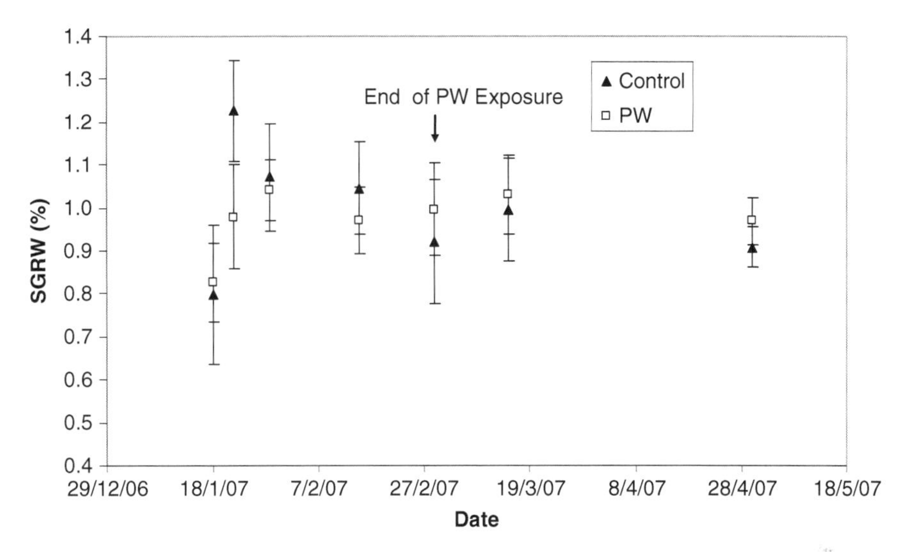

Fig. 17.5 Specific growth rate by weight (%) of juvenile cod exposed for 45 days to 0.05% PW from Hibernia (2006), then transferred to untreated water for 60 days; each data point represents the mean value from a sample of eight fish

Sturve et al. (2006) reported that chronic exposure of juvenile cod to North Sea oil (0.5 ppm) resulted in a significant elevation of EROD activity in livers. When alkylphenols were added to the mixture the level of EROD induction was significantly reduced. These authors suggest alkylphenols present in PW are capable of interfering with the oil-mediated CYP1A response in cod. PW contains significant amounts of alkylphenols (Hasselburg et al. 2004). Analysis of raw PW from Hibernia in 2005 and 2006 showed dramatically different levels of phenols. The average total phenol concentration in 2005 PW was 11.934 $mg \cdot L^{-1}$ of which nonylphenol represented 276 $ng \cdot L^{-1}$ (data not shown). In 2006 the raw PW had an average phenol concentration of 3.962 $mg \cdot L^{-1}$ and no detectable levels of nonylphenol. Alkylphenols represented about 50% of the total phenols and were made up of methyl, ethyl and propyl phenols (Table 17.2). It is possible, therefore, that the higher concentrations of alkylphenols in the 2005 samples may have had an inhibitory effect on EROD activity in the mummichog. This remains speculative as experiments were not conducted with mummichog in 2006 and no cod were tested in 2005.

Growth of cod, as indicated by the specific growth rate, was unaffected ($P>0.05$) by exposure to Hibernia PW for 45 days (Fig. 17.5).

3.3 Juvenile Cod VTG

There was no significant difference ($P > 0.05$) in plasma VTG between control male cod and male cod exposed to PW for 45 days (data not shown). This result is consistent with the low level of alkylphenols in Hibernia PW collected in 2006.

McLeese et al. (1981) reported that Atlantic salmon (*Salmo salar*) bioaccumulate alkylphenols. While this study was conducted with freshwater fish, it shows the potential for fish to accumulate alkylphenols and suggests the potential for direct endocrine effects within the organism and for bio-magnification in the food chain. Scott et al. (2006) reported a strong relationship between VTG concentration in male cod and size. They suggest larger (>5 kg) fish are picking up endocrine disrupting compounds in their diet (other fish) as opposed to benthic invertebrates which constitute the main diet items for smaller cod or from water.

4 Conclusion

It is clear from the data presented that PW from the Hibernia platform has no effect on the species tested at environmentally relevant concentrations. The endpoints investigated: EROD induction, growth and embryological abnormalities were only observed at PW concentrations of 1% or greater. These data show that indicators of exposure and of effects are only observed at high concentrations relative to expected environmental exposure. Hamoutene et al. (2007) state that dispersion models predict that PW will be diluted by at least 240 times within 50–100 m of the PW discharge and by 9000 times 20 km from the discharge. Other authors suggest dilution factors of 1000:1 within 50 m of the discharge (Furuholt 1996). Regardless of the models used, it is clear that the concentrations of PW that produced measurable effects in the endpoints investigated in this chapter would only be observed very close to the PW discharge meaning that the risks associated with these endpoints are very small. While our results indicate low risk for the endpoints investigated, other biomarkers have been identified that do appear to be affected by exposure to PW or to representative compounds. Holth et al. (2009) have shown formation of hepatic DNA adducts in cod chronically exposed to low levels of PAHs and alkylphenols. Bohne-Kjersem et al. (2009) exposed cod to 0.01% PW from the North Sea and using proteomics found that a large number of proteins were affected. These proteins play a role in immune systems, fertility, metabolism, morphology and perception.

Acknowledgements The authors thank Dr. Peter Hodson and members of his staff at Queen's University in Kingston, ON, for their analytical support and for training provided to members of our research team. Mr. Tom King of the Bedford Institute of Oceanography (BIO) provided analytical data and Ms. Susan Cobanli, also of BIO, helped in arranging research cruises and collection of the PW. Mr. David Wong and Ms. Tammy Blair of the St. Andrews Biological Station provided analytical and statistical support. The project was supported by the Panel of Energy Research and Development (PERD) and the Department of Fisheries and Oceans.

References

Aas E, Klungsøyr J (1998) PAH metabolites in bile and EROD activity in North Sea fish. Mar Environ Res 46:229–232

Abrahamson A, Brandt I, Brunström B, Sundt RC, Jørgensen EH (2008) Monitoring contaminants from oil production at sea by measuring gill EROD activity in Atlantic cod (*Gadus morhua*). Environ Poll 153:169–175

Allen Y, Matthiessen P, Scott AP, Haworth S, Feist S, Thain JE (1999) The extent of oestrogenic contamination in the UK estuarine and marine environments – further surveys of flounder. Sci Total Environ 233:5–20

Azetsu-Scott K, Yeats P, Wohlgeschaffen G, Dalziel J, Niven S, Lee K (2007) Precipitation of heavy metals in produced water: Influence on contaminant transport and toxicity. Mar Environ Res 63:146–167

Billiard SM, Querbach K, Hodson P (1999) Toxicity of retene to early life stages of two freshwater fish species. Environ Toxicol Chem 18:2070–2077

Bohne-Kjersem A, Skadsheim A, Goksøyr A, Einar-Grøsvik B (2009) Candidate biomarker discovery in plasma of juvenile cod (Gadus morhua) exposed to crude North Sea oil, alkyl phenols and polycyclic aromatic hydrocarbons (PAHs). Mar Environ Res 68:2 68–77

Boudreau M, Courtenay SC, MacLatchy DL, Bérubé CH, Parrott JL, Van Der Kraak GJ (2004a) Utility of morphological abnormalities during early-life development of the estuarine mummichog, *Fundulus heteroclitus*, as an indicator of estrogenic and antiestrogenic endocrine disruption. Environ Toxicol Chem 23:415–425

Boudreau M, Courtenay SC, MacLatchy DL, Bérubé CH, Hewitt LM, Van Der Kraak GJ (2005) Morphological abnormalities during the early-life development of the estuarine mummichog, *Fundulus heteroclitus*, as an indicator of androgenic and anti-androgenic endocrine disruption. Aquat Toxicol 71:357–369

Boudreau M, Sweezey MJ, Lee K, Hodson PV, Courtenay SC (2009) Toxicity of Orimulsion-400 to early life stages of Atlantic herring (*Clupea harengus*) and mummichog (*Fundulus heteroclitus*). Environ Toxicol Chem 28:1206–1217

Boudreau M, Courtenay SC, Van Guelpen L, MacLatchy D, Robertson D, Freeborn M, Bérubé C, Van Der Kraak G (2004b) Are developmental abnormalities a useful endpoint for the wild fish survey element of the Canadian EEM program for pulp and paper mills? In: Borton D (ed) Proceedings, 5th int. conf. on fate and effects of pulp mill effluents. DEStech Publications, Lancaster, PA, pp 491–505

Brinkworth LC, Hodson PV, Tabash S, Lee P (2003) CYP1A induction and blue sac disease in early developmental stages of rainbow trout (*Oncorhynchus mykiss*) exposed to retene. J Toxicol Environ Health 66:627–646

Burnett K, Bain L, Baldwin W, Callard G, Cohen S, De Giulio R (2007) *Fundulus* as the premier teleost model in environmental biology: opportunities for new insights using genomics. Comp Biochem Physiol D2:257–286

Busacker GP, Adelman IR, Goolish EM (1990) Growth. In: Schreck CB and Moyle PB (eds) Methods for fish biology. American Fisheries Society, Bethesda, MD, pp 363–389

Canada-Newfoundland and Labrador Offshore Petroleum Board. Accessed 23 Nov 2009. http://www.cnlopb.nl.ca/stat_rm.shtml

Couillard CM (2002) A microscale test to measure petroleum oil toxicity to mummichog embryos. Environ Toxicol 17:195–202

Furuholt E (1996) Environmental effects of discharge and reinjection of produced water. In: Reed M, Johnsen S (eds) Produced water 2. Environmental issues and mitigation technologies. Plenum Press, New York, NY, pp 275–288

Hamoutene D, Mabrouk G, Samuelson S, Volkoff H, Parrish C, Mansour A, Mathieu A, Lee K (2007) Effect of produced water on cod (*Gadus morhua*) immune responses. In: Kidd KA, Allen Jarvis R, Haya K, Doe K, Burridge LE (eds) Proceedings of the 34th annual aquatic toxicity workshop: September 30 to October 3, 2007, Halifax, Nova Scotia. Can Tech Rep Fish Aquat Sci 2793:59

Hasselburg L, Meir S, Svardal A (2004) Effects of alkylphenols on redox status in first spawning Atlantic cod (*Gadus morhua*). Aquat Toxicol 69:95–105

Helder T (1980) Effects of 2,3,7,8-tetrachlorodibenzo-p-dioxin (TCDD) on early life stages of the pike (*Esox lucius* L.). Sci Total Environ 14:255–264

Hodson PV, Efler S, Wilson JY, El-Shaarawi A, Maj M, Williams TG (1996) Measuring the potency of pulp mill effluents for induction of hepatic mixed function oxygenase activity in fish. J Toxicol Environ Health 49:101–128

Hodson PV, Kloepper-Sams PJ, Munkittrick KR, Lockhart WL, Metner DA, Luxon L, Smith IR, Gagnon MM, Servos M, Payne JF (1991) Protocols for measuring mixed function oxygenases of fish liver. Can Tech Rep Fish Aquat Sci 1829:49 p

Hodson PV, Khan C, Saravanabhavan G, Clarke L, Brown S, Hollebone B, Wang Z, Short J, King T, Lee K (2008) Alkylphenanthrenes, fluorenes and naphthobenzothiophenes are the main components of crude oil that are chronically toxic to fish. In: Kidd KA, Allen Jarvis R, Haya K, Doe K, Burridge LE (eds) Proceedings of the 34th annual aquatic toxicity workshop: September 30 to October 3, 2007, Halifax, Nova Scotia. Can Tech Rep Fish Aquat Sci 2793:62

Holth TF, Beylich BA, Skarphédinsdóttir H, Liewenborg B, Grung M, Hylland K (2009) Genotoxicity of environmentally relevant concentrations of water-soluble oil components in cod (Gadus morhua). Environ Sci Technol. 2009 May 1;43(9):3329–34

Kime DE (1998) Endocrine disruption in fish. Kluwer, Norwell, MA, 356 pp

King TL, Lee K. 2004 Assessment of sediment quality based on toxic equivalent benzo[*a*]pyrene concentrations. In: Proceedings of the 27th arctic and marine oilspill program (AMOP), Edmonton, Alberta, Canada, June 8–10, 2004. Environmental Science and Technology Division, Environment Canada, Ottawa, Ontario, Canada, pp 793–806

Lee RF, Anderson JW. 2005 Significance of cytochrome P450 system responses and levels of bile fluorescent aromatic compounds in marine wildlife following oil spills. Mar Poll Bull 50: 705–723

Lee K, Azetsu-Scott K, Cobanli SE, Dalziel J, Niven S, Wohlgeschaffen G, Yeats P (2005) Overview of potential impacts from produced water discharges in Atlantic Canada. In: Armsworthy SL, Cranford PJ, Lee K (eds) Offshore oil and gas environmental effects monitoring approaches and technologies. Battelle Memorial Institute, Columbus, OH, pp 319–342.

McCarty LS, Power M, Munkittrick KR (2002) Bioindicators versus biomarkers in ecological risk assessment. Human Ecol Risk Assess 8:159–164

McLeese DW, Zitko V, Sergeant DB, Burridge L, Metcalfe CD (1981) Lethality and accumulation of alkylphenols in aquatic fauna. Chemosphere 10:723–730

Neff JM, Johnsen S, Frost TK, Utvik TIR, Durell GS (2006) Oil well produced water discharges to the North Sea. part II: comparison of deployed mussels (*Mytilus edulis*) and the DREAM model to predict ecological risk. Mar Environ Res 62:224–246

Schnell S, Schiedek D, Schneider R, Balk L, Vuorinen PJ, Karvinen H, Lang T (2008) Biological indications of contaminant exposure in Atlantic cod (*Gadus morhua*) in the Baltic Sea. Can J Fish Aquat Sci 65:1122–1134

Scott AP, Katsiadaki I, Wittames PR, Hylland K, Davies IM, McIntosh AD Thain J (2006) Vitellogenin in the blood plasma of male cod (*Gadus morhua*): A sign of oestrogenic endocrine disruption in the open sea. Mar Environ Res 61:149–170

Sturve J, Hasselburg L, Fälth H, Celander M, Förlin L (2006) Effects of North Sea oil and alkylphenols on biomarker responses in juvenile Atlantic cod (*Gadus morhua*). Aquat Toxicol 78S:S73–S78

Thériault MH, Courtenay SC, Munkittrick KR, Chiasson AG (2007) The effect of seafood processing plant effluent on sentinel fish species in coastal waters of the southern Gulf of St Lawrence, New Brunswick. Water Qual Res J Can 42:172–183

von Westernhagen H (1988) Sublethal effects of pollutants on fish eggs and larvae. In: Hoar WS, Randall DJ (eds) Fish physiology, vol 11-the physiology of developing fish part a eggs and larvae. Academic, San Diego, CA, pp 253–346

Walker MK, Spitsbergen JM, Olsen JR, Peterson RE (1991) 2,3,7,8-tetrachlorodibenzo-p-dioxin toxicity in early life stage development of lake trout (Salvelinus namaycush). Can J Fish Aquat Sci 48:875–883

Whyte JJ, Jung RE, Schmitt CJ, Tillitt DE (2000) Ethoxyresorufin-*O*-deethylase (EROD) activity in fish as a biomarker of chemical exposure. Crit Rev Toxicol 30: 347–570

Zhao L, Chen Z, Lee K (2008) A risk assessment model for produced water discharge from offshore petroleum platforms-development and validation. Mar Poll Bull 56:1890–1897

Chapter 18
Microbial Community Characterization of Produced Water from the Hibernia Oil Production Platform

C. William Yeung, Kenneth Lee, and Charles W. Greer

Abstract The Hibernia production platform is the largest oil producing platform off the east coast of Canada. The produced water is the major source of contamination from the platform into the ocean. A comprehensive study on the potential impact of the produced water discharge is needed. Microorganisms can rapidly respond to change, whether negative or positive, and at the population level, are powerful indicators of change in their environment. The objective of this study was to characterize the indigenous microbial community structure, by denaturing gradient gel electrophoresis (DGGE), in the produced water and in seawater around the production platform, and to determine whether the release of produced water is impacting the natural ecosystem. The DGGE results showed that the production water did not have a detectable effect on the bacterial populations in the surrounding water. Cluster analysis showed a >90% similarity for all near surface water (2 m) samples, ~86% similarity for all the 50 m and near bottom (NB) samples, and ~78% similarity for the whole water column from top to bottom across a 50 km range, based on two consecutive yearly sampling events. However, there were distinct differences in the composition of the bacterial communities in the produced water compared to seawater near the production platform (~50% similar), indicating that the effect from produced water may be restricted to the region immediately adjacent to the platform. Specific microorganisms (*Thermoanaerobacter* for eubacteria and *Thermococcus* and *Archaeoglobus* for archaea) were detected as significant components of the produced water. These particular signature microorganisms may be useful as markers to monitor the dispersion of produced water into the surrounding ocean.

1 Introduction

The Hibernia production platform is the largest oil producing platform off the east coast of Canada. The geological formation contains about 1 billion barrels of oil and is currently producing ~200,000 barrels of crude oil per day. The produced

---

C.W. Greer (✉)
National Research Council of Canada, Montreal, QC, Canada, H4P 2R2

K. Lee, J. Neff (eds.), *Produced Water*, DOI 10.1007/978-1-4614-0046-2_18,

water, which contains minor amounts of natural organic (petroleum hydrocarbons, organic acids, alkylphenols) and inorganic (heavy metals, radionuclides) components from the subsurface geologic formation and the chemical amendments that aid in oil production, is the major source of contamination from the platform into the ocean. It is currently discharged into the surrounding marine environment under strict regulation (Canada's revised Offshore Waste Treatment Guideline 2002). In general, there is no evidence of harmful effects on the marine environment from produced water due to rapid dilution. However, the uncertainty of long-term effects still remains. Therefore, a comprehensive study on the potential impact of produced water discharged from the Hibernia production platform is needed.

A number of studies in the North Sea, that deployed fish (Abrahamson et al. 2008; Hylland et al. 2008) and shellfish (Hylland et al. 2008) to monitor the long-term effects of the produced water in the surrounding environment, have shown that the exposure levels were low and caused minor environmental impact at the deployment locations. Therefore, there is a need to develop methods that are sufficiently sensitive to components in produced water at the levels found in the surrounding ecosystem. Microorganisms are typically the first organisms to encounter changes in their environment, and with their short generation times and relatively large population densities, can rapidly respond negatively or positively to change. In light of their sensitivity and rapid response time, we hypothesized that monitoring the microbial community structure may be used to define the extent of impact zones around offshore platforms.

Culture-independent surveys of rRNA genes have greatly expanded knowledge of the microbial community structure and species content. Denaturing gradient gel electrophoresis (DGGE), a method to separate PCR-amplified DNA fragments based on their base composition (Muyzer et al. 1993), is an extensively used rRNA gene screening method for fast assessment of the dominant microbial community members' diversity and dynamics and allows high sample throughput with DNA-based phylogenetic resolution for an entire target community. Because DGGE can handle larger numbers of samples, throughput is often more extensive than clone libraries for comparing microbial assemblages over space and time (Casamayor et al. 2002). DGGE can also provide a quick semi-quantitative analysis of community diversity. Each band on the DGGE gel represents one of the major species in the community. Using the DGGE method to explore variability of the microbial community structure may provide important data on the potential impact of produced water on indigenous marine microbial populations.

The objectives were to

(i) identify the impact of produced water discharges on the environment, if any;
(ii) compare microbial community structure changes related to released produced water;
(iii) identify unique microorganisms from the produced water for tracking in the surrounding seawater.

The results of this research will provide insight for determining the acceptable disposal limits for produced water to minimize the environmental impact, while taking into account the need and cost of produced water treatment and/or disposal by re-injection. The present study used DGGE to compare the microbial community composition by depth and distance, from around 500 m to 50 km, around the production platform.

2 Materials and Methods

2.1 Site Description and Sample Collection

Seawater samples were collected in July 2005 and June 2006 from a number of locations and depths from the Hibernia platform. All seawater depth profile samples were collected using the Seabird Niskin rosette frame (24 × 10 L bottles) containing a Seabird conductivity, temperature, depth detector (CTD). For 2005, depth profile samples (of 2 m, 5 m, and Near Bottom (NB – 0.5 m off the bottom)) were collected at a 500 m location (500 m) and at a reference seawater location 50 km (R50k) from the Hibernia production platform. Also, two surface water samples (<500 mA and <500 mB) were collected at locations within the 500 m exclusion zone using only the Seabird CTD. Similarly, for 2006, depth profile samples (2 m, 5 m, and NB (0.5 m off the bottom)) were collected at two 500 m locations (500 mA and 500 mB) and at the same reference seawater location 50 km (R50k) from the Hibernia production platform. In addition, a sample of fresh produced water was kindly provided by the personnel of the Hibernia production platform. All containers used in the filtration were rinsed three times with the sample water. About 4 L of seawater samples and 2 L of the produced water sample were immediately filtered through sterile 0.22 μm GSWP (Millipore) filters. Following filtration all filters were transferred to sterile 50 mL Falcon tubes and were stored at –20°C until analyzed.

2.2 DNA Extraction

The recovery of total community DNA was performed using a modified version of Fortin's method (Fortin et al. 1998). Prior to lysis treatment, 4.5 mL of sterilized distilled water was added to each 50 mL Falcon tube containing the filter paper. A 500 μL aliquot of 250 mM Tris–HCl (pH 8.0) containing 50 mg lysozyme was added and the samples were incubated for 1 h at 37°C with gentle orbital mixing. Fifty microliters of proteinase K (20 mg/mL) was added to the samples, and they were incubated for 1 h at 37°C with gentle orbital mixing. The lysis treatment was completed with the addition of 500 mL of 20% SDS and 30 min of incubation at 85°C with gentle inversion mixing every 10 min. The filter paper was then removed from the 50 mL Falcon tube. Supernatants were treated with one-half volume of 7.5 M ammonium acetate, incubated on ice for 15 min to precipitate proteins, and centrifuged for 15 min at 4°C (9,400 × *g*). The supernatants were transferred to a sterilized 50 mL Falcon tube and treated with one volume of cold 2-propanol. The DNA was precipitated overnight at –20°C. Samples were centrifuged at 4°C for 30 min (9,400 × *g*). Pellets were washed with 70% cold ethanol and air-dried. The

DNA was re-suspended in 250 μL of Tris–EDTA (pH 8.0). DNA concentrations were estimated by running 5 μL of purified material against the Lambda *Hind*III DNA ladder (Amersham Biosciences, Piscataway, NJ) standard on a 1.4% agarose gel with SYBR Safe (Molecular Probes, Eugene, OR, USA).

2.3 PCR Amplification of the 16S rRNA Gene

For PCR amplification of the 16S rRNA gene, the bacteria-specific forward primer U341F (5′-CCTACGGGAGGCAGCAG-3′) and the reverse primer U758R (5′-CTACCAGGGTATCTAATCC-3′) were used. These primers, complementary to conserved regions of 16S rRNA gene, were used to amplify a 418-bp fragment corresponding to positions 341–758 in the *Escherichia coli* sequence (Muyzer et al. 1993) and covered the variable regions V3 and V4. The bacteria-forward primer used for DGGE possessed a GC clamp (5′-GGCGGGGCGGGGGCACGGGGGGCGCGGCGGGCGGGGCGGGGG-3′) at the 5′ end. This GC-clamp stabilized the melting behavior of the amplified fragments (Sheffield et al. 1989). The archaea-specific forward primer ARC344F (5′-ACGGGGYGCAGCAGGCGCGA-3′) and the reverse primer ARC915R (5′-GTGCTCCCCCGGCAATTCCT-3′) were used to amplify archaea DNA. These primers amplified a 572-bp fragment corresponding to positions 344–915 in the *E. coli* sequence. The archaea-forward primer used for DGGE possessed another GC clamp (5′-CGCCCGCCGCGCCCCGCGCCCGTCCCGCCGCCCCCGCCCG-3′) at the 5′ end. Each 50 μL PCR mixture contained 1 μL of the template DNA (undiluted, 10^{-1} or 10^{-2}), 25 pmol of each oligonucleotide primer, 200 μM of each dNTP, 1 mM $MgCl_2$ and 2.5 units of Taq polymerase (Amersham Biosciences, Piscataway, NJ, USA) in 1x Taq polymerase buffer (10 mM Tris–HCl pH 9.0, 50 mM KCl, 1.5 mM $MgCl_2$). Briefly, after an initial temperature of 96°C for 5 min and thermocycling at 94°C for 1 min, the annealing temperature was set to 65°C (for bacterial PCR) or 60°C (for archaeal PCR) and was decreased by 1°C at every cycle for 10 cycles, and 1 min (bacteria) or 3 min (archaea) elongation time at 72°C. Additional cycles (15–20) were performed with annealing temperatures of 55°C for bacterial and 50°C for archaeal. PCR products were loaded onto a 1% agarose gel with SYBR Safe (Molecular Probes, Eugene, OR, USA) with a 100-bp ladder (MBI Fermentas, Amherst, NY, USA) to determine the presence, size, and quantity of the PCR products.

2.4 Denaturing Gradient Gel Electrophoresis (DGGE)

The 16S rRNA gene PCR products from eight PCR reactions were combined for each sample and concentrated by ethanol precipitation for DGGE analysis. For both eubacteria and archaea DGGEs, about 650 ng of 16S rRNA gene PCR product from each sample was applied to a lane, and was analyzed on 8% poly-acryalmide gels containing gradients of 35–65% denaturant (7 M urea and 40% deionized formamide were considered to be 100% denaturant). DGGE was done with a DCode Universal Mutation Detection System (Bio-Rad). Electrophoresis was run at a constant voltage of 80 V for 16 h at 60°C in 1x TAE (40 mM Tris–acetate, 1 mM EDTA, pH 8.3) running buffer. The gels were then stained with VistaGreen (Amersham Biosciences, Piscataway, NJ, USA). The gels were imaged with the FluoroImager System Model 595 (Molecular Dynamics, Sunnyvale, CA, USA). The gel images were analyzed with GelCompar II v4.6 (Applied Maths, Sint-Martens-Latem, Belgium) to generate dendrogram profiles. The genotypes were visually detected based on presence or absence of bands in the different lanes. A band was defined as "detected" if the ratio of its peak height to the total peak height in the profile was >1%. After conversion and normalization of gels, the degrees of similarity of DNA pattern profiles were computed using the Dice similarity coefficient (Dice 1945), and dendrogram patterns were clustered by the Unweighted Pair Group Method using Arithmetic average (UPGMA) groupings with a similarity coefficient (S_{AB}) matrix. Also, some of the individual bands from produced water DGGEs were excised, eluted, and re-amplified with the same set of primers without the GC-clamp. The PCR products from DGGE band re-amplification were sequenced at the Université Laval Plate-forme d'analyses biomoléculaires using a model ABI Prism 3130XL (Applied Biosystems, Foster City, CA, USA) with their respective primers. The sequences were edited and were tentatively identified using NCBI BLASTN search (http://ncbi.nlm.nih.gov/blast/).

3 Results and Discussion

Two eubacterial DGGEs for samples from 2005 and 2006 and one archaeal DGGE for samples from 2005 were analyzed.

From both 2005 and 2006 eubacterial DGGEs, the cluster analyses showed a >90% similarity for all the near surface water (2 m) samples from within 500 m to the 50 km reference seawater location, and a high similarity (~86%) for deeper seawater (50 m and NB) from the 500 m and 50 km reference locations (Figs. 18.1 and 18.2). There was also a high similarity (>78%) of eubacterial community structure for the whole water column from top to bottom (Figs. 18.1 and 18.2). However, the composition of the eubacterial community in the produced water was only about 50% similar to the seawater around the production platform (Figs. 18.1 and 18.2), even though the major component of the produced water comes from the injected surrounding seawater.

The 2005 archaeal DGGE also revealed a high similarity of archaeal community structure in the surrounding seawater, with 79% for the surface seawater (2 m) and 67.7% for the bottom seawater (50 m & NB), from within 500 m to the 50 km reference location (Fig. 18.3). Like the eubacterial DGGE results, there is a higher similarity (34%) value for archaeal community structure from top to bottom in the

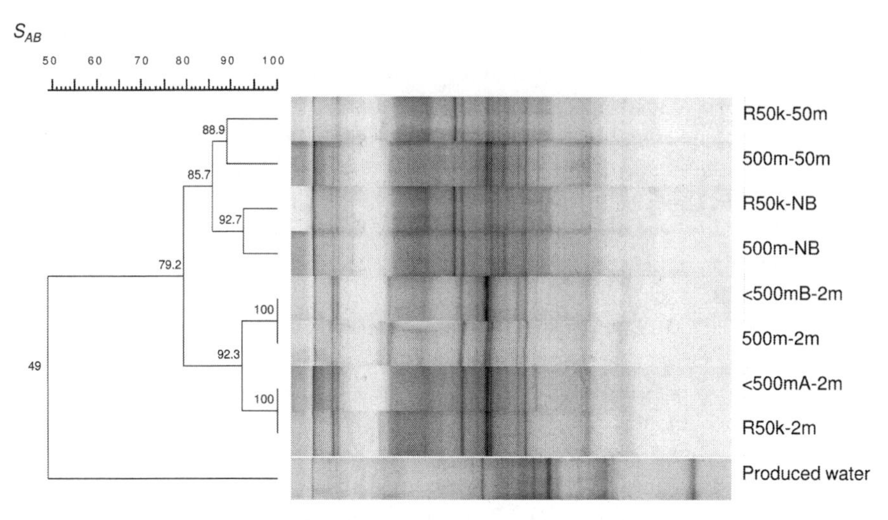

Fig. 18.1 Cluster analysis of Hibernia 2005 surrounding seawater and Hibernia 2006 produced water eubacterial DGGE banding patterns based on band positions using UPGMA of a S_{AB} matrix

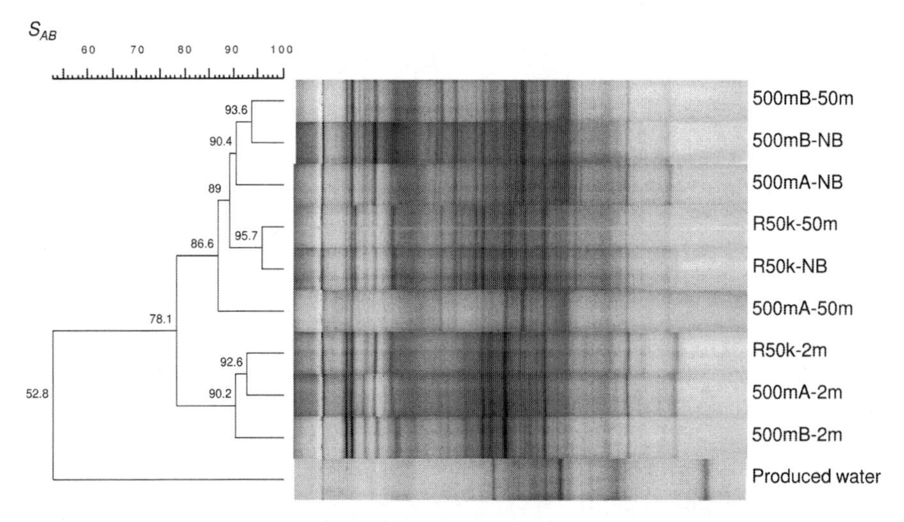

Fig. 18.2 Cluster analysis of Hibernia 2006 surrounding seawater and produced water eubacterial DGGE banding patterns based on band positions using UPGMA of a S_{AB} matrix

water column in comparison to the produced water archaeal community structure (9.4%) (Fig. 18.3).

The results from both eubacterial and archaeal DGGEs showed a high similarity for the water column both horizontally and vertically, suggesting that there is a fairly stable microbial community in the surrounding seawater. Therefore, any abnormal

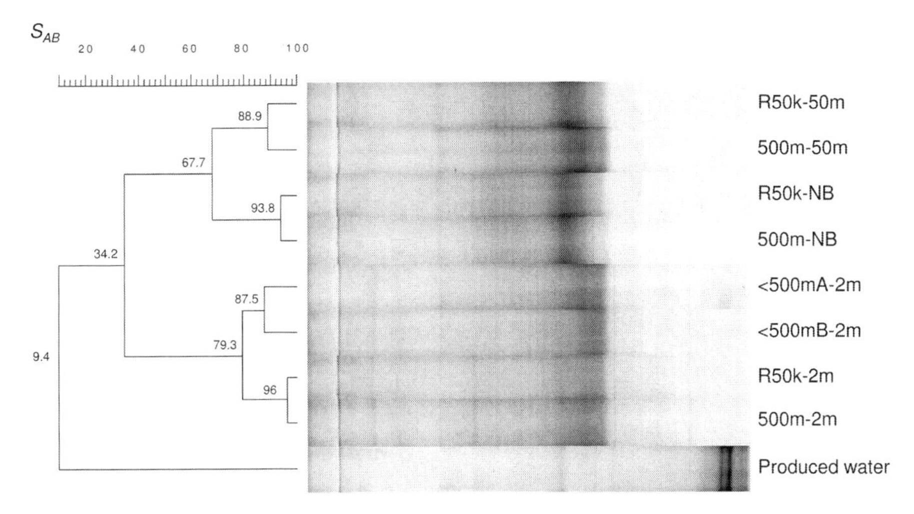

Fig. 18.3 Cluster analysis of Hibernia 2005 surrounding seawater and Hibernia 2006 produced water archaeal DGGE banding patterns based on band positions using UPGMA of a S_{AB} matrix

changes in the microbial community in the water column could potentially be used as an indicator of foreign input. Furthermore, the DGGE pattern from produced water indicated that the microbial community structure was distinctly different from the seawater microbial community structure near the production platform, suggesting that the microbial community structure in the produced water is unique. Thus, the findings of stable microbial communities in the surrounding seawater may indicate that any effects from the produced water are restricted to the region immediately adjacent to the platform or on the sediments.

From the eubacterial and archaeal produced water DGGEs, a number of species (*Thermoanaerobacter sp.* for eubacteria and *Thermococcus sp.* and *Archaeoglobus sp.* for archaea) were identified through band excision and sequencing. The anaerobic and thermophilic nature of these species was unique to the produced water and strongly suggests that they originate from the geological formation. Furthermore, these species were detected as significant components of the produced water, but appeared to be below detection limits in the surrounding water outside the 500 m exclusion zone (data not shown). The uniqueness of these produced water specific species may be useful as targets to monitor the dispersion of produced water in the surrounding marine ecosystem. The discovery of these produced water specific species has enabled us to design species-specific 16S rRNA gene primers for the quantitative PCR (q-PCR) of target microorganisms in the marine environment following the discharge of the produced water.

Acknowledgments The authors thank Susan Cobanli, Nathalie Fortin, Sylvie Sanschagrin, and Marc Auffret for their excellent technical assistance. This research was supported by Fisheries and Oceans Canada and the National Research Council of Canada.

References

Abrahamson A, Brandt I, Brunström B, Sundt R, Jørgensen E (2008) Monitoring contaminants from oil production at sea by measuring gill EROD activity in Atlantic cod (*Gadus morhua*). Environ Pollut 153:169–175

Casamayor E, Massana R, Benlloch S, Øvreås L et al (2002) Changes in archaeal, bacterial and eukaryal assemblages along a salinity gradient by comparison of genetic fingerprinting methods in a multipond solar saltern. Environ Microbiol 4:338–348

Dice L (1945) Measures of the amount of ecologic association between species. Ecology 26: 297–302

Fortin N, Fulthorpe R, Allen D, Greer C (1998) Molecular analysis of bacterial isolates and total community DNA from kraft pulp mill effluent treatment systems. Can J Microbiol 44:537–546

Hylland K, Tollefsen K, Ruus A et al (2008) Water column monitoring near oil installations in the North Sea 2001–2004. Mar Pollut Bull 56:414–429

Muyzer G, De Waal E, Uitterlinden A (1993) Profiling of complex microbial populations by denaturing gradient gel electrophoresis analysis of polymerase chain reaction-amplified genes coding for 16S rRNA. Appl Environ Microbiol 59:695–700

Sheffield V, Cox D, Lerman L, Myers R (1989) Attachment of a 40-base-pair G + C-rich sequence (GC-clamp) to genomic DNA fragments by the polymerase chain reaction results in improved detection of single-base changes. Proc Natl Acad Sci USA 86:232–236

Chapter 19
Application of Microbiological Methods to Assess the Potential Impact of Produced Water Discharges

Kenneth Lee, Susan E. Cobanli, Brian J. Robinson, and Gary Wohlgeschaffen

Abstract Microbial production and activity in produced water directly recovered from the discharge stream of offshore oil and gas production facilities off the east coast of Canada were examined before and after aeration in a series of concentrations to determine the effect of dilution at sea. Aeration and dilution resulted in reduced toxicity due to volatilization and oxidation of the lighter hydrocarbons including polycyclic aromatic hydrocarbons (PAHs), alkylated PAHs, benzene, toluene, ethylbenzene, xylene, and short-chain alkanes (C_{10}–C_{14}). A fraction of the detrimental effects on microbial productivity and activity could also be attributed to the elevated salinity associated with produced water. These results suggest that caution should be used in the manipulation of produced water samples used for toxicity/risk assessment studies.

1 Introduction

Produced water represents the largest volume waste stream in oil and gas production operations on most offshore platforms (Chapter 1, this volume; Krause 1995) It typically contains inorganic compounds (trace metals, nutrients), volatile aromatic compounds such as BTEX (benzene, toluene, ethylbenzene, xylenes), polycyclic aromatic hydrocarbons (PAHs such as naphthalene), phenols, organic acids, and additives (Manfra et al. 2007). There is concern that the ocean discharge of produced water with its associated manufactured and naturally occurring chemicals may pose adverse impacts to the marine ecosystem (Din and Abu 1992; Krause et al. 1992; Stromgren et al. 1995; Stagg and McIntosh 1996; Holdway 2002; Querbach et al. 2005; Hamoutene et al. 2010; Perez-Casanova et al. 2010). Despite predicted and measured high rates of dilution for the produced water plume following discharge (Neff 1987), a level of concern over its discharge at sea still remains for several reasons including the following: (1) the volume of produced water discharge typically increases as reservoirs become depleted, and (2) recent studies have suggested that some components in produced water such as metals, high molecular weight aromatic

K. Lee (✉)
Centre for Offshore Oil, Gas and Energy Research (COOGER), Fisheries and Oceans Canada, Bedford Institute of Oceanography, Dartmouth, NS, Canada B2Y 4A2

K. Lee, J. Neff (eds.), *Produced Water*, DOI 10.1007/978-1-4614-0046-2_19,

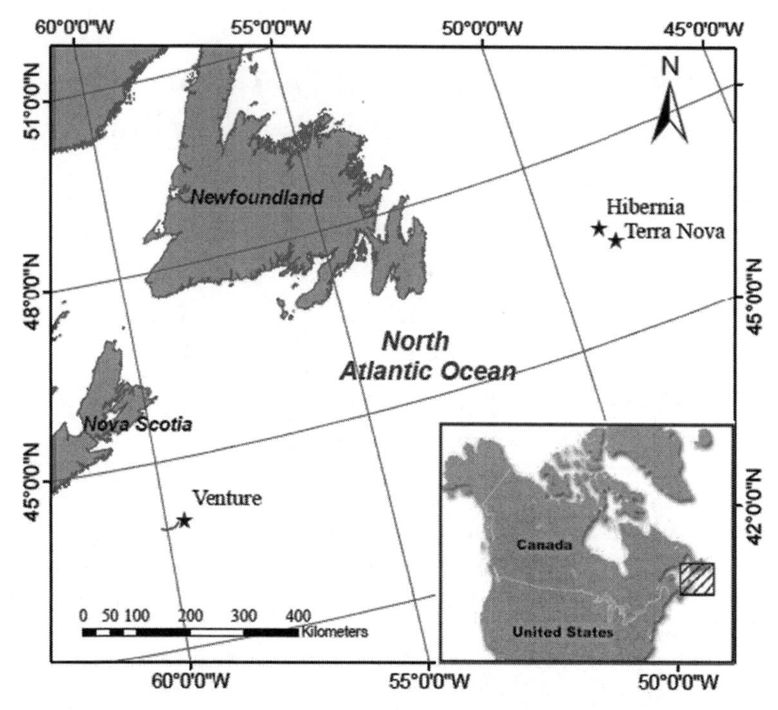

Fig. 19.1 Locations of offshore oil and gas production sites for sample collection off the east coast of Canada

hydrocarbons and saturated hydrocarbons may accumulate in the sediments and the surface micro-layer as a result of natural physico-chemical processes (Neff 2002; Marty and Saliot 1976; Lee et al. 2005; Azetsu-Scott et al. 2007).

Site specific studies are needed to assess the impact of produced waters as its chemical constituents can vary, depending on the location of its extraction within various parts of the oil and gas reserve and the waste treatment technologies used prior to its discharge (Collins 1975). To determine the environmental risk of produced water discharges in waters off the east coast of Canada, changes in the microbial production and activity before and after aeration were investigated in samples of produced water collected from the Hibernia Gravity Based Structure (GBS), the Terra Nova Floating Production, Storage, and Offloading (FPSO) unit on the Grand Banks, and the Venture platform on the Scotian Shelf of Canada (Fig. 19.1, Table 19.1). These experiments were done to elucidate the effect of dilution on the toxicity of produced water when discharge occurs at sea.

2 Methods

2.1 Preparation of Produced Water Sample Containers

Produced water was collected in new, high density polyethylene (HDPE) 4 L wide mouth bottles, 10 L HDPE Nalgene®jerricans, and 4 L amber glass bottles (Fisher

Table 19.1 Volume (m^3) of produced water discharged by platforms during the sample collection period (C-NLOPB 2011; CNSOPB 2011)

Platform	Hibernia	Hibernia	Terra Nova	Venture
Month	April	July	July	July
Year	2004	2008	2008	2009
For month	312,493	592,101	374,783	8,449
For year	4,882,460	7,175,520	3,692,715	135,730
Cumulative as of 31 Aug 2010	45,075,904		20,325,220	1,551,264

Scientific). The wide mouth plastic bottles, which were specifically designed to study the influence of chemical kinetics on produced water toxicity, were fitted with conical ultra high molecular weight polyethylene lid inserts to eliminate a headspace with the closure of the bottles. The bottles and jerricans were cleaned by filling with 1 M HCl for 48–72 h, rinsing five times with de-ionized, distilled water, air drying, and sealing. Amber glass bottles (4 L) were solvent rinsed with acetone, hexane and dichloromethane and air dried prior to closure.

2.2 Collection of Produced Water

Produced water sample collection was performed onboard Hibernia, Terra Nova, and Venture by personnel associated with each production facility. Samples were drawn from a point in the process stream after treatment for regulatory compliance and immediately prior to discharge. The bottles were filled to overflowing to eliminate any headspace and returned to the laboratory as soon as logistically possible. Samples were maintained at 4°C in the dark prior to testing. Dilution experiments commenced from the day of collection to as long as one week after collection. Details of collection dates and experimental commencements are provided in Table 19.2.

2.3 Sub-sampling of Produced Water

Sealed samples from the field were opened under nitrogen within a glove box (Hibernia 2004 samples) or within a fume hood (Hibernia 2008, Terra Nova and Venture samples) for the distribution of subsamples immediately prior to analysis for

Table 19.2 Produced water collection, date of experimental manipulation, and test conditions

Source	Sampling date	Salinity (ppt)	pH	Expt. start date – fresh PW	Expt. start date – aerated PW	Aeration (h)	Incubation temp (°C)
Hibernia	16 Apr 04	46.5	7.4	20 Apr 04	3 May 04	68	2.4–3.2
Hibernia	13 Jul 08	45	8	13 Jul 08	14 Jul 08	25	15.2–16.5
Terra Nova	7 Jul 08	45	8	8 Jul 08	11 Jul 08	70.75	15.2–16.5
Venture	9 Jul 09	201.3	6.5	16 Jul 09	18 Jul 09	46	15–16

- microbiology dilution experiments (500 mL Nalgene® bottles),
- nutrients (2 × 60 mL acid rinsed plastic bottles, frozen at –20°C),
- salinity (WTWLF 197 conductivity meter),
- pH (Orion® 230A pH meter or Colorphast EM Reagents test strips),
- dissolved metals (1 × 1 L Teflon bottle),
- organics (1 L for PAH and hydrocarbons, 1 L for phenols, in a 2.3 L solvent rinsed amber glass bottle, acidified to pH <2 with 6N HCl),
- benzene, toluene, ethylbenzene and xylene (BTEX) analysis (2 × 40 mL purge and trap vials, refrigerated at 4°C and analyzed within 14 days of collection).

2.4 Microbiological Experiments

2.4.1 Aeration Experiments

Upon contact with oxygenated seawater, produced water is oxidized, resulting in the transformation of complex hydrocarbons into simpler molecules and the precipitation of metals out of solution, forming flocs (Azetsu-Scott et al. 2007). To examine the differences between "fresh" un-oxidized produced water and that which has reached equilibrium at sea (i.e. on mixing with oxygenated seawater during discharge), subsamples of the produced water were aerated by the addition of compressed air through an air-stone (Table 19.2). Measurements of pH and salinity were made immediately following the aeration period and subsamples were collected for bacterial productivity (^{3}H thymidine) and activity (^{14}C glutamic acid) experiments.

2.4.2 Dilution Experiments

Dilution experiments were conducted using the freshly collected produced water recovered from the Hibernia, Terra Nova, and Venture production sites to examine the effects on bacterial productivity (^{3}H thymidine) and activity (^{14}C glutamic acid). Experiments were conducted in the laboratory (2004, 2009) and at sea (2008).

Freshly dispensed produced water was mixed with fresh, unfiltered seawater collected immediately prior to the start of each experiment, to prepare the following concentration series (as % produced water) in 2004: 0, 0.25, 0.5, 1, and 10%. In 2008 and 2009, concentrations of 2.5, 5, and 20% were added to the test series. Due to the high dilution rate following discharge (SOEP 1996; Neff 2002), these concentrations represented worst-case conditions expected to be encountered in close proximity to the platform.

2.4.3 Assessing the Influence of Salinity

Salinity has been recognized as a component of produced water that can cause or contribute to toxic effects on marine organisms (Neff 2002) including indigenous bacteria. The salinity of Venture produced water was over 200 parts per thousand (ppt) – almost seven times greater than that of the surrounding seawater. To determine the effect of salinity on the bacterial response, a dilution experiment was run concurrently with Venture produced water, whereby sodium chloride was added to the seawater as an additional treatment to match the salinity of each dilution of produced water.

2.4.4 Assessing Bacterial Productivity

Bacteria contain one chromosome. When a cell grows and divides, every new chromosome that is synthesized represents a new bacterial cell; hence, there is a direct correlation between rates of DNA synthesis and cell division. Bacterial productivity was measured by quantifying the rates of ^{3}H-thymidine incorporation into DNA (Fuhrman and Azam 1982) using protocols adapted from Bell (1993) and Li et al. (1993).

For each produced water test concentration, 4 replicates of 32 mL were measured into 50 mL culture tubes and high specific activity ^{3}H thymidine (Methyl-^{3}H, Perkin Elmer, 72.2–87.3 Ci/mmol, cat # NET-027Z) was immediately added for a final concentration of 5 nM. Tubes were capped and vortexed.

Triplicate aliquots of 10 mL from two replicate samples were filtered immediately ("zero time filtration blanks") onto 0.2 μm Nuclepore® polycarbonate membrane filters, rinsed with freshly filtered (0.2 μm Millipore®) seawater, then folded in half, placed in a glassine envelope and immediately frozen at –20°C. After 3 h, the filtration process was repeated for the remaining 2 tubes which had been incubated in the dark at ambient water temperature (Table 19.2). All filters in glassine envelopes were maintained frozen until extraction of total radiolabeled macromolecules.

Individual frozen polycarbonate filters were subsequently placed into an 18 × 150 mm glass test tube in an ice bath at 0°C to thaw for 15 min and then re-suspended in 5 mL of ice cold 5% (v/v) trichloroacetic acid (TCA) for 20 min to precipitate DNA. Then the contents were filtered onto a Whatman® GF/f filter, rinsed twice with 5% TCA followed by 1 mL of 95% ethanol to remove thymidine adsorbed to cell wall lipids, but not incorporated into DNA. Both the GF/f and polycarbonate filters were placed into 20 mL glass scintillation vials and were digested by addition of 0.25 mL of hyamine hydroxide, for 30 min at 55°C. After cooling, the pH was adjusted by adding 50 μL of glacial acetic acid. This was followed by the addition of 10 mL of liquid scintillation cocktail (Beckman Ready Safe). Samples were maintained in the dark until radioactivity was measured using a Beckman-Coulter LS 6500 liquid scintillation counter and data expressed as disintegrations per minute (DPM).

Equations from Bell (1993) were used to calculate productivity as the rate of thymidine uptake in moles/L/h. Conversion factors provided by Ducklow and Carlson (1992) and Li et al. (1993) were used to calculate this rate in terms of cells (cells/L/h) and grams of carbon (gC/L/h). Data were normalized and expressed as a percent of the control value (unfiltered seawater collected on the day of the experiment taken to be 100%).

2.4.5 Assessing Bacterial Activity

Since produced water contains organic acids, changes in heterotrophic uptake may be a suitable means to monitor the potential biological effect of produced water discharges. In the current study, the relative rates of heterotrophic activity in seawater were determined by the amount of radio-labeled organic substrate (^{14}C-glutamic acid) added at one substrate concentration, and taken up in a unit of time (Griffiths et al. 1977). For each dilution of produced water, eight replicates of 10 mL were dispensed into glass test tubes (8 mm $\times$ 150 mm), and 0.1 μmol of $^{14}C(U)$-glutamic acid (Perkin Elmer, Specific activity 250–278 mCi/mmol, cat # NEC-290E) was added to each tube. Blanks were prepared by adding 1 mL of 6N H_2SO_4 to two of the replicates prior to isotope addition. Tubes were capped with respiration traps consisting of a rubber stopper fitted with a plastic center well cup (Kontes) containing a Whatman® 25 mm GF/c filter folded in an accordion-like manner. All samples were incubated in the dark for 4 h at ambient seawater temperatures (Table 19.2). The incubation period was terminated and $^{14}CO_2$ expelled by injecting 1 mL of 6N H_2SO_4 into each tube using a disposable 23G needle fitted to a 1 mL syringe. Another 1 mL syringe fitted with a 23G needle was used to inject 0.1 mL of β-phenethylamine onto the GF/c filter in order to absorb the expelled $^{14}CO_2$.

After 12 h, the rubber stopper with the well was removed from the test tube, and the GF/c respiration filter taken with forceps and placed in a glass scintillation vial along with 10 mL of liquid scintillation cocktail (Beckman Ready Safe). The remaining sample in the tube was used to measure uptake. It was filtered onto Millipore® nitrocellulose MF filters (0.22 μm, 25 mm), rinsed three times with 5 mL fresh, filtered (0.22 μm Millipore®) seawater and placed in a glass scintillation vial along with 1 mL of ethyl acetate to dissolve the filter before addition of 10 mL of cocktail. Radioactivity of these uptake and respiration samples was measured on a Beckman-Coulter LS 6500 liquid scintillation counter, and results expressed in DPM.

Uptake and respiration rates were calculated as ng/L/h (Parsons et al. 1984) and summed to determine gross uptake. Data were normalized and expressed as a percent of the control value (unfiltered seawater collected on the day of the experiment taken to be 100%).

2.5 Chemistry Analysis

2.5.1 PAH and Aliphatic Hydrocarbons

PAH and aliphatic hydrocarbons were analyzed using a modified version of EPA Method 8270. The 1 L produced water sample was spiked with a surrogate standard containing a series of deuterated aliphatic and aromatic hydrocarbons, and extracted with 3 $\times$ 50 mL of dichloromethane in a separatory funnel. The solvent was concentrated on a TurboVap and the extract purified on a silica gel column. The purified extract was exchanged into isooctane and spiked with internal standards. Samples were analyzed using an Agilent 6890 Gas Chromatograph (GC) coupled to a 5975 Mass Spectrometer (MS). The column was a Supelco MDN-5s of 30 m length × 250 μm internal diameter × 0.25 μm film thickness, with a 1 m retention gap of deactivated fused silica. A 1 μL aliquot was injected using the oven track mode. Helium was the carrier gas with a flow rate of 1.0 mL/min. The oven temperature program was set to hold at 85°C for 2 min, followed by a ramp to 280°C at 4°C/min held for 20 min. Total run time was 70.75 min. The MS was operated in the selected ion monitoring (SIM) mode with specific ions and retention windows applied for each compound. Samples were calibrated against a seven-point curve containing a mixture of aliphatic hydrocarbons as well as parent and alkyl PAH. For some of the alkyl PAH where standards were not available, the response of the parent PAH was used for quantification.

2.5.2 Alkylated and Nonyl Phenols

Phenols were processed according to a modified version of EPA Method 8041. The 1 L produced water sample was acidified with 6N HCl to a pH <2 and extracted in a separatory funnel with dichloromethane, concentrated on a TurboVap, the extract exchanged into hexane and spiked with an internal standard of deuterated phenol. Samples were analyzed with an Agilent 6890 GC setup as before (Section 2.5.1), except that the oven temperature was held at 55°C for 2 min followed by a 10°C/min ramp to 100°C held for 2 min, a 1°C/min ramp to 115°C held 2 min, and a 20°C/min ramp to 220°C held 4 min for 34.75 min total run time. The MS was set up as before (Section 2.5.1) and calibration was done with a ten-point curve.

2.5.3 BTEX

The 40 mL purge and trap samples for BTEX were analyzed using a modified version of EPA Method 8240. The analytical system consisted of a Teledyne Tekmar purge and trap system, coupled to an Agilent 6890 GC and a 5973N MS. An auto sampler was used to dispense 5 mL into the purge chamber. The sample was purged with helium for 11 min, followed by a desorption time of 2 min. The GC column was set up (Section 2.5.1) and run in split/splitless mode with a ratio of 50:1. Helium was the carrier gas at 1.0 mL/min. The oven was set to hold 50°C (8 min) followed by a 4°C/min ramp to 60°C and then a 40°C/min ramp to 280°C (total run = 18 min). The MS was run in SIM mode and samples were calibrated against a nine-point curve.

2.5.4 Inorganics

Inorganics in fresh produced water were quantified by RPC Science and Engineering (Fredericton, NB, Canada) by inductively coupled plasma mass spectrometry analysis of a highly diluted sample using Standards Council of Canada (SCC) approved procedures 4.M01 and 4.M29. Mercury was analyzed using cold vapor atomic fluorescence spectroscopy (SCC approved procedure 4.M52).

Nutrients (ammonia, silicate, nitrate, and nitrite) were analyzed by segmented flow analysis (Technicon II). The determination of soluble silicates (Technicon Industrial Method No. 186–72W released March 1973, adapted from Strickland and Parsons 1972) in seawater was based on the reduction of a silicomolybdate in acidic solution to "molybdenum blue" by ascorbic acid which is read colorimetrically at 660 nm. Oxalic acid is introduced to the sample stream before the addition of ascorbic acid to eliminate interference from phosphate.

The determination of nitrate/nitrite followed Technicon Industrial Method No. 158–71W released December 1972 (adapted from Armstrong et al. 1967; Grasshoff 1969; U.S. Department of the Interior 1969). The method was based on the principle of nitrate reduction to nitrite by a copper–cadmium reductor column. The nitrite ion then reacts with sulphanilamide under acidic conditions to form a diazo compound. This compound then couples with N-1-naphlylethylenediamine dihydrochloride to form a reddish-purple azo dye, which is read colorimetrically at 550 nm. Nitrite is determined with identical chemistry but omitting the copper–cadmium column from the sample stream.

The method for ammonia (Kerouel and Aminot 1997) was based on the reaction of ammonia with ortho-phthalaldehyde (OPA) and sulfite to form an intense fluorescent product. Determination is done fluorometrically with excitation at 370 nm and emission at 418–700 nm.

3 Results

3.1 Microbiology

3.1.1 Aeration and Dilution Experiments

In the 2004 Hibernia sample, microbial productivity rates at 85–100% of control values were only slightly depressed in the aerated samples over a concentration range of 0.25–1% produced water. However, activity was significantly suppressed to 20% of the control value in fresh and aerated test samples at a 10% produced water concentration level (Fig. 19.2a). Despite higher rates of bacterial productivity in the aerated samples at produced water concentrations <1%, analysis of the data showed that very little glutamic acid was metabolized. This might have been because of the abundance of alternative or preferred substrate present in the produced water.

In terms of relative heterotrophic activity, while significant suppression was also observed at the higher experimental concentration (10% produced water), at lower concentrations (<1% produced water) there was little effect on microbial response (Fig. 19.2b).

In the 2008 Hibernia sample, both bacterial productivity and relative heterotrophic activity rates remained close to control levels or were positively stimulated in the aerated samples in produced water concentrations up to 10%

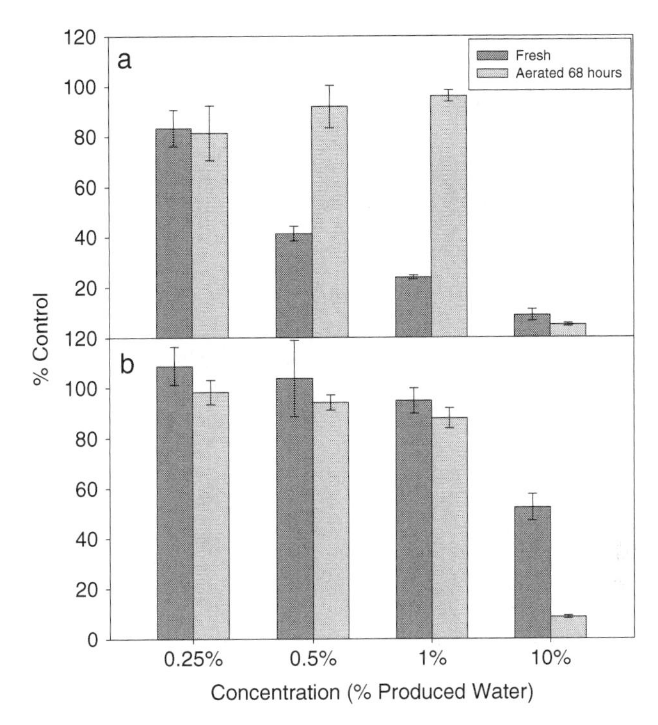

Fig. 19.2 Bacterial productivity (**a**) and activity (**b**) in fresh and aerated Hibernia 2004 produced water

(Fig. 19.3). For fresh produced water, the average microbial productivity rates fluctuated between about 55 and 95% of the control values (Fig. 19.3a), while relative heterotrophic activity rates remained at about 75% of the control value in the 0.25–5% concentration range and declined at 10 and 20% (Fig. 19.3b).

In the aerated samples there was no significant suppression in bacterial productivity and glutamic acid metabolism rates at concentrations below 20% produced water.

The trend was even clearer and more striking in the Terra Nova 2008 fresh sample compared with the aerated (Fig. 19.4). Bacterial productivity rates were sustained at 90–120% of the control values for seawater samples containing fresh produced water at a concentration of 0.25–5%. While an inhibitory effect was observed in microbial productivity and heterotrophic activity at higher concentrations (10–20%) of fresh produced water, in the aerated samples microbial productivity was clearly stimulated and continued to increase to a maximum average of about 380% of the control value in the 10% concentration, then declined to about 280% of the control value at 20% produced water (Fig. 19.4a).

Activity in the fresh sample was maintained at an average of about 75–105% of the control value over a concentration range of 0.25–5% produced water before

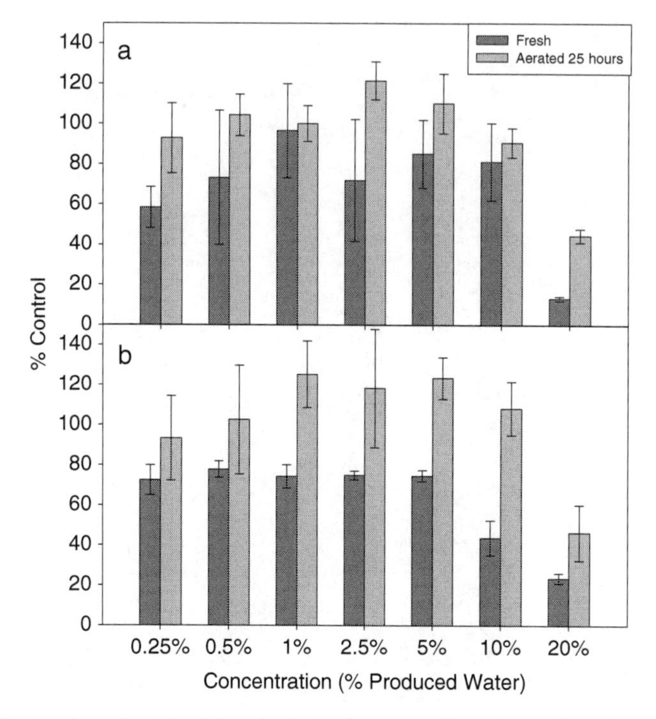

Fig. 19.3 Bacterial productivity (**a**) and relative heterotrophic activity (**b**) in fresh and aerated Hibernia 2008 produced water

showing an inhibitory effect (Fig. 19.4b). In contrast, activity in the aerated sample remained at a value within 80% of the control rates (Fig. 19.4b).

For the 2009 Venture sample, dilution experiments showed an inhibitory effect on bacterial productivity and heterotrophic uptake potential in seawater containing fresh produced water at concentrations as low as 0.5% (Fig. 19.5). Over the lower experimental concentration range (up to 2.5% produced water), aeration significantly reduced the inhibitory effect.

3.1.2 Assessing the Influence of Salinity

As the salinity values for produced water from the Venture platform were extremely high at 200 ppt, an effort was made to elucidate the influence of salinity on the results of microbial assays. This was accomplished by adding additional treatments to the dilution experiment (a sample set with NaCl additions only) to account for the salinity toxicity associated with produced water.

Under the experimental test conditions (Fig. 19.6), salinity associated with produced water from Venture could account for a fraction of the observed inhibition in bacterial productivity and activity at concentrations >2.5%.

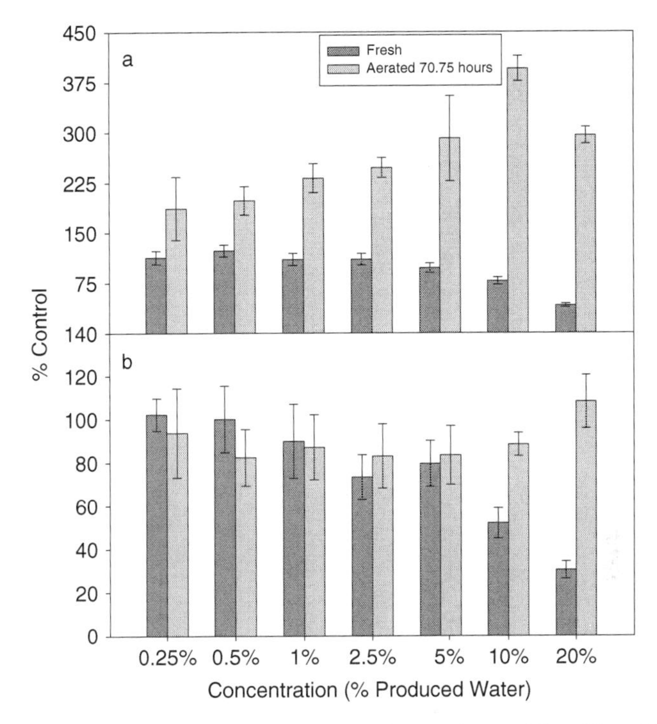

Fig. 19.4 Bacterial productivity (**a**) and activity (**b**) in fresh and aerated Terra Nova 2008 produced water

3.2 Chemistry

3.2.1 Organics

The organics composition of the produced water collected from the three different rigs was similar (Fig. 19.7), with BTEX and phenols dominating. Alkane concentrations were also similar among the three locations, while PAH concentrations followed the order of Venture > Terra Nova > Hibernia. The PAH content of Terra Nova and Hibernia produced water consisted of 50–75% naphthalene (parent + alkylated), and Venture was 85%.

Samples were aerated to simulate the effects of chemical kinetic reactions after produced water discharged into the open sea reaches equilibrium. The loss of organics from aeration (Fig. 19.8) followed the order of their relative vapor pressures with BTEX almost completely removed, followed by PAHs (dominated by highly volatile naphthalene) and their alkylated homologs, phenols, and longer-chain alkanes (C_{15}–C_{30}).

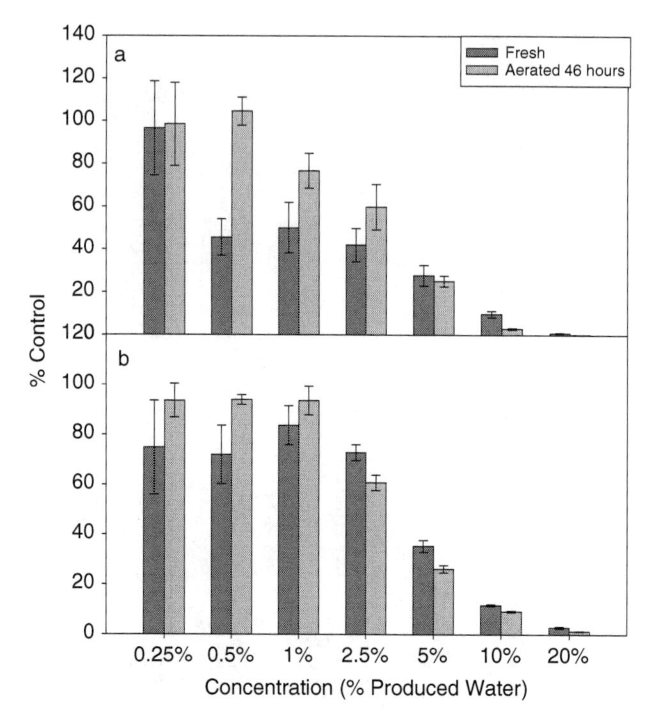

Fig. 19.5 Bacterial productivity (**a**) and activity (**b**) in fresh and aerated Venture 2009 produced water

3.2.2 Inorganics

Inorganic constituents were relatively similar among the three platforms (Table 19.3), except that Venture produced water had concentrations of ammonia that were an order of magnitude greater than Hibernia and Terra Nova. Venture produced water was also much more saline than Hibernia or Terra Nova.

Concentrations of dissolved metals (Table 19.4) were also similar between Hibernia and Terra Nova, while Venture had levels of barium, manganese, iron, strontium, and zinc elevated by 1–4 orders of magnitude. Compared to Hibernia and Terra Nova, the concentration of sulfur at Venture was approximately 4 orders of magnitude lower. As expected, aeration had little or no effect on the concentration of dissolved metals.

4 Discussion

Microorganisms were chosen for toxicity test analysis in this study as they react very quickly to their surrounding environment and are responsible for basic metabolic processes of environmental significance in the ocean such as nutrient regeneration, carbon fixation, contaminant biodegradation, and biotransformation.

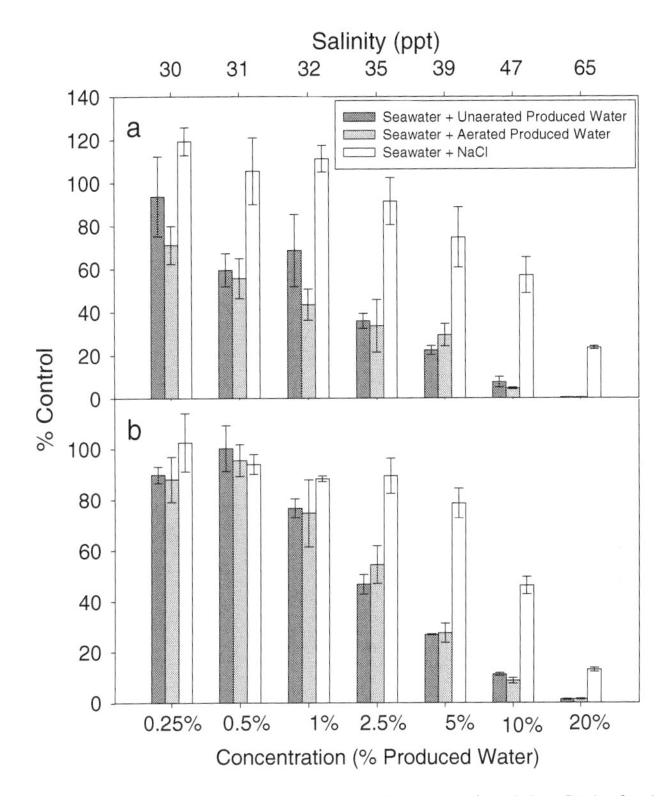

Fig. 19.6 Influence of salinity on bacterial productivity (**a**) and activity (**b**) in fresh and aerated Venture 2009 produced water

The results of the dilution experiments showed a decline in the rates of bacterial production and relative heterotrophic activity at concentrations >5–10% produced water. This is in general agreement with a reported effective toxicity concentration (EC_{50}) for the marine bacterium, *Vibrio fischeri* (formerly *Photobacterium phosphoreum*) of 3.5–6.3% produced water from the North Sea oil platforms, Clyde, Forties Charlie, Brent Delta and Brae Alpha (Stagg et al. 1996).

Chemical kinetic reactions that occur following the release of anoxic produced water and its subsequent dilution in the open ocean upon discharge have been found to alter its toxicity over time (Lee et al. 2005). The significance of this process was clearly illustrated in the controlled dose–response aeration–dilution experiments using natural microbial populations as the test organisms. A typical toxicity dose-response curve, with initial increase in productivity at low concentrations of produced water due to addition of nutrients followed by inhibition above a threshold value, was observed with fresh produced water. Following aeration for 44 h, to simulate equilibration in the ocean following discharge, produced water of the same concentration gradient elicited a stimulatory response. The difference associated with sample aeration can be attributed to the loss of predominantly volatile

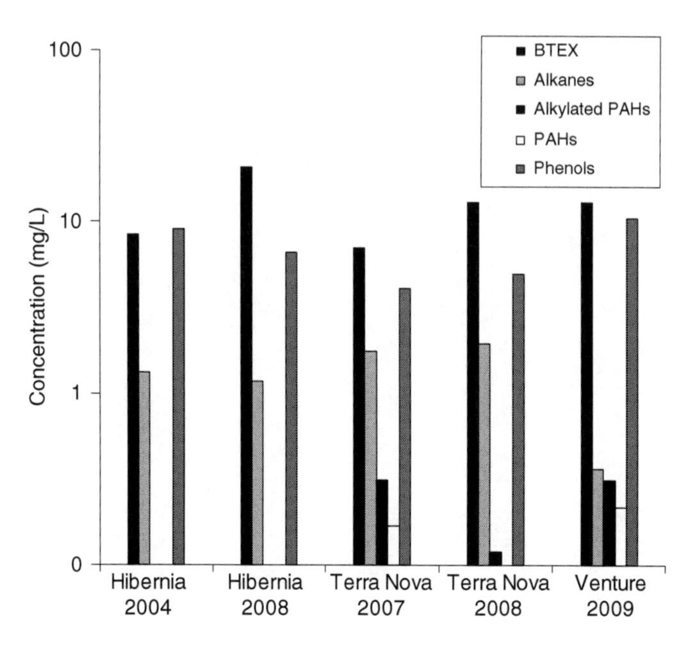

Fig. 19.7 Comparison of organic composition of all fresh produced water samples

low molecular weight hydrocarbons and sequestering of hydrolysis metals as precipitates or in suspended organic matter, thereby reducing the toxicity of these constituents. Under favorable environmental conditions where nutrients are not limited and toxicity is not an issue, indigenous bacteria have the capacity to metabolize the labile organic compounds associated with produced water.

In this study, phenol was one of the major constituents of toxic concern found in all the produced water samples. This compound is a known toxicant to the marine bacterium, *V. fischeri*, and is, in fact, used as a reference toxicant in the Microtox assay (Microbics 1992). Dehydrogenase catalyzes the oxidation of organic chemicals, and its activity is a measure of bacterial growth and respiration. At about 300 mg/L, phenol has been shown to cause a 50% reduction in bacterial community dehydrogenase activity after 24 h exposure, and dehydrogenase activity for *Bacillus* sp. and *Escherichia* sp. was cut 50% within 24 h at phenol concentrations of 700–1400 mg/L (Nweke and Okpokwasili 2010). Numerical modelling studies have suggested that the stimulatory effect of nutrients associated with produced water discharges on plankton growth may alter ecosystem trophic level dynamics (Rivkin et al. 2000; Khelifa et al. 2003). However, high concentrations of ammonia (>400 mM) are also known to be toxic to bacteria (Sprott and Patel 1986). Our experiments showed that ammonia concentration (about 2–20 mM) was not reduced by aeration, and it is possible that its high concentration in the produced water from the Venture facility could account for the consistent inhibitory response generated for growth and metabolism both before and after aeration.

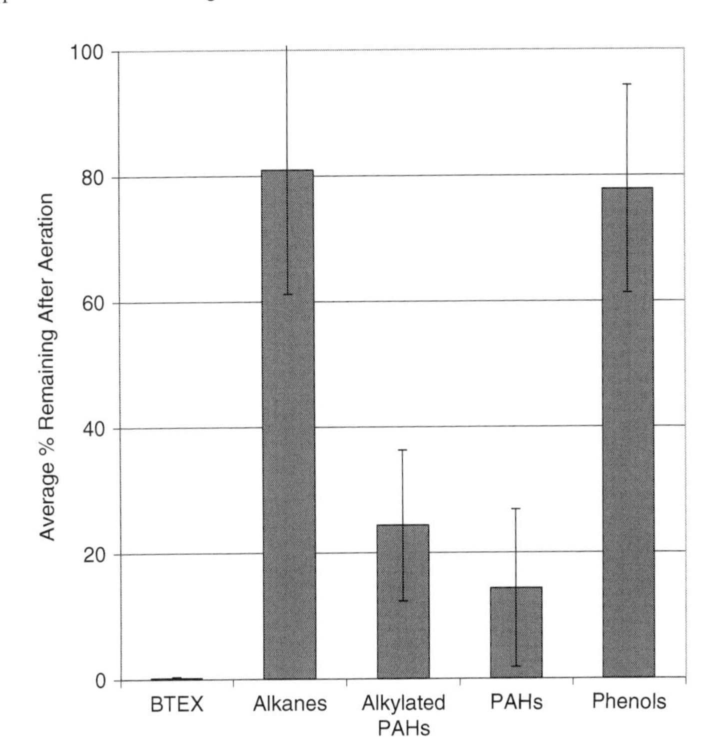

Fig. 19.8 Effect of aeration on organic constituents of produced water from the three platforms

Table 19.3 Salinity, and nutrients (silicate, phosphate, nitrate, nitrite, and ammonia) for fresh and aerated produced water samples

	Hibernia				Terra Nova				Venture	
	2004		2008		2007		2008		2009	
Parameter	Raw	Aerated	Raw	Aerated	Raw	Aerated	Raw	Aerated	Raw	Aerated
Salinity (ppt)	46.4	46.3	37.3	–	47.9	–	45.7	–	203.5	198.0
Silicate (μM)	742.9	861.6	526.5	785.9	539.3	916.0	694.9	823.5	349.7	455.5
Phosphate (μM)	37.5	34.2	1.1	3.0	0.5	1.0	0.6	0.6	–	–
Nitrate (μM)	0.4	0.4	0.5	0.6	0.1	0.2	0.3	0.5	1.3	1.6
Ammonia (μM)	2,033	2055	1152	1315	3444	2647	2497	2115	20,828	21,790
Nitrite (μM)	0.30	0.25	0.14	0.14	0.15	0.17	0.27	0.30	1.72	2.13

Two distinct strategies can be employed to discern the ecological effects of produced water effluents: predictive and observational (Middleditch 1984). The predictive approach, as described in our field sampling and laboratory experiments, is based on determining the identification and concentrations of components within the produced water and their toxicity. The observational approach, although site

Table 19.4 Analysis of metals and other inorganics for fresh and aerated produced water

Parameter	Hibernia		Venture	
	2004		2009	
(μg/L)	Fresh	Aerated	Fresh	Aerated
Aluminum	2.0	4.3	100	120
Antimony	0.1	0.1	< 2	< 2
Arsenic	< 20	< 20	< 50	< 50
Barium	478	481	1240,000	1260,000
Beryllium	0.1	0.1	1	1.4
Bismuth	–	–	< 0.5	< 0.5
Boron	15,800	15,850	29,000	29,000
Cadmium	0.04	0.02	2.00	2.30
Calcium	885,000	889,000	21800,000	22200,000
Chromium	< 1	< 1	< 10	< 10
Cobalt	< 1	< 1	< 10	< 10
Copper	2.5	1.3	< 10	< 10
Iron	925	1310	137,000	137,000
Lanthanum	0.0	0.0	16	17
Lead	1.9	0.2	27	35
Lithium	1870	1865	36,000	35,000
Magnesium	815,000	832,500	1,380,000	1,410,000
Manganese	447	459	24,100	24,500
Mercury	–	–	0.1	–
Molybdenum	0.8	0.7	1	1.1
Nickel	76.0	80.5	< 20	< 20
Phosphorus	5340	5030	70	< 50
Potassium	274,000	276,000	1,110,000	1,050,000
Rubidium	315	310	4400	4300
Selenium	< 10	< 10	< 50	< 50
Silicon	21,700	21,850	25,600	25,900
Silver	–	–	1	0.5
Sodium	15,350,000	15,300,000	49,500,000	50,700,000
Strontium	62,750	63,100	2,410,000	2,460,000
Sulfur	670,000	673,500	460	300
Tellurium	–	–	< 2	< 2
Thallium	2.7	2.8	140	150
Thorium	< 0.005	< 0.005	< 0.2	< 0.2
Tin	< 0.05	< 0.05	< 0.5	< 0.5
Titanium	< 20	< 20	< 1	2
Uranium	< 0.002	< 0.002	< 0.005	0.058
Vanadium	0.9	1.6	< 5	< 5
Zinc	7.5	6.6	2400	2400

specific, will provide direct and unequivocal information relating to “real-world” effects. In this regard, field observations have also been made within the study region. Lee et al. (2005) noted no significant differences for ^{3}H-thymidine uptake rates in water column depth profiles in transect lines from 0.5 to 20 km away from the Hibernia offshore platform. Using advanced techniques in genomics to assess changes in microbial community structure and function, Yeung et al. (Chapter 8, this volume) determined that the impact zone from produced water discharge from the Hibernia offshore platform (with a much higher produced water discharge rate than the Venture offshore platform on the Scotian Shelf) was limited to within 500 m of the discharge point. On the Scotian Shelf, the analysis of potential environmental effects of produced water discharges using an integrated modelling approach indicated that the soluble benzene and naphthalene components reach chronic no-effect concentration levels at a distance of only 1.0 m from the discharge point (Berry and Wells 2004). The limited persistence of these compounds that resulted in a low rank for the risk of adverse environmental effects, was attributed to advection processes linked to tidal currents within the region.

5 Conclusions

The concentrations of organic compounds in the Hibernia, Terra Nova and Venture fresh produced water samples under study were well within the range of values reported worldwide (Neff 2002). The lower molecular weight compounds, BTEX and naphthalene, are less influenced by the efficiency of the oil–water separation process during cleanup before discharge than the higher molecular weight PAHs, and are not measured by standard regulatory oil and grease analytical methods (Argonne National Laboratories 2004). However, as shown in the dilution studies which simulated natural weathering processes, these compounds have limited environmental persistence. Under current regulatory practices, the hydrocarbons in produced water (especially the PAHs) represent the organic compounds of greatest environmental concern (Neff 2002). Nevertheless, as suggested by the results of our aeration and dilution studies, natural chemical kinetic reactions following the discharge of produced water can effectively reduce the toxicity to a point at which indigenous microbes are capable of biodegrading or biotransforming the remaining residual organic compounds.

The results of this scientific study clearly highlight the importance of standardizing protocols for sample collection, treatment, and storage for environmental impact assessment of produced water discharged at sea. Furthermore, as the composition of produced water may vary from one geological formation to another, consideration must be made on a case-to-case basis. While the characteristics of the produced water samples in this study from offshore oil and gas production sites were similar to other produced waters worldwide, specific anomalies must be taken into account when interpreting toxicity data. For example, the Venture produced water had a higher salinity and was higher in ammonia, but lower in sulfur than the other samples.

Ecological relevance must also be considered. Toxic responses are directly linked to both concentration and exposure time. When produced water is discharged at sea, numerous scientific studies have suggested that natural physical oceanographic processes would result in rapid dilution by factors of 10–100-fold within tens of meters of the platform (Brandsma and Smith 1996; Flynn et al. 1996; Stromgren et al. 1995; Somerville et al. 1987). In a comprehensive North Sea study with field monitoring data from mussels and semi-permeable membrane devices (SPMDs) and the application of the Dose-related Risk and Effect Assessment Model, Durell et al. (2006) reported that surface water total PAH concentrations ranged from 25 to 350 ng/L within 1 km of the platform discharge point, and reached background levels of 4–8 ng/L within 5–10 km of the discharge – a 100,000-fold dilution of the PAH in the discharge water. In terms of the impact zone from produced water discharges, Burns et al. (1999) used data on bioaccumulation from bivalves and microbial growth inhibition studies from a shallow tropical marine ecosystem on the Northwest Shelf of Australia, to validate chemistry and model predictions of a potential biological impact area which extended 900 m from the point of discharge. Yet they also concluded that dispersion and degradation processes were fast enough to mitigate any long-term build up of contaminants within the system.

Our controlled dilution studies in the laboratory with aerated produced water (to simulate equilibrium conditions following discharge into the sea) showed that toxic effects were only elicited at concentrations exceeding 0.5% produced water. Analysis of the data clearly showed the reduction of toxicity attributed to chemical kinetic reactions. In addition, as noted by Flynn et al. (1996), a large proportion of the organic material in produced water samples are highly biodegradable, as standard laboratory tests showed the biodegradation of >90% of phenols and PAHs in addition to the reduction of BTEX to concentrations below the detection limit (<0.5 ppb). Considering the limited volume of produced water released, and the high rates of dilution following discharge, the results of our microbiological studies showed that deleterious effects occur only within the immediate vicinity of the discharge point. Based on these findings, there would be minimal environmental risks associated with the discharge of produced water from oil and gas activities off the east coast of Canada.

Acknowledgments Funding for this research was provided by Fisheries and Oceans Canada (DFO), the Program of Energy Research and Development (PERD), and the Environmental Studies Research Funds (ESRF) managed by Natural Resources Canada (NRCan). We would like to thank Carol Anstey (DFO) for the nutrient analyses.

References

Argonne National Laboratory (2004) A white paper describing produced water from production of crude oil, natural gas, and coal bed methane. Report under contract no. W-31-109-Eng-38. pp 1–87

Armstrong FAJ, Sterns CR, Strickland JDH (1967) The measurement of upwelling and subsequent biological processes by means of the Technicon Autoanlyzer and associated equipment. Deep-Sea Res 14(3):381–389

Azetsu-Scott K, Yeats P, Wohlgeschaffen G, Dalziel J, Niven S, Lee K (2007) Precipitation of heavy metals in produced water. Influence on contaminant transport and toxicity. Mar Environ Res 63:146–167

Bell RT (1993) Estimating production of heterotrophic bacterioplankton via incorporation of tritiated thymidine. In: Kemp, PF, Sherr BF, Sherr EB, Cole JJ (eds) Handbook of methods in aquatic microbial ecology. Lewis Publishers, Boca Raton, FL, pp 495–503

Berry JA, Wells PG (2004) Integrated fate modeling for exposure assessment of produced water on the Sable Island Bank (Scotian Shelf, Canada). Environ Toxicol Chem 23:2483–2493

Brandsma MG, Smith JP (1996) Dispersion modeling perspectives on the environmental fate of produced water discharges. In: Reed M, Johnsen S (eds) Produces water 2: environmental issues and mitigation technologies. Plenum Press, New York, NY, pp 215–224

Burns KA, Codi S, Furnas M, Heggie D, Holdway D, King B, McAllister FA (1999) Dispersion and fate of produced formation water constituents in an Australian Northwest Shelf shallow water ecosystem. Mar Pollut Bull 38:593–603

C-NLOPB (2011) Statistics, Resource Management Statistics, Monthly Production Summary for the Hibernia Field. Canada – Newfoundland and Labrador Offshore Petroleum Board, http://www.cnlopb.nl.ca/stat_rm.shtml. Accessed Jan 2011

CNSOPB (2011) Production data, Canada-Nova Scotia Offshore Petroleum Board, http://www.cnsopb.ns.ca/production.php. Accessed Jan 2011

Collins AG (1975) Geochemistry of oilfield waters. Elsevier, , New York, NY, 496 pp

Din ZB, Abu AB (1992) Sublethal effects of produced water from crude oil terminals on the clam Donax faba. In: Ray JP, Engelhardt FR (eds) Produced water: technical/environmental issues and solutions (Environmental Science Research vol 46). Plenum Press, New York, NY, pp 445–454

Ducklow HW, Carlson CA (1992) Oceanic bacterial production. In: Marshall KC (ed) Advances in microbial ecology, vol 12. Plenum Press, New York, NY, pp 113–181

Durell G, Utvik TR, Johnsen S, Frost T, Neff J (2006) Oil well produced water discharges to the North Sea. Part I: comparison of deployed mussels (Mytilus edulis), semi-permeable membrane devices, and the DREAM model predictions to estimate the dispersion of polycyclic aromatic hydrocarbons. Mar Environ Res 62:194–223

Flynn SA, Butler EJ, Vance I (1996) Produced water composition, toxicity, and fate. In: Reed M, Johnson S (eds) Produced water 2: environmental issues and mitigation technologies. Plenum Press, New York, NY, pp 69–80

Fuhrman JA, Azam F (1982) Thymidine incorporation as a measure of heterotrophic bacterioplankton production in marine surface waters: evaluation and field results. Mar Biol 66:109–120

Grasshoff K (1969) Technicon International Congress, June 1969. Technicon Industrial Systems, Tarrytown, New York, NY

Griffiths RP, Hayasaka SS, McNamara TM, Morita R (1977) Comparison between two methods of assaying relative microbial activity in marine environments. Appl Environ Microbiol 34(6):801–805

Hamoutene D, Samuelson S, Lush L, Burt K, Drover D, King TL, Lee K (2010) In vitro effect of produced water on cod, *Gadus morhua*, sperm cells and fertilization. Bull Environ Contam Toxicol 84:559–563

Holdway DA (2002) The acute and chronic effects of wastes associated with offshore oil and gas production on temperate and tropical marine ecological processes. Mar Pollut Bull 44: 185–203

Kerouel R, Aminot A (1997) Fluorometric determination of ammonia in sea and estuarine waters by direct segmented flow analysis. Mar Chem 57:265–275

Khelifa A, Pahlow M, Vezina A, Lee K, Hannah C (2003) Numerical investigation of impact of nutrient inputs from produced water on the marine planktonic community. In: Proceedings of the 26th Arctic and marine oilspill program (AMOP) technical seminar, Victoria, BC, June 10–12, 2003, pp 323–334

Krause PR, Osenberg CW, Schmitt RJ (1992) Effects of produced water on early life stages of a sea urchin: stage-specific responses and delayed expression. In: Ray JP, Engelhardt FR (eds) Produced water – technological/environmental issues and solutions. Plenum Press, New York, NY, pp 431–444

Krause PR (1995) Spatial and temporal variability in receiving water toxicity near an oil effluent discharge site. Arch Environ Contam Toxicol 29:523–529

Lee K, Azetsu-Scott K, Cobanli SE, Dalziel J, Niven S, Wohlgeschaffen G, Yeats P (2005) Overview of potential impacts of produced water discharges in Atlantic Canada. In: Armsworthy SL, Cranford PJ, Lee K (eds) Offshore oil and gas environmental effects monitoring: approaches and technologies. Battelle Press, Columbus, OH, pp 319–342

Li WKW, Dickie PM, Harrison WG, Irwin BW (1993) Biomass and production of bacteria and phytoplankton during the spring bloom in the western North Atlantic Ocean. Deep-Sea Res II 40(1–2):307–327

Manfra L, Moltedo G, Virno Lamberti C, Maggi C, Finoia G, Giuliani S, Onorati F, Gabellini M, Di Mento R, Cierco AM (2007) Metal content and toxicity of produced formation water (PFW): Study of the possible effects of the discharge on marine environment. Arch Environ Contam Toxicol 53:183–190

Marty JC, Saliot A (1976) Hydrocarbons (normal alkanes) in the surface microlayer of seawater. Deep-Sea Res 23:863–873

Microbics (1992) Detailed basic test protocol. In: Microtox manual – a toxicity testing handbook (vol 2). Microbics Corporation, Carlsbad, CA, pp 101–127

Middleditch BS (1984) Ecological effects of produced water effluents from offshore oil and gas production platforms. Ocean Manage 9:191–316

Neff JM (1987) Biological effects of drilling fluids, drill cuttings and produced waters. In: DF Boesch, N.N. Rabalais (eds) Long-term environmental effects of offshore oil and gas development. Elsevier Applied Science Publishers Ltd., London, pp 469–538

Neff JM (2002) Bioaccumulation in marine organisms. Effect of contaminants from oil well produced water. Elsevier, Oxford, 452 pp

Nweke CO, Okpokwasili GC (2010) Influence of exposure time on phenol toxicity to refinery wastewater bacteria. J Environ Chem Ecotoxicol 2(2):020–027

Parsons TR Maita Y, Lalli CM (1984) A manual of chemical and biological methods for seawater analysis. Pergamon Press, Oxford. 173 pp

Perez-Casanova JC, Hamoutene D, Samelson-Abbott S, Burt K, King TL, Lee K (2010) The immune response of juvenile Atlantic cod (*Gadus morhua* L.) to chronic exposure to produced water. Mar Environ Res 70(1):26–34

Querbach K, Maillet G, Cranford PJ, Taggart C, Lee K, Grant J (2005) Potential effects of produced water discharges on the early life stages of three resource species. In: Armsworthy S, Cranford PJ, Lee K (eds) Offshore oil and gas environmental effects monitoring (approaches and technologies). Battelle Press, Columbus, OH, pp 343–372

Rivkin RB, Tian R, Anderson MR, Payne JF, and Deibel D (2000) Ecosystem level effects of offshore platform discharges – identification, assessment and modeling. In: Proceedings of the 27th annual aquatic toxicity workshop: October 1–4, 2000, St. John's, Newfoundland, Penny KC, Coady KA, Murdoch MH, Parker WR, Niimi AJ (eds) Can Tech Rep Fish Aquat Sci 2331:3–12

SOEP (Sable Offshore Energy Project) (1996) Volume 3. Environmental Impact Statement. Prepared by MacLaren Plansearch (1991) Ltd

Somerville HJ, Bennett D, Davenport JD, Holt MS, Lynes A, Mahieu A, McCourt B, Parker JG, Stephenson RR, Watkinson RJ, Wilkinson TG (1987) Environmental effect of produced water from North Sea oil operations. Mar Pollut Bull 18: 549–558

Sprott GD, Patel GB (1986) Ammonia toxicity in pure cultures of methanogenic bacteria. Syst Appl Microbiol 7(2–3):358–363

Stagg R, Gore DJ, Whale GF, Kirby MF, Blackburn M, Bifield S, McIntosh AD, Vance I, Flynn SA, Foster A (1996) Field evaluation of toxic effects and dispersion of produced water discharges from North Sea oil platforms, implications for monitoring acute impacts in the environment. In: Reed M, Johnson S (eds) Produced water 2: environmental issues and mitigation technologies. Plenum Press, New York, NY, pp 81–100

Stagg RM, Mclntosh A (1996) Hydrocarbon concentrations in the northern North Sea and effects on fish larvae. Sci Total Environ 186(3):189–201

Strickland JDH, Parsons TR (1972) A practical handbook of sea-water analysis, 2nd edn, Series 167. Fisheries Research Board of Canada, Ottawa, ON, 311pp

Stromgren T, Sorstrom SE, Schou L, Kaarstad I, Aunaas T, Brakstad OG, Johansen O (1995) Acute toxic effects of produced water in relation to chemical composition and dispersion. Mar Environ Res 40:147–169

US Department of the Interior (1969) FWPCA methods for chemical analysis of water and wastes November 1969, Federal Water Pollution Control Administration, Cincinnati, 280 pp

Chapter 20
Studies on Fish Health Around the Terra Nova Oil Development Site on the Grand Banks Before and After Discharge of Produced Water

Anne Mathieu, Jacqueline Hanlon, Mark Myers, Wynnann Melvin, Boyd French, Elisabeth M. DeBlois, Thomas King, Kenneth Lee, Urban P. Williams, Francine M. Wight, and Greg Janes

Abstract Bioindicators or health effect indicators have the potential to identify adverse health conditions in fish in advance of effects on populations. A commercially valuable species, American plaice (*Hippoglossoides platessoides*), was chosen by the oil industry in consultation with Fisheries and Oceans Canada as an important species for Environmental Effects Monitoring (EEM) programs on the Grand Banks. We report here on fish health studies carried out at the Terra Nova Offshore Oil Development site before and after discharge of produced water, which began in 2003. These studies constitute one component of the overall Terra Nova EEM program. Fish were collected in the near vicinity of the Terra Nova Development site as well at a Reference site located approximately 20 km southeast of the development. Approximately 500 fish were studied in total over 5 survey years from 2000 to 2006. The health effect indicators studied included fish condition, visible skin and organ lesions, ethoxyresorufin *O*-deethylase (EROD) activity, haematology (differential cell counts) and a variety of histopathological indices in liver (e.g. nuclear pleomorphism, megalocytic hepatosis, foci of cellular alteration, macrophage aggregation, neoplasms) and gill (e.g. epithelial lifting, oedema, fusion and aneurysms). Although a slight elevation of EROD activity was observed in fish from the Development site in 2002, before discharge of produced water, and in 2006, the suite of other health bioindicators were found to be generally absent or similar between the Development and Reference sites. Overall, on the basis of the various indicators studied, the results support the hypothesis of no significant project effects on the health of American plaice.

A. Mathieu (✉)
Oceans Ltd., St. John's, NL, Canada

K. Lee, J. Neff (eds.), *Produced Water*, DOI 10.1007/978-1-4614-0046-2_20,

1 Introduction

Produced water from offshore oil and gas platforms contains a complex mixture of chemicals and while it is generally accepted that acute toxicological effects associated with produced water may be reduced to acceptable regulatory limits by dilution within a short distance from the point of release, there is ongoing concern over the long-term effects generated by constant low level exposure (e.g. Lee et al. 2005). On-site Environmental Effects Monitoring (EEM) programs are thus needed to provide an early warning of any undesirable change resulting from offshore development operations.

The EEM program associated with the Terra Nova offshore oil development being carried out by Petro-Canada on the Grand Banks was formulated in a fairly integrated manner with guidance from regulators, nationally and internationally recognized experts and the fishing industry. Provision for input by the public was also a key element in program design. The final program accepted by the Canada Newfoundland and Labrador Offshore Petroleum Board contained a number of components including fish health; fish and shellfish contamination and tainting; water chemistry, including chlorophyll as an indicator of primary productivity; sediment chemistry and toxicity; and benthic community structure. We report here on the fish health component which was recommended by Fisheries and Oceans Canada and generally recognized to be of key importance from a fisheries perspective.

Given that population level effects in fish would be both highly expensive to investigate and likely all but impossible to detect in the open ocean in the absence of major impacts, the fish health component included the use of early warning bioindicators (e.g. Adams 1990; Hugget et al. 1992). Such bioindicators can provide an appreciation of the degree and severity of any potentially impending fish health problems, or otherwise provide a reconnaissance tool for assessing if effects are occurring, and if so, whether they might be of regulatory or socioeconomic importance. The value of using bioindicators in environmental monitoring and assessment studies was especially spearheaded by the International Commission for the Exploration of the Seas (ICES) in the 1990s and is presently recognized by various agencies including the European Environmental Agency, the Oslo-Paris Commission, the Helsinki Commission, the Mediterranean Action Plan and the National Oceanic and Atmospheric Administration.

During the development of the Terra Nova EEM program, American plaice was identified as the species of primary interest for studies on fish health. There is a particular commercial interest in plaice since it was the most abundant flatfish on the Grand Banks, but stocks have undergone a major decline which has been attributed to overfishing and natural mortality (Dwyer et al. 2009). Plaice also displays favourable biological characteristics for assessing chronic toxicity, namely a long life span and frequent contact with sediments where contaminants can accumulate affording bioavailability through either fish contact with fines or ingestion of contaminated benthic prey. Plaice also undergoes feeding related movements into the water column in addition to its tendency for sediment contact (Beamish 1966; Pitt 1967). This enhances its potential for integrating contaminant exposure via the water column (through both feeding and water contact) as well as through interaction with sediments. Another consideration for use of a flatfish species is the relatively large number of studies, both laboratory and field, detailing the various types of pathologies that can be associated with chemical contamination (e.g. Myers and Fournie 2002; Stentiford et al. 2003; Koehler 2004; Lyons et al. 2004). American plaice is also regarded as a fairly sedentary species compared to more migratory species such as cod (Pitt 1969; Morgan 1994). It is reasonable to suggest that American plaice are resident for periods of weeks to months in specific areas with early warning indicators of health, such as EROD induction and selected gill and liver lesions being causally linked to these areas. Furthermore, the Terra Nova EEM program has involved multiyear surveys.

Given that fish are well known to be attracted to oil and gas platforms (e.g. Love et al. 2000), this could also be happening to some extent around the Terra Nova Development site. However, fish might move to some extent in and out of potential contaminant zones very close to offshore petroleum development sites, such that exposure would be insufficient to result in deleterious effects. Overall, either the presence or absence of health effects under natural ecological conditions of exposure is important to assess from an environmental management and fisheries perspective.

The bioindicators studied for the Terra Nova EEM program included fish condition, visible lesions on skin and internal organs, liver and gill histopathology, hepatic EROD activity and haematology (Mathieu et al. 2005). Externally visible fish abnormalities have been recommended as bioindicators for biological monitoring programs at the national and international level (ICES 1999; USEPA 2000; Lang 2002; OSPAR 2004; Lehtonen et al. 2006), and commonly used in survey work in different countries. Fish liver and gill histopathology is increasingly being used in biological monitoring and assessment programs (e.g. Hinton et al. 1992; Myers and Fournie 2002; Stentiford et al. 2003) while MFO enzymes such as ethoxyresorufin *O*-dealkylase (EROD), or its cytochrome linked component, CYPIA, have found extensive use in field studies over the past two decades (e.g. Whyte et al. 2000; Payne et al. 2003). Although fewer field studies exist for gill than liver histopathology, gill histopathology was also assessed since its epithelia are quite sensitive to certain classes of chemicals including low levels of hydrocarbons (Solangi and Overstreet 1982; Kiceniuk and Khan 1987).

A haematological component, namely inspection for changes in various types of blood cells, was also included since changes in blood cell counts, can provide insight into potential effects on immune function and resistance to disease (e.g. Weeks et al. 1992). Moreover, numerous environmental pollutants have been shown to interact with components of the fish immune system (e.g. Rice and Arkoosh 2002; Tierney et al. 2004; Tort et al. 2004). Changes in white blood cell profiles have also recently been observed in laboratory studies with fish exposed to wastewaters from oil-sand refining operations (Farrell et al. 2004) and produced waters (Payne et al. 2005). Differential white blood cell counts are currently used as bioindicators by the US Geological Survey in their Biomonitoring of Environmental Status and Trends (Best) Program (Jenkins 2003, 2004).

It is noted that the various indices studied could potentially encompass different time frames of exposure. For instance, induction of EROD and gill lesions (depending on type) could occur in a matter of hours or days, while liver lesions might take from days to months or years with tumours probably requiring the longest time for expression. As indicated previously, given its long life-span, American plaice has potential for addressing chronic long-term as well as more acute toxicological conditions.

There are sources of general information dealing with the potential risks of produced water (e.g. Holdway 2002; Neff 2002). However, there are few studies which have investigated effects on fish under natural ecological conditions of exposure in the near vicinity of offshore petroleum development sites. Also, these studies, as discussed hereafter, have mainly dealt with observations on CYP1A or CYP1A linked enzyme activity and not on higher level responses such as fish condition, gross skin and organ abnormalities and detailed liver and gill histopathology.

2 Materials and Methods

2.1 Project Area Description

The Terra Nova oil field is situated on the Grand Banks of Newfoundland (Canada), approximately 350 km east-southeast of St. John's (Fig. 20.1). The oil field is being developed using a floating production storage and offloading (FPSO) facility with periodic support from a semi-submersible drilling rig.

The discharges from the Terra Nova Development site principally include produced water, drill cuttings, sewage, cooling water and deck drainage while the burning of fuel for the generation of electricity might result in a low level of

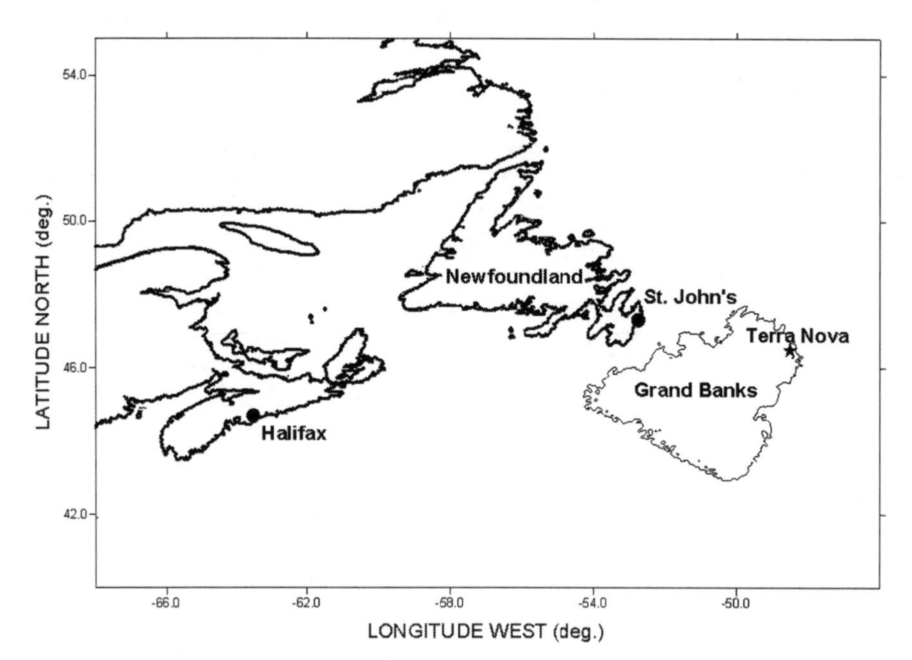

Fig. 20.1 Location of Terra Nova oil field

combustion hydrocarbons entering the nearby area through atmospheric deposition. Produced water began to be released at the site in 2003 approximately 3 years after drilling began. Transport and fishing vessels would also be expected to be a minor source of contaminants for the general area from hydrocarbons derived from fuel combustion and release of bilge waters. Sewage is passed through a secondary treatment prior to discharge, while bilge water and deck drainage are passed through an oil and water separator. The cooling water uses an electric anti-fouling system and is treated to a minimum chlorine residual before discharge.

The Development site is situated sufficiently far to the west of Flemish Pass to be out of the influence of the main Labrador Current. Currents on the Grand Banks are generally weak (< 10 cm/s) and southwards and are dominated by wind-induced and tidal current variability (Petrie and Anderson 1983). Water depth at the site is ~90 m.

The Reference site is situated ~20 km downstream of the study site, a distance where dilution of produced water discharges would be expected to be considerable. One of the main difficulties in plume modelling is related to accuracy of predictions beyond the near field (e.g. Zhao et al. 2008). However, a number of near field modelling studies which have been carried out worldwide suggest dilution rates of 1,000–10,000 are typical at a distance of 0.5–1.0 km from the point of discharge (e.g. OGP 2002, 2005). Plume studies carried out at the Terra Nova site indicated similar results, with a dilution ranging from 400 to 800 at 0.25 km from the source of the discharge (Lorax Environmental 2006).

Thus, the levels of any contaminants at the Reference (far field) site some ~20 km away would be expected to be very low. Gray (2002) has specifically cautioned that model predictions beyond the near field can be problematic, due to difficulties in measuring the very low concentrations of contaminants expected to be present.

Both water-based and synthetic-based drilling muds have been used at the site during drilling operations. The water-based muds were typically muds enriched in barite and bentonite with lesser amounts of other constituents. Since the beginning of drilling to August 2006, Petro-Canada reported cumulative water-based mud discharges of 41,793 m^3 and synthetic-based mud on cuttings discharges of 4,832 m^3. The synthetic-based fluids have been reported to have negligible acute toxicity to fish larvae, copepods and ctenophores (Payne et al. 2001).

Produced water represents the major reportable discharge stream from the FPSO with the first discharges occurring in April 2003. The total produced water discharged to June 2006 was 4,127,005 m^3. Monthly discharges are given in Fig. 20.2. Release has always been under the approved discharge limit of 18,300 m^3/day.

2.2 Chemical Composition of Produced Water

Neat produced water was analysed for organic compounds in four samples collected at Terra Nova in 2007. Samples were preserved with 6 M HCl. Polycyclic aromatic hydrocarbons (PAHs) were determined by solid-phase microextraction and gas chromatography/mass spectrometry according to U.S. EPA method 8272,

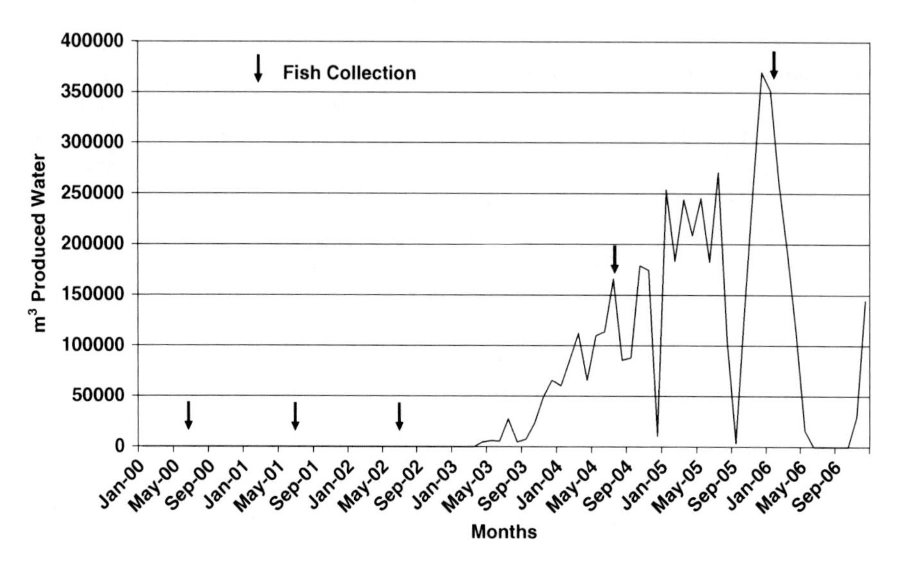

Fig. 20.2 Monthly discharges of produced water

benzene, toluene, ethylbenzene and xylenes (BTEX) by trap capillary column gas chromatography according to EPA method 502, and phenols by gas chromatography according to EPA method 8041.

2.3 Fish Collection and Necropsy

Adult American plaice were collected by otter trawl from two areas on the Grand Banks, one in the vicinity of the Terra Nova Development (~1–3 km from the Fisheries Exclusion Zone) and the other, at a Reference site (~20 km southeast of the rig). Three surveys (2000, 2001, 2002) were carried out before release of produced waters, whereas the 2004 and 2006 surveys were carried out after release of produced water. Regarding a Reference site, it is reasonable to suggest that any contaminant conditions of a background nature in a given region in the open sea should be more or less similar, and one Reference site was considered to be reasonable and cost effective for this type of study with fish. However, it is recognized that use of more than one reference site can be of value in an EEM program (e.g. Green 2005) with any decision on a reasonable number being related to the indices being measured and the potential for dissimilarities in background conditions in the region, either natural or anthropogenic. For instance, the requirement for fish health could be different than for benthic community structure. It is also noted that collection of fish at the Reference site still involved trawling over a substantial area which would be expected to provide fairly representative samples. Furthermore, as previously noted, the Terra Nova EEM program has involved multiyear surveys.

Upon capture, blood was drawn from a dorsal blood vessel near the tail and blood smears were prepared. Total length and weight were recorded. Observations for gross pathology were carried out under the general framework described by Goede and Barton (1990), except no observations were made on the thymus. Fish were dissected and sexed and maturity stage recorded according to procedures used by Fisheries and Oceans Canada in the Newfoundland Region. Otoliths were collected for determination of age.

A 3–4 mm section from the centre portion of each liver (along the longitudinal axis) and the first arch of the right gill were fixed in 10% buffered formalin for histopathological analysis. The remaining liver tissue was frozen on Dry Ice until returning to port when it was placed in a –65°C freezer for EROD analysis.

2.4 Fish Condition

Fish condition was expressed as Fulton's condition factor and calculated as 100 × body gutted weight/length^3. However, since use of this index assumes that body weight is proportional to the cube of length (which is not always the case), log–log regressions of weight on length were further tested by ANCOVA.

2.5 Tissue Histopathology

Liver and gill samples fixed in formalin were processed for paraffin embedding by standard histological methods (Lynch et al. 1969), sectioned at 6 μm and stained with hematoxylin and eosin (Luna 1968). Histological slides were examined by microscopy at ×25, ×63 and ×250 magnification. Histological examination was conducted by the same investigators in all years. Information on the site of capture was not revealed until final diagnoses were made.

Liver lesions previously identified as having a putative chemical aetiology in flat fish (e.g. Myers et al. 1987; ICES 2004; Lyons et al. 2004) were recorded for each fish as present or absent. The percentage of fish affected by each type of lesion or prevalence of lesion was then calculated. Macrophage aggregation was recorded on a relative scale from 0 to 7 and prevalence was calculated for fish showing a moderate to high aggregation (3 or higher on the scale).

Secondary gill lamellae were examined for the presence of gill lesions associated with chemical toxicity (Mallat 1985). This included observations for epithelial lifting (separation of the epithelial layer from the basement membrane), aneurysms (dilation of blood vessel at the tip of the secondary lamellae), fusion (fusion of two or more adjacent secondary lamellae) or oedema (swelling between or within cells). Any other observations, such as presence of X-cells were also recorded. A semi-quantitative examination was carried out where the total number of secondary lamellae and the secondary lamellae presenting the lesions were counted on four primary lamellae. Results for each fish were expressed as the percentage of lamellae presenting the condition in relation to the total number of lamellae counted (up to

800 lamellae per fish). The degree of oedema was also recorded on a 0 (absent) to 3 (pronounced) relative scale.

2.6 Hepatic EROD Activity

Liver samples were thawed within 3 weeks of storage at –65°C and homogenized in 4 volumes of 50 mM Tris buffer, pH 7.5, using 10 passes of a glass Ten Broek hand homogenizer. Homogenates were centrifuged at $9000 \times g$ for 15 min at 4°C and the post mitochondrial supernatant (S9 fraction) frozen at –65°C until assayed. Assays were carried out within 3 weeks of storage but essentially no loss of activity has been noted in homogenates stored for up to 9 months.

EROD activity was assessed as 7-ethoxyresorufin *O*-deethylase (EROD) according to the fluorimetric method of Pohl and Fouts (1980) as modified by Porter et al. (1989). Enzyme activity was assayed in a final volume of 1 ml containing 50 mM Tris buffer, pH 7.5, 2 μM ethoxyresorufin (Sigma) dissolved in dimethyl sulphoxide, 0.15 mM NADPH and S9 protein. After 15 min incubation at 27°C, the reaction was stopped with 2 ml of methanol (HPLC grade) and samples were centrifuged ($3600 \times g$ for 5 min) to remove the protein precipitate. The fluorescence of resorufin formed in the supernatant was measured at an excitation wavelength of 550 nm and an emission wavelength of 580 nm using a Perkin-Elmer LS-5 fluorescence spectrophotometer. Protein concentration was determined using the Lowry protein method (Lowry et al. 1951) with bovine serum albumin as standard. The rate of enzyme activity in pmol/min.mg protein was obtained from the regression of fluorescence against standard concentrations of resorufin.

2.7 Haematology

Blood smears were fixed in methanol and stained with Giemsa for identifying different types of cells (Ellis 1976). A differential blood cell count was performed on lymphocytes, neutrophils and thrombocytes and expressed as a percentage of each type of cell on 200 cells counted per slide, using the Exaggerated Battlement Differential counting technique (Lynch et al. 1969).

2.8 Statistical Analysis

EROD activity as well as arcsine square root-transformed percentages of blood cells and gill lamellae were analysed by the unpaired *t*-test or the Mann–Whitney Rank Sum test, when the groups were not normally distributed. Fish condition was assessed by analysis of covariance (ANCOVA). Degree of oedema was analysed by the Mann–Whitney Rank Sum test and prevalence of gross pathology and liver lesions by the Fisher exact test. Comparisons between sites having a $p < 0.05$ were considered to be statistically significant.

3 Results and Discussion

3.1 Chemical Composition of Produced Water

The concentration range (μg/L) for various chemicals found in 4 samples of neat produced water collected at Terra Nova in 2007 is given for 29 alkanes (Table 20.1), 55 PAHs (Table 20.2) as well as for 16 phenols and BTEX (Table 20.3).

Average concentrations of total compounds were 1,766 μg/L for alkanes, 311 μg/L for methylated PAHs, 169 μg/L for PAHs, 4,080 μg/L for phenols and 7.3 mg/L for BTEX. Concentrations were below or in the range of the values reported for other oil developments in the North Sea and Brazil (Flynn et al. 1995; OGP 2002; Utvik and Gartner 2006; Dorea et al. 2007).

Measurement of concentrations of PAH in produced water is of interest in relation to their historical importance in toxicology in general, especially in relation to mutagenesis and carcinogenesis (e.g. Hylland 2006) while alkanes and BTEX are of interest in relation to their value as tracers (alkanes) or potential acute toxicity in the near field (BTEX). The concentrations of PAH measured in 4 samples of produced water from the Terra Nova Development were in the range of those found in produced water from 18 fields in the Norwegian sector of the North Sea (OGP 2002) with many of the compounds found in the 4 samples from Terra Nova being present in the lower part of the range.

Preliminary information is also provided on concentrations of a number of alkylphenols which are found in produced water (as well as petroleum).

The low level of alkylphenol compounds present in produced water has recently drawn interest since some compounds (viz higher alkylated forms) may act as weak estrogens or potentially present other forms of toxicity (Thomas et al. 2004; Martin-Skilton et al. 2006; Sturve et al. 2006; Meier et al. 2007; Boitsov et al. 2007).

Table 20.1 Concentrations of alkanes in four samples of produced water from the Terra Nova oil field (2007)

Alkanes	Range (μg/L)	Alkanes	Range (μg/L)
n-Decane	9.8–19.2	Tricosane	56.1–115.6
Undecane	17.0–35.8	Tetracosane	50.6–103.8
Dodecane	28.9–43.1	Pentacosane	44.9–92.9
Tridecane	48.1–111.3	Hexacosane	40.9–84.8
Tetradecane	71.0–175.2	Heptacosane	36.6–76.1
Pentadecane	89.0–205.3	Octacosane	30.3–69.0
Hexadecane	93.4–187.3	*n*-Nonacosane	27.1–63.1
Heptadecane	97.0–186.8	Tricontane	17.5–44.8
2,6,10,14-TMPdecane (pristane)	36.3–71.4	*n*-Heneicontane	15.5–40.1
Octadecane	86.1–168.8	Dotriacontane	14.1–34.8
2,6,10,14-TMHdecane (phytane)	45.9–89.3	Tritriacontane	18.4–41.2
Nonadecane	78.6–154.9	Tetratriacontane	10.6–23.6
Eicosane	63.0–127.2	*n*-Pentatriacontane	13.9–28.8
Heneicosane	62.4–127.5	17β (H), 21α (H)-hopane	0.2–0.4
Docosane	58.2–119.4		

Table 20.2 Concentrations of PAHs in four samples of produced water from the Terra Nova oil field (2007)

PAHs	Range (μg/L)	PAHs	Range (μg/L)
Naphthalene	120.1–166.4	Trimethylphenanthrene	5.0–9.3
1-Methylnaphthalene	88.5–135.1	Tetramethylphenanthrene	3.0–5.7
Methylnaphthalene	99.4–152.8	Fluoranthene	0.4–0.5
d10-Methylnaphthalene	6.7–11.1	Pyrene	0.4–0.6
2,6-Dimethlynaphthalene	13.8–25.5	Methylpyrene	1.2–2.1
Dimethylnaphthalene	48.2–83.4	Trimethylpyrene	1.7 –3.0
2,3,5-Trimethylnaphthalene	9.4–18.7	Tetramethylpyrene	1.5–2.7
Trimethylnaphthalene	22.7–41.1	Dimethylpyrene	1.6–3.0
Tetramethylnaphthalene	14.2–25.2	Naphthobenzothiophene	0.7–0.8
Acenaphthene	1.1–1.5	Methylnaphthobenzothiophene	1.6–3.0
Fluorene	5.0–7.7	Dimethylbenzothiophene	6.7–12.8
Methylfluorene	5.6–9.4	Benz[a]anthracene	0.5–0.6
Dimethylfluorene	1.1–1.6	Chrysene	0.9–1.2
Dibenzothiophene	4.7–7.5	Methylchrysene	1.2–1.9
Methyldibenzothiophene	5.7–10.2	Dimethylchrysene	1.2–1.9
Dimethyldibenzothiophene	4.7–8.9	Trimethylchrysene	1.4–2.7
Trimethyldibenzothiophene	3.0–5.6	Benzo[b]fluoranthene	0.3–0.4
Tetramethyldibenzothiophene	1.9–3.5	Benzo[e]pyrene	0.3–0.4
Phenanthrene	7.8–12.7	Benzo[a]pyrene	0.5–0.6
Methylphenanthrene	10.0–18.1	Dibenz[a,h]anthracene	0.3–0.4
2-Methylphenanthrene	2.8–4.9	Benzo[ghi]perylene	0.3–0.4
Dimethylphenanthrene	8.0–14.9	Trimethylphenanthrene	5.0–9.3
3,6-Dimethylphenanthrene	1.1–1.7		

Table 20.3 Concentrations of phenols and BTEX in four samples of produced water from the Terra Nova oil field (2007)

Phenols	Range (μg/L)	BTEX	Range (mg/L)
Phenol	1340–1782	Benzene	3.3–4.2
o-Cresol	729–901	Toluene	1.9–2.7
m- and *p*-Cresol	827–971	Ethyl Benzene	0.1–0.2
2,6-Dimethylphenol	39–50	*p*-Xylene	0.2–0.4
2-Ethylphenol	72–80	*m*-Xylene	0.05–0.07
2,4- and 2,5-Dimethylphenol	147–246	*o*-Xylene	0.3–0.4
3- and 4-Ethylphenol	250–271		
2,3-Dimethylphenol	33–39		
2-Isoproplyphenol	47–53		
2-Proplyphenol	15–17		
3- and 4-Isopropylphenol	115–137		
2-sec-Butylphenol	11–12		
3- and 4-Tert butylphenol	34–35		
4-sec-Butylphenol	1–14		
4-Isopropyl-3-methylphenol	5–6		
4-Nonylphenol	nd		

A useful biomarker for estrogenic effects is the elevation (beyond very low basal levels normally) of an egg yolk precursor protein, vitellogenin, in the serum of male fish, which if produced in sufficiently high concentrations may result in the formation of eggs in testes (Sumpter and Johnson 2005).

The low levels of alkylphenols in produced water would be expected to undergo considerable dilution upon release, especially under open sea conditions, and may not present ecological concerns for fish reproduction. There is some preliminary information available in this regard. A field study in the North Sea found a slight elevation of vitellogenin in the serum of male codfish, but this was in fish caged within 500 m of the Statfjord B platform for 5 weeks (Scott et al. 2006). In the meantime, a slight depression, not elevation, of vitellogenin was found in the serum of female fish in a second caging trial (cited in Hylland et al. 2008).

Vitellogenin was not recommended as a biomarker for industry to use in their monitoring programs on the Grand Banks. It is also noted that although Payne et al. (2005) detected an high level of vitellogenin induction in male cunners (*Tautogalabrus adspersus*) exposed to the model estrogen β-estradiol, no induction was observed in fish chronically exposed for as long as 3 months to a relatively high level of produced water from the Grand Banks.

3.2 Gross Pathology

Of the approximately 500 fish examined over 5 years (250 from the Reference site and 250 from the Development site), no lesions were observed on the external surface of the liver, gonad, digestive tract, body cavity or spleen (Table 20.4).

Given the importance accorded to gross pathology in monitoring programs at the national and international levels (e.g. ICES 1999; USEPA 2000; Lang 2002; OSPAR 2004; Lehtonen et al. 2006), it is of interest that fish from the Development site as well as the Reference site were essentially free of signs of gross pathology.

Table 20.4 Percentage of fish affected by visible abnormalities in American plaice before and after discharge of produced water

							Discharge of produced water ⟷			
	2000		2001		2002		2004		2006	
	REF	DEV	REF	DEV	REF	DEV	REF	DEV	REF	DEV
Number of fish	49	44	48	50	50	50	49	51	50	50
Growth on fins	0	0	2	0	0	0	0	0	0	0
Lymphocystis	0	0	2	0	0	0	0	0	0	0
Parasites on gills	0	0	0	0	0	0	4	0	0	0
White gills	0	0	0	0	0	0	0	2	0	2
Lesions on internal organs	0	0	0	0	0	0	0	0	0	0

REF = reference site, DEV = development site

3.3 Fish Condition

Fulton's condition factor relates body weight to length and although the index requires interpretation with caution (see below), it is a common measurement carried out in fisheries science for assessing overall fish health.

Inter-site comparisons were carried out on fish of comparable age and length, but it is noted that sample size was often relatively small for calculation of condition factors. There were no statistically significant differences ($p < 0.05$) for Fulton's condition factor in male fish for the years 2001, 2002 and 2004 (Fig. 20.3). However, fish taken at the Development site in 2006 displayed a slight but significantly lower condition factor. Further analyses of adjusted means by ANCOVA indicated a marginally significant difference ($0.05 < p < 0.10$). In the case of female fish (Fig. 20.4), there were no statistically significant differences in years 2002 and 2006 but fish from the Development site displayed a higher condition factor in 2001 and in 2004.

As noted, measurement of fish condition (especially in isolation) as an indicator of fish health needs to be interpreted with caution since in wild fish populations, food availability, competition and other factors may interact synergistically to produce a general decline in fish health (Goede and Barton 1990; Barton et al. 2002). Thus it may not be possible to elucidate pollutant effects or condition due to masking by other biotic or abiotic variables. Pollutants may therefore decrease the condition factor or have no apparent effect.

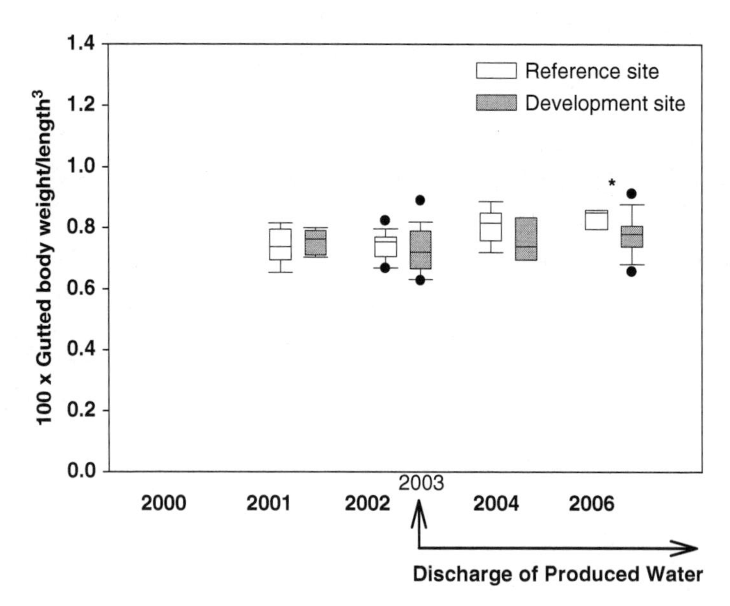

Fig. 20.3 Fulton's condition factor in male American plaice, calculated as 100 × gutted body weight/length³; data plotted are median (*line in the box*), 25th and 75th percentiles (*bottom* and *top of box*) and 10th and 90th percentiles (*error bars*); data points beyond the 5th and 95th percentiles are also displayed; *significantly different ($p < 0.05$)

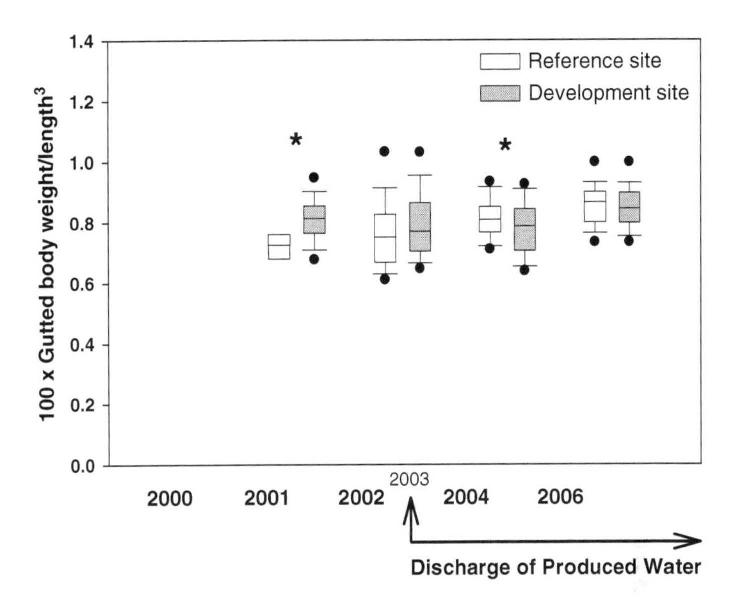

Fig. 20.4 Fulton's condition factor in female American plaice, calculated as 100 × gutted body weight/length^3; data plotted are median (*line in the box*), 25th and 75th percentiles (*bottom* and *top of box*) and 10th and 90th percentiles (*error bars*); data points beyond the 5th and 95th percentiles are also displayed; *significantly different ($p < 0.05$)

Although fish condition is an important index to assess in monitoring programs, given the consideration noted above and the results obtained in the 5 survey years at the Terra Nova site, caution is warranted in drawing conclusions on the basis of a single year data, with analysis of trends over time being the preferred approach. It is also noted in this regard that growth variability which may be linked to fish condition has been well documented in studies of American plaice on the Grand Banks and adjacent waters (Brodie 2002; Morgan et al. 2005), reinforcing the importance of obtaining trend data at any particular development site. To this end, it is of interest to note that the present 5 years of results from the Terra Nova site do not indicate a "trend" for effects on fish condition index.

3.4 Liver Histopathology

Detailed histopathological studies were carried out on liver tissues with a focus on various lesions that have been associated with chemical toxicity in flatfish (e.g. Myers et al. 1987; ICES 2004). This included but was not limited to nuclear pleomorphism, megalocytic hepatosis, basophilic, eosinophilic and clear cell foci, carcinoma, cholangioma, cholangiofibrosis, moderate to high macrophage aggregation and hydropic vacuolation.

Results, expressed as percentage of fish affected by each type of lesion/observation in the Reference and Development sites before and after discharge of produced water, are summarized in Table 20.5.

Table 20.5 Percentage of fish affected by hepatic lesions in American plaice before and after discharge of produced water

	2000		2001		2002		2004 (Discharge of produced water →)		2006	
	REF	DEV	REF	DEV	REF	DEV	REF	DEV	REF	DEV
Number of fish	49	44	48	50	50	50	49	51	50	49
Nuclear pleomorphism	0	2.3	0	0	0	2.0	0	0	0	0
Megalocytic hepatosis	0	2.3	0	0	0	0	0	0	0	0
Cell foci alterations	0	0	0	0	0	0	2.0	0	0	2.0
Carcinoma	0	0	0	0	0	0	0	0	0	0
Cholangioma	0	0	0	0	0	0	0	0	0	0
Hydropic vacuolation	0	0	0	0	0	0	0	0	0	0
Fibrillar inclusions	0	0	0	0	0	0	0	0	0	0
Peri-cholangiofibrosis	0	0	0	0	0	0	2.0	0	0	0
Macrophage aggregation[a]	4	2.3	0	0	0	0	0	0	0	0

REF = reference site, DEV = development site
[a]moderate to high aggregation (3 or higher rating on a 1–7 relative scale)

Results on the approximately 500 fish examined indicated that they were relatively free of many liver lesions commonly associated with chemical toxicity with no indication of differences between Reference and Development sites.

Although detailed histopathological studies of the kind reported here have not (to our knowledge) been carried out on fish collected around petroleum development sites in the North Sea, there have been a few observations on possible metabolic disturbance in caged or wild fish around petroleum development sites. Smolders et al. (2006) noted in caging trials with cod in the BECPELAG program that fish from the Statfjord field site contained significantly lower levels of hepatic lipid and protein than fish caged at the Reference site. Cellular energy allocation was also significantly higher at the Reference site than at sites close to the oil field. Some cellular alterations were also reported in studies of liver tissues of wild saithe (*Pollachius vivens*) carried out within the framework of the BECPELAG workshop (Bilbao et al. 2006). Lysosomal membrane stability was 3-fold more stable in fish from the reference station compared to three stations at 0.5, 2 and 10 km on the Statfjord transect. Lysosomal density was also significantly increased at the 0.5 km station, while peroxisomal palmitoyl-CoA oxidase activity was significantly elevated at all stations along the Statfjord transect compared to the reference site.

Histopathological effects were observed in earlier studies on fish and shellfish in association with some oil developments in the Gulf of Mexico (Gallaway et al. 1981; Grizzle 1986; Wilson-Ormond et al. 1994), but equivocal results have also been obtained (Menzie 1982). However, relatively high levels of lesions were often noted in fish from reference sites, and such pathological effects may be background in nature and linked to general water quality conditions in the Gulf. This would be expected to confound resolution of any specific rig-related effects.

Regarding the present study, it is of interest to note that a large variety of histopathological effects that have been observed in fish liver tissues in association with chemical contamination were absent or very rare at both the Reference and Terra Nova Development sites on the Grand Banks.

3.5 Gill Histopathology

Epithelial lifting, aneurysms and fusion of secondary lamellae were absent or mostly below 0.5% in all surveys before and after release of produced water with no distinction between the Reference and Development sites (Table 20.6). A representative photograph of a section of a normal gill is given in Fig. 20.5a. X-cells which are

Table 20.6 Percentage of gill lamellae displaying pathological conditions in American plaice before and after discharge of produced water

	2000		2001		2002		2004 (Discharge of produced water)		2006 (Discharge of produced water)	
	REF	DEV	REF	DEV	REF	DEV	REF	DEV	REF	DEV
Number of fish	48	43	47	50	49	50	47	48	47	49
Epithelial lifting [a]	0	0	0	0	0	0	0	0	0	0
Aneurysm [a]	0.3	0.1	0.5	0	<0.1	0	<0.1	0	0	0
Fusion [a]	0.5	0.6	<0.1	<0.1	0	0	0.2	0.2	<0.1	<0.1
X-cells [a]	0	0	0	0	0	0	0	2.1	0	0
Oedema [b]	1.4	1.4	1.3	1.0*	1.2	1.3	1.1	1.2	0.9	1.2

REF = reference site, DEV = development site
[a]Mean percentage of lamellae presenting the condition (in relation to the total number of lamellae counted)
[b]Mean of rating on a relative 0–3 scale
*Significantly different ($p < 0.05$) after the Mann–Whitney Rank Sum test analysis

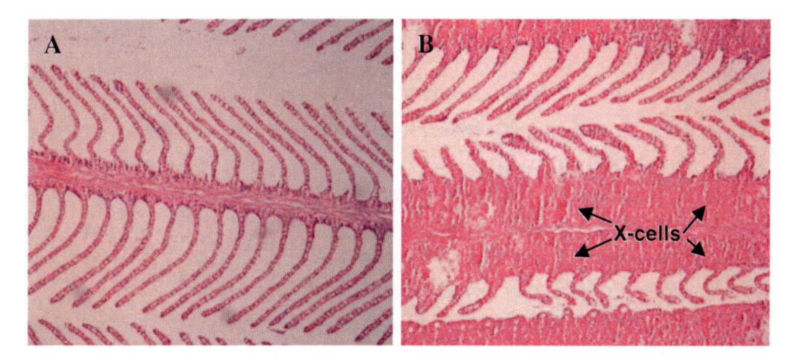

Fig. 20.5 Microscopic appearance of American plaice gill (H&E, 63×); (**a**) normal gill, (**b**) X-cells between gill secondary lamellae (*arrow*)

believed to be due to infection by protozoans (Miwa et al. 2004) were observed in one fish from the Development site in 2004 (Fig. 20.5b). A low level of oedema, putatively of a background nature, was common to all fish from both the Reference and Development sites either before or after the discharge of produced water. There were no significant differences between sites except in 2001 when fish from the Development site exhibited a slightly lower level of gill oedema than fish from the Reference site.

Fusion of gill lamellae has been reported in turbot (*Scophthalmus maximus*) exposed in the laboratory for 3 weeks to 1% produced water from a platform on the United Kingdom Continental Shelf (Stephens et al. 2000). The fusion recorded was quite severe, approximately 50% of the lamellae and it is not known if lesser pathologies, such as mild fusion, mucus cell production or aneurysms might be produced in chronic exposures with much lower levels of produced water. An increase of mucus cells was recently observed in the gills of cod chronically exposed for 3 months in the laboratory to low levels of produced water from the Hibernia field on the Grand Banks (Hamoutene et al. 2007).

Regarding the present study at the Terra Nova site, as for liver histopathological indices, it is of interest to note the absence of significant gill lesions at both the Reference and Development sites.

3.6 EROD Activity

Since basal levels of EROD activity can vary between males and females of the same species (e.g. Mathieu et al. 1991), results were analysed separately for each sex. Except for a low level of induction noted in males from the Development site in 2002, before release of produced water, and in females from the Development site in 2006, after release of produced water (which commenced in 2003), there were no other inter-site differences (Figs. 20.6 and 20.7).

Caution is also warranted should there be temperature differences, especially major differences, between sites being compared, since some enzyme activities can in theory follow temperature compensation mechanisms such that "basal" levels could increase at lower temperature, or conversely, decrease at higher temperature (Ankley et al. 1985). However, in the Terra Nova EEM program, temperature differences have generally not been a factor for consideration due to the near uniformity of temperature between the two sites. Bottom water temperatures were measured at each site over the 5 year monitoring period and intersite variability in any given year was $< 1.0^{\circ}C$.

Induction of EROD (or its counterpart such as CYP1A) can be a quite sensitive response to selected aromatic compounds (e.g. planar PAH) and has been used extensively in laboratory and field studies including in association with oil spills and around offshore petroleum development sites (e.g. Payne et al. 1987; Whyte et al. 2000). Induction of EROD is also of special interest due to its association with various toxic effects in animals including fish (discussed briefly below).

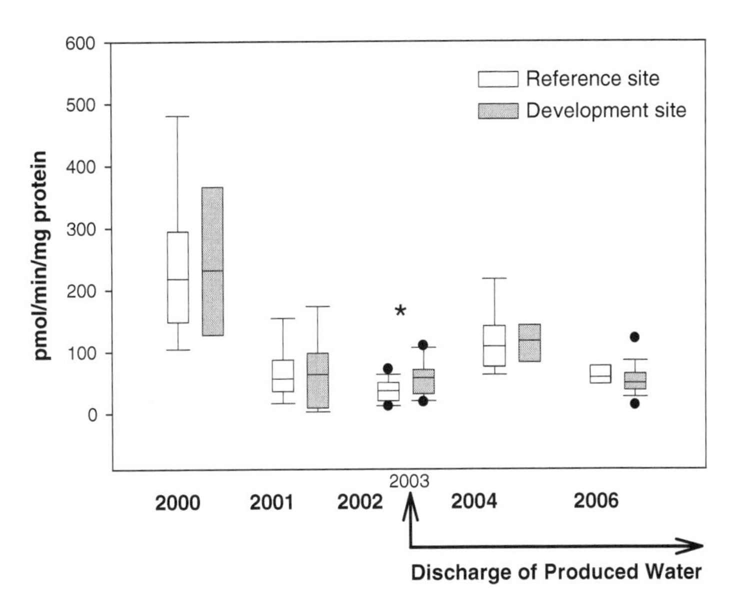

Fig. 20.6 Hepatic EROD activity in female American plaice; data plotted are median (*line in the box*), 25th and 75th percentiles (*bottom* and *top of box*) and 10th and 90th percentiles (*error bars*); data points beyond the 5th and 95th percentiles are also displayed; *significantly different ($p < 0.05$)

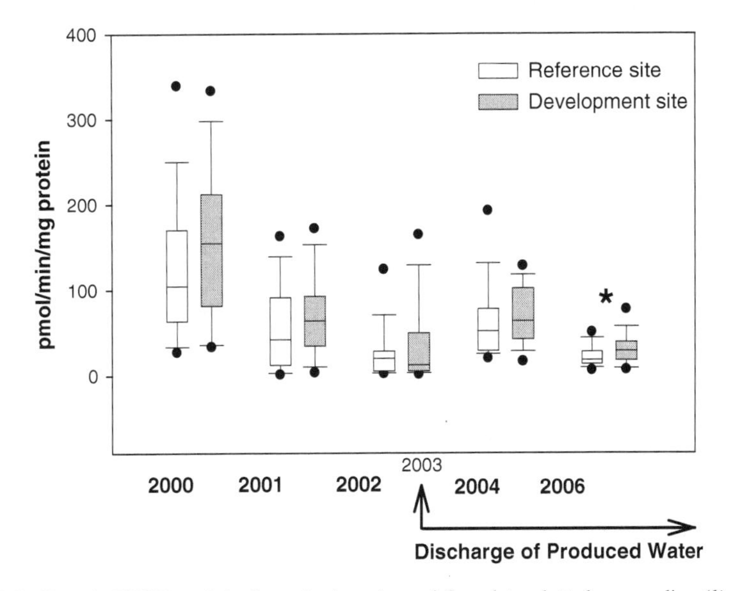

Fig. 20.7 Hepatic EROD activity in male American plaice; data plotted are median (*line in the box*), 25th and 75th percentiles (*bottom* and *top of box*) and 10th and 90th percentiles (*error bars*); data points beyond the 5th and 95th percentiles are also displayed; *significantly different ($p < 0.05$)

Although produced waters can be highly variable in content there is some information available from laboratory studies on their potential for induction of CYP1A or CYP1A linked enzyme activity in fish. Exposure of cod for 14 days to environmentally relevant concentrations of produced water from the Osberg C platform in the North Sea using a flow through system resulted in a concentration-dependent induction of gill EROD (Abrahamson et al. 2008). Gene expression has also been studied in gill and liver tissues of zebrafish exposed to produced water from the Osberg C platform (Olsvik et al. 2007). Of the large variety of genes altered in the gills of exposed fish, CYP1A displayed the greatest upregulation. Induction of CYP1A1 has similarly been reported to occur in the liver tissues of mosquito fish (*Gambusia affinis*) exposed in the laboratory to produced water from a platform in the Mediterranean Sea (Casini et al. 2006). Exposure of juvenile Atlantic cod (*Gadus morhua*) to produced water from the Hibernia oil field on the Grand Banks resulted in EROD enzyme induction in gill, heart and liver tissues with liver displaying the highest level of induction (Payne et al. 2005). Evidence to support the presence of a large variety of CYP1A enzyme inducing compounds in produced water, namely in waters from the United Kingdom Continental Shelves, has also been provided by Hurst et al. (2005) through studies on agonists for the cellular hydrocarbon receptor involved in induction.

There are also a few studies, mostly from the North Sea, in which induction of CYP1A or CYP1A linked enzyme activity has been investigated in fish caged around platform sites. Early studies by Aas and Klungsøyr (1998) reported a lack of induction in fish caged at a distance of 500 m from some platforms in the North Sea. As part of a special workshop on biological effects of contaminants in pelagic systems (BECPELAG), Forlin and Hylland (2006) caged Atlantic cod at various distances from the Statfjord production area as well as in the German Bight. No oil field related effects on hepatic CYP1A activity were noted. Enzyme activity was, however, higher in cod caged near the oil field compared to cod caged in the German Bight. No significant differences in gill EROD activity were found in cod caged for 6 weeks at reference sites and 500–10,000 m from the Troll and Statfjord A platforms (Abrahamson et al. 2008). However, the activity in cod caged 1,000 m from the Statfjord A platform was significantly higher compared to cod caged 10,000 m from the platform. An increase in CYP1A was also observed in liver tissue of stripey seaperch (*Lutjanus carponotatus*) caged near the Harriet A platform on the Northwest shelf of Australia for 10 days (Zhu et al. 2008). This study also made an interesting observation with respect to an induction of two other CYP forms, namely CYP2M1 and CYP2K1.

Regarding observations on natural field populations, which as noted above can have an additional degree of ecological relevance, an early study noted aryl hydrocarbon hydroxylase induction in 3 species, namely cod (*Gadus morhua*), haddock (*Melanogrammus aeglefinnus*) and whiting (*Merlangius merlangus*) taken around production platforms in the North Sea compared to nearby reference sites (Davies et al. 1984). Interestingly, the species in which induction was observed are generally considered to be much more migratory than flatfish. Stagg et al. (1995) also documented considerably higher levels of EROD induction in a flatfish (*Limanda limanda*) (4-fold in males and 16–20-fold in females) around a group of oil platforms in the northern North Sea. Furthermore, elevated levels of EROD enzymes were also observed in fish larvae around the same group of platforms (Stagg and McIntosh 1996). However, it is not known to what extent discharges of produced waters versus discharges of diesel base muds may have been responsible, whole or in part, for these earlier observations on enzyme induction in wild fish. EROD induction was not observed in fish collected around 3 platforms in the Gulf of Mexico (McDonald et al. 1996). However, reference fish as well as fish collected near the platforms contained quite high levels of PAH metabolites in their bile, suggesting extensive background contamination by hydrocarbons at the reference sites – which may be related to the various major inputs of contaminants into the Gulf of Mexico.

A more recent study of wild fish around the Harriet A and Harriet C platforms on the Northwest Shelf of Australia (King et al. 2005) documented induction of CYP1A protein in two species, gold-spotted trevally (*Carangoides fulvoguttatus*) and bar-cheeked coral trout (*Plectropomus maculatus*).

In the Terra Nova EEM program, EROD induction was observed on two occasions in wild fish. Such differences might be a reflection of natural variation but a contaminant related effect cannot be ruled out. If contamination related, it is noted that the induction observed was quite low (less than 2-fold) with the level observed after discharge of produced water being similar to that observed before hand. However, although the toxicity of some PAH could be independent of the aryl hydrocarbon receptor (Incardona et al. 2006), should any future monitoring indicate a trend towards high levels of induction, it could generate greater interest since induction in juvenile and adult fish as well as fish larvae has been associated with a variety of metabolic, cellular, organ and developmental disturbances (Spies et al. 1988; Au et al. 1999; Carls et al. 2005; Colavecchia et al. 2007; Morales-Caselles et al. 2007).

3.7 Haematology

With respect to potential for indicating potential immunological effects and disease status or susceptibility, there were no significant differences in the percentage of lymphocytes, neutrophils and thrombocytes between fish from the Reference and Development sites (Table 20.7). There were also no apparent qualitative differences in morphology or staining characteristics of red blood cells.

Thus, although produced water (Payne et al. 2005) as well as similar wastewaters (Farrell et al. 2004) have been shown to have potential for affecting white blood cell profiles in fish, no apparent development related effects have been noted in relation to the Terra Nova Development site.

Table 20.7 Percentage of blood cell types in American Plaice before and after discharge of produced water; all data are expressed as mean percentage ± standard deviation of each type of cell on 200 white blood cells counted

	2000		2001		2002		2004		2006	
							Discharge of produced water ⟷			
	REF	DEV	REF	DEV	REF	DEV	REF	DEV	REF	DEV
Number of fish	50	44	10	10	50	50	49	51	50	50
Lymphocytes	ns	ns	77.9 ± 2.6	78.4 ± 2.3	76.3 ± 4.5	76.2 ± 4.7	ns	ns	88.9 ± 3.5	88.9 ± 4.1
Neutrophils	ns	ns	3.7 ± 1.2	3.7 ± 0.7	5.0 ± 1.8	4.9 ± 2.1	ns	ns	0.7 ± 1.6	0.5 ± 0.7
Thrombocytes	ns	ns	18.4 ± 2.5	17.9 ± 2.0	18.7 ± 4.2	18.9 ± 4.3	ns	ns	10.4 ± 3.2	10.6 ± 4.2

REF = reference site, DEV = development site, ns = not suitable for carrying out reliable differential cell counts

4 Conclusions

Fish health studies have been carried out on approximately 500 American plaice over 5 survey years for the Terra Nova EEM program. Three surveys were carried out before release of produced water and two thereafter. The health effect indicators studied included fish condition, visible skin and organ lesions, hepatic EROD activity, haematology (differential cell counts) and a variety of histopathological indices in liver (e.g. nuclear pleomorphism, megalocytic hepatosis, foci of cellular alteration, macrophage aggregation, neoplasms) and gill (e.g. hyperplasia, oedema, fusion and aneurysms). These indicators have been extensively used in laboratory and field investigations with various fish species. Although a slight elevation of EROD enzyme activity was observed in fish from the Development site in 2002, before release of produced water, and in 2006, after release of produced water, other health bioindicators were generally absent or similar between the Reference and Development sites. Overall, on the basis of the various indicators studied, the results support the hypothesis of no significant project effects on the health of American plaice.

Acknowledgements Special thanks to the many sea-going personnel who work offshore on the various EEM programs.

References

Aas E, Klungsøyr J (1998) PAH metabolites in Bile and EROD activity in North Sea fish. Mar Environ Res 46(1–5):229–232

Abrahamson A, Brandt I, Brunström B, Sundt RC, Jorgensen EH (2008) Monitoring contaminants from oil production at sea by measuring gill EROD activity in Atlantic cod (*Gadus morhua*). Environ Pollut 153:169–175

Adams SM (eds) (1990) Biological indicators of stress in fish. American Fisheries Symposium, Bethesda, Maryland

Ankley GT, Reinert RE, Wade AE, White RA (1985) Temperature compensation in the hepatic mixed-function oxidase system of bluegill. Comp Biochem Physiol 81C:125–129

Au DWT, Wu RSS, Zhou BS, Lam PKS (1999) Relationship between ultrastrucutral changes and EROD activities in liver of fish exposed to benzo(a)pyrene. Environ Pollut 104:235–247

Barton BA, Morgan JD, Vijayan MM (2002) Physiological and condition-related indicators of environmental stress in fish. In: Adams M (ed) Biological indicators of aquatic ecosystem stress. American Fisheries Society, Bethesda, MD, pp 111–148

Beamish FWH (1966) Vertical migration by demersal fish in the Northwest Atlantic. J Fish Res Bd Can 23(1):109–139

Bilbao E, Soto M, Cajaraville MP, Cancio I, Margomez I (2006) Cell-and tissue-level biomarkers of pollution in wild pelagic fish, herring (*Clupea harengus*) and saithe (*Pollachius virens*) from the North Sea. In: Hylland K, Lang T, Vethaak Dick (eds) Biological effects of contaminants in marine pelagic ecosystems. SETAC Press, Pensacola, FL, USA pp 121–142

Boitsov S, Mjoes SA, Meier S (2007) Identification of estrogen-like alkylphenols in produced water from offshore oil installations. Mar Environ Res 64(5):651–665

Brodie WB (2002) American plaice *Hippoglossoides platessoides* on the Grand Bank (NAFO Divisions 3LNO) – a review of stock structure in relation to assessment of the stock. NAFO SCR Doc. 02/36, Serial No N4647, 19p

Carls MG, Heintz RA, Marty GD, Rice SD (2005) Cytochrome P4501A induction in oil-exposed pink salmon *Oncorhynchus gorbuscha* embryos predicts reduced survival potential. Mar Ecolog Prog Ser 301:253–265

Casini S, Marsili L, Fossi MC, Mori G, Bucalossi D, Porcelloni S, Caliani I, Stefanini G, Ferraro M, Alberti di Catenaja C (2006) Use of biomarkers to investigate toxicological effects of produced water treated with conventional and innovative methods. Mar Environ Res 62(Suppl 1):S347–S351

Colavecchia MV, Hodson PV, Parrott JL (2007) The relationships among CYP1A induction, toxicity, and eye pathology in early life stages of fish exposed to oil sands. J Toxicol Environ Health 70(18):1542–1555

Davies JM, Bell JS, Houghton C (1984) A comparison of the levels of hepatic aryl hydroxylase in fish caught close to and distant from North Sea oil fields. Mar Environ Res 14:23–45

Dorea H, Bispo, JR, Aragao, KA, Cunha BB, Navickiene S, Alves JP, Romao LP, Garcia CA (2007) Analysis of BTEX, PAHs and metals in the oilfield produced water in the State of Segipe, Brazil. Microchemic J 85(2):234–238

Dwyer KS, Morgan MJ, Maddock PD, Brodie WB, Healey BP (2009) Assessment of American plaice in NAFO. Div. LNO. NAFO SCR Doc. 09/35

Ellis AE (1976) Leucocytes and Related Cells in the Plaice *Pleuronectes platessa.* J Fish Biol 8:143–156

Farrell AP, Kennedy CJ, Kolok A (2004) Effects of wastewater from an oil-sand refining operation on survival, haematology, gill histology, and swimming of fathead minnows. Can J Zool 82:1519–1527

Flynn SA, Butler EJ, Vance I (1995) Produced water composition, toxicity and fate: a review of recent BP North Sea studies. In: Reed M, Johnsen S (eds) Produced water 2 – Environmental issues and mitigation technologies. Plenum Press, London. Environ Sci Res 52:69–80

Forlin L, Hylland K (2006) Hepatic cytochrome P4501A concentration and activity in Atlantic cod caged in two North Sea pollution gradients. In: Hylland K, Lang T, Vethaak D (eds) Biological effects of contaminants in marine pelagic ecosystems. SETAC Press, Pensacola, FL, USA, pp 253–260

Gallaway BJ, Martin LR, Howard RL, Boland GS, Dennis GD (1981) Effects on artificial reef and demersal fish and macrocrustacean communities. In: Middleditch BS (ed), Environmental effects of offshore oil production: the Buccaneer gas and oilfield study, Mar. Sci. 14, Plenum, New York, pp 237–299

Goede RW, Barton BA (1990) Organismic indices and an autopsy-based assessment as indicators of health and condition of fish. In: Adams SM (eds) Biological indicators of stress in fish. American fisheries symposium 8, Bethesda, Maryland, pp 93–108

Gray JS (2002) Perceived and real risks: produced water from oil extraction. Mar Poll Bull 44:1171–1172

Green RH (2005) Marine coastal monitoring: designing an effective offshore oil and gas environmental effects monitoring program. In: Armsworthy SL, Cranford PJ, Lee K (eds) Proceedings of the offshore oil and gas environmental effects monitoring workshop: approaches and technologies, Bedford Institute of Oceanography, Dartmouth, Nova Scotia, May 26–30, 2003, Battelle Press, Columbus, OH, pp 373–397

Grizzle JM (1986) Lesions in fishes captured near drilling platforms in the Gulf of Mexico." Mar Environ Res 18:267–276

Hamoutene D, Mabrouk G, Samuelson S, Volkoff H, Parrish C, Mansour A, Mathieu A, Lee K (2008) Effect of produced water on cod (*Gadus morhua*) immune responses. Proceedings of the 34th annual aquatic toxicity workshop, Halifax, Nova Scotia, Can Tech Rep Fish Aquat Sci 2793:59.

Hinton DE, Baumann PC, Gardner GR, Hawkins WE, Hendricks JD, Murchelano RA, Okihiro MS (1992) Histopathologic biomarkers. In Hugget RJ, Kimerle RA, Mehrle PM, Bergman HL (eds) Biomarkers: biochemical, physiological and histological markers of anthropogenic stress. Society of Environmental Toxicology and Chemistry (SETAC) Special Publication Series, Lewis Publishers, Chelsea, MI, pp 155–209

Holdway DA (2002) The acute and chronic effects of wastes associated with offshore oil and gas production on temperate and tropical marine ecological processes. Mar Pollut Bull 44: 185–203

Huggett RJ, Kimerle RA, Mehrle PM, Bergman HL (eds) (1992) Biomarkers: biochemical, physiological, and histological markers of anthropogenic stress. Society of Environmental Toxicology and Chemistry (SETAC) Special Publication Series, Lewis Publishers, Chelsea, MI, 347 pp

Hurst MR, Chan-Man YL, Balaam J, Thain E, Thomas KV (2005) The stable aryl hydrocarbon receptor agonist potency of United Kingdom Continental Shelf (UKCS) offshore produced water effluents. Mar Poll Bull 50:1694–1698

Hylland K (2006) Polycyclic aromatic hydrocarbon (PAH) ecotoxicology in marine ecosystems. J Toxicol Environ Health A69:109–123

Hylland K, Tollefsen KE, Ruus A, Jonsson G, Sundt RC, Sanni S, Utvik TI, Johnsen S, Nilssen I, Pinturier L, Balk L, Barsiene J, Marigomez I, Feist SW, Borseth JF (2008) Water column monitoring near oil installations in the North sea 2001–2004. Mar Pollut Bull 56:414–429

ICES (1999) Report of the ICES Advisory Committee on the Marine Environment, Copenhagen, Denmark, 31 May-5 June 1999. ICES Coop Res Report # 239

ICES (2004) Biological effects of contaminants: Use of liver pathology of the European flatfish dab (*Limanda limanda*) and flounder (*Platichthys flesus*) for monitoring. By SW Feist, T Lang, GD Stentiford, A Kohler. ICES Tech Mar Environ Sci 38:42p

Incardona JP, Day HL, Collier TK, Scholz NL (2006) Developmental toxicity of 4-ring polycyclic aromatic hydrocarbons in zebrafish is differentially dependent on AH receptor isoforms and hepatic cytochrome P450 1A metabolism. Toxicol Appl Pharmacol 217:308–321

Jenkins JA (2003) Pallid sturgeon in the lower Mississippi Region: haematology and genome Information. USGS Open-File Report 03-406, 32p

Jenkins JA (2004) Fish bioindicators of ecosystem condition at the Calcasieu Estuary, Louisiana: USGS Open-File Report 2004-1323, 47p

Kiceniuk J, Khan RA (1987) Effect of petroleum hydrocarbons on Atlantic cod, *Gadus morhua*, following chronic exposure. Can J Zool 65:490–494

King SC, Johnson JE, Haasch ML, Ryan DA, Ahokas JT, Burns KA (2005) Summary results from a pilot study conducted around an oil production platform on the Northwest Shelf of Australia. Mar Pollut Bull 50:1163–1172

Koehler A (2004) The gender-specific risk to liver toxicity and cancer of flounder (*Platichthys flesus*) at the German Wadden Sea coast. Aquat Toxicol 70(4):257–276

Lang T (2002) Fish disease surveys in environmental monitoring: the role of ICES. ICES Marine Science Symposia, 215:202–212

Lee K, Azetsu-Scott, Cobanli SE, Dalziel J, Niven S, Wohlgeschaffen G, Yeats P (2005) Overview of potential impacts from produced water discharges in Atlantic Canada. In: Armsworthy SL, Cranford PJ, Lee K (eds) Proceedings of the offshore oil and gas environmental effects monitoring workshop: approaches and technologies, Bedford Institute of Oceanography, Dartmouth, Nova Scotia, May 26–30, 2003. Battelle Press, Columbus, OH, pp 297–317

Lehtonen KK, Schiedek D, Kohler A., Lang T, Vuorin P, Forlin L, Barsiene J, Pempkowiak J, Gercken M (2006) The BEEP project in the Baltic Sea: overview of results and outline for a regional biological effects monitoring strategy. Mar Poll Bull 53:523–537

Lorax Environmental (2006) Calibration and validation of a numerical model of produced water dispersion at the Terra Nova FPSO. Prepared for Petro-Canada, St. John's

Love MS, Caselle JE, Snook L (2000) Fish assemblage around seven oil platforms in the Santa Barbara Channel Area. Fish Bull 98:96–117

Lowry OH, Rosebrough NJ, Fan AL, Randall RJ (1951) Protein measurement with the folin phenol reagent. J Biol Chem 193:265–275

Luna LG (1968) Manual of histological staining methods of the Armed Forces Institute of Pathology. McGraw-Hill, New York, NY

Lynch M, Raphael S, Mellor L, Spare P, Inwood M (1969) Medical laboratory technology and clinical pathology. W. B. Saunders Company, Philadelphia, PA

Lyons BP, Stentiford GD, Green M, Bignell J, Bateman K, Feist SW, Goodsir F, Reynolds WJ, Thain JE (2004) DNA adduct analysis and histopathological biomarkers in European flounder (*Platichthys flesus*) sampled from UK estuaries. Mutat Res 552(1–2):177–186

Mallatt J (1985) Fish gill structure changes induced by toxicants and other irritants: a statistical review. Can J Fish Aquat Sci 42:630–648

Martin-Skilton R, Lavado R, Thibaut R, Minier C, Porte C (2006) Evidence of endocrine alteration in the red mullet, *Mullus barbatus*, from the NW Mediterranean. Environ Pollut 141(1):60–68

Mathieu A, Lemaire P, Carriere S, Drai P, Giudicelli J, Lafaurie M (1991) Seasonal and sex linked variations in hepatic and extra hepatic biotransformation activities in striped mullet (*Mullus barbatus*). Ecotox Environ Saf 22:45–57

Mathieu A, Melvin W, French B, Dawe M, DeBlois EM, Power F, Williams U (2005) Health Effect Indicators in American plaice (*Hippoglossoides platissoides*) from the Terra Nova Development Site on the Grand Banks. In: Armsworthy SL, Cranford PJ, Lee K (eds) Proceedings of the offshore oil and gas environmental effects monitoring workshop: approaches and technologies, Bedford Institute of Oceanography, Dartmouth, Nova Scotia, May 26–30, 2003. Battelle Press, Columbus, OH, pp 297–317

McDonald SJ, Willett KL, Thomsen JT, Beatty KB, Connor K, Narasimhan TR, Erickson CM, Safe SH (1996) Sublethal detoxification responses to contaminant exposure associated with offshore production platforms. Can J Fish Aquat Sci 53:2606–2617

Meier S, Andersen TE, Norberg B, Thorsen A, Taranger GL, Kjesbu OS, Dale R, Morton HC, Klungsøyr J, Svardal A (2007) Effects of alkylphenols on the reproductive system of Atlantic cod (*Gadus morhua*). Aquat Toxicol 81(2):207–218

Menzie CA (1982) The environmental implications of offshore oil and gas activities. Environ Sci Tech 16(8):454–472

Miwa S, Nakayasu C, Kamaishi T, Yoshiura Y (2004) X-cells in fish pseudotumors are parasitic protozoans. Dis Aquat Org 58:165–170

Morales-Caselles C, Jimenez-Tenorio S, Riba I, Sarasquete C, DelValls A (2007) Kinetic of biomarker responses in juveniles of the fish *Sparus aurata* exposed to contaminated sediments. Environ Monit Assess 131:211–220

Morgan MJ, Brodie WB, Shelton PA (2005) An assessment of American plaice in NAFO subdivision 3PS. Can Sci Advis Secret, Research Document 2005/069, 42p

Morgan MJ (1994) Preliminary results of tagging experiments on American plaice in NAFO Divs. 3LNO. NAFO SCR Doc 96/61, 13p

Myers MS, Fournie JW (2002) Histopathological biomarkers as integrators of anthropogenic and environmental stressors. In: Adams M (ed) Biological indicators of aquatic ecosystem stress. American Fisheries Society, Bethesda, MD, pp 221–287

Myers MS, Rhodes LD, McCain BB (1987) Pathologic anatomy and patterns of occurrence of hepatic neoplasms, putative preneoplastic lesions, and other idiopathic hepatic conditions in English sole (*Parophrys vetulus*) from Puget Sound, Washington. J Nat Cancer Inst 78(2): 333–363

Neff JM (2002) Bioaccumulation in marine organisms: effect of contaminants from oil well produced water. Elsevier, New York, NY, p 468

OGP (International Association of Oil and Gas Producers) (2002) Aromatics in produced water: occurrence, fate and effects, and treatment. Report No1.20/324, 24p

OGP (International Association of Oil and Gas Producers) (2005) Fate and effects of naturally occurring substances in produced water on the marine environment. OGP Report, No 363: 35 pp

Olsvik PA, Lie KK, Stavrum AA, Meier S (2007) Gene-expression profiling in gill and liver of zebrafish exposed to produced water. Int J Environ Anal Chem 87(3):195–210

OSPAR (2004) OSPAR Coordinated Environmental Monitoring Programme (CEMP), OSPAR Commission Ref. No 2004-16

Payne JF, Andrews C, Guiney J, Gagnon S, Fancey L, Lee K (2005) Production water releases on the Grand Banks: potential for endocrine and pathological effects in fish. Proceedings of the 32nd annual aquatic toxicity workshop. Can Tech Rep Fish Aquat Sci 2617:24

Payne JF, Mathieu A, Collier TK (2003) Ecotoxicological studies focusing on marine and freshwater fish. In: Douben P (eds) PAHs: an ecotoxicological perspective, John Wiley & Sons Ltd., Chichester, UK pp 191–224

Payne JF, Fancey L, Andrews C, Meade J, Power F, Lee K, Veinott G, Cook A (2001) Laboratory exposures of invertebrate and vertebrate species to concentrations of IA-35 (Petro-Canada) drill mud fluid, production water, and Hibernia drill mud cuttings. Can Man Rep Fish Aquatic Sci 2560:1–27

Payne JF, Fancey L, Rahimtula A, Porter E (1987) Review and perspective on the use of mixed-function oxygenase enzymes in biological monitoring. Comp Biochem Physiol 86C(2): 233–245

Petrie B, Anderson C (1983) Circulation on the Newfoundland continental shelf. Atmos Ocean 21:207–226

Pitt TK (1967) Diurnal variation in the catches of American plaice, *Hippoglossoides platessoides*, from the Grand Bank. ICNAF Res Bull 4:53–58

Pitt TK (1969) Migrations of American Plaice on the Grand Bank and in St. Mary's Bay, 1954, 1959, and 1961. J Fish Res Board Can 26(5):1301–1319

Pohl RJ, Fouts JR (1980) A rapid method for assaying the metabolism of 7-ethoxyresorufin by microsomal subcellular fractions. Anal Biochem 107:150–155

Porter E, Payne JF, Kiceniuk JW, Fancey L, Melvin W (1989) Assessment of the potential for mixed-function oxygenase enzyme induction in the extrahepatic tissues of cunners during reproduction. Mar Environ Res 28:117–121

Rice CD, Arkoosh MR (2002) Immunological indicators of environmental stress and diseases susceptibility in fishes. In: Adams M (ed) Biological indicators of aquatic ecosystem stress. American Fisheries Society, Bethesda, MD, pp 187–220

Scott AP, Kristiansen SI, Katsiadaki I, Thain J, Tollefsen KE, Goksoyr A, Barry J (2006) Assessment of oestrogen exposure in cod (*Gadus morhua*) and saithe (*Pollachius virens*) in relation to their proximity to an oilfield. In: Hylland K, Lang T, Vethaak D (eds) Biological effects of contaminants in marine pelagic ecosystems SETAC Press, Pensacola, FL, pp 329–340

Smolders R, Cooreman K, Roose P, Blust R, De Coen W (2006) Cellular energy allocation in caged and pelagic free-living organisms along two North Sea pollution gradients. In: Ketil Hylland, Thomas Lang, Dick Vethaak (eds) Biological effects of contaminants in marine pelagic ecosystems. SETAC Press, Pensacola, FL, pp 235–251

Solangi MA, Overstreet RM (1982) Histopathological changes in two estuarine fishes, *Menidia beryllina* (Cope) and *Trinectes maculatus* (Bloch and Schneider), exposed to crude oil and its water-soluble fractions. J Fish Dis 5:13–35

Spies RB, Rice DW, Felton J (1988) Effects of organic contaminants on reproduction of the starry flounder *Platichthys stellatus* in San Francisco Bay. Hepatic contamination and mixed-function oxidase (MFO) activity during the reproductive season. Mar Biol 98:181–189

Stagg RM, McIntosh A, Mackie P (1995) Elevation of hepatic monooxygenase activity in the dab (*Limanda limanda L.*) in relation to environmental contamination with petroleum hydrocarbons in the northern North Sea. Aquat Toxicol 33:245–264

Stagg RM, McIntosh A (1996) Hydrocarbon concentrations in the Northern Sea and effects on fish larvae. Sci Total Environ 186:189–201

Stentiford GD, Longshaw M, Lyons BP, Jones G, Green M, Feist SW (2003) Histopathological biomarkers in estuarine fish species for the assessment of biological effects of contaminants. Mar Environ Res 55:137–159

Stephens SM, Frankling SC, Stagg RM, Brown A (2000) Sub-lethal effects of exposure of juvenile turbot to oil produced water. Mar Poll Bull 40(11):928–937

Sturve J, Hasselberg L, Falth H, Celander M, Forlin L (2006) Effects of North Sea oil and alkylphenols on biomarker responses in juvenile Atlantic cod (*Gadus morhua*). Aquat Toxicol 78:S73–S78

Sumpter JP, Johnson AC (2005) Lessons from endocrine disruption and their application to other issues concerning trace organics in the aquatic environment. Environ Sci Technol 39:4321–4332

Thomas KV, Balaam J, Hurst MR, Thain JE (2004) Identification of in vitro estrogen and androgen receptor agonists in north sea offshore produced water discharges Environ Toxicol Chem 23(5):1156–1163

Tierney K, Stockner E, Kennedy CJ (2004) Changes in immunological parameters and disease resistance in juvenile coho salmon (*Oncorhynchus kisutch*) in response to dehydroabietic acid exposure under varying thermal conditions. Water Qual Res J Can 39(3):175–182

Tort L, Balash JC, MacKenzie S (2004) Fish health challenge after stress. Indicators of immunocompetence. Contrib Sci 2(4):443–454

USEPA (2000) A framework for an integrated and comprehensive monitoring plan for the estuaries of the Gulf of Mexico. EPA, Washington, DC, 87 pp

Utvik T, Gartner, L (2006) Concentration of polycyclic aromatic hydrocarbons in seawater: Comparison of results from dispersion modeling with measured data from blue mussels (*Mytilus edulis*) and SPMD residues. In: Hylland K, Lang T, Vethaak D (eds) Biological effects of contaminants in marine pelagic ecosystems SETAC Press, Pensacola, FL, pp 29–42

Weeks BA, Anderson DP, DuFour AP, Fairbrother A, Goven AJ, Lahvis GP, Peters G (1992) Immunological biomarkers to assess environmental stress. In Huggett RJ, Kimerle RA, Mehrle PM, Bergman HL (eds) Biomarkers: biochemical, physiological, and histological markers of anthropogenic stress. Society of Environmental Toxicology and Chemistry (SETAC) Special Publication Series, Lewis Publishers, Chelsea, MI, pp 211–233

Whyte JJ, Jung RE, Schmitt CJ, Tillitt DE (2000) Ethoxyresorufin-O-deethylase (EROD) activity in fish as a biomarker of chemical exposure. Critical Rev Toxicol 30(4):347–570

Wilson-Ormond EA, Ellis MS, Powel EN (1994) The effect of proximity to gas producing platforms on size, stage of reproductive development and health in shrimp and crabs. National Shellfisheries Association, Charleston, South Carolina, Abstracts, 1994 Annual Meeting, April 24–28:306

Zhao L, Chen Z, Lee K (2008) A risk assessment for produced water discharge from offshore petroleum platforms-development and validation. Mar Pollut Bull 56:1890–1897

Zhu S, King SC, Haasch ML (2008) Biomarker induction in tropical fish species on the Northwest Shelf of Australia by produced formation water. Mar Environ Res 65:315–324

Chapter 21
Risks to Fish Associated with Barium in Drilling Fluids and Produced Water: A Chronic Toxicity Study with Cunner (*Tautogolabrus adspersus*)

Jerry F. Payne, Catherine Andrews, Linda Fancey, Boyd French, and Kenneth Lee

Abstract Barite is composed largely of barium sulfate, and the high density of this particular compound lends support to the use of barite as a weighting agent in oil well drilling fluids, which are often released into the environment in conjunction with petroleum exploration and development. The discharge of produced water can also be an important source of barium sulfate around petroleum development sites. Barium sulfate is practically insoluble in seawater and is commonly indicated to have little toxicity potential. However, there is limited information on the chronic toxicity potential of the compound. We have carried out a long-term chronic toxicity study on the effects of barium sulfate, as barite, on fish health. Cunners (*Tautogolabrus adspersus*) were exposed on a weekly basis for 10 months to dispersions of 200 g of barite in a tank containing 1800 L of water. A range of health effect indicators were investigated, including visible skin and organ lesions, fish and organ condition indices, histopathological alterations in liver, gill, and kidney tissues, and levels of ethoxyresorufin *O*-deethylase (EROD), a catalytic activity associated with induction of cytochrome P4501A1. No differences were noted between the control and experimental groups other than a slight induction of EROD. Given the prolonged and high-level exposure to barite, results support the hypothesis that the relatively insoluble barium sulfate associated with the disposal of drilling fluids and produced waters does not pose a significant toxicity risk to finfish.

1 Introduction

Large amounts of barite are used as a weighing agent in drilling muds, particularly when drilling deep wells or penetrating strata containing potentially problematic geopressures. Although composed largely of barium sulfate, barite often contains small amounts of minerals such as limestone and dolomite as well as several metals in the form of insoluble sulfides. The most abundant metal in barite is barium sulfate,

J.F. Payne (✉)
Science Branch, Fisheries and Oceans Canada, Northwest Atlantic Fisheries Centre, St. John's, NL, Canada

K. Lee, J. Neff (eds.), *Produced Water*, DOI 10.1007/978-1-4614-0046-2_21,

and the high density (4.1–4.5 g/cm^3) of this particular compound contributes to the use of barite as a weighting agent in drilling muds (NRC 1983; Neff 2002).

Although interest in barium principally revolves around its use in drilling muds, particularly water-based muds, which are often deposited in association with cuttings near rig sites, barium is also commonly found in produced water with concentrations frequently exceeding those in seawater by a factor of 1000 (Neff 2002). Produced water is composed of ancient "fossil" water common to oil well reservoirs or water injected into reservoirs to displace oil and enhance its flow. Seawater is enriched in sulfate, and soluble salts of barium rapidly precipitate as barium sulfate in the presence of sulfate (Trefry et al. 1995; Monnin et al. 1999). The barium in highly saline reservoir waters containing sulfate could exist in the form of barium sulfate. We have noted in this regard that barium chloride, which is quite soluble in water, instantly formed a precipitate upon addition to a sample of produced water from the Grand Banks. Barium sulfate would also be expected to form upon injection of seawater into oil well reservoirs having low sulfate content or upon the discharge of produced water itself into the ocean. Thus, barium is also important from a produced water perspective. Moreover, produced water commonly disperses over broad scale areas and can be discharged in relatively high volumes at a particular development site for a number of years.

Regarding toxicity, particulate barium sulfate is only slightly soluble in seawater. The solubility of barium in seawater in equilibrium with barium sulfate has been reported to be in the range of 37–52 μg/L (Chow and Goldberg 1960; Church 1979). Barium sulfate also has a low acute toxicity potential, with median lethal concentrations being reported at concentrations exceeding approximately 7000 mg/L (NRC 1983).

Embryonic forms are often more sensitive to contaminants than adult forms, and barium (as barite) was toxic to embryos of the crab *Cancer anthonyi* at concentrations greater than 1000 mg/L (Macdonald et al. 1988). Since this concentration greatly exceeds the solubility of barium in seawater, the effect could have been physical in nature. Barite dispersed in water in high concentrations was also not acutely toxic to capelin (*Mallotus villosus*) and snow crab (*Chionecetes opilio*) larvae, or planktonic jellyfish exposed for 24 h to 200, 2000, and 1000 mg/L respectively (Payne et al. 2006). Regarding potential effects on behavior, the swimming activity of Dungeness crab (*Cancer magister*) and dock shrimp larvae (*Pandalus danae*) was inhibited by suspensions of barite in the range of 16–71 mg/L (Carls and Rice 1984). Given the very high exposure, it is again likely that the effect was physical in nature.

It is important to note that when discussing the potential toxicity of barium, it is barium sulfate which is relevant in relation to discharges by the oil and gas industry. By comparison, soluble salts of barium, such as barium chloride and acetate, are well-known toxicants for plants and animals (e.g., Wang and Wang 2008 and references therein). Relatively low concentrations of barium acetate (EC median of 189 μg/L) have been reported to inhibit embryonic development in mussel larvae (*Mytilus edulis*) (Spangenberg and Cherr 1996). This may have been due to a relatively slow rate of precipitation of barium complexed with acetate since higher concentrations of barium acetate, which resulted in precipitation, were less inhibitory. Relatively high concentrations of barium chloride (100 mg/L) were lethal to early life stages of yellow crab (*Cancer anthonyi*) (Macdonald et al. 1988). Although exposures in small chambers are valuable for obtaining a general appreciation of toxicity, mixing and dilution would be expected to be slower than in the ocean, where a decrease in exposure to the hydrated barium ion, the bioavailable form of barium, would be anticipated.

A few studies have dealt with chronic effects. Exposure of grass shrimp (*Palaemonetes pugio*) to a substrate of barite for approximately 100 days resulted in gut lesions which could have been caused by barite abrasion (Conklin et al. 1980). Gill damage was observed in a suspension feeding bivalve *Cerastoderma edule* and the deposit feeder *Macoma balthica* treated with high daily doses of barite, namely levels that would result in 1, 2, and 3 mm depth equivalents on the ocean bottom (Barlow and Kingston 2001). Cranford et al. (1999) carried out a detailed study on the effects of drilling fluid, including specific ingredients such as barite, on sea scallop *Placopecten magellanicus*, a species of major commercial importance on the Eastern seaboard of Canada and the United States. This study, as well as a companion modelling study (Hannah et al. 2006), has been of particular value for assessing concentrations that could potentially affect growth and reproductive success in scallop under relatively long-term conditions of exposure. Although the effects observed might be primarily linked to a physical cause such as tissue abrasion and/or effects on feeding, a role for some degree of toxicity cannot be ruled out.

To our knowledge, no studies have been carried out investigating the potential for barium sulfate to adversely affect the health of fish under chronic conditions of exposure. Such a study was considered to be of interest for further evaluation of environmental risks that could be associated with the discharge of drill cuttings as well as the discharge of produced waters, which have been the subject of special attention for some time (Ray 1996). This study reports on the results obtained with cunner exposed to weekly suspensions of barite in water for approximately 10 months.

2 Materials and Methods

Cunners caught in Portugal Cove in costal Newfoundland were acclimatized in the laboratory for 2 weeks before the start of the experiment. Two groups of 50 fish of approximately the same size were then constituted and placed in two flow-through 1800 L seawater tanks having a water flow rate of approximately 20 L/min at ambient water temperature, which ranges from approximately 0.0–2.0°C from December to May and rising to a peak temperature of approximately 12–14°C in August. During the first month of exposure in August, fish were fed food contaminated with barite, however, there was considerable washout of barite from the food and during the next 10 months of exposure (September to June), fish were exposed to barite through dispersing 200 g of barite onto the surface of the water on a weekly basis. From November to June, cunners are normally in torpor, during which time they do not feed.

In order to obtain an appreciation of the solubility potential for barium, a leachate test was carried out on the barite used in the study. Two grams of dry barite was added to 50 mL of reagent grade water. Following a 2 h extraction period, the sample was filtered through a 0.45 μm filter. The filtrate was then analyzed for barium and other metals by ICP-MS (EPA Method 6020A).

Ovary tissues of 10 control and 10 exposed fish were likewise analyzed for barium and other metals by ICP-MS. Tissues were digested in concentrated nitric acid, and 1–2% solutions of the acid were used for injections. Recovery of matrix spike parameters were generally greater than 90% with established QC limits of 75–125%.

Barite was also analyzed for polycyclic aromatic hydrocarbons by GC-MS (EPA Method 8270C) to account for any extensive contamination of the ore by this class of chemicals, which might for instance occur during mining and transport operations.

All chemical analyses were carried out by an accredited analytical laboratory, Maxxam Analytics, Inc.

2.1 Necropsy

A total of 95 cunners, 48 control and 47 exposed to barite, were autopsied. Fish were examined for gross pathology (skin, fins, gills, and internal organs) under the general framework described by Adams et al. (1993). Fork length, whole weight, gutted weight, sex, and liver weight were recorded. A slice of liver, the first right gill arch, and kidney tissue were dissected and fixed in 10% buffered formalin for histopathological analysis. A piece of liver and gonad was stored at –65°C for EROD and barium analysis respectively.

2.2 Tissue Histopathology

Liver, gill, and kidney tissues of female cunner were processed for histological analysis (Lynch et al. 1969) using a Tissue-Tek® VIP Processor. Sections were cut at 6 μ on a Leitz microtome, floated on a 47°C water bath containing gelatin, and picked up on labeled microscope slides. After air drying, the slides were fixed at 60°C for approximately 2 h to remove most of the embedding media and allow the sections to adhere properly to the slide. Tissue sections were stained using Mayers Haematoxylin and Eosin (H&E) method (Luna 1968). Coverslips were applied using Entellan® and the slides were left to air dry and harden overnight.

Tissue sections were examined under different magnifications by transmission light microscopy (Wild Leitz Aristoplan bright field microscope). To minimize interpretive bias, a "blind" system in which the examiner is not aware of the treatment of the specimen was used. This is accomplished by using a "pathology" number on the slide label generated from a random number table matched with the actual specimen number.

2.2.1 Liver

Liver sections were assessed for the presence of various lesions that have been associated with chemical toxicity including nonspecific necrosis, nuclear pleomorphism, megalocytic hepatosis, various foci of cellular alteration, hydropic vacuolation, and macrophage aggregation (e.g., Myers et al. 1987; Hinton et al. 1992; Myers and Fournie 2002; Lyons et al. 2004; ICES 2004).

2.2.2 Gill

Gill sections were assessed for the presence of major gill lesions including telangiectasis, fusion, epithelial lifting, severe hyperplasia, or clubbing as well as extensive edema (Mallatt 1985).

2.2.3 Kidney

Kidney sections were assessed for any structural differences in interstitial tissues and renal tubules.

2.3 Mixed-Function Oxygenase (MFO)

MFO activity was assessed in liver samples of cunner as 7-ethoxyresorufin *O*-deethylase (EROD), a cytochrome P450 linked catalytic activity, according to the method of Pohl and Fouts (1980) as modified by Porter et al. (1989).

2.4 Statistical Analyses

Comparisons between the control and exposed groups were carried out using Sigma-Stat 3.0 for the analysis of variance (ANOVA) and SAS for the analysis of covariance (ANCOVA). Length, whole and gutted weight, liver weight, condition factor, hepato-somatic index, and EROD activity were analyzed by ANOVA (unpaired *t*-test or the Mann–Whitney Rank Sum test, when the groups were not normally distributed). Log–log regressions of gutted weight versus length and liver weight versus gutted weight were compared by ANCOVA. Comparisons between groups having a $p<0.05$ were considered to be statistically significant.

3 Results

Regarding the potential for barium to leach from the barite used in the study, the high concentrations of salts in seawater precluded examination by ICP-MS of any low levels of barium as well as possible trace levels of other dissolved metals that might be present in the barite as relatively insoluble sulfides. However, leachate studies carried out on barite with reagent grade water demonstrated that although the leachate was essentially free of a variety of heavy metals (i.e., if present were below the limit of detection), some leaching of barium did occur (Table 21.1). The results were generally consistent with other studies indicating that barium in barite, which is largely composed of barium sulfate, has limited potential for bioavailability and thus toxicity due to its very low aqueous solubility. It is also noted that the analysis was carried out on a distilled water leachate of barite and lower concentrations would be expected in seawater leachate.

Notwithstanding the lack of potential for cunners to be exposed to significant levels of dissolved barium, ingestion of particulate barite could also be an important route of exposure since saltwater fish ingest a considerable volume of water on a daily basis. We are unaware of any specific information on cunners, but studies on a large number of marine teleosts indicate ingestion rates in the order of 1–5 mL/kg/h (Marshall and Grosell 2006). Additional exposure in the environment for many benthic fish species (such as cunners) could come from ingestion of sediments

Table 21.1 Elements measured in distilled water leachate of barite

Elements	Units	Barite leachate	RDL
Total mercury	μg/L	ND	0.02
Elements (ICP-MS)			
Dissolved aluminum	μg/L	95	10
Dissolved antimony	μg/L	ND	2
Dissolved arsenic	μg/L	ND	2
Dissolved barium	μg/L	210	5
Dissolved beryllium	μg/L	ND	2
Dissolved bismuth	μg/L	ND	2
Dissolved boron	μg/L	ND	5
Dissolved cadmium	μg/L	ND	0.3
Dissolved chromium	μg/L	ND	2
Dissolved cobalt	μg/L	ND	1
Dissolved copper	μg/L	ND	2
Dissolved iron	μg/L	ND	50
Dissolved lead	μg/L	ND	0.5
Dissolved manganese	μg/L	ND	2
Dissolved molybdenum	μg/L	ND	2
Dissolved nickel	μg/L	ND	2
Dissolved selenium	μg/L	ND	2
Dissolved silver	μg/L	ND	0.5
Dissolved strontium	μg/L	640	5
Dissolved thallium	μg/L	ND	0.1
Dissolved tin	μg/L	ND	2
Dissolved uranium	μg/L	ND	0.1
Dissolved vanadium	μg/L	ND	2
Dissolved zinc	μg/L	ND	5

ND = not detectable
RDL = reportable detection limit

during feeding, and the acidity of gut fluids might result in a small dissolution of barite entering the gut from either water drinking or feeding.

Considering an environmental context, it is also of interest that various zooplankton species are known to contain barium sulfate granules (Gooday and Nott 1982; Collier and Edmond 1984). Thus, some fish species residing in the near vicinity of oil development sites could have an enhanced exposure to barium, through the ingestion of zooplankton or their predators. However, any such exposure of zooplankton to barium could be trivial compared to their potential for exposure by the ingestion of phytoplankton, such as diatoms, which commonly contain barium (from natural sources) in their skeletons (e.g., Sternberg et al. 2005 and references therein).

Although fish movement throughout the water column as well as the turbulence associated with fish movement presented problems in trying to determine the levels of particulates to which the fish were exposed in the study (through the ingestion of water), fine visible particulates were observed for up to 10 h after dispersing 200 g of barite onto the water surface.

Calculations have been carried out on the settling rates of crushed ores such as barite and are presented for general interest (Gibbs et al. 1971). Assuming a modal grain size of 25 μm and a specific gravity of 4.5 g/cm^3, the settling rate was calculated to be approximately 0.1 cm/s for undisturbed water. This would translate into a settling time of approximately 4 h in our tank which was approximately 1.5 m in depth. However, it is noted that this calculation is on the basis of a modal grain size of 25 μm and smaller particulates would be expected to settle more slowly. Also, in addition to turbulence generated by the movement of fish, as noted above, the current induced in the tank by the inflowing water ($\sim$ 20 L/min) would be expected to result in a reduction of settling rate. Detailed studies dealing with settling rates of drilling fluid particulates, including observations in the field, can be found in Muschenheim and Milligan (1996) and Milligan and Hill (1998).

An analysis was carried out to determine whether barium or other metals had accumulated to some extent in the ovary – with the understanding that there might be some low level of exposure via the water column or through internal leaching of barium sulfate in the digestive tract.

Any evidence of bioaccumulation in the ovary was considered to be of interest from a reproductive viewpoint. Furthermore, since barium is chemically similar to calcium, which is typically found in high concentrations in ovary tissues (Hellou et al. 1992), it might "substitute" to some extent for calcium in the sense that Ba may move through Ca ion channels more readily than Ca itself (e.g., O'Donnell 1988). Samples of ovary tissues from control (10) and exposed (10) fish were analyzed, but there was no evidence for bioaccumulation of barium other than in one exposed fish in which the level detected (2.1 mg/kg) was just above the detection limit (Table 21.2). A low level of aluminum and strontium was also found in one exposed fish each. Low levels of copper, arsenic, iron, manganese, selenium, and zinc were detected in the ovary, but there were no statistically significant differences between the groups.

Table 21.2 Concentrations (mean wet weight) of elements in cunner ovary following exposure to barite

Elements (ICP-MS)	Units	10 individuals per treatment		RDL	Elements (ICP-MS)	Units	10 individuals per treatment		RDL
Aluminum	mg/kg	C	ND	2.5	Lithium	mg/kg	C	ND	0.50
		E	3.2				E	ND	
Antimony	mg/kg	C	ND	0.50	Manganese	mg/kg	C	0.50–0.62	0.50
		E	ND				E	0.53–0.57	
Arsenic	mg/kg	C	0.53–0.58	0.50	Molybdenum	mg/kg	C	ND	0.50
		E	0.65				E	ND	
Barium	mg/kg	C	ND	1.5	Nickel	mg/kg	C	ND	0.50
		E	2.1				E	ND	
Beryllium	mg/kg	C	ND	0.50	Selenium	mg/kg	C	0.72–1.04	0.50
		E	ND				E	0.65–0.88	
Boron	mg/kg	C	ND	1.5	Silver	mg/kg	C	ND	0.12
		E	ND				E	ND	
Cadmium	mg/kg	C	ND	0.05	Strontium	mg/kg	C	ND	1.5
		E	ND				E	2.4	
Chromium	mg/kg	C	ND	0.50	Thallium	mg/kg	C	ND	0.02
		E	ND				E	ND	
Cobalt	mg/kg	C	ND	0.20	Tin	mg/kg	C	ND	0.50
		E	ND				E	ND	
Copper	mg/kg	C	0.90–1.45	0.50	Uranium	mg/kg	C	ND	0.02
		E	0.94–1.22				E	ND	
Iron	mg/kg	C	17–59	15	Vanadium	mg/kg	C	ND	0.50
		E	27–37				E	ND	
Lead	mg/kg	C	ND	0.18	Zinc	mg/kg	C	86.8–110	1.5
		E	ND				E	82.8–99.9	

ND = not detectable
RDL = reportable detection limit
C = control samples
E = experimental samples

3.1 Gross Pathology

No abnormalities were observed upon necropsy on the gills, fins, or skin of fish or on the external surface of internal organs including the liver, gonad, digestive tract, body cavity, spleen, and kidney.

3.2 Morphometrics and Fish Condition

Biological characteristics (length, whole and gutted weight, and liver weight) were analyzed separately for each sex and compared between the control group and the group exposed to barite using the unpaired *t*-test or the Mann–Whitney Rank Sum test, when the groups were not normally distributed.

Fish condition, which can be defined as a state of physical fitness, was assessed by calculating two conditions indices: Fulton's condition factor (100 $\times$ whole or gutted body weight/length^3) and the hepato-somatic index (100 $\times$ liver weight/gutted body weight). These condition indices are commonly used and are presented for general interest with comparisons between groups being carried out with the unpaired *t*-test or the Mann–Whitney Rank Sum test. However, since these indices assume that body weight is proportional to the cube of length, and liver weight is linearly related to gutted weight, log–log regressions of adjusted means of gutted weight (predicted mean of gutted weight at overall mean of length) and liver weight (predicted mean of liver weight at overall mean of gutted weight) were also compared between the two groups by analysis of covariance (ANCOVA).

Biological characteristics and fish condition measured at the end of the experimental period are summarized in Table 21.3 (males) and Table 21.4 (females), respectively. There were no significant differences in biological characteristics (ANOVA) and fish condition (ANCOVA) in males from the control and exposed groups. With respect to females, differences in length, whole and gutted weight, and liver weight were marginally significant ($0.05<p<0.11$) between the control and exposed groups. Fish in the control group were slightly smaller, however, condition factor, hepato-somatic index as well as adjusted means of gutted weight and liver weight were similar between the two groups (ANOVA or ANCOVA).

3.3 Histopathology

Histopathological studies were carried out on female cunner of similar size (16.8–19.7 cm in length): 20 from the control group and 19 from the group exposed to barite. Histopathology involves the analysis of organs for the presence of cellular and tissue damage and is especially well suited for assessing chronic as well as acute exposure to toxicants. Also, being a higher level biological response, it can provide indication of the potential for more serious health impairment in organisms.

Table 21.3 Biological characteristics and condition of male cunner

	Control group	Exposed group	*p*-value
Fish number	19	18	
Length (cm)	17.6 ± 1.2	17.7 ± 1.5	0.879 (AN[f])
Whole body weight (g)	68.8 ± 18.2	70.7 ± 20.8	0.682 (AN[f])
Gutted body weight (g)	63.6 ± 17.0	65.3 ± 19.1	0.682 (AN[f])
Liver weight (g)	0.78 ± 0.27	0.80 ± 0.29	0.891 (AN[f])
Condition factor (whole body weight)[a]	1.240 ± 0.067	1.251 ± 0.069	0.648 (AN[f])
Condition factor (gutted body weight)[b]	1.147 ± 0.059	1.156 ± 0.015	0.648 (AN[f])
Hepato-somatic index[c]	1.212 ± 0.215	1.205 ± 0.221	0.923 (AN[f])
Adjusted mean of gutted weight[d]	62.3	62.7	0.678 (ANC[g])
Adjusted mean of liver weight[e]	0.748	0.741	0.870 (ANC[g])

All data are means ± standard deviations, except for adjusted means; [a]condition factor = 100 × whole body weight/length^3; [b]condition factor = 100 × gutted body weight/length^3; [c]100 × liver weight/gutted body weight; [d]adjusted mean of gutted weight is predicted mean gutted weight at overall mean length; [e]adjusted mean of liver weight is predicted mean liver weight at overall gutted weight; [f]*p* value obtained after ANOVA analysis (unpaired *t*-test or Mann–Whitney Rank Sum test); [g]*p* value obtained after ANCOVA analysis of log–log regression of gutted body weight on length or liver weight on gutted body

Table 21.4 Biological characteristics and condition of female cunner

	Control group	Exposed group	*p*-value
Fish number	29	29	
Length (cm)	17.4 ± 1.1	17.9 ± 1.4	0.106 (AN[f])
Whole weight (g)	66.1 ± 13.5	74.3 ± 17.9	0.053 (AN[f])
Gutted weight (g)	60.2 ± 12.3	67.6 ± 16.4	0.057 (AN[f])
Liver weight (g)	0.80 ± 0.23	0.91 ± 0.27	0.107 (AN[f])
Condition factor (whole weight)[a]	1.243 ± 0.099	1.270 ± 0.071	0.241 (AN[f])
Condition factor (gutted weight)[b]	1.133 ± 0.091	1.155 ± 0.066	0.297 (AN[f])
Hepato-somatic Index[c]	1.310 ± 0.228	1.338 ± 0.212	0.756 (AN[f])
Adjusted mean of gutted weight[d]	61.7	62.8	0.356 (ANC[g])
Adjusted mean of liver weight[e]	0.809	0.817	0.820 (ANC[g])

All data are means ± standard deviations, except for adjusted means; [a]condition factor = 100 × whole body weight/length^3; [b]condition factor = 100 × gutted body weight/length^3; [c]100 × liver weight/gutted body weight; [d]adjusted mean of gutted weight is predicted mean gutted weight at overall mean length; [e]adjusted mean of liver weight is predicted mean liver weight at overall gutted weight; [f]*p* value obtained after ANOVA analysis (unpaired *t*-test or Mann–Whitney Rank Sum test); [g]*p* value obtained after ANCOVA analysis of log–log regression of gutted body weight on length or liver weight on gutted body

3.3.1 Liver

The structure of the liver was similar in control fish and fish exposed to barite. There were no lesions typically associated with chemical toxicity (e.g., Myers et al. 1987; Hinton et al. 1992; Myers and Fournie 2002; Lyons et al. 2004; ICES 2004). These included necrosis, nuclear pleomorphism, megalocytic hepatosis, eosinophilic foci, basophilic foci, clear cell foci, carcinoma, cholangioma, cholangiofibrosis, steatosis, and hydropic vacuolation. A representative photo micrograph is provided in Fig. 21.1.

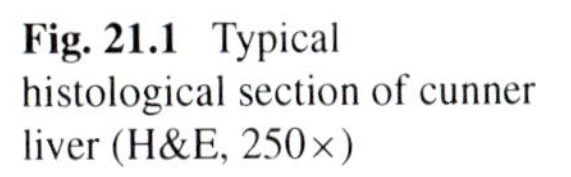
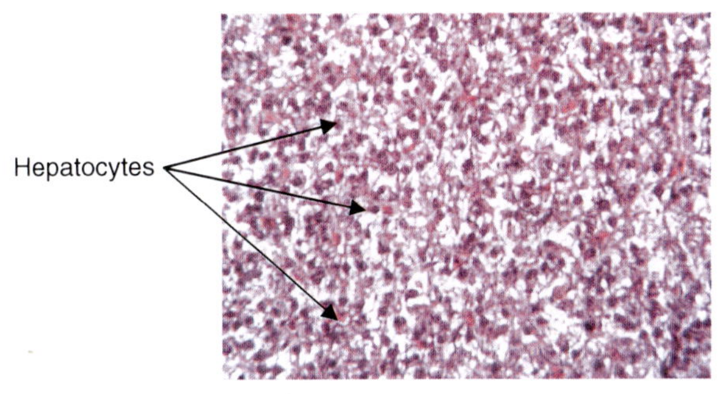

Fig. 21.1 Typical histological section of cunner liver (H&E, 250×)

3.3.2 Gill

Particular attention was given to effects on gills since barium is known to accumulate at least in the gills of some tuna species to a greater degree than other organs (Patterson and Settle 1977; Bassari 1994). However, the structure of primary and secondary gill lamellae appeared normal in control fish and fish exposed to barite. There were a few observations of telangiectasis, fusion, and hyperplasia, but the cases were very mild and observed in fish from both groups. Parasites were also observed in most of the samples examined. A representative photomicrograph is provided in Fig. 21.2. Given the known sensitivity of fish gill lamellae to acute and chronic exposure to a variety of chemicals and other irritants (e.g., Mallatt 1985),

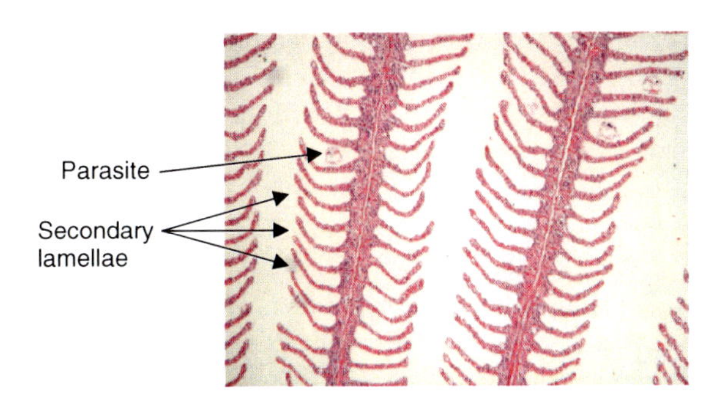

Fig. 21.2 Typical histological section of cunner gill (H&E, 250×)

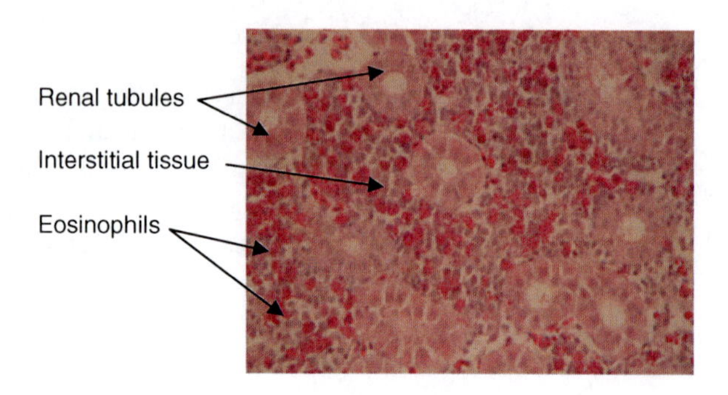

Fig. 21.3 Typical histological section of cunner kidney (H&E, 250×)

it is of interest that no effects were noticed under chronic conditions of exposure of fish to suspensions of barite for 10 months.

3.3.3 Kidney

The structure of the kidney appeared similar in control cunner and cunner exposed to barite. It was composed of renal tubules surrounded by interstitial tissue which contained variable amounts of eosinophils. A representative photomicrograph is provided in Fig. 21.3.

3.4 MFO Activity

It is of interest that a small induction of EROD, a cytochrome P450 linked catalytic activity, was observed in both male and female fish exposed to barite (Fig. 21.4). The cytochromes P450 belong to a super-family of hemoproteins, of which EROD is commonly associated (but not exclusively) with cytochrome P450 1A1 (or CYP1A1). A variety of environmentally important contaminants including many PAH, PCBs, and dioxins having appropriate stereochemical properties are relatively strong activators of a cytosolic receptor, (aryl hydrocarbon receptor or AHR) which upon movement into the nucleus results in an increased transcription of cytochrome linked mRNA and enzyme induction (Denison and Nagy 2003).

Such enzyme induction has led to the wide use of CYP1A1 or a related catalytic activity such as EROD, in environmental monitoring and assessment programs with fish (e.g., Whyte et al. 2000; Payne et al. 2003). It is also worth noting that the cytosolic receptor, which plays an initial role in enzyme induction interacts with a variety of cellular pathways, and both the significance of receptor interaction and induction itself are topics of considerable interest in developmental biology and toxicology (Timme-Laragy et al. 2007 and references therein). Induction itself has also been associated with a variety of metabolic, cellular, and other organ disorders in fish (Spies et al. 1988; Au et al. 1999; Carls et al. 2005; Colavecchia et al. 2007; Morales-Caselles et al. 2007).

Since the process of induction is mediated by organic compounds having appropriate stereo-chemical binding properties for the cytosolic receptor, why are fish displaying elevated levels of MFO upon exposure to barite?

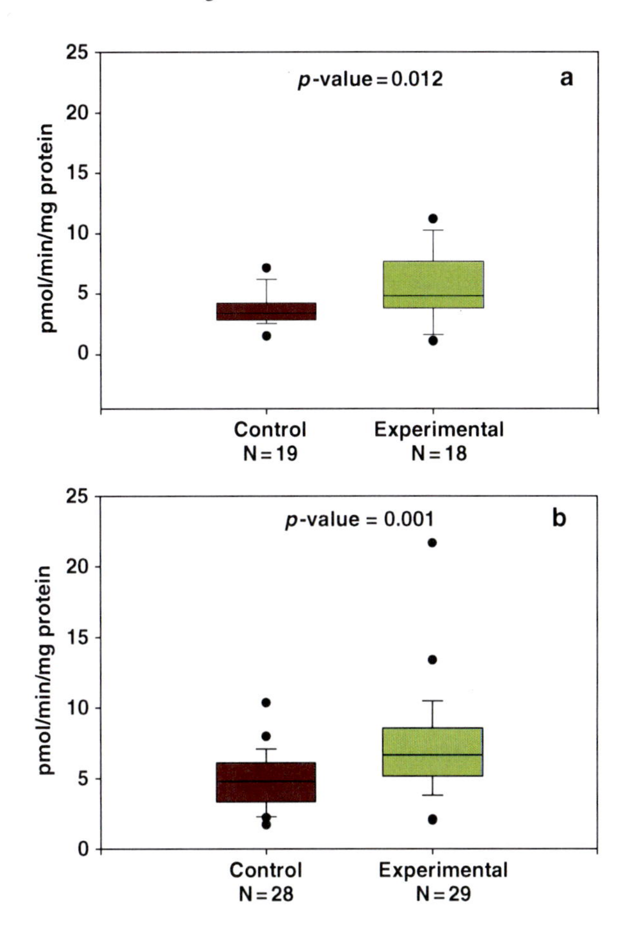

Fig. 21.4 EROD activity (pmol resorufin/min/mg protein) in **a** male and **b** female cunner; data plotted are median (*line in the box*), 25th and 75th percentiles (*bottom* and *top of box*), and 10th and 90th percentiles (*error bars*) with points beyond the 5th and 95th percentiles also displayed

Since many PAH are well-known inducers of MFO in fish, the stock of barite used in the study was analyzed for this particular class of compounds. PAHs were considered since the open pit mining of barite might provide opportunity for PAH contamination from a source such as combustion hydrocarbons from transport trucks.

Barite was solvent extracted for analysis of PAH by GC-MS but none of a variety of compounds was detected at a detection limit of 0.05 mg/kg. The compounds measured included 1-methylnaphthalene, 2-methylnaphthalene, acenaphthene, acenaphthylene, anthracene, benzo(a)anthracene, benzo(a)pyrene, benzo(b)fluoranthene, benzo(g, h, i) perylene, benzo(k)fluoranthene, chrysene, dibenz(a, h)anthracene, fluorene, indeno(1, 2, 3-cd)pyrene, naphthalene, perylene, phenanthrene, and pyrene.

Returning to the question of which compounds might be linked to the increased EROD levels found in fish exposed to barite, it is known from laboratory studies that acute exposure to relatively high levels of some heavy metals results (not unexpectedly) in enzyme inhibition (Viarengo et al. 1997; Bozcaarmutlu and Arinc 2004; Oliveira et al. 2004). There is also a case history indicating that heavy metals may have been responsible for the low enzyme levels found in fish at a field site in the Mediterranean (Romeo et al. 1994). Enzyme inhibition has also been a common feature in studies with mammalian models exposed to relatively high levels of heavy metals (Kaminsky 2006). However, an increase in P450 mRNA has been reported on occasion in mammalian models exposed to some concentration of a metal (Kadiiska et al. 1985; Elbekai and El-Kadi 2007; Korashy and El-Kadi 2008). Although earlier studies reported on the inhibitory effects of copper on fish EROD (Viarengo et al. 1997; Oliveira et al. 2004), a recent study has also noted an increase in EROD activity in a fish species exposed to copper (Henczová et al. 2008). Instead of such increase in enzyme activity being mediated via ligand binding to the aryl hydrocarbon receptor, and represent "true" induction by way of gene transcription, it can be speculated that the increase in enzyme activity may be linked to some posttranscription mechanism.

Little can be said at this time about the increase in EROD in fish exposed to barite other than the fact that it may generate some interest on the mechanism involved and whether it might be of some toxicological importance.

4 Relevant Field Studies for Comparison

Given the results obtained in this chronic toxicity study, the question can be asked on whether there are field studies around oil development sites discharging barium sulfate which have examined similar health effect indicators in fish as studied here. Multiyear environmental effects monitoring (EEM) studies carried out on American plaice (*Hippoglossoides platessoides*) around the Terra Nova site on the Grand Banks of Newfoundland have reported comparable results between development and reference sites (Mathieu et al. 2005, 2010).

5 Conclusions

Barium sulfate is discharged into the marine environment in association with exploration and development activities for oil and gas and the disposal of drilling fluids and produced waters. Ionic barium is slightly toxic to marine organisms but precipitates rapidly as nearly insoluble barium sulfate in seawater, substantially reducing its bioavailability and toxicity. This was supported in a chronic toxicity study with cunner in which gross pathology, fish and organ condition, histopathology, and a cytochrome P450 linked catalytic activity, EROD, were studied. No differences were observed between the control and exposed fish other than a slight induction of EROD. Given the high and prolonged level of exposure, results support the hypothesis that the discharge of barium sulfate poses no significant toxicity risk to fish. Multiyear field studies conducted on fish around a development site on the Grand Banks also support this viewpoint.

Acknowledgments This study was principally supported through funding provided by the Program on Energy Research and Development (PERD), Natural Resources Canada. We also thank Hugh Bain, Science Branch, Department of Fisheries and Oceans, Ottawa, for facilitating the study through his office.

References

Adams SM, Brown AB, Goede RW (1993) A quantitative health assessment index for rapid evaluation of fish condition in the field. Trans Amer Fish Soc 122:63–73

Au DWT, Wu RSS, Zhou BS, Lam PKS (1999) Relationship between ultrastrucutral changes and EROD activities in liver of fish exposed to benzo(a)pyrene. Environ Pollut 104: 235–247

Bassari A (1994) A study on the trace elements concentrations of *Thunnus thynnes*, *Thunnes obesus* and *Katwuwonus pelamis* by means of ICP-AES. Toxicol Environ Chem 44:193–198

Barlow MJ, Kingston PF (2001) Observations on the effects of barite on the gill tissue of the suspension feeder *Cerastoderma edule* (Linné) and the deposit feeder *Macoma balthica* (Linné). Mar Poll Bull 42(1):71–76

Bozcaarmutlu A, Arinc E (2004) Inhibitory effects of divalent metal ions on liver microsomal 7-ethoxyresorufin *O*-deethylase (EROD) activity of leaping mullet. Mar Environ Res 58:521–524

Carls MG, Heintz RA, Marty GD, Rice SD (2005) Cytochrome P4501A induction in oil exposed pink salmon *Oncorhynchus gorbuscha* embryos predicts reduced survival potential. Mar Ecolog Prog Ser 301:253–265

Carls MG, Rice SD (1984) Toxic contributions of specific drilling mud components to larval shrimp and crabs. Mar Environ Res 12:45–62

Chow TJ, Goldberg ED (1960) On marine geochemisty of barium. Geochim Cosochim Acta 20:192–198

Church TM (1979) Marine Barite. In: Burns RG (ed) Marine minerals. Mineralogical Society of America Short Course Notes. vol 6, pp 175–209

Colavecchia MV, Hodson PV, Parrott JL (2007) The relationship among CYP1A induction, toxicity, and eye pathology in early life stages of fish exposed to oil sands. J Toxicol Environ Health 70(18):1542–1555

Collier RW, Edmond J (1984) The trace element geochemistry of marine biogenic particulate matter. Prog Oceanog 13:113–199

Conklin PJ, Doughtie DG, Rao KR (1980) Effects of barite and used drilling muds on crustaceans, with particular reference to the grass shrimp Palaemonetes pugio. In: Symposium on research on environmental fate and effects of drilling fluids and cuttings. Courtesy Associations, Washington, DC, pp 912–943

Cranford PJ, Gordon DC Jr, Lee K, Armsworthy SL, Tremblay G-H (1999) Chronic toxicity and physical disturbance effects of water- and oil-based drilling fluids and some major constituents on adult sea scallops (*Placopecten magellanicus*). Mar Environ Res 48:225–256

Denison MC, Nagy SR (2003) Activation of the aryl hydrocarbon receptor by structurally diverse exogenous and endogenous chemicals. Annu Rev Pharmacol Toxicol 43:309–334

Elbekai RH, El-Kadi AO (2007) Transcriptional activation and posttranscriptional modification of Cyp1a1 by arsenite, cadmium, and chromium. Toxicol Lett 172(3):106–119

Gibbs RJ, Mathews MD, Link DA (1971) The relationship between sphere size and settling velocity. J Sediment Petro 41:7–18

Gooday AJ, Nott JA (1982) Intracellular barite crystals in two xenophyophores, *Aschemonella ramuliformis* and *Galatheammina sp.* with comments on the taxonomy of *A. ramuliformis*. J Mar Biol Ass UK 62:545–605

Hannah CG, Drozdowski A, Loder J, Muschenheim K, Milligan T (2006) An assessment model for the fate and environmental effects of offshore drilling mud discharges. Estuar Coast Shelf Sci 70:577–588

Hellou J, Warren WG, Payne JF, Belkhode S, Lobel P (1992) Heavy metals and other elements in three tissues of cod, *Gadus morhua*, from the Northwest Atlantic. Mar Poll Bull 24(9):452–458

Henczová M, Deér AK, Filla A, Komlósi V, Mink J (2008) Effects of Cu^{2+} and Pb^{2+} on different fish species: Liver cytochrome P450-dependant monooxygenase activities and FTIR spectra. Comp Biochem Physiol Part C 148:53–60

Hinton DE, Baumann PC, Gardner GR, Hawkins WE, Hendricks JD, Murchelano RA, Okihiro MS (1992) Histopathologic biomarkers. In: Hugget RJ, Kimerle RA, Mehrle PM, Jr, Bergman HL (eds) Biomarkers: biochemical, physiological and histological markers of anthropogenic stress. Lewis Publishers, Chelsea, MI, pp 155–209

ICES (International Council for the Exploration of the Sea) (2004) Biological effects of contaminants: Use of liver pathology of the European flatfish dab (*Limanda limanda*) and flounder (*Platichthys flesus*) for monitoring. By: SW Feist, T Lang, GD Stentiford, A Kohler. ICES Tech. Mar Environ Sci 38:42

Kaminsky L (2006) The role of trace metals in cytochrome P4501 regulation. Dru Metab Rev 38(1–2):227–234

Kadiiska M, Stoytchev T, Serbinova E (1985) On the mechanism of the enzyme-inducing action of some heavy metal salts. Arch Toxicol 56(3):167–169

Korashy HM, El-Kadi AO (2008) Modulation of TCDD-mediated induction of cytochrome P450 1A1 by mercury, lead, and copper in human HepG2 cell line. Toxicol In Vitro 22(1):154–158

Luna LG (1968) Manual of histological staining methods of the Armed Forces Institute of Pathology. McGraw-Hill, New York, NY, 258 pp

Lynch M, Raphael S, Mellor L, Spare P, Inwood M. (1969) Medical laboratory technology and clinical pathology. Saunders Company, Bradenton, FL, p 1359

Lyons BP, Stentiford GD, Green M, Bignell J, Bateman K, Feist SW, Goodsire F, Reynolds WJ, Thain JE (2004) DNA adduct analysis and histopathological biomarkers in European flounder (*Patichthys flesus*) sampled from UK estuaries. Mutat Res 552(1–2):177–186

Marshall WS, Grosell M (2006) Ion transport, osmoregulation, and acid-base balance. In: Evans DH, Claiborne JB (eds) The physiology of fishes, Third Edition. CRC Press, Boca Ration, FL, pp 177–230

Mathieu A, Hanlon J, Myers M., Melvin W, French B, DeBlois EM, King T, Lee K, Williams UP, Wight FM, Janes G (2011) Studies on fish health around the Terra Nova development site on the Grand Banks before and after release of produced water. In: Lee K, Neff J (eds) Environmental risks and mitigation technologies for produced water. Springer, New York, NY (in press)

Mathieu A, Melvin W, French B, Dawe M, DeBlois EM, Power F, Williams U (2005) Health effects indicators in American plaice (*Hippoglossoides platissoides*) from the Terra Nova development site on the Grand Banks. In: Armsworthy SL, Cranford PJ, Lee K (eds) Proceedings of the offshore oil and gas environmental effects monitoring workshop: approaches and technologies. Bedford Institute of Oceanography, Dartmouth, NS, May 26–30, 2003. Battelle Press, Columbus, OH, pp 297–317

Macdonald JM, Shields JD, Zimmer-Faust RK (1988) Acute toxicities of eleven metals to early life-history stages of the yellow crab Cancer anthonyi. Mar Biol 98:201–207

Mallatt J (1985) Fish gill structure changes induced by toxicants and other irritants: a statistical review. Can J Fish Aquat Sci 42:630–648

Milligan TG, Hill PS (1998) A laboratory assessment of the relative importance of turbulence, particle composition, and concentration in limiting maximal floc size and settling behaviour. J Sea Res 39:227–241

Monnin C, Jeandel C, Cattaldo T, Dehairs F (1999) The marine barite saturation state of the world's oceans. Mar Chem 65:253–261

Morales-Caselles C, Jimenez-Tenorio S, Riba I, Sarasquete C, DelValls A (2007) Kinetic of biomarker responses in juveniles of the fish *Sparus aurata* exposed to contaminated sediments. Environ Monit Assess 131:211–220

Muschenheim DK, Milligan TG (1996) Flocculation and accumulation of fine drilling waste particles on the Scotian Shelf. Mar Poll Bull 32:740–745

Myers MS, Rhodes LD, McCain BB (1987) Pathologic anatomy and patterns of occurrence of hepatic neoplasms, putative preneoplastic lesions, and other idiopathic hepatic conditions in English sole (*Parophrys vetulus*) from Puget Sound, Washington. J Nat Cancer Inst 78(2): 333–363

Myers M.S, Fournie JW (2002) Histopathological biomarkers as integrators of anthropogenic and environmental stressors. In Adams M (ed) Biological indicators of aquatic ecosystem stress, Bethesda, MD, pp 221–287

Neff JM (2002) Barium in the ocean. Chapter 4 in: bioaccumulation in marine organisms. Effect of contaminants from oil well produced water. Elsevier, Amsterdam, pp 79–87
NRC (National Research Council). (1983) Drilling discharges in the marine environment. National Academy Press, Washington, DC
O'Donnell MJ (1988) Potassium channel blockers unmask electrical excitability of insect follicles. J Exp Zool 245:137–143
Oliveira M, Santos MA, Pacheco M (2004) Glutathione protects heavy metal-induced inhibition of heptic microsomal ethoxyresorufin *O*-deethylase activity in *Dicentrarchus labrax* L. Ecotox and Environ Saf 58:379–385
Patterson C, Settle D (1977) Comparative distributions of alkaline earths and lead among major tissues of tune *Thunna alalunga*. Mar Biol 39:289–295
Payne J, Andrews C, Guiney J, Whiteway S (2006) Risks associated with drilling fluids at petroleum development sites in the offshore: evaluation of the potential for an aliphatic hydrocarbon based drilling fluid to produce sedimentary toxicity and for barite to be acutely toxic to plankton. Can Tech Rep of Fish and Aquat Sci 2679:28p
Payne JF, Mathieu A, Collier TK (2003) Ecotoxicology studies focusing on marine and freshwater fish. In: Douben P (ed) PAHs: an ecotoxicological perspective. John Wiley and Sons, Chichester, pp 191–224
Pohl RJ, Fouts JR (1980) A rapid method for assaying the metabolism of 7-ethoxyresorufin by microsomal subcellular fractions. Analyt Biochem 107:150–155
Porter EL, Payne JF, Kiceniuk J, Fancey L, Melvin W (1989) Assessment of the potential for mixed-function oxygenase enzyme induction in the extrahepatic tissues of cunners during reproduction. Mar Environ Res 28:117–121
Ray JP (1996) Session summary. In: Reed M, Johnsen S (eds) Produced water 2. Environmental issues and mitigation technologies. Plenum Press, New York, NY, pp 3–5
Romeo M, Mathieu A, Gnassia-Barelli M, Romana A, Lafaurie M (1994) Heavy metal content and biotransformation enzymes in two fish species from the NW Mediterranean. Mar Ecol Prog Ser 107:15–22
Spangenberg JV, Cherr GN (1996) Developmental effects of barium exposure in a marine bivalve (*Mytilus californianus*). Environ Toxicol Chem 15(10):1769–1774
Spies RB, Rice DW, Felton J (1998) Effects of organic contaminants on reproduction of the starry flounder *Platichthys stellatus* in San Francisco Bay. Hepatic contamination and mixed function oxidase (MFO) activity during the reproductive season. Mar Biol 98:181–189
Sternberg E, Tang D, Ho TY, Jeandel C, Morel FMM (2005) Barium uptake and adsorption in diatoms. Geochim Cosmochim Acta 69(11):2745–2752
Timme-Laragy AL, Cockman CJ, Matson CW, Giulio RT (2007) Synergistic induction of AHR regulated genes in developmental toxicity from co-exposure to two model PAHs in zebrafish. Aquat Toxicol 85:241–250
Trefry JH, Naito KL, Trocine RP, Metz S (1995) Distribution and bioaccumulation of heavy metals from produced water discharges to the Gulf of Mexico. Wat Sci Technol 32:31–36
Viarengo A, Bettela E, Fabbri R, Bruno B, Lafaurir M (1997) Heavy metal inhibition of EROD activity in liver microsomes from the bass *Dicentrarchus labrax* exposed to organic xenobiotics: role of GSH in the reduction of heavy metal effects. Mar Environ Res 44:1–11
Wang DY, Wang Y (2008) Phenotypic and behavioral defects caused by barium exposure in nematode *Caenorhabditis elegans*. Arch Environ Contam Toxicol 54:447–453
Whyte JJ, Jung RE, Schmitt CJ, Tillitt DE (2000) Ethoxyresorufin-*O*-deethylase (EROD) activity in fish as biomarkers of chemical exposure. Critical Rev Toxicol 30(4):347–570

Part V
Monitoring Technologies

Chapter 22
Historical Perspective of Produced Water Studies Funded by the Minerals Management Service

Mary C. Boatman

Abstract The Minerals Management Service (MMS) is the bureau within the US Department of the Interior that offers for lease areas of the Outer Continental Shelf for mineral extraction and regulates the offshore oil and gas industry. As part of its mission, the agency collects information about the environmental impacts from these regulated activities through a program of funded research. Since its inception in 1973, the MMS Environmental Studies Program has collected baseline information about the marine environment and examined the potential impacts of the industry, such as the effects of produced water. The Gulf of Mexico is the primary area for offshore oil and gas activities, contributing about 27% of the domestically produced oil and 17% of the natural gas. In Southern California, 23 platforms are currently operating. Over the years, MMS has funded studies examining the impacts from this produced water in coastal areas and offshore. Generally, studies have focused on the contamination in the sediments as the ultimate sink rather than measurements in the water column. A few studies have examined the genotoxicity on specific species or life stages. One recent study examined the contribution of produced water to the hypoxic zone that forms off the coast of Louisiana. The information gathered through these studies is used by MMS in the preparation of Environmental Impact Statements, which support both offshore leasing decisions and permits promulgated by the US Environmental Protection Agency.

1 Introduction

The Minerals Management Service (MMS) is the bureau within the US Department of the Interior whose primary responsibilities include managing development of the energy and mineral resources of the 1.7 billion acres comprising the Outer Continental Shelf (OCS). The OCS extends from approximately 3 miles offshore to the edge of the Exclusive Economic Zone at 200 miles. As part of this responsibility, today MMS oversees the extraction of 27% of the oil and 15% of the natural

M.C. Boatman (✉)
U.S. Department of the Interior, Bureau of Ocean Energy Management, Regulation, and Enforcement, Herndon, VA 20170, USA

K. Lee, J. Neff (eds.), *Produced Water*, DOI 10.1007/978-1-4614-0046-2_22,

gas domestically produced, along with the collection and disbursement of more than $10 billion collected as royalties and revenues annually. Existing activities on the OCS include initial resource assessments and lease offerings for oil and gas extraction through exploration, production, and decommissioning. The MMS is tasked with ensuring that these activities are conducted in a safe and environmentally sound manner. As such, the MMS coordinates with other responsible Federal agencies to regulate the offshore industry.

The regulatory framework for discharges from oil and gas operations is led by the US Environmental Protection Agency (EPA), which regulates all discharges including the two major sources: drilling discharges and produced water. These discharges are regulated under the Clean Water Act and permitted through the National Pollutant Discharge Elimination System (NPDES) process. National guidelines are established for discharges and adopted within EPA Regions, which issue the permits. The MMS works with EPA by conducting inspections on platforms. The NPDES limitations for produced water address oil and grease, toxicity, and visual sheen. These limits were established in 1993 after extensive evaluation of both the environmental impacts of discharges and the economic impacts of control technology. The MMS worked with the EPA during this evaluation and continues this relationship by reevaluating the impacts through the National Environmental Policy Act process.

While EPA established regulations for discharges on the OCS, MMS is still responsible for evaluating the potential impacts from those discharges, both individually and cumulatively. Through its Environmental Studies Program (ESP), the MMS contracts with consulting firms, universities, and other Federal agencies to conduct research and provide information to decision makers regarding current issues. The studies are used to gather baseline information, provide input for NEPA documents, establish mitigation measures, summarize existing information, identify areas for further research, and provide monitoring.

Since its inception in 1973, the ESP has mostly funded studies that review and summarize existing information with regard to the discharge of produced water and the potential impacts. However, a few studies involving original research were funded to evaluate nearshore discharges; to characterize the chemical constituents, particularly identifying the bioactive compounds, chemical toxicity on various life stages, effects on genetic diversity and genotoxicity; and to evaluate the extent of the discharge plume. Specific studies are discussed in the following sections. This review focuses only on studies of produced water that are original research and were funded by MMS through the ESP and, therefore, is not an exhaustive presentation of produced water research.

2 Discharges in Coastal Areas of the Gulf of Mexico

In the Gulf of Mexico, offshore oil and gas activities have operated for more than 60 years with the first production on the OCS occurring in 1947. Nearshore discharges were not always processed on the platform, but rather the oil and water were piped to shore and processed at coastal facilities. The treated water was then discharged into the local environment or reinjected into disposal wells. By the late 1980s, there were over 1,200 locations where produced water was discharged into coastal environments of Louisiana and Texas (coastal discharges were not authorized in Alabama, Mississippi, or Florida). However, only 15 of these locations processed produced water generated from OCS facilities. The MMS first funded a study of these discharges in 1987, which evaluated the chemical and biological effects from three locations in southeast Louisiana (Boesch et al. 1989). The sites examined were in relatively shallow environments that were tidally influenced. The produced water was sampled at the discharge point and analyzed for hydrocarbons and trace metals. Sediments and benthic macroinfauna were sampled along the discharge plume to approximately 1 km from the discharge site. Oysters and ribbed mussels were collected and evaluated for body burden of selected chemicals. In addition, sediment cores were collected near one discharge site.

As expected, close to the discharge site, fine grained sediments were heavily contaminated with hydrocarbons and trace metals. At the most heavily contaminated sites, the benthic macroinfauna were nonexistent, particularly where high concentrations of polynuclear aromatic hydrocarbons were observed. The contamination extended for several hundred meters to greater than 1 km downstream from the discharge site. Oysters and mussels collected near discharge sites exhibited concentration factors of 3 for total polynuclear aromatic hydrocarbons and total saturated hydrocarbons relative to reference sites. Trace metal contamination of the sediments was much lower than hydrocarbons.

A follow-on study extended the work of Boesch et al. (1989) to better characterize the discharges and the spatial and temporal extent of the produced water (Rabalais et al. 1991). At that point in time, an estimated 250,000 barrels/day (bbl/day) of produced water with OCS origins were brought to shore, treated, and discharged into coastal waters. The sites analyzed ranged in discharge rates from 3,000 to 106,000 bbl/day. The influence of the outfalls extended up to 1,300 m from the discharge site, impacting the benthic macroinfauna at most sites. Oysters were affected at locations up to 1 km from the discharge point. The discharges extended away from the discharge point as a brine layer overlying the sediment. Along with elevated salinity, the brine contained elevated levels of volatile hydrocarbons, sulfides, and total radium activity.

One location examined by Rabalais et al. (1991) was revisited in 1996 to evaluate the temporal changes to a location where discharges of produced water had ceased in 1987 (Rabalais et al. 1998). This location was initially sampled in 1989, 2 years after cessation of discharges. New analytical techniques were used in 1996, making some of the comparisons problematic. As expected, surficial sediments did not exhibit the levels of contamination observed 7 years prior, and the benthic macroinfauna abundance and species richness were much improved. However, sediment cores did show heavily weathered petroleum residue as expected. In 1999, the EPA banned the discharge of produced water in the coastal environments of Louisiana and Texas, including State waters.

Concern about the fate and effects of produced water constituents in wetland and coastal ecosystems led to an examination of the biodegradation of aromatic heterocyclic compounds in these environments (Catallo 2000). The study used microcosms to simulate environmental conditions in wetlands, which include both static and dynamic hydrology. Sediments were exposed to concentrations of N-, O-, and S-heterocycles that commonly occur in crude oils and other sources. The results indicated that tidal situations, where there is a cycle of flooding and draining, removed these chemicals from the environment through enhanced degradation. From these results, it is suggested that an optimal remediation strategy would be to leave sediments in place where tidal forces can clean the system as opposed to dredging and depositing contaminated sediments in an inland landfill.

The chronic effects of hydrocarbons in produced water are a concern but often difficult to evaluate. Various methodologies were examined to assess biochemical responses in marine organisms exposed to produced water discharges in coastal Louisiana and reported as a compilation of eight separate studies (Winston and Means 1994). The induction of 7-ethoxyresorufin-*O*-deethylase (EROD) was evaluated in the marine worm *Nereis virens* and the killifish *Fundulus grandis*. For *Nereis*, initial studies indicated that exposure to dissolved and sediment-bound normal, alkylated, and heterocyclic aromatic hydrocarbons induced EROD, but subsequent tests demonstrated that the results were not reproducible. However, in killifish, EROD was induced in embryos, larvae, and adults, suggesting that this may be a good methodology for assaying petroleum contaminants. This methodology was tested on fish collected from sites near oil platforms in the Gulf of Mexico. While induction of EROD was observed in some of the fish collected, no clear pattern was discernable. The authors concluded that too many variables including species, sex, gonadal status, season, temperature, and nutrition can affect the results. After determining that EROD induction may not be a good assay technique, the authors examined superoxide dismutase and catalase in fish and concluded that neither is a good indicator of oxidative stress from contamination due to variable results. Another assay technique using the measurement of induced P450 was evaluated for various fish. Winston and Means (1994) concluded that toxicological assessment on field samples was not possible even though the assays were successful in laboratory experiments.

In addition to biochemical responses, genotoxicity and bioaccumulation of hydrocarbons from produced water were evaluated by Winston and Means (1994). Genotoxicity was evaluated on fish from coastal Louisiana using the *umu* mutagenicity assay. In each case, no gene induction was observed. Bioaccumulation was evaluated in the oyster *Crassostrea virginica* using gas chromatograph mass spectrometry to measure 60 individual aromatic petroleum hydrocarbon compounds. The oysters were exposed to various concentrations of sediment contaminated with aromatic hydrocarbons up to 26 parts per million. Mortality was high for oysters exposed to the higher concentrations. Bioaccumulation of aromatic compounds was observed after 3 days of exposure, and levels continued to increase, for the most part, through the entire 21 days of exposure.

The Gulf of Mexico Offshore Operations Monitoring Experiment (GOOMEX, Kennicutt 1995) examined sediments from around five offshore oil and gas platforms in the western Gulf of Mexico. Sediment samples were collected in a radial pattern extending out to greater than 3,000 m from the platform. Biological samples from the water column and sediment surface were collected using an otter trawl in the near field and far field. Within the sediments, biological studies measured meiofaunal life history, reproduction, and genetic diversity and abundance, and diversity and community structure of meiofauna and megafauna while chemical analysis focused on contaminant distribution in sediments and pore water toxicity by bioassay. Sediment contamination for both metals and hydrocarbons was primarily linked to drilling discharges of muds and cuttings which overwhelmed any chronic contamination from produced water discharges. Fish collected in the water column were evaluated using in vivo assays to determine CYP1A induction in livers. In vitro bioassays were conducted to evaluate the toxicities and potencies of contaminant extracts from invertebrate tissues. Comparison of near field and far field samples did not provide statistically significant differences in response to contamination. The GOOMEX study used a statistically rigorous study design to support the scientific findings. This type of rigor is critical when evaluating chronic impacts.

3 Discharges in Offshore Areas of the Gulf of Mexico

The GOOMEX study results suggested that exposure to produced water resulted in a reduction in genetic diversity in meiobenthic copepods. This result was further investigated by Fleeger et al. (2001) to determine whether a reduction in genetic diversity from contamination was due to genotypic selection or the result of interspecific variation of cryptic species. To evaluate their hypothesis, harpacticoid copepods (*Cletocamptus deitersi*) from four different locations were used. The results of 96-h bioassays revealed that *C. deitersi* from two locations had similar tolerances to polycyclic aromatic hydrocarbons (phenanthrene) but different tolerances to heavy metals, a mix of zinc, lead, and cadmium based on ratios present in produced water. *C. deitersi* is a cryptic species that could be used to evaluate the implications of contamination on genetic diversity since there is a differential response to trace metal contamination.

In 2004, the EPA raised concerns that produced water discharged in the Gulf of Mexico was contributing to the hypoxic events that occur each summer off the coast of Louisiana. Rabalais (2005) evaluated the contributions from produced water relative to the inputs of carbon and nitrogen from the Mississippi and Atchafalaya river systems. The organic compounds in produced water may add to the other organic material that is using oxygen during natural degradation processes. In addition, nitrogen compounds may be fueling primary production, which also consumes oxygen during degradation of phytoplankton after a bloom. The report summarized and evaluated existing information, which included a limited amount of existing data on the produced water constituent load. However, initial conclusions determined that contributions to hypoxia from discharged produced water were minimal, with the recommendation that more data be collected.

4 BACIPS and Associated Studies

In California, coastal oil and gas activities have occurred since the early 1900s. Today, 23 platforms are operating between Point Conception and Palos Verdes in Federal waters. In 1989, a series of studies was initiated through the ESP to evaluate the impacts of produced water in a coastal marine environment. The research team attempted to apply Before After Control Impact Paired Series (BACIPS) to one location along the California Coast and examined discharges from another during and after cessation. Studies were conducted to characterize the discharge including both organic and inorganic constituents with particular interest in bioactive compounds. An assessment of chronic toxicity from produced water discharges on urchin larvae, kelp gametes, and mussels was also conducted.

An evaluation of the impacts of produced water using the BACIPS assessment design to obtain a good time series of data prior to the discharge of produced water was attempted beginning in 1988 (Osenberg 1999). The challenge was to collect data for several years prior to the initiation of produced water discharge at a location along the California Coast. Sampling continued until 1995, when it was finally determined that the discharge would never occur. While this was not successful in the initial quest, a 7-year time series of data was collected and evaluated for environmental variability. The field program focused on the enumeration of benthic infauna, epifauna, and demersal fishes; growth and tissue production of mussels transplanted to the study site; and characterization of a variety of chemical and physical attributes (e.g., temperature, sedimentation rates, grain size of sediments, and trace metal concentrations in the water column and sediments). The statistical evaluation of the resulting data set suggested that individual-based parameters could be used to evaluate the effects of contamination, but relatively few of the population and chemical–physical parameters could provide adequate power given time constraints of most studies.

Concurrent with the study using BACIPS, another location where produced water discharge was occurring was sampled focusing on the control-impact aspects (Osenberg and Schmitt 1999). During the course of the study, the produced water discharge ceased, so the study was extended to evaluate the recovery of the location. Analysis of the complete data set, collected both during discharge and after cessation of discharge, revealed that spatial patterns in mussel performance, shell barium content, and infaunal density were almost entirely due to the effects of produced water. Shortly after cessation of produced water discharge, spatial variation in these parameters declined considerably.

Produced water is a complex mixture of both organic and inorganic constituents. Typically, analyses focus on the petroleum hydrocarbons and trace metal components. Higashi et al. (1997; extended in Higashi and Jones 1997), as a part of the larger evaluation of produced water at two locations in southern California, examined a range of constituents in order to systematically characterize and identify key chemicals that are of concern because of bioeffects. The purpose of this evaluation was to reduce the list of produced water constituents to those of critical concern.

The study examined produced water directly drawn from a discharge pipe, marine water, and sediments. An analytical scheme was devised to separate the produced water into seven fractions. A variety of analytical techniques were used to identify the components of each fraction. As expected in the produced water, hydrocarbons and alkylphenols were the major organic constituents, but organosulfur compounds, in particular thiocarboxylic acids and thiopyranones, were also found. Inorganic forms of sulfur, including sulfides, thiosulfates, and polysulfides, were identified. The presence of these types of sulfur compounds is important because of their potential bioeffects due to the strong interactions of sulfur compounds with metal ions as well as the bioactivity of the compounds themselves. The use of pyrolysis gas chromatography mass spectrometry allowed for the identification of polysaccharidic material that was not cellulosic in produced water from one location.

While acute toxicity is easily measured and recognized, the effects of sublethal chronic toxicity are always elusive. In an effort to elucidate the cause and effect relationships amid the background of variability, several techniques were evaluated on a range of organisms. Cherr et al. (1993) evaluated the effects of exposure to low levels of produced water on the early life stages of the purple sea urchin (*Strongylocentrotus purpuratus*), the giant kelp (*Macrocystis pyrifera*), and the California mussel (*Mytilus californianus*). A new methodology employing nuclear magnetic resonance spectroscopy was studied for the ability to conduct non-invasive analyses of the effects of exposure to contaminants in order to monitor long-term metabolic changes. From the analyses, it appeared that produced water may perturb the later stages of embryonic development, as expressed in the development of abnormal spicules. While for kelp, effects from produced water exposure were not observed, in mussel embryos, produced water may inhibit normal shell calcification. This observation was further studied by Cherr and Fan (1997) and possibly linked to elevated levels of barium. Laboratory exposure to produced water also perturbed the ovarian energy balance and caused ovarian degeneration in mussels, resulting in possible effects on reproduction. Cherr and Fan (1997) continued the evaluation of the toxic effects of produced water on mussel embryos, determining that there was no effect prior to hatching. Embryos were exposed continuously to 1–2% produced water as well as a pulsed exposure where fresh seawater was used to wash the embryos after 15 h of exposure. Stage-specific effects were observed suggesting that gastrulation and cell differentiation were the most sensitive stages.

Mussel outplants were studied at one location in California (Cherr and Fan 1997) using adults placed near the outfall and at successive distances. Mussels planted 1 m from the outfall showed the highest concentration of barium in the tissues. Those planted at 5 m from the outfall displayed poorer growth and reproductive performance. Effects dropped off quickly with distance from the outfall. Since barium in seawater rapidly precipitates as the relatively insoluble barium sulfate, barium concentrations are relatively low; therefore, the process for the incorporation of barium into mussels is not direct. Cherr and Fan (1997) made some initial investigations into the uptake of barium in phytoplankton as a potential source of barium in mussels that ingest the plankton.

A later study by Raimondi and Boxshall (2002) revisited the effects of produced water on invertebrate larvae by examining the effects of laboratory exposure of bryozoans *Watersipora* spp, *Bugula neritina, Schizoporella unicornis*, the red abalone *Haliotis rufescens*, the sea star *Asterina miniata*, and the ascidian *Botrylloides* spp. The larvae were exposed at an early stage and then evaluated as they grew into adults. Maximum effects were observed when the larvae were exposed to a 10% mixture of produced water in filtered seawater including strong sublethal effects on growth, competitive ability, or reproductive output. The higher concentration was used to determine the potential effects at lower concentrations. Effects at 1 and 0.1% dilutions were not observable. Of course, produced water is rapidly diluted from the discharge point because of the use of diffusers, making even the lowest dilutions only applicable very close to the discharge point.

5 Plume Dynamics

The direction and extent of the plume is another factor to consider when evaluating the potential environmental impacts from the discharge of produced water. A study of the plume extent through evaluation of the density field around the discharge point was conducted in Southern California (MacIntyre and Washburn 1996). Conductivity, temperature, and depth measurements were taken over a 18-month period to delineate the stratification of the area around the discharge point. Current measurements were also taken. The discharge occurred in relatively shallow waters that were influenced by tidal action. Two models were evaluated: the Morton–Taylor–Turner model, which is an integral model, and the Roberts–Snyder–Baumgartner model, which is based on dimensional analysis. Plume dynamics are difficult to evaluate because of the rapid dilution of the plume, the variability of the current regime at the discharge point, and the effects of local stratification.

Plume dynamics was studied later at four platforms off the coast of southern California (Diener and Jones 2004). Field sampling was conducted using a towed sampling array equipped with a conductivity, depth, and temperature profiler; transmissometer; chlorophyll fluorometer; two Rhodamine fluorometers; and a colored dissolved organic matter sensor to map the plume. Natural tracers failed to show evidence of the discharge plume, even at distances of 25–50 m from the discharge point. Primarily, this was because the parameters within the discharge were not significantly different than the surrounding marine waters. Rhodamine dye was injected into the produced water before the discharge point and measured with a towed fluorometer. The measured initial dilution of the plumes ranged from 216:1 to 3,536:1, depending on the discharge rate of the effluent. The observed plumes extended out from the platform in a horizontal direction, and the dye was detectable at distances ranging from 400 to 1,500 m, again dependent on the discharge volume. The observed plume trajectory was compared to model results using EPA's VISUAL PLUMES model with good results for relatively high discharge rates, but poorer results for low and intermittent rates.

6 Summary and Conclusions (Challenges)

This summary has focused entirely on the technical reports that are products of MMS funded studies. All of these reports are available on the MMS website through the Environmental Studies Program Information System.

While the MMS has only funded a few studies to evaluate the environmental impacts, it is clear that the challenges of conducting produced water studies are many. There is no control over the discharge parameters including when they will occur, the rate of discharge, and their intermittency. The lack of control over these parameters makes it difficult, if not impossible, to apply BACIPS because it requires a number of years of data prior to initiation of a discharge. The measurable effects are very close to the outfall, with a rapid decrease in concentration due to dilution. Choosing the best species for monitoring effects is difficult. Constituents that may be of greatest concern are not necessarily metals or petroleum hydrocarbons. Other chemicals may be present including treatment chemicals, sulfur compounds, and polysaccharides. Produced water observations are site specific, and constituents vary depending on the composition of the formation oil and water. Chronic effects are difficult to observe and document, and are highly species dependent as well as dependent on life stage. In laboratory experiments, relatively high concentrations are often needed to elicit an observable response and are often not representative of actual field experience. A statistically rigorous study design to evaluate chronic effects is probably the best way to address the post-discharge impacts.

References

Boesch DF, Rabalais NN (eds) (1989) Produced waters in sensitive coastal habitats: an analysis of impacts, Central Coastal Gulf of Mexico. OCS Report MMS 89-0031, U.S. Dept. of the Interior, Minerals Management Service, Gulf of Mexico OCS Region, New Orleans, Louisiana, 157 pp

Catallo WJ (2000) Biodegradation of aromatic heterocycles from petroleum-produced water and pyrogenic sources in marine sediments. OCS Study MMS 2000-060. U .S. Dept. of the Interior, Minerals Management Service, Gulf of Mexico OCS Region, New Orleans, Louisiana, 28 pp

Cherr GN, Fan TW–M (1997) Chronic toxicological effects of produced water on reproduction and development in marine organisms. MMS OCS Study 97-0024. Coastal Research Center, Marine Science Institute, University of California, Santa Barbara, California. MMS Cooperative Agreement Number 14-35-0001-30471. 71 pp

Cherr GN, Higashi RM, Shenker JM (1993) Assessment of chronic toxicity of petroleum and produced water components to marine organisms. Coastal Research Center, Marine Science Institute, University of California, Santa Barbara, California. MMS Cooperative Agreement Number 14-35-0001-30471. 130 pp

Diener D, Jones B (2004) Produced water discharge plumes from Pacific offshore oil and gas platforms. Report Prepared for MMS Pacific OCS Region under Contract 1435-01-02-CT-85136. 43 pp

Fleeger JW, Foltz DW, Rocha-Olivares A (2001) How does produced water cause a reduction in the genetic diversity of Harpacticoid copepods? Final report. OCS Study MMS 2001-078. U.S. Dept. of the Interior, Minerals Management Service, Gulf of Mexico OCS Region, New Orleans, Louisiana, 35 pp

Higashi RM, Jones AD (1997) Identification of bioactive compounds from produced water discharge/characterization of organic constituent patterns at a produced water discharge site. OCS Study MMS 97-0023. Coastal Research Center, Marine Science Institute, University of California, Santa Barbara, California. MMS Cooperative Agreement Number 14-35-0001-30761. 43 pp

Higashi RM, Jones AD, Fan TW–M (1997) Characterization of organic constituent patterns at a produced water discharge site/barium relations to bioeffects of produced water. OCS Study MMS 97-0022. Coastal Research Center, Marine Science Institute, University of California, Santa Barbara, California. MMS Cooperative Agreement Number 14-35-0001-30761. 43 pp

Kennicutt MC II (ed) (1995) Gulf of Mexico offshore operations monitoring experiment, phase i: sublethal responses to contaminant exposure, Final Report. OCS Study MMS 95-0045. U.S. Dept. of the Interior, Minerals Management Service, Gulf of Mexico OCS Region, New Orleans, Louisiana, 748 pp

MacIntyre S, Washburn L (1996) Spatial scale of produced water impacts as indicated by plume dynamics. OCS Study MMS 97-0024. Coastal Research Center, Marine Science Institute, University of California, Santa Barbara, California. MMS Cooperative Agreement Number 14-35-0001-30471. 39 pp

Osenberg CW (1999) Long-term monitoring of biological parameters at a proposed produced water discharge: application of a BACIPS assessment design. OCS Study MMS 99-0062. Coastal Research Center, Marine Science Institute, University of California, Santa Barbara, California. MMS Cooperative Agreement Numbers 14-35-0001-30471 and 14-35-0001-30761. 52 pp

Osenberg CW, Schmitt RJ. (1999) Ecological responses to, and recovery from, produced water discharge: application of a BACIPS assessment design. OCS Study MMS 99-0061. Coastal Research Center, Marine Science Institute, University of California, Santa Barbara, California. MMS Cooperative Agreement Number 14-35-0001-30761. 82 pp

Rabalais NN (2005) Relative contribution of produced water discharge in the development of hypoxia. OCS Study MMS 2005-044. U.S. Dept. of the Interior, Minerals Management Service, Gulf of Mexico OCS Region, New Orleans, Louisiana, 56 pp

Rabalais NN, McKee BA, Reed DJ, Means JC (1991) Fate and effects of nearshore discharges of OCS produced waters. Volume 2. Technical Report. OCS Study MMS 91-0005. U.S. Dept. of the Interior, Minerals Management Service, Gulf of Mexico OCS Region, New Orleans, Louisiana, 337 pp

Rabalais N N, Smith LE, Henry CB, Jr, Roberts PO, Overton EB (1998) Long-term effects of contaminants from OCS-produced water discharges at Pelican Island Facility, Louisiana. OCS Study MMS 98-0039. U.S. Dept. of the Interior, Minerals Management Service, Gulf of Mexico OCS Region, New Orleans, Louisiana, 91 pp

Raimondi P, Boxshall A (2002) Effects of produced water on complex behavior traits of invertebrate larvae. OCS Study MMS 2002-050. Coastal Research Center, Marine Science Institute, University of California, Santa Barbara, California. MMS Cooperative Agreement Number 14-35-0001-30758. 38 pp

Winston GW, Means JC (1994) Bioavailability and genotoxicity of produced water discharges associated with offshore drilling operations. OCS Study MMS 95-0020. A Final Report by the Louisiana Universities Marine Consortium for the U.S. Dept. of the Interior, Minerals Management Service, Gulf of Mexico OCS Region, New Orleans, Louisiana, 124 pp

Chapter 23
Water Column Monitoring of Offshore Oil and Gas Activities on the Norwegian Continental Shelf: Past, Present and Future

Ingunn Nilssen and Torgeir Bakke

Abstract As a result of aging offshore fields discharging higher volumes of produced water, and after the discharge of oil contaminated cuttings was terminated on the Norwegian Continental Shelf (NCS) in 1993, oil originating from produced water has become the dominant contributor of the Norwegian E&P industry to hydrocarbon and chemical input to the marine environment. To reflect the present situation, the environmental monitoring programmes have also changed. The Norwegian requirements related to offshore environmental monitoring are given in the HSE regulations for the petroleum activity. The water column monitoring consists of two parts: the condition monitoring and the effects monitoring. The programmes have developed through dialogue between the authorities, the scientific community, consultants, and the E&P industry, but will still be subjected to revision and improvement in the years to come.

1 Introduction

On the NCS, environmental monitoring related to the petroleum industry has been carried out since the mid-1970s. The primary focus has been on the potential impact of discharges of drill cuttings to the seabed, but after the discharges of oil contaminated cuttings was terminated on NCS in 1993 and the increase in produced water discharges during the 1990s, water column monitoring is now receiving equal attention as sediment monitoring. Water column monitoring has undergone significant development during the past decade (Røe 1998; OLF 2006). Continuous changes in the discharge and emission patterns (e.g. substitution into more environmental friendly chemicals (Nilssen and Johnsen 2008)) along with technology development has led to less impact on the environment. As a result, the environmental monitoring methodology is constantly challenged when it comes to measuring possible impacts. The development of the water column monitoring has taken place in parallel with the introduction of the "Zero discharge" initiative launched by Norwegian authorities in 1996 (Norwegian Ministry of Environment 1997; Røe 1998). In

I. Nilssen (✉)
Statoil Research Centre, NO-7005 Rotvoll, Trondheim, Norway

K. Lee, J. Neff (eds.), *Produced Water*, DOI 10.1007/978-1-4614-0046-2_23,

this process, the zero harm principle was introduced in addition to the existing requirement of zero discharge of especially hazardous chemical compounds.

The changes described above together with the cooperation regime that has developed between the Climate and Pollution Agency (Klif), scientific institutions, and the E&P industry in Norway have led to the present situation where the oil industry collaborates internally, the results are discussed among authorities, consultants, NGOs, and industry in regular forums, and the programmes are constantly adjusted based on previous experience and new knowledge.

The environmental monitoring should reflect the actual discharges, and for the E&P industry, it should also acts as feedback to mitigation measures (e.g. implementation of BAT), and as validation of the predictive risk assessment tools. In this chapter we focus on one important aspect in such a holistic environmental management, the water column monitoring.

2 Regulations

The main objectives in the Norwegian Pollution Control Act are the precautionary principle and the polluter pays principle (Pollution Control Act 1996), which means that the offshore operators have the burden of proof. It is the operators who must show that their activities do not cause unacceptable harm to the environment. Although the cost of obtaining this knowledge lies with the operators, it is of interest to all parties that the monitoring is conducted in a cost-effective way.

The requirements for environmental monitoring on the NCS are covered in the HSE regulations for the offshore activities (Petroleum Safety Authority Norway 2009). The monitoring requirements are expressed in several articles in the offshore regulations, e.g. The Management Regulations (Petroleum Safety Authority Norway 2004) and The Activities Regulations chapter X-I. Operators are required to cooperate on the planning of environmental monitoring of both the sediments and the water column according to the Activities Regulations (Petroleum Safety Authority Norway 2010).

The purpose (Petroleum Safety Authority Norway 2010) of the offshore monitoring is to get;

- an overview and control of pollution from the offshore activity, including environmental impacts, and
- an overview of the general condition and development around the various facilities and in the regions (trends).

The results from the monitoring will among other things be used for;

- early warning of deterioration of the environmental situation;
- the development of forecasts for the expected environmental condition;
- evaluation of the risk for environmental damage and ecological effects;
- verification of models for calculating the environmental risk as a function of the existing and expected discharges from the offshore industry;
- verification of laboratory-based research.

The programmes have a common procedural platform to ensure spatial and temporal comparability, but these are adjusted on the basis of results obtained from earlier surveys, new knowledge, and discharge patterns. Each year the programmes are endorsed by the Klif after prior discussion with the operators.

3 Water Column Monitoring Strategy

The water column monitoring programme has two prime elements: the *environmental effects monitoring* (EEM) and the *environmental condition monitoring* (ECM). While the EEM utilises caged marine organisms (fish and blue mussels) to assess fate and effects of discharges from specific installations, the ECM is based on regional sampling and analysis of wild fish.

The water column monitoring (EEM and ECM) emphasises methods measuring evidence of exposure to hydrocarbons, particularly di- and polyaromatic hydrocarbons (NDP/PAH) and alkylphenols which are component groups occurring naturally in produced water. Produced water is a complex mixture, and to intercept the total exposure load from the discharge, general stress indicators are also included in the suite of methods.

As described above, the purpose of the EEM is to measure potential effects from a produced water discharge by the use of biomarkers. Biomarkers are used as early warning signals of biological effects, and the practice of caging experimental organisms initiates a worst case scenario for observing possible effects related to a gradient being emitted from a point source of produced water. The philosophy behind this is that if no effects are measured in the caged organisms, the likelihood of effect to occur in wild organisms is small.

The major objective of the ECM, on the other hand, is to document to what extent, if any, discharges from the E&P industry have effects on the quality of fish, one of Norway's other main exports. The ECM is based on measurements of hydrocarbons in the muscle of wild fish caught in several regions on the NCS.

Since the EEM covers the possible effects from point sources, and the ECM should detect a diffuse increase in hydrocarbons on a regional scale, these two programmes are complementary, however, since these two programmes were introduced in the late 1990s, they have moved towards each other in that biomarkers are now included in the ECM, and wild fish have been analysed as a part of the EEM.

4 Environmental Effects Monitoring (EEM)

During the first years when environmental monitoring of the water column was carried out (1997–2000), the focus was on measuring concentrations of selected substances around discharge points (OLF 2008) in order to assess the range in contamination levels and geographical scale of impacts, and to validate the dilution and dispersion module of the Dose-related Environmental Assessment Model (DREAM, Nilssen and Johnsen 2008).

Since 2001 the EEM has been carried out either as field surveys (one field or area each year) or as supporting laboratory/desktop studies focusing on measurement of local exposure and potential biological effect (biomarkers) from produced water discharges on caged fish and mussels. The cages are deployed at given distances from the source for 4–6 weeks during spring. Spring is chosen as a worst-case scenario since a shallow thermocline and stable stratification reduce plume dilution as compared to other seasons. In order to minimise stress on the fish prior to caging, they have been transported to the site in the cages onboard a live-fish carrier. Figure 23.1 gives a sketch of a cage setup with buoys and moorings.

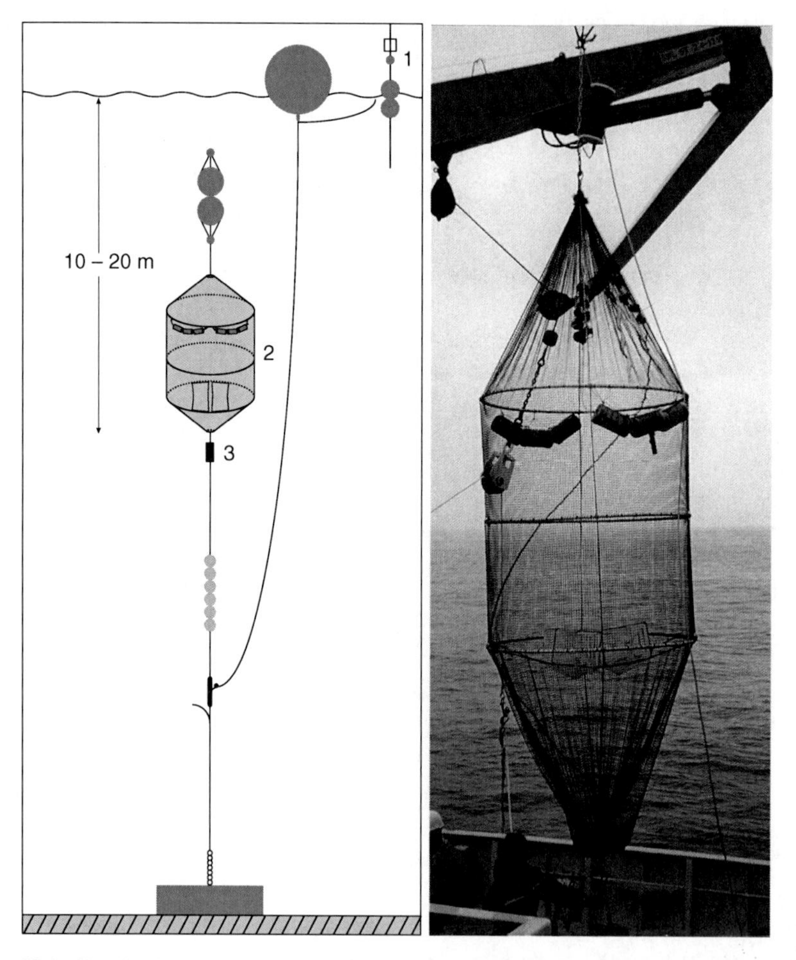

Fig. 23.1 Sketch of a cage (*left*) setup with buoys and moorings and photo (*right*) of cage (source Bjørn Serigstad). 1 Surface marker buoy with reflector and/or light or submersed rig with acoustic release for retrieval; 2 Cage with mussels and/or fish and/or passive sampling device; 3 Physical/chemical sensors, e.g. measurement of currents (directions and strength)

Table 23.1 Biological and chemical end points for Atlantic cod, *Gadus morhua* (Effects Monitoring Programme 2008)

Method	Matrix
Cytochrome P450 1A proteins (CYP1A)	Liver
Glutathione-*S*-transferase (GST)	Liver
Vitellogenin (precursor of egg yolk protein) (VTG)	Blood plasma
Zona radiate protein (egg shall protein) (ZRP)	Blood plasma
PAH metabolites, Fixed Fluorescence	Bile
PAH metabolites, GC/MS	Bile
Alkylphenol metabolites (AP met)	Bile
DNA adducts	Liver

Table 23.2 Biological and chemical analyses of mussels, *Mytilus edulis* (Effects Monitoring Programme 2008)

Method	Matrix
Pyrene hydroxylase	Digestive gland
Lysosomal stability	Haemocytes
Lipofuscin accumulation	Digestive gland
Neutral lipid accumulation	Digestive gland
PAH concentration	Soft tissue
Lipid content	Soft tissue
Micronucleus	Haemocytes

The end points for the analysis of the caged organisms (Tables 23.1 and 23.2) have been selected on the basis of an international workshop (Hylland et al. 2002) and later EEM surveys.

The main findings from 12 years with EEM on the NCS are comforting in that the results have not revealed any serious impact on the endpoints included.

A challenge in most of the surveys has been to predict the produced water plume behaviour reliably enough to ensure that the cages cover an exposure gradient when the number of cages used catching the plume is uncertain; this is of particular concern in areas with strong currents, or at times when currents deviate from their typical pattern (Hylland et al. 2008).

The first phase of water column monitoring (1997–2000) has formed the basis for a good understanding of the dilution and fate of produced water discharges. By using PAH as tracing compounds, dilution models have been verified and validated (OLF 2006; Nilssen and Johnsen 2008). Increased concentrations of PAHs originating from produced water have been detected at various distances from different production fields, and up to several kilometres away from the discharge source (Røe 1998), however, concentrations are generally lower than the observed threshold levels for toxic effects on marine organisms. The early phase of the monitoring programme has also established natural background levels of hydrocarbons in specific areas of the North Sea.

Results from the EEM surveys (2001–2007) confirm that the caged fish and blue mussel have been exposed to low concentrations of PAH at the near field stations. PAH residues are found in mussels and PAH bile metabolites in cod, but the increased levels have all been close to the analytical detection limits (Røe 1998;

Hylland et al. 2002). General stress indicators such as lysosomal membrane stability, and formation of micronuclei in mussels support the pattern seen for the hydrocarbons and alkylphenols; i.e. a slight increase near the discharge point (Røe 1998; Hylland et al. 2002).

A laboratory support study replacing the EEM field experiment in 2007 showed that the lowest exposure levels of produced water that induced enhanced vitellogenin levels in fish (indicative of alkylphenol exposure) were at least one order of magnitude higher than levels recorded in field exposed fish (Sundt et al. 2008).

The biomarker responses verify exposure and changes in genetic, biochemical, and anatomical properties of the organism, but to what extent these changes influence organism function and survival, and hence populations, are still uncertain. The challenge of linking risk assessment and biomarker responses is followed up through a 3-year project financed by The Research Council of Norway (2007–2010).

5 Condition Monitoring

The condition monitoring process, ECM, was established to meet the concern of possible contamination of fish in areas which are exposed to large produced water discharges. The focus of ECM is on measurement of selected hydrocarbons in commercially important fish species (Petroleum Safety Authority Norway 2010). Selected biomarkers have also been included in later surveys (Klungsøyr et al. 2003; Grøsvik et al. 2006) in an attempt to link the ECM and EEM. The ECM is carried out every 3 years with each survey covering several regions of the NCS. Table 23.3 shows the methods applied for the survey in 2008 and Fig. 23.2 shows the areas for fish sampling in 2005.

Also, for the ECM, the end points included in each survey are selected on the basis of previous experience. Results so far have not shown any increase in polyaromatic hydrocarbons (PAHs and NPDs) in fish filets or in the livers of cod (*Gadus morhua*), haddock (*Melanogrammus aeglefinus*), saithe (*Pollachius virens*) and herring (*Clupea harengus*) residing in the NCS.

Table 23.3 Overview of parameters, species and number per species, from the four areas: Tampen, Haltenbanken, the Barents Sea and reference in the North Sea (Condition Monitoring programme 2008)

Method	Matrix	Species
PAH metabolites GC/MS	Bile	Haddock
PAH metabolites GC/MS	Bile	Long rough dab
PAH metabolites GC/MS	Bile	Saithe
DNA adducts	Liver	Haddock
DNA adducts	Liver	Long rough dab
Alkylphenol GC/MS	Bile	Saithe
Alkylphenol GC/MS	Bile	Haddock
Vitellogenin	Blood	Cod (M)
Lipid/fatty acid composition		Haddock
Histology	Gonad	Haddock
Histology	Gonad	Long rough dab

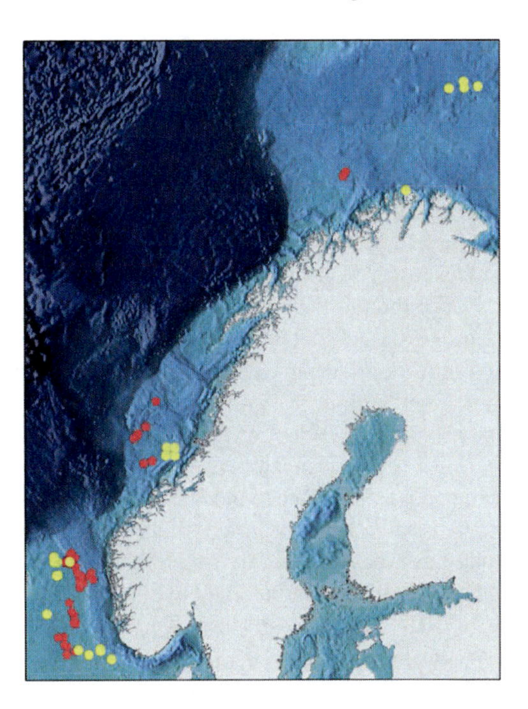

Fig. 23.2 Stations for fish sampling in 2005 (Grøsvik et al. 2006)

The ECM biomarkers have generally been in accordance with the levels of hydrocarbon contamination observed, i.e. no effects signals. However, in 2002 and 2005, slightly elevated levels of DNA adducts were measured in haddock from the North Sea (Klungsøyr et al. 2003; Grøsvik et al. 2006; Beyer et al. 2004). The source for this is still unknown, but the ECM carried out in 2008 (not yet reported to the authorities) was adjusted to investigate this more closely. Since haddock feed near the sediment, there is a possibility that the increased levels of DNA adducts were caused by hydrocarbons associated with sediments and cutting piles. To evaluate sediment as the hydrocarbon source, the flatfish, long rough dab (*Hippoglossoides platessoides*), replaced saithe in the 2008 ECM.

6 Conclusions and the Way Forward

More than 10 years with WCM on the NCS have contributed to a better understanding of the fate and potential impact of produced water discharges to the marine environment of the North Sea. Dilution models have been validated and improved, and a set of biomarker methods that seem sufficiently reliable for the purpose has been established. The results from the EEM show evidence of hydrocarbon exposure in the vicinity of a produced water effluent source. The ECM has also shown increased levels of DNA adduct in bottom-dwelling fish. The source for this is still unknown, and the DNA adduct measurements have been followed up during the ECM in 2008.

The EEM demands quite extensive investment in a number of cages in order to cover the area of possible influence and describe spatial exposure gradients: the risk of losing equipment is relatively high. The translation of responses observed in caged organisms to the natural environment is uncertain, but the caging approach is useful to detect worst-case impacts. Since even these impacts have been found to be modest, the risk of impact to wild fish and possibly other pelagic organisms is low.

Even though the water column monitoring programmes are continuously being adjusted based on experience, there is an ongoing search for improved methodologies, more sensitive methods to measure the real exposure levels of produced water, and a more realistic approach to monitor the impact of such exposure on the natural pelagic community. One of the primary challenges with this type of monitoring is to distinguish between impacts originating from the E&P industry, other anthropogenic sources, and natural variations. Technology development has now made on-line monitoring possible, so one solution that may meet this demand is long-term real-time monitoring of physical, chemical, and biological properties. At present, seabed observatory network programmes have been initiated both in Europe (ESONET 2010) and in North America (NEPTUNE 2010) where a range of different sensors are deployed in nodes connected via fiberoptic cables. The latest E&P industry field developments are subsea units which posses fiberoptic connections to shore or to an offshore field centre. Utilising the existing industry infrastructure for environmental monitoring, and introducing organisms as one of the sensors, e.g. Biota Guard (2010), one may be presented with a tool capable of generating important knowledge useful in interpreting results from the WCM; in addition, when feasible, these two monitoring approaches could be beneficially merged.

References

Beyer J, Balk L, Hylland K (2004) Extra analyses of fish from the Tampen, Sleipner and Egersund offshore fields. Report AM-2004/010 RF-Akvamiljø

Biota Guard (2010) http://www.biotaguard.no/

ESONET (2010) http://mars-srv.oceanlab.iu-bremen.de/

Grøsvik B, Einar SM, Westerheim K, Skarphéðinsdóttir H, Liewenborg B, Balk L, Klungsøyr J (2006) Condition monitoring in the water column 2005: oil hydrocarbons in fish from Norwegian waters. http://www.olf.no/miljoerapporter/vannsoeyleovervaaking-tilstandsovervaaking-condition-monitoring-2005-article1946-247.html

Hylland K, Becker G, Klungsøyr J, Lanf T, McIntosh A, Serigstad B, Thain JE, Thomas KV, Utvik TIR, Vethaak D, Wosniok W (2002) An ICES workshop on biological effects in pelagic ecosystems (BECPELAG): summary of results and recommendations. ICES ASC 2002 CM 2002/X:13

Hylland K, Tollefsen K-E, Ruus A, Jonsson G, Sundt RC, Sanni S, Utvik TIR, Johnsen S, Nilssen I, Pinturier L, Balk L, Baršienė J, Marigòmez I, Feist SW, Børseth JF (2008) Water column monitoring near oil installations in the North Sea 2001–2004. Mar Pollut Bull 56: 414–429

Klungsøyr J, Balk L, Berntssen MHG, Beyer J, Melbye AG, Hylland K (2003) NFR project No. 152231/720 – Contamination of fish in the North Sea by the offshore oil and gas industry. Summary report to NFR. 30 pp

NEPTUNE (2010) Transforming ocean science http://www.neptunecanada.ca/

Nilssen I, Johnsen S (2008) Holistic environmental management of discharges from the oil and gas industry – combining quantitative risk assessment and environmental monitoring. Society of Petroleum Engineers HSE conference Nice 111586

Norwegian Ministry of Environment (1997) Stortingsmelding 58 "Miljøvernpolitikk for en bærekreftig utvikling; dugnad for framtida". The Norwegian Ministry of Environment, Oslo. (1996–97)

OLF (2006) "Water column monitoring – summary report 2005" OLF, Statvanger, Norway http://www.olf.no/miljoerapporter/water-column-monitoring-summary-report-2005-article1941-247.html

OLF (2008) "Environmental Report 2007" OLF, Stavanger, Norway http://www.olf.no/miljoerapporter/olf-miljoerapport-2007-article1959-247.html

Petroleum Safety Authority Norway (2004) Regulations relating to management in the petroleum activities (The Management Regulations) http://www.ptil.no/management/category401.html

Petroleum Safety Authority Norway (2009) HSE regulations for the Norwegian Continental Shelf http://www.ptil.no/regulations/category216.html

Petroleum Safety Authority Norway (2010) Regulations relating to conduct of activities in the petroleum activities (The Activities Regulations). http://www.ptil.no/activities/category399.html

Pollution Control Act (1996) Act of 13 March 1981 No.6 concerning protection against pollution and concerning waste (the Pollution Control Act), most recently amended by Act of 12 June 1996 No.36 http://www.regjeringen.no/nb/dokumentarkiv/Regjeringen-Brundtland-III/md/260597/260604/t-1300_pollution_control_act.html?id=260605

Røe, Toril I (1998) Produced water discharges to the North Sea: a study of bioavailability of organic produced water compounds to marine organisms. Ph.D. Thesis, Faculty of Chemistry and Biology, Norwegian University of Science and Technology.

SFT (1998) Zero discharge report. SFT Oslo, November 1998

Sundt RC, Beyer J, Meier S (2008) Exposure levels of alkylphenols causing xenoestrogenic effects in laboratory PW exposed cod, versus realistic levels in field exposed fish. Report IRIS – 2008/055. http://www.olf.no/publikasjoner/miljorapporter/

Sundt RC, Ruus A, Grung M, Pampanin D, Baršienė J, Skarphédinsðóttir H (2006) "Water Column Monitoring 2006". IRIS Akvamiljø Report AM 2006/013 http://www.olf.no/miljoerapporter/vannsoeyleovervaaking-effektovervaaking-wcm-2006-article1947-247.html

Chapter 24
Bioaccumulation of Hydrocarbons from Produced Water Discharged to Offshore Waters of the US Gulf of Mexico

Jerry Neff, T.C. Sauer, and A.D. Hart

Abstract At the request of the U.S. Environmental Protection Agency (USEPA), the Gulf of Mexico Offshore Operators Committee sponsored a study of bioconcentration of selected produced water chemicals by marine invertebrates and fish around several offshore production facilities discharging more than 731 m^3/day of produced water to outer continental shelf waters of the western Gulf of Mexico. The target chemicals, identified by USEPA, included five metals (As, Cd, Hg, ^{226}Ra and ^{228}Ra), three volatile monocyclic aromatic hydrocarbons (MAH), benzene, toluene, and ethylbenzene, and four semivolatile organic chemicals (SVOC), phenol, fluorene, benzo(a)pyrene, and di(2-ethylhexyl)phthalate (DEHP). Additional MAH (*m*-, *p*-, and *o*-xylenes) and a full suite of 40 parent and alkyl-PAH and dibenzothiophenes were analyzed in produced water, ambient water, and tissues at some platforms.

Concentrations of MAH, PAH, and phenol were orders of magnitude higher in produced water than in ambient seawater. All MAH and phenol were either not detected (>95% of tissue samples) or were present at trace concentrations in all invertebrate and fish tissue samples, indicating a lack of bioconcentration of these relatively soluble, low molecular weight produced water chemicals. Concentrations of several petrogenic PAH, including alkyl naphthalenes and alkyl dibenzothiophenes, were slightly, but significantly higher in some bivalve molluscs, but not fish, from discharging than from non-discharging platforms. These PAH could have been derived from produced water discharges or from tar balls or small fuel oil spills. Concentrations of individual and total PAH in mollusc, crab, and fish tissues in this study are well below concentrations that might be harmful to the marine animals themselves or to humans who might collect them for food at offshore platforms.

1 Introduction

Produced water is the largest volume waste associated with oil and gas production (Neff 2002). Approximately 1.1 m^3 of produced water is generated for each m^3 of oil or gas equivalent produced worldwide (OGP 2004; Clark and Veil 2009). Most

J. Neff (✉)
Neff & Associates, LLC, Duxbury, MA, USA

K. Lee, J. Neff (eds.), *Produced Water*, DOI 10.1007/978-1-4614-0046-2_24,

produced water generated from offshore production facilities is treated to remove dispersed oil and discharged to the sea (Clark and Veil 1009; OSPAR 2009). The remainder is reinjected either into the producing formation for enhanced oil recovery or into another formation for disposal (Neff 2002). In 2007, approximately 334 million m^3 of produced water was discharged to federal marine waters of the United States (Clark and Veil 2009) and approximately 402 million m^3 was discharged to offshore waters of the northeast Atlantic, most of it to the United Kingdom and Norwegian Sectors of the North Sea (OSPAR 2009).

Produced water is fossil water from the hydrocarbon-bearing formation or a mixture of fossil water and water injected into the formation to enhance fossil fuel recovery. It often contains high concentrations of salts, metals, hydrocarbons, phenols, and organic acids derived from the geologic formations in which it has resided for millions of years. Produced water also may contain water soluble additives use to improve oil/gas/water separation and water treatment. The chemicals of greatest environmental concern in produced water, because of their potential for bioaccumulation and toxicity, are metals, hydrocarbons, and some additives.

Petroleum hydrocarbons are present in produced water in solution or as dispersed oil droplets. Most countries regulate ocean discharge of produced water by placing limits on the concentration of total petroleum hydrocarbons in the produced water destined for ocean disposal. The current limit for total oil and grease (dispersed and dissolved oil, measured by gravimetric or infrared analysis – see Chapter 2) in treated produced water destined for disposal in offshore waters of Canada is 60 mg/L (ppm) daily maximum and 30 ppm monthly average; the limits for discharges to US federal waters are 42 ppm daily maximum and 29 ppm monthly average. The regulatory limit for total oil and grease in produced waters discharged to most offshore waters of the North Sea, the Mediterranean Sea, the Arabian Gulf, and most of Asia is 40 ppm (Neff 2002; Veil et al. 2004).

Modern produced water treatment technology removes most of the dispersed oil but is inefficient in removing dissolved hydrocarbons, metals, and additives. The average concentration of total petroleum hydrocarbons in produced water discharged offshore in Europe in 2007 was 23.9 mg/L (OGP 2009).

The US Environmental Protection Agency, Region 6, included in a modified general National Pollutant Discharge Elimination System (NPDES) permit for discharges from oil and gas platforms in federal waters of the western Gulf of Mexico a requirement for site-specific monitoring of the bioconcentration in indigenous marine animals of several metals and organic compounds of concern from all platforms discharging more than 731 m^3/d (4,600 barrels/day) of produced water to outer continental shelf waters of the western Gulf of Mexico. The target chemicals, identified by USEPA, included five metals (As, Cd, Hg, ^{226}Ra, and ^{228}Ra), three volatile monocyclic aromatic hydrocarbons (MAH), benzene, toluene, and ethylbenzene, and four semivolatile organic chemicals (SVOC), phenol, fluorene, benzo(a)pyrene, and di(2-ethylhexyl)phthalate).

In response, the Offshore Operators Committee (an industry organization composed of 107 member and associate member companies that collectively account for approximately 97% of oil and gas production in the Gulf of Mexico) proposed an industry-wide bioconcentration study. USEPA Region 6 approved the industry-wide bioconcentration study, which consisted of a Definitive Component and a Platform Survey Component. The Definitive Component consisted of intensive, statistically designed sampling to determine if marine bivalve molluscs and fish at two pairs of platforms, each representing a produced water discharging and non-discharging platform pair, were bioconcentrating target chemicals from produced water. The Platform Survey Component met USEPA requirements for sampling of tissue residues of target chemicals in marine bivalves, crustaceans, and fish at several platform pairs in different geographic regions of the outer continental shelf of the central and western Gulf of Mexico.

We summarize here the results of the Definitive and Platform Survey Components of the monitoring program for the seven MAH and SVOC stipulated in the permit. Additional MAH (*m*-, *p*-, and *o*-xylenes, C_3- and C_4-benzenes) and a full suite of 40 parent and alkylated petroleum PAH and dibenzothiophenes also were analyzed in water and tissue samples in the Definitive Component. Our study was designed to comply with USEPA requirements and to test the null hypothesis that resident bivalve molluscs, crustaceans, and fish do not bioconcentrate volatile and semi-volatile organic chemicals from produced water discharges to concentrations higher than those in tissues of the same species from non-discharging platforms on the outer continental shelf of the Gulf of Mexico.

The final technical report for this study was completed in 1997 (CSA International, Inc. 1997) and preliminary reports of some of the results for PAH were published elsewhere (Neff et al. 2001).

2 Materials and Methods

2.1 Study Sites

The offshore oil and gas structures database from Minerals Management Service (MMS) Gulf of Mexico OCS Region and the EPA Region VI Discharge Monitoring Report (DMR) database were reviewed to identify candidate platform pairs, each consisting of a platform discharging more than 731 m^3/day (4,600 bbl/day) of produced water and a nearby non-discharging platform, for the Definitive Component and the Platform Survey Component of the Gulf of Mexico Produced Water Bioaccumulation Study. A screening survey was performed during the fall of 1994 on four candidate Definitive Component platform pairs to determine if the platform pairs were suitable and had sufficient resident fish and invertebrates for sampling. Two platform pairs were selected for the Definitive Component and 12 platform pairs, including the four candidate platform pairs evaluated in the screening survey for the Definitive Component, were selected for the Platform Survey Component (Table 24.1). Platforms are identified by lease block area and block number. At the request of EPA, the Platform Survey Component included at least two platform pairs each in four areas of the Gulf of Mexico:

Table 24.1 Discharging and non-discharging platforms in the Gulf of Mexico where samples of resident biota were collected as part of the Definitive and Platform Survey Components of the Gulf of Mexico Produced Water Bioaccumulation Study; produced water and ambient seawater samples also were collected at the Definitive Component platform pair; the mean produced water discharge rates from discharging platforms and the water depths at all platforms are included; platforms are identified by lease block area and block number

Discharging platform			Non-discharging platform	
Platform	Water depth (m)	Discharge rate (m^3/day)	Platform	Water depth (m)
Definitive component				
East Breaks 165A[a]	262	1,750	High Island 356A[b]	91
Green Canyon 19A[c]	229	1,130 (20.6)[d]	Eugene Island 361A	91
Platform survey component				
Definitive component candidate sites				
Vermilion 214A[c]	35	1,080	East Cameron 229A	35
Ship Shoal 277A[c]	67	827	Eugene Island 360C	91
Off Mississippi River Delta				
Mississippi Canyon 194A	305	2,226	Mississippi Canyon 280A	305
South Timbalier 130	49	827	South Timbalier 128X	31
High-density platform area				
Vermilion 245E	37	843	South Marsh Island 72C	37
Eugene Island 330C	76	1,050	Eugene Island 352B	91
Offshore Texas				
High Island 376A	101	763	West Cameron 587B	58
High Island 382F	104	1,304	High Island 553A	79
Water depth <10 m				
Eugene Island 100C	6.5	2,226	Ship Shoal 100DA	6.5
Vermilion 31A	6.5	1,304	South Marsh Island 229C	6.5

[a]Included in the Offshore Texas area in the Platform Survey Component
[b]A low-volume produced water discharge, averaging 4.6 m^3/day, was initiated at HI356A shortly before the spring 1995 field survey
[c]Included in the High-Density Platform Area of the Platform Survey Component
[d]The average produced water discharge rate at a secondary discharge point at GC19A at the time of the fall 1995 field survey

- influenced by the Mississippi River;
- high platform density;
- less than 10 m of water;
- off the Texas coast.

Figure 24.1 shows the locations of the platform pairs in the Definitive and Platform Survey Components of the study. The screening survey was performed in the fall (October–November) of 1994. Data for MAH and SVOC concentrations in fish from the four candidate platform pairs collected during the screening survey were used in the Platform Survey Component. Field surveys for the Definitive

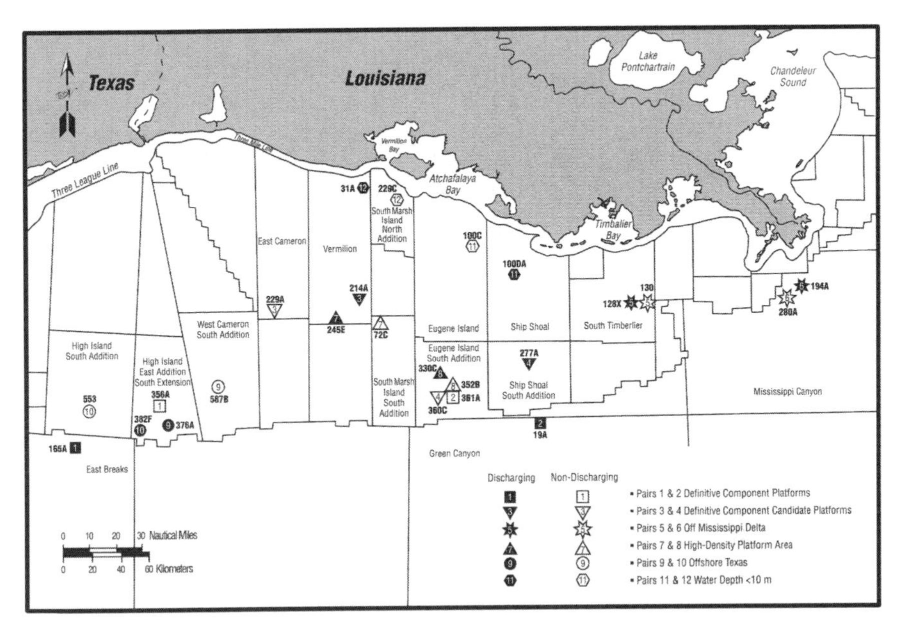

Fig. 24.1 Produced water-discharging and non-discharging (reference) platforms sampled during the Definitive and Platform Components of the Gulf of Mexico Produced Water Bioconcentration Study

Component and the Platform Survey Component were performed in the spring (April–May) and fall (September–December) of 1995.

2.2 Sample Collection

Eight replicate produced water (from the two discharging platforms) and ambient seawater samples were collected at each platform of the two Definitive Component platform pairs during the screening field survey (two replicates) and the two definitive surveys (three replicates). Produced water samples were collected from a spigot located near the discharge end of the produced water discharge pipe. Ambient sea water samples were collected at a depth of 5 m, 2000 m up-current from each of the four platforms.

Produced water and seawater samples for SVOC analysis were collected in 2.5-L precleaned amber glass (I-CHEM®) bottles with Teflon®-lined lids, preserved with reagent grade methylene chloride, and stored at 4°C. Water samples for MAH analysis were collected in triplicate 40-mL precleaned volatile organic analysis (VOA) vials with Teflon®/silicone septa, preserved to a pH <2 with HCl, and stored at 4°C.

Six replicate samples of each of two species of bivalve molluscs and two or more species of platform-associated fish were collected at each of the four platforms during the two definitive field surveys. Two species of fish were collected at 11 of the Platform Survey Component platform pairs. One species each of bivalve mollusc, crustacean, and fish were collected from the other Platform Survey Component platform pair (Vermilion 31A/South Marsh Island 229C in less than 10 m of water).

Two species of bivalve molluscs that are abundant on submerged platform structures in the Gulf of Mexico were collected for analysis at the four platforms in the Definitive Component: the American thorny oyster (*Spondylus americanus*) and the jewel box (*Chama macerophylla*). American oysters (*Crassostrea virginica*) and blue crabs (*Callinectes sapidus*) were collected from platform legs at the Vermilion 31A/South Marsh Island 229C platform pair as part of the Platform Survey Component. Bivalves and crustaceans were collected by divers between the surface and 20 m from submerged surfaces of all six platforms.

Species of bivalves or crustaceans collected at a platform were randomly divided into three replicate samples of 15–30 individuals and stored frozen. Soft tissues of each replicate were removed from the shell, combined, and refrozen. To prevent the loss of MAH during processing, frozen composite tissue was homogenized in a Tekmar® Tissumizer and subsampled immediately for MAH and SVOC analyses. Each of the three composite sample homogenates for each species was divided into two subsamples, yielding six replicates for each species from each platform and survey. A MAH subsample was taken first from the homogenate and immediately returned to the freezer to minimize handling time. Although there may be some loss of volatiles during homogenization of composite bivalve or crustacean samples, analysis of individual specimens would not have been appropriate to represent the entire sample. A small piece of homogenized tissue was retained for determination of water content.

Fish were collected by hook-and-line, spearfishing, and traps from the upper water column within 100 m of each platform. Creole-fish (*Paranthias furcifer*) and yellow chub (*Kyphosus incisor*) were collected at all four Definitive Component platform sites, and one or both species were collected at eight of the Platform Survey Component platforms. Six additional fish species were collected at one or more platforms on one or more field surveys: red snapper (*Lutjanus campechanus*); gray triggerfish (*Balistes capriscus*); spadefish (*Chaetodipterus faber*); rock-hind (*Epinephelus adscensionis*); hardhead catfish (*Arius felis*); and sheepshead (*Archosargus probatocephalus*).

Fish were rinsed in seawater, weighed, measured, wrapped in precleaned aluminum foil and stored frozen until analysis. Groups of one or more (depending on fish size) specimens of each fish species were randomly separated into three replicate composite samples and filleted. Equal amounts of fillet (muscle tissue) from each specimen in a composite group were diced into approximately 0.6 cm cubes and placed immediately into a VOA vial that was stored frozen until analysis of MAH. To minimize the potential loss of volatile compounds, fish fillets for analysis of MAH were not homogenized. The rest of the fillets in each composite sample was homogenized in a Tekmar® Tissumizer equipped with a stainless steel probe. The fish tissue homogenate was then subsampled for SVOC analysis. As dictated by the statistical design, two subsamples, designated as 1 and 2, of each composite sample (A, B, and C) were analyzed. A portion of each sample was retained for determination of water content.

2.3 Sample Analysis

2.3.1 Monocyclic Aromatic Hydrocarbons (MAH)

MAH were measured in water and tissue samples by purge and trap gas chromatography/mass spectrometry (GC/MS), a modification of EPA Method 8260. Target MAH were benzene, toluene, ethylbenzene, *m*-, *p*-xylenes, *o*-xylene, C_3-benzenes, and C_4-benzenes (Table 24.2). Only benzene, toluene, and ethylbenzene were measured in tissues collected for the Platform Survey Component. Water sample size usually was 25 mL and tissue sample size was 15–24 g wet wt. Water samples were transferred to the sparge vessel, fortified with surrogate internal standards (SIS) and recovery internal standards (RIS), and analyzed. Tissue samples were prepared for

Table 24.2 Target monocyclic aromatic hydrocarbons (MAH), polycyclic aromatic hydrocarbons (PAH), and other semivolatile organic compounds (SVOC) analyzed in water and tissue samples in the Definitive Component; only EPA-specified target chemicals ([a]) were analyzed in tissues collected in the Platform survey component

Monocyclic aromatic hydrocarbons (MAH)	
Benzene[a]	*o*-Xylenes
Toluene[a]	C_3-Benzenes
Ethylbenzene[a]	C_4-Benzenes
m/p-Xylenes	
Polycyclic aromatic hydrocarbons (PAH)	
Naphthalene (N)	C_1-Phenanthrenes/anthracenes (C1P)
2-Methylnaphthalene	C_2-Phenanthrenes/anthracenes (C2P)
1-Methylnaphthalene	C_3-Phenanthrenes/anthracenes (C3P)
2,6-Dimethylnaphthalene	C_4-Phenanthrenes/anthracenes (C4P)
2,3,5-Trimethylnaphthalene	Fluoranthene (FL)
C_1-Naphthalenes (C1N)	Pyrene (PY)
C_2-Naphthalenes (C2N)	C_1-Fluoranthenes/pyrenes (C1FL/PY1)
C_3-Naphthalenes (C3N)	C_2-Fluoranthenes/pyrenes (C2FL/PY2)
C_4-Naphthalenes (C4N)	Benz[a]anthracene (BaA)
Acenaphthylene (ACY)	Chrysene (C)
Acenaphthene (ACE)	C_1-Chrysenes/Benz(a)anthracenes (C1C)
Biphenyl (BiP)	C_2-Chrysenes/Benz(a)anthracenes (C2C)
Fluorene[a] (F)	C_3-Chrysenes/Benz(a)anthracenes (C3C)
C_1-Fluorenes (C1F)	C_4-Chrysenes/Benz(a)anthracenes (C4C)
C_2-Fluorenes (C2F)	Benzo[b]fluoranthene (BbF)
C_3-Fluorenes (C3F)	Benzo[k]fluoranthene (BkF)
Dibenzothiophene (D)	Benzo[e]pyrene (BeP)
C_1-Dibenzothiophenes (C1D)	Benzo[a]pyrene[a] (BaP)
C_2-Dibenzothiophenes (C2D)	Perylene (PER)
C_3-Dibenzothiophenes (C3D)	Indeno[1,2,3-c,d]pyrene (IcdP)
Anthracene (A)	Dibenz[a,h]anthracene (DBA)
Phenanthrene (P)	Benzo[g,h,i]perylene (BghiP)
1-Methylphenanthrene	
Other semivolatile organic compounds (SVOC)	
Phenol[a]	Di(2-ethylhexyl)phthalate (DEHP) [a]

[a]Target organic chemicals specified by EPA Region 6 in the NPDES permit

MAH analysis by the standard EPA method of Hiatt (1981) and Easley et al. (1981). Frozen tissue samples were quickly crushed, placed into a sparge vessel, weighed, and fortified with SIS and anti-foam agent. Water was then added to the container to eliminate headspace. The container was sealed and sonicated in a cold water bath for approximately 15 min, fortified with RIS, and analyzed.

Water and tissue samples were purged and trapped on a Tekmar® purge and trap/desorption unit (purge temperature: 40°C). Target compound separation and identification was achieved with a Hewlett-Packard Model 5890 gas chromatograph equipped with a 30 m × 0.25 mm internal diameter DB-624 column and a Hewlett-Packard Model 5970 Mass Selective Detector. Samples were quantified for MAH by the method of internal standards, using SIS.

2.3.2 Semivolatile Organic Compounds (SVOC)

Water and tissue samples collected in the Definitive Component were extracted and analyzed for 40 PAH and PAH isomer groups, phenol, and di(2-ethylhexyl)phthalate (DEHP; Table 24.2). Fish, blue crab, and oyster samples collected in the Platform Survey Component were analyzed for fluorene, benzo(a)pyrene, phenol, and DEHP. After extraction, seawater samples were analyzed directly by GC/MS for all target analytes; produced water extracts were fractionated by high performance liquid chromatography (HPLC) and then analyzed by GC/MS for all target analytes. Tissue sample extracts were passed though silica gel cleanup columns, fractionated by HPLC, and split in half: one half was analyzed for phenols, and the other half was further cleaned up by passing through alumina before analysis for PAH and DEHP. The methods for extraction, cleanup, and instrumental analysis followed procedures described by Douglas et al. (1992), USEPA (1993), and Page et al. (1995), with some modifications for analysis of phenol and DEHP as described below.

Water Extraction

Two-litre water samples were extracted for SVOC by separatory funnel liquid–liquid extraction. The water samples were acidified with clean 6 N HCl prior to sample processing to enable the extraction of phenol along with the PAH and DEHP. All samples were spiked with surrogate compounds and extracted with three 120-mL volumes of ultrapure methylene chloride. Extracts were combined, dried with sodium sulfate, concentrated, spiked with the appropriate internal standards in preparation for fractionation by HPLC or instrumental analysis.

Tissue Extraction

Sodium sulfate was added to each 25 g wet wt sample and the sample was spiked with surrogate compounds. The samples were then extracted three times by maceration in 100 mL volumes of methylene chloride and the extracts were combined and concentrated. The lipid weight (extract weight) was determined for each sample. The extracts were cleaned up by HPLC fractionation before instrumental analysis.

Sample Cleanup

Prior to HPLC fractionation, all tissue extracts were passed through a silica gel cleanup column in order to remove any interfering highly polar compounds. The extracts were eluted with 9:1 methylene chloride:ethyl ether through a chromatography column packed with sodium sulfate and 5% deactivated silica gel in a slurry of 9:1 methylene chloride:ethyl ether.

The extracts were fractionated by HPLC on a size exclusion chromatography column (Envirosep® Gel Permeation Column) with methylene chloride elution. The HPLC fraction collection windows were calibrated with a solution containing phenol, 4,4′-dibromo-octafluorobiphenyl, DEHP, benzo(g,h,i)perylene, and the primary sample interferences, lipid (as corn oil) and sulfur, prior to the analysis of a sample sequence. The PAH, phenol, and DEHP fractions of the extract were collected by an automated fraction collector programmed to collect the specified fraction based on the HPLC calibration solution.

One half of the sample extract was spiked with phenol recovery standard and analyzed by GC/MS for phenol. The other half of the sample extract was processed through an alumina clean-up column packed with 7% deactivated alumina, and eluted with methylene chloride. The eluate was analyzed for PAH and DEHP by GC/MS.

Instrumental Analysis

Target compound separation and identification was achieved with a Hewlett-Packard Model 5890 gas chromatograph equipped with a 30 m × 0.25 mm i.d. DB-5 column and a Hewlett-Packard Model 5970 Mass Selective Detector. The GC/MS analysis was conducted in the selected ion monitoring mode. The GC/MS was tuned and calibrated with a minimum of a five-point PAH/phenol/phthalate calibration spanning the linear range of the mass spectrometer prior to the analysis of a sample set. The PAH, phenol, and DEHP target analytes were quantified versus the surrogate compounds added to the samples prior to extraction.

Data Quality

Method detection limits (MDLs) for the MAH and SVOC were determined by the EPA seven standard deviation method (Fed. Reg. 1984). MDLs for MAH in water and tissue samples ranged from 90 to 320 ng/L and 2.4 to 4.1 ng/g dry wt, respectively. MDLs for phenol in water and tissue samples were 1.8 ng/L and 38 to 270 ng/g dry wt and for DEHP were 90 ng/L and 140 ng/g dry wt. MDLs for individual PAH or alkyl-isomer groups in water and tissue ranged from 1 to 11 ng/L and

1.3 to 16 ng/g dry wt. MDLs for different alkyl-PAH isomer groups were assigned the value of the parent compound.

The practical quantification level (PQL), which is the lowest concentration that can be achieved within specified limits of precision and accuracy during routine laboratory operating conditions (50 FR 46902-46933, November 13, 1985), is defined in this investigation as five times the MDL. Concentrations between the MDL and PQL are considered estimates because there is reduced confidence in the precision and accuracy of the value.

The data quality for the MAH and SVOC analysis met data quality objectives (DQOs) and the results achieved very low MDLs. The occasional sample that was out of the range of the relevant acceptance criteria was reanalyzed. Because the analytical methods used in this study achieved very low MDLs in water (in the low parts per trillion range), phenol and DEHP were consistently detected in the laboratory procedural blanks for the water samples (phenol: range from 25 to 260 ng/L; DEHP: range from 1.0 to 350 ng/g). Procedural blanks for some tissue sample batches contained phenol and DEHP concentrations higher than the MDL (designated not detected: ND) although extraordinary measures were implemented to minimize contamination during processing and analysis, (e.g., covering all surfaces with baked aluminum foil, contact of fillet with only fresh foil). Phenol procedure blank concentrations ranged from 180 to 400 ng/g dry wt (MDL = 38 ng/g), and DEHP concentrations in blanks ranged from 300 to 670 ng/g dry wt (MDL = 140 ng/g). The concentrations of phenol and DEHP in tissue blanks were in the range observed in many field-collected tissue samples.

2.4 Statistical Analysis

Tissue residue data collected in the Platform Component were not evaluated statistically. Each of the two platform pairs in the Definitive Component was treated separately by organism and by chemical for statistical analysis. Several of the target chemicals were present in some produced water, ambient sea water, or animal tissue samples at concentrations below the method detection limit (MDL). All concentrations below the MDL (ND) were set to ½ their MDL value for statistical analysis. The statistical tool used to test the null hypothesis depended on the number of nondetectable (ND) values in the data set based on the following modifications to the approach recommended by the USEPA (1989): (1) when the number of ND values for a particular chemical was <15% of the total, two-way analysis of variance was used to test for differences between mean concentrations for discharging and reference platforms within surveys; (2) when the number of ND values was ≥15% and <50%, data were transformed to ranks and the Friedman test with a correction for ties (Conover 1980) was used to test for equality of concentrations between platform pairs and within survey differences; (3) when the number of ND values was ≥50% and <90%, the data were transformed to binary values, either ND or not ND, and Fisher's exact test (Conover 1980) of proportions was used; and (4) when the number of ND values was >90%, no statistical tests were performed because too few data were available.

3 Results

3.1 Study Sites

The two discharging/non-discharging (D/R) platform pairs in the Definitive Component are East Breaks 165A (EB165A)/High Island 356A(HI356A) and Green Canyon 19A (GC19A)/Eugene Island 361A (EI361A) (Table 24.1). EB165A (D) and HI356A (R) are located approximately 70 km apart southeast of the Texas coast (Fig. 24.1) at water depths of 262 and 91 m, respectively. GC19A (D) and EI361A (R) are located approximately 69 km apart south of the Louisiana coast (Fig. 24.1) at water depths of 229 and 91 m, respectively.

Produced water from EB165A and GC19A is discharged approximately 2.4 m above the sea surface and approximately 1.8 m below the sea surface, respectively. Mean produced water discharge rates at the two discharging platforms during the two field surveys were 1,750 and 1,130 m^3/day (Table 24.1). There was a low volume produced water discharge at the reference platform HI356A at the time of the first Definitive Component field survey and a secondary low volume discharge at GC19A at the time of the second Definitive Component field survey.

The 12 platform pairs in the Platform Survey Component are widely spread over the outer continental shelf of the western and central Gulf of Mexico (Fig. 24.1) in 6.5 m to 305 m of water (Table 24.1). The platform pairs include two each off the Mississippi River, in a high-density platform area, offshore Texas, and at a water depth less than 10 m. In addition, the Definitive Component platform pair EB165A/HI356A is in the offshore Texas area and the other Definitive Component platform pair GC19A/EI361A and the two Definitive Component candidate sites, VR214A/EC229A and SS277A/EI360C, are in the high-density platform area. Mean produced water discharge rates from the 12 discharging platforms at the time of the three field surveys ranged from 763 to 2,226 m^3/day (Table 24.1).

3.2 Monocyclic Aromatic Hydrocarbons (MAH)

Benzene, toluene, ethylbenzene, and xylenes (BTEX) and total C_3- and C_4-benzenes (MAH) were analyzed in produced water from the two discharging platforms, in ambient seawater near the four platforms, and in tissues of bivalves and fish collected near the four Definitive Component platforms (Table 24.2). Tissues of marine animals (mostly fish) collected near 12 platform pairs in the Platform Survey Component were analyzed for benzene, toluene, and ethylbenzene.

3.2.1 MAH in Produced Water

Mean total MAH concentrations in eight replicate produced water samples collected during the screening survey and two definitive surveys from the two discharging platforms, EB165A and GC19A, were 1,934 μg/L and 1,825 μg/L (parts per billion), respectively. The mean concentration of MAH in three replicate samples from a low-volume secondary discharge from GC19A at the time of the second definitive survey was 1,025 μg/L (Table 24.3). The most abundant MAH in produced water

Table 24.3 Mean concentrations in μg/L (parts per billion) of total MAH (BTEX plus C_3- and C_4-benzenes), total PAH (sum of 40 PAH and alkyl-PAH isomer groups), phenol, and DEHP in produced water from discharging platforms and ambient seawater from discharging and reference platforms in the Definitive Component, and in blanks; produced water ($n = 8$ for each discharging platform) and ambient seawater samples ($n = 6$ for each platform pair) were collected during the screening field survey and the two Definitive Component surveys; three replicate samples of produced water also were collected from a secondary, low volume discharge at GC19A during the spring 1995 field survey

Platform	Produced water (mean ± SD)	Ambient seawater (Range)	Lab/field blanks
Total monocyclic aromatic hydrocarbons (MAH) (MDL = 0.09–0.31 μg/L)			
EB165A (D)	1,934 ± 145	ND – 0.63 J	
HI356A (R)	–	ND – 0.50 J	
GC19A (D)	1,825 ± 641	ND – 0.53 J	ND – 3.1
GC19A (D)[a]	1,025 ± 22	–	
EI361A (R)	–	ND – 0.16 J	
Total Polycyclic Aromatic Hydrocarbons (PAH) (MDL = 0.001–0.011 μg/L)			
EB165a (D)	37.2 ± 5.3	0.004–0.013 J	
HI356A (R)	–	ND – 2.26	
GC19A (D)	73.5 ± 67.2	ND – 0.075	ND – 0.01 J
GC19A (D)[a]	136 ± 11.3	–	
EI361A (R)	–	ND – 0.013 J	
Phenol (MDL = 0.002 μg/L)			
EB165A (D)	438 ± 114	ND – 0.25	
HI356A (R)	–	ND – 0.08	
GC19A (D)	435 ± 162	0.03 J – 0.11	ND – 0.26
GC19A (D)[a]	367 ± 23.6	–	
EI361A (R)	–	0.03 J – 0.24	
Di(2-ethylhexyl)phthalate (DEHP) (MDL = 0.09 μg/L)			
EB165a (D)	0.18 ± 0.16 J	ND – 0.18 J	ND – 0.35 J
HI356A (R)	–	ND – 1.8	
GC19A (D)	0.10 ± 0.07 J	ND – 0.20 J	
GC19A (D)[a]	0.41 ± 0.41 J	–	
EI361A (R)	–	0.09 J – 0.68	

ND = Concentration below the MDL
J Qualifier indicating that the value is between the method detection limit (MDL) and the practical quantification level (PQL), defined as five times the MDL
[a]Produced water from a temporary 20.6 m^3/day secondary discharge at the time of the spring 1995 field survey

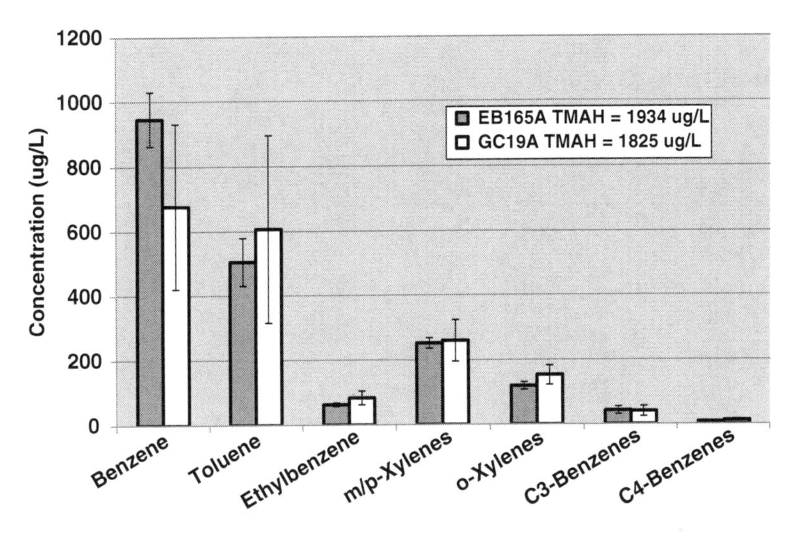

Fig. 24.2 Mean (± standard deviation: $n = 8$) concentrations of BTEX and C_3- and C_4-benzenes concentrations in produced water from the two discharging platforms in the Definitive Component

was benzene, followed by toluene and total xylenes (Fig. 24.2). Concentrations of ethylbenzene, C_3- and C_4-benzenes were low.

3.2.2 MAH in Ambient Seawater

Ambient seawater up-current of discharging and reference platforms was essentially devoid of MAH. Toluene was detected at 0.15–0.50 μg/L in single water samples collected on the screening and spring field surveys at the two reference platforms. Benzene (0.28, 0.62 μg/L) and toluene (0.53, 0.13 μg/L) were detected in single water samples collected at discharging platforms, GC19A and EB165A, respectively, during the screening and fall 1995 field surveys, respectively. Laboratory and field blanks sometimes contained up to about 3 μg/L total MAH, indicating that the low concentrations measured in ambient seawater may be artifacts.

3.2.3 MAH in Tissues: Definitive Component

Approximately 96% of the tissue samples did not contain any target MAH at concentrations above the MDL. Benzene was detected in the tissues of the two species of bivalve molluscs and in yellow chub collected during the spring 1995 survey from two discharging and one reference platform at concentrations of 7.3– 9.3 ng/g dry wt (parts per billion) (Table 24.4).

Toluene was detected in tissues of the two bivalves and three species of fish collected during the fall 1995 survey. Toluene concentrations were higher in bivalves from the two reference platforms than from the two discharging platforms (Table 24.4). Low concentrations were detected in fish from one discharging platform pair (GC19A/EI361A) but not in fish from the other platform pair

Table 24.4 Range of concentrations in ng/g dry wt (parts per billion) of total MAH in tissues of bivalve molluscs and fish collected during the spring and fall 1995 field surveys at the two platform pairs in the Definitive Component; only benzene (MDL 3.2 ng/g dry wt) or toluene (MDL 3.8 ng/g dry wt) was detected above the MDL in the spring and fall surveys, respectively – no other MAH were detected in tissues at concentrations above the MDL

Platform	Species	Spring survey	Fall survey
EB165A (D)	Jewel box	ND – 7.3 J	10 J – 16 J
	Thorny oyster	ND	ND – 22
	Yellow chub	ND	ND
	Creole fish	ND	ND
	Rockhind	ND	ND
HI356A (R)	Jewel box	ND	18 J – 43
	Thorny oyster	ND – 9.3 J	8.3 J – 37
	Yellow chub	ND	ND
	Creole fish	ND	ND
	Rockhind	ND	ND
	Sergeant major	–	ND
GC19A (D)	Jewel box	ND	5.8 J – 20
	Thorny oyster	ND	18 J – 46
	Yellow chub	ND – 7.7 J	ND
	Gray triggerfish	ND	ND – 11 J
	Creole fish	ND	ND
EI361A (R)	Jewel box	ND	40 – 64
	Thorny oyster	ND	20 – 68
	Yellow chub	ND	7.5 J – 11 J
	Creole fish	ND	ND – 15 J
	Gray triggerfish	ND	5.8 J – 10 J

ND Concentration below the MDL
J Qualifier indicating that the value is between the method detection limit (MDL) and the practical quantification level (PQL), defined as five times the MDL

(EB165A/HI356A). No other MAH were detected in any tissues at concentrations above the MDL.

3.2.4 MAH in Tissues: Platform Survey Component

A total of 460 fish muscle samples from seven species, 16 blue crab samples, and 20 oyster samples from the 12 platform pairs in the Platform Survey Component were analyzed for the three target MAH, benzene, toluene, and ethylbenzene. Blue crabs and oysters were collected at one platform pair in <10 m of water, VR31A/SS100DA. Concentrations of benzene, toluene, and ethylbenzene were below the MDL in 97–99% of the tissue samples (Table 24.5). Most measured concentrations were between the MDL and the PQL, indicating that the concentrations are estimates. The highest concentrations of the three MAH in fish and invertebrate tissues ranged from 25 to 37 ng/g dry wt. Toluene was detected most frequently. Concentration ranges of the three MAH were similar in fish and invertebrates from discharging and reference platforms.

Table 24.5 MAH, PAH, phenol, and DEHP in tissues of marine animals (mostly fish) collected at 12 platform pairs in the Platform Survey Component (in ng/g dry wt or parts per billion)

Chemical	MDL	N[a]	%<MDL	Concentration range above MDL
Benzene	3.2	496	97	3.2 J – 25
Toluene	3.8	496	97	5.9 J – 37
Ethylbenzene	2.5	496	99	6.7 J – 26
Fluorene	0.66	500	89	0.99 J – 140
Benzo(a)pyrene	0.60	500	97	1.9 J – 19
Phenol[b]	270	500	85	280 J – 2,000
DEHP[b]	140	490	90	141 J – 1,700

J Qualifier indicating that the value is between the method detection limit (MDL) and the practical quantification level (PQL), defined as five times the MDL
[a]Number of replicate samples analyzed
[b]Phenol and DEHP were detected in some procedural blanks at concentrations in the range of those detected in tissues

MAH were detected most frequently in tissues of fish, crabs, and oysters collected at the two platform pairs in less than 10 m of water. One or more blue crab, oyster, hardhead catfish, and sheepshead samples from the two discharging platforms in shallow water contained benzene, toluene, and/or ethylbenzene at concentrations above the MDL. One sample each of hardhead catfish and sheepshead from the corresponding reference platforms contained ethylbenzene or toluene. Between one and seven fish samples from the other platform pair categories contained benzene, toluene, or ethylbenzene at concentrations above the MDL.

3.3 Semivolatile Organic Compounds (SVOC)

The SVOC analyzed in produced water, ambient seawater, and tissues of bivalves and fish in the Definitive Component include 40 PAH and alkyl-PAH isomer groups, phenol, and DEHP (Table 24.2). The SVOC analyzed in tissues of marine animals (mostly fish) collected in the Platform Survey Component are fluorene, benzo(a)pyrene, phenol, and DEHP. The same samples were analyzed for both MAH and SVOC.

3.3.1 SVOC in Produced Water

PAH

Mean + SD ($n = 8$ for each discharge) total PAH (TPAH) concentrations in produced water samples collected at EB165A and GC19A during the fall 1994 screening survey and the spring and fall 1995 field surveys were 37.2 $\pm$ 5.34 and 73.5 $\pm$ 67.2 μg/L, respectively (Table 24.3). Mean TPAH concentrations were similar in produced water collected at EB165A during the three surveys. However, TPAH concentrations in produced water from GC19A were higher in the fall of

1994 than in produced water collected at the time of the two subsequent surveys. The produced water samples collected at GC19A during the screening survey contained some high molecular weight PAH, indicating that oil droplets were present. A low-volume produced water discharge from GC19A at the time of the spring 1995 field survey contained a higher mean concentration of TPAH (136 ± 11.3 μg/L, $n = 3$) than the primary produced water discharge did (24.2 ± 0.51 μg/L, $n = 3$) at the time of the spring survey. The estimated mass emission rates of total PAH from these discharges are 65 g/day at EB165A, 83 g/day at the primary discharge at GC19A, and 2.8 g/day at the secondary discharge at GC19A.

The NPDES permit required monitoring of two PAH, fluorene and benzo(a)pyrene in tissue samples. Therefore, special attention was paid to the concentrations of these PAH in produced water and ambient water samples. The range of fluorene concentrations in the 19 replicate produced water samples from the two discharging platforms was 0.059–1.10 μg/L. Benzo(a)pyrene was detected in three of the eight replicate produced water samples collected from the primary discharge at GC19A during the three field surveys and in all three replicate produced water samples collected during the third field survey from the secondary discharge at GC19A. The range of benzo(a)pyrene concentrations in the six produced water samples was 0.003–0.14 μg/L. Benzo(a)pyrene was not detected in any produced water samples from EB165a.

The most abundant PAH in all produced water samples, except those from the secondary discharge from GC19A, were naphthalene and C_1- through C_4-alkyl naphthalenes (Fig. 24.3). Total naphthalenes represented 87% (EB165A) to 43% (secondary discharge from GC19A) of the total PAH in the produced water samples. Alkyl PAH, except the alkyl naphthalenes, were more abundant than the corresponding parent PAH. This PAH profile is typical of a water soluble fraction of crude oil (Neff 2002). The produced water from the secondary source at GC19A contained relatively high concentrations of alkyl phenanthrenes, dibenzothiophenes, and chrysenes, which have a very low solubility, indicating that this produced water probably contained some dispersed oil.

Phenol

Mean phenol concentrations in eight replicate produced water samples from each of the two discharging platforms were similar; 438 ± 114 μg/L (EB165A) and 435 ± 162 μg/L (GC19A). The average phenol concentration was lower in the low-volume secondary discharge at GC19A (Table 24.3).

DEHP

Mean concentrations of DEHP, which is not a natural component of or additive to oil, gas, or produced water (Hudgins 1989, 1991, 1992), in produced water from the two discharging platforms ranged from 0.10 ± 0.07 μg/L in the primary discharge at GC19A to 0.41 ± 0.41 μg/L in the secondary discharge at GC19A (Table 24.3). However, up to 0.35 μg/L DEHP was detected in field and laboratory blanks, so

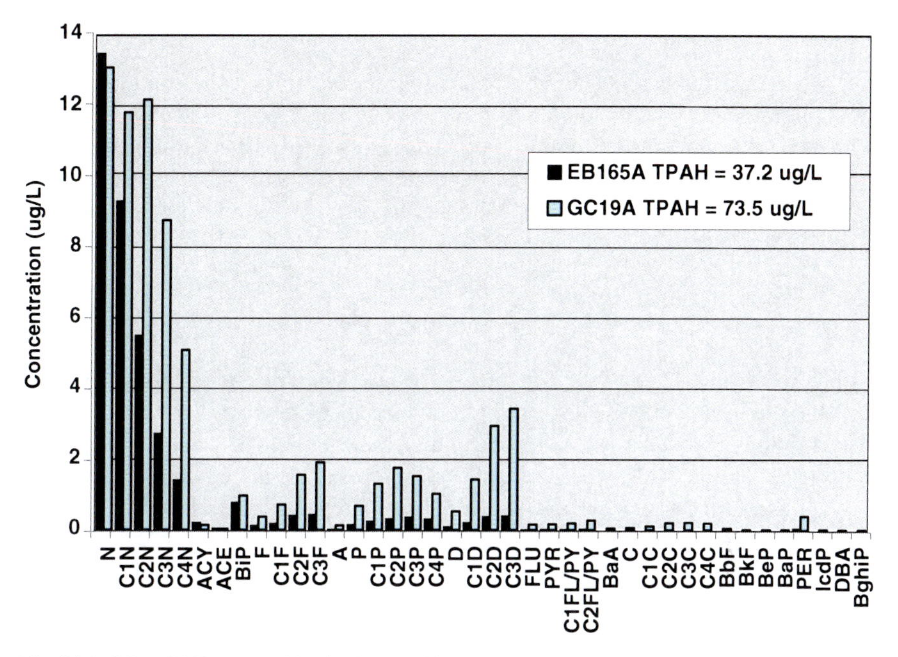

Fig. 24.3 Mean PAH composition in eight replicate produced water samples collected during three surveys from the two discharging platforms in the Definitive Component

these concentrations may be higher than the actual concentrations in the produced water.

3.3.2 SVOC in Ambient Seawater

PAH

Thirty-two ambient seawater samples were collected from near the surface, up-current from the four platforms in the Definitive Component during the three field surveys. The TPAH concentration in 31% of the samples was below the MDL. The remaining samples contained 0.004–2.26 μg/L TPAH (Table 24.3). The only PAH detected in most ambient seawater samples were naphthalene and occasionally methyl naphthalene. However, one replicate sample collected on the spring and fall 1995 field surveys at reference platform HI356A contained 0.75 and 2.26 μg/L TPAH. These samples contained low concentrations of alkyl fluorenes, phenanthrenes, dibenzothiophenes, and chrysenes. Ambient sea water collected during the third field survey at discharging platform GC19A also contained low concentrations of alkyl naphthalenes and phenanthrenes. These low-solubility PAH in ambient seawater samples near a discharging and a reference platform probably were derived from dispersed oil droplets or a tar ball. No fluorene or BAP were detected in any ambient seawater samples up-current of discharging or reference platforms.

Phenol

Phenol was detected in all but three ambient seawater samples; concentrations above the MDL ranged from 0.03 to 0.25 μg/L at all platforms (Table 24.3). There was no difference in phenol concentrations in ambient seawater near discharging and reference platforms. Phenol was detected in laboratory and field blanks at concentrations up to 0.26 μg/L, indicating that phenol concentrations in seawater were lower than those reported here.

DEHP

All but six of the 32 replicated ambient seawater samples collected during three field surveys near the two discharging/reference platform pairs contained DEHP concentrations above the MDL. Concentrations above the MDL ranged from 0.09 to 1.8 μg/L (Table 24.3). Because DEHP is ubiquitous in the environment, due to its wide use as a physical plasticizer in plastics (Lopezavila et al. 1990; Sullivan and Carty 1994), and because it was detected in field and laboratory procedural blanks at concentrations similar to those in ambient seawater, concentrations reported here probably are substantially higher than actual concentrations in ambient seawater.

3.3.3 SVOC in Tissues

Concentrations of 40 PAH and alkyl-PAH isomer groups, phenol, and DEHP were measured in 240 tissue samples from two species of bivalve molluscs and five species of fish collected at the two platform pairs in the Definitive Component. Concentrations of fluorene, benzo(a)pyrene, phenol, and DEHP were measured in tissues of 470–480 samples of fish and 16–20 samples each of blue crabs and oysters collected on the two Platform Survey Component field surveys.

3.3.4 PAH in Tissues

Definitive Component

TPAH concentrations were below the MDL (1.3–16 ng/g dry wt for individual PAH) in 13.3% of the 240 samples of molluscs and fish collected at the two platform pairs during the two definitive field surveys. One replicate sample of thorny oysters and 31 replicate fish muscle samples did not contain any PAH above the MDL. TPAH concentrations greater than the MDL in individual tissue samples were highly variable within and among species and platforms, resulting in high standard deviations for the site means. Mean ± SD ($n = 6$ with values less than the MD recorded as 0) TPAH concentrations for the spring 1995 field survey ranged from 0.83 ± 1.86 ng/g in creole fish collected from discharging platform EB165A to 498 ± 76.2 ng/g in thorny oysters collected at discharging platform GC19A (Table 24.6). Mean concentrations were higher (usually by an order of magnitude

Table 24.6 Mean ± SD ($n = 6$ for each survey) TPAH concentrations (ng/g dry wt) in two species of bivalve molluscs and five species of fish collected during the spring and fall 1995 Definitive Component field surveys at two discharging (D)/reference (R) platform pairs

	Spring 1995 field survey			
Species	EB165A (D)	HI356A (R)	GC19A (D)	EI361A (R)
Jewel box	60.2 ± 19.8	78.6 ± 30.8	207 ± 98.1	78.1 ± 17.0
Thorny oyster	40.1 ± 13.7	10.6 ± 7.1	498 ± 76.2	24.4 ± 13.4
Creole fish	0.83 ± 1.86	4.70 ± 3.59	4.93 ± 3.98	1.47 ± 3.26
Gray triggerfish	NS	NS	<MDL	2.88 ± 3.16
Rockhind	7.57 ± 6.82	5.93 ± 7.03	NS	NS
Yellow chub	7.30 ± 4.22	29.8 ± 39.6	23.7 ± 7.63	21.52 ± 4.02
	Fall 1995 field survey			
	EB165A (D)	HI356A (R)	GC19A (D)	EI361A (R)
Jewel box	120 ± 55.9	147 ± 57.5	113 ± 36.2	104 ± 15.4
Thorny oyster	124 ± 34.9	419 ± 275	290 ± 82.1	109 ± 30.3
Creole fish	4.03 ± 0.52	14.4 ± 8.1	2.78 ± 2.03	8.97 ± 5.78
Gray triggerfish	NS	NS	5.90 ± 6.24	7.42 ± 2.61
Yellow chub	6.07 ± 2.24	3.90 ± 0.50	4.38 ± 1.22	21.62 ± 8.12
Sergeant major	5.20 ± 0.83	4.12 ± 0.66	NS	NS

NS, no sample collected. MDL, mean concentration below the MDL (1.3–16 ng/g) dry wt

or more) in soft tissues of the two mollusc species than in muscle of the five fish species.

Mean TPAH concentrations were significantly higher in soft tissues of the two mollusc species from discharging platform GC19A than in those from the paired reference platform, EI361A, at the time of the spring field survey and in thorny oysters at the time of the fall field survey (Table 24.6). Thorny oysters also contained significantly higher mean TPAH concentrations at discharging platform EB165A than at the paired reference platform, HI356A, at the time of the spring 1995 field survey and at the reference platform than at the discharging platform at the time of the fall 1995 field survey.

TPAH concentrations in individual fish muscle samples were low and highly variable. Differences between samples from discharging and reference platforms were not meaningful, because the individual PAH contributing to the TPAH differed for each species between discharging and reference platforms and most concentrations of individual PAH were below the PQL. The highest concentrations of TPAH were in the muscle of yellow chub collected at discharging platform GC19A and reference platforms HI356A and EI361A in the spring 1995 field survey and at reference platform EI361A in the fall 1995 field survey (Table 24.6).

None of the mollusc or fish tissue samples contained the full suite of 40 PAH and alkyl-PAH isomer groups analyzed in the Definitive Component. The mollusc samples contained more target PAH than the fish muscle samples did. Individual

Table 24.7 Concentration ranges in ng/g dry weight (parts per billion) of individual PAH in soft tissues of thorny oysters from platform pair EB165A (D)/HI365A (R); PAH are listed only if one or more values is above the method detection limit (MDL), shown with a < for each PAH; ([a]) denotes concentration ranges for which one or more values exceeded the PQL; mean concentrations (with <MDL values set at 1/2MDL for statistical analysis) that are significantly higher than the mean concentration in the paired sample are highlighted

	Spring 1995 survey		Fall 1995 survey	
PAH	EB165A (D)	HI365A (R)	EB165A (D)	HI365A (R)
C_1-Naphthalenes	<2.7	<2.7	<2.7–9.2	<2.7 - 10
C_2-Naphthalenes	<5.1	<5.1	9.2–28[a]	9.4–57[a]
C_3-Naphthalenes	<5.1	<5.1	9.2–32[a]	**12–140[a]**
C_4-Naphthalenes	<5.1	<5.1	<5.1	**18–170[a]**
C_1-Fluorenes	<3.6–5.6	<3.6	<3.6–24[a]	<3.6
Anthracene	**<2.3–4.8[a]**	<2.3	**7.4–9.3[a]**	<2.3
C_1-Phenanthrenes	<16	<16–12	<16–19	<16–80[a]
C_2-Phenanthrenes	<16	<16	<16–44	34–120[a]
Dibenzothiophene	<1.3	<1.3	<1.3	<1.3–6.0
C_1-Dibenzothiophenes	<1.3	<1.3	<1.3	<1.3–36
C_2-Dibenzothiophenes	<1.3–11[a]	<1.3	<1.3–33	12–70[a]
C_3-Dibenzothiophenes	**5.6–30[a]**	<1.3–6.7[a]	<1.3–26	7.5–47[a]
C_1-Fluoranthenes/Pyrenes	<3.3	<3.3	<3.3	<3.3–17[a]
Chrysene	<2.2–3.9	<2.2	<2.2	<2.2
C_1-Chrysenes	<2.2–8.9	<2.2	<2.2	<2.2
C_2-Chrysenes	<2.2–11	<2.2	<2.2	<2.2
Benzo(e)pyrene	4.1–5.8	<2.3–4.1	<2.3	<2.3
Benzo(a)pyrene	5.6–8.9	<2.8–5.7	<2.8	<2.8
Perylene	<3.3	<3.3–4.8	<3.3	<3.3
Total PAH range	**26–61**	ND – 20	79–183	**113–805**

thorny oyster samples contained up to 24 PAH or PAH isomer groups at concentrations above the MDL (Tables 24.7 and 24.8). Jewel boxes contained fewer individual PAH, in most cases, at lower concentrations. Most PAH concentrations were below the PQL, indicating that the values are estimates, decreasing confidence of any significant differences observed between concentrations of individual PAH in discharge site and reference bivalve molluscs (Tables 24.7 and 24.8).

Mean concentrations of anthracene in thorny oysters from discharging platform EB165A were significantly higher than in those from reference platform HI365A at the time of the spring and fall Definitive Component field surveys; however, all anthracene concentrations were below the PQL (Table 24.7). C_3-dibenzothiophenes concentrations were significantly higher in thorny oysters from discharging platform EB165A than in those from reference platform HI365A at the time of the spring Definitive Component field survey. Concentrations of C_3- and C_4-naphthalenes were significantly higher in thorny oysters from reference platform HI356A than in those from discharging platform EB165A at the time of the fall Definitive Component survey.

Table 24.8 Concentration ranges in ng/g dry weight (parts per billion) of individual PAH in soft tissues of thorny oysters from platform pair GC19A (D)/EI361A (R); PAH are listed only if one or more values is above the method detection limit (MDL), shown with a < for each PAH; ([a]) denotes concentration ranges for which one or more values exceeded the PQL; mean concentrations (with <MDL values set at 1/2MDL for statistical analysis) that are significantly higher than the mean concentration in the paired sample are highlighted

	Spring 1995 survey		Fall 1995 survey	
PAH	GC19A (D)	EI361A (R)	GC19A (D)	EI361A (R)
Naphthalene	<8.6–17	<8.6	<8.6	<8.6
C_1-Naphthalenes	**4.9–11**	<2.7–4.4	<2.7–5.2	**5.1–8.1**
C_2-Naphthalenes	14–24	<5.1–8.8	<5.1–12	8.3–15
C_3-Naphthalenes	**22–40[a]**	<5.1	<5.1–15	<5.1–9.0
C_4-Naphthalenes	**41–54[a]**	<5.1	<5.1–38[a]	<5.1
Biphenyl	<2.3–6.8	<2.3–19[a]	<2.3	<2.3
C_1-Fluorenes	<3.6–23[a]	<3.6	13–23[a]	12–22[a]
Anthracene	<2.3–3.6	<2.3	3.9–9.5	<2.3–6.4
C_1-Phenanthrenes	**<16–16**	<16	<16	<16
C_2-Phenanthrenes	**39–50**	<16	<16–57	26–40
C_3-Phenanthrenes	**51–60**	<16	**<16–180[a]**	<16
C_4-Phenanthrenes	**46–55**	<16	**<16–60**	<16
Dibenzothiophene	<1.3–1.8	<1.3	<1.3	<1.3
C_1-Dibenzothiophenes	**<1.3–16[a]**	<1.3	<1.3	<1.3
C_2-Dibenzothiophenes	**47–60[a]**	<1.3–9.1[a]	**25–40[a]**	<1.3–22[a]
C_3-Dibenzothiophenes	**<1.3–100[a]**	<1.3–14[a]	**41–65[a]**	<1.3–25[a]
Fluoranthene	<2.6–3.6	<2.6	<2.6	<2.6–11
Pyrene	<3.3	<3.3	<3.3	<3.3–11
C_2-Fluoranthenes/Pyrenes	**12–16**	<3.3	<3.3	<3.3–6.4
C_1-Chrysenes	**<8.5–12[a]**	<2.2	<2.2–5.9	<2.2–5.5
C_2-Chrysenes	**17–27[a]**	<2.2	<2.2	<2.2
Benzo(e)pyrene	7.1–11	<2.3–4.0	<2.3–3.3	<2.3
Benzo(a)pyrene	6.6–12	<2.8–5.4	<2.8	<2.8
Perylene	**26–35[a]**	<3.3	**4.6–7.7**	<3.3
Benzo(g,h,i)perylene	**<2.9–5.7**	<2.9	<2.9	<2.9
Total PAH range	**402–593**	9.0–49	**185–412**	62–144

Mean concentrations of C_1-, C_3-, and C_4-naphthalenes, C_1-, C_2-, C_3-, and C_4-phenanthrenes, C_1-, C_2-, and C_3-dibenzothiophenes, C_2-fluoranthenes/pyrenes, C_1- and C_2-chrysenes, perylene, and benzo(g,h,i)perylene were significantly higher in thorny oysters from discharging platform GC19A than in those from reference platform EI136A at the time of the spring Definitive Component field survey (Table 24.8). The highest concentration of seven of these PAH was above the PQL. Mean concentrations of C_3-phenanthrenes, C_2- and C_3-dibenzothiophenes, and perylene were significantly higher in thorny oysters from GC19A than in those from EI136A at the time of the fall Definitive Component field survey. The mean concentration of C_1-naphthalenes was higher in thorny oysters from reference platform EI136A than in those from discharging platform GC19A at the time of the fall Definitive Component Survey. Concentrations of C_1-naphthalenes, C_4-phenanthrenes, and perylene were below the PQL.

The concentrations of only C_2- and C_3-dibenzothiophenes were significantly higher in jewel boxes from discharging platform GC19A than in those from reference platform EI136A at the time of the spring Definitive Component survey. There were no significant differences in concentrations of any PAH in jewel boxes from the EB165A (D)/HI356A (R) platform pair or in jewel boxes collected at the other platform pair during the fall Definitive Component survey.

The PAH profile in tissues of most thorny oyster and jewel box samples collected from both discharging and reference platforms were dominated by alkyl-PAH, particularly alkyl-naphthalenes, alkyl-phenanthrenes, and alkyl-dibenzothiophenes (Fig. 24.4a, b), indicating that the molluscs were being exposed to petrogenic PAH. Several high molecular weight PAH, usually indicative of a pyrogenic source, also were present in some thorny oysters and jewel boxes from the two discharging platforms (Fig. 24.4a, b). These PAH included benz(a)anthracene, C_1- and C_2-chrysenes, benzo(e)pyrene, benzo(a)pyrene, and perylene and were present in some produced water samples collected during the screening and first Definitive Component surveys, but not the second Definitive Component survey. The resemblance of the PAH profiles in produced water from GC19A and in the two mollusc species from this discharging platform is a strong indication that some of the PAH in the mollusc tissues were bioconcentrated from the produced water discharge.

The target PAH identified by EPA for monitoring are fluorene and benzo(a)pyrene. Fluorene was not present at a concentration above the MDL (3.6 ng/g dry wt) in any of the 240 mollusc and fish samples collected in the Definitive Component. Benzo(a)pyrene was not present at a concentration above the MDL (2.8 ng/g dry wt) in any fish muscle samples. Benzo(a)pyrene was present in 67% of the jewel box and thorny oyster samples at concentrations ranging from 4.2 to 21 ng/g. Sixty-nine percent of benzo(a)pyrene concentrations in jewel boxes were above the PQL; none of the benzo(a)pyrene concentrations in thorny oysters was above the PQL.

Platform Survey Component

Fluorene was detected in 5–33% of fish collected at discharging platforms and in 0–17% of the fish collected at reference platforms in four production areas of the Gulf of Mexico (Table 24.9). Flourene also was detected in 70% and 50% of the blue crabs and eastern oysters collected at the discharging and reference platforms (VR31A/SMI229C), respectively, in less than 10 m of water off Louisiana. Concentrations of fluorene above the MDL in fish and invertebrate tissues ranged from 0.99 to 140 ng/g at the discharging platforms and from 1.1 to 52 ng/g at the reference platforms. The highest concentrations were in oysters from a discharging platform in less than 10 m of water and in gray triggerfish from a reference platform in the high-density platform area. Differences in detection of fluorene in fish tissues in the Definitive Component and Platform Survey Component samples can be

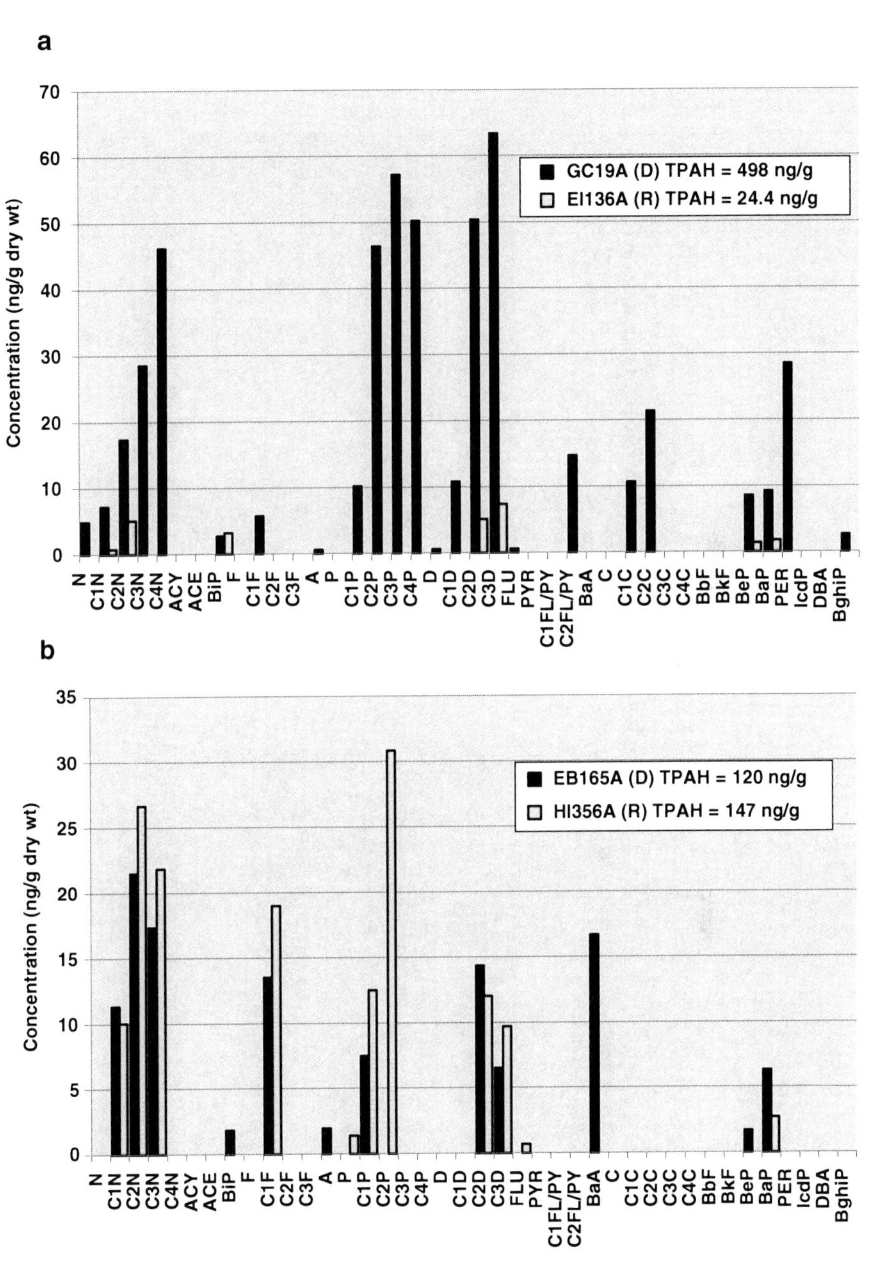

Fig. 24.4 Mean PAH composition in tissues for the Definitive Component field survey of (**a**) thorny oysters ($n = 6$) collected during the spring 1995 at discharging and reference platform pair, GC19A (D)/EI136A (R), and (**b**) jewel boxes ($n = 6$) collected during the fall 1995 at discharging and reference platform pair, EB165A (D)/HI365A (R)

Table 24.9 Percent SVOC concentrations below the MDL[a] and range of concentrations (ng/g dry wt) above the MDL in fish and invertebrates (blue crabs and eastern oysters) collected during two field surveys at 12 discharging/reference platform pairs in different regions of the central and western Gulf of Mexico in the Platform Survey Component

	Discharging platforms		Reference platforms	
SVOC	% < MDL	Range > MDL	% < MDL	Range > MDL
High density platform area (5 Platform Pairs)				
Fluorene	94	0.99 J – 8.2	97	2.9 J – 52
Benzo(a)pyrene	99	3.3	98	2.2 J – 12
Phenol	83	280 J – 990 J	90	300 J – 1,000 J
DEHP	85	141 J – 780	80	150 J – 1,700
Off Texas Coast (3 Platform Pairs)				
Fluorene	92	1.8 J – 18	100	–
Benzo(a)pyrene	98	2.0 J	98	4.1
Phenol	88	340 J – 600 J	87	300 J – 790 J
DEHP	98	470 J	93	160 J – 170 J
Off Mississippi River (2 Platform Pairs)				
Fluorene	95	1.2 J – 4.6	98	1.1 J
Benzo(a)pyrene	100	–	100	–
Phenol	85	280 J – 570 J	88	300 J – 910 J
DEHP	90	180 J – 400 J	92	150 J – 440 J
In less than 10 m of water				
Two species of fish (2 Platform Pairs)				
Fluorene	67	1.3 J – 5.3	83	1.6 J – 24
Benzo(a)pyrene	100	–	100	–
Phenol	73	300 J – 630 J	90	650 J – 1,300 J
DEHP	87	160 J – 350 J	100	–
Blue Crabs and Eastern Oysters (1 Platform Pair)				
Fluorene	30	1.6 J – 140	50	7.0–18
Benzo(a)pyrene	70	1.9 J – 19	85	2.6 J – 4.9
Phenol	80	290 J – 1,300 J	75	470 J – 2,000
DEHP	90	160 J – 350 J	100	–

[a]MDLs for Platform Survey Component: Fluorene = 0.66 ng/g; Benzo(a)pyrene = 0.60 ng/g; Phenol = 270 ng/g; DEHP = 140 ng/g

attributed to differences in the MDL for fluorene in tissues in the two components (3.6 ng/g and 0.66 ng/g, respectively).

Benzo(a)pyrene was detected in only 3% of tissue samples collected during the two Platform Survey Component surveys (Table 24.9). The highest concentrations were in the same American oyster and gray trigger fish samples that contained the highest concentrations of fluorene. As with fluorene, differences in detection of benzo(a)pyrene in muscle tissues of fish collected in the two components of the study can be attributed to differences in the MDL for benzo(a)pyrene in the two components (2.8 ng/g in the Definitive Component and 0.60 ng/g in the Platform Component).

3.3.5 Phenol in Tissues

Definitive Component

Phenol concentrations were highly variable in tissues of the two species of molluscs and the five species of fish collected at the four Definitive Component platforms and in the procedural blanks. Procedural blanks ranged from less than the MDL (38 ng/g dry wt) to 170 ng/g phenol. Phenol was not detected at concentrations above the MDL in 17–50% of the jewel boxes, creole fish, and yellow chub collected at discharging platform EB165A, and in 17–42% of the creole fish collected at the other three platforms (Table 24.10). Forty-seven percent of the mollusc samples and 14% of the fish samples contained phenol concentrations equal to or higher than the PQL (190 ng/g). The range of phenol concentrations above the MDL was similar in tissues of the two species of molluscs (68–610 ng/g) and in the five species of fish (44–1,000 ng/g) from discharging and reference platforms (Table 24.10). The highest tissue concentration measured was 1,000 ng/g in a sample of gray triggerfish from discharging platform GC19A. Thus, it is unlikely that any of the tissue phenol came from the produced water discharges.

Table 24.10 Percent phenol concentrations below the MDL[a] and range of concentrations (ng/g dry wt) above the MDL in soft tissues of two species of molluscs and muscle of five species of fish collected from paired discharging and reference platforms during the spring and fall 1995 surveys of the Definitive Component; $N = 12$ for each species and platform, except for rockhind and sergeant major for which $N = 6$ at each platform

	EB165A (D)		HI356A (R)	
Species	% < MDL	Range above MDL	% < MDL	Range above MDL
Jewel box	33	160 J – 450	0	83 J – 280
Thorny oyster	0	100 J – 610	0	68 J – 210
Creole fish	50	54 J – 110 J	17	54 J – 180 J
Rockhind	0	81 J – 150 J	0	50 J – 190
Yellow chub	17	70 J – 200	8	52 J – 170 J
Sergeant major	0	54 J – 94 J	0	70 J – 89 J
	GC19A (D)		EI361A (R)	
Species	% < MDL	Range above MDL	% < MDL	Range above MDL
Jewel box	0	150 J – 470	0	140 J – 800
Thorny oyster	0	120 J – 520	0	75 J – 220
Creole fish	42	66 J – 280	25	49 J – 160 J
Yellow chub	0	70 J – 550	0	44 J – 180 J
Gray triggerfish	0	310 – 1,000	0	72 J – 280

J Qualifier indicating that the value is between the method detection limit (MDL) and the practical quantification limit (PQL), defined as five times the MDL
[a]MDL for tissue phenol in the Definitive Component = 38 ng/g dry wt

Platform Survey Component

Phenol concentrations were below the MDL in 73–90% of fish and invertebrate tissue samples collected during the Platform Survey Component (Table 24.9), as well as in some procedural blanks. The high percentage of phenol concentrations below the MDL can be attributed to the high MDL for tissues collected in the Platform Component (270 ng/g). Phenol concentrations above the MDL ranged from 290 to 2,000 ng/g dry wt (Table 24.9). Highest concentrations were in red snapper from a reference platform in 90 m of water, hard head catfish from a reference platform in 6.5 m of water, and blue crabs from a discharging/reference platform pair in 6.5 m of water. Thus, there is no evidence that fish from the vicinity of discharging platforms in the four regions of the Gulf of Mexico are bioaccumulating phenol from produced water discharges.

3.3.6 DEHP in Tissues

Definitive Component

Concentrations of DEHP were highly variable in tissues of the two species of molluscs and five species of fish sampled in the Definitive Component. Seventy percent of the 240 tissue samples contained DEHP concentrations less than the MDL (140 ng/g dry wt). Some procedural blanks contained DEHP at concentrations greater than the MDL. DEHP concentrations above MDL were similar in molluscs and fish from the discharging and reference platforms and ranged from 140 ng/g in yellow chub to 3,600 ng/g in thorny oysters both from reference platform EI361A.

Platform Survey Component

DEHP concentrations were below the MDL (140 ng/g) in 80 to 100% of tissue samples from the four platform regions sampled in the Platform Component (Table 24.9), as well as in some procedural blanks. DEHP concentrations above the MDL ranged from 140 to 780 ng/g at discharging platforms and 150–1,700 ng/g at reference platforms. Little DEHP was detected in produced water or ambient seawater (Table 24.3). Thus, there is no indication from either component of this survey that any species bioaccumulated DEHP from produced water. The sources of DEHP and the reasons for the large differences in concentrations in different species is unexplained, but may be due to variable contamination of samples with this ubiquitous contaminant during sampling, processing, or analysis.

4 Discussion

4.1 MAH and SVOC in Produced Water and Ambient Seawater

Produced water generated during offshore production of oil and gas usually is treated on the platform to remove dispersed and some dissolved hydrocarbons before discharge to the sea or reinjection into a disposal well (Clark and Veil 2009). Much of the BTEX and low molecular weight alkanes and phenols are lost to the atmosphere during produced water treatment and discharge, particularly if the discharge is above the sea surface. If properly treated, produced water contains very little dispersed oil droplets. Most of the hydrocarbons and other organic compounds remaining in the produced water after treatment are in solution or colloidal suspension in the water phase. Higher molecular weight SVOC (e.g., 3- through 6-ring alkyl-PAH, DEHP, and >C_{16} saturated hydrocarbons (e.g., pristane, phytane) are present in produced water mainly associated with dispersed oil droplets (Faksness et al. 2004); their presence can be used as an indication of the presence of dispersed oil in produced water.

The most abundant hydrocarbons in solution in treated produced water usually are MAH, particularly the more soluble benzene, C_1- and C_2-alkyl benzenes (BTEX) (Neff 2002; OGP 2005, Dórea et al. 2007). BTEX (benzene, toluene, ethylbenzene and xylene, or MAH) usually represent more than 95% (and PAH less than 5%) of the total aromatic hydrocarbons in produced water (Neff 2002; OGP 2005). Solubilities of aromatic hydrocarbons decrease with increasing molecular weight. Thus, concentrations of benzene and toluene usually are highest and concentration decreases with increasing molecular weight and ring number, and decreasing aqueous solubility. Alkyl PAH often are more abundant than the corresponding parent PAH, reflecting their relative abundances in the crude oil (Neff 2002). Phenol and alkyl phenols usually are not abundant in crude oils and their concentrations in produced water usually are low. DEHP and other phthalate esters are not natural ingredients of produced water and are not production treatment chemicals.

Saline produced waters dilute rapidly upon discharge to well-mixed marine waters. Brandsma and Smith (1996) modeled the fate of produced water discharged under typical Gulf of Mexico conditions. Produced water discharged at a rate of 115 m^3/d is diluted by 500-fold within 10 m from the outfall and by 1,000-fold within 100 m from the outfall. They predicted that the 1,130 m^3/d produced water discharge plume from Platform GC19A penetrated to a total depth of about 30 m and was diluted by a factor of 100-fold within a few meters of the outfall under stratified water column conditions. A further ten-fold dilution had occurred by the time the plume had drifted 100 m from the outfall. The larger volume (1,750 m^3/d), more saline discharge plume from Platform EB165A descended about 5 m deeper following discharge and diluted nearly as rapidly as the produced water plume from Platform GC19A. The rapid dilution of the produced water plume limits the opportunities for marine animals to encounter elevated concentrations of produced water chemicals and bioconcentrate them in their tissues.

4.1.1 Monocyclic Aromatic Hydrocarbons (MAH)

The concentration and composition of the BTEX fraction in the two produced waters was similar to that in produced water discharged offshore elsewhere in the world (Neff 2002: OGP 2005). The BTEX concentration in produced water discharged offshore worldwide ranges from about 70 μg/L to nearly 600,000 μg/L, with most concentrations in recent years in the lower part of this range (Neff 2002). The BTEX concentration usually is higher in produced water from gas wells than that from oil wells. For example, the mean BTEX concentration in produced water from oil and gas platforms in the Norwegian Sector of the North Sea is 8,600 and 26,000 μg/L, respectively (Johnsen et al. 2004). BTEX concentrations are much lower in produced water from platforms on the outer continental shelf off California, with median concentrations of benzene, toluene, and ethylbenzene, of 93.5, 127, and 23 μg/L, respectively (Marine Research Specialists 2005). Benzene or toluene is most abundant. Total C_3- plus C_4-benzenes concentrations usually represent only 0.5–2.5% of total MAH in produced waters from the US, North Sea, and Indonesia (Neff 2002).

Because MAH are highly volatile, they are lost rapidly by evaporation from the produced water plume following discharge to the ocean, particularly if the discharge is above the sea surface, as was the case for the produced water discharge from EB165A. Dilution/dispersion plays a small role in decreasing the concentration of MAH in the produced water plume. There was a 14,900-fold dilution of BTEX in surface waters 20 m down-current from an 11,000 m^3/day produced water discharge from a production platform in the Bass Strait off southeast Australia, from 6,410 μg/L in the produced water to 0.43 μg/L in the ambient seawater (Terrens and Tait 1996). In the present investigation, no MAH were detected at concentrations above those in associated field and laboratory blanks in ambient seawater collected 2,000 m up-current from the four platforms in the Definitive Component. Only benzene or toluene was detected at concentrations above the MDL (0.13 and 0.09 μg/L, respectively) in a few ambient seawater samples. BTEX concentrations in the upper water column near production platforms in shallow waters off Louisiana ranged from 0.008 to 0.33 μg/L (Sauer 1980), compared to background concentrations of 0.009 to 0.10 in surface waters of the outer continental shelf off Texas and Louisiana (Sauer et al. 1978). These low ambient seawater concentrations are typical for open waters of the Gulf of Mexico (Sauer 1980). Because of their volatility, MAH do not persist long in seawater, even near point sources such as produced water outfalls (Neff 2002; National Research Council 2003).

4.1.2 Polycyclic Aromatic Hydrocarbons (PAH)

The concentrations of TPAH (40 individual PAH and PAH isomer groups) and the relative concentrations of individual PAH in produced water from the two discharging platforms in the Definitive Component were in the lower part of the range for produced water discharged offshore elsewhere in the world (Neff 2002; OGP 2005). Concentrations of TPAH in individual produced water samples collected during three field surveys from discharging platforms EB165A and GC19A ranged from 5.3 to 246 μg/L. Produced water collected from four other platforms in the US Gulf of Mexico in 1993 contained 37–580 μg/L TPAH (Neff and Sauer 1996a). TPAH concentrations in produced waters from offshore platforms throughout the world range from 40 to 3,000 μg/L, with most concentrations in the lower part of this range in recent years (Neff 2002). Concentrations of TPAH in produced water sampled between 1997 and 2000 from 27 production facilities in the Tampen, Ekofisk, and Sleipner regions of Norwegian Sector of the North Sea ranged from 240 to 3,830 μg/L (Durell et al. 2004).

Mean estimated mass emission rates for TPAH in the three produced water discharges from the two platforms monitored in this study ranged from 2.8 to 83 g/day. By comparison, PAH emission rates to the North Sea in produced water from the 27 production facilities in the Norwegian Sector of the North Sea range from 5.5 to 39,000 g/day. The reasons for the higher PAH emission rates from most North Sea platforms than from Gulf of Mexico platforms are the higher produced water discharge rates (3– 28,200 m^3/day to the North Sea versus 20–2,225 m^3/day to the Gulf of Mexico) and the higher TPAH concentrations in produced water from 27 platforms in the North Sea (240–3,830 μg/L; Durell et al. 2004) than in produced water from seven platforms in the Gulf of Mexico (40–600 μg/L).

The dominant PAH compounds in produced water from the two discharging platforms monitored in this study as well as offshore produced waters analyzed in detail by others are naphthalene and alkyl naphthalenes. Naphthalenes represented 40–91% of the TPAH in the produced waters analyzed in this study, and 80–96% of the TPAH in produced water from 21 North Sea platforms analyzed by Neff et al. (2006). The PAH profile in some produced water samples from GC19A was enriched with higher molecular weight (4- to 5-ring) PAH, indicating that these samples probably contained higher concentrations of dispersed oil droplets than the other samples (Faksness et al. 2004).

Concentrations of PAH decrease rapidly in the produced water plume by dilution and dispersion. As discussed above, the predicted dilution of the produced water plume from EB165A and GC19 is about 1,000-fold within 100 m down-current of the discharge (Brandsma and Smith 1996). Terrens and Tait (1996) measured dilution of naphthalene (2-rings), phenanthrene (3-rings), and fluoranthene (4-rings) in a produced water plume discharged at a rate of 11,000 m^3/day from a production platform in the Bass Strait, Australia. Naphthalene was diluted 11,000-fold and phenanthrene and fluoranthene were diluted by 6,000- and 4,000-fold, respectively, 20 m down-current from the discharge. The more rapid dilution of naphthalene indicated that some of it evaporated from the plume.

4.1.3 Phenol

Phenol and alkylphenols in produced water discharged offshore are of concern because some more highly alkylated phenols are potent endocrine disruptors (Thomas et al. 2004; Boitsov et al. 2007; Balaam et al. 2009). Phenol and alkylphenols probably are derived from natural biodegradation of oil in the reservoir (Taylor et al. 2001) and their concentrations are highly variable in produced water. Phenol, methylphenols (cresols) and dimethylphenols (xylenols) usually are most abundant and concentration decreases with increasing alkylation (Neff 2002; Boitsov et al. 2007). Concentrations of total C_0- through C_9-alkylphenols, some of which are weak endocrine disruptors (Balaam et al. 2009), in produced water from several platforms in the Tampen, Ekofisk, and Sleipner Regions of the North Sea range from

264 to 55,000 μg/L (Durell et al. 2004; Boitsov et al. 2007). Between 7 and 36% of total phenols is phenol. Total C_8- and C_9-phenols (the strongest phenol endocrine disruptor: Routledge and Sumpter 1997) usually represent less than 0.1% of total phenols.

Concentrations of phenol in the produced water samples analyzed in this study ranged from 84 to 670 μg/L compared to 19 to 2,2300 μg/L in the 11 North Sea produced water samples analyzed by Boitsov et al. (2007). Concentrations of the more toxic C_2- and C_9-alkyl phenols were not measured in the Gulf of Mexico produced water samples, but they probably are much lower than concentrations in the North Sea samples analyzed by Boitsov et al. (2007).

Phenol dilutes rapidly in the produced water plume through evaporation, photolysis, biodegradation, and dilution/dispersion. Under summer conditions, microbial and photolysis transformation rate constants for phenol in seawater are 0.03/h (half-life = 28 h) and 0.016/h (half-life 43 h) (Hwang et al. 1986). Phenol in the produced water from a production facility in the Statfjord Field was diluted in the receiving waters by 10,000-fold within 10 m and by 20,000-fold at a distance of 500 m from the discharge (Riksheim and Johnson 1994). The concentration of phenol in surface waters near the platform ranged from 0.22 to 0.49 μg/L; higher than the phenol concentration in surface waters up-current from the two platform pairs monitored in this study (<0.002 to 0.25 μg/L).

4.1.4 Di(2-ethylhexyl)phthalate (DEHP)

DEHP is a synthetic organic chemical that is used primarily as a physical plasticizer in several types of plastics. It softens the plastic resins without reacting with them. DEHP, added to plastic resins in amounts of 20–67% by weight, leaches slowly from the plastics during use and evaporates directly into the air from which it spreads throughout the environment (Giam et al. 1980). Large amounts of DEHP also may be released to the environment during incomplete incineration of plastics or by leaching directly into water (IPCS 1992). Because of the large amounts manufactured and its facile release to the environment, DEHP has become a ubiquitous trace contaminant in the environment.

Because phthalates are so widely distributed in the environment and because they have a high affinity for adsorption to surfaces (Sullivan et al. 1982; Mackintosh et al. 2006), they are very difficult to analyze at trace levels in environmental samples, including petroleum products, without laboratory contamination (Giam et al. 1975; Lopezavila et al. 1990; Sullivan and Carty 1994; Rogers et al. 2001). Therefore, in reviewing the data on the concentrations of DEHP in environmental samples, it is important to recognize that much of the data may provide anomalously high measures of actual DEHP concentrations in the natural environment. Although elaborate precautions were taken in the present investigation to minimize laboratory contamination, DEHP was detected at variable concentrations in most laboratory and field blanks. Actual concentrations in most samples probably are overestimates.

DEHP is not a known ingredient or an intentional additive in oil production and produced water treatment. Mean concentrations of DEHP in produced water from the two discharging platforms were in the range of 0.1–0.4 μg/L, comparable to concentrations in ambient seawater and blanks. Thus, DEHP is not being discharged to the Gulf of Mexico in produced water.

4.2 MAH and SVOC Bioconcentration from Produced Water in Marine Animals

The USPA (1991) defines bioconcentratable nonpolar organic compounds in aquatic environments as those with a log octanol/water partition coefficient (log K_{ow}) greater than 3.5. BTEX, C_0- through C_4-phenols, and naphthalene have log K_{ow} values lower than 3.5 (Eastcott et al. 1988; Neff 2002) and, so, are not expected to bioaccumulate in marine food webs. Although bioconcentrated rapidly from water, MAH, lighter phenols, and naphthalene also are lost very rapidly from animal tissues when concentrations in the ambient seawater decline (Struhsaker 1977; Neff 2002; Sundt et al. 2009). Therefore, one would not expect to encounter high concentrations of these organic compounds in tissues of marine animals collected from well-mixed marine environments, even those near point sources of hydrocarbons.

Log K_{ow} values for C_3- and C_4-alkylbenzenes, PAH from methylnaphthalene to indeno(123-cd)pyrene, and DEHP range from 3.58 to 7.5 (Eastcott et al. 1988; Neff 2002) and, so, are expected to be bioconcentratable. However, fish and most invertebrates have a mixed function oxygenase enzyme system (CYP1A) in several tissues that rapidly metabolizes and excretes accumulated MAH, PAH, and less methylated phenols (Stegeman and Hahn 1994). Bivalve molluscs excrete PAH slowly and, so, bioaccumulate PAH to concentrations higher than fish do (Meador et al. 1995). DEHP also can be metabolized by most marine animals (OSPAR 2006). There are no published studies of bioconcentration and metabolism of C_3- and C_4-alkylbenzenes by marine animals. DEHP does not bioaccumulate to high concentrations in marine food webs; its high log K_{ow} and low solubility ensure that concentrations in solution (the most bioavailable form) are very low (Mackintosh et al. 2004).

4.2.1 MAH

There are very few published values for the concentrations of BTEX, and none for C_3- and C_4-alkylbenzenes, in the tissues of marine organisms (Whelan et al. 1982; Gossett et al. 1983), and none involving species from the Gulf of Mexico. BTEX concentrations in benthic and demersal marine invertebrates and fishes from the vicinity of the outfall for the Palos Verdes sewage treatment plant in the Southern California Bight ranged from <5 to 260 ng/g (Gossett et al. 1983).

We detected individual MAH in bivalve mollusc and fish tissues from discharging and reference platforms at concentrations of 7–68 ng/g in the Definitive Component and 3–37 ng/g in the Platform Survey Component. Despite the relatively high MAH concentrations in produced water, there was no difference in MAH concentrations in marine animals from discharging and reference platforms. Thus, marine animals near offshore produced water discharges do not bioconcentrate MAH from the produced water.

4.2.2 Phenol

Phenol is a soluble, ionizable aromatic compound that is difficult to extract quantitatively from tissues. Therefore, MDLs are high, ranging from 37 ng/g dry wt to 270 ng/g in the Definitive and Platform Survey Components of this study. Concentrations in mollusc, crab, and fish tissues from discharging and non-discharging platforms ranged from less than the MDL to 2000 ng/g; the different taxa bioaccumulated similar amounts of phenol. Phenol and alkylphenols are natural products that are abundant in many marine plants and are released to the ambient seawater when the organisms die. Atlantic cod rapidly bioconcentrated C_4- through C_7-alkylphenols during an 8-day exposure to dissolved phenols in the water, but metabolized and excreted them rapidly. Little alkylphenol was bioaccumulated from the food (Sundt et al. 2009). Much of the phenol residues in marine animal tissues probably comes from natural biogenic sources, not from produced water.

4.2.3 DEHP

As discussed above, DEHP is a ubiquitous contaminant in the environment. The MDL for DEHP in tissues was 140 ng/g dry weight (PQL = 700 ng/g) for both components of this program. DEHP was found in several procedural blanks at concentrations of 300–670 ng/g. Therefore, tissue DEHP concentrations below the PQL probably are overestimates. DEHP concentrations were below the MDL in 25–100% of the invertebrate and fish tissues from the two survey components.

In this study, DEHP concentrations above the MDL ranged from 160 to 3,600 ng/g dry wt in invertebrate tissues and from 140 to 2,800 ng/g in fish muscle. There was no difference between discharging and reference platforms in tissue concentrations of DEHP. Thus, this survey found little evidence that marine invertebrates and fish are bioconcentrating DEHP from produced water.

4.2.4 PAH

PAH are considered the most toxic naturally-occurring organic compounds in produced water (Neff 2002). In this investigation, TPAH concentrations above the MDL (1.3–16 ng/g for different PAH) ranged from 6 to 805 ng/g dry wt in two species of bivalve molluscs and from 3.0 to 117 ng/g in muscle tissue of five species of platform-associated fish collected at two discharging and two reference platforms in the Definitive Component.

By comparison, concentrations of TPAH in tissues of bivalve molluscs and fish collected at four produced water-discharging platforms in 6–122 m of water off the Louisiana coast were in the range of 187–12,000 ng/g and <0.1–1,900 ng/g, respectively (Neff and Sauer 1996a). The highest concentration in fish muscle was in a single replicate sample of Atlantic croaker (a demersal feeder) collected at the platform in 6 m of water; the platform was discharging an average of 2,353 m^3 of produced water containing an average of 158 μg/L TPAH (Neff and Sauer 1996b), both in the range for the two discharging platforms monitored in this study. The PAH profile in the fish tissues was dominated by alkyl naphthalenes, indicating that the fish may have bioaccumulated the PAH from produced water or a discharge of light fuel oil. The PAH profile in the bivalve tissues resembled that of crude oil and may have come from the produced sand that also was discharged from the platforms (produced sand is no longer permitted for discharge to the Gulf of Mexico).

Most bivalves sampled in the present study contained TPAH at concentrations in the same range as the medians for mussels (*Mytilus edulis* and *M. californianus*) and oysters (*Crassostrea virginica*) sampled in US coastal waters in the National Status and Trends Mussel Watch program. The annual median concentrations of TPAH in mussels and oysters sampled in the Mussel Watch Program from 1988 through 2003 ranged from 140 to 500 ng/g dry wt (O'Connor and Lauenstein 2006).

By comparison, mussels (*Mytilus edulis*), deployed for a month in cages within 2 km down-current of high-volume produced water discharges at the Statfjord and Ekofisk production facilities in the North Sea, contained 3,450–9,290 ng/g TPAH (Durell et al. 2006; Neff et al. 2006), up to an order of magnitude higher than observed for thorny oysters and jewel boxes in this study. The greater bioconcentration of PAHs by the mussels can be attributed to the exposure regime (artificially confining the mussels in the mean down-current direction from the discharge), the large volumes of the produced water discharges (74,100 m^3/d from three discharges in the Statfjord complex and 5,420 m^3/day from four discharges in the Ekofisk complex) and higher concentrations of TPAH in the produced water (Statfjord, 990–1, 480 μg L^{-1}; Ekofish, 520–1, 170 μg/L). The two Gulf of Mexico platforms monitored in this investigation discharged 1,130–1,750 m^3d^{-1} of produced water containing 24–136 μg L^{-1} PAHs.

The NPDES permit required monitoring the bioconcentration of two PAH, fluorene and benzo(a)pyrene from platforms in different regions of the western Gulf of Mexico in fish, bivalve molluscs, and crustaceans. Fluorene was not detected in muscle of any fish or bivalve samples collected in the Definitive component; it was detected in 5–33% of the fish samples from the four production areas and in 70 and 50% of the blue crab and American oyster samples collected from one of the shallow-water platform pairs in the Platform Component. Fluorene concentrations above the MDL in fish muscle ranged from 0.99 to 52 ng/g and in American oysters ranged from 7 to 140 n/g.

Benzo(a)pyrene was detected at a concentration above the MDL in 3% of the fish and oyster samples collected in the Platform Survey Component. It was not detected in any blue crabs collected in the Platform Survey Component or in any of the fish collected in the Definitive Component. The highest concentrations were in the same American oysters (19 ng/g) and in the same gray triggerfish (12 ng/g) that contained the highest concentrations of fluorene.

The concentration range of total and individual PAH in muscle of the platform-associated fish sampled in the two components of this study is in the lower part of the range reported for fish from other areas of the world (Hellou et al. 2002; Neff 2002; Neff et al. 2009). For example, TPAH concentrations in muscle of eight species of fish collected in the oil development area of the US Beaufort Sea between 2004 and 2006 ranged from 1.5 to 92 ng/g (Neff et al. 2009). Reported concentrations in muscle tissue of marine fish from throughout the world range from <2 to as much as 23,000 ng/g TPAH, <0.1 to 100 ng/g fluorene, and <0.04 to 250 ng/g benzo(a)pyrene (Neff 2002).

This study has shown that some bivalve molluscs, but not fish, from produced water discharging platforms are bioaccumulating alkyl PAH, probably derived from the produced water. PAH metabolism in fish probably is rapid enough that it would be unlikely that elevated concentrations of PAH would be detected in fish muscle tissue. Fluorene and benzo(a)pyrene, because of their low concentrations in produced water, are not suitable for monitoring PAH bioaccumulation from produced water.

5 Summary

Concentrations of MAH, PAH, and phenol are orders of magnitude higher in produced water than in ambient seawater. DEHP is not a normal constituent of produced water and its concentration is similar in produced and ambient water, probably reflecting laboratory contamination.

There is no evidence that MAH or phenol are being bioconcentrated from produced water discharges to four production areas of the western Gulf of Mexico. Concentrations of several petrogenic PAH, including alkyl naphthalenes and alkyl dibenzothiophenes, were slightly higher in some bivalve mollusc samples from discharging than from non-discharging platforms. These PAH could have been derived from produced water discharges or from tar balls or small fuel oil spills. Concentrations of individual and total PAH in mollusc and fish tissues in this study are well below concentrations that might be harmful to the marine animals themselves or to humans who might collect them for food at offshore platforms.

None of the seven organic chemicals recommended by EPA for monitoring bioconcentration from produced water is suitable for this purpose because of their physical behavior in the marine environment and concentrations in produced water. Better candidates for this purpose are alkyl naphthalenes and phenanthrenes.

References

Balaam JL, Chan-Man Y, Roberts PH, Thomas KV (2009) Identification of nonregulated pollutants in North Sea produced water discharges. Environ Toxicol Chem 28:1159–1167

Boitsov S, Mjøs SA, Meier S (2007) Identification of estrogen-like alkylphenols in produced water from offshore oil installations. Mar Environ Res 64:651–665

Brandsma MG, Smith JP (1996) Dispersion modeling perspectives on the environmental fate of produced water discharges. In: Reed M, Johnson S (eds) Produced water 2: environmental issues and mitigation technologies, Plenum Press, New York, NY, pp 215–224

Clark CE, Veil JA (2009) Produced water volumes and management practices in the United States. Report ANL/EVS/R-09/1. Argonne National Laboratory, Chicago, IL, 64pp

Conover WJ (1980) Practical nonparametric statistics, 2nd edn. Wiley, New York, NY

CSA International, Inc. (1997) Gulf of Mexico produced water bioaccumulation study. Definitive component. Technical Report. Report to the Offshore Operators Committee, New Orleans, LA from Continental Shelf Associates, Jupiter, FL

Dórea HS, Bispo JRL, Aragão KAS, Cunha BB, Navickiene S, Alves JPH, Romão LPC, Garcia CAB (2007) Analysis of BTEX, PAHs and metals in the oilfield produced water in the State of Sergipe, Brazil Microchem J 85:234–238

Douglas GS, McCarthy KJ, Dahlen DT, Seavy JA, Prince RC, Elmendorf DL (1992) The use of hydrocarbon analyses for environmental assessment and remediation. J Soil Contam 1:197–216

Durell G, Johnsen S, Røe-Utvik T, Frost T, Neff J (2004) Produced water impact monitoring in the Norwegian Sector of the North Sea: Overview of water column surveys in the three major regions. SPE 86800. In: Paper presented at the seventh SPE international conference on health, safety, and environment in oil and gas exploration and production, Calgary, Alberta, Canada, 15pp

Durell G, Johnsen S, Røe Utvik T, Frost T, Neff J (2006) Oil well produced water discharges to the North Sea. Part I: comparison of deployed mussels (*Mytilus edulis*), semi-permeable membrane devices, and the DREAM Model predictions to estimate the dispersion of polycyclic aromatic hydrocarbons. Mar Environ Res 62:194–223

Easley DM, Kleopfer RD, Carasea AM (1981) Gas chromatographic-mass spectrometric determination of volatile organic compounds in fish. J Assoc Off Anal Chem 64:653–656

Eastcott L, Shiu WY, Mackay D (1988) Environmentally relevant physical-chemical properties of hydrocarbons: a review of data and development of simple correlations. Oil Chem Pollut 4:191–216

Faksness LG, Grini PG, Daling PS (2004) Partitioning of semi-soluble organic compounds between the water phase and oil droplets in produced water. Mar Pollut Bull 48:731–742

Federal Register (1984) Appendix B to Part 136—Definition and procedure for the determination of the method detection limit—Revision 1.11. Federal Register, Vol. 49 (209), October 26, 1984, 198–199

Giam CS, Atlas E, Chan HS, Neff GS (1980) Phthalate esters, PCB and DDT residues in the Gulf of Mexico atmosphere. Atmos Environ 14:65–69

Giam CS, Chan HS, Neff GS (1975) Sensitive method for determination of phthalate ester plasticizers in open-ocean biota samples. Anal Chem 47:2225–2229

Gossett RW, Brown DA, Young DR (1983) Predicting the bioaccumulation of organic compounds in marine organisms using octanol/water partition coefficients. Mar Pollut Bull 14:387–392

Hellou J, Leonard J, Anstey C (2002) Dietary exposure of finfish to aromatic contaminants and tissue distribution. Arch Environ Contam Toxicol 42:470–476

Hiatt MH (1981) Analysis of fish and sediment for volatile priority pollutants. Anal Chem 53:1, 541–543

Hudgins CM Jr (1989) Chemical treatments and usage in offshore oil and gas production systems. Report to the American Petroleum Institute, Washington, DC, 52pp

Hudgins CM Jr (1991) Chemical usage in North Sea oil and gas production and exploration operations. Report to Oljeindustriens Landsforening (OLF). Petrotech Consultants, Inc., Houston, TX

Hudgins CM (1992) Chemical treatment and usage in offshore oil and gas production systems. J Petrol Technol May 1992:604–611

Hwang H-M, Hodson RE, Lee RF (1986) Degradation of phenol and chlorophenols by sunlight and microbes in estuarine water. Environ Sci Technol 20:1002–1007

IPCS (International Programme on Chemical Safety) (1992) Environmental health criteria 131. Diethylhexyl Phthalate. World Health Organization, Geneva, 141pp

Lopezavila V, Milanes J, Constantine F, Beckert WF (1990) Typical phthalate ester contamination incurred using EPA Method 8060. J Assoc Offic Anal Chem 73:709–720

Johnsen S, Røe Utvik TI, Garland E, de Vals B, Campbell J (2004) Environmental fate and effects of contaminants in produced water. SPE 86708. Paper presented at the seventh SPE international conference on health, safety, and environment in oil and gas exploration and production, Calgary, Alberta, Canada, 29–31 March, 2004. Society of Petroleum Engineers, Richardson, TX

Lopezavila V, Milanes J, Constantine F, Beckert WF (1990) Typical phthalate ester contamination incurred using EPA Method 8060. J Assoc Offic Anal Chem 73:709–720

Mackintosh CE, Maldonato JA, Honfwu J, Hoover N, Chong A, Ikonomou MG, Gobas FAPC (2004.) Distribution of phthalate esters in a marine aquatic food web: comparison to polychlorinated biphenyls. Environ Sci Technol 38:2011–2020

Mackintosh CE, Maldonado JA, Ikonomou MG, Gobas FAPC (2006) Sorption of phthalate esters and PCBs in a marine ecosystem. Environ Sci Technol 40:3481–3488

Marine Research Specialists (2005) The effect of produced-water discharges on federally managed fish species along the california outer continental shelf. Report to the Western States Petroleum Association, Santa Barbara, CA, 221pp

Meador JP, Stein JE, Reichert WL, Varanasi U (1995) Bioaccumulation of polycyclic aromatic hydrocarbons by marine organisms. Rev Environ Contam Toxicol 143:79–165

National Research Council (2003) Oil in the Sea III. Inputs, Fates, and Effects. The National Academies Press, Washington, DC, 265pp

Neff JM (2002) Bioaccumulation in marine organisms. Effects of contaminants from oil well produced water. Elsevier, Amsterdam, 452pp

Neff JM, Johnsen S, Frost T, Røe Utvik T, Durell G (2006) Oil well produced water discharges to the North Sea. Part II: comparison of deployed mussels (*Mytilus edulis*) and the DREAM Model to predict ecological risk. Mar Environ Res 62:224–246

Neff JM, Sauer TC Jr (1996a) An ecological risk assessment for polycyclic aromatic hydrocarbons in produced water discharges to the western Gulf of Mexico. In: Reed M, Johnson S (eds) Produced water 2: environmental issues and mitigation technologies. Plenum Press, New York, NY, pp 355–366

Neff JM, Sauer TC Jr (1996b) Aromatic hydrocarbons in produced water: bioaccumulation and trophic transfer in marine food webs. In: Reed M, Johnson S (eds) Produced water 2: environmental issues and mitigation technologies. Plenum Press, New York, NY, pp 163–176

Neff JM, Sauer TC, Hart A (2001) Monitoring polycyclic aromatic hydrocarbon (PAH) bioavailability near offshore oil wells. In: Greenberg BM, Hull RN, Roberts MH Jr, Gensemer RW (eds) Environmental toxicology and risk assessment: science, policy and standardization – implications for environmental decisions (tenth volume), ASTM STP 1403. American Society for Testing and Materials, Philadelphia, PA, pp 160–180

Neff JM, Trefry JH, Durell G (2009) Task 5. Integrated biomonitoring and bioaccumulation of contaminants in Biota of the cANIMIDA study area. Final Report. US Dept. of the Interior, Minerals Management Service, Alakska OCS Office, Anchorage, AK, 182pp

O'Connor TP, Lauenstein GG (2006) Trends in chemical concentrations in mussels and oysters collected along the US coast: update to 2003. Mar Environ Res 62:261–285

OGP (International Association of Oil & Gas Producers) (2004) Environmental performance in the E&P industry, 2003 data. Report No. 359, December, 2004. OGP, London, UK

OGP (International Association of Oil & Gas Producers) (2005) Fate and effects of naturally occurring substances in produced water on the marine environment. Report No. 364, February 2005. OGP, London, UK

OSPAR (The Convention for the protection of the marine environment of the North-East Atlantic) (2006) Phthalates. Hazardous substances series. Publication No. 70/2006. OSPAR Commission, London, UK

OSPAR (The Convention for the protection of the marine environment of the North-East Atlantic) (2009) Discharges, spills and emissions from offshore oil and gas installations in 2007 Including Assessment of Data Reported in 2006 and 2007. OSPAR Commission, London, UK, 58pp

Page DS, Boehm PD, Douglas GS, Bence AE (1995) Identification of hydrocarbon sources in benthic sediments of Prince William Sound and the Gulf of Alaska following the *Exxon Valdez* oil spill. In: Wells PG, Butler JN, Hughes JS (eds) Exxon valdez oil spill: fate and effects in alaskan waters. STP 1219. American Society for Testing and Materials, Philadelphia, PA, pp 41–83

Riksheim H, Johnsen S (1994) Determination of produced water constituents in the vicinity of offshore production fields in the North Sea. SPE Paper 27150. In: Proceedings of the annual meeting of the society of petroleum engineers, Djakarta, Indonesia. Society of Petroleum Engineers, Richardson, TX

Rogers RP, Hendrickson CL, Emmett MR, Marshall AG, Greaney M, Qian K (2001) Molecular characterization of petroporphyrins in crude oil by electrospray ionization Fourier transform ion cyclotron resonance mass spectrometry. Can J Chem 79:546–551

Routledge EJ, Sumpter JP (1997) Structural features of alkylphenolic chemicals associated with estrogenic activity. J Biol Chem 272:3280–3288

Sauer TC Jr (1980) Volatile liquid hydrocarbons in waters of the Gulf of Mexico and Caribbean Sea. Limnol Oceanog 25:338–351

Sauer TC Jr, Sackett WM, Jeffrey LM (1978) Volatile liquid hydrocarbons in the surface coastal waters of the Gulf of Mexico. Mar Chem 7:1–16

Stegeman JJ, Hahn ME (1994) Biochemistry and molecular biology of monooxygenases: current perspectives on forms, functions and regulation of cytochrome P-450 in aquatic species. In: Malins DC, Ostrander GK (eds) Aquatic toxicology: molecular, biochemical and cellular perspectives. Lewis Publishers, Boca Raton, FL, pp 87–206

Struhsaker JW (1977) Effects of benzene (a toxic component of petroleum) on spawning Pacific herring, *Clupea harengus pallasi*. Fish Bull 75:43–49

Sullivan B, Carty D (1994) Is it real contamination or is it just plastic? Soils Jan–Feb. 1994:6–8

Sullivan KF, Atlas EL, Giam CS (1982) Adsorption of phthalic acid esters from seawater. Environ Sci Technol 16:428–432

Sundt RC, Baussant T, Beyer J (2009) Uptake and tissue distribution of C_4-C_7 alkylphenols in Atlantic cod (*Gadus morhua*): relevance for biomonitoring of produced water discharges from oil production. Mar Pollut Bull 58:72–79

Taylor P, Bennett B, Jones M, Larter S (2001) The effect of biodegradation and water washing on the occurrence of alkylphenols in crude oils. Org Geochem 32:341–358

Terrens GW, Tait RD (1996) Monitoring ocean concentrations of aromatic hydrocarbons from produced formation water discharges to Bass Strait, Australia. SPE 36033. In: Proceedings of the international conference on health, safety & environment. Society of Petroleum Engineers, Richardson, TX, pp 739–747

Thomas KV, Balaam JL, Hurst MR, Thain JE (2004) Identification of in vitro estrogen and androgen receptor agonists in North Sea offshore produced water discharges. Environ Toxicol Chem 23:1156–1163

USEPA (U.S. Environmental Protection Agency) (1989) Statistical analysis of ground-water monitoring data at RCRA facilities. Interim Final Guidance. EPA/530/SW-89/026. EPA, Office of Solid Waste, Washington, DC

USEPA (U.S. Environmental Protection Agency) (1991) Assessment and control of bioconcentratable contaminants in surface waters. Draft. EPA, Office of Water, Washington, DC. 290pp

USEPA (U.S. Environmental Protection Agency) (1993) Guidance for assessing chemical contaminant data for use in fish advisories. Volume 1: fish sampling and analysis. Office of Science and Technology, Office of Water, Washington DC. EPA 823-R-93. 254pp + appendices

Veil JA, Puder MG, Elcock D, Redweik RJ Jr (2004) A white paper describing produced water from production of crude oil, natural gas, and coal bed methane. Report to the U.S. Dept. of Energy, National Energy Technology Laboratory. Argonne National Laboratory, Washington, DC, 79pp

Whelan JK, Tarafa ME, Hunt JM (1982) Volatile C_1-C_8 organic compounds in macroalgae. Nature, Lond. 299:50–52

Part VI
Risk Assessment and Management

Chapter 25
Offshore Environmental Effects Monitoring in Norway – Regulations, Results and Developments

Torgeir Bakke, Ann Mari Vik Green, and Per Erik Iversen

Abstract The first oil on the Norwegian continental shelf was found at the Ekofisk field in the southern North Sea in 1969. Several new discoveries were made in the years after, and from 1973, the Norwegian Pollution Control Authority (SFT from 2010 Climate And Pollution Agency (Klif)) required that all the licensed companies should submit annual reports on the environmental conditions in their impact areas. This requirement was one of the conditions for discharge permits, and guidance was given on the minimum scope and content of the environmental surveys to be performed. The annual reports quickly demonstrated that there was a need for better harmonisation of the monitoring methods to be used. In 1987 SFT and the offshore operators jointly hosted a 2-day workshop to agree on a common strategy and methodology for offshore baseline and monitoring surveys. On the basis of the workshop outcome, an expert group appointed by SFT developed a guideline document for sediment monitoring that was subsequently discussed in an open forum with the offshore operators. The guidelines were put into force in 1988 and in the same year they were adopted by the Paris Commission for use in the convention waters (PARCOM 1989). In 1991 the guidelines (SFT 1990) were made mandatory for monitoring around Norwegian fields. In 1997 the guidelines were revised. A concept of regional monitoring was introduced, and guidelines for monitoring of the water column were included (SFT 1997). The latter was a response to the change in impact focus from discharge of drilling waste to produced water (PW). In 1993 strong restrictions ended regular discharges of oily drill cuttings. At the same time, national prognoses estimated an increase in PW discharges from around 25 million m^3/yr in 1993 to more than 250 million m^3/yr in 2009. Subsequent minor guideline revisions were made in 1999, (SFT 1999) and 2010 (Klif 2009).

T. Bakke (✉)
Norwegian Institute for Water Research (NIVA), NO-0349 Oslo, Norway

K. Lee, J. Neff (eds.), *Produced Water*, DOI 10.1007/978-1-4614-0046-2_25,

1 Purpose

The offshore environmental monitoring programme serves several proposes. It provides knowledge on the general environmental conditions at the offshore installations and how the conditions might change in time and space. Through combined analysis of environmental factors, contaminant exposure and biological conditions, it generates information on dose–response relationships between cuttings deposits and benthic organisms and between PW effluents and pelagic organisms. It also provides a feedback mechanism that allows the Authorities to assess the sufficiency of their regulations to help ensure that impact from regular offshore operations does not exceed what was predicted in environmental impact assessments (EIAs) made prior to the development of the offshore fields.

2 How Does Norway Regulate?

The HSE regulations are based on the "polluter pays" principle. These regulations require that the offshore operators perform environmental monitoring at their own cost and according to the monitoring requirements in the regulations (http://www.ptil.no/regulations/category87.html). SFT may also require additional environmental investigations and expects the operators to actively improve the monitoring strategies and procedures. The strongest example of the latter is the funding of a 2-year international field-going workshop (2001–2002) to develop appropriate water column monitoring procedures (Hylland et al. 2002).

3 Timing of Environmental Assessments

The regulations address the whole lifetime of an offshore field (Fig. 25.1). Prior to the opening of a new offshore region or area for exploration, the Authorities prepare an EIA for the region describing the expected petroleum developments and the natural environment that may be affected. The EIA identifies all potential sources of impact from exploration, development, production and decommissioning and evaluates impact patterns and significance. After the exploration phase of a field and prior to its development, the operator is requested to produce a similar, but site-specific EIA, addressing the development of that particular field. At this stage, the operator performs an environmental baseline survey of the sea floor sediments according to the regulations. A corresponding description of the water column baseline conditions is not required since the focus here is on spatial (i.e. distance from a PW source) rather than temporal trends. For fields in new areas not formely investigated or in particularly sensitive areas, the sediment baseline survey is to be performed before the exploration phase. During development, production period and decommissioning of a field, regular environmental monitoring surveys

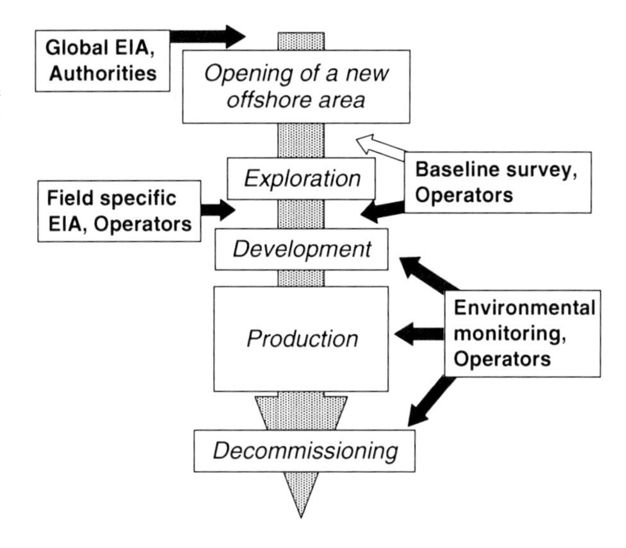

Fig. 25.1 Norwegian environmental assessment requirements and responsible entities at different phases of an offshore development; *white arrow*: requirement in sensitive areas

of the water column and sediments are to be performed according to the frequency and procedures specified in the regulations. SFT may then decide if there is a need to continue the monitoring after decommissioning.

4 Baseline and Monitoring Strategies

The monitoring strategy has taken a regional approach. The Norwegian continental shelf is divided into 11 geographical regions according to latitude (Fig. 25.2). Sediment monitoring surveys are performed in the six regions with ongoing petroleum activities. The water column monitoring has two independent elements: Environmental Condition Monitoring (ECM) and Environmental Effects Monitoring (EEM). The ECM focuses on health and contamination status of commercial fish stocks and the EEM covers effects measurements around specific PW effluents. The ECM is at present performed in 10 of the regions, including 2 reference regions with no offshore activity and the EEM is performed in 4 regions with ongoing production.

4.1 Sediment Monitoring

Sediment monitoring is performed every third year in each region, and the region is treated as an entity where predefined stations around each installation and several regional reference stations are sampled in one common field survey. The total number of stations per survey ranges from about 140 to more than 250 in a region. Replicate sampling from each station is done by use of 0.1 m^2 grabs. The basic principles for the sediment monitoring are

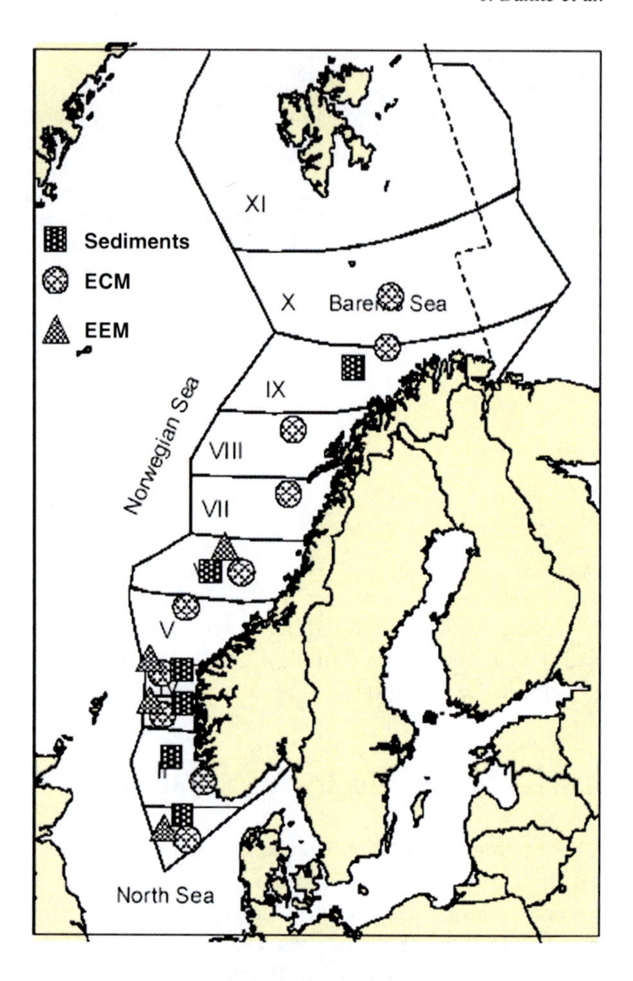

Fig. 25.2 Regions for offshore sediment and water column monitoring on the Norwegian shelf

- synoptic physical, chemical and macrofauna characterisation of the same sediment stations
- surveys performed during the same time window each year
- standard operating procedures, uniform across all fields and regions, in all phases of a survey from planning of sampling design to final report format
- emphasis on multivariate statistical analyses of environmental and biological data
- requirement of formalised quality assurance procedures and documentation by the consultants performing the work

All sediment results since 1990 have been gathered in a common database where they are subjected to quality control. This has ensured a standardisation that is as complete as possible for such a large-scale, multi-year sampling programme. The database is available to the general public.

4.2 Water Column Monitoring

The ECM started in 1994 and since 2000 has been performed every 3 years. Each year a subset of the 10 regions is monitored. The focus is on body burdens of selected petroleum derived contaminants as well as selected biological effects parameters (biomarkers) in commercial pelagic fish species, e.g. cod (*Gadus morhua*), saithe (*Pollachius virens*) and haddock (*Melanogrammus aeglefinus*), with an option for including other species as well. The surveys are done as an element of the regular fish stock assessment surveys performed by the Institute of Marine Research (IMR). The main purpose of the ECM is to provide a quality assessment of fish stocks that may have been exposed to PW or drilling effluents. The regulations do not specify the analyses to be done, but both the selection of regions and the measurement programme must be approved by SFT prior to each survey.

The EEM is done annually and in one of the regions, the first survey occurred in 1996. One field in the region is selected as the monitoring target, and the purpose is to assess biological effects of the PW effluents from that field. The EEM covers measurements of exposure and effects biomarkers in caged cod (*G. morhua*) and blue mussel (*Mytilus edulis*) deployed at set distances from the PW outlets (Fig. 25.3). Cage positioning is based on plume dispersion modelling (Johnsen et al. 2000; Durell et al. 2006). The high level principles for the EEM are laid down in the regulations. A survey protocol detailing the site to be monitored, the field

Fig. 25.3 Combined cod and blue mussel cage for the water column EEM prior to deployment at the Ekofisk field (Photo C. Harman, NIVA)

sampling and experimental procedures and the selection of biomarkers is developed for each year. The survey protocol is approved by SFT prior to a tender procedure.

5 Monitoring Results

5.1 Sediments

The sediment monitoring has created time series of sediment conditions around each of the Norwegian offshore installations and for many of them from before their development started. Some of these series cover a time span of more than 25 years (e.g. the Ekofisk and Statfjord fields). The results have revealed not only extensive sediment contamination and faunal effects out to 3–5 km from installations discharging oil-based drill cuttings (Olsgard and Gray 1995; Gray et al. 1999), but also a clear reduction in the impacted areas at most of these fields after such discharges ceased around 1993 (Bakke and Nilssen 2004; Renaud et al. 2008). An example is the development of faunal effects at Gyda (Fig. 25.4) showing an increase in total hydrocarbon (THC) contamination until 1993 and a subsequent rapid decrease within 3 years. The extent of faunal impact follows a similar pattern, but with a delayed recovery probably regulated by the recruitment rate of the fauna. Renaud et al. (2008) showed that the total area contaminated with hydrocarbons from drill cuttings and the area with impacted fauna decreased during 1996–2006 in three of the five regions with available time series data (monitoring in the Barents region has recently started). The 2006 surveys concluded that in one of the regions, no faunal impact could be seen at any station (UNIFOB 2007), implying that any remaining

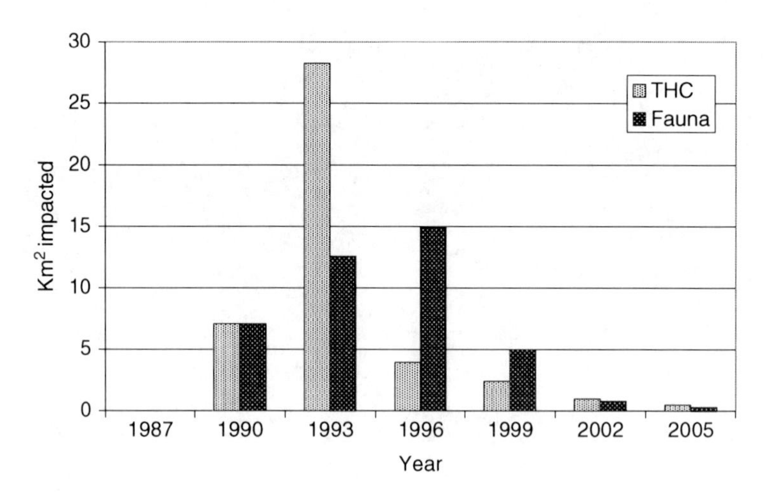

Fig. 25.4 Temporal change in estimated areas (km^2) of THC contamination and faunal effects around the Gyda field, southern North Sea, from 1987 (baseline survey) to 2005

effects are confined to within the 250 m periphery around the installations, which is the closest distance covered by the monitoring. The time series from the regional reference stations have also provided insight into the natural variation of the shelf benthic fauna over time. Renaud et al. (2008) showed that a systematic long-term shift in the community structure had occurred in all regions during 1996–2006. A typical feature is that the natural change in community structure from one year to another is so strong that it most often overrides the differences between stations within the same year.

5.2 Water Column Monitoring

None of the ECM surveys has detected any elevation in the tissue levels of compounds that can be linked to PW discharges. The EEM monitoring has generated data on a large range of parameters for genetic, biochemical, physiological and anatomical effects in the caged cod and mussel (e.g. Sundt et al. 2006), but only a few responses have been linked convincingly to PW exposure. One of the main problems has been to deploy cages in positions that ensure a gradient in exposure to the PW plume, and also to quantify the real exposure level that the cages endure for 1–2 months. During the surveys in 2003 and 2004, no effects were detected, but at the Ekofisk field in 2006 the levels of PAHs in mussel, PAH metabolites in cod bile, as well as two biomarkers (lysosomal membrane stability in mussel and CYP1A in cod) could be linked reliably to distance from the PW outlets (Sundt et al. 2006). The results suggested that effects were only detectable close to the outlets, which was in line with earlier risk analyses of these PW effluents (Neff J, unpublished).

6 Use of Results

The Authorities have used the results to obtain an overview of the environmental status and temporal changes in sediments, both regionally and in the vicinity of the petroleum installations. The results are also important to ensure sufficiency of the national regulations. In 1988, an assessment of the monitoring data from the Statfjord field suggested that much greater bottom areas than predicted by the oil companies showed biological effects of oil-based drill cuttings (Reiersen et al. 1988). Although the assessment was much debated (Gray et al. 1999), the response of the Authorities was to implement legislation regarding the amount of oil allowed in cuttings to be discharged, resulting in the effective ban on discharges of oil-based cuttings in 1993. A later comprehensive analysis of the available monitoring data by Olsgard and Gray (1995) showed that the legislation was justified. Later the monitoring results revealed that discharges of cuttings from drilling with synthetic mud, which replaced oil-based muds after 1993, also resulted in effects on the benthic fauna that were considered unacceptable. Discharges of synthetic cuttings are now prohibited from Norwegian installations. The ‘precautionary’ principle and also the comprehensive knowledge of the benthic impact from drilling discharges that has been generated from the North Sea and the Mid Norway shelf has elicited a political "zero discharge" policy for future petroleum exploration and production in Norwegian Arctic regions, implying no operational discharges of oil or environmentally harmful substances.

Application of the water column monitoring data to assess sufficiency of legislation has hitherto been far more difficult than for sediments. The ECM serves its function as regular control of the quality of fish harvest going to the market, but cannot support regulation of discharges from individual fields. The present EEM biomarker approach is sufficiently sensitive to function as an early warning signal for significant effects on pelagic populations, and the cages may be seen as "watch dogs" to ensure that there are no unexpected impacts from PW discharges. Still the history of water column monitoring is, as yet, too short to confirm the impression that the effects are only local.

An important use of the monitoring data that has not yet been fully utilised is testing of compliance with impact predictions given in the site-specific EIAs. As long as the monitoring tries to answer questions such as "Are there effects on the ecosystem?" value judgements will be needed which are open to constant argument (Gray et al. 1999). Thus it is essential that EIAs make quantitative predictions that can be assessed against agreed acceptance criteria. When the predictions are acceptable, the main purpose of the environmental monitoring will be to control whether the predictions hold or not.

7 Future Development of the Norwegian Offshore Environmental Monitoring

The present knowledge on the sediment ecosystem response to the offshore petroleum activities generated through nearly 25 years of harmonised monitoring is strong. The sediment monitoring reveals few surprises today. This alone calls for a revision of the existing requirements and probably a simplification of the surveys. During 2008 this was evaluated by SFT with the help of a scientific expert group, and several modifications to the monitoring guidelines are included in the current version (Klif 2009).

The change from drilling with oil and synthetic fluids to new generations of water-based muds calls for a change in sediment parameters to be measured. Oil-related parameters will become less important. Recent experiments have shown that water-based muds may also have significant effects on benthic fauna (Schaanning et al. 2008). Oxygen depletion is one important effect, but chemical toxicity or physical impact from cuttings particles cannot be ruled out (Cranford et al. 1999). Hence it is still too early to identify the optimal set of monitoring parameters that should be included to cover such impacts.

Since more fields are moving into a late life stage with less drilling activity, the frequency of the sediment surveys in one region could be more flexible than every

3 years. The frequency will probably be tailored to each field individually, based on such things as technical configuration, discharge pattern and local impact trends seen.

As the areas of impact are decreasing, the monitoring emphasis will be on near-zone sites, and there is a need for moving closer to the installations than the present 250 m. This will often be in conflict with the safety zone, but one will also have to abandon the present radial station design (c.f., Gray et al. 1999) and tailor the station layout to the configuration of the subsea installations. Approaching the platforms may also call for other sampling techniques than grabs, such as ROV-operated point samplers, photo and video recording, and automatic instrument vehicles deployed onto the sediment.

Increased petroleum activity in the Barents Sea region is also a challenge. The regulations apply to the whole Norwegian shelf, but the "zero discharge" requirement for Arctic developments[1] could justify simplification of the environmental monitoring, especially regarding chemical contamination. The benthic fauna monitoring should however be kept at the present level until one knows whether the impact pattern seen in the shelf regions further south is representative of the Arctic benthic ecosystems (Olsen et al. 2007).

The water column EEM is still in a developmental stage and is expected to change as the research on biomarkers advances. The core set of biomarkers used is considered sufficiently sensitive for the purpose, and others may become candidates if their procedural variability can be reduced. The caging approach is a worst-case exposure scenario, since pelagic organisms by nature are not forced to stay in fixed positions relative to a pollution source, but it seems to be the only viable way to cover exposure gradients. There is a need to improve methods to estimate the real exposure, and also to ensure that caged organisms are actually exposed to PW plumes, for example by increasing the number of cages. The latter is an economic issue, but could be overcome by putting more emphasis on caging only mussels and less on combined mussel and cod cages.

8 Conclusions

The offshore monitoring programme established for the oil and gas industry is by far the most comprehensive marine environmental monitoring activity in Norway today and may be the most comprehensive programme related to the offshore industry anywhere in the world. Procedural harmonisation, strict quality assurance requirements, and open access to the data have provided authorities, operators and the scientific community with reliable knowledge on the impact of the petroleum industry. Furthermore, these monitoring activities have generated a very good data set for research on the sediment ecology of the Norwegian shelf. Even after revision and

[1]A white paper from 2011 has withdrawn the specific requirements for Arctic developments and the same "zero discharge" policy is now applicable for the entire Norwegian continental shelf.

simplification of the requirements, it is certain that the monitoring will continue to add to the valuable long-term time series of data.

Acknowledgements This chapter is based on a large number of survey reports produced by numerous consulting companies on contract from the offshore operators. The ones referred directly to are cited, and the others are hereby thanked collectively for their contribution. This chapter has been prepared under contract 4008001 from the Norwegian Pollution Control Authority (SFT) to the Norwegian Institute for Water Research.

References

Bakke T, Nilssen I (2004) Harmonised monitoring of offshore drilling waste effects in Norway. In: Armsworthy SL, Cranford PJ, Lee K (eds) Proceedings of the offshore oil and gas environmental effects monitoring workshop: approaches and technologies. Battelle Press, Columbus, OH

Cranford PJ, Gordon Jr DC, Lee K, Armsworthy SL, Tremblay G-H (1999) Chronic toxicity and physical disturbance effects of water- and oil-based drilling fluids and some major constituents on adult sea scallops (*Placopecten magellanicus*). Mar Envir Res 48:225–256

Durell G, Utvik TR, Johnsen S, Frost T, Neff J (2006) Oil well produced water discharges to the North Sea. Part I: Comparison of deployed mussels (Mytilus edulis), semi-permeable membrane devices, and the DREAM model predictions to estimate the dispersion of polycyclic aromatic hydrocarbons. Mar Environ Res 62:194–223

Hylland K, Becker G, Klungsøyr J, Lanf T., McIntosh A, Serigstad B, Thain JE, Thomas KV, Utvik TIR, Vethaak D, Wosniok W (2002) An ICES workshop on biological effects in pelagic ecosystems (BECPELAG): summary of results and recommendations. ICES ASC 2002 CM 2002/X:13

Gray JS, Bakke T, Beck HJ, Nilssen I (1999) Managing the environmental effects of the Norwegian oil and gas industry: from conflict to consensus. Mar Pollut Bull 38:525–530

Johnsen S, Frost TK, Hjelsvold M, Utvik TR (2000) The environmental impact factor – a proposed tool for produced water impact reduction, management and regulation. SPE paper 61178. SPE Int Conf, Stavanger, Norway

Klif 2009. Guidelines for environmental monitoring. Climate and Pollution Agency Report TA-2586/2009 (in Norwegian)

Olsen GH, Carrol ML, Renaud PE, Ambrose WG Jr, Olssøn R Carrol J (2007) Benthic community response to petroleum-associated components in Arctic versus temperate marine sediments. Mar Biol 151:2167–2176

Olsgard F, Gray JS (1995) A comprehensive analysis of the effects of offshore oil and gas exploration and production on the benthic communities of the Norwegian continental shelf. Mar Ecol Prog Ser 122:277–306

PARCOM (1989) Guidelines for monitoring methods to be used in the vicinity of platforms in the North Sea. Chameleon Press, London

Reiersen LO, Gray JS, Palmork KH, Lange R (1988). Monitoring in the vicinity of oil and gas platforms: results from the Norwegian sector of the North Sea and recommended methods for forthcoming surveillance. In: Engelhardt FR, Ray JP, Gillam AH (eds) Drilling waste. Elsevier, London

Renaud PE, Jensen T, Wassbotten I, Mannvik HP, Botnen H (2008) Offshore sediment monitoring on the Norwegian shelf. A regional approach 1996–2006. Report no 3487-003. Akvaplan-niva AS, Tromsø

Schaanning MT, Trannum HC, Øxnevad S, Carroll J, Bakke T (2008) Effects of drill cuttings on biogeochemical fluxes and macrobenthos of marine sediments. J Exp Mar Biol Ecol, 361:49–57

SFT (1990) Manual for monitoring surveys around petroleum installations in Norwegian marine areas (in Norwegian). SFT 90:01 TA-699/1990. Norwegian Pollution Control Authority, Oslo

SFT (1997) Guidelines for Environmental Monitoring of the Petroleum Activity on the Norwegian Shelf (in Norwegian). SFT 97:01 TA-1424/1997. Norwegian Pollution Control Authority, Oslo

SFT (1999) Environmental monitoring of the petroleum activity on the Norwegian shelf (in Norwegian). SFT 99:01 TA-1641/1999. Appendix 1 to the Information Duty Regulations Requirements for Reporting from Offshore Petroleum Activities. Norwegian Pollution Control Authority, Oslo. http://www.sft.no/arbeidsomr/petroleum

Sundt RC, Ruus A, Grung M, Pampanin DM, Barsiene J, Skarphedinsdottir H (2006) Water column monitoring 2006. Summary report. Report AM 2006/013. Akvamiljø, Stavanger

UNIFOB (2007) Environmental monitoring of oil and gas fields in Region II, 2006. UNIFOB, Bergen (in Norwegian with English summary)

Chapter 26
Fuzzy-Stochastic Risk Assessment Approach for the Management of Produced Water Discharges

Zhi Chen, Lin Zhao, and Kenneth Lee

Abstract In recent years, the large volumes of produced water discharges from offshore petroleum production activities have been identified to be an issue of concern by regulators and environmental groups. In the majority of the existing risk assessment research of produced water discharges, local environmental guidelines have been used as the evaluation criteria. However, these guidelines are mostly impractical and cannot be implemented. In this chapter, a fuzzy-stochastic risk assessment approach has been developed to predict the risks and to examine the uncertainties associated with produced water discharge and the related regulated values for the heavy metal, lead, in the marine environment. Specifically, the concept of fuzzy membership is established to reflect the suitability of local standards for produced water risk assessment. The Monte Carlo method has been used to address system uncertainties and provides stochastic simulation of pollutant dispersion associated with produced water discharges. The developed risk assessment approach is applied to an offshore petroleum facility located on the Grand Banks of Canada. As an extension of the previous risk assessment studies, the proposed approach will contribute to the development of effective decision tools for the assessment and management of produced water discharges in the marine environment.

1 Introduction

Offshore oil and gas production is often accompanied by the production of large quantities of wastewater. This "produced water" may contain a number of toxic contaminants, including petroleum hydrocarbons, heavy metals, radionuclides, and treatment process chemicals (Hodgins 1993), which, when discharged into the ocean, may have adverse effects on the marine ecosystem.

In recent years, environmental impacts associated with produced water discharges have been of great concern to regulators, the public, environmental interest groups, and the oil industry. To aid in assessing the environmental risks of produced water discharges, Thatcher et al. (1999, 2001) introduced the Chemical

Z. Chen (✉)
Department of Building Civil and Environmental Engineering, Concordia University, Montreal, PQ, Canada H3G 1M8

K. Lee, J. Neff (eds.), *Produced Water*, DOI 10.1007/978-1-4614-0046-2_26,

Hazard Assessment and Risk Management (CHARM) model. This uses a set of rules to calculate the ratio of predicted contaminant concentrations to local environmental standards. Reed et al. (2001) developed the Dose-related Risk and Effect Assessment Model (DREAM) by coupling the contaminant dispersion model PROVANN (Reed et al. 1996) and CHARM. Using CHARM in concert with a Monte Carlo simulation method, Mukhtasor et al. (2004) evaluated toxicity risks arising from produced water generated by an offshore platform. Similarly, Meinhold et al. (1996a) used Monte Carlo simulations to assess the human health risks of radium and lead in produced water.

Most risk assessment studies have used local environmental guidelines or standards as evaluation criteria, without addressing their practicability or the uncertainties inherent to them. Current environmental surveys and regulations suggest the need for a quantitative risk-based approach in developing environmental protection policies. The evaluation of existing standards, and of the scale of potential deleterious impacts on ecosystems and human health, requires a consultative risk assessment approach. Indeed, risk assessment, the process of assigning magnitudes and probabilities to the adverse effects of anthropogenic activities or natural catastrophes (Suter 1993), should be an integral part of the decision-making and planning processes involved in framing a given region's environmental management plan (Meinhold et al. 1996b).

However, for uncertain parameters that cannot be expressed as probability distributions, stochastic risk assessment methods are inapplicable, so other risk assessment approaches must be sought. Fuzzy set theory, widely used to handle uncertainties associated with discrete or imprecise characteristics (Bardossy et al. 1991), can produce results of moderate acceptability (suitability) (Klir and Yuan 1995; Zimmermann 2001). For example, a fuzzy membership function can be established to quantify uncertainties associated with evaluation criteria.

The objective of this study was to use a fuzzy-stochastic risk assessment approach to predict the risks associated with produced water discharges and to examine uncertainties in discharge parameters and related environmental standards. A numerical dispersion approach based on an integration of ocean hydrodynamic and pollutant dispersion models was used to provide simulations of the dispersion of lead within the produced water effluent stream. The Monte Carlo method was used to provide a stochastic simulation of pollutant dispersion associated with produced water discharges. The developed risk assessment approach was validated with data from an offshore petroleum facility located on the Grand Banks of Newfoundland, Canada.

2 The Integrated Modelling and Risk Assessment System

2.1 Numerical Contaminant Transport Model

A numerical dispersion approach, based on the integration of ocean hydrodynamic and pollutant dispersion models, was used to provide simulations of the dispersion of ocean contaminants. The Random Walk Model was used for the pollutant dispersion simulation, while the Princeton Ocean Model (POM) was embedded in the dispersion model and provided the three component velocities of prevailing ocean currents.

2.1.1 Ocean Circulation Model – POM

As a three-dimensional ocean circulation model with a vertical sigma coordinate and curvilinear horizontal grids, POM can address large-scale and long-term phenomena according to basin size and grid resolution. The POM's continuity and momentum equations are as follows (Mellor 2004):

$$\frac{\partial DU}{\partial x} + \frac{\partial DV}{\partial y} + \frac{\partial \omega}{\partial \sigma} + \frac{\partial \eta}{\partial t} = 0 \tag{26.1}$$

$$\begin{aligned}&\frac{\partial UD}{\partial t} + \frac{\partial U^2 D}{\partial x} + \frac{\partial UVD}{\partial y} + \frac{\partial U\omega}{\partial \sigma} - fVD + gD\frac{\partial \eta}{\partial x}\\ &\quad + \frac{gD^2}{\rho_0}\int_{\sigma}^{0}\left[\frac{\partial \rho'}{\partial x} - \frac{\sigma'}{D}\frac{\partial D \partial \rho'}{\partial x \partial \sigma'}\right]d\sigma' = \frac{\partial}{\partial \sigma}\left[\frac{K_{\mathrm{m}}}{D}\frac{\partial U}{\partial \sigma}\right] + F_x\end{aligned} \tag{26.2}$$

$$\begin{aligned}&\frac{\partial VD}{\partial t} + \frac{\partial UVD}{\partial x} + \frac{\partial V^2 D}{\partial y} + \frac{\partial V\omega}{\partial \sigma} + fUD + gD\frac{\partial \eta}{\partial y}\\ &\quad + \frac{gD^2}{\rho_0}\int_{\sigma}^{0}\left[\frac{\partial \rho'}{\partial y} - \frac{\sigma'}{D}\frac{\partial D \partial \rho'}{\partial y \partial \sigma'}\right]d\sigma' = \frac{\partial}{\partial \sigma}\left[\frac{K_{\mathrm{m}}}{D}\frac{\partial V}{\partial \sigma}\right] + F_y\end{aligned} \tag{26.3}$$

where f is the Coriolis parameter (s^{-1}); g is the gravitational acceleration ($m\ s^{-2}$); t is time (s); x, y are horizontal Cartesian coordinates (m); $D \equiv H + \eta$ is the total elevation of the surface water (m); H is the bottom topography (m); η is the surface elevation (m); F_x, F_y are the horizontal diffusion terms ($m^2\ s^{-2}$); K_{m} is the vertical kinematic viscosity ($m^2\ s^{-1}$); U and V are the horizontal velocities ($m\ s^{-1}$); ρ' is the density after subtraction of the horizontally averaged density, a strategy that redness reduces error (Mellor et al. 1994, 1998); σ is the sigma vertical coordinate (m); and ω is the velocity component normal to sigma surfaces ($m\ s^{-1}$).

2.1.2 Dispersion Model – The Random Walk Model

The Random Walk Model, based on a particle tracking method, served as the dispersion model for this study. Its governing equations are (Riddle et al. 2001):

$$\begin{aligned}\Delta x &= U\Delta t + \sqrt{2K_{\mathrm{h}}\Delta t}R_x\\ \Delta y &= V\Delta t + \sqrt{2K_{\mathrm{h}}\Delta t}R_y\\ \Delta z &= W\Delta t + \sqrt{2K_z\Delta t}R_z\end{aligned} \tag{26.4}$$

where Δx, Δy, and Δz represent the three-dimensional movements of a particle occurring within a model timestep (m); Δt is the timestep duration (s); drawn from the POM model embedded in the contaminant transport model, U, V, and W are the horizontal and vertical velocity components at time t (m s^{-1}), and these represent hydrodynamic ocean circulation; R_x, R_y, and R_z are random numbers from a standard normal distribution; K_h and K_z are the horizontal and vertical mixing coefficients (m^2 s^{-1}), uncertain parameters which can be quantified based on Eq. (26.5). Each particle represents a fixed mass of effluent. It is assumed that no contaminant degradation occurs.

2.2 Integrated Fuzzy-Stochastic Risk Assessment Approach

2.2.1 Monte Carlo Method for Quantifying System Uncertainty

Risk assessment requires that both model and data uncertainty be taken into account. This can be accomplished using a Monte Carlo simulation to quantify system uncertainties, with model outputs serving as the basis for risk quantification.

Given the uncertainty of parameters in the contaminant transport model, the K_h and K_z mixing coefficients are the simulations' key input variables. As these coefficients tend to follow a normal distribution (Riddle et al. 2001), a Monte Carlo approach can generate coefficient values from a uniform distribution. The normal generators can be expressed as follows:

$$x = N(\sigma_x, \mu_x) \tag{26.5}$$

where x represents either K_h or K_z; $N(\sigma_x, \mu_x)$ represents a normal distribution function of σ_x and μ_x; σ_x is the standard deviation of x; and μ_x is the mean value of x.

The results of the uncertainty analysis depend directly on the distributions assumed for each of the parameters. Hence, good representative data are required to obtain reliable estimates of uncertainty.

After generating sets of random values for each parameter, the distribution of predicted concentrations for each grid square can be calculated using the transport model. The resultant distribution can then be used to define 5th and 95th percentile concentrations (95% of all sample concentrations are less than this concentration), which reflect system uncertainties.

2.2.2 Probabilistic Risk Assessment

To account for uncertainties, presumed to owe their existence to random processes within spatial systems (Schuhmacher et al. 2001), a probabilistic risk assessment modelling approach was used. A range of model scenarios was evaluated by repeatedly picking uncertain variables' values from a probability distribution and using these values in the model. Such probabilities were propagated through the model,

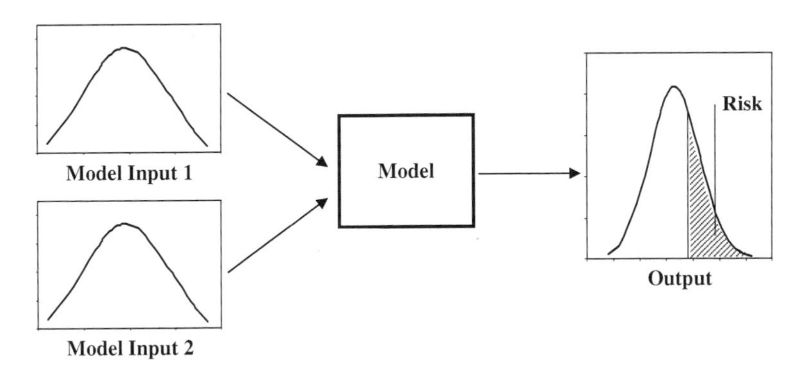

Fig. 26.1 Stochastic risk assessment

and an output distribution describing the probability of various outcomes was generated (Fig. 26.1).

More specifically, environmental risks associated with the discharge of produced water could be expressed as the probability of a random pollutant's concentration, L, exceeding local environmental guidelines (denoted as C_0), i.e., $R = P(L > C_0)$, where R denotes risk. Thus, the risk can be quantified as follows (Chen et al. 1998):

$$R = P(L > C_0) = \int_{C_0}^{\infty} f_L(L)\, dL \tag{26.6}$$

where $f_L(L)$ is the probability density function and R is the risk level quantified as the probability of system failure.

2.2.3 Construction of Fuzzy Membership Functions for Evaluation Criteria

In most risk assessment studies, local environmental standards have been used as evaluation criteria. However, such standards can vary widely from one location to another and are often overly conservative and thus impractical. For example, while the Netherlands mandate a maximum lead concentration [Pb] in surface waters of 10 μg L^{-1}, the Canadian Water Quality Guidelines' (CWQG 2006) protection of aquatic life criteria for 4-day mean [Pb] are 1, 2, 4, or 7 μg L^{-1} for *freshwater* bearing 0–60, 60–120, 120–180, or >180 mg $CaCO_3$ L^{-1} (hardness), respectively (Table 26.1). However, Canadian guidelines offer no [Pb] criteria for seawater.

Table 26.1 Lead criteria for the protection of aquatic life

Origin (type of water)	Lead concentration (μg L^{-1})
Canada (freshwater)	1–7
Netherlands (freshwater)	10
USEPA (seawater)	5.6
California (seawater)	2

While the USEPA has established a seawater protection standard for aquatic life of 5.6 μg Pb L^{-1} (4-day mean; Eisler 2000), the State of California (COP 2001) has mandated a more stringent standard of 2 μg Pb L^{-1} (6-month median).

Given the wide variability in the protection of aquatic life standards (Table 26.1), their practicability must be further addressed. Uncertainties associated with a standard's suitability involve imprecise concepts that cannot be solved through probability theory, but can be quantified using fuzzy logic. When a standard is suitable, it has a good chance of being adopted without significant modifications (Leung 1988). Therefore, the concept of membership grade can be established to reflect the suitability of the standard and its associated uncertainties. In this study, a triangle membership function was formulated by an analysis of chronic toxicity to aquatic organisms.

In order to formulate the triangle membership function, three concentrations have to be determined in advance. The first concentration is the maximum tolerable lead [Pb] concentration ($C_{\max}$). Many studies of the toxic effects of lead in aquatic ecosystems have been conducted. California's environmental protection agency (COP 2001) estimated chronic lead toxicity to occur at a [Pb] of 22 μg L^{-1}. Any water quality objectives should be below this value. From mortality rates observed in a life-cycle study of the snail *Lymnaea palustris*, Borgmann et al. (1978) established a chronic lead toxicity concentration of 19 μg L^{-1}. Rivkin (1979) showed that in a marine environment 12 days under a [Pb] of 10 μg L^{-1} caused complete growth inhibition of the diatom *Skeletonema costatum*. A waterborne [Pb] above 10 μg L^{-1} is expected to cause increasingly severe long-term effects on fish and fisheries (DeMayo et al. 1982; Ruby et al. 1993). As it would be too risky to chose any greater a [Pb] as our standard, a [Pb] of 10 μg L^{-1} was chosen as a completely unsuitable level, with a suitability score of 0 (i.e., membership grade $\mu\,(C_{\max}) = 0$). Lower Pb concentrations would be progressively more suitable.

The second concentration is the most suitable standard acceptable concentration (C_{optimal}). It is obvious that the lower the standard concentration, the stricter the environmental regulation. However, too strict a regulation may become impractical and may simply not be implemented. Lead adversely affects the survival, growth, reproduction, development, and metabolism of most species under controlled conditions, but its effects are substantially modified by numerous physical, chemical, and biological variables. Surveying previous studies of Pb toxicity to aquatic organisms, Borgmann et al. (1978) noted no deaths of the snail *L. palustris* when exposed to a [Pb] of 3.8 μg L^{-1} over its entire lifetime. Eisler (2000) reported adverse effects of Pb on daphnia reproduction at a [Pb^{2+}] of 1.0 μg L^{-1}. Therefore, taking a precautionary approach, we consider the most suitable standard level as 0.5 μg Pb L^{-1}, with membership grade $\mu\left(C_{\text{optimal}}\right) = 1$.

The last concentration for the formulation of the triangle membership function is the minimum possible lead concentration ($C_{\min}$). For the minimum possible lead concentration, we consider an extreme (although impractical) situation as a standard, which is, $C_{\min} = 0\,\mu\text{g L}^{-1}$ with $\mu\,(C_{\max}) = 0$.

Based on the above analysis, a membership function of fuzzy evaluation criteria pertaining to the "suitability" of [Pb] standards for the protection of aquatic life was obtained (Fig. 26.2):

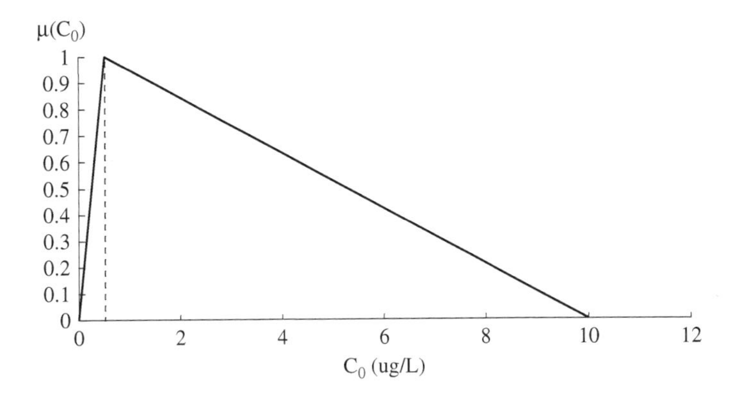

Fig. 26.2 Fuzzy evaluation criteria for Pb

$$\mu\,(C_0) = 2.0\,C_0 \text{ when } 0 \leq C_0 \leq 0.5 \tag{26.7a}$$

$$\mu\,(C_0) = 1.053 - 0.105\,C_0 \text{ when } 0.5 < C_0 \leq 10 \tag{26.7b}$$

2.2.4 Integration of Transport Simulation, Uncertainty Analysis, and Risk Assessment

Lowrance (1976) defined risk as a measure of the probability and severity of adverse effects. In order to quantify risks, it is necessary to specify the spatial and temporal distribution of contaminants in the environment, the uncertainties in the system, and the method of risk evaluation. Generally, uncertainties can be reflected using stochastic, fuzzy, and interval analysis techniques. Among these, stochastic methods can effectively describe probabilistic information, while fuzzy methods are used to describe the uncertainties related to the imprecision or fuzziness of the system (Leung 1988).

Figure 26.3 illustrates the detailed computational process for the integrated simulation and risk assessment model proposed. This system comprises a numerical dispersion model, the Monte Carlo method, and an integrated risk assessment module. Based on the application of known regulatory toxic threshold limits, the following steps were used in quantifying the level of environmental risk associated with the release of produced water into marine water.

1. A random field is generated from site-investigation data (i.e., a random number subroutine was called to generate probabilistic distributions of the horizontal and vertical mixing coefficients).
2. A numerical dispersion model is used to simulate pollutant transport for each set of coefficients generated in Step 1.
3. Predicted contaminant concentrations generated for each modelling point in the study area (Step 2) were stored until the chosen number of simulation runs was reached.

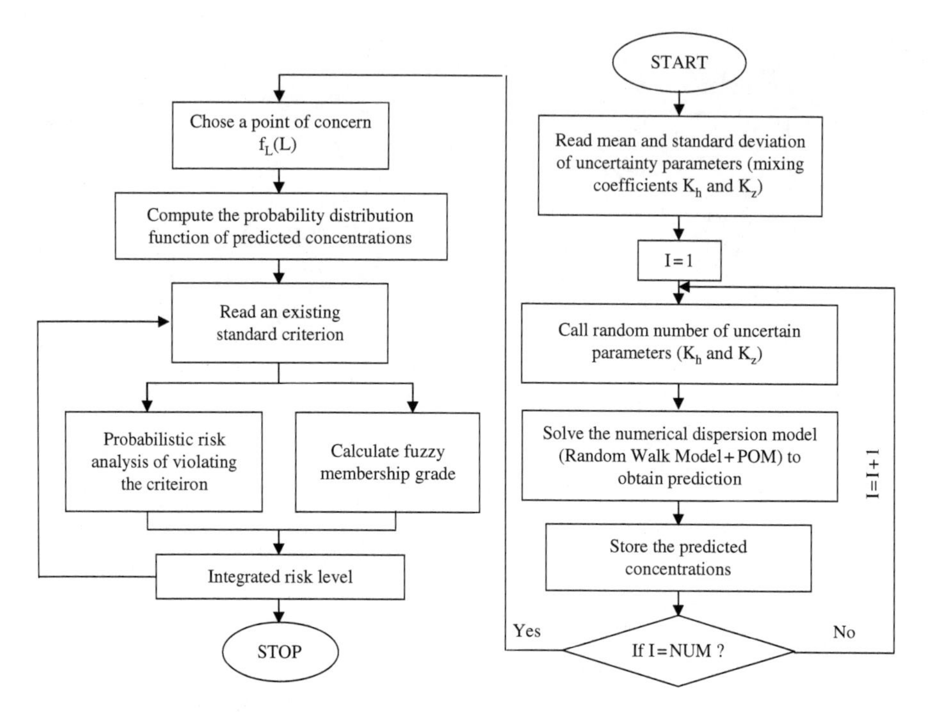

Fig. 26.3 Integrated fuzzy-stochastic risk assessment approach processes

4. A point of concern is chosen. The statistics of simulation outputs is generated for this point based on the results from Step 3.
5. Chemical toxicity threshold standards are identified. The probabilistic risk level of violating each standard and the corresponding fuzzy membership grade based on the fuzzy evaluation formula (Eq. (26.7) and Fig. 26.2) are calculated.
6. A complete set of integrated risk levels is obtained.

3 Case Study

3.1 Overview of the Study Site

A study site around an offshore oil platform on the Grand Banks of Newfoundland, Canada, was chosen for the risk assessment analysis. The Hibernia production platform, a Gravity Base Structure (GBS), is located above 80 m depth of sea and accounts for the greatest volume of produced water currently discharged into the waters of Atlantic Canada (CAPP 2001). A roughly 50 km × 50 km grid with the Hibernia GBS at its center is shown in Fig. 26.4.

Since there are no existing current monitoring stations around the study area's four boundaries, numerical radiation open boundary conditions (OBCs) were

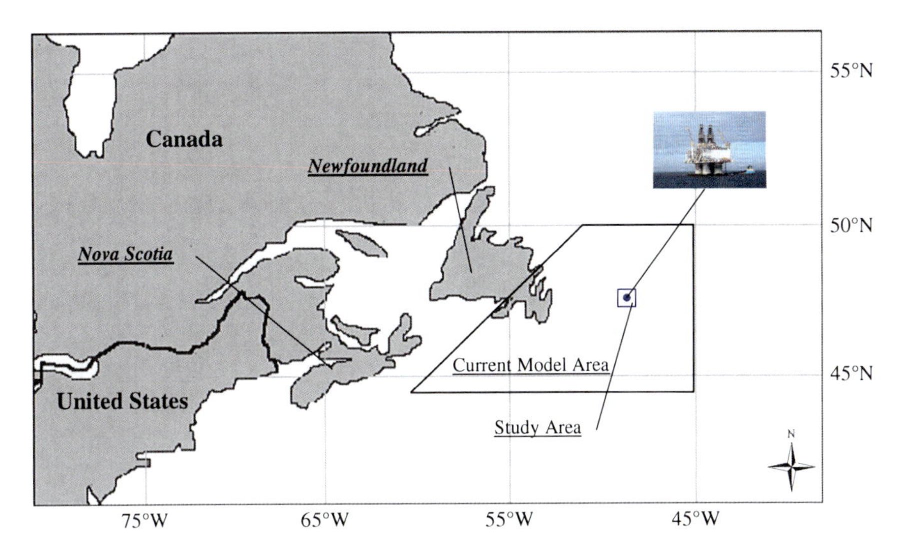

Fig. 26.4 Location of the study site

performed in the ocean circulation model (POM) to simulate the ocean current. However, no matter which type of OBCs were chosen, numerical errors could exist that would create an unrealistic flow across the boundary, thus affecting simulation results. To eliminate such errors, ocean currents were simulated for an area almost 200-fold that of the study area (Fig. 26.4). Only the partial results for the middle of the larger area were used for the simulation of pollutant dispersion.

The numerical dispersion model was initialized over a virtual 30-day period using June 2005 climatological data, including monthly mean temperature and water salinity as well as hourly averaged wind data. These data were obtained from Fisheries and Oceans Canada's Oceanographic database, Environment Canada's climatic database, the National Data Buoy Centre's database, and PAL Environmental Services (2005).

A produced water release port 40 m below the surface at the Hibernia platform was identified and considered as the origin of waterborne Pb. A background [Pb] of 0 for Atlantic Canada seawater was assumed. Samples of produced water and ambient seawater were collected in close proximity to the Hibernia platform during a research expedition in June 2005. These were analyzed at the COOGER (Centre for Offshore Oil and Gas Environmental Research) and TAF (Trace Analysis Facility) at the University of Regina.

3.2 Monte Carlo Simulation for the Study Area

Over the life of a producing field, the volume of the produced water can exceed by ten times the volume of the hydrocarbons produced. During the later stages of production, it is not uncommon to find that produced water can account for as much as 98% of the extracted fluids (Stephenson 1992; Shaw et al. 1999). Petro-Canada (2005) has estimated that the Hibernia site still has a production life of from 20 to 25 years. Therefore, using known trends in production from 1997 to 2006 (BASIN 2007), future risks in the Hibernia area were assessed for the end stage of production and an emission rate of 5.1×10^7 m^3 yr^{-1}.

Based on Eq. (26.5) the values of mixing coefficients K_h and K_z are normally distributed (Riddle et al. 2001). The K_h was assigned a uniform distribution between 50 and 120 m^2 s^{-1}, while K_z was assigned a normal distribution with a mean and standard deviation of 0.007 and 0.003 m^2 s^{-1}, respectively. The selection of parameter distributions is discussed later. Since K_h and K_z are uncertain, the generation of 200 pairs of K_h and K_z values was based on the assumption of a normal distribution (Eq. (26.5)).

The ocean's physical conditions and the [Pb] in produced water discharges were assumed to remain unchanged over the period of 2005–2025. Using the multiple pairs of K_h and K_z values, three-dimensional [Pb]-distribution patterns were generated through a Monte Carlo approach for 2025, thereby allowing the mapping of the uncertainty in the concentration predictions and risks for 2025. A three-dimensional distribution map of [Pb] generated through a Monte Carlo run with $K_h = 64.3\,\text{m}^2\,\text{s}^{-1}$ and $K_z = 0.006\,\text{m}^2\,\text{s}^{-1}$ showed that the highest [Pb] (2.19 μg L^{-1}) exists around the releasing source, due to its continuous emissions (Fig. 26.5).

After Monte Carlo results were generated for each of the grid squares, 95th and 5th percentile concentration contour maps were prepared (Fig. 26.6). The three-dimensional plots of 95th percentile [Pb] concentration (Fig. 26.6a) do not necessarily represent an instantaneous area-wide concentration distribution, because the parameters contributing to one grid location's 95th percentile [Pb] may differ significantly from those prevailing at another location; consequently, the 95th percentile [Pb] distribution plots may represent a "worst-case" prediction (Riddle et al. 2001). Predicted 95th percentile [Pb] distribution plots were of the order of 0.2–2.4 μg L^{-1} and may be much greater at the produced water outlet. The corresponding 5th percentile [Pb] was of the order of 0.05–0.85 μg L^{-1}.

3.3 Integrated Risk Assessment for the Study Area

Risk assessment scenarios were investigated for each of four criteria of the protection of aquatic life: (i) Environment Canada, (ii) the Netherlands, (iii) USEPA, and (iv) the State of California. Criteria (i) and (ii) are for freshwater, while (iii) and (iv) are for seawater (Table 26.2). Probabilistic risk levels were generated by means of a stochastic risk assessment approach (Eq. (26.6)). The membership grade representing the suitability of using these standards as risk assessment criteria was calculated by using Eq. (26.7). This analysis revealed the ties between the strictness and the applicability of environmental standards.

Under Scenario (i), the risk assessment criterion was based on regulations which took into account water hardness (Table 26.2). For a [Pb] criterion of 1 μg L^{-1} and

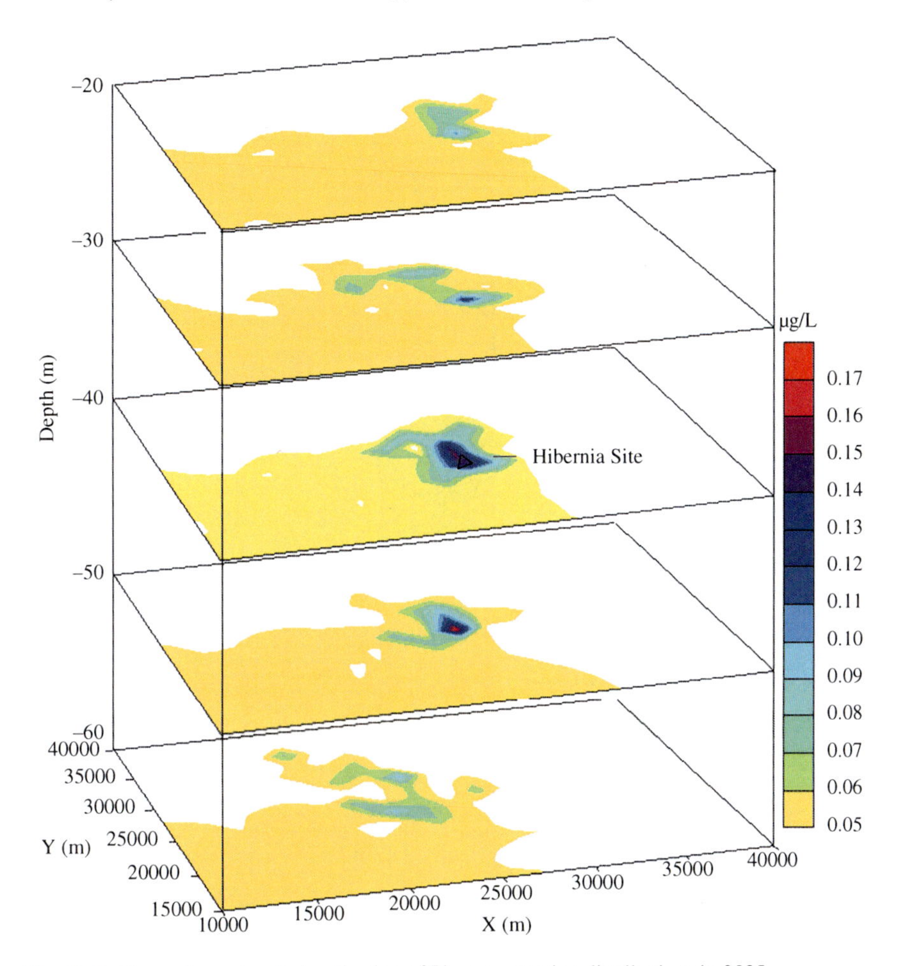

Fig. 26.5 Three-dimensional visualization of Pb concentration distributions in 2025

a water hardness of 0–60 mg $CaCO_3$ L^{-1}, the model calculated a probabilistic risk level of 0.94, representing a 94% probability that the criterion would be violated. Based on Eq. (26.7), the corresponding membership grade is 0.95, indicating a high suitability of the guideline used. Similarly, for the criterion of 2 μg L^{-1} for a water hardness of 60–120 mg $CaCO_3$ L^{-1}, the probabilistic risk level was 0.25 with a corresponding membership grade of 0.84, indicating a slightly lower suitability than the 1 μg Pb L^{-1} criterion. For the criteria of 4 and 7 μg L^{-1} for hardness values of 120–180 and >180 mg $CaCO_3$ L^{-1}, the probabilistic risk levels were zero, indicating that the standards would never be violated. The corresponding membership grades were 0.63 and 0.32, respectively, showing the standards' low applicability.

Under Scenario (ii), using a [Pb] guideline of 10 μg Pb L^{-1} issued by the Netherlands led to an integrated risk level of 0 and a membership grade of 0, indicating a completely unsuitable standard. Scenario (iii), using the USEPA's aquatic [Pb]

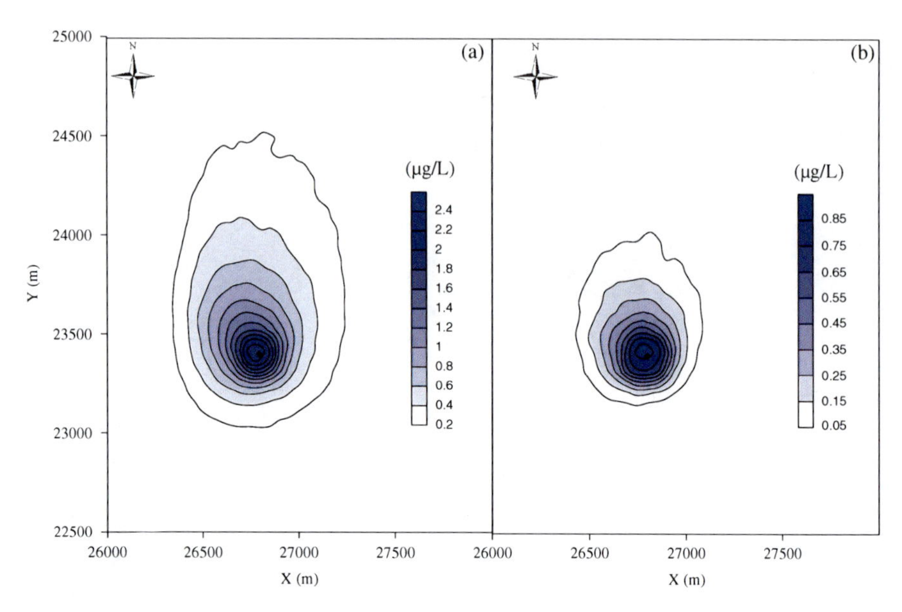

Fig. 26.6 Uncertainty predictions showing (**a**) 95th percentile lead concentrations and (**b**) 5th percentile lead concentrations for produced water dispersion

Table 26.2 Summary of fuzzy-stochastic risk assessment results

Issuer	Regulated value ($\mu g\ L^{-1}$)	Integrated risk level	
		Stochastic risk	Membership grade
Canada (freshwater)			
Water hardness (mg $CaCO_3$ L^{-1})			
0–60	1	0.94	0.95
60–120	2	0.25	0.84
120–180	4	0	0.63
>180	7	0	0.32
Netherlands (freshwater)	10	0	0
USEPA (seawater)	5.6	0	0.47
California (seawater)	2	0.25	0.84

criterion of 5.6 μg Pb L^{-1} also led to a risk level of 0, indicating that the standard would never be violated. The corresponding membership grade of 0.47, indicated the standard's relatively low applicability.

Under Scenario (iv), using the State of California's [Pb] guideline of 2 μg Pb L^{-1} led to an integrated risk level of 0.25 with a membership grade of 0.84, indicating the standard's high applicability. For the four criteria of protection of aquatic life standards investigated, probabilistic risk levels obtained through a fuzzy-stochastic risk assessment approach ranged from 0 to 0.94, and corresponding membership grades from 0 to 0.95.

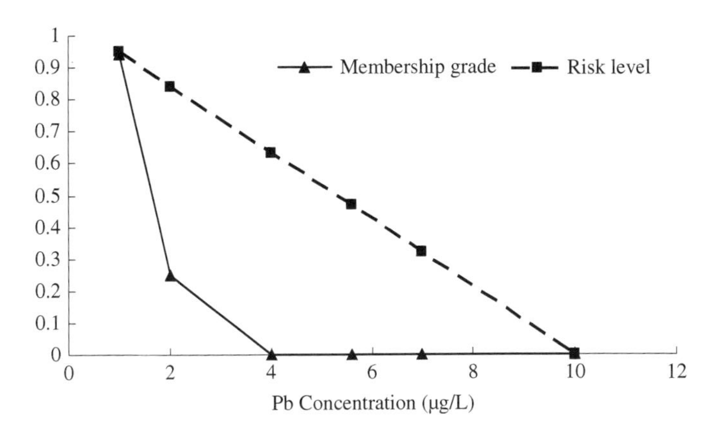

Fig. 26.7 Results of the integrated risk assessment

Compared with traditional risk-quantifying methods, the fuzzy-stochastic risk assessment approach is not only capable of quantifying probabilistic risk levels based on numerical simulation but also capable of reflecting the applicability of the evaluation criteria. The ties between the suitability (applicability) and relevant risk levels are further illustrated in Fig. 26.7. In general, for the protection of aquatic life in seawater, the scenario based on the guideline issued by the State of California has a higher membership grade than the others, indicating that the California criterion is more applicable in regulating [Pb] in the marine environment than the other criteria studied.

In order to provide a simple and effective way to visually illustrate the level of risks associated with production discharges around an offshore production platform, a severity scale map was plotted using the 2 μg Pb L^{-1} California criterion, which has a fuzzy membership grade of 0.84 (Fig. 26.8). The risk zone with a stochastic risk level above 0.2 (i.e., mean [Pb] >1.6 μg L^{-1}) is approximately 25,000 m^2, bounded within a 90-m radius around the Hibernia GBS platform. The majority of the study area had a zero probability of violating the adverse effects threshold value selected.

4 Discussion

Estimating the range of mixing coefficients K_h and K_z in the ocean is difficult and complicated by the spatial variability in water velocity and turbulence in such a large expanse as the study area. Riddle et al. (2001), working in the North Sea, used K_h values ranging from 10 to 50 m^2 s^{-1} and a mean, standard deviation, and minimum of 0.05, 0.004, and 1×10^{-4} m^2 s^{-1}, respectively, for K_z. However, as the physical characteristics of the North Sea and Atlantic Canada are very different, their mixing coefficient would be too. Figure 26.9a shows the probability distribution function at the Hibernia site in 2025 under Riddle's range of the mixing coefficients. A large tail is present in the probability density function's distribution and much higher

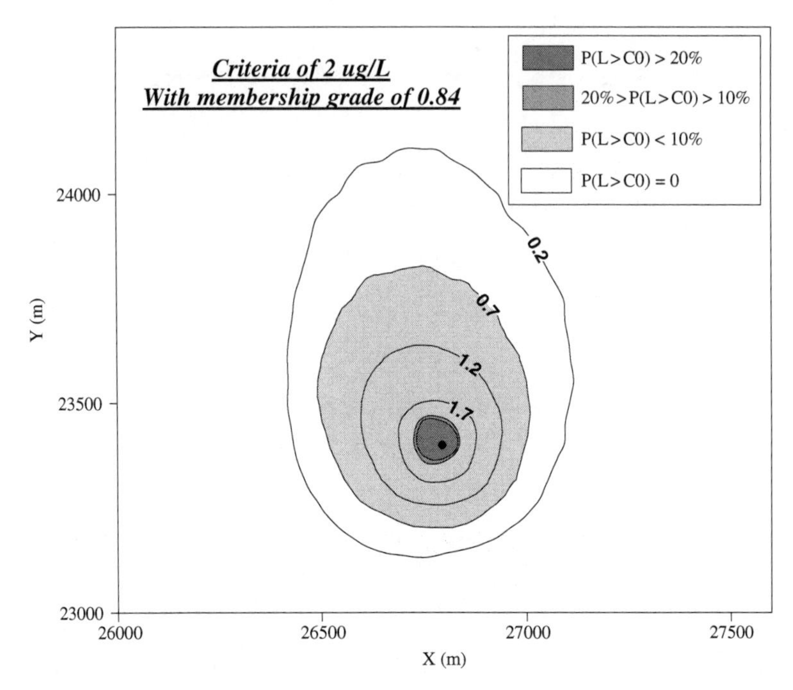

Fig. 26.8 Mean lead concentration distribution and severity scale map for integrated risks with a membership grade of 0.84 based on the criterion issued by the State of California (2 μg/L) at 40-m depth in 2025

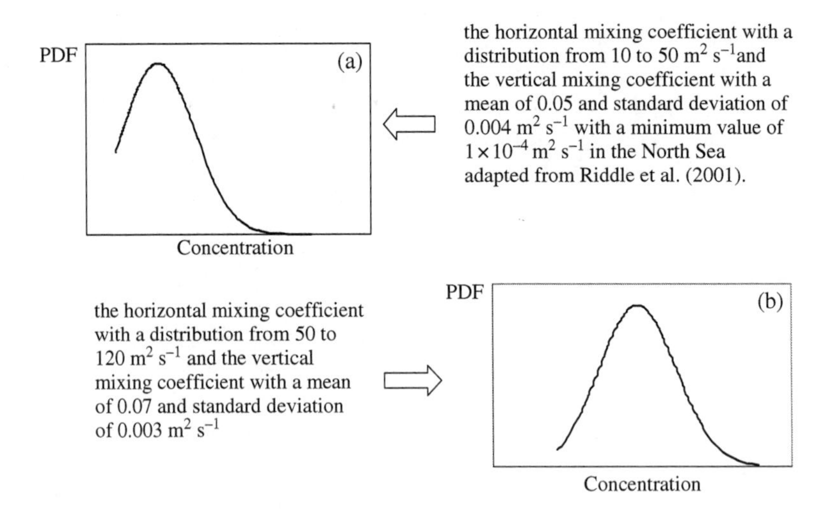

Fig. 26.9 Simulation of probability density function for 2025 at the Hibernia site

[Pb] are generated, indicating little mixing and a high [Pb] accumulation under this distribution of the parameters. In order to avoid such an influence of low mixing coefficients on the risk analysis, local conditions at the Hibernia platform were monitored and a calibration undertaken. The results of both support the use of a uniform distribution of K_h from 50 to 120 $m^2\ s^{-1}$, and of a normal distribution of K_z, with a mean and standard deviation of 0.007 and 0.003 $m^2\ s^{-1}$, respectively (Fig. 26.9b).

This study characterized uncertainties related to environmental standards based on toxicity studies in marine environments, through the use of an integrated risk modelling framework. A triangle fuzzy membership function was formulated, which was used to estimate the suitability of a standard for risk assessment and thereby address the uncertainty of the standard. Other forms of membership function, such as the *L–R* nonlinear membership function (Leung 1988), can be used to address these complexities in the future. However, more computational and experimental efforts are required.

5 Conclusions

A fuzzy-stochastic risk assessment approach was developed to reflect uncertainties associated with produced water discharges and related regulated pollution criteria for the marine environment. A triangle fuzzy membership function was established to reflect the toxicity effects of waterborne Pb and the suitability of local Pb standards. To generate a numerical dispersion model, an ocean hydrodynamic model (POM) was coupled with a pollutant dispersion model (Random Walk Model). Using the Monte Carlo method, the resultant model was used to conduct a stochastic simulation of pollutant dispersion. This stochastic simulation served as the basis for the risk assessment. The quantification of risks associated with Pb in produced water discharges was based on combined fuzzy/stochastic information.

This developed risk assessment approach was applied to an offshore oil production facility located on the Grand Banks of Canada. Reasonable results were generated, thereby demonstrating the utility of this approach to support the effective management of produced water discharges in the future.

The proposed fuzzy-stochastic risk assessment approach is an extension of previous risk assessment studies, and it reflects the uncertainties associated with not only the modelling system but also those of regulatory criteria. This approach provides a quantitative means for the examination of the uncertainties and risks resulting from produced water discharges on a regional spatial scale. With its expanded evaluation dimensions, the developed risk assessment approach can be used as an effective risk assessment protocol for the management of produced water discharges in the marine environment.

Acknowledgments We would like to acknowledge the professional sample analysis work by Prof. Renata Bailey at the University Regina and Mrs. Susan Cobanli from Fisheries and Oceans Canada. This study was funded in part by the Natural Science and Engineering Research Council Canada (NSERC), Fisheries and Oceans Canada (DFO), and the Program of Energy Research and Development (PERD).

References

Bardossy A, Bogardi I, Duckstein L (1991) Fuzzy set and probabilistic techniques for health-risk analysis. Appl Math Comput 45:241–268

BASIN. (2007) Offshore Eastern Canada hydrocarbon database. Natural Resources Canada, <http://basin.gsca.nrcan.gc.ca/index_e.php> (April, 2007)

Borgmann U, Kramar O, Loveridge C (1978) Rates of mortality, growth, and biomass production of Lymnaea palustris during chronic exposure to lead. J Fish Res Board Can 35: 1109–1115

CAPP (2001) Offshore produced water waste management, Canadian Association of Petroleum Producers (CAPP) East Coast Committee. Alberta, Nova Scotia, and Newfoundland, Canada

Chen Z, Huang GH, Chakma A, Tontiwachwuthikul P (1998) Environmental risk assessment for aquifer disposal of carbon dioxide. In: Proceeding of 4th international conference on greenhouse gas control technologies, Interlaken, Switzerland

COP (2001) California Ocean plan – ocean quality control plan: ocean waters of California, State Water Resources Control Board, Resolution No. 2000-108. California Environmental Protection Agency, Sacramento, CA, USA

CWQG (2006) Canadian water quality guidelines for the protection of aquatic life, 2006, Canadian environmental quality guidelines, Canadian Council of Ministers of the Environment. The guidelines were originally published in CCREM 1987, Chapter 3

DeMayo A et al (1982) Toxic effects of lead and lead compounds on human health, aquatic life, wildlife, plants, and livestock. CRC Crit Rev Environ Control 12:257–305

Eisler R (2000) Handbook of chemical risk assessment: health hazards to humans, plants and animals: VI metals. Lewis Publishers, New York

Hodgins DO (1993) Hibernia effluent fate and effects modeling, report prepared for Hibernia management and Development Company, Seaconsult Marine Research Ltd. (revised January 1994). Sidney, BC, Canada

Klir GJ, Yuan B (1995) Fuzzy sets and fuzzy logic: theory and applications. Prentice-Hall, Englewood Cliffs, NJ

Leung Y (1988) Spatial analysis and planning under imprecision. Elsevier, New York

Lowrance WW (1976) Of acceptable risk. William Kaufmann Inc., Los Altos, CA

Meinhold AF, Holtzman S, DePhillips MP (1996a) Risk assessment for produced water discharges to Open Bays in Louisiana. In: Reed M, Johnsen S (eds) Produced water 2: environmental issues and mitigation technologies. Plenum Press, New York, pp 395–409

Meinhold AF, Holtzman S, DePhillips MP (1996b) Produced water discharges to the Gulf of Mexico: background information for ecological risk assessments. Brookhaven National Laboratory, New York

Mellor GL (2004) Users guide for a three-dimensional, primitive equation, numerical ocean model. Princeton University, Princeton, NJ

Mellor GL, Ezer T, Oey L-Y (1994) The pressure gradient conundrum of sigma coordinate ocean models. J Atmos Oceanic Technol 11:1126–1134

Mellor GL, Oey L-Y., Ezer T (1998) Sigma coordinate gradient errors and the seamount problem. J Atmos Oceanic Technol 12:1122–1131

Mukhtasor, Husain T, Veitch B, Bose N (2004) An ecological risk assessment methodology for screening discharge alternatives of produced water. Hum Ecol Risk Assess 10: 505–524

PAL Environmental Services (2005) Hibernia annual environmental data summary (2005), Prepared for Exxon Mobil, PAL Environmental Services, Newfoundland, Canada

Petro-Canada (2005) Canadian east coast oil: hibernia. Petro Canada, <http://www.petro-canada.ca/en/about/717.aspx> (April, 2007)

Reed M, Johnsen S, Rye H, Melbye A (1996) PROVANN: a model system for assessing exposure and bioaccumulation of produced water components in marine pelagic fish, eggs, and larvae.

In: Reed M, Johnsen S (eds) Produced water 2: environmental issues and mitigation technologies. Plenum Press, New York. Proceedings of the 1995 International Produced Water Seminar, Trondheim, Norway, Plenum Press, New York, pp 317–332

Reed M, Hetland B, Ditlevsen MK, and Ekrol N (2001) DREAM Version 2.0. Dose related Risk Effect Assessment Model. Users manual, SINTEF Applied Chemistry, Environmental Engineering, Trondheim, Norway.

Riddle AM, Beling EM, Murray-Smidth RJ (2001) Modeling the uncertainties in predicting produced water concentrations in the North Sea. Environ Model Softw 16:659–668

Rivkin R (1979) Effects of lead on growth of the marine diatom, Skeletonema costatum. Mar Biol 50:239–247

Ruby MV, Davis A, Link TE, Schoof R, Chaney RL, Freeman GB, Bergstrom P (1993) Development of an in vitro screening test to evaluate the in vivo bioaccessibility of ingested mine-waste lead. Environ Sci Technol 27(13):2870–2877

Schuhmacher M, Meneses M, Xifro A, Domingo JL (2001) The use of Monte-Carlo simulation techniques for risk assessment: study of a municipal waste incinerator. Chemosphere 43: 787–799

Shaw DG, Farrington JW, Connor MS, Trippm BW, Schubel JR (1999) Potential environmental consequences of petroleum exploration and development on grand banks, New England Aquarium Aquatic Forum Series Report 99-3, Boston, p 64

Stephenson MT (1992) A survey of produced water studies. In: Ray JP, Engelhardt RF (eds) Produced water: technological/environmental issues and solutions. Plenum Press, New York, pp 1–11

Suter GW, II (1993) Defining the field. In: Suter GW (ed) Ecological risk assessment. Lewis Publishers, Boca Raton, FL, pp 3–20

Thatcher M, Robson M, Henriquez LR (1999) A CIN revised CHARM III report. A user guide for the evaluation of chemicals used and discharged offshore, version 1.0, Netherlands Ministry of Transportation, The Netherlands

Thatcher M, Robson M, Henriquez LR, Karman CL (2001) A user guide for the evaluation of chemicals used and discharged offshore: version 1.2. CIN revised CHARM III Report, Charm Implementation Network (CIN), European Oilfield Speciality Chemicals Association (EOSCA)

Zimmermann HJ (2001). Fuzzy set theory and its application, 4th edn. Kluwer, Norwell, MA

Chapter 27
Application of Quantitative Risk Assessment in Produced Water Management – the Environmental Impact Factor (EIF)

Ståle Johnsen and Tone K. Frost

Abstract The Dose-related Risk and Effect Assessment Model (DREAM) was developed through a JIP in the period 1997–2000 and was implemented for produced water (PW) management in the Norwegian sector of the North Sea as a part of the 'Zero discharge work', 2000–2005. The initial version of DREAM included two approaches to PW management, the Environmental Impact Factor (EIF) and a body burden related risk assessment model focusing on selected PW compounds. The EIF, addressed in the present chapter, has found broad application in the North Sea and has also been used in other offshore production areas by different companies. The produced water EIF is based on the risk assessment principles described in the EU Technical Guidance Document (TGD), comparing the Predicted Environmental Concentration (PEC) and the Predicted No Effect Concentration (PNEC) of PW compounds. The quantitative risk element in the model is represented by the water volume where PEC exceeds PNEC, including the combined risk of all major PW constituents, both naturally occurring compounds and industry-added chemicals. The EIF is used as a management tool, primarily to identify and perform cost–benefit analyses of PW mitigation measures and best available technology (BAT). The method enables the operator to identify the compounds posing the most significant environmental risk in PW, and further to rank different PW discharges with respect to environmental significance and risk. This chapter describes the EIF method and focuses on examples of application of the tool on specific offshore production fields. A description of how the EIF fits into Statoil's environmental management system is also given, including the link between risk assessment, selection of BAT and field validation through environmental monitoring.

1 Introduction

Produced water (PW) discharges to the marine environment became a major focus area from an environmental point of view as some of the major oil producing fields in the Norwegian Continental Shelf (NCS) matured and their water production

S. Johnsen (✉)
Statoil Research Centre, N-7005, Trondheim, Norway

K. Lee, J. Neff (eds.), *Produced Water*, DOI 10.1007/978-1-4614-0046-2_27,

increased significantly in the early 1990s. The primary approach of the Norwegian operators and authorities was initially to increase the understanding of the potential environmental harm of these discharges by chemical and toxicological characterisation of PW from representative offshore fields (Brendehaug et al. 1992; Ray and Engelhart 1992). These early studies in general showed that PW contains a number of potential environmentally harmful compounds, varying from field to field. However, dilution with a factor of 1000–10,000 will normally bring these toxicants below their toxicity threshold concentration level. These observations led to several studies on the fate and dilution of PW in the recipient by field studies and model development (Riksheim and Johnsen 1994; Johnsen et al. 1998; Utvik and Johnsen 1999; Utvik et al. 1999). A conclusion from this phase was, as could be expected, that dilution of PW discharges is varying significantly from case to case, depending on the environmental conditions in the fields' location. Thus, a risk assessment approach to PW management would seem appropriate, taking into account the nature, time and location of the discharge.

The first attempt to identify the potential harm of PW to the environment by risk assessment approaches was presented in 1995 (Reed and Johnsen 1996), and this led to the development of the Dose-related Risk and Effect Assessment Model (DREAM) and the introduction of the Environmental Impact Factor (EIF) in the period 1996–1999 (Johnsen et al. 2000; Reed and Hetland 2002; Smit et al. 2003). The need for a quantitative risk assessment tool for PW management had then become evident through the Norwegian authorities' White Paper 58 (1996) followed by the 'Zero discharge report', a joint venture from the operators in the NCS and the Norwegian Pollution Control Authority (SFT 1998). The EIF has since been implemented as a management tool for risk reduction by all operators in the NCS and is today commonly accepted and used for PW management and performance improvement.

2 Dose-related Risk and Effect Model (DREAM)

The development of DREAM was initiated based on a desire to build a risk and effect assessment model being able to quantify potential environmental harm of PW compounds by determining internal (tissue) concentrations of selected compounds in marine organisms and comparing these to observed critical body burdens for chronic effects. Indeed, the DREAM model can do this for a limited number of PW compounds, primarily PAHs and alkyl phenols. The data requirements to perform this exercise are so comprehensive that it is practically impossible to operate DREAM for all PW compounds. This led to the development of the Environmental Impact Factor which is based on the simpler PEC/PNEC approach, the internationally agreed principles for risk assessment as described by the European Commission's Technical Guidance Document (EU-TGD) (EC 1996, 2003). The EIF (Johnsen et al. 2000; Smit et al. 2003) is based on comparing the environmental abundance of discharged PW compounds, expressed by their predicted environmental concentration (PEC), with the predicted no effect concentration (PNEC) derived from available toxicity data on the respective compounds. These two variables are used to calculate the Risk Characterisation Ratio (RCR) which is the basis for the risk management principles described in the EU-TGD (Eq. (27.1)). While the EU approach addresses the risk assessment of single compounds in the environment based on their toxicity properties, the EIF has been developed to handle complex mixtures and multiple environmental stressors depending on the actual discharge and compartment (Smit et al. 2008).

$$\mathrm{RCR} = \mathrm{PEC/PNEC} \tag{27.1}$$

The PEC variable in Eq. (27.1) can be derived from dilution modelling of the discharges or from measurements of environmental concentrations through field monitoring. Due to the complex nature of PW, modelling is normally the only realistic approach. However, validation of dilution models through field studies is a crucial element in a holistic approach to environmental management. An example of a concentration field of PW discharges given by the DREAM dilution model is shown in Fig. 27.1. The total concentration of PW compounds is given as a time-variable, three-dimensional image, projected to the sea surface in a two-dimensional presentation. The composition of the produced water will be field

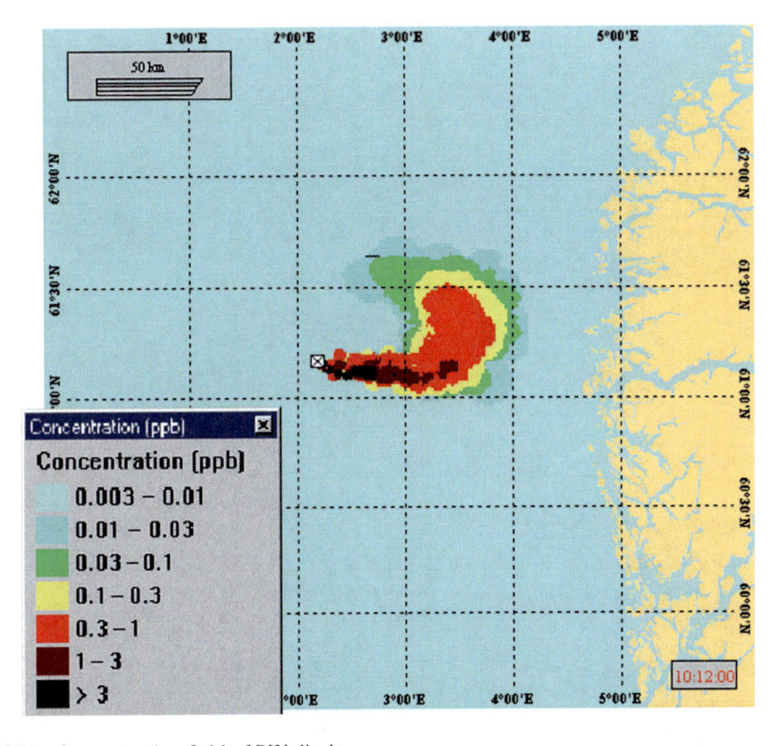

Fig. 27.1 Concentration field of PW discharge

dependent and the input data (presence and abundance of PW compounds) will decide the nature of the concentration field.

The PNEC values are derived from available toxicity data on PW compounds. Due to the variable quality and availability of such data, ranging from comprehensive chronic toxicity values on natural compounds like PAH and heavy metals to limited acute toxicity data on process chemicals, the EU-TGD defines a set of assessment factors to be used in the determination of PNEC. This may lead to relatively conservative PNEC values for PW compounds with limited toxicity data available. In addition, that time-variable exposure is not considered in the TGD approach and chronic exposure is assumed. The EIF therefore represents a conservative risk estimate. As a consequence, the EIF should not be used as a descriptive tool (describing ecosystem status) but for management purposes only with the aim of assisting in the selection of cost-effective mitigation measures. This is illustrated in the examples presented below. Combining the PEC and PNEC values (Eq. (27.1)) enables the DREAM model to image the RCR as a reflection of the water volume in the recipient where PEC exceeds PNEC. The selected unit for the EIF is the recipient water volume of 100 m × 100 m × 10 m (100,000 m^3). An EIF of 10 represents a water volume of 1,000,000 m^3. This water volume is the quantitative element in the risk assessment (Fig. 27.2).

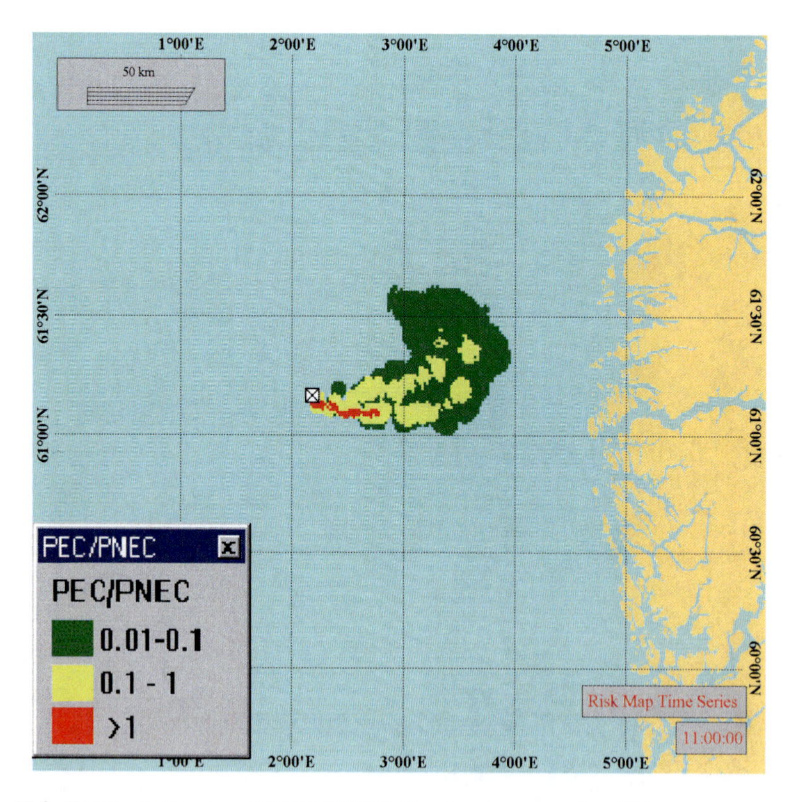

Fig. 27.2 Water volume where RCR (PEC/PNEC) ratio >1, representing EIF

3 Practical Applications

Figure 27.3 gives EIF values determined for all Statoil offshore production fields in the NCS (2002 values) and illustrates how EIF can be used to rank the importance of the different fields with respect to potential environmental risks. In addition to a quantitative image of the risk, the EIF calculations will provide a pie chart for each scenario, showing PW compounds that are contributing to the risk (example included in the figure).

This information will enable the operator to identify mitigation measures to reduce the risk of PW discharges at each installation and to document the cost effectiveness of each mitigation measure considered, by recalculating the EIF using the expected efficiency of a given measure on the PW discharges and composition.

A specific example on how EIF calculations are used to quantify potential risk reduction from different mitigation measures considered for a specific PW discharge scenario is shown in Fig. 27.4. This example is taken from the Statfjord field and illustrates the work to identify mitigation measures inside the 'Zero

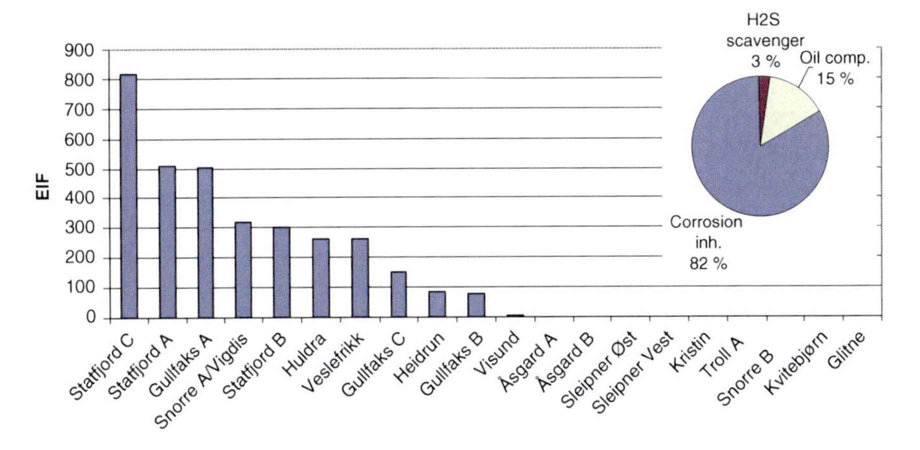

Fig. 27.3 EIFs for Statoil operated fields in the NCS (2002)

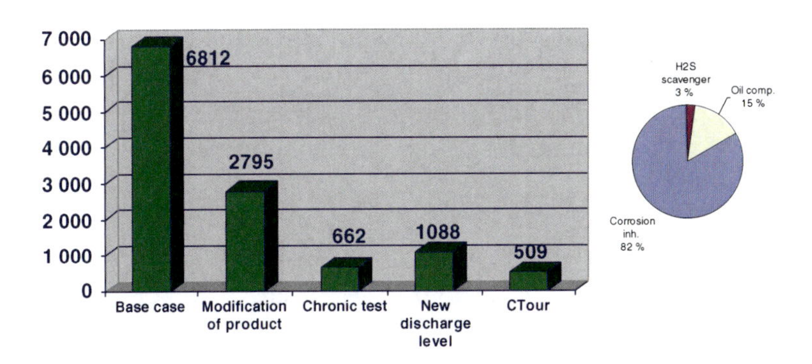

Fig. 27.4 Using EIF for selecting BAT for a specific PW discharge scenario

discharge' initiative for this field. The figure shows that both specific discharge reducing technologies and improving the quality of input data, e.g. toxicity of discharged compounds, can contribute to a better understanding and management of environmental risks (Knudsen et al. 2004).

The first bar in Fig. 27.4 represents the field's 'base case' EIF, e.g. the EIF prior to any risk reducing actions taken. As seen from the pie chart included in the figure, the main contributing discharge compound to the risk is a corrosion inhibitor applied at the installation. The chemical has been used at the field for more than 10 years with an excellent protection of the flowlines and production equipment. A closer study of this chemical revealed that the active components included were poorly biodegradable and accounted for a major part of the toxicity

A cooperation between the operator and the chemical supplier was initiated to modify the chemical to a more environmental friendly product, e.g. replacing one of the active components with a less toxic component. Thereby, a significant reduction in the EIF can be achieved (Fig. 27.4, bar 2). As for a majority of process chemicals, only acute toxicity data were available for the corrosion inhibiter used at this production site. In agreement with the EU-TGD requirements, this resulted in applying the highest assessment factor in the PNEC determination for that chemical, which may result in an unrealistically conservative focus on the particular chemical. To reduce this uncertainty, long-term toxicity tests were carried out for the corrosion inhibitor and the PNEC value was recalculated. Using the new PNEC value in the PEC/PNEC comparison led to a significant reduction in the contribution from the chemical to the overall EIF, and thus a reduction in the total EIF value (Fig. 27.4, bar 3). However, the corrosion inhibitor still showed up as a major contributor in the EIF pie chart.

To further reduce the uncertainty in the risk assessment, a field study to validate the oil/water distribution of the chemical between the oil phase and the water phase was conducted. This showed that the partition model applied for calculation of the concentration of the chemical following the water phase and discharge to sea had underestimated this value, leading to an increase in the overall EIF (Fig. 27.4, bar 4). Next, several alternative risk reducing mitigation approaches were evaluated, including improving material quality (steel in pipelines and process equipment), PW re-injection and several water treatment technologies. Based on a cost–benefit evaluation the C-Tour technology was selected for this actual PW discharge. The expected impact of this is shown in bar 5 of Fig. 27.4. C-Tour removes both dispersed oil and oil-soluble components from the produced water. Tests at this field have shown that the technology reduces the discharge of dispersed oil to the sea by 50–70%, and in addition a 30–40% reduction of the active component of the corrosion inhibitor.

4 Holistic Environmental Management

Figure 27.5 illustrates how Statoil integrates Quantitative Risk Assessment (QRA) represented by the EIF and environmental monitoring, together with identification of BAT, in a holistic management circle for securing Best Environmental

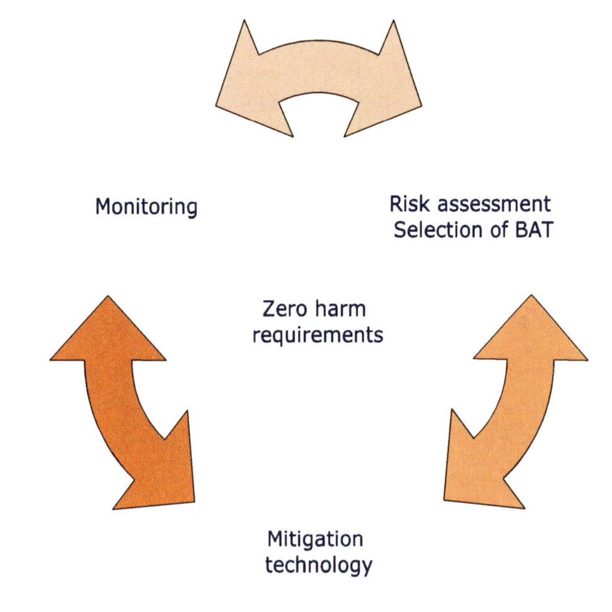

Fig. 27.5 Holistic management circle

Practice (BEP) during a given activity. The purpose of this system is to identify major contributions to risk for a given discharge or emission, identify cost effective measures to reduce potential environmental risk, identify needs for mitigation technology and to document the benefit/effect of implemented measures.

The fundament of the holistic approach to environmental management is sufficient knowledge of the influenced ecosystems and the nature of the actual discharges and emissions. This must be achieved prior to any activity by establishing the environmental context through reviewing existing information, identifying knowledge gaps and closing these by performing baseline surveys of the area. For a given activity, application of BAT and BEP will define the expected discharges and emissions.

QRA is, in addition to other risk assessment tools, used as a decision support tool in the early phase of any planned activity or development. In new developments, QRA is a part of the Environmental Impact Assessment (EIA) to design the monitoring program and is essential in the selection of an appropriate holistic management approach. QRA is also utilised in the operational phase when action is needed to reduce discharges or emissions from a process or activity. The identified actions may be a result of an observed need to reduce the potential harm to the environment, regulatory requirements or meeting the goal of continuous improvement. Finally, QRA is used to identify cost-effective mitigation measures and the need for development and qualification of new discharge or emission-reducing technology. Environmental monitoring is used to establish the characteristics of the area (baseline studies), to identify any changes or potential impact to the environment and to document the effect/benefit of mitigation measures.

5 Conclusions

The DREAM/EIF method has proven to be a powerful management tool for PW mitigation and is presently applied by all operators in the Norwegian Continental Shelf. Risk assessment represented by the EIF is accepted by national authorities. However, as the EIF identifies the most cost-effective measures to reduce environmental risk based on precautionary principles, it does not reflect the condition of the ecosystem. The EIF should always be used together with other assessment techniques such as environmental monitoring as the EIF does not identify the need for reducing environmental risk from PW discharges, but it is a powerful tool to find the best approach if such an action is desired.

Acknowledgments The DREAM model and the produced water EIF were developed through a joint industry project financed by Statoil, Norsk Hydro, Total and ENI. Several scientific institutions were involved in the development, with TNO and SINTEF playing a key role.

References

Brendehaug J, Johnsen S, Bryne K.H, Gjøse AL, Eide TH, Aamot E (1992) Toxicity testing and chemical characterisation of produced water – a preliminary study. In: Ray JP, Engelhart FR (eds) Produced water: technological/environmental issues and solutions. Plenum Press, New York, pp 245–256

European Communities (1996) Technical guidance document in support of commission directive 93/67/EEC on risk assessment for new notified substances and commission regulation (EC) No. 1488/94 on risk assessment for existing substances. Part I to IV, Office for official publications of the European Communities. ISBN 92-827-8011-2

European Communities (2003) Technical guidance document on risk assessment (TGD). 2nd edn, Part II. In support of Commission Directive 93/67/EEC on Risk Assessment for new notified substance, Commission Regulation (EC) No 1488/94 on Risk Assessment for existing substances and Directive 98/8/EC of the European Parliament and of the Council concerning the placing of biocidal products on the market. European Commission Joint Research Centre. EUR 20418 EN/2, 203 pp + App

Johnsen S, Frost TK, Hjelsvold M, Utvik TR (2000) The environmental impact factor – a proposed tool for produced water impact reduction, management and regulation. SPE 61178. Society of Petroleum Engineers

Johnsen S, Røe TI, Durell G, Reed M (1998) Dilution and bioavailability of produced water compounds in the Northern North Sea. A combined modeling and field study. SPE paper SPE 46269. Presented at the HSE meeting, Caracas, Venezuela

Knudsen BL, Hjeldsvold M, Frost TK, Svarstad MBE, Grini P, Willumsen CF, Torvik H (2004) Meeting the zero discharge challenge for produced water. Soc Petrol Eng J SPE 86671

Ray JP, Engelhart FR (1992) Produced water – technology/environmental issues and solutions. Plenum Press, New York, p 616

Reed M, Hetland B (2002) "DREAM": a dose-related exposure assessment model. Technical description of physical-chemical fates components. SPE paper No. 73856

Reed M, Johnsen S (1996) Produced water – environmental issues and mitigation technologies. Plenum Press, New York, p 536

Riksheim H, Johnsen S (1994) Determination of produced water constituents in the vicinity of offshore production fields in the North Sea. SPE 27150, SPE HSE, Jakarta

SFT (1998) Zero discharge report. SFT Oslo, November 1998

Smit MGD, Jak RG, Holthaus KIE, Karman CC (2003) An outline of the DREAM project and development of the environmental impact factor for produced water discharges. TNO-report 2003/376, TNO, Den Helder, The Netherlands

Smit MGD, Jak RG, Rye H, Frost TK, Singsaas I, Karman CC (2008) Assessment of environmental risks from toxic and non-toxic stressors; a proposed concept for a risk-based management tool for offshore drilling discharges. Integ Env Assess Manag 4:173–183

Utvik TIR, Durell GS, Johnsen S (1999) Determining produced water originating PAH in North Sea waters: comparison of sampling techniques. Mar Pollut Bull 38:977–989

Utvik TIR, Johnsen S (1999) Bioavailability of PAH in the North Sea. Environ Sci Technol 33:12

Chapter 28
Challenges Performing Risk Assessment in the Arctic

Gro Harlaug Olsen, JoLynn Carroll, Salve Dahle, Lars-Henrik Larsen, and Lionel Camus

Abstract Increasing offshore oil and gas activities in the European Arctic have raised concerns of the potential anthropogenic impact of oil-related compounds on polar marine ecosystems. For the Barents Sea, the Norwegian government has therefore enforced a zero discharge policy which does not allow any discharges to sea. This policy poses some challenges to routine operations, and it has been questioned whether this is the overall best environmental strategy. An alternative could be to handle the Barents Sea in the same way as the rest of the Norwegian shelf, which is to apply the zero harmful discharge strategy. This strategy involves performing risk assessments of harm made to the environment by petroleum-related activities. However, risk quantification procedures for petroleum operations were originally established on scientific knowledge derived from investigations mainly performed on temperate living organisms. Before risk calculations can be performed in Arctic areas, basic knowledge of sensitivity of Arctic species has to be in place, and risk assessment procedures need to be adapted to Arctic environments. This chapter describes current procedures and discusses key challenges to performing risk assessments in the Arctic, with special focus on the Barents Sea. For Arctic organisms there is a general lack of information concerning possible effects of oil-related compounds at all levels of biological organization. Further research is needed to understand sub-lethal and long-term impacts of acute and chronic exposure to petroleum compounds as well as to increase basic ecosystem understanding.

1 Introduction

Since the discovery of major oil and gas reservoirs in the North Sea in the 1970s, the petroleum industry in Norway has operated under a Norwegian government regulatory policy known as the "zero harmful discharge" policy. The policy allows the discharge of some chemicals, while discharges of other more harmful chemicals are prohibited. To demonstrate compliance, petroleum operators use an environmental risk-based approach, providing a quantitative measure of the impact of chemical discharges on the biota present in the marine environment.

G.H. Olsen (✉)
Akvaplan-niva, FRAM Centre, N-9296 Tromsø, Norway

K. Lee, J. Neff (eds.), *Produced Water*, DOI 10.1007/978-1-4614-0046-2_28,

This risk assessment approach is coupled to ongoing monitoring programs to verify that environmental impacts from ongoing operations result in "zero harm" to the environment.

In 1980 the Norwegian sector of the southern Barents Sea was opened for oil and gas exploration. Petroleum resources in the Norwegian and Russian Arctic are believed to be substantial, making the region a new frontier for development. The region also contains major fisheries resources. Fishing is the second largest industry in Norway, representing 1–2% of the gross national product while the petroleum industry accounts for over 20% of the gross national product. In order to protect from adverse impacts both commercial and non-commercial biological resources of the Barents Sea, the Norwegian government introduced the concept of an Integrated Management Plan to be developed for the seas extending from Lofoten to the Barents Sea. This concept uses an ecosystem-based management approach and calls for a number of scientific studies to be in place before decision on opening an area for oil and gas activities can be made (Olsen et al. 2007). Until sufficient knowledge is in place, a strict policy should be adopted, known as the "zero discharge policy"' – a policy that prohibits all discharges of chemicals, oil and waste water to the sea. Several Norwegian fields are currently under development under this more stringent policy directive, e.g. Snøhvit and Goliath, and a number of exploratory wells have been drilled.

It has been questioned if the zero discharge policy is the best overall environmental policy because it leads to other increases of potentially more serious waste streams (e.g. increased CO_2 emissions and land storage of wastes) as well as greater occupational risks for humans from accidents. This policy also entails challenges in developing monitoring and assessment methodologies that can unequivocally establish environmental compliance to regulations. On the other hand, if a traditional risk-based management approach such as in the North Sea is to be applied in the Arctic, there are several challenges yet to be addressed. The risk assessment procedures currently used by the oil industry are developed for temperate areas and may therefore need to be adapted to Arctic species and Arctic environments.

Here we describe current risk assessment procedures for North Sea petroleum operations. Our purpose is to discuss the key challenges in adapting these procedures for use in Arctic environments. We ask the following questions:

- Current risk assessment procedures – are they appropriate for the Arctic seas?
- What are the main effects of petroleum activities on Arctic ecosystems?
- Have we sufficient ecosystem knowledge to perform risk assessments in Arctic seas?

The main challenges are related to lack of toxicity data on Arctic species and reformulation of the regulations. Herein, we focus the discussion on risk calculations for operational discharges (chronic effects), although acute discharges will be discussed briefly. We further present the status of ongoing research efforts toward addressing the identified challenges and suggest what remains to be done.

2 Current Risk Assessment Procedures

To comply with a "zero harmful discharge" policy for the North Sea marine environment, the petroleum industry of Norway developed a methodology to calculate the risk of harmful effects to the environment from petroleum operations. The approach involves a science-based prediction of the potential ecological effects of discharges of produced water and drilling mud. The model, known as the Environmental Impact Factor (EIF), provides a quantitative estimate of possible ecological risks on a regional-scale (Johnsen et al. 2000), following the procedures described in the EU-Technical Guidance Document (EU-TGD). The objective is to keep discharges within regulatory guidelines specified in permits issued by the Norwegian State Pollution Control Authority (SFT). These limits are based on the results from toxicity testing (acute or chronic).

A key step in the EIF impact assessment is to calculate a risk characterization ratio, where the predicted environmental concentration (PEC) of any pollutant is compared with the predicted no effect concentration (PNEC) (Fig. 28.1). The PEC value is based on modelling of chemical fate and exposure while PNEC is derived from toxicity data for different species. The PNEC value is the LC_{50} value (the concentration of a chemical which kills 50% of a sample population) of the most sensitive species divided by an assessment factor. Typically, LC_{50} values are

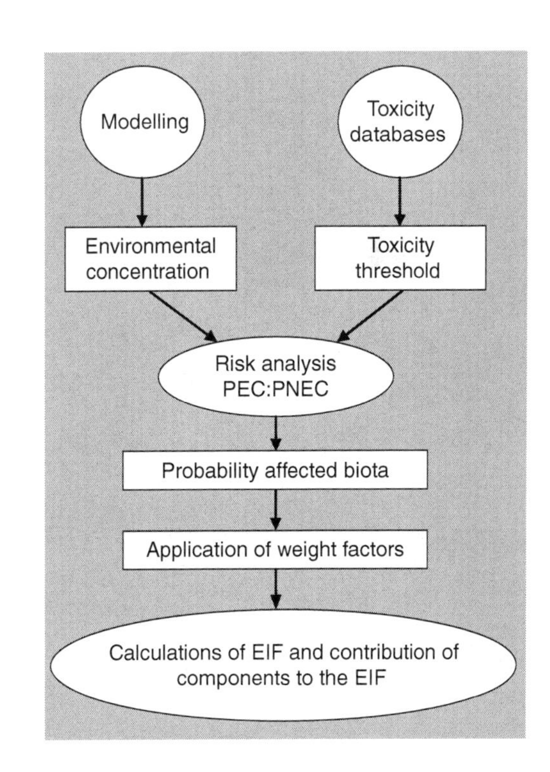

Fig. 28.1 Steps in the EIF produced water approach to risk assessment

derived from available toxicity test results and the assessment factor varies from 10 to 10,000 depending on the quality of the available toxicity data. If the PEC/PNEC is lower than 1, no risk reduction measures are required. If the ratio is higher than 1, the risk of effects to organisms is unacceptable and risk reduction measures must be implemented by the operator. While the toxicity assessment in the EIF-model has traditionally been based on the LC_{50} for the most sensitive species, a more recent development has been to apply a Species Sensitivity Distribution (SSD) curve in the toxicity assessment. In the SSD approach, EC_{50} (the concentration of a chemical which has effects on 50% of a sample population) or PNEC values for a number of species from various taxonomic and functional groups (at least 8) are needed. Therefore, a large toxicity database consisting of long-term tests (preferably 10–15) must be available to derive PNECs for risk assessment.

When enough toxicity data are available, the SSD approach is considered a more ecosystem relevant approach. Although in recent years much research has been performed toward developing biomarkers as more sensitive early warning signals of the impact of chemicals in the environment, biomarkers have as of yet not been incorporated into the EIF risk assessment model. In some cases internal body burden concentrations of toxicants are used, e.g. alkylated phenols and some PAHs. Finally, PEC/PNEC values are incorporated into a fate and exposure model that defines an environmental impact factor. The model requires site-specific information on physical factors (currents, wind, etc.) and the physical and chemical properties of discharged chemicals in order to calculate chemical transport, dilution and fate in the environment. The EIF derived from the fate and exposure model represents the volume of seawater containing concentrations of produced water chemicals that exceed the pre-determined risk criterion (PEC/PNEC > 1). As long as the PEC/PNEC ratio is less than 1, no risk reduction measures are required by the operator.

These risk assessment tools have been developed for general use. However application in new regions, such as the Arctic, requires additional site-specific information both for the toxicity assessment used to derive PNEC values and for the fate and exposure modelling to derive the EIF factor.

3 Effects of Petroleum Activities on Arctic Ecosystems

In 2003, Chapman and Riddle published a paper in Marine Pollution Bulletin titled *Missing and Needed: Polar Marine Ecotoxicology*, which highlighted the lack and need of toxicity data for polar marine organisms. However, a number of studies published in the 1970s were not included in their review (Wells 2004). Nonetheless, their review highlighted that the bulk of available data are not suitable for risk assessment nor does the data provide a good understanding of short- and long-term effects of contaminants in Arctic ecosystems. To put this in perspective, the USEPA Ecotox database includes over 220,000 records for aquatic species alone, including data from tests on more than 4000 species and in excess of 7000 chemicals (Chapman and Riddle 2005). A recent review of Arctic toxicity studies (Camus et al. 2008) identified only ~38 peer-reviewed publications most of which reported on experiments measuring the biological effects of oil compounds (Table 28.1). The literature includes toxicity tests with 46 species from 14 different taxonomic groups. The dominant groups were crustaceans (20 species), amphipods (10 species), bivalues (5 species) and algae (9 species), as well as three studies on benthic communities. Most of these studies report on LC_{50} tests or sub-lethal tests of effects on growth and reproduction. Generally, LC_{50} values for Arctic species are lower than that for non-Arctic species. Sub-lethal effects were demonstrated on most of the test species (amphipods decapods, Arctic fish, molluscs and Arctic benthic communities), and early life stages seem to be more vulnerable to oil than adults (Table 28.1).

The data available in the published literature are of significant interest toward assessing the environmental risks to Arctic organisms, but the data are not directly comparable due to differences in methodologies, exposure scenarios and endpoints. The relevance of these data for EIF risk evaluation is presently unknown and therefore it is timely to extract and evaluate these data for this purpose. To extrapolate and understand potential cascading effects from individual organisms to the whole ecosystem as required in the ecosystem approach to environmental management adopted by the Norwegian authorities, further requires a substantial amount of new information on the structural and functional characteristics of Arctic ecosystems.

Table 28.1 Available peer-reviewed literature addressing ecotoxicology of dominant groups of Arctic organisms

Dominant groups	References
Amphipods	Busdosh and Atlas (1977), Camus and Olsen (2008), Percy (1977), Olsen et al. (2007b, 2008a, b), Percy (1976), Busdosh (1981), Carls and Korn (1985), Percy and Mullin (1977), Aunaas et al. (1991)
Decapods	Riebel and Percy (1990), Perkins et al. (2005), Brodersen (1977), Camus et al. (2002a, 2004)
Algae	Petersen and Dahllöf (2007), Hsiao (1970)
Benthic communities	Gulliksen and Taasen (1982), Olsen et al. (2008a, b)
Bacteria	Gerdes et al. (2005), Knowles and Wishart (1977)
Fish	Carls and Korn (1985), George et al. (1995), Christiansen and George (1995), Christiansen et al. (1996), Jonsson et al. (2003), Ingebrigsten et al. (2000), Aas et al. (2003), Christiansen (2000)
Benthos	Camus et al. (2000, 2003, 2002a, b), Olsen et al. (2007b), Mageau et al. (1987), Aarset and Zachariassen (1982)
Mysid	Chapman and McPherson (1993), Riebel and Percy (1990), Carls and Korn (1985)
Cnidaria, Isopods, Copepods, Gastropods	Percy and Mullin (1977), Hjorth and Dahllöf (2008), Strand et al. (2006)

4 Arctic Ecosystems

The Barents Sea is the most biologically productive among Arctic shelf seas, contributing approximately 49% of total primary production for the Pan-Arctic shelf environment (Stein and McDonald 2004). This productivity supports large populations of zooplankton, a rich benthic community, large numbers of migratory birds, large fish stocks and communities of marine mammals. The zooplankton species, *Calanus finmarchicus* and *Calanus glacialis*, together with capelin and herring are the keystone prey organisms in the Barents Sea, creating the basis for a rich assemblage of higher trophic level organisms and facilitating one of the world's largest fisheries. Atlantic cod, seals, whales, seabirds and man compete for a small proportion (< 1%) of the harvestable energy of this sea, and 80% of the harvestable energy is channelled through deep-water communities and benthos (Wassmann et al. 2006). Benthic crustaceans and decorator crabs, as well as amphipods (mainly northern areas), are very important in the diet of cod (key species in the Barents Sea), for example, while the diet of haddock consists of ~50% benthic organisms. The most numerous benthic faunal groups in the Barents Sea are bivalves, gastropods, polychaetes, echinoderms and crustaceans. Previously, benthic communities in Arctic regions were thought to have a lower diversity than benthic communities in temperate regions, but there is actually a moderate increase in diversity with latitude (Renaud et al. 2009).

The Barents Sea is a relatively shallow shelf area (average depth is 230 m), but with variable water depths, from deep troughs in the Hopen Trench system in the northeast to shallower areas (e.g. southwest of Novaya Zemlya) (Fig. 28.2). During winter, sea ice may cover up to 90% of the Barents Sea, although the southern Barents Sea remains ice-free year-round. Variable inflow of water masses of Atlantic and Arctic origin causes marked inter-annual and cyclic variabilities in water characteristics (temperature, salinity, turbidity, etc.) as well as in the extent of ice cover. Wind and temperature variations control water mass mixing and stratification in the Barents Sea. Winter convection and ice-free conditions result in a well-mixed water column to depths as great as 200–300 m; in spring, solar radiation heats the sea surface, leading to stratification of the upper water column by May–June (AMAP, Arctic Pollution 2002). A special hydrographical feature of the Barents Sea is the Polar Front. This is a region where relatively warm and saline water of Atlantic origin meets colder and less saline Arctic water (Fig. 28.2). The area is characterized by formation of eddies and by upwelling of deep water with high nutrient concentrations. The availability of nutrients supports high concentrations of plankton which in turn attract fish, sea birds and mammals. The precise location of the Polar Front varies with season and from year to year due to variations in the inflow of Atlantic water.

The dynamics and interplay of these complex physical processes in the Barents Sea are further linked to some special biological characteristics of organisms and their relationships to one another, e.g. behaviour, reproduction, species diversity, and food web interactions. Arctic species are generally longer-lived, and grow and reproduce slower than temperate species (AMAP, Arctic Pollution 2002). Seasonal

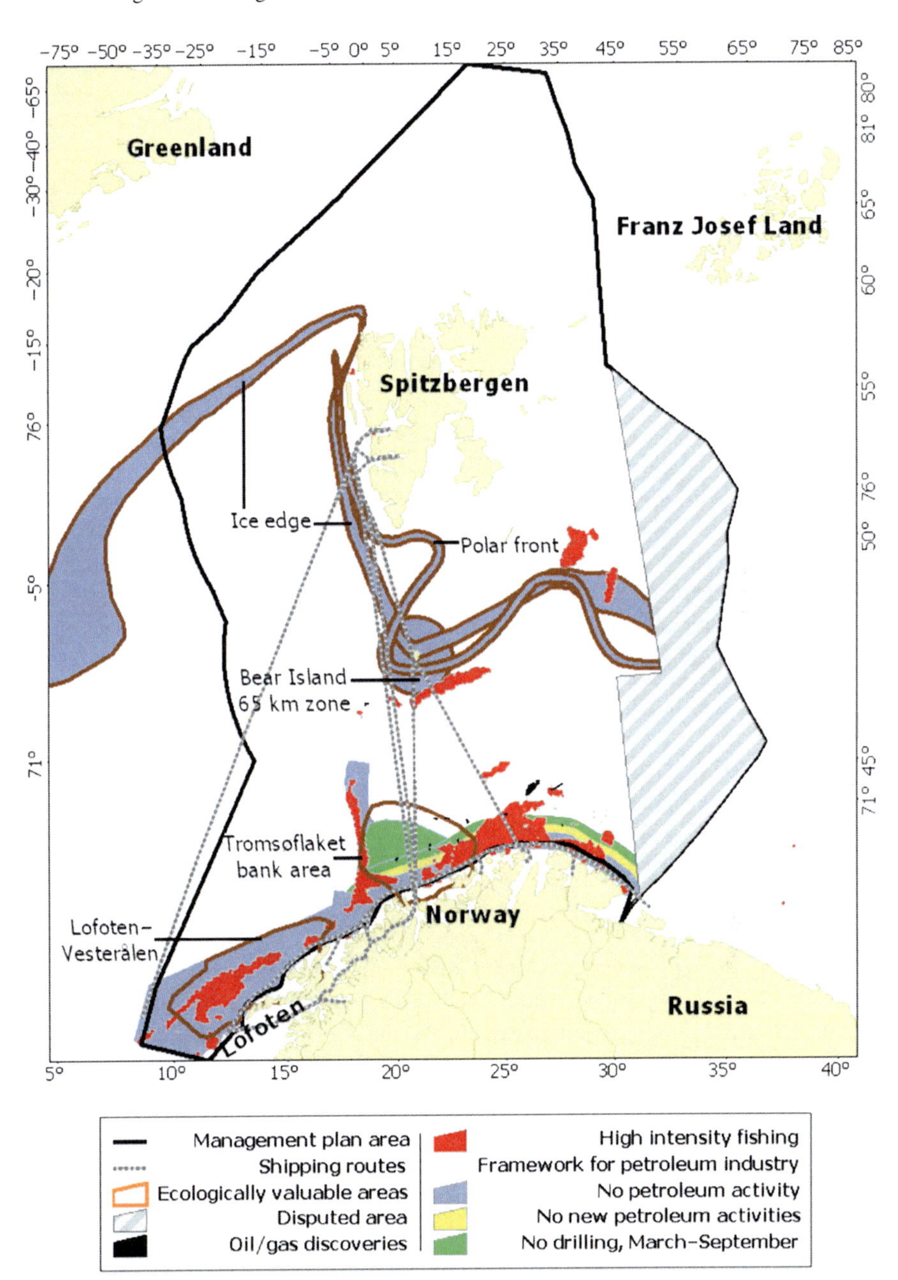

Fig. 28.2 Diverse interests and activities in the Barents Sea (ICES 2007)

variations in temperature and light in the Arctic result in short, intense summers where large numbers of resident and migratory animals undergo critical stages of their life cycle (e.g. nesting, rearing of young, moulting) within a relatively short period of time. This short period is for some species the only time during the year when reproduction occurs; other lower trophic level species (e.g. copepods) have several life cycles during a year (Hansen et al. 1996). The energy flux through the food web of the Barents Sea, from algae to marine mammals, is lipid based. High lipid levels, an adaptation to cold temperatures, are characteristic of Arctic organisms. Lipids are used for energy reserves, for insulation (mammals), and for buoyancy. These characteristics play a role in determining the uptake, transfer and effects of chemicals in Arctic food webs.

Climate variability and change, differences in recruitment, variable resource availability, harvesting restrictions and management schemes influence the changing state of the biological resources of the Barents Sea. Fish stocks, in particular, are already heavily influenced by anthropogenic activity, with fishing as the most important, and many of the stocks have shown large-scale population fluctuations over the last 25 years (IMR/PINRO Joint report series 3 2006). In context with future industrial development, research coupled to monitoring will be necessary as a basis for assessing ecosystem health and to ensure that both the causes and consequences of major ecosystem alterations are not misidentified.

5 Discussion

5.1 Current Risk Assessment Procedures – Are They Appropriate for the Arctic?

The risk assessment tools described previously have been shown to be a robust methodology to assess impacts from petroleum operations in the North Sea. However application in new regions, such as the Arctic, requires additional site-specific information for toxicity assessments and for modelling chemical fate and exposure. As pointed out by Chapman and Riddle (2005), the bulk of available data is not on Arctic species nor does the data provide an adequate foundation to evaluate short- and long-term effects of contaminants in Arctic ecosystems.

5.1.1 Acquire LC_{50} Data for Arctic Organisms

As the basis for the calculation of PNEC values are LC_{50} values for different chemicals for the most sensitive species, a major research effort is needed to derive these data. Although SSD curves exist for temperate regions, none have been developed for the Arctic. Therefore, toxicity assessments performed for Arctic areas based on the SSD approach require similar LC_{50} data but for a much larger number of

Table 28.2 Relevant test species for the Barents Sea; selection criteria are based on the SSD approach, that is, species from different taxonomic groups, different habitats and feeding modes

Taxon	Classification	Habitat	Trophic level
Calanus glacialis	Arthropoda	Ice edge, pelagic	Grazer
Gammarus wilkitzkii	Arthropoda	Ice edge	Grazer, scavenger
Boreogadus saida	Vertebrata	Ice edge, pelagic	Planktivore
Skeletonema costatum	Vertebrata	Pelagic	Primary consumer
Calanus hyperboreus	Chrysophyta	Pelagic	Grazer
Gadus morhua	Arthropoda	Pelagic	Predator
Mya truncata	Vertebrata	Benthic	Filter feeder
Liocyma fluctuosa	Mollusca	Benthic	Deposit feeder
Ophiura/Ophioctin	Echinodermata	Benthic	Suspensioin/deposit feeder
Myoxocephalus scorpea	Vertebrata	Benthic	Predator/scavenger
Paralithodes camtschatica	Arthorpoda	Benthic	Predator/scavenger

Arctic species. Recommended species must represent different taxonomic groups, habitats and different feeding modes to achieve an integrated ecosystem approach for the Arctic. Based on these criteria, Arctic test species for determination of LC_{50} values and calculations of Arctic SSD curves are presented in Table 28.2. The species chosen are important components of the Arctic ecosystem and can be widely found in the Barents Sea. They have wide species distributions, available on a pan-Arctic geographic scale, making them suitable test species for the Arctic as a whole.

5.1.2 Incorporate a Benthic Compartment into Toxicity and Fate Models

When risk calculations are currently performed, EIF values are based on the modelling of chemical transport, dilution and fate in the surface water column layer only. However, in the Arctic, there is strong benthic–pelagic coupling because primary consumers (herbivores) are not able to consume all the organic matter that is produced during the intense spring bloom. A large fraction of organic matter reaches the sea bottom to provide food to benthic populations and 80% of the harvestable primary production of the Barents Sea is channelled through deep-water communities and benthos (Wassmann et al. 2006). Hence, when adapting risk assessment tools to Arctic areas, it may be necessary to perform EIF calculations for deeper water layers. Furthermore, Arctic benthic invertebrates are especially relevant from a monitoring perspective as well (e.g. Ambrose and Renaud 1995; Wassmann et al. 2006). In low-temperature environments, oil compounds will adsorb more readily to particles. Higher material fluxes and a strong affinity of oil for particles suggests that in the Arctic, a larger proportion of oil compounds will be transferred to the sea floor, facilitating exposure of benthic organisms.

5.2 *What Do We Still Need to Know About Effects of Petroleum Activities on Arctic Organisms and Ecosystems?*

Chapman and Riddle (2005) outlined future research needs for producing polar ecotoxicology data which summarizes well, to our knowledge, the general research objectives needed for the Arctic. The main research challenges are as follows.

5.2.1 Direct, Immediate Acute Impacts (Physical and Toxicological)

Among the points raised by Chapman and Riddle (2005), the "LC_{50} test of 96 h exposure period is not appropriate for polar organisms which tend to have longer acute response times." Indeed in deriving LC_{50} values for Arctic organisms, a number of difficulties has arisen (Honkanen et al. 2008). For example, LC_{50} values are not always achieved when applying the standard test protocol (i.e. 100% mortality is not achieved). Performing tests at low water temperatures reduce chemical solubility, while the low metabolism of polar organisms slows down chemical uptake. Thus LC_{50} may not be an appropriate endpoint for risk assessment in the Arctic. Mean Survival Time (MST), which is the time needed at a respective concentration of a chemical to kill 50% of organisms, may be a more appropriate endpoint (Carls and Korn 1985). The difficulty to define LC_{50} values for Arctic organisms calls for a need to define a common toxicity protocol for acute tests on polar animals. Since most risk and impact assessment models/tools are based on LC_{50} data, it is critical and of high importance that guidelines for toxicity tests be established and accredited at the international level.

5.2.2 Direct but Delayed Sub-lethal Impacts Compromising Life History

Current risk assessments are performed mainly on adult organisms. While adult Arctic species have been shown to be relatively robust to oil exposure, the health and growth of early life stages and subsequent reproduction appear to be more sensitive (Peterson et al. 2003). Although the PNEC calculation indirectly accounts for uncertainty in the risk measure through inclusion of an assessment factor, the use of assessment factors is not linked to specific causes of uncertainty. If further experiments indicate that early life stages are predominantly the most sensitive life forms, their use rather than adults as test species for LC_{50} values will be a more ecologically relevant approach to establishing a conservative measure of risk in the assessment procedure. Comparison of LC_{50} values among adults and their early life stages (Table 28.2) could serve as a starting point for such an evaluation.

5.2.3 Indirect Effects on Trophic Structure

Subtle long-term effects such as trophic and interaction cascades (e.g. loss of organisms that serve as or create important physical structure in the environment), leading to postponed recovery of populations may occur after an acute oil spill (Peterson et al. 2003). These potential effects are currently not accounted for in oil spill risk assessments which consider death as the only endpoint. Incorporating sub-lethal effects could involve adding a parameter to account for sub-lethal effects to the estimation of ecosystem recovery time. Another approach would be to include biomarkers of effects into risk assessment procedures. To reach this step, more research on Arctic organisms is needed to evaluate sub-lethal effects, for example, related to species' growth and reproduction. A range of different Arctic species with different ecology, feeding mode and habitat must be tested and evaluated for their applicability in risk assessment procedures. If done properly, a more ecologically relevant risk assessment approach will be achieved (Depledge and Galloway 2005). Some preliminary steps in this direction have been taken for the EIF (produced water) model where it is possible to include some long-term effects for alkylated phenols and selected PAHs.

5.3 Have We Enough Basic Knowledge About Arctic Ecosystems?

Arctic ecosystems exhibit a number of key differences from temperate ecosystems. These differences are better known today due in part to a number of large multi-disciplinary ecosystem investigations carried out in recent years (e.g. Wassmann et al. 2008; Deming et al. 2002; Carmack et al. 2006; Grebmeier et al. 2006; Fortier and Cochran 2008). While new knowledge has been acquired, attempts to assess individual species' and ecosystem responses to a variety of human impacts still suffer from a lack of some key information. This limits the pace of development for risk assessment tools for the Arctic. As a corollary to the ecosystem understanding, we have today (section 4), here we highlight three relevant needs.

5.3.1 Acquire More Information on Basic Life History Traits of Arctic Species

Arctic organisms are specially adapted to their harsh environmental conditions, such as cold temperatures and long periods of starvation. Little is known about how these and other adaptations influence the sensitivity of Arctic biota to anthropogenic chemicals in their environment. For example, polar marine surface waters are characterized by high levels of dissolved oxygen, seasonally intense UV irradiance and high levels of dissolved organic carbon. A recent study found evidence that compounds in the body of an amphipod protect the organism from UV irradiation (R. Krapp, personal communication) and furthermore that amphipod species possess a stronger antioxidant capacity during summer than winter (Krapp et al. 2009). Low food availability in polar marine waters may result in organisms diverting some energy into the maintenance of high anti-oxidant (AOX) defence systems and that high AOX may explain the relatively long lifespan of most polar organisms (Camus et al. 2005). However, UV exposure has been shown to inhibit the total oxyradical scavenging capacity in some amphipod species (R. Krapp, personal communication). Seasonal differences in an organism's resistance to oxyradical stress will have consequences for oil toxicity associated with exposure at different times of the year. However, in general few studies have been carried out on this issue. A solid knowledge base on the relationships between polar adaptations and sensitivity to chemicals in Arctic environments is still lacking.

5.3.2 Understand What Changes Are Occurring in the Arctic

There has been a spectacular recent acceleration of Arctic ice loss in summer 2007, and indeed in all years since the 1970s (Stroeve et al. 2007). There is also mounting evidence that ecosystem response to certain types or magnitudes of extrinsic pressures (climate, human impacts, etc.) is often abrupt and non-linear, leading to a significant reorganization of system properties and processes: regime shifts (Scheffer et al. 2001). Regime shifts can result in alterations of the most basic ecosystem parameters, including food web structure, the flow of organic matter and nutrients through the ecosystem or the patterns of space occupation, leading to a cascade of changes in the ecosystem. Because the Arctic is warming about three times faster than the global rate (ACIA 2004), Arctic ecosystems are likely to encounter climate-driven thresholds and tipping points leading to abrupt changes much sooner than in other ecosystems. Thus our current understanding is transitory. Continuing investigation of ecosystem structure and function is necessary to ensure that risk assessment procedures are environmentally relevant and that changes caused by other impact factors are not erroneously assigned to the petroleum industry. The Barents Sea, with its balance between Arctic and Atlantic water and its sharp Polar Front, is particularly susceptible to regime shifts in response to climate changes.

5.3.3 Scale Up to a Pan-Arctic Ecosystem Understanding

A number of Arctic ecosystem campaigns have been conducted during the last decade (e.g. Deming et al. 2002; Carmack et al. 2006; Grebmeier et al. 2006; Fortier and Cochran 2008). These have highlighted key differences in the structure and function of the Arctic shelf seas. There is a continuing need to support efforts toward integrating knowledge across regions, disciplines and national boundaries, into a holistic, pan-Arctic understanding of ecosystem functioning (Carmack and Wassmann 2006; Wassmann 2006). This must be achieved through support for the coordination of multi-year, field-based, multi-disciplinary, process studies in key regions such as the Seasonal Ice Zone and the central Arctic. Such efforts will yield dividends for the petroleum industry through the acquisition of new ecosystem knowledge that can be used to develop more flexible Arctic risk assessment tools with general application, and thereby eliminate the need to "reinvent the wheel" when moving into new areas of exploration and development. For example, indicators of responses to anthropogenic impacts that are representative for the ecosystem that can be applied similarly throughout the Arctic must be defined and developed through research. Benthic invertebrates are especially relevant because of strong pelagic–benthic coupling of Arctic food webs (e.g. Ambrose and Renaud 1995; Wassmann et al. 2006).

6 Conclusion

The major challenge facing the petroleum industry toward adapting risk assessment tools from the North Sea to the Arctic is the acquisition of sufficient LC_{50} data for Arctic species, both for the assessment of toxicity in sensitive species and for the development of SSD curves. When such information exists, it is possible to calculate PEC/PNEC ratios to evaluate harmful effects concentrations using the EIF risk assessment approach. If EIF assessments indicate little risk of harmful effects, then the basis for implementing a zero harmful strategy similar to what is in use in the North Sea today should be reconsidered. However, as the Barents Sea is one of the most biologically diverse and productive marine ecosystems within the Arctic, there is a need for special considerations when oil-related activities are being established in the area. Further research is needed to understand sub-lethal and long-term impacts of acute and chronic exposure to petroleum compounds as well as to increase the available knowledge on species life history and the understanding of Arctic ecosystem processes.

References

Aarset AV, Zachariassen KE (1982) Effects of oil pollution on the freezing tolerance and solute concentration of the blue mussel *Mytilus edulis*. Mar Biol 72(1):45–51

Aas E, Liewenborg B, Grosvik BE, Camus L, Jonsson G, Borseth JF, Balk L (2003) DNA adduct levels in fish from pristine areas are not detectable or low when analysed using the nuclease P1 version of the P-32-postlabelling technique. Biomarkers 8(6):445–460

ACIA (2004) Arctic climate impact assessment. Impacts of a warming Arctic: Arctic climate impact assessment. Cambridge University Press, Cambridge

AMAP (2002) Arctic pollution 2002, persistent organic pollutants, heavy metals, radioactivity, human health, changing pathways. AMAP, Oslo, Norway

Ambrose WG Jr, Renaud PE (1995) Benthic response to water column productivity patters- evidence for benthic-pelagic coupling in the Northeast Water Polynya. J Geophys Res Ocean 100(C3):4411–4421

Aunaas T, Olsen A, Zachariassen KE (1991) The effects of oil and oil dispersants on the amphipod *Gammarus oceanicus* from arctic waters. Polar Res 10(2):619–630

Brodersen CC, Rice SD, Short JW, Mecklenburg TA Karinen JF (1977) Sensitivity of larval and adult Alaskan shrimp and crabs to acute exposures of the water- soluble fraction of Cook inlet crude oil; (IN) 1977 Oil Spill Conf. (Prevention, Behavior, Control, Cleanup), API, EPA, and USCG, Washington, DC, pp 575–578, RU 0072

Busdosh M, Atlas RM (1977) Toxicity of oil slicks to arctic amphipods. Arctic 30(2):85–92

Busdosh M (1981) Long-term effects of the water-soluble fraction of Prudhoe Bay crude-oil on survival, movement and food search success of the arctic amphipod B*oeckosimus* (=*Onisimus*) *affinis*. Mar Environ Res 5(3):167–180

Camus L, Birkely SR, Jones MB, Børseth JF, Grosvik BF, Gulliksen B, Lonne OJ, Regoli F, Depledge MH (2003) Biomarker responses and PAH uptake in *Mya truncata* following exposure to oil-contaminated sediment in an Arctic fjord (Svalbard). Sci Total Environ 308(1–3):221–234

Camus L, Grosvik BE, Borseth JF, Jones MB, Depledge MH (2000) Stability of lysosomal and cell membranes in haemocytes of the common mussel (Mytilus edulis): effect of low temperatures. Mar Environ Res 50(1–5):325–329

Camus L, Davies PE, Spicer JI, Jones MB (2004) Temperature-dependent physiological response of Carcinus maenas exposed to copper. Mar Environ Res 58(2–5):781–785

Camus L, Gulliksen B, Depledge MH, Jones MB (2005) Polar bivalves are characterized by high antioxidant defences. Polar Res 24(1–2):111–118

Camus L, Jones MB, Borset JF (2002a) Heart rate, respiration and sum oxyradical scavenging capacity of the Arctic spider crab *Hyas araneus*, following exposure to polycyclic aromatic compounds via sediment injection. Aquat Toxicol 62(1–2):1–13

Camus L, Jones MB, Børseth JF, Grosvik BE, Regoil F, Depledge MH (2002b) Total oxyradical scavenging capacity and cell membrane stability of haemocytes of the Arctic scallop, *Chlamys islandicus*, following benzo(a)pyrene exposure. Mar Environ Res 54(3–5):425–430

Camus L, Olsen GH, Carroll M (2008) Arctic ecotoxicological studies review. APN3975, 54 pp

Carls MG, Korn S (1985) Sensitivity of Arctic marine amphipods and fish to petroleum hydrocarbons. Can Tech Rep Fish Aquat Sci 1368:11–26

Carmack E, Barber D, Christensen J, Macdonald R, Rudels B, Sakshaug E (2006) Climate variability and physical forcing of the food webs and the carbon budget on Pan-Arctic shelves. Prog Oceanogr 71:145–182

Carmack E, Wassmann P (2006) Food webs and physical-biological coupling on pan-Arctic shelves: comprehensive perspectives, unifying concepts and future research. Prog Oceanogr 71:446–477

Chapman P, McPherson C (1993) Comparative zinc and lead toxicity tests with Arctic marine invertebrates and implications for toxicant discharges. Polar Record 29(168):45–54

Chapman P, Riddle JM (2005) Polar marine toxicology—future research needs. Mar Pollut Bull 50(9):905–908

Christiansen JS, Dalmo RA, Ingebrigsten K (1996) Xenobiotic excretion in fish with aglomerular kidneys. Mar Ecol Prog Ser 136:303–304

Christiansen JS, George SG (1995) Contamination of food by crude oil affects food selection and growth performance, but not appetite, in an Arctic fish, the polar cod (Boreogadus saida). Polar Biol 15:277–281

Christiansen JS (2000) Sex differences in ionoregulatory responses to dietary oil exposure in polar cod. J Fish Biol 57:167–170

Deming JW, Fortier L, Fukuchi M (2002) The international North water Polynya study (NOW): a brief overview. Deep Sea Res II 49:4887–5295

Depledge MH, Galloway TS (2005) Healthy animals, healthy ecosystem. Front Ecol Environ 3(5):251–258

Fortier L, Cochran JK (2008) Introduction to a special section on annual cycles on the Arctic Shelf. J Geophys Res 113:C03S00. doi:10.1029/2007JC004457

Greenhill L (2007) On the relevance of the Norwegian Zero discharge policy with particular reference to the Norwegian Barents Sea. Master thesis Heriot-Watt University School of Life Science

George SG, Christiansen JS, Killie B, Wright J (1995) Dietary crude oil exposure during sexual maturation induces hepatic mixed function oxygenase (CYP 1A) activity at very low environmental temperatures in polar cod Boreogadus saida. Mar Ecol Prog Ser 122: 307–312

Gerdes B, Brinkmeyer R, Dieckmann G, Helmke E (2005) Influence of crude oil on changes of bacterial communities in Arctic sea-ice. Fems Microbiol Ecol 53(1):129–139

Grebmeier JM, Cooper LW, Feder HM., Sirenko BI (2006) Ecosystem dynamics of the Pacific-influenced Northern Bering and Chukchi Seas in the Amerasian Arctic. Prog Oceanogr 71: 331–361

Gulliksen B, Taasen JP (1982) Effect of an oil-spill in Spitzbergen in 1978. Mar Pollut Bull 13(3):96–98

Hansen B, Christiansen S, Pedersen G (1996). Plankton dynamics in the marginal ice zone of the central Barents Sea during spring: Carbon flow and structure of the grazer food chain. Polar Biol 16:115–128

Hjorth M, Dahllöf I (2008) A harpacticoid copepod Microsetella spp. from sub-Arctic coastal waters and its sensitivity towards the polyaromatic hydrocarbon pyrene. Polar Biol 31(12):1437–1443

Honkanen JO, Källkvist T, Camus L (2008). Acute toxicity of glutaraldehyde and 3,5-dichlorophenol in the Polar cod (Boreogadus saida) at two temperatures. In: Akvaplan-niva AS (ed). Tromsø, Norway, p 20

Hsiao SIC (1978) Effects of crude oils on growth of arctic marine-phytoplankton. Environ Pollut 17(2):93–107

Ingebrigtsen K, Christiansen JS, Lindhe Ö, Brandt I (2000) Disposition and cellular binding of 3H-benzo(a)pyrene at subzero temperatures: studies in an aglomerular Arctic teleost fish – the polar cod (Boreogadus saida) Polar Biol 23:503–509

Johnsen S, Frost TK, Hjelsvold M, Utvik TR (2000) The environmental impact factor– a proposed tool for produced water impact reduction, management and regulation. SPE paper 61178. SPE International Conference, 26–28 June 2000, Stavanger, Norway

Jonsson G, Beyer J, Wells D, Ariese F (2003) The application of HPLC-F and GC-MS to the analysis of selected hydroxy polycyclic hydrocarbons in two certified fish bile reference materials. J Environ Monit 5(3):513–520

Knowles R, Wishart C (1977) Nitrogen-fixation in arctic marine-sediments – effect of oil and hydrocarbon fractions. Environ Pollut 13(2):133–149

Krapp RH, Bassinet T, Berge J, Pampanin DM, Camus L (2009) Antioxidant responses in the polar marine sea-ice amphipod *Gammarus wilkitzkii* to natural and experimentally increased UV levels. Aquat Toxicol 94:1–7

Mageau C, Engelhardt FR, Gilfillan ES, Boehm PD (1987) Effects of short-term exposure to dispersed oil in Arctic invertebrates. Arctic 40:162–171

Olsen E, Gjøsæter H, Røttingen I, Dommasnes A, Fossum P, Sandberg P (2007) The Norwegian ecosystem-based management plan for the Barents Sea. Ices J Mar Sci 64:599–602

Olsen GH, Carroll ML, Renaud PE, Ambrose WG Jr, Olssøn R, Carroll J (2007a) Benthic community response to petroleum-associated components in Arctic versus temperate marine sediments. Marine Biol 151:2167–2176

Olsen GH, Carroll J, Berge J., Sva E, Camus L (2008a) Energy budget in the Arctic sea ice amphipod *Gammarus wilkitzkii* exposed to the water soluble fraction of oil. Mar Environ Res 66:213–214

Olsen GH, Carroll J, Sva E, Camus L (2008b) Cellular energy allocation in the Arctic sea ice amphipod *Gammarus wilkitzkii* exposed to the water soluble fractions of oil. Mar Environ Res 66(1):215–216

Olsen GH, Sva E, Carroll J, Camus L, De Coen W, Smolders R, Øveraas H, Hylland K (2007b) Alterations in energy budget of Arctic benthic species exposed to oil related compounds. Aquat Toxicol 83:85–92

Percy JA (1976) Responses of Arctic marine crustaceans to crude oil and oil-tainted food. Environ Pollut 10(2):155–162

Percy JA (1977) Responses of Arctic marine benthic crustaceans to sediments contaminated with crude-oil. Environ Pollut 13(1):1–10

Percy JA, Mullin TC (1977) Effects of crude oil on the locomotery activity of Arctic marine invertebrates. Mar Pollut Bull 8(2):35–40

Perkins RA, Rhoton S, Behr-Andres C (2005) Comparative marine toxicity testing: a coldwater species and standard warm-water test species exposed to crude oil and dispersant. Cold Reg Sci Technol 42(3):226–236

Peterson CH, Rice SD, Short JW, Esler D, Bodkin JL, Ballachey BE, Irons DB (2003) Long-term ecosystem response to the *Exxon Valdez* oil spill. Science 302:2082–2086

Petersen DG, Dahllof I (2007) Combined effects of pyrene and UV-light on algae and bacteria in an arctic sediment. Ecotoxicology 16(4):371–377

Pinro/IMR report on the state of the Barents Sea ecosystem (2005/2006) Joint report series nr. 3. 2006

Renaud PE, Webb TJ, Bjørgesaeter A, Karakassis I, and others (2009) Continental-scale patterns in benthic invertebrate diversity: insights from the MacroBen database. Mar Ecol Prog Ser 382:239–252

Riebel PN, Percy JA (1990) Acute toxicity of petroleum hydrocarbons to the Arctic shallow-water mysid, mysis oculatat (Fabricus). Sarsia 75:223–232

Scheffer M, Carpenter S, Foley JA, Folke C, Walker B (2001) Catastrophic shifts in ecosystems. Nature 413:591–596

Stein R, Macdonald RW (2004) Organic carbon budget: Arctic vs. global ocean. In: Stein R, Macdonald RW (eds) The organic carbon cycle in the Arctic Ocean. Springer, Berlin, Heidelberg, pp 315–322

Strand J, Glahder CA, Asmund G (2006) Imposex occurrence in marine whelks at a military facility in the high Arctic. Environ Pollut 142(1):98–102

Stroeve J, Holland MM, Meier W, Scambos T, Serreze M (2007) Arctic sea ice decline: Faster than forecast. Geophys Res Lett 34:L09501

Wassmann P (2006) Structure and function of contemporary food webs on Arctic shelves: an introduction. Prog Oceanogr 71(2–4):123–128

Wassmann P, Carroll J, Bellerby RGJ (2008) Carbon flux and ecosystem feedback in the northern Barents Sea in an era of climate change: an introduction. Deep-Sea Res II 55(20–21): 2143–2453

Wassmann P, Reigstad M, Haug T, Rudels B, Carroll ML, Hop H, Gabrielsen GW, Falk-Petersen S, Denisenko SG, Arashkevich E, Slagstad D, Pavlova O (2006) Food webs and carbon flux in the Barents Sea. Prog Oceanogr 71(2–4):232–287

Wells PG (2004) The editorial 'Missing and needed: polar marine ecotoxicology' by P.M. Chapman and M.J. Riddle. Mar Pollut Bull 48(5–6):604–605

Chapter 29
Produced Water Management Options and Technologies

John A. Veil

Abstract Produced water is by far the largest volume by-product or waste stream associated with oil and gas exploration and production. Because of the large volumes involved, management of produced water presents important costs to the industry. This chapter describes the broad range of options that may be used to manage produced water. In some situations, technologies can be employed to reduce the volume of water that is managed within the well or at the platform deck. In other situations (primarily at onshore wells), produced water can be treated and then reused for various purposes. Produced water can be disposed using discharge, injection, removal to an offsite disposal facility, and evaporation. These management options are described along with numerous technologies currently used by the international oil and gas industry for treating produced water. Because this book focuses primarily on offshore produced water, this chapter will emphasize technologies for treating and managing offshore produced water. However, the chapter will include summary descriptions of other management practices and technologies used for onshore wells, too.

1 Introduction

Much of the information presented in this chapter is based on a 2004 white paper prepared by Argonne National Laboratory (Argonne) for the US Department of Energy (DOE) (Veil et al. 2004). Several years later, Argonne expanded the information from the white paper and created the Web-based Produced Water Management Information System (PWMIS), now housed on DOE's website at http://www.netl.doe.gov/technologies/PWMIS/. PWMIS provides separate fact sheets on 25 different produced water management practices and technologies in the Technology Description module. In a separate interactive Technology Identification Module, users are asked a series of questions, most of which have "yes/no" answers. Depending on how the user answers the questions, PWMIS can suggest a short list of technologies that are applicable to the user's well location. Although outside the scope of this chapter, another part of PWMIS, the Regulatory Module, provides

J.A. Veil (✉)
Veil Environmental, LLC, Annapolis, MD, USA

K. Lee, J. Neff (eds.), *Produced Water*, DOI 10.1007/978-1-4614-0046-2_29,

online access to the produced water regulations for federal and state agencies in the United States.

Although this chapter describes many produced water management technologies and practices, it does not include drawings or photographs of the different technologies. Readers are encouraged to visit PWMIS, which contains many images that help in understanding the nature of the produced water management technologies.

1.1 What Is Produced Water?

Produced water is water trapped in underground formations that is brought to the surface along with oil or gas. Because the water has been in contact with the hydrocarbon-bearing formation for centuries, it contains some of the chemical characteristics of the formation and the hydrocarbon itself. It may include water from the formation, water injected into the formation, and any chemicals added during the production and treatment processes. Produced water may also be referred to as "brine" or "salt water." The major constituents of concern in produced water are

- salt content (salinity, total dissolved solids, electrical conductivity)
- oil and grease (this is a measure of the organic chemical compounds)
- various natural inorganic and organic compounds or chemical additives used in drilling and operating the well
- naturally occurring radioactive material

Produced water is not a single commodity. The physical and chemical properties of produced water vary considerably depending on the geographic location of the field, the geological host formation, and the type of hydrocarbon product being produced. Produced water properties and volume can even vary throughout the lifetime of a reservoir.

1.2 How Much Produced Water Is Generated?

Produced water is a very large volume by-product or waste stream associated with oil and gas exploration and production. According to a recent data compilation, 867,853 wells were producing crude oil worldwide as of December 31, 2008 (Oil & Gas Journal 2008), nearly 500,000 of which are found in the United States. Hundreds of thousands of other natural gas wells also generate produced water. Approximately 21 billion barrels (bbl)[1] of produced water are generated each year in the United States (Clark and Veil 2009). Khatib and Verbeek (2003) estimate that, in 1999, an average of 210 million bbl of water was produced each day worldwide. This volume represents about 77 billion bbl of produced water for that entire year.

[1] 1 bbl = 42 US gallons, or about 0.16 m^3

Early in the life of an oil well, the oil production is high and water production is low. Over time, the oil production decreases and the water production increases. Another way of looking at this is to examine the ratio of water to oil.

- Worldwide estimate – 2:1 to 3:1
- US estimate – 7:1, because many US fields are mature and past their peak production
- Many older wells in the United States and other countries have ratios >50:1

At some point, the cost of managing the produced water exceeds the profit from selling the oil. When this point is reached, the well is shut in.

Coal bed methane wells are developed and produced in a much different manner. A well is drilled into a coal seam, then the ground water present in the coal seam is pumped out of the ground as rapidly as possible. Eventually, the hydrostatic pressure in the coal seam changes to allow the methane to be released from the coal matrix and move to the well for collection. Initially the well produces a large volume of water, but the volume declines over time. Methane production starts low, builds to a peak, and then decreases.

2 Principles of Produced Water Management

Historically, produced water was managed in ways that were the most convenient or least expensive. Today, many companies recognize that water can be either a cost or a value to their operations. For example, Shell has established a formal Water-to-Value program through which the company attempts to minimize the production of water, reduce the costs of water treatment methods, and look for ways in which existing facilities can handle larger volumes of water (Khatib and Verbeek 2003). Greater attention to water management allows production of hydrocarbons and the concomitant profits to remain viable.

Produced water management typically differs between onshore and offshore facilities. This is partly due to the space and weight restrictions at most offshore sites. Also, the primary contaminant of concern typically differs between onshore discharges (salt content) and offshore discharges (oil and grease level).

This chapter describes produced water management technologies and strategies in terms of a three-tiered water management or pollution prevention hierarchy (i.e., minimization, recycle/reuse, and disposal). The PWMIS website follows the same approach. Companies are encouraged to first evaluate practices that reduce the volumes of produced water entering the well or being handled at the surface. In this water minimization tier, processes are modified, technologies are adapted, or products are substituted so that less water is generated. When feasible, water minimization can often save money for operators and results in greater protection of the environment.

For the water that is still produced following water minimization, operators move next to the second tier, in which water is reused in a beneficial manner or recycled. In some situations, these more desirable management options may not be practical, cost effective, or permitted by the regulatory agencies. Then water must be disposed of by injection, discharge, evaporation, or removal offsite to a commercial water disposal facility.

Prior to reusing or disposing produced water, companies often must first treat the produced water. As part of the discussion of the disposal tier, this chapter provides information on a variety of technologies that can be used to treat different components of the produced water. In particular, the options are grouped into technologies for removing (a) salts and other inorganic chemicals and (b) oil and grease and other organic chemicals.

3 Basic Separation

When a well has been completed in a productive formation, a mixture of reservoir fluids enters the wellbore and is brought to the surface, either pushed by internal reservoir pressure or lifted with some type of pump. The mixture contains three types of fluids (crude oil, natural gas, and produced water) and often some solids. As produced fluids are brought to the surface, the first step in managing produced water is basic separation of fluids (oil, gas, water) and solids from one another. Once the materials have been separated, companies can use a variety of technologies and practices to manage the water.

Segregation of fluids is typically accomplished by gravity separation in a horizontal or vertical separator. Gas rises to the top of the tank where it can be collected. The oil layer floats on top of the water layer. Strategically placed outflows can collect each fluid layer separately. A common type of separator is the "free-water knockout tank." The separated oil stream may contain some water, and the water stream may contain additional dissolved hydrocarbons or emulsified oil.

Emulsions are droplets of one liquid suspended in another liquid. Since in most cases they cannot be removed through basic gravity separation, emulsions typically require additional treatment. One common oil field treatment method used to break emulsions is the application of heat generated by burning gas and passing the hot exhaust gas through a pipe running through the middle of a tank known as a "heater-treater." This treatment helps to break emulsions, thereby allowing gravity separation to take place. Other approaches used to break emulsions involve electrostatic precipitation and emulsion-breaking chemicals (demulsifiers).

Chemicals are used in many different steps of produced water management. They are an important part of optimizing the performance of many of the technologies described elsewhere in this chapter. Proper chemical selection, the point of chemical application within the water handling system, and chemical dosage contribute to the effectiveness of the water treatment. This chapter does not include any detailed discussion of the use of chemicals.

At some point, the oil is either sent to a refinery or stored in tanks until it can be collected by truck. Likewise, the produced water is either sent directly for management or stored in tanks. Many onshore fields contain a series of tanks to hold and store crude oil and produced water. These are the "tank batteries." Certain sites have one tank battery per well. More commonly, however, one tank battery serves several wells in the same area.

Any solids carried into the separators and storage tanks settle to the bottom and form a layer of sediment or sludge. The solids can be sand particles from the formation, solids left over from the well stimulation, scale or corrosion products, cement particles from well construction, or other materials. The solids – "tank bottoms" – accumulate and eventually must be removed, usually by a vacuum truck hauling the material offsite for recycling or disposal. Sediment and sludge from separators and tanks are considered to be exploration and production waste.

4 Water Minimization Technologies

Each additional barrel of produced water that is generated increases the costs of pumping it to the surface and storing, treating, and managing it. An optimal solution for reducing costs is to reduce the amount of water being handled. Two situations are described in the following sections – reduction of the volume of water entering the well and reduction of the volume of water reaching the surface.

4.1 Technologies to Reduce the Volume of Water Entering the Well

Two primary technologies can be used to restrict water from entering the wellbore: mechanical blocking devices and chemicals that shut off water-bearing channels or fractures within the formation, preventing water from making its way into the well.

4.1.1 Mechanical Blocking Devices

Various mechanical and well construction techniques have been used to block water from entering the well. Seright et al. (2001) provide a good discussion of the following techniques.

- Straddle packers
- Bridge plugs
- Tubing patches
- Cement
- Wellbore sand plugs
- Well abandonment
- Infill drilling

- Pattern flow control
- Horizontal wells

Not all types of reservoir or well construction problems can be effectively mitigated by mechanical devices. When considering water reduction approaches, operators should first diagnose the specific cause of the increased water production. Failure to determine the cause can lead to selecting a solution that does little to correct the problem.

4.1.2 Water Shutoff Chemicals

Most water shutoff chemicals are polymer gels or their pre-gel forms (gelants). In the process of selectively entering the cracks and pathways that water follows, gel solutions displace the water. Once the gels are in place in the cracks, they block most of the water movement to the well while allowing oil to flow to the well. The specifics of gel selection and deployment are driven by the type of water flow being targeted. Some of the key factors recommended for consideration with respect to gel treatment designs and operations include the following.

Component ingredients

- Type of gel polymer (in most cases a polyacrylamide polymer; microbial products and lignosulfonate have also been applied)
- Type of crosslinking agent (metal ion or organic)
- Fluid used to mix the gel

Properties of the gel (subject to variation in different stages throughout gel treatment)

- Concentration of polymer
- Molecular weight of polymer
- Degree of crosslinking
- Viscosity (affects the size of cracks or fractures that can be penetrated at a given pressure and can inject as pre-mixed gel or as gelant)
- Density (if too dense, gel can sink too far into the water layer and lose effectiveness)
- Set-up time (influences how far into the cracks or fractures the gel will penetrate)

Treatment procedure

- Preparation of well before treatment
- Volume of gel used
- Injection pressure
- Injection rate

Reynolds and Cole (2003) suggest using the following criteria for selecting candidate wells for gel treatment.

- Wells already shut in or near the end of their economic life
- Significant remaining mobile oil in place
- High water–oil ratio
- High producing fluid level
- Declining oil and flat water production
- Wells associated with active natural water drive
- High-permeability contrast between oil- and water-saturated rock

The results of many successful gel treatment jobs have been reported in the literature. Seright et al. (2001) describe 274 gel treatments conducted in naturally fractured carbonate formations. The median water-to-oil ratio was 82 before the treatment, then fell to 7 shortly after the treatment, and stabilized at 20 a year or two after treatment. On average, those wells produced much less water after the treatment. Following gel treatment, the oil production increased and remained above pretreatment levels for 1–2 years.

4.2 Technologies to Reduce the Volume of Water Reaching the Surface

Remote separation of water can be performed inside the wellbore using downhole oil/water or gas/water separators or through dual-completion wells. At offshore locations, remote separation can be performed using large sea floor separators. These are described below.

4.2.1 Downhole Separators

Downhole oil/water separation (DOWS or DHOWS) technology is installed in the bottom of an oil well. It separates oil and water in the wellbore. The oil-rich stream is pumped to the surface, while the water-rich stream is pumped directly to an injection formation without ever coming to the surface. This can lower costs and improve environmental protection. DOWS technology has two primary components – an oil/water separation component and one or more pumps. Two basic methods of separation (hydrocyclones and gravity separation) have been employed in commercial units.

Hydrocyclones use centrifugal force to separate fluids of different specific gravity. This does not involve any moving parts. A mixture of oil and water enters the hydrocyclone at a high velocity from the side of a conical chamber. The subsequent swirling action causes the heavier water to move to the outside of the chamber and exit through one end, while the lighter oil remains in the interior of the chamber and exits through a second opening. The water fraction containing a low concentration of oil (typically less than 500 mg/L) can then be injected, and the oil fraction along with some water is pumped to the surface. Hydrocyclone-type DOWS systems have been designed with electric submersible pumps, progressive cavity pumps (also called progressing cavity pumps), and rod pumps.

Gravity separator-type DOWS systems are designed to allow the oil droplets that enter a wellbore through the perforations to rise and form a discrete oil layer in the well. Most gravity separator tools are vertically oriented and have two intakes – one in the oil layer and the other in the water layer. This type of gravity separator-type DOWS system uses rod pumps. As the sucker rods move up and down, the oil is lifted to the surface and the water is injected.

Argonne described DOWS technology and developed an extensive database including many DOWS installations throughout the world (Veil et al. 1999; Veil and Quinn 2004). Most DOWS installations were found in North America (34 in Canada and 14 in the United States). Six were identified in Latin America, two in Europe, two in Asia, and one in the Middle East. All trials were conducted at onshore facilities, except for one in China. Two-thirds of the installations used gravity-separation-type DOWS systems.

DOWS technology has been used almost exclusively onshore. Only one offshore installation of DOWS technology (in China) was identified. The system worked for a few weeks, but subsequently failed. When the DOWS unit was lifted to the surface, the company discovered that a worker had not properly tightened a bolt, leading to equipment failure. Nevertheless, since the cost of drilling or working over an offshore well is so high, oil companies have been reluctant to try additional DOWS installations.

Although a few of the DOWS installations were very successful and remained in service for extended periods, many other installations did not meet expectations. Either their performance was deemed not adequate or they stopped working after only a few weeks or months of operation. As a result of the inconsistent performance of the installed DOWS systems, few, if any, installations have been made in recent years. Veil and Quinn (2004) updated an international database of DOWS installations in 2004 and identified that only two installations were made after 2001. The literature since 2004 has not reported on any new DOWS installations. The technology concept is sound, but improved quality control on the part of the developers and better well selection (with a good understanding of the formation to which injection will be made) are needed before the technology regains favor.

Downhole gas/water separation (DGWS) technology is installed in the bottom of a gas well. It separates gas and water in the wellbore. A report prepared in 1999 by Radian International for the Gas Research Institute (GRI 1999) offers a very comprehensive discussion of DGWS technology. Much of the information in this section is based on that report. DGWS technologies can be classified into four main categories: bypass tools, modified plunger rod pumps, electric submersible pumps, and progressive cavity pumps. There are tradeoffs among the various types, depending on the depth involved and the specific application. Both produced water rates and well depth determine which type of DGWS tool is appropriate for deployment in a specific case.

Veil and Quinn (2004) include a database of DGWS installations. Data on 48 of the DGWS installations was distilled from GRI (1999). Thirty-four of the installations were in the United States, with Oklahoma (16) and Kansas (11) heading the list. Fourteen installations were in Alberta, Canada. More than 60% of the installations (30) used modified plunger rod pump systems.

Before a well is produced, the interface between the oil and water layers in the formation is relatively horizontal and flat. Once production of the oil layer begins, near-well flow patterns tend to disrupt the horizontal interface and form a cone around the production perforations. Increasing portions of the water layer can enter into the well, thereby limiting the volume of oil that can be produced. This situation can be reversed and controlled by completing the well with two separate tubing strings and pumps. The primary completion is made at a depth corresponding to strong oil production, and a secondary completion is made lower in the interval at a depth with strong water production. The two completions are separated by a packer. The oil collected above the packer is produced to the surface, while the water collected below the packer is injected into a lower formation (Shirman and Wojtanowicz 2002). Pumping rates are adjusted to maintain a horizontal interface between the oil and water layers. This technology is called a "dual-completion well" or "downhole water sink." In another version of the process, the water can be separately produced to the surface, for management there.

4.2.2 Sea Floor Separators

Operators can still minimize water production with technologies that do not impede produced water from entering the well, but instead reduce the volume of water brought to the land surface or the platform. This type of technology is predicated upon separating oil and water remotely. Lifting water to the platform represents a substantial expense for operators. Platform space and weight constraints further restrict treatment options and increase costs. Sea floor (also called seabed) separation involves a large module that sits on the sea floor. Fluids from one or more wells are sent there for separation. The oil is lifted to a platform or to a floating production, storage, and offloading vessel, while the water is typically pumped directly to an injection well.

Prior to the end of 2007, only one sea floor separation unit has been used in full-scale operations. A Norwegian company developed a subsea separation and injection system (SUBSIS) that separates the produced fluids from an offshore well at a treatment module located on the sea floor. The SUBSIS module weighs 400 tons, is 17 m long and wide, and is 6 m high. The SUBSIS began full operation in August 2001 at Norsk Hydro's Troll field, about 4 km from the Troll C platform. It operated at a water depth of 350 m. Initial results indicated that 23,000 bbl/day (bbl/d) of produced fluids was separated into 16,000 bbl/d of oil and gas and 7,000 bbl/d of water. The water was injected directly from the SUBSIS unit into a dedicated injection well (Wolff 2000; Offshore 2000).

Von Flatern (2003) reports the results of a year-long trial of the SUBSIS. The SUBSIS handled a maximum flow of 60,000 bbl/d and a typical flow of 20,000 bbl/d. The oil concentration in the separated water stream dropped from an initial level of about 600 to 15 ppm. Because the water injected from the SUBSIS did not need to come to the surface at the Troll platform and occupy some of its water handling capacity, the Troll platform was able to produce an additional 2.5 million bbl of oil during the year-long trial (von Flatern 2003).

Another sea floor separator, located at Statoil-Hydro's TORDIS project, began operation in late 2007. It is located at about 200 m water depth. The sea floor separator is expected to extend the life of the TORDIS project by an additional 15–17 years and produce an incremental 35 million bbl of oil from the field (Gjerdseth et al. 2007).

Petrobras continues development of a subsea unit called the vertical annular separation and pumping system (VASPS). A VASPS unit was used successfully to separate gas from other fluids at the Marimba field in the Campos basin (Vale et al. 2002). Additional technology improvements are under way to allow oil/water separation. The system was designed to operate in 2008 at water depths ranging from 800 to 1,500 m, at the Marlim field in the Campos basin. The VASPS's main features included a modular configuration, small size, and reduced weight, to allow handling by a wider range of installation vessels. During the Marlim field tests, the system was installed at 872 m, and as close as possible to the wellhead to minimize heat loss in the flow line (Offshore 2006).

The challenges posed by sea floor systems include the following.

- Subsea systems are costly to implement.
- The technology is new, which increases implementation risks.
- In light of the costs, they are better suited for use in relatively young reservoirs.

5 Recycle and Reuse Technologies

This section describes several ways in which produced water can be recycled or reused. Some of the processes can use the water regardless of its chemical constituents. Other processes require water to be quite clean before it can be reused. In some coal bed methane fields, the produced water comes to the surface with very limited salt content. However, most produced water is quite salty and must be treated before reuse. Most of these applications are relevant to onshore rather than offshore wells. Therefore, with the exception of injection for enhanced recovery, the recycle and reuse technologies are discussed only briefly.

5.1 Injection for Enhanced Recovery

Reinjection into an underground formation represents the most commonly used approach for onshore management of produced water. Some produced water is injected solely for disposal. Most produced water is injected to maintain reservoir pressure and hydraulically drive oil toward a producing well. This practice can be referred to as enhanced oil recovery (EOR), water flooding, or, if the water is heated to make steam, as steam flooding. The growing interest in producing heavy oil formations around the world may increase the extent of steam flooding or other steam uses like steam-assisted gravity drainage for oil sands. In the context of improving oil recovery, produced water becomes a reusable resource rather than a waste product.

Significant volumes of produced water are injected in the United States. In 2003, Argonne interviewed staff from oil and gas agencies in three large oil- and gas-producing states (California, New Mexico, and Texas) to determine the number of injection wells in each state and what percentage were used for EOR. The numbers of wells and water volumes injected are estimated. Nevertheless, they highlight the importance of injection and EOR as produced water management options.

California has nearly 25,000 produced water injection wells. The annual injected volume is approximately 1.8 billion bbl, distributed as follows: disposal wells – 360 million bbl; water flood – 900 million bbl; and steam flood – 560 million bbl (Stettner 2003).

New Mexico has 903 permitted disposal wells, 264 of which are active. The state has an additional 5,036 wells permitted for enhanced recovery, 4,330 of which are active. The approximate volume of produced water injected for disposal is 190 million bbl. The volume injected for enhanced recovery is about 350 million bbl (Stone 2003).

Texas has 11,988 permitted disposal wells, 7,405 of which are active. It has an additional 38,540 wells permitted for enhanced recovery; 25,204 of those are active. The approximate volumes of produced water injected in 2000 were 1.2 billion bbl disposed into nonproducing formations, 1 billion bbl disposed into producing formations, and 5.3 billion bbl injected for enhanced recovery (Ginn 2003).

In sum, operators in just these three states inject more than 7 billion bbl of produced water per year for reuse through water or steam flooding. The ratio of produced water volume injected for water and steam flooding to the volume injected for disposal ranges from 1.8:1 to 4.0:1 for these three states.

5.1.1 Offshore Produced Water Injection for Enhancing Recovery

Historically, produced water has rarely been injected for EOR at offshore platforms. Several factors account for this divergence from the onshore pattern. First, most US platforms (as well as many international platforms) are authorized to discharge produced water to the ocean following treatment. Surface discharge represents the preferred option for operators in most cases. Second, at some point in the life of a field (when pressure maintenance is needed), the offshore wells at a platform do not generate sufficient produced water to meet the volumetric needs for water flooding. Third, platforms have ready access to virtually unlimited supplies of seawater. As discussed in the following section, operators must ensure that the water injected for EOR does not clog the pores of the producing formation. Seawater is nearly always cleaner than produced water, and it requires less pretreatment before injection. Therefore, seawater generally provides the preferred source of injection water for EOR.

In recent years, particularly in the North Sea region, where produced water discharge standards are becoming more stringent, offshore operators are beginning to inject produced water combined with seawater or to replace seawater where it has been used previously. Studies have found some problems of reservoir souring (i.e., hydrogen sulfide production) when produced water is injected to a formation that had received seawater previously (Jenneman et al. 2004). Another consideration with using produced water for reinjection is that produced water is likely to be considerably warmer than seawater, thereby losing some of the potential for thermal fracturing that cool seawater can offer.

While some produced water generated from US offshore activities in 2007 was injected, the vast majority of produced water was managed through discharge. In 2007, injection for enhanced recovery managed 48,673,102 bbl and injection for disposal managed 1,298,417 bbl of produced water. The remaining produced water (537,381,327 bbl) was treated and then discharged (Clark and Veil 2009).

5.1.2 Treatment Before Reinjection

It is important to ensure that the water being injected is compatible with the formation's receiving the water, to prevent premature plugging of the formation or other damage to equipment. It may therefore be necessary to treat the water to control excessive solids, dissolved oil, corrosion, chemical reactions, or growth of microbes.

Solids are usually treated by gravity settling or filtration. Residual amounts of oil in the produced water not only represent lost profit for producers, but can also contribute to plugging of formations receiving the injectate. Various treatment chemicals are available to break emulsions or make dissolved oil more amenable to oil removal treatment.

Corrosion can be exacerbated by various dissolved gases, primarily oxygen, carbon dioxide, and hydrogen sulfide. Oxygen scavengers and other treatment chemicals are available to minimize levels of undesirable dissolved gases.

The water chemistry of a produced water sample is not necessarily the same as that of the formation that will receive the injected water. Various substances dissolved in produced water may react with the rock or other fluids in the receiving formation and have undesirable consequences. Before beginning a water flood operation, it is important to analyze the constituents of the produced water for the purpose of avoiding chemical reactions that form precipitates. If necessary, treatment chemicals can minimize undesirable reactions.

Bacteria, algae, and fungi can be present in produced water. They can also be introduced during water handling processes at the surface. Bacteria, algae, and fungi are generally controlled by adding biocides or by filtration.

5.2 Injection for Future Use

Produced water can potentially be stored underground in shallow aquifers for future use. This process is known as aquifer storage and recovery (ASR). Most produced water is very salty. The cost of removing salinity from produced water poses a barrier to injecting it into an aquifer. However, some types of produced water are relatively fresh, and hence can be used directly with little or no treatment. This has been particularly true for produced water from some coal bed methane fields.

Although this is not a common use, one example can be found in Wellington, Colorado, USA (Stewart 2006). An oil company is treating produced water from oil wells as a raw water resource that will be used to augment shallow water aquifers to ensure adequate water supplies for holders of water rights. The oil company is embarking on this project to increase oil production. A separate company will then purchase and utilize this water as an augmentation water source. This water will eventually be used to allow the Town of Wellington and northern Colorado water users to increase their drinking water supplies by 300%.

Before selecting ASR as the preferred technology for managing produced water, oil and gas operators must evaluate various relevant factors.

- The availability of an aquifer suitable for recharge. Suitability considerations involve the areal extent, thickness, depth of the aquifer, and the presence and size of confining layers or aquitards.
- The hydrogeology of the aquifer formation, including its porosity, permeability, transmissivity, hydraulic conductivity, flow direction, and velocity.
- The chemical characteristics of the water already residing in the aquifer and the incoming produced water. These should be compared to ensure that undesirable chemical reactions are not likely to occur in the aquifer.
- The willingness of the jurisdictions ultimately using the water to emplace produced water in the aquifer. Operators must have the ability to obtain the necessary permits. Further, they must know the types and extent of monitoring that will subsequently be required.
- The cost of treating the produced water to ensure compliance with all applicable injection standards.

5.3 *Use for Hydrological Purposes*

In addition to its value as 'water,' produced water is a fluid that can be used to occupy space or resist earth or fluid movement (i.e., for hydrological purposes). Potential hydrological uses of injected produced water include

- controlling surface subsidence in the wake of large withdrawals of ground water or oil and gas,
- blocking salt water intrusions in aquifers in coastal environments, and
- augmenting the regional ground water or local stream flows.

5.3.1 Subsidence Control

The most likely hydrological use of produced water is to control subsidence of the ground surface. Subsidence control wells are injection wells designed to reduce or remediate the loss of land surface elevation due to the removal of ground water, oil, or gas. In their natural state in the formation, these fluids provide physical support to the soil and rock layers above them. When large quantities of the fluids are extracted, the upper formations can compress or collapse. Surface subsidence can cause damage to building foundations, roadways and railways, water wells, and pipelines. Land subsidence control is achieved by injecting water into an underground formation to maintain fluid pressure and avoid compaction.

One of the most compelling examples of subsidence resulting from oil and gas extraction involves the Wilmington oil field in Long Beach, California. Since the 1930s, more than 1,000 wells withdrew about 2.5 billion bbl of oil. Between the 1940s and the 1960s, this field experienced a total of 29 ft of subsidence, caused primarily by the withdrawal of hydrocarbons (Colazas et al. 1987). Subsidence in the Wilmington oil field caused extensive damage to Long Beach port industrial and naval facilities. A massive repressurization program, based on the injection of salt water into the oil reservoirs, reduced the subsidence area from approximately 50 to 8 km^2. Approximately 2.3 billion bbl of water was reinjected through 1969. The rate of subsidence at the historic center of the bowl has been reduced from a maximum of 28 in. per year in 1952 to 0 in 1968. A small surface rebound has occurred in areas of heaviest water injection (Mayuga and Allen 1969).

5.3.2 Other Hydrological Uses of Produced Water

Other potential hydrological uses of injected produced water include blocking salt water intrusions in aquifers in coastal environments and augmenting the regional ground water or local stream flows. Liske (2005) and Ouellette et al. (2005) report on a project to evaluate possible new uses for produced water generated in the San Ardo field in central California. Among the options being considered is control of salt water intrusion in the Salinas River valley. This area has overdrawn ground water for domestic and agricultural uses, resulting in the salt water/fresh water interface moving 6 miles upstream. In this example, the treated produced water is not expected to be injected. Instead, it would be discharged to the Salinas River or used locally for irrigation, thereby avoiding ground water withdrawal and reducing the driving force of the salt water intrusion.

Produced water can potentially be used to augment stream flows. Where discharges are permitted, treated produced water meeting applicable discharge standards could be directly discharged to surface water bodies. Produced water could also be injected into formations exhibiting hydrologic interconnection with surface water bodies, or allowed to infiltrate to the water table through holding ponds. When used for stream augmentation, the quality of produced water must be controlled to avoid impairment of the surface water quality pursuant to the criteria adopted by the host state. Moreover, the quantity of the added produced water must not increase excessive erosion or damage to stream channels.

5.4 Agricultural Use

Produced water meeting the water quality requirements of agricultural users offers the potential to supplement and replace existing water supplies. Ayers and Westcot (1994) provide an excellent reference on water quality requirements for agricultural uses. This section describes the use of produced water for irrigation, livestock watering, wildlife watering, and managed wetlands.

Perhaps the most significant barrier to using produced water for agricultural purposes involves the salt content of the water. Most crops do not tolerate much salt, and sustained irrigation with salty water can damage soil properties. In addition, if livestock drink water containing too much salt, they can develop digestive disorders. However, not all produced water is equally salty. For example, some of the coal bed methane fields in Wyoming's Powder River Basin generate relatively fresh water. However, in addition to the salt content, the relative proportion of sodium to other ions is important because excessive sodium is harmful to soils. Soil scientists use the term "sodium adsorption ratio" (SAR) to characterize the ionic proportions. The SAR is defined as the milliequivalent weight of sodium divided by the square root of the sum of the milliequivalent weights of calcium and magnesium, divided by 2, or in equation format:

$$\text{SAR} = \text{Na}^{+1} / \left[\left(\text{Ca}^{+2} + \text{Mg}^{+2} \right) / 2 \right] 0.5, \text{ with Na, Ca, and Mg expressed as meq/L.}$$

ALL (2003) summarizes crop irrigation water quality requirements, noting that the three most critical parameters include salinity, sodicity, and toxicity. Salinity is expressed as electrical conductivity in units of microSiemens per cm (μS/cm; previously, mmhos/cm). Crops exhibit varying susceptibility to salinity. When salinity rises above a species-specific threshold, crop yields decrease. Excess sodium can damage soils. Higher SAR values lead to soil dispersion and loss of soil infiltration capability. When sodic soils are wet, they become sticky, and when dry, they form a crusty, nearly impermeable layer. Some trace elements in produced water can cause harmful effects to plants when present in sufficient quantities. ALL (2003) also offers several case examples of produced water use to irrigate crops, provide water to livestock, and create wildlife watering and habitat impoundments. These examples are found in the Rocky Mountain region of the United States.

Although most irrigation projects using produced water are found in the Rocky Mountain region, Brost (2002) describes a complex system used by ChevronTexaco to treat produced water in the Kern River field in central California. The treatment system provides about 480,000 bbl/d of water for irrigating fruit trees and other crops and recharging shallow aquifers. Another 360,000 bbl/d of water is further purified and used to make steam at a cogeneration facility.

During a speech at a September 2008 carbon sequestration conference, a representative of the Brazilian oil company, Petrobras, reported on several pilot tests being conducted to manage carbon dioxide emissions. One of the examples described an experiment that combined produced water with carbon dioxide from exhaust stack emissions in a bioreactor tank. The goal of the experiment is to grow microalgae that could eventually become a feedstock for biofuels. The Petrobras official did not have any results or written papers to share at that time, but relayed this information in a conversation with this chapter's author during the conference (Murce 2008).

5.5 Industrial Use

Many parts of the western United States are characterized by limited supplies of potable water. In light of the increasing demands, the costs of identifying and treating new water supplies continue to climb. Yet, many of these arid regions are home to oil and gas production, yielding significant amounts of produced water. In areas where traditional surface and ground water resources are scarce, produced water could become a significant replacement resource in many industrial processes, as long as the quality of the produced water remains adequate. The degree of prior treatment required depends on the quality of the produced water and the intended use.

5.5.1 Oil Field Use

Produced water can be substituted in oil field operations where other types of water are used. Peacock (2002) describes a program in New Mexico where produced water is treated to remove hydrogen sulfide. It is then used in drilling operations. This beneficial reuse of produced water saves more than 4 million bbl per year of local ground water.

Many natural gas wells must be hydraulically fractured to enhance production operations. Each "frac job" requires thousands to hundreds of thousands of bbl of water. In areas where natural gas fields are expanding rapidly (e.g., the Barnett Shale in Texas, the Fayetteville Shale in Arkansas, and the Marcellus Shale in the northern Appalachian region), local water supplies may not be adequate to meet the demand for frac water. Produced water or "flowback water" – the water returning from the formation following a frac job – can be treated and reused for new frac jobs. Veil (2008) describes a technology that is being employed in Texas for treating frac flowback water for reuse.

5.5.2 Use for Power Production

In 2003, the DOE funded a group of researchers led by the Electric Power Research Institute to evaluate the feasibility of using coal bed methane produced water to meet up to 25% of the cooling water needs at the San Juan Generating Station in northwestern New Mexico. The plant requires nearly 500,000 bbl/d of makeup water to replace water lost to evaporation or blowdown. The researchers prepared a series of reports, which are all available at http://www.netl.doe.gov/technologies/coalpower/ewr/water/pp-mgmt/epri.html. The economics of using produced water at that specific plant does not appear favorable. Therefore, the utility is not moving forward with implementation.

In at least one case, produced water is used to supply water for steam generation. After softening, about 360,000 bbl/d of produced water from a ChevronTexaco facility in central California is sent to a cogeneration plant where it serves as a source of boiler feed water (Brost 2002).

When produced water comes to the surface with sufficiently high temperature (typically at least 95°C), it may be possible to extract electricity through geothermal power processes. Although many researchers have studied geological formations for their geothermal potential, petroleum-producing formations rarely provide hot enough produced water to economically generate geothermal power, and therefore have not been studied extensively. The first successful production of power from heated produced water occurred in October 2008 at DOE's Rocky Mountain Oilfield Testing Center (RMOTC 2008).

5.5.3 Dust Control

In most oil fields, the lease roads are unpaved and can create substantial dust. Some oil and gas regulatory agencies allow operators to spray produced water on dirt roads to control the dust. This practice is generally controlled so that produced water is not applied beyond the road boundaries or within buffer zones around stream crossings and near buildings.

Coal bed methane produced water may be generated in areas with active surface coal mining operations. Surface mining, processing, and hauling are inherently dusty activities. Produced water can be used for dust suppression at those locations. Murphree (2002) outlines plans to use coal bed methane produced water for dust suppression at the North Antelope/Rochelle Complex in Campbell County, Wyoming, the world's largest coal mine. IOGCC and ALL (2006) describe several coal mines in Wyoming's Powder River Basin that use coal bed methane produced water.

5.5.4 Fire Control

Fires often break out during the driest portions of the year. Areas experiencing drought conditions are particularly vulnerable. In many cases, only limited surface and ground water resources are available for firefighting. Although application of large volumes of saline produced water can adversely impact soils, this is far less devastating than a large fire. ALL (2003) reports that firefighters near Durango, Colorado, used coal bed methane produced water impoundments as sources of water to fill air tankers (i.e., helicopters spraying water onto fires) during the summer of 2002.

5.6 Use as Drinking Water

In the past, the treatment costs to reduce salinity and remove other constituents from produced water for purposes of meeting US Environmental Protection Agency

(EPA) drinking water standards were prohibitively high. However, in recent years, costs to develop and deploy treatment technology have dropped. At the same time, communities running out of water are willing to pay higher prices for clean water. Treatment costs are approaching water prices in some cases. These developments provide the crucial incentive for many water treatment technology developers deciding to enter the marketplace. A related but important issue involves managing the concentrated by-product stream that results from treating the produced water (Veil et al. 2006).

Texas A&M University developed a portable produced water treatment system that can be moved into oil fields to convert produced water to potable water. This can be used to augment scarce water supplies in arid regions, while also providing economic paybacks to operators in the form of prolonged productive lives of their wells (Burnett et al. 2002; Burnett and Veil 2004). During the past few years, the desalination trailer developed by the university conducted pilot tests using produced water from several locations in Texas. The water treated by the trailer met the applicable drinking water standards (Veil et al. 2006).

In addition to these tests, a sample of produced water from an oil field in Grimes County, Texas, was treated during a membrane desalination workshop held at Texas A&M in August 2006. Some brave attendees drank the water, including the author of this chapter. According to staff of the Texas A&M University, the laboratory analysis of the water showed, "the input total dissolved solids [TDS] was 13,320 mg/L, and the output was 323 mg/L. Sodium input was 4,490 mg/L, and output was 127 mg/L. Chlorides input was 7,494 mg/L, and the output was 184 mg/L. Potassium input was 76 mg/L, and the output was 1.2 mg/L. This is better than our city water here in the Bryan/College Station area." Additional information and data can be viewed on the website of the Texas A&M Global Petroleum Research Institute at http://www.pe.tamu.edu/gpri-new/home/ConversionBrine.htm.

6 Disposal Technologies

At least four methods can be used to dispose of produced water. One common disposal method (discharge) typically requires treatment before the produced water can be released. A second method (injection) may or may not require treatment. The other two methods (evaporation and commercial disposal) can directly dispose of produced water without undertaking any treatment.

6.1 Discharge to Surface Waters

In many parts of the world, discharge to surface water bodies is the principal option for disposing of produced water from offshore facilities. Many countries allow produced water discharge subject to regulatory discharge standards for "oil and grease" and other constituents. Oil and grease is a key constituent in produced water. Different countries use different oil and grease discharge standards.

Offshore wells are not the only wells that use discharge for produced water management. Although US discharge standards prohibit surface discharges from most onshore oil and gas wells, they do not currently apply to coal bed methane wells. Many US coal bed methane well operators prefer to discharge produced water to surface water bodies, where allowed by the regulatory agency.

6.2 Injection for Disposal

Section 5.1 describes injection for enhanced recovery. Most onshore operators reinject produced water for enhanced recovery. However, not all produced water is needed for enhanced recovery. A large volume of produced water is injected into non-producing formations solely for disposal.

Operators injecting for disposal will typically seek formations that exhibit the right combination of permeability, porosity, injectivity, and other geologic features enabling the injected water to enter the formation under pressures lower than fracture pressure. However, because produced water that is handled at the surface prior to injection is likely to be cooler than the injection formation temperature, the possibility of small-scale thermal fracturing exists during injection operations. Injection formations are geologically isolated from any underground source of drinking water and from hydrocarbon-producing formations. Operators also typically avoid areas with excessive faulting, fractures that extend vertically, or improperly cemented wellbores.

6.3 Evaporation

Evaporation is a natural process for transforming water in liquid form to water vapor in the air. Evaporation depends on local humidity, temperature, and wind. Drier climates generally favor evaporation as a waste management technique. The simplest approach to evaporation involves placing produced water in a pond, pit, or lagoon with a large surface area. Water can then passively evaporate from the surface. As long as evaporation rates exceed inflows (including precipitation), produced water will be removed.

One potential problem posed by evaporation ponds stems from their attractiveness to migratory waterfowl. If evaporation ponds contain oil or other hydrocarbons on the surface, birds landing in the ponds could become coated with oil and suffer harm. Covering ponds with netting helps to avoid this problem.

Evaporation rates can be enhanced by spraying the water through nozzles. The nozzles create many small droplets with increased surface areas. In some arid parts of the western United States, this has been done through portable misting towers. The water is sprayed into the air and evaporates before hitting the ground. However, this practice can lead to salt damage to soil and vegetation. Even in more humid climates, equipment similar to artificial snow-making equipment has been used to create many fine water droplets that can evaporate.

In cold climates, evaporation can be coupled with natural freezing and thawing in the Freeze/Thaw Evaporation (FTE®) process (Boysen et al. 1999). When the ambient temperature drops below 0°C, produced water is pumped from a holding pond and sprayed onto a freezing pad. As the spray freezes, an ice pile forms. When the temperature warms, ice on the freezing pad melts. The highly saline brine, identified by its high electrical conductivity, is separated and pumped to a pond where it can be utilized as an additive for drilling fluids. The remaining purified water is then pumped from the freezing pad to a holding pond where it can be stored prior to beneficial reuse or discharge. Unless artificial refrigeration is employed, the FTE process is limited to cooler climates during the colder times of the year. However, ponds can be used for spray evaporation in warmer months.

6.4 Offsite Commercial Disposal

Most onshore produced water is managed onsite by the operators using injection. Under certain circumstances, however, operators prefer to send their produced water offsite to a commercial disposal facility. This is typically accomplished by having a truck periodically visit the well locations, remove the accumulated water, and haul it away to the destination facilities. Offsite commercial disposal becomes the option of choice when small producers do not want to have the responsibility for constructing, operating, and closing onsite facilities. Onshore operators who do not have access to nearby formations deemed suitable for accepting produced water through injection wells may also look to offsite disposal.

In 2006, Argonne published a report and database describing the network of offsite commercial disposal facilities that accept different types of exploration and production wastes, including produced water (Puder and Veil 2006). The report provides information on the location of the disposal facilities, what types of wastes they accept, the disposal methods they employ, and the cost for disposal.

7 Technologies for Treating Salt and Other Inorganic Chemicals

This section describes four categories of technologies for removing salt and other inorganic chemicals from produced water. These technologies are used almost exclusively for treating onshore produced water before discharge or reuse.

7.1 Membrane Technologies

Salt is the key parameter that determines how produced water is managed onshore. If produced water is injected underground for enhanced recovery or disposal, the salt content does not pose a major issue. However, reuse and discharge to an onshore surface water body require that the salt content is low enough to prevent salt-related problems. This section describes several processes that remove salt and other inorganic chemicals from produced water by using membranes. Membrane processes are the most commonly used treatment technology for removing salt from produced water.

7.1.1 Filtration

Filtration is common in many industrial applications. Filters can be designed of different materials and in different configurations. The filtration process involves passing liquids through a membrane that has a minimum pore size. Suspended and dissolved particles that are larger than the membrane pore size are blocked by the membrane, while the water and smaller particles pass through. Filtration membranes and processes are subdivided by pore size ranges. The various categories, from largest to smallest pore size, include microfiltration, ultrafiltration, nanofiltration, and reverse osmosis. Various authors offer side-by-side comparisons of the four filtration classes. It should be noted that while the relative relationships between membrane filtration technologies are not subject to much controversy in the literature, pore size cutoffs and other details vary by author. Table 29.1 represents a

Table 29.1 Comparison of classes of membrane filtration

Parameter	Microfiltration	Ultrafiltration	Nanofiltration	Reverse osmosis
Pore size	0.01–1.0 μm	0.001–0.01 μm	0.0001–0.001 μm	<0.0001 μm
Molecular weight cutoff	>100,000	1,000–300,000	300–1,000	100–300
Operating pressure	<30 psi	20–100 psi	50–300 psi	225–1,000 psi
Membrane materials	Ceramics, polypropylene, polysulfone, polyvinylidenedifluoride	Ceramics, polysulfone, polyvinylidenedifluoride, cellulose acetate, thin film composite	Cellulose acetate, thin film composite	Cellulose acetate, thin film composite, polysulfonated polysulfone
Membrane configuration	Tubular, hollow-fiber	Tubular, hollow-fiber, spiral-wound, plate and frame	Tubular, spiral-wound, plate and frame	Tubular, spiral-wound, plate and frame
Types of materials removed	Clay, bacteria, viruses, suspended solids	Proteins, starch, viruses, colloidal silica, organics, dyes, fats, paint, solids	Starch, sugar, pesticides, herbicides, divalent anions, organics, biochemical oxygen demand (BOD), chemical oxygen demand (COD), detergents	Metal cations, acids, sugars, aqueous salts, amino acids, monovalent salts, BOD, COD

composite summary distilled from Wagner (2001), IOGCC and ALL (2006), and Cartwright (2006).

As membrane pore size decreases, the energy required to push the water solution through the membrane increases. In addition, the tendency to foul the membrane increases as the pore size decreases. Hence, membrane filtration is often conducted in stages. A pretreatment module removes the larger constituents before they reach the membranes. The pretreatment stage can include many types of processes. This depends on the different materials in the wastewater influent. If reverse osmosis is required in the final treatment step for removing salt or metals, the pretreatment module could include microfiltration or ultrafiltration. In the pretreatment stage, operators typically add different chemicals to the treatment system for cleaning, anti-fouling, or other process control purposes.

7.1.2 Electrodialysis

Hayes (2004) and IOGCC and ALL (2006) report on electrodialysis, another type of membrane process for removing salt from produced water. Electrodialysis is a separation process using a stack of alternating anion- and cation-selective membranes separated by spacer sheets. Water passes through the stack of membranes. Electrical current is applied to the cell, causing the anions to migrate in one direction and the cations in the other direction. As the migrating ions intersect the selectively permeable membranes, alternating cells of concentrated and diluted solutions are produced in the spaces between the membranes.

A related process – electrodialysis reversal – operates in a similar fashion. In contrast to conventional electrodialysis, however, the polarity of the electrodes is reversed several times per hour, and the flows are simultaneously switched so the brine channel becomes the clean water channel and vice versa. The reversal feature is useful in breaking up films, scales, and other deposits, and flushing them out of the process before they can foul the membranes.

Electrodialysis consumes less energy than reverse osmosis because it occurs at lower pressures (IOGCC and ALL 2006). In practice, conventional electrodialysis and electrodialysis reversal can reduce salt concentrations to less than 200 mg/L of TDS before the internal resistance of the solution to current flow – and therefore, power demand – rapidly increases (Hayes 2004).

7.2 Ion Exchange

Home water softeners have been used for decades to remove iron and manganese from tap water. Water softeners use ion exchange columns filled with a specific type of resin. The resin is charged with sodium ions by passing a concentrated salt solution through the column. As water loaded with iron and manganese moves through the column, those ions are attracted to the resins, which prefer to bind with the iron and manganese. Sodium is released when the resin attaches to the iron and manganese. Once stripped of the undesirable iron and manganese, the water is piped into the home water supply. Periodically, the resin-exchange sites are all used. Then, the column is backwashed and regenerated with fresh salt solution.

The water softener example is provided to explain the principle of ion exchange. Exchange resins can be designed to selectively remove different types of chemicals. For produced water ion exchange, sodium is the target for removal. For that application, resins are chosen that are regenerated with hydrogen ions. As the produced water passes through the column, sodium ions are removed from the water. They replace the hydrogen on the resin's exchange sites. Because the treated water contains more hydrogen ions, its pH drops. Often the pH is raised by contact with calcium carbonate.

One version of ion exchange commonly used for treatment of coal bed methane produced water in the Rocky Mountain region is the Higgins Loop (Dennis 2006; IOGCC and ALL 2006). The Higgins Loop uses a continuous countercurrent ion exchange contactor for liquid phase separations of ionic components. The contactor consists of a vertical cylindrical loop, which contains a packed bed of resin separated into four operating zones by butterfly or "loop" valves. These operating zones – adsorption, regeneration, backwashing, and pulsing – function like four separate vessels.

Produced water containing high sodium levels is fed to the adsorption zone within the Higgins Loop. There, it contacts a strong acid cation resin, which accepts sodium ions in exchange for hydrogen ions. Treated water containing less than 10 mg/L sodium then exits the loop. In the lower section of the Higgins Loop, resin filled with sodium is regenerated with either hydrochloric or sulfuric acid. This generates a small, concentrated spent brine stream. Regenerated resin is rinsed with water prior to reentry into the adsorption zone to remove acid from its pores. As resin in the upper layer of the adsorption zone becomes loaded with sodium, the flows to the Higgins Loop are momentarily interrupted. This allows advancement of the resin bed (pulsing) through the loop in the opposite direction of liquid flow. Liquid flows are restarted after resin pulsing is complete.

Beagle (2008) reports that more than 20 Higgins Loop ion exchange units were operating in Montana, Wyoming, and Colorado during 2008 to treat produced water. The volume of water treated in this fashion has steadily increased to 425,000 bbl/d. Other equipment vendors provide alternative versions of ion exchange technology for treating produced water.

The backwash stream from ion exchange treatment of produced water contains high levels of sodium. Therefore, disposal requirements for the concentrated brine should be considered when selecting ion exchange treatment technologies.

7.3 Capacitive Deionization

Capacitive deionization is based on an electrostatic process operating at low voltages and pressures. Produced water is pumped through an electrode assembly, and ions in the water are attracted to the oppositely charged electrodes. This concentrates the ions at the electrodes, while reducing the concentration of the ions in the water. The cleaned water then passes through the unit. When the electrodes' capacity is reached, the water flow is stopped and the polarity of the electrodes is reversed. This causes the ions to move away from the electrodes where they had previously accumulated. The concentrated brine solution is then purged from the unit.

Capacitive deionization offers more cost-effective performance than other types of salt removal technology (e.g., reverse osmosis), especially when the produced water is not highly salty or the treatment goal is not geared toward achieving drinking water quality. Capacitive deionization is just beginning to be used to treat produced water, so very few data are available. Atlas (2007) reports on a test to treat coal bed methane water at flows of 5 gpm (18.9 L/min) with two 5 gpm systems in series. Solids were filtered using a 30 μm filter. The incoming feed water had a TDS of 2,500 ppm. Following treatment in the unit, the TDS was 270 ppm (85% recovery). The SAR was reduced from 24 to 3.

7.4 Thermal Distillation

Thermal distillation has been used for decades to desalinate seawater for drinking water. Historically, thermal distillation has not been cost effective for treating produced water. A newer version of thermal distillation (AltelaRain^{SM}) has been developed for treating produced water. Since the technology recaptures the energy previously used to evaporate water, energy costs fall to approximately 30% of comparable distillation/evaporation processes (Godshall 2007).

Ambient-temperature air is brought into the bottom of the tower on the evaporation side of a heat transfer wall. After entering the evaporation side at the top of the tower, the produced water spreads over and coats the heat transfer wall in a thin film. As the air moves from the bottom to the top of the tower, low-temperature heat is transferred into the evaporation side through the heat transfer wall. This raises air temperatures and evaporates water from the brine coating the wall. Water, now highly concentrated in contaminants, leaves from the bottom of the tower, while warm saturated air rises to the top of the tower. Steam is added to further heat the warm air. This hotter saturated air is then sent back down through the tower on the condensation side of the heat transfer wall. Since the evaporation side of the tower is slightly cooler than the condensation side, the air cools and transfers the latent heat from the condensation to the evaporation side. Meanwhile, pure distilled water condensate leaves the condensation side of the tower at the bottom of the tower.

The technology is modular. A cluster of units placed inside of a large metal shipping container are able to process approximately 100 bbl/d of produced water. As of late 2008, the AltelaRain process has been used in New Mexico, Colorado, and Alberta. Water quality test results show significant removals of produced water contaminants.

A different type of thermal distillation system has been used in Texas to treat dirty water resulting from a frac job so that it can be used again on a subsequent frac job. The Aqua-Pure® system is considerably larger than the AltelaRain unit. It uses a mechanical vapor recompression evaporation process. Each mobile evaporator unit is capable of processing 2,300 bbl/d of contaminated water and returning approximately 2,000 bbl/d as fresh water (Veil 2008). Although the Aqua-Pure system has not been used to treat produced water, it could be adapted for that purpose.

8 Technologies for Treating Oil and Grease and Other Organic Chemicals

This section describes six categories of technologies for removing oil and grease and other organic chemicals from produced water. These technologies are used primarily for treating offshore produced water before it is discharged, although some of the physical separation technologies are used to pretreat water that is reinjected.

8.1 Background on Oil and Grease

Not all produced waters contain the same constituents, even if they have the same oil and grease content. Oil and grease is made up of at least three forms:

- free oil (this is in the form of large droplets that are readily removable by gravity separation methods),
- dispersed oil (this is in the form of small droplets that are more difficult to remove), and
- dissolved oil (these are hydrocarbons and other similar materials that are dissolved in the water stream; they are often challenging to remove).

The most appropriate individual technology or sequence of technologies should be determined after evaluating the nature of the compounds making up the oil and grease content of the untreated produced water. For example, take two untreated produced water samples, both of which contain 100 mg/L of oil and grease. Produced water sample A has primarily free oil, whereas produced water sample B has primarily dissolved oil. In order to meet the US maximum discharge limit of 42 mg/L, the types of treatment processes and the cost of those processes would be vastly different. This is the challenge faced by offshore operators.

Oil and grease does not occur as a single chemical compound. Rather, as an "indicator pollutant," it is a measure of many different types of organic materials that respond to a particular analytical procedure. Different analytical methods will measure different organic fractions and compounds. Therefore, the specific analytical method used is important in determining the magnitude of oil and grease measurements (see Chapter 2).

This is particularly important due to the phasing out over the past decade of Freon-113 as an extraction solvent (under EPA Method 413.1). That longstanding standard-approved method (which was used to collect all the effluent data used in establishing the statistically derived discharge standards for oil and grease) has been replaced by EPA Method 1664, which uses *n*-hexane as the extraction solvent. Raia and Caudle (1999) describe a study sponsored by the American Petroleum Institute, comparing the results of the two methods. The standard deviations of the results were the same order of magnitude or were larger than the means, thereby making it difficult to determine if the results were comparable. Most of the samples showed higher values when measured by the new method. This raises some concern over compliance. For example, if the new method measures 44 mg/L, the old method measures 40 mg/L for the same sample, this is the difference between compliance and noncompliance, with the maximum discharge limit for oil and grease of 42 mg/L. It is also worth noting that the oil and grease measuring method approved for North Sea discharges was changed in 2007 to the use of gas chromatography with flame ionization detection, but it still differs from the two methods introduced above (Yang 2007).

8.2 Physical Separation

As a first step in removing oil and grease from produced water, companies often choose physical separation methods that rely on gravity, centrifugal force, or pore-size trapping. At onshore sites, this generally involves some form of oil/water separator or free water knockout vessel (for separation of the free oil). In offshore settings, oil/water separators and skim piles are deployed to remove oil droplets greater than 150 μm in diameter. More physical separation steps can be added to remove any remaining free oil and some dispersed oil. Additional treatment iterations may be required to achieve compliance with all applicable discharge limits. Examples of technologies in this category include advanced separators, hydrocyclones, filters, and centrifuges.

Physical separation methods can generally capture much of the free oil and some of the dispersed oil. They are not very effective at removing dissolved oil. Table 29.2 (based on Frankiewicz 2001) provides guidance for selecting treatment equipment based on the size of the oil droplets that need to be removed.

Table 29.2 Droplet size removal capabilities (Frankiewicz 2001)

Technology	Removal capacity by droplet size (in μm)
API gravity separator	150
Corrugated plate separator	40
Induced gas flotation without chemical addition	25
Induced gas flotation with chemical addition	3–5
Hydrocyclone	10–15
Mesh coalescer	5
Media filter	5
Centrifuge	2
Membrane filter	0.01

8.2.1 Advanced Separators

Separators rely on the difference in specific gravity between oil droplets and produced water. The lighter oil rises at a rate dependent on the droplet diameter and the fluid viscosity (Stokes Law). Smaller diameter droplets rise more slowly. If insufficient retention time is provided, the water exits the separator before the small droplets have risen through the water to collect as a separate oil layer. Advanced separators contain additional internal structures that shorten the path followed by the oil droplets before they are collected (e.g., inclined plate separators, corrugated plate separators). This gives smaller oil droplets the opportunity to reach a surface before the produced water overflows and exits the separator.

8.2.2 Hydrocyclones

Hydrocyclones, which do not contain any moving parts, apply centrifugal force to separate substances of different densities. Hydrocyclones can separate liquids from solids or liquids from other liquids. The liquid/liquid type of hydrocyclone is used for produced water treatment. Produced water is pumped tangentially into the conical portion of the hydrocyclone. Water, the heavier fluid, spins to the outside of the hydrocyclone and moves toward the lower outlet. The lighter oil remains in the center of the hydrocyclone before being carried toward the upper outlet. Hydrocyclones have been used for surface treatment of produced water for several decades.

8.2.3 Filtration

Membrane filters are discussed in Section 7.1.1. They are typically used for removing salt and inorganic chemicals. However, other types of filtration devices, primarily vessels containing granular or particulate media, are used for offshore produced water treatment. Many types of solid media are used, including sand, anthracite, crushed walnut shells, as well as multimedia filters. Filters can be operated in various ways, such as upflow, downflow, and fluidized bed. Media filters operate until they reach a predetermined level of solids loading, then they are taken offline and backwashed to remove the collected material.

8.2.4 Centrifuges

Sellman (2007) describes centrifuges that remove oil and solids. Like hydrocyclones, centrifuges separate oil from water by centrifugal force. However, centrifuges use a rapidly spinning bowl and generate much stronger forces than hydrocyclones. Hence, centrifuges are capable of removing oil droplets with smaller diameters. In produced water centrifuges, the spinning axis is vertically positioned. Centrifuges are often used to help achieve compliance with strict oil and grease discharge standards, but because of their greater cost, are not used as frequently as hydrocyclones.

8.3 Coalescence

Coalescing technologies cause small oil droplets to join together to form larger ones that are more amenable to removal. According to Stokes Law, oil rises at a rate dependent on the droplet diameter and the fluid viscosity. Smaller diameter droplets rise more slowly. One way of increasing the rate at which droplets rise is to increase the droplet diameter. Once the drop size has increased, other technologies can be used to remove the oil.

Coalescers provide surfaces on which oil droplets can congregate and merge. Most coalescers use fiberglass, polyester, metal, or Teflon® media, which are arranged in a mesh, co-knit, or irregular "wool" format. Finer meshed media are more capable of capturing and coalescing smaller droplets. However, tighter mesh becomes more vulnerable to fouling with solids. If solids are likely to pose a problem, filtering devices or other solids-removal equipment should be employed before running the coalescer.

Another form of coalescer consists of a bundle of oleophilic polypropylene fibers inside a cartridge positioned along a flow line just upstream of another separation device (e.g., hydrocyclone, filter). The fibers serve to aggregate small oil droplets for easier downstream removal. The coalescence occurs rapidly (within 2 s). Tulloch (2003) reports that oil droplet growth was enhanced by increasing either the length of the fibers or the number of fibers packed into the cartridge.

The Total Oil Remediation and Recovery process technology is based on a multistage adsorption and separation system with the capacity of multiphase separation of large and small (free-floating and emulsified) oil droplets in produced water (Plebon et al. 2005; Saad et al. 2006). This is accomplished by means of a patented reusable petroleum adsorbent media coalescing agent, which is polyurethane-based, oleophilic, hydrophobic, and nontoxic. The oily water passes through multiple vessels containing the media and a recovery chamber. The media continuously adsorb, coalesce, and desorb the very small droplets in oil emulsions. This process creates larger oil droplets. In the recovery chamber, oil droplets float to the top where the final separation of oil, gas, and the water occurs.

Controlled application of ultrasound in a standing wave pattern has been shown to coalesce oil droplets. As of early 2007, the technology was being tested and showed promise (Sinker 2007).

8.4 Flotation

Flotation technologies introduce bubbles of air or other gas into the bottom of a sealed tank. As the bubbles rise, they attach to oil droplets and solid particles and lift them to the surface where they can be skimmed off. Gas flotation technology is subdivided into dissolved gas flotation (DGF) and induced gas flotation (IGF). The two technologies differ by the method used to generate gas bubbles and the resultant bubble sizes. In DGF units, gas (usually air) is fed into the flotation chamber, which is filled with a fully saturated solution. Inside the chamber, the gas is released by applying a vacuum or by creating a rapid pressure drop. IGF technology uses mechanical shear or propellers to create bubbles that are introduced into the bottom of the flotation chamber.

DGF units create smaller gas bubbles than IGF systems. However, they require more space than IGF systems and more operational and maintenance oversight. Because space and weight are at a premium on offshore platforms, IGF systems are more commonly used at most offshore facilities.

Many IGF systems use multiple cells in series to enhance the hydraulic characteristics and improve oil and solids removal. Chemicals are often added to aid the flotation process. They can break emulsions, improve aggregation of particles, and serve other functions. Cline (2000) offers a useful overview of flotation technology. The author discusses the advantages and disadvantages of different types of flotation equipment offered by various vendors.

8.5 Combined Physical and Chemical Processes

Several technologies combine physical processes with chemical extraction and adsorption to provide excellent oil and grease removal. These devices are able to remove most of the dispersed oil and some of the dissolved oil and organics. If applicable discharge standards require oil and grease removal to low levels, these technologies can be employed.

8.5.1 EPCON

The EPCON compact flotation unit (CFU) consists of a vertical vessel acting as a three-phase water/oil/gas separator. Centrifugal forces and gas flotation contribute to the separation process. The oil droplets are made to agglomerate and coalesce to produce larger oil drops. This eventually creates a continuous oil or emulsion layer at the upper liquid level of the flotation chamber. Internal devices in the chamber and simultaneous gas flotation effects triggered by the release of residual gas from the water facilitate the separation process. In some cases, process optimization can be achieved by introducing external gas and/or specific flocculating chemicals. The resultant oil and gas deposits are removed in a continual process through separate outlet pipes. Treatment through a single CFU separation step has proven to reduce the oil-in-water content to 15–25 mg/L, while simultaneously degassing the water (Jahnsen and Vik 2003). When more than one CFU is used in series, oil and grease removal performance further improves to levels often below 10 mg/L.

8.5.2 CTour

The CTour process system enhances the traditional hydrocyclone technology. It uses gas condensate to extract hydrocarbons from water. The condensate is injected into the produced water stream before being routed through existing hydrocyclone systems. The condensate functions as a solvent, which draws dissolved hydrocarbons out of the water phase and into the condensate. In addition, the condensate helps to coalesce the small dispersed oil droplets, which then form larger oil droplets before being removed in the hydrocyclones. The CTour process system is capable of removing many of the dissolved organics from the produced water.

Pilot studies at Statoil's Statfjord C platform initially demonstrated 99% removal of dispersed oil, 2–3 ring polycyclic aromatic hydrocarbons (PAHs), 4–6 ring PAHs, and phenols of C6 or greater (Grini et al. 2003). A subsequent study showed that the performance of the hydrocyclone had improved from 70% removal (pre-CTour) to 85–90% removal (post-CTour). Dispersed oil was measured as low as 1–5 mg/L (Torvik et al. 2005).

8.6 Solvent Extraction

Macro porous polymer extraction (MPPE) is a fluid extraction technology that removes dissolved organics from produced water. The MPPE process takes place in the pores of polymer particles. Water contaminated with hydrocarbons is passed through a column packed with porous polymer beads, which contain a proprietary extraction liquid that readily removes dissolved oil and organics from produced water. Due to their heightened affinity for the extraction liquid, hydrocarbons are removed. The purified water can be reused or discharged. Periodical in situ regeneration makes use of low-pressure steam to strip the hydrocarbons of the extraction liquid. The stripped hydrocarbons are condensed and then separated from the water phase by gravity. The resulting almost 100% pure hydrocarbon phase is recovered, removed from the system, and readied for subsequent use, reuse, or disposal. Meanwhile, the condensed aqueous phase is recycled within the system (Meijer and Kuijvenhoven 2002). MPPE is a technology that can be employed if the applicable discharge standards require oil and grease removal to lower than 5 mg/L.

8.7 Adsorption

Adsorption technologies can quite effectively remove most organic materials from produced water. Adsorption is generally utilized as a polishing step, rather than a primary or secondary treatment. This avoids rapid loading of the adsorption media, which fill up at a rate proportional to the concentration of organic compounds in the influent.

Most produced water adsorption is done using organoclay cartridges. Other types of adsorption media that have been used include activated carbon and zeolite.

Organoclays are manufactured by modifying bentonite clay with quaternary amines. Organoclay mixtures are effective at attracting and adsorbing a wide range of hydrocarbons. They can be designed to attract and adsorb more than their own weight in organic compounds (Ali et al. 1999). Several companies supply cartridge filters containing different types of granular organoclays.

Organoclay cartridges offer many advantages, among which are that the technology

- uses an adsorption process that does not require chemicals,
- has a performance record that is unaffected by droplet size,
- does not require a power supply,
- increases oil recovery rates and ensures maximum platform productivity, and
- removes total petroleum hydrocarbons and other soluble components.

Different levels of organoclay treatment are used, depending on the discharge standard. The cost of treating is somewhat proportional to the amount of oil removed. As an example, the US produced water discharge standard for oil and grease in the Gulf of Mexico is 29 mg/L. If an operator can achieve 35 mg/L using other treatment devices, it may employ organoclay to remove incremental levels of oil and grease. It is not necessary to achieve 15 or 10 mg/L; the cost to reach those levels may be prohibitive. The operator may treat using enough organoclay to reach 20 or 25 mg/L. This allows compliance plus some cushion to account for variation.

Once organoclay cartridges become fully loaded, they must be brought back to shore for disposal. At present, cost-effective methods for recycling or reusing the cartridges are not available.

9 Summary

Produced water is a high-volume by-product of oil and gas production. The cost of managing large volumes of produced water can be an important consideration in determining the viability of oil and gas projects. Fortunately, many technologies are available to manage produced water. Some technologies can minimize the amount of water that is handled at the surface, and other technologies can reuse water for different purposes. Many types of water treatment processes have been developed. Some of those that have been used traditionally in the oil and gas industry are also described in this chapter, segregated by the primary group of pollutants the technology is designed to remove (i.e., inorganics or organics).

In recent years, many new technologies or sequences of technologies designed to treat produced water have been introduced. Some of these technologies are completely new, while others have been adapted from other industrial or municipal wastewater treatment or process operations. Based on the author's experience over the past several years, the produced water market is attracting a great deal of investment and innovation. New technologies get introduced quite frequently. Because of the "clearinghouse value" of information products such as PWMIS, the author of this chapter receives frequent contacts from new companies that want to know how to get their produced water technologies featured and better known. The advice offered by the author is to conduct pilot or full-scale tests using actual produced water, then publish or present the data in technical venues. Many of the successful companies have followed that advice.

In addition to the technology descriptions contained in this chapter and in PWMIS, several other sources provide valuable information on produced water management and treatment technologies. Two reports prepared by ALL Consulting provide extensive information and examples on methods for managing produced water from coal bed methane production (ALL 2003; IOGCC and ALL 2006). Both of these were cited in several places throughout this chapter, and they contain a great deal of other useful information.

During the 14th International Petroleum Environmental Conference (November 2007), the author organized a Produced Water Technology Symposium that featured technology presentations on many of the technologies described in this chapter. Speakers during the morning session described technologies used to remove salt and inorganic chemicals, while the afternoon speakers described technologies used to remove oil and grease and organics. All of the presentations can be viewed at the conference website (http://ipec.utulsa.edu/Conf2007/2007agenda.html). The symposium sessions were held on Thursday, November 8.

Acknowledgments Mr. Veil's produced water studies over the past 15 years were sponsored by the US Department of Energy's Office of Fossil Energy and National Energy Technology Laboratory under Contract DE-AC02-06CH11357.

References

Ali SA, Henry LR, Darlington JW, Occhipinti J (1999) Novel filtration process removes dissolved organics from produced water and meets federal oil and grease guidelines. In: Presented at the 9th annual produced water seminar, Houston, TX, Jan. 21–22

ALL (ALL Consulting) (2003) Handbook on coal bed methane produced water: management and beneficial use alternatives, prepared by ALL Consulting for the Ground Water Protection Research Foundation, U.S. Department of Energy, and U.S. Bureau of Land Management, July

Atlas R (2007) Purification of brackish water using hybrid CDI-EDI technology. In: Presented at the 14th international petroleum environmental conference, Houston, TX, Nov. 5–9. Available at http://ipec.utulsa.edu/Conf2007/Papers/Atlas_77a.pdf

Ayers RS, Westcot DW (1994) Water quality for agriculture, Food and Agriculture Organization of the United Nations Irrigation and Drainage Paper, 29 Rev. 1 (reprinted 1989, 1994). Available at http://www.fao.org/DOCREP/003/T0234E/T0234E00.htm#TOC

Beagle D (2008) CBM-produced water management in the Powder River Basin of Wyoming and Montana. In: Presented at the 15th international petroleum and biofuels environmental conference, Albuquerque, NM, Nov. 10–13

Boysen JE, Harju JA, Shaw B, Fosdick M, Grisanti A, Sorensen JA (1999) The current status of commercial deployment of the freeze thaw evaporation treatment of produced water, SPE 52700. In: Presented at SPE/EPA 1999 exploration and production environmental conference, Austin, TX, March 1–3

Brost DF (2002) Water quality monitoring at the kern river field. In: Presented at the 2002 ground water protection council produced water conference, Colorado Springs, CO, Oct. 16–17. Available at http://www.gwpc.org/meetings/special/PW%202002/Papers/Dale_Brost_PWC2002.pdf

Burnett D, Fox WE, Theodori GL (2002) Overview of Texas A&M's program for the beneficial use of oil field produced water. In: Presented at the 2002 ground water protection council produced water conference, Colorado Springs, CO, Oct. 16–17. Available at http://www.gwpc.org/meetings/special/PW%202002/Papers/David_Burnett_PWC2002.pdf

Burnett DB, Veil JA (2004) Decision and risk analysis study of the injection of desalination by-products into oil-and gas-producing zones. SPE 86526. In: Presented at the SPE formation damage conference, Lafayette, LA, Feb. 13–14

Cartwright PS (2006) Water recovery and reuse – a technical perspective. In: Presented at the 2nd annual desalination workshop, Texas A&M University, College Station, TX, Aug. 6–8

Clark CE, Veil JA (2009) Produced water volumes and management practices in the United States, ANL/EVS/R-09/1. In: Prepared for the U.S. Department of Energy, National Energy Technology Laboratory, September, 64 pp. Available at http://www.ead.anl.gov/pub/dsp_detail.cfm?PubID=2437

Cline JT (2000) Survey of gas flotation technologies for treatment of oil & grease. In: Presented at the 10th produced water seminar, Houston, TX, Jan. 19–21

Colazas XC, Strehle RW, Bailey SH (1987) Subsidence control wells, wilmington oil field, Long Beach, California. In: Proceedings of the international symposium on class V injection well technology, Washington, DC, Sept. 22–24, Underground Injection Practices Council Research Foundation

Dennis R (2006) Coal industry turns to ion exchange technology for wastewater minimization. Industrial WaterWorld, PennWell Corporation, Tulsa, OK, Sept. Available at http://ww.pennnet.com/Articles/Article_Display.cfm?Section=ARTCL%26;PUBLICATION_ID=41%26;ARTICLE_ID=276178%26;C=colmn

do Vale OR, Garcia JE, Villa M (2002) VASPS installation and operation at Campos Basin, OTC paper 14003. In: Presented at the offshore technology conference, Houston, TX, May 6–9

Frankiewicz T (2001) Understanding the fundamentals of water treatment, the dirty dozen – 12 common causes of poor water quality. In: Presented at the 11th produced water seminar, Houston, TX, Jan. 17–19

Ginn R (2003) Personal communication between Ginn, Railroad Commission of Texas, Austin, TX, and J. Veil, Argonne National Laboratory, Washington, DC, Feb. 14

Gjerdseth AC, Faanes A, Ramberg R (2007) The Tordis IOR project, OTC 18749. In: Presented at the 2007 offshore technology conference, Houston, TX, April 30–May 3

Godshall N (2007) AltelaRain[SM] produced water treatment technology: making water from waste. In: Presented at the 14th international petroleum environmental conference, Houston, TX, Nov. 5–9. Available at http://ipec.utulsa.edu/Conf2007/Papers/Godshall_44a.pdf

GRI (Gas Research Institute) (1999) Technology assessment and economic evaluation of downhole gas/water separation and disposal tools, GRI-99/0218, prepared by Radian Corporation for the Gas Research Institute (now Gas Technology Institute), Oct. Available at http://www.gastechnology.org/webroot/app/xn/xd.aspx?xd=10AbstractPage%5C12154.xml

Grini PG, Clausen C, Torvik H (2003) Field trials with extraction based produced water purification technologies. In: Presented at the 1st NEL produced water workshop, Aberdeen, Scotland, March 26–27

Hayes T (2004) The electrodialysis alternative for produced water management, GasTIPS, Summer, pp. 15–20. Available at http://media.godashboard.com/gti/4ReportsPubs/4_7GasTips/Summer04/TheElectrodialysisAlternativeForProducedWaterManagement.pdf

IOGCC and ALL (Interstate Oil and Gas Compact Commission and ALL Consulting) (2006) A guide to practical management of produced water from onshore oil and gas operations in the United States, prepared for U.S. Department of Energy, National Energy Technology Laboratory, by the Interstate Oil and Gas Compact Commission and ALL Consulting, Oct. Available at http://www.all-llc.com/IOGCC/ProdWtr/ProjInfo.htm

Jahnsen L, Vik EA (2003) Field trials with EPCON technology for produced water treatment. In: Presented at the produced water workshop, Aberdeen, Scotland, March 26–27

Jenneman GE, Webb RH, Dinning AJ, Voldum K, Bache O (2004) Evaluation of nitrate and nitrite for control of biogenic sulfides in Ekofisk produced water. In: Presented at the 11th international petroleum environmental conference, Albuquerque, NM, Oct. 11–15

Khatib Z, Verbeek P (2003) Water to value – produced water management for sustainable field development of mature and green fields. J Petrol Technol, Jan: 26–28

Liske B (2005) Recovery of more oil-in-place at lower production costs while creating a beneficial water resource. In: Presented at the DOE/PERF water program review, Annapolis, MD, Nov. 1–4. Available at http://www.perf.org/pdf/liske.pdf

Mayuga MN, Allen DR (1969) Subsidence in the wilmington oil field, Long Beach, California, USA, IAHS-AISH Publication 88

Meijer DT, Kuijvenhoven CAT (2002) Field-proven removal of dissolved hydrocarbons from offshore produced water by the macro porous polymer-extraction technology. In: Presented at the 12th produced water seminar, Houston, TX, Jan. 16–18

Murce T (2008) Personal communication between Murce, Petrobras Research and Development Center, Rio de Janeiro, Brazil, and J.A. Veil, Argonne National Laboratory, Argonne, IL, Sept. 9

Murphree PA (2002) Utilization of water produced from coal bed methane operations at the North Antelope/Rochelle Complex, Campbell County, Wyoming, presented at the 2002 Ground Water Protection Council Produced Water Conference, Colorado Springs, CO, Oct. 16–17. Available at http://www.gwpc.org/meetings/special/PW%202002/Papers/Phil_Murphree_PWC2002.pdf

Offshore (2000) ABB looking to progress subsea processing into ultra-deepwater. Offshore 60(8). Aug. 1

Offshore (2006) Subsea separation, reinjection system solves problem of produced water, Offshore 66(9). Sept

Oil & Gas Journal (2008) Worldwide look at reserves and production. Oil Gas J Dec. 22

Ouellette R, Ganesh R, Leong LYC (2005) Overview of regulations for potential beneficial use of oilfield produced water in California. In: Presented at the 12th international petroleum environmental conference, Houston, TX, Nov. 8–11. Available at http://ipec.utulsa.edu/Conf2005/Papers/Ouellette_Overview.pdf

Peacock P (2002) Beneficial use of produced water in the Indian Basin Field: Eddy County, NM. In: Presented at the 2002 Ground Water Protection Council produced water conference, Colorado Springs, CO, Oct. 16–17. Available at http://www.gwpc.org/meetings/special/PW%202002/Papers/Paul_Peacock_PWC2002.pdf

Plebon MJ, Saad M, Fraser S (2005) Further advances in produced water de-oiling utilizing a technology that removes and recovers dispersed oil in produced water 2 microns and larger. In: Presented at the 12th international petroleum environmental conference, Houston, TX, Nov. 8–11. Available at http://ipec.utulsa.edu/Conf2005/Papers/Plebon_Further_Advances.pdf

Puder MG, Veil JA (2006) Offsite commercial disposal of oil and gas exploration and production waste: availability, options, and cost, prepared for U.S. Department of Energy, National Energy Technology Laboratory, Aug., 148 pp. Available at http://www.ead.anl.gov/pub/dsp_detail.cfm?PubID=2006

Raia JC, Caudle DD (1999) Methods for the analysis of oil and grease and their application to produced water from oil and gas production operations. In: Presented at the 9th produced water seminar, Houston, TX, Jan. 21–22

Reynolds RR, Kiker RD (2003) Produced water and associated issues – a manual for the independent operator, Oklahoma Geological Survey Open File Report 6-2003, prepared for the South Midcontinent Region of the Petroleum Technology Transfer Council

RMOTC (Rocky Mountain Oilfield Testing Center) (2008) Geothermal electrical generation holds promise for older oil fields, News Release, Oct. 18. Available at http://www.rmotc.doe.gov/newsevents/ormat.html

Saad M, Plebon MJ, Fraser S (2006) Fundamental approach to produced water treatment: validation of an innovative technology. In: Presented at the 16th produced water seminar, Houston, TX, Jan. 18–20

Sellman E (2007) Produced water deoiling using disc stack centrifuges. In: Presented at the 17th produced water seminar, Houston, TX, Jan. 17–19

Seright RS, Lane RH, Sydansk RD (2001) A strategy for attacking excess water production, SPE 70067. In: Presented at the SPE permian basin oil and gas recovery conference, Midland, TX, May 15–16

Shirman EI, Wojtanowicz AK (2002) More oil using downhole water-sink technology: a feasibility study, SPE 66532. SPE Production and Facilities, Nov

Sinker A (2007) Less oil in, less oil out: a holistic approach to enhanced produced water treatment. In: 17th produced water seminar, Houston, TX, Jan. 17–19

Stettner M (2003) Personal communication between Stettner, California Department of Conservation, Sacramento, CA, and J. Veil, Argonne National Laboratory, Washington, DC, Feb. 13

Stewart DR (2006) Developing a new water resource from production water. In: Presented at the 13th international petroleum environmental conference, San Antonio, TX, Oct. 23–27. Available at http://ipec.utulsa.edu/Conf2006/Papers/Stewart_18.pdf

Stone B (2003) Personal communication between Stone, New Mexico Oil Conservation Division, Santa Fe, NM, and J. Veil, Argonne National Laboratory, Washington, DC, Feb. 14

Torvik H, Bergersen L, Paulsen C (2005) One year of operational experience with CTour at Statfjord C. In: Presented at the 3rd NEL produced water workshop, Aberdeen, Scotland, April 20–21

Tulloch SJ (2003) Development & field use of the Mare's Tail® Pre-Coalescer. In: Presented at the produced water workshop, Aberdeen, Scotland, March 26–27

Veil JA (2008) Thermal distillation technology for management of produced water and frac flow-back water, water technology brief #2008-1. In: Prepared for U.S. Department of Energy, National Energy Technology Laboratory, May 13, 12 pp. Available at http://www.ead.anl.gov/pub/dsp_detail.cfm?PubID=2321

Veil JA, Burnett D, Grunewald B (2006) Disposal of concentrate from treatment of water for beneficial reuse. In: Presented at the 13th international petroleum environmental conference, San Antonio, TX, Oct. 23–27. Available at http://ipec.utulsa.edu/Conf2006/Papers/Veil_concentrate.pdf

Veil JA, Langhus BG, Belieu S (1999) Feasibility evaluation of downhole oil/water separation (DOWS) technology, prepared by Argonne National Laboratory, CH2M-Hill, and Nebraska Oil and Gas Conservation Commission for the U.S. Department of Energy, Office of Fossil Energy, National Petroleum Technology Office, Jan. Available at http://www.ead.anl.gov/pub/dsp_detail.cfm?PubID=416

Veil JA, Puder MG, Elcock D, Redweik RJ Jr (2004) A white paper describing produced water from production of crude oil, natural gas, and coal bed methane, prepared by Argonne National Laboratory for the U.S. Department of Energy, National Energy Technology Laboratory, Jan

Veil JA, Quinn JJ (2004) Downhole separation technology performance: relationship to geological conditions, prepared by Argonne National Laboratory for the U.S. Department of Energy, National Energy Technology Laboratory, Nov. Available at http://www.ead.anl.gov/pub/dsp_detail.cfm?PubID=1783

von Flatern R (2003) Troll pilot sheds light on seabed separation, Oil Online, May 16

Wagner J (2001) Membrane filtration handbook – practical tips and hints, Second Edition, Revision 2, Osmonics, Inc., Minnetonka, MN, Nov., 129 pp. Available at http://www.gewater.com/pdf/1229223-%20Lit-%20Membrane%20Filtration%20Handbook.pdf

Wolff EA (2000) Reduction of emissions to sea by improved produced water treatment and subsea separation systems, SPE#61182. In: Presented at the Society of Petroleum Engineers international conference on health, safety, and environment, Stavanger, Norway, June 26–28

Yang M (2007) Oil in produced water measurement –an overview of regulatory requirements and current practices. In: Presented at the II international seminar on oilfield water management, Rio de Janeiro, Brazil, October 16–19

Chapter 30
Decision-Making Tool for Produced Water Management

Abdullah Mofarrah, Tahir Husain, Kelly Hawboldt, and Brian Veitch

Abstract Produced water (PW) is the most significant source of waste discharged in the production phase of oil and gas operations. The management of PW provides distinct challenges for the oil and gas industry. There are number of technologies to treat and manage the PW, but selection of the best alternative often involves competing criteria and needs sophisticated decision-making tools. This chapter introduces a decision-making tool for selecting the best PW management option by utilizing a multi-criteria decision-making (MCDM) approach. The methodology introduces several important concepts, and definitions in decision analysis related to PW management. These are the trade-offs among technical feasibility, environment, cost and health and safety. The Analytical Hierarchy Process (AHP) and additive value model is integrated with MCDM to enhance the decision-making process. The proposed methodology is applied to a hypothetical example, and its efficacy is demonstrated through an application dealing with the selection of PW management technology for oil and gas operations.

1 Introduction

Produced water (PW) is the most significant source of waste discharged in the production phase of oil and gas operations (Patin 1999). Once discharged into the ocean, PW may introduce toxicity and bioaccumulation (i.e., PAHs and metals) in marine organisms (CAPP 2001; OGP 2005; Neff 2002). Treatment of produced water has become an important issue in the petroleum industry. Selection of the best PW treatment facilities is a complex process and depends on specific conditions. The aim of this chapter is to provide a methodology for the selection of PW management technologies that will guide the decision makers. The decision-making tool attempts to minimize the time as well as the conflicts that occur from various opinions during the selection process. A multi-criteria decision-making (MCDM) process is used as a basic structure to develop this methodology. MCDM techniques

A. Mofarrah (✉)
Faculty of Engineering and Applied Science, Memorial University of Newfoundland,
St. John's, NL, Canada, A1B 3X5

K. Lee, J. Neff (eds.), *Produced Water*, DOI 10.1007/978-1-4614-0046-2_30,

Table 30.1 Fundamental scales of importance (adapted from Saaty 1980)

How important is *A* relative to *B*?	Preference index assigned
Equally important	1
Moderately more important	3
Strongly more important	5
Very strongly more important	7
Overwhelmingly more important	9
Intermediate values (need to judge two)	2,4,6,8

deal with the problems whose alternatives are predefined and the decision-maker ranks the available alternatives (Tesfamariam and Sadiq 2006). There are a variety of MCDM methods: one of them classified under utility theory is the analytical hierarchy process (AHP) developed by Saaty (1977, 1980). The analytical hierarchy process (AHP) is one of the most commonly used MCDM methods, which integrates subjective and personal preferences in performing analyses. AHP works on a premise that can handle complex problems by structuring them into a simple and comprehensible hierarchical structure (Tesfamariam and Sadiq 2006; Khasnabis et al. 2002; Holguin-Veras 1993; Uddameri 2003; McIntyr et al. 1999). The AHP uses objective mathematics to process the subjective and personal preferences of an individual or a group in decision making (Saaty 1980).

In conventional AHP, the judgment matrix is developed by means of a pairwise comparison between any two criteria. The pairwise comparison is established using a 9-point scale (Table 30.1) which converts the human preferences between available alternatives. In the preferences scale, the dimensionless value "1" represents that two criteria are equally important, while the other extreme "9" represents that one criterion is absolutely more important than the other (Saaty 1980).

The primary objective of this chapter is to present a methodology to guide the decision making in selecting the best PW management technology. This task is carried out by structuring a multi-criteria decision-making problem with the help of integrating the Analytical Hierarchy Process (AHP) and additive value model.

2 Methodology

2.1 Criteria Evaluation

An integrated multi-criteria decision-making model is developed for this study to assess and compare the PW treatment alternatives. Figure 30.1 shows the structure of the study. The first step of the evaluation is to identify the most important criteria and establish a hierarchic tree. A hierarchic tree is a way of representing the hierarchical nature of the problem in a graphical form (Tesfamariam and Sadiq 2006; Saaty 1997, 1980). In this study, the hierarchic tree is formed by selecting

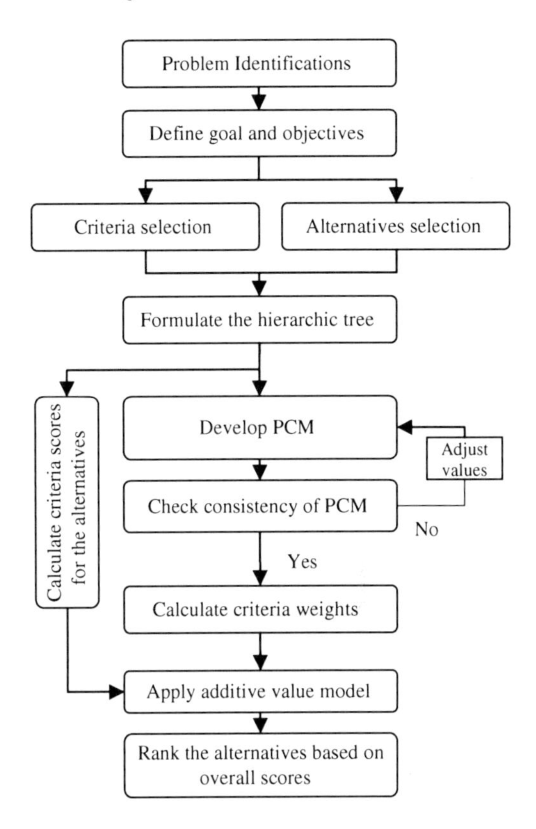

Fig. 30.1 Structure of the study

the produced water management alternatives and the criteria under which the alternatives will be evaluated. A four-level hierarchical tree is constructed in this case. The first level of the hierarchy corresponds to the objective or goal, and the last level corresponds to the different produced water management options, whereas the intermediate levels correspond to the criteria and sub-criteria. The criteria are the controlling factors under which the options will be scored.

In this study, researchers used two distinct groups of criteria: one threshold criterion and another decision-making criterion. The threshold criterion is used to screen out inappropriate technologies. The technology that is unable to meet the regulatory oil and grease discharge standard limit described in Table 30.2 will be rejected and not considered for further evaluation.

The decision-making criteria are used to evaluate and compare the technologies. Four important factors are considered as decision-making criteria to develop this methodology. These principle criteria categories are technical feasibility, cost, environment, and health and safety. For detailed investigation, principle criteria are subdivided into sub-criteria. Figure 30.2 shows the criteria hierarchical structure for the study.

Table 30.2 Offshore PW discharged standards (CAPP 2001)

Country	Effluent limits	Exception thresholds	Routine reporting
USA	29 mg/L monthly avg. 42 mg/L daily max.	Any exception	Annual
UK	40 ppm monthly avg. 30 ppm annual avg.	>100 ppm	Monthly O&G Annual Comprehensive
Norway	40 ppm monthly avg.	> 40 ppm monthly avg.	Quarterly O&G Annual Comprehensive
Canada	40 ppm 30 day avg. 80 ppm 2 day avg.	Any exception	Monthly

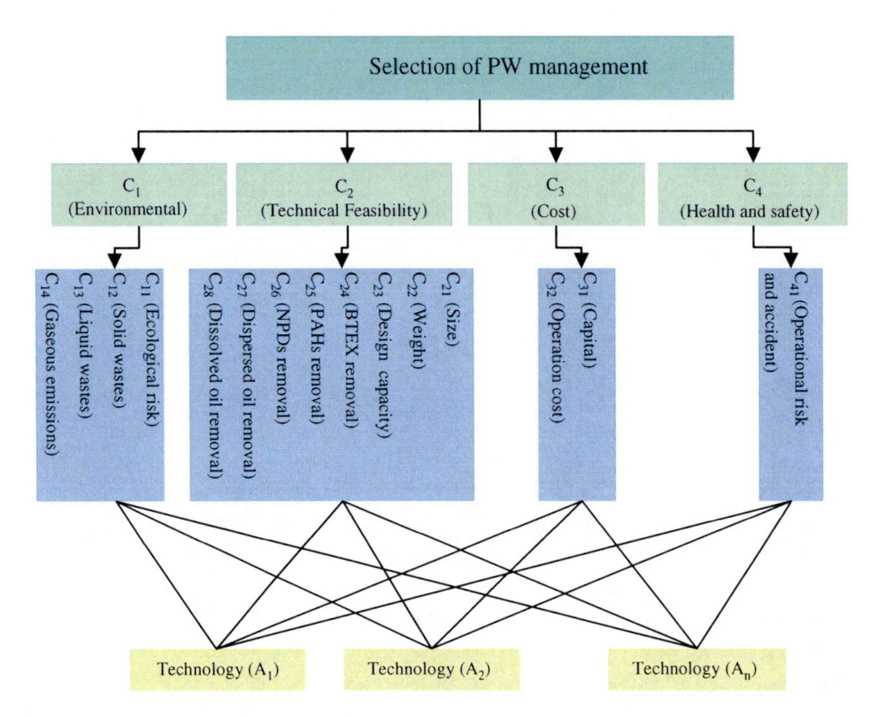

Fig. 30.2 Criteria and sub-criteria of the case study

2.2 Scoring Scheme

A scoring scheme is introduced based on the performance of the technologies under each criterion. Quantitative data such as the weight and footprint are normalized using linear functions as shown in Eq. (30.1). The dimensionless scores are used to measure the performance of the alternatives. As this type of data may have either increasing or decreasing value, two ranges of normalized scores were used. Positive scores range from 0 to 1 and negative scores range from –1 to 0. For this study, the positive scores are treated as beneficial scores. For example, the pollutants removal efficiency of the technology, the higher is better. On the other hand, high cost, large foot print, heavy weight etc., which are drawbacks of the technology are assigned negative scores.

$$r_i = \pm \frac{C_i}{\sum_{i=1}^{n} C_i} \tag{30.1}$$

where, r_i is the normalized value of the criteria C_i.

2.3 Subjective Judgments

In environmental and social studies, most of the information is imprecisely defined due to the unquantifiable nature of data or lack of proper knowledge (Turban and Meredith 1991). Experts often use linguistic scales to express the existing scenarios. Where quantitative data are not available, subjective rankings are used in this study to measure the option. The subjective rankings are then converted into numbers using the scale range of 0–1 shown in Fig. 30.3. Linguistic terms such as low (L), moderate (M), high (H), and extremely high (EH) are used to capture the subjective ranking.

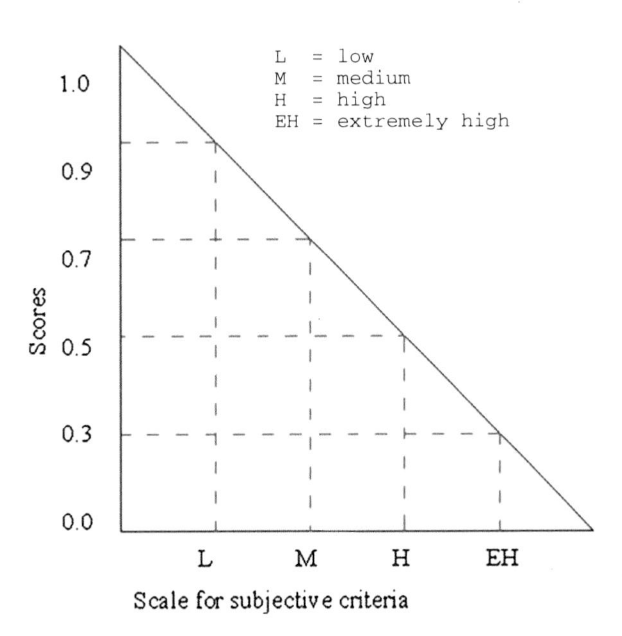

Fig. 30.3 The conversion scale of linguistic variables

2.4 Weights Calculation

To measure the relative importance of the criteria, weighting factors are assigned to the criteria following a pairwise comparison. To calculate the weights, pairwise comparison matrices (PCM) are formed between the elements at each level of the hierarchy. AHP is used to solve the PCM. A simple example to demonstrate the AHP analysis is illustrated here. Considering the criteria C_{11}, C_{12}, C_{13}, and C_{14}, a PCM, A_{C_1} is developed by using the pairwise comparison technique. According to the theory (Saaty 1980), each element of the lower triangle in the PCM is reciprocal to the upper triangle.

It is necessary to estimate the consistency of the PCM. The following equations can be used to check the consistency of the PCM (Saaty 1980; Modarres 2006):

$$\mathrm{CI} = \frac{\lambda_{\max} - n}{n - 1} \tag{30.2}$$

$$\mathrm{CR} = \frac{\mathrm{CI}}{\mathrm{RI}} \tag{30.3}$$

where CI = consistency index, CR = consistency rate, $\lambda_{\max}$ maximum eigenvalue, n = number of parameters in the matrix, and RI = random index. The suggested RI values are given in Table 30.3.

When CR ≤ 0.10, the judgment matrix can be considered to have a satisfactory consistency (Saaty 1977, 1980; Modarres 2006) which means the weight vector is reliable. Otherwise, we must reconstruct the judgment matrix. In this case, the maximum eigenvalue of A_{C_1} was $\lambda_{\max} = 4.0069$ and the corresponding eigenvector is shown in Table 30.3.

$$\begin{matrix} C_{11} & C_{12} & C_{13} & C_{14} \end{matrix}$$

$$A_{C_1} = \begin{bmatrix} C_{11} \\ C_{12} \\ C_{13} \\ C_{14} \end{bmatrix} \begin{bmatrix} 1 & 1.267 & 0.914 & 1.036 \\ 0.789 & 1 & 0.790 & 1.087 \\ 1.093 & 1.266 & 1 & 1.272 \\ 0.965 & 0.92 & 0.786 & 1 \end{bmatrix}$$

The consistency ratio (CR) found for matrix A_{C_1}, is 0.002 (by Eqs. (30.2) and (30.3)) which is less than 0.1, so according to the assumptions the PCM is consistent. The eigenvectors of the PCM is considered the weight factors of the corresponding criteria (Saaty 1980). The weighting factors of this matrix are calculated from the corresponding eigenvector and normalized as shown in Table 30.4.

Table 30.3 Random index values (Saaty 1980)

n	1	2	3	4	5	6	7	8	9	10
RI	0.0	0.0	0.58	0.9	1.12	1.24	1.32	1.41	1.45	1.49

Table 30.4 Example for weight factors calculation

Criteria	Eigenvector	Weights factors
C_{11}	1.042	$= (1.042 \times 0.12)/4.0 = 0.031$
C_{12}	0.903	$= (0.930 \times 0.12)/4.0 = 0.027$
C_{13}	1.144	$= (1.144 \times 0.12)/4.0 = 0.034$
C_{14}	0.910	$= (0.910 \times 0.12)/4.0 = 0.027$
Sum	4.000	

3 Ranking of the Alternatives

The overall scores are dimensionless numbers used to represent the alternative's final ranking (Chen et al. 1992). The higher overall score represents the better performance of the alternative. Additive mathematical models are used to calculate overall scores. The overall score V_i for the ith alternative can be computed as follows:

$$\text{Overall values } V_i = \sum_{i}^{n} W_i C_{ij} \quad \text{for } i = 1, 2, \ldots, n \tag{30.4}$$

where W_i represents the weighting factors of ith criteria, and C_{ij} represents the criteria scores.

4 Sensitivity Analysis

To determine the sensitivity of the evaluation results, the criteria weights are varied to observe new alternative ranks. This analysis assures that the weights used in the evaluation are well defined among the criteria. In order to carry it out, one criterion weight is varied by 75% increase to 50% decrease and the other criteria weights are adjusted proportionally, so that in all cases, the sum of all criteria remains one.

5 Application

To demonstrate the proposed methodology, a simple hypothetical case study was considered with four PW treatment technologies, namely a gas flotation unit (A_1), a high pressure water condensate separator (A_2), coalescence technology (A_3), and an advanced oxidation process (A_4).

5.1 Gas Flotation Unit (A_1)

In this process, a gas is finely distributed in the PW. Gas bubbles and oil form foam in the water, which is skimmed, by mechanical means. The foam and part of the water is skimmed off an overflow. Gas flotation technology can be divided into dissolved gas flotation (DGF) and induced gas flotation (IGF). The two technologies differ by the method used to generate gas bubbles and the resultant bubble sizes. In DGF units, gas (usually air) is fed into the flotation chamber, which is filled with a fully saturated solution. Inside the chamber, the gas is released by applying a vacuum or by creating a rapid pressure drop (OSPAR 2002). IGF technology uses mechanical shear or propellers to create bubbles that are introduced into the bottom of the flotation chamber. Dissolved particles such as benzene and heavy metals are not removed, although gas injection may remove some volatile components (OSPAR 2002). Sometimes, air is used instead of gas, in which case a major part of BTEX is also removed from the PW.

5.2 High Pressure Water Condensate Separator (A_2)

On gas platforms, the dispersed and dissolved oil content in PW can be reduced by a high pressure (HP) water condensate separator, which operates at approximately the same pressure as the primary production separator (OSPAR 2006). In this process, the gas condensate from the high pressure separator flows through a throttling control valve to a low pressure separator. The reduction in pressure across the control valve causes condensation of the mixture. The formation of small condensate droplets in water (emulsion) in the regulating valve is prevented by separating the mixture and by releasing pressure in separate valves (OSPAR 2006). With this technology, acceptable oil concentrations are achieved using relatively simple add-on treatment equipment. The technique may also be used for condensate–water mixtures from the gas filter/separator and high pressure scrubbers.

5.3 Coalescence Technology (A_3)

This PW treatment technology includes the installation of a pre-coalescence step before entering the hydrocyclone liners. The coalescence media provides sufficient surface area on which oil droplets collect and merge. Most coalescers use fiberglass, polyester, metal, or Teflon media, which are arranged in a mesh, co-knit, or irregular "wool" format. Finer meshed media are more capable of capturing and coalescing smaller droplets. However, tighter mesh becomes more vulnerable to fouling with solids (Tulloch 2003). When PW flows through a permeable pack of fibrous material, the small droplets convert into larger ones and improve the oil removal efficiency. The installation of coalescence activity can boost the performance of the downstream de-oiling hydrocyclones and reduce the oil in water concentration in the discharge stream by up to 80% (OSPAR 2006). Solid materials present in the PW may reduce the performance of this technology. But, employing filtering devices or other solids-removal equipment before running the coalescer improves its efficiency.

5.4 Advanced Oxidation Process (A_4)

Advanced oxidation process (AOP) degrades organic species by utilizing the hydroxyl radical (OH^-). This results in degradation of the organics to carbon dioxide, water, and inorganic salts. Ozone (O_3) may be injected into the waste-water stream as part of an airstream. The ozone starts to react with the water to form hydroxyl radicals, but this process is enhanced in the presence of ultraviolet (UV) light. The UV/O_3 process is the best-developed AOP method that is currently available to the industry. This technique removes dissolved oil (BTEX), aliphatic hydrocarbons, and other pollutants from produced water (OSPAR 2006).

5.5 Score Calculations

Quantitative data such as the weight and size of the treatment technology were collected from different sources, including OSPAR (2002, 2006) and Ekins et al. (2005), and reported in Table 30.5. The data were then normalized to unity by

Table 30.5 Data matrix for the study; subjective judgments were assigned based on the performance of the technology, and numerical data was compiled from OSPAR (2002, 2006) and Ekins et al. (2005)

Criteria		Technology			
		Gas flotation unit (A_1)	High pressure water condensate separator (A_2)	Coalescence technology (A_3)	Advanced oxidation process (A_4)
C_1	Environmental	–	–	–	–
C_{11}	Ecological risk	M	L	L	L
C_{12}	Solid wastes	H	L	M	L
C_{13}	Liquid wastes	H	L	M	L
C_{14}	Gaseous emissions	H	L	L	L
C_2	Technical feasibility	–	–	–	–
C_{21}	Size (m^3)	75	0	0	6
C_{22}	Weight ave. (tons)	45	4	0.1	3
C_{23}	Design capacity ave. (m^3/h)	175	6	5	2
C_{24}	Ave. BTEX removal	20	30	50	75
C_{25}	Ave. PAHs removal	20	30	50	75
C_{26}	Ave. NPDs removal	20	30	0	75
C_{27}	Ave. dispersed oil removal	75	20	99	75
C_{28}	Ave. dissolved oil removal	20	30	50	75
C_{29}	Ave. metal removal	0.000	0.000	0.000	0.000
C_3	Cost	–	–	–	–
C_{31}	Capital (new) ave. (€)	250,000	86,000	5000	40,000
C_{32}	Operation cost ave. (€/y)	185,000	3400	Negligible	Negligible
C_4	Health and safety	–	–	–	–
C_{41}	Operational risk and accident	M	L	M	L

Eq. (30.1). For example, the normalized values of criteria C_{22} for A_1 can be calculated as

Normalized score $= -45/(45 + 4 + 0.1 + 3) = -0.8637$. The negative sign (–) was assigned to the criteria because it was assumed as cost criteria. Similarly, all normalized scores were calculated and reported in Table 30.6.

Where quantitative data were not found, subjective rankings were used to measure the option. The linguistic terms described in the earlier section were used for subjective judgment. The detailed judgments for different alternatives are reported in Table 30.5. The linguistic terms were directly mapped with Fig. 30.3 to calculate the corresponding criteria score. The scores for different criteria are reported in Table 30.6.

Criteria weights were calculated by conducting pairwise comparisons at different hierarchy levels. The weights of the principle criteria (i.e., C_1, C_2, C_3, and C_4) were distributed into the sub-criteria, assuming the equal importance among the sub-criteria under principle criteria. The complete weight scenarios for this study are shown in Table 30.7.

Table 30.6 Normalized scores of alternatives

Criteria		Normalized scores			
		Gas flotation unit (A_1)	High pressure water condensate separator (A_2)	Coalescence technology (A_3)	Advanced oxidation process (A_4)
C_{11}	Ecological risk	0.70	0.90	0.90	0.90
C_{12}	Solid wastes	0.50	0.90	0.70	0.90
C_{13}	Liquid wastes	0.50	0.90	0.70	0.90
C_{14}	Gaseous emissions	0.50	0.90	0.90	0.90
C_{21}	Size (m^3)	−0.9259	0.0000	0.0000	−0.0741
C_{22}	Weight ave. (tons)	−0.8637	−0.0768	−0.0019	−0.0576
C_{23}	Design capacity ave. (m^3/h)	0.9309	0.0319	0.0266	0.0106
C_{24}	Ave. BTEX removal	0.1143	0.1714	0.2857	0.4286
C_{25}	Ave. PAHs removal	0.1143	0.1714	0.2857	0.4286
C_{26}	Ave. NPDs removal	0.1600	0.2400	0.0000	0.6000
C_{27}	Ave. dispersed oil removal	0.2788	0.0743	0.3680	0.2788
C_{28}	Ave. dissolved oil removal	0.1143	0.1714	0.2857	0.4286
C_{29}	Ave. metal removal	0.0000	0.0000	0.0000	0.0000
C_{31}	Capital (new) ave. (€)	−0.6562	−0.2257	−0.0131	−0.1050
C_{32}	Operation cost ave. (€/y)	−0.9818	−0.0180	0.0000	−0.0001
C_{41}	Operational risk and accident	0.7000	0.90	0.70	0.90

Table 30.7 Weighting factors of criteria and sub-criteria

$C_1 = 0.20$				$C_2 = 0.50$									$C_3 = 0.20$		$C_4 = 0.10$
C_{11}	C_{12}	C_{13}	C_{14}	C_{21}	C_{22}	C_{23}	C_{24}	C_{25}	C_{26}	C_{27}	C_{28}	C_{29}	C_{31}	C_{32}	C_{41}
0.0500	0.0500	0.0500	0.0555	0.0555	0.0555	0.0555	0.0555	0.0555	0.0555	0.0555	0.0555	0.0555	0.1000	0.1000	0.1000

6 Results and Discussion

The overall scores V_i for each alternative were computed by using Eq. (30.4). The overall scores V_i are dimensionless numbers which represent the ranking of the alternatives. The higher overall scores reflect better performance of the technology. By comparing $V_{i(s)}$, decision makers can identify the best alternatives. The alternative ranking for this study is shown in Fig. 30.4. The overall scores of the gas flotation unit (A_1), high pressure water condensate separator (A_2), coalescence technology (A_3), and advanced oxidation process (A_4) are 0.012, 0.289, 0.298, and 0.373, respectively.

The result shows that the positive scores for all technologies were not significantly different except for technology (A_4), which means the technical characteristics such as pollutants removal efficiency, occupational health and hazards are relatively equivalent for all other technologies. Coalescence technology (A_3) has very few negative scores compared to others. This means that the alternative A_3 is most cost effective and compact in shape. Comparing the overall scores, the advanced oxidation process has the highest score (0.373) and was ranked highest overall.

To determine the sensitivity of the evaluation results, we varied the criteria weights to observe new overall values. Four sets of analyses were conducted by increasing the weight for one criterion or a group of criteria and proportionally

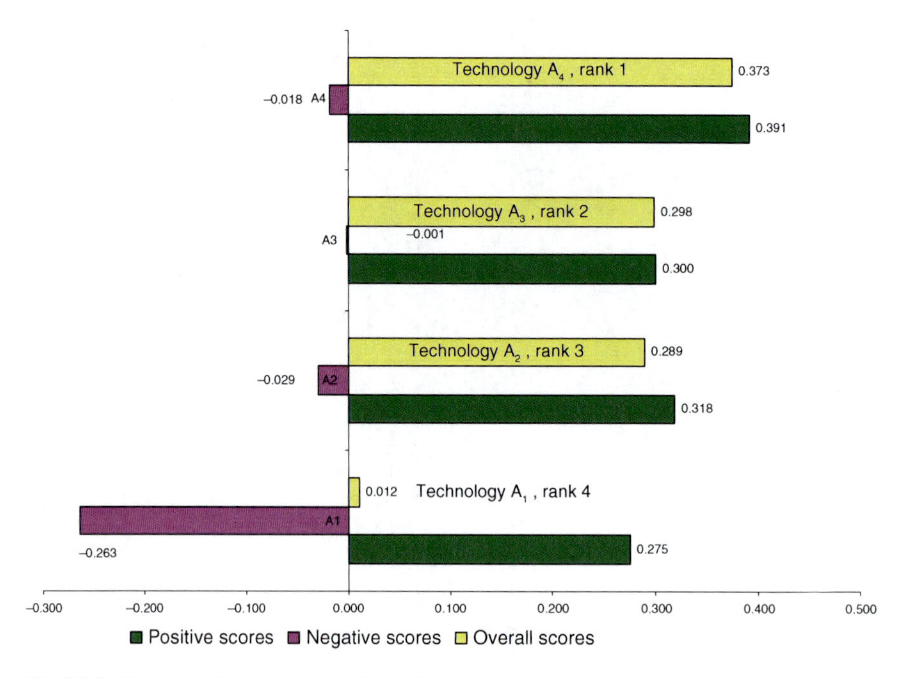

Fig. 30.4 Final overall scores and ranking of the alternatives

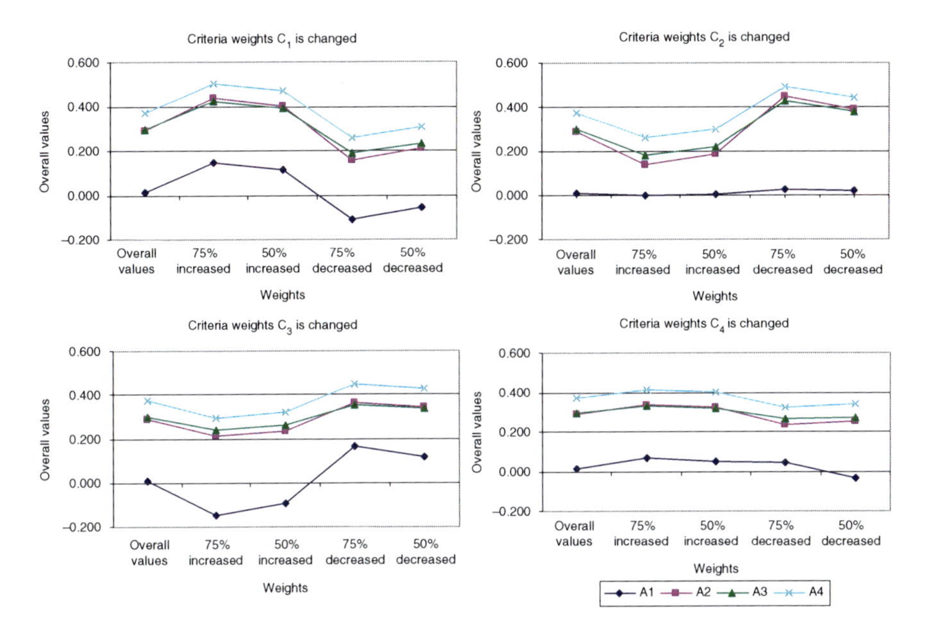

Fig. 30.5 Sensitivity analysis results

decreasing the weights of the other criteria to keep the total weights one. The sensitivity analysis shown in Fig. 30.5 indicates that the ranking of the evaluation was not significantly changed due to the redistribution of criteria weights. The sensitivity analysis has shown that there is a significant effect of assigned weights on the final ranking. However, for this study the ranks of the alternatives were found unchanged in most cases.

The sensitivity analysis assures that the weights used in this evaluation were well defined among the criteria.

7 Conclusions

In this chapter, we presented a decision-making process for evaluation of produced water management alternatives. AHP has been integrated with an additive value model to provide reliable and effective decision support. The major advantage of the AHP approach is that it gives a relative importance to the criteria involved in the assessment of alternatives. Many parameters influence the validity of the evaluation results, including the availability of data, subjective scoring, and the distribution of weights. Therefore, a sensitivity analysis was conducted to authenticate the results of the evaluation. This evaluation was structured to provide a simple and comprehensive methodology to assess produced water management technologies. Although advanced oxidation was ranked highest in the example, the outcome of the methodology will depend on the site specifics. In many ways, this represents the advantage of this approach, the ability to factor in site specifics. With some modification of the hierarchy structures, this methodology would be able to the select the best produced water management technology from available options with minimum effort.

Acknowledgments The research support provided by the Petroleum Research Atlantic Canada (PRAC) and Natural Sciences and Engineering Research Council's Collaborative Research and Development (NSERC CRD) grant is acknowledged.

References

CAPP (Canadian Association of Petroleum), August (2001) Technical report: offshore produced water waste management. http://www.capp.ca. Accessed 20 July 2007

Chen S-J, Hwang C-L, in collaboration with Hwang F-P (1992) Fuzzy multiple attribute decision making: methods and applications. Springer, New York, NY. ISBN: 0387549986 (U.S.)

Ekins P, Vanner R, Firebrace J (2005, March) Management of produced water on offshore oil installations: a comparative assessment using flow analysis final report of PSI. http://www.psi.org.uk/docs/2005/UKOOA/ProducedWater-Workingpaper.pdf. Accessed 20 Feb 2008

Holguin-Veras J (1993) Comparative assessment of AHP and MAV in highway planning: case study. J Trans Eng 121(2):191–200

Khasnabis S, Alsaidi E, Liu L and Ellis RD (2002) Comparative study of two techniques of transit performance assessment: AHP and GAT. J Transp Eng 128(6):499–508

McIntyre C, Kirschenman M, Seltveit S (1999) Applying decision support software in selection of division director. J Manage Eng 15(2):86–92

Modarres M (2006) Risk analysis in engineering technique, tools and trends. CRC press, Taylor & Francis, New York.

Neff JM (2002) Bioaccumulation in marine organisms: effects if contaminants in oil well produced water. Elsevier, New York, ISBN: 0-080-43716-8

OGP (International Association of Oil & Gas producers) (2005) Fate and effects of naturally occurring substances in produced water on the marine environment. February 2005, Report No.364. http://www.ogp.org.uk/pubs/364.pdf. Accessed 20 July 2007

OSPAR (2002) Background document concerning techniques for the management of produced water from offshore installations, OSPAR Commission 2002. http://www.ospar.org/v_publications/download.asp?v1=p00162. Accessed 10 July 2007

OSPAR (2006) Background document concerning techniques for the management of produced water from offshore installations, OSPAR Commission 2006. http://www.ospar.org/v_publications/download.asp?v1=p00162/2002. Accessed 10 July 2007

Patin S (1999) "Environmental impact of the offshore oil and gas industry" (trans: Cascio, E). Eco-monitoring Publishing, East Northport, NY

Saaty TL (1977) A scaling method for priorities in hierarchical structure. J Math Psychol (15): 234–281

Saaty TL (1980) The analytic hierarchy process. McGraw- Hill, New York

Tesfamariam S, Sadiq R (2006) Risk-based environmental decision-making using fuzzy analytic hierarchy process (F-AHP). Stoch Environ Res Risk Assess 21: 35–50

Tulloch SJ (2003) Development & field use of the Mare's Tail® Pre-Coalescer. In: Presented at the produced water workshop, Aberdeen, Scotland, March 26–27

Turban E, Meredith JR (1991) Fundamentals of management science, 5th edn. Richard D. Irwin, New York

Uddameri V (2003) Using the analytic hierarchy process for selecting an appropriate fate and transport model for risk-based decision-making at hazardous waste sites. Pract Period Hazardous Toxic Radioactive Waste Manage 7(2):139–146

Index

K. Lee, J. Neff (eds.), *Produced Water*, DOI 10.1007/978-1-4614-0046-2,

E

I

J

K

W

Z